2023 IEEE/ACM International Symposium on Low Power Electronics and Design (ISLPED 2023)

Vienna, Austria
7-8 August 2023

IEEE Catalog Number: CFP23LOW-POD
ISBN: 979-8-3503-1176-1

IEEE Catalog Number: CFP23LOW-POD
ISBN (Print-On-Demand): 979-8-3503-1176-1
ISBN (Online): 979-8-3503-1175-4

Additional Copies of This Publication Are Available From:

Curran Associates, Inc
57 Morehouse Lane
Red Hook, NY 12571 USA
Phone: (845) 758-0400
Fax: (845) 758-2633
E-mail: curran@proceedings.com
Web: www.proceedings.com

TABLE OF CONTENTS

Energy-Efficient Machine Learning Acceleration: From Technologies to Circuits and Systems

Chukwufumnanya Ogbogu[1], Madeleine Abernot[2], Corentin Delacour[2], Aida Todri-Sanial[2,3], Sudeep Pasricha[4], Partha Pratim Pande[1].
[1]School of EECS Washington State University, Pullman WA, USA. [2]LIRMM, University of Montpellier, CNRS, France. [3]Eindhoven University of Technology, Netherlands. [4]Dept. of Electrical and Computer Engg., Colorado State University, Fort Collins, Colorado, USA
[1]{c.ogbogu, pande}@wsu.edu, [2]{madeleine.abernot, corentin.delacour}@lirmm.fr, [3]a.todri.sanial@tue.nl, [4]sudeep@colostate.edu.

***Abstract*—Advanced computing systems have long been enablers for breakthroughs in Machine Learning (ML) algorithms either through sheer computational power or form-factor miniaturization. However, as ML algorithms become more complex and the size of datasets increase, existing computing platforms are no longer sufficient to bridge the gap between algorithmic innovation and hardware design. With the rising needs of advanced algorithms for large-scale data analysis and data-driven discovery, and significant growth in emerging applications from the edge to the cloud, we need energy-efficient, low-cost, high- performance, and reliable computing systems targeted for these applications. This paper presents the latest developments in oscillatory neural networks, optical computing, and memristive processing-in-memory (PIM) to address the various challenges in designing efficient computing systems specifically targeting ML applications.**

Keywords—optical neural networks, silicon photonics, oscillatory neural networks, ReRAM, GNNs

I. Introduction

Deep machine learning (ML) algorithms are employed in a wide variety of real-world applications, e.g., self-driving cars, medical diagnosis, network security, and industrial automation. Both training and inferencing of these deep ML models are computationally demanding tasks and are typically accomplished on the cloud. However, there is a growing necessity to implement deep learning on edge platforms due to privacy and security concerns, the need for user-specific customization, low latency, and real-time requirements (such as in augmented/virtual reality applications). However, implementing these applications on edge devices is challenging due to area and energy constraints. Addressing this necessitates suitable high-performance and energy efficient hardware support. In this paper, we present the salient features and design challenges with three emerging hardware paradigms, viz., oscillatory analog computation, optical computing, and memristive processing-in-memory (PIM).

Computing with phase dynamics of coupled oscillators enables not only signal voltage amplitude reduction but brings massive parallelism allowing for fast computation with energy efficiency. We believe that analog computing based on coupled oscillatory neural networks might be what is needed for certain ML tasks running on edge devices that have latency and power constraints. Using phase dynamics allows for solving associative memory and combinatorial optimization types of problems with simple circuits that can enable more versatile edge AI functions.

Optical computing has become an attractive substrate for accelerating emerging ML workloads due to the continued miniaturization of silicon photonics devices that also possess compatibility with CMOS fabrication. This paradigm combines light speed latencies for communication with light-speed computational operations, such as matrix vector multiplications and summations, which can significantly improve performance and energy-efficiency when executing various types of ML algorithms.

Resistive random-access memory (ReRAM) based processing-in-memory (PIM) architectures have been used to accelerate deep ML applications such as CNNs, RNNs, GNNs, transformers, etc. ReRAM-based systems are more area-efficient compared to their GPU counterparts and do not require expensive off-chip memory access due to the "in-memory" nature of the ReRAM-based computation. The crossbar structure of ReRAM-based architectures enables efficient Matrix-Vector Multiplication (MVM) operations, which are ubiquitous in modern ML tasks.

II. Computation with Oscillatory Neural Networks

In response to the challenges introduced by cloud computing, current research focuses on enabling ML at the edge, bringing ML capabilities closer to the data source, to reduce latency and minimize energy requirements.

A solution takes inspiration from biology to design neuromorphic computing techniques, like Spiking Neural Networks [1]. In this work, we focus on another promising neuromorphic paradigm with Oscillatory Neural Networks (ONN) [2] inspired by brain oscillations. ONNs are networks of coupled oscillators computing with inherent parallel phase synchronisation of coupled oscillators. Phase computing encodes information in the phase relationship among oscillators. It allows to reduce power consumption by limiting the voltage amplitude. Low power and fast parallel ONN computing makes it attractive for edge AI [3].

Using phase-computing ONNs, information is encoded in the phase relationship among oscillators. For example, for binary information, a logic '0' is encoded with a 0° phase, while a logic '1' is encoded with a 180° phase. ONN computation starts by initializing phases of each oscillator in the network from input data. Then, phases evolve depending on the coupling between oscillators, and stabilizes to a final phase state. The evolution of phases corresponds to the minimization of an intrinsic energy parameter, like in attractor networks [4]. The final phase state gives the network output information. Thus, the coupling among oscillators,

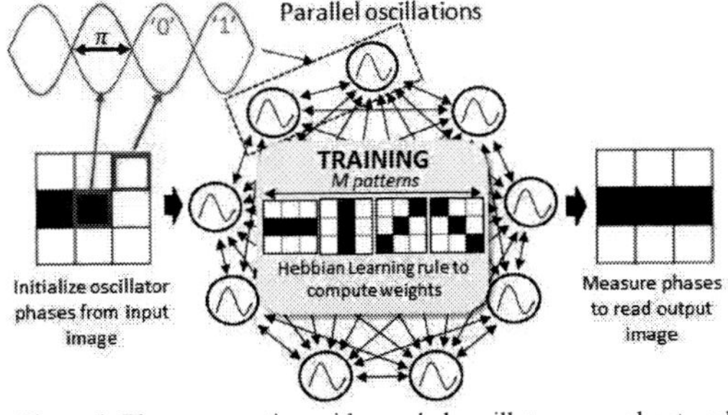

Figure 1. Phase computing with coupled oscillatory neural networks.

defined by learning, is the main parameter to solve specific tasks efficiently.

In the current state-of-the-art, ONNs are typically implemented using a fully connected (FC) architecture, utilizing unsupervised learning for auto-associative memory tasks similar to Hopfield networks, as illustrated in Fig. 1. This configuration allows the network to memorize patterns based on its connections and converge to a learned pattern when initialized with a corrupted input. FC-ONNs configured for pattern recognition have found applications in various domains. However, the FC-ONN architecture necessitates a large number of coupling elements, which increases quadratically with the number of neurons. For a network of N neurons, the number of synaptic elements required is given by N(N-1)/2, making large-scale implementation challenging. Additionally, while associative memory tasks are intriguing, the capacity of the network is often limited by unsupervised learning, restricting the number of patterns that can be efficiently learned and retrieved. Therefore, there is a pressing need to explore alternative architectures and applications that can be applied to ONNs.

Recently, researchers have introduced two-layer ONN architectures that incorporate bidirectional and feedforward connections between layers while excluding connections among oscillators within the same layer [5]. These two-layer ONNs have been successfully utilized for image edge detection in the respective studies. Notably, the implementations in these works were digitally executed on FPGA platforms. However, it is important to highlight that achieving a feedforward ONN architecture using analog implementation is challenging, as the default nature of coupling among oscillators is bidirectional.

A. Analog ONNs and combinatorial optimization

As ONNs are dynamical systems, continuous-time analog architectures that follow the "let physics compute" principle, they are well suited for efficient ONN implementation. Analog ONNs naturally perform gradient descent of their energy function through time, without any algorithm and clock. This intrinsic minimization mechanism is appealing for many optimization tasks, especially for combinatorial optimization problems that are often Nondeterministic Polynomial time hard (NP-hard) and require the exploration of a solution space that scales exponentially. Thus, accelerating the search is crucial for large instances. ONN high parallelism is particularly promising for this task and can provide up to four orders of magnitude runtime improvement compared to CPUs [6]. As a matter of fact, all NP-complete problems can be reduced to finding the ground states of a corresponding Ising Hamiltonian, which in principle can be mapped to the ONN energy by relaxing the initial discrete variables to continuous phases [7].

In practice, various CMOS ONN solvers have been demonstrated [8] [9], the largest being a 1968-node ONN based on ring oscillators oscillating at 1 GHz [10]. Other approaches focus on reducing the oscillator and synapse footprint using novel devices such as magnetic tunnel junctions (MTJ), memristors, or transition metal oxide devices (TMO). Vanadium dioxide (VO2) is an example of TMO device that has a hysteresis switching behaviour and produces oscillations at room temperature [11]. When scaled down to submicron dimensions, VO2-based oscillators are expected to reach very low energy consumptions down to 10 fJ at 100 MHz [3], but yet remains to be demonstrated.

Just like any other neural network, implementing densely connected ONNs is challenging. Most of current hardware limit the synaptic connections to nearest neighbours which overall necessitate more physical neurons than the initial network after mapping [10]. Moreover, finding the optimal network mapping to the hardware can cause an important computational overhead compared to the initial problem to solve [12]. Another interesting ONN capability is to emulate all-to-all connectivity using the injection of a modulated external signal [13]. This approach replaces the connectivity complexity by a modulation scheme which is also challenging to design, but could advantageously be performed off-chip.

B. Digital ONN implementation for edge applications

With all challenges related to analog ONN implementations, researchers also developed digital-based ONN implementations. [8] introduced a mixed signal implementation using digital oscillators with additional analog components, and more recently, a fully-digital ONN was implemented on FPGA to demonstrate ONN computing paradigm for AI edge applications [14].

The ONN on FPGA first demonstrated interesting properties to solve pattern recognition tasks based on the fully-connected recurrent architecture from Hopfield. In particular, a 10x6 fully-connected ONN was trained with images of digits and implemented inside the FPGA performing real-time pattern recognition from a camera stream [14]. An additional system architecture has also been proposed to allow on-chip unsupervised learning with the digital ONN design, being able to solve pattern recognition with on-chip Hebbian or Storkey learning implementation. Then, two cascaded fully-connected ONNs were used to perform obstacle avoidance on mobile robots. A first ONN detects obstacles from proximity sensor data, and a second ONN uses the detected obstacles to define a novel direction allowing real-time obstacle avoidance from proximity sensors dataflow [15].

Later, the digital ONN implementation has been adapted to fit with layered architectures, considering bidirectional or feedforward synaptic connections, and was applied to the image edge detection application. Finally, recently, the ONN for image edge detection was proposed as an accelerator of the SIFT feature detection algorithm [16].

Comparable to analog ONN, dense digital ONN implementation is challenging. The recently introduced layered architectures permits the implementation of larger scale ONNs, the investigation of novel general learning methods, and the exploration of novel edge applications.

III. Optical Computing for ML Acceleration

The gradual plateauing of improvement in performance-per-watt with electronic ML accelerators in recent years has led to a widespread and urgent search for alternative computing technologies. One very promising technology that has emerged from this search is silicon photonics (SiPh), which involves the use of light to transmit data [17] [18]. Increasingly, SiPh components are being used to perform computations with data. Optical computing systems built with SiPh components are not constrained by the limits of electron movement that hamper today's electronic platforms. Instead, they rely on speedy photons for information transfer and manipulation. Such systems have the potential to revolutionize data-hungry ML application acceleration by providing ultra-fast data transfers and processing (e.g., multiply and accumulate, or Fast Fourier Transform operations) in the analog optical domain while consuming much less energy than traditional electronic computing [19].

A. Photonic Devices and Circuits

Non-coherent optical computing involves the use of multiple wavelengths that can perform parallel data transfers as well as operations such as matrix-vector operations concurrently [20]. Both data transfers and computation require the use of various optoelectronic components that can be fabricated in CMOS-compatible foundries.

Lasers (off-chip or on-chip) are used to generate optical signals necessary for computation and communication in optical circuits. *SiPh* waveguides made of a core (Si) and a cladding (SiO_2) material provide high-refractive-index contrast and allow for total internal reflection and hence optical signal confinement and transmission. Microring Resonators (MRs) involve a ring-shaped waveguide that is designed to be sensitive to a particular wavelength, referred to as the MR's resonant wavelength. MRs are used for modulation and filtering data carrying wavelengths from optical interconnects. These same devices can also be used to perform computation via amplitude modulation on different wavelengths using a tuning circuit (which can be electro-optic (EO) or thermo-optic (TO) [20]) that modifies the operational characteristics of the MRs. As an example, two parameters in an ML model that need to be multiplied can be deployed on MRs with the same resonant wavelength, along the same input waveguide. The first MR modulates the signal to represent the first parameter. The parameter specific amplitude modulation with the second MR results in a multiplication operation [21]. Lastly, photodetectors are used to detect optical signals and convert them to electrical signals.

B. Low Power Optical Computing Design Challenges

To achieve energy-efficient and low-power non-coherent optical computing, several challenges must be addressed. Fabrication-process variations (FPVs) and thermal variations create unintended changes in fabricated optoelectronic components, requiring power-hungry tuning circuits to mitigate their effects [22]. Tuning circuits also have high energy overheads due to their high latencies of operation. Thermal crosstalk arising due to such tuning circuits further reduces MR stability and increases energy consumption. Crosstalk between multiple wavelengths requires increasing laser power to overcome degradation in signal-to-noise ratio (SNR) [23]. Lastly, electro-optic conversions require power

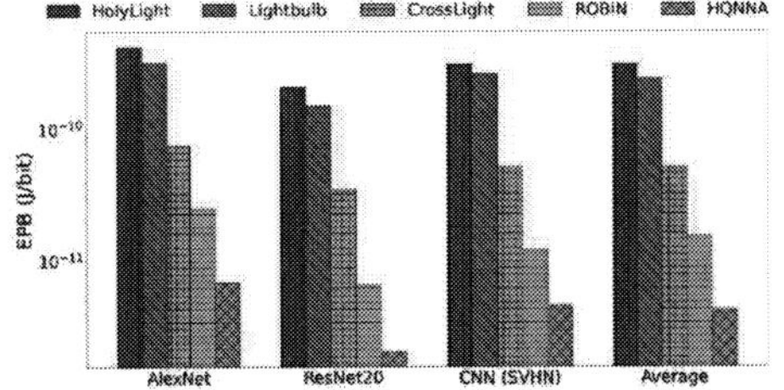

Figure 2: EPB for CNN models, across optical-domain accelerators

hungry analog-to-digital converters (ADCs) and digital-to-analog converters (DACs). All of these factors must be addressed during the design of an optical computing accelerator for ML.

C. Cross-Layer Design for Optical Computing

In [24], we proposed *CrossLight*, the first cross-layer optimized optical computing accelerator for ML workloads that included CNNs and DNNs.

At the device level, we fabricated a 1.5×0.6 mm^2 chip with high-resolution Electron Beam (EBeam) lithography and performed a comprehensive design-space exploration of MRs to compensate for FPVs while improving MR device insertion loss and Q-factor. We found that in an MR design of any radii and gap, when the input waveguide is 400 nm wide and the ring waveguide is 800 nm wide at room temperature (300 K), the undesired resonant wavelength shift due to FPVs can be reduced from 7.1 to 2.1 nm (70% reduction). This is a significant result, as these engineered MRs require less compensation for FPV-induced resonant wavelength shifts, which reduces the power consumption of architectures using such MRs.

At the circuit level, we attempted to reduce the reliance on thermo-optic (TO) tuning circuits that create thermal crosstalk. We developed a hybrid tuning circuit where both thermo-optic (TO) and electro-optic (EO) tuning are used to compensate for resonant wavelength shifts due to FPVs and thermal variations. The hybrid tuning approach supports faster operation of MRs with fast EO tuning to compensate for small wavelength shifts and, using TO tuning when large wavelength shifts need to be compensated. This significantly reduces the energy overheads associated with MR operation. We further used a commercial 3D heat transport simulation EDA tool for *SiPh* devices (Lumerical HEAT) to determine phase crosstalk ratio to determine optimal inter-MR distance to minimize power consumption in TO microheaters.

At the architecture level, we designed vector dot product (VDP) with banks of MRs for optical matrix multiplication. The VDP units were clustered into two groups: one to support convolution (CONV) layer acceleration and the other to support fully connected (FC) layer acceleration. We focused on these two types of layers as they are the most widely used and consume the most latency and power in computational platforms that execute CNNs and DNNs. We also optimized laser reuse across VDP units, optimized the dimensions and cluster sizes of VDPs, and proposed dataflow mechanisms to map computations to VDP units.

Table 1 shows the energy-per-bit (EPB) and performance-per-watt (in terms of thousands of image frames per second processed, per watt) for *CrossLight* compared to other *SiPh*-based accelerators (HolyLight, DEAP_CNN) and electronic platforms. The cross-layer optimizations allow *CrossLight* to outperform the state-of-the-art ML accelerators, highlighting the promise of optical computing.

D. Sparse and Quantized Optical Computing

To limit the DAC and ADC power consumption, while sustaining model accuracy, in [25] we proposed the *ROBIN* optical ML accelerator for binarized neural networks (BNNs). BNNs use 1-bit weights, but activations use more bits (e.g., 4 or 8) to preserve accuracy. As BNNs only require simple switching circuits for weight parameter representation and

Table 1: Average EPB and kiloFPS/Watt values across accelerators

Accelerator	Avg. EPB (pJ/bit)	Avg. kiloFPS/watt
P100	971.31	24.9
IXP 9282	5099.68	2.39
AMD-TR	5831.18	2.09
DaDianNao	58.33	0.65
Edge TPU	697.37	17.53
Null Hop	2727.43	4.48
DEAP_CNN	44453.88	0.07
Holylight	274.13	3.3
CrossLight	28.78	52.59

simpler DAC circuitry for activations, they lead to better energy efficiency. In [26], we proposed the *HQNNA* optical accelerator that supported heterogeneous quantization, where each layer in CNN/DNN models could be quantized uniquely between 2 to 8 bits. By allocating more bits to layers that are critical for preserving model accuracy, heterogeneous layer-wise quantization can better balance model accuracy with energy costs. We improved computation orchestration via time division multiplexing and bit-slicing, and optimized laser power to aggressively reduce energy overheads in HQNNA. Fig. 2 shows how these optimizations allow HQNNA to achieve lower EPB than ROBIN as well as *CrossLight*.

Another approach to reduce power involves leveraging sparsification. Sparsity in a DNN model refers to removal of weights (i.e., replacing them with 0s) which do not contribute to the overall accuracy of the model. This technique is often used to reduce memory footprint for ML models. But sparse models incur unwanted operations which lead to a 0 output because of the 0 valued weights involved. This leads to wasted energy consumption. Our work in [27] tackled this challenge by combining a dataflow mechanism for compressing sparse matrices to dense matrices along with better VDU design to handle any residual sparsity. These hardware/software codesign optimization helped achieve 27.6× lower EPB than state-of-the-art electronic and optical-domain accelerators.

E. Emerging Use Cases for Optical Computing

There are many opportunities to utilize optical computing to accelerate ML models beyond CNNs and DNNs.

In [28], we proposed RecLight, the first optical accelerator for sequence learning tasks (e.g., network anomaly detection [29]) that involve Recurrent Neural Networks (RNNs). RNNs, which can include Gated Recurrent Units (GRUs) and Long Short-Term Memory (LSTM) cells, are challenging to accelerate due to the recursive nature of these models and the compute-intensive operations required for large-dimensional sequence data. We developed custom optical computing units that supported accelerating GRU and LSTM cell operations, as well as efficient optical-domain implementations of non-linear activation functions such as sigmoid and tanh.

Large Language Models (LLMs), such as those used in OpenAI's ChatGPT and Google's Bard rely on transformer neural networks. These unique neural networks combine attention layers, feed-forward layers, and normalization layers to learn context and meaning by tracking relationships in data sequences. However, the complex structure of these models creates challenges for accelerating their execution. In [30] we proposed the first optical computing platform to accelerate a broad family of language-based and vision-based transformer models such as BERT and Vision Transformers.

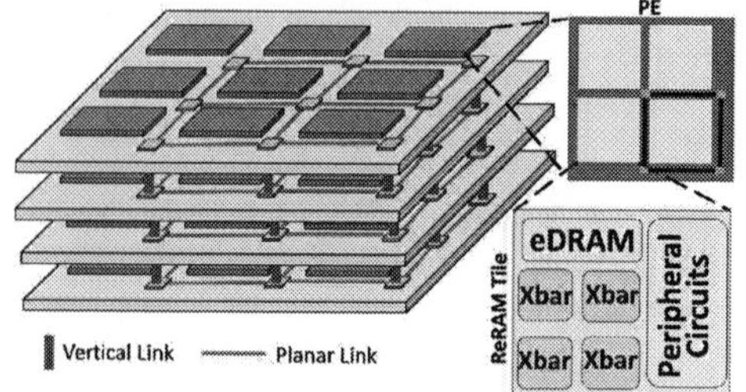

Figure 3: Illustration of ReRAM-based 3D PIM Architecture [41].

We adapted many of the cross-layer optimizations described earlier to the unique requirements of transformer models.

IV. GNN Training on ReRAM-Based PIM Architecture

Training machine learning (ML) models at the edge (training on-chip or on embedded systems) can address many pressing challenges, including data privacy/security, increase the accessibility of ML applications to different parts of the world by reducing the dependence on the communication fabric and the cloud infrastructure, and meet the real-time requirements of emerging applications like augmented/virtual reality (AR/VR) applications. However, existing edge platforms do not have sufficient capabilities to support on-device training of ML models such as Convolutional Neural Networks (CNNs), Graph Neural Networks (GNNs) etc. Moreover, it is estimated that training a single unpruned neural network on conventional compute platforms, such as GPUs, can cost over $10,000 and emit as much carbon as five cars over their lifetimes [31]. Resistive random-access memory (ReRAM) based processing-in-memory (PIM) architectures are a promising solution to address this problem. ReRAM-based PIM systems have been proposed to accelerate both CNN and GNN computation [32]. The crossbar structure of ReRAM enables efficient Matrix-Vector Multiplication (MVM), which is ubiquitous in modern ML tasks including CNN/GNN training and inferencing. In this section, we principally focus on discussing the advantages and challenges of designing ReRAM-based accelerators for GNN training. However, by considering area, power and storage, the overall architecture needs to be divided into multiple ReRAM tiles with bounded crossbar size. Hence, despite the PIM capability, when we design a ReRAM-based manycore architecture for large-scale CNN/GNN computation, it gives rise to a substantial amount of on-chip traffic that creates performance bottlenecks if not addressed appropriately. Moreover, ReRAM crossbar arrays suffer from low write endurance [33] [34]. By incorporating model and graph pruning, we can reduce the on-chip traffic and storage requirements and improve the endurance of the ReRAM-based architectures significantly. ReRAM-based PIM architecture for GNN training.

Recently, we have proposed a ReRAM-based 3D PIM architecture called ReMaGN, tailored for on-chip training of GNNs [32]. We adopt a 3D architecture to enable high degree of integration [35]. This architecture consists of multiple PEs stacked vertically across four layers as shown in Fig. 3. Each PE contains multiple ReRAM crossbar arrays for executing MVM operations. The salient features of the ReMaGN architecture are:

1. To effectively utilize the high-throughput computation provided by ReRAM-based PEs, the overall architecture needs to be supported with a high-performance and efficient communication backbone. In this architecture, we utilize a 3D mesh network-on-chip (NoC) as the interconnection backbone for communication between PEs during GNN training. We have reduced the NoC traffic by incorporating DropEdge and Dropout.
2. ReMaGN employs a pipelined training methodology. Training GNNs on one big monolithic graph is often impractical due to memory concerns. In addition, training on large graphs does not exploit the benefits of ReRAM-based architectures, which rely on a pipelined implementation [32]. Pipelining reduces the number of ReRAM writes (which are slow), leading to higher overall throughput. However, this strategy is not amenable to GNNs if the entire graph needs to be processed altogether. Graph clustering/partitioning is used in ReMaGN to address this problem.
3. Typically, ReRAMs compute using 16-bit fixed-point, which has significantly less representation capability than 32-bit floating point used by traditional GPUs. We have incorporated stochastic rounding to successfully address the accuracy loss due to reduced precision.
4. ReMaGN outperforms conventional GPUs by up to 9.5X (on average 7.1X) in terms of execution time, while being up to 42X (on average 33.5X) more energy efficient without sacrificing accuracy.

A. Model and Data Pruning for GNN Training

To improve the performance and energy efficiency of ReRAM-based architectures without loss of accuracy, we proposed crossbar-aware pruning techniques to simultaneously prune the GNN model weights and the input data graphs. Model pruning for neural networks helps reduce redundant computations [36]. Several crossbar-aware model pruning techniques have been proposed to exploit the ReRAM crossbar structure to reduce area and improve energy efficiency without compromising accuracy [37]. However, all these methods prune pre-trained DNN models for inferencing purposes, hence they are not suited for training. Moreover, existing crossbar-aware methods prune weights along rows/columns only, which leads to marginal energy and area savings. Here, we generally refer to these crossbar row/column-based pruning methods as 'RCP'. Fig. 4(a) illustrates how the weights when pruned using RCP are mapped to ReRAM crossbars. However, despite the structured row-/column-wise pruning in RCP, this only yields marginal energy savings as an entire crossbar is still activated for computation [38].

Recently, an iterative model pruning method known as the Lottery Ticket Hypothesis (LTH) was proposed to obtain highly sparse DNN models for training [39]. An LTH-inspired unified graph sparsification (UGS) has also been used to prune GNN models and graph adjacency matrices [40]. UGS jointly prunes GNN weights and graph adjacency matrices using trainable masks to reduce the number of matrix-vector-multiplication (MVM) operations associated with GNN training. However, the graph pruning in UGS only removes the edges from the graph adjacency matrix and thus the overall size of the input (i.e., the number of input subgraphs) still remains unchanged. Additionally, UGS introduces huge storage overhead for mask parameters, which are proportional to the GNN weights and number of adjacency matrix entries. Moreover, existing LTH-based methods don't take the ReRAM crossbar structure into consideration. As a result, there is no significant area or power savings. As an illustration, Fig. 4(b) shows how the GNN weights pruned using the standard LTH-based methods are mapped to ReRAM crossbars. The unstructured pruning nature of the LTH method only prunes individual weights and yields no significant advantage from a hardware standpoint.

We have proposed an LTH-inspired crossbar-aware pruning method for GNN training on ReRAM-based architectures. Our method enhances existing RCP methods by taking the overall crossbar structure into consideration and implements the pruning in an iterative manner to ensure that the pruned GNN models can be trained from scratch without loss in accuracy. Fig 4(c) illustrates how weights are mapped to the ReRAM crossbars after they are pruned using our proposed crossbar-aware method. Unlike RCP, we can see that entire (c×c) crossbars are pruned out, which leads to significant area and energy savings. We complement the crossbar-aware weight pruning with an optimized non-zero-storage mechanism for the graph adjacency matrix to further reduce the on-chip crossbar storage overhead as shown in Fig. 4(d) [41]. The optimized graph storage is achieved by dividing the adjacency matrix into non-overlapping segments based on the crossbar size. The size of the sliding window is determined by the crossbar size (c × c) to decompose the N×N graph adjacency matrix into "valid" and "invalid" segments (as shown in green and red boxes in Fig. 4(d) respectively) for storing on ReRAM crossbar arrays. A valid segment contains at least a non-zero element and is stored on the crossbar. However, an invalid segment consists of all-zeros and is discarded, as an MVM operation with zeros

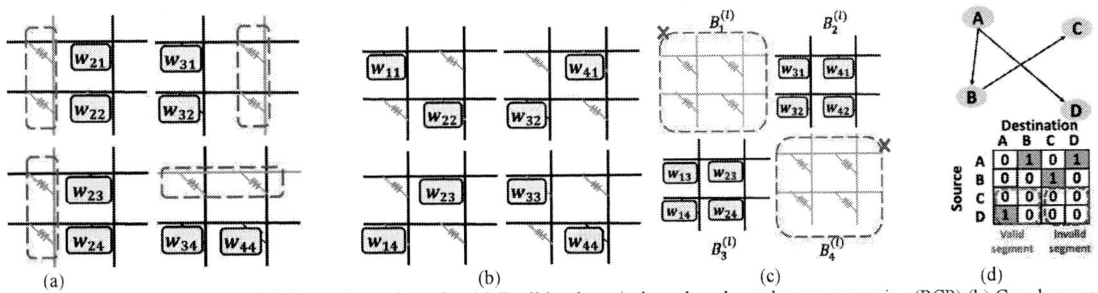

Figure 4. Mapping weights to ReRAM crossbars after using (a) Traditional row/column-based crossbar-aware pruning (RCP) (b) Crossbar-unaware pruning (LTH and UGS) (c) DietGNN pruning. (d) optimized graph adjacency matrix storage technique in DietGNN [41].

mapped to ReRAM crossbars is redundant and only leads to zeros. As a result, this non-zero-storage optimization mechanism reduces the number of crossbars required for the adjacency matrix while maintaining the graph connectivity. We refer to our crossbar aware model pruning method and the optimized graph storage mechanism together as 'DietGNN' henceforth (as proposed in [41]). Overall DietGNN enables energy-efficient GNN training and inferencing on ReRAM-based PIM architectures.

In addition to DietGNN, we can further prune the input data graphs for GNN training on ReRAM-based architectures. Large training datasets pose a challenge for resource-constrained platforms (such as edge devices) as they require high memory and processing power [42]. Recent work has proposed methods to reduce the amount of data required for training, generally referred to as data pruning (DP). An importance score-based data pruning methodology for CNNs was proposed to greatly reduce the amount of input data [43]. Graph Early Bird (GEB) prunes the edges of the input graphs to reduce MVM operations for GNN training very early during the training process [44]. Thus, the size of the input feature vector remains unchanged as only graph edges are pruned. In contrast to existing data pruning techniques for GNNs, our proposed method intelligently prunes large portions of an input graph (both nodes and edges) to reduce the end-to-end pipeline depth of GNN training on ReRAM-based architectures.

It is well known that ReRAM crossbar arrays suffer from low write endurance [33] [34]. ReRAM cells become highly prone to permanent faults after a limited number of writes (typically 106 - 109 writes), thus limiting their lifetime. As graph pruning reduces the number of input subgraphs, it also helps to reduce the number of weight updates occurring during training. Hence, incorporating data pruning helps to preserve the lifetime of the ReRAM-based architecture and is complementary to existing reliability enhancement techniques such as gradient sparsification [34].

For a comprehensive evaluation of DietGNN+DP, we have considered three diverse GNN models, namely: Graph Convolution Networks (GCN), Graph Attention Networks (GAT), and Graph Sample and Aggregate (GSA); and six benchmark real-world graph datasets: PPI, Reddit (RDT), Amazon2M (A2M), Flickr (FKR), Yelp (YLP), and Open Graph Benchmarks-Proteins (OGP) [45]. In Fig. 5(a) and (b), we show the accuracy and achievable sparsity respectively of DietGNN+DP method relative to other pruning techniques (LTH, UGS and RCP). As an example, in Fig 5(a) and (b), we consider the GCN model pruned using each method, and mapped to the ReRAM-based PIM architecture for training. We note that we observe similar accuracy and sparsity trends with the other GNN models (GAT and GSA) across multiple datasets. As shown in Fig. 5(a), UGS is unable to prune a lot of weights without experiencing accuracy loss due to its deletion of graph edges hence it achieves lower sparsity. Overall, Fig. 5 shows that except for UGS, all the pruned models are very sparse (~90%), and they can be trained from scratch with very minimal accuracy (as shown in Fig. 5(b)) loss compared to their unpruned counterparts.

Finally, we carry out a performance evaluation of the DietGNN+DP-enabled GNN training on the ReRAM-based PIM architecture. Fig 6(a) and (b) compare the DietGNN+DP method with; the unpruned counterpart, and other pruning approaches (UGS, RCP, and DietGNN alone) in terms of execution time and energy consumption. Here, we assume an iso-area scenario for this analysis, i.e., the number of ReRAM crossbars available is the same for all the cases. As shown in Fig. 6(a) and 6(b), the DietGNN+DP training improves execution time and energy consumption by ~57% and ~73% respectively on average compared to the unpruned model running on an iso-area ReRAM-based PIM architecture. Overall, the DietGNN+DP-enabled training achieves low energy- and storage-efficient GNN computation. The key highlights of the DietGNN+DP framework are summarized as follows:

1.) DietGNN demonstrates that it is possible to prune more than 90% of GNN weights for diverse GNNs and real-world graph datasets.
2.) The pruned GNNs enable significant energy savings (that is, crossbar diet) and performance improvements without sacrificing accuracy.
3.) The experimental results demonstrate that DietGNN+DP accelerates GNN training by up to 4.5 × while using 6.6 × less energy on average when compared to its unpruned counterpart on ReRAM-based 3D PIM architectures.

B. Limitations of Existing ReRAM-based PIM Architectures

Reliability of ReRAMs: The ReRAM fabrication process is not as mature as conventional CMOS fabrication [33]. As a result, ReRAMs are prone to many types of hardware faults and noise. For instance, hard faults prevent the resistance of a ReRAM cell from being updated, resulting in write failures. Moreover, the limited endurance of ReRAM cells makes them suffer from short programming cycles before they suffer permanent write failures. The write endurance of ReRAM chips typically ranges from 106 to 1012 writes before they fail [33]. However, training of state-of-the-art large-scale deep neural networks (DNN) usually demands numerous weight updates which result in multiple ReRAM cell programming cycles [46]. As a result, this limits the lifetime of ReRAM devices. Hence, enhancing the lifetime of ReRAM crossbars is important to facilitate their widespread

(a)

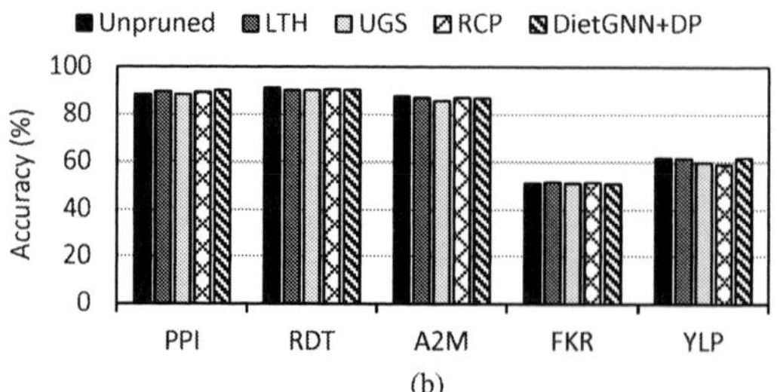

(b)

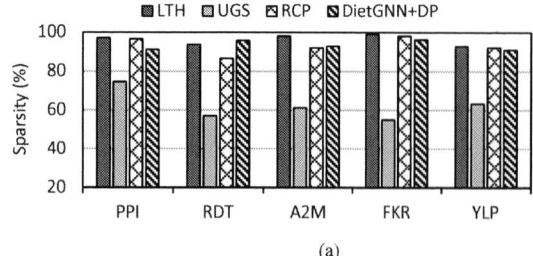

Figure 5: (a) Sparsity and (b) Accuracy of pruned GNN models (winning tickets) obtained using different methods. All models are trained on ReRAM crossbars.

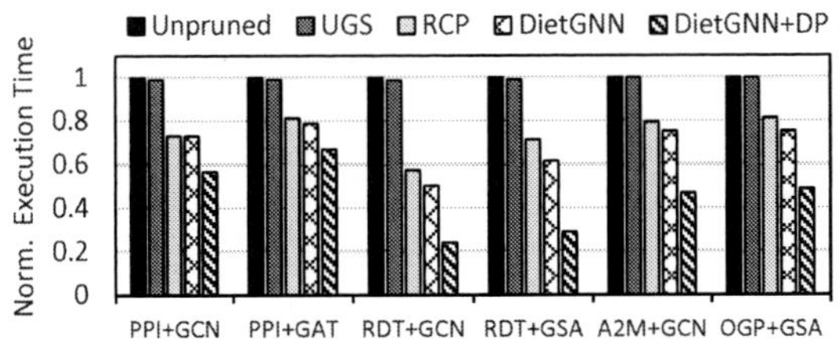

(a)

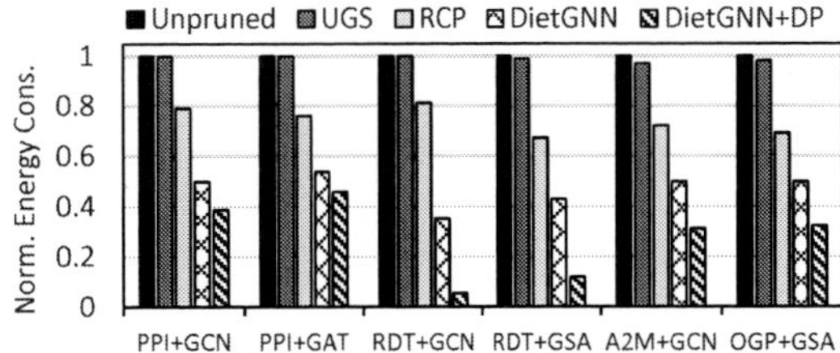

(b)

Figure 6: Normalized (a) execution time and (b) energy consumption for the unpruned, UGS, RCP, DietGNN, and DietGNN+DP-enabled GNN (normalized with respect to the execution of the unpruned model on the ReRAM-based PIM architecture).

adoption as hardware accelerators for DNN training. Recent efforts have tried to address the write endurance issue of ReRAM devices some of which include device-level optimizations and software-level approaches. However, most device-level methods aim at reducing only the per-write energy and do not reduce the actual number of writes. Meanwhile, software techniques focus solely on reducing the number of write operations due to weight updates; and do not consider ReRAM cell writes due to activations or intermediate products. Overall, these methods often introduce area, power, and performance overheads.

Write Latency and Energy of ReRAMs: In addition to the write endurance challenge of ReRAM cells, they also suffer from high write energy (~2 nJ) and latency (~100 ns) compared to other Non-volatile Memory (NVM) technologies for PIM architectures. As a result, most ReRAM-based PIM platforms require traditional memory hierarchies and a pipelined implementation to efficiently hide the memory latency due to slow ReRAM writes. Researchers have explored parallel writes via bit-slicing as a technique to mitigate the expensive write operations during ML training, thus enabling ReRAM-based training architectures.

Accelerating Large Language Models (LLMs): LLMs have achieved significant success in a wide variety of natural language processing (NLP) tasks. However, language tasks are increasingly becoming complex, and are fast outpacing the computing capabilities of traditional platforms such as GPUs and CPUs. This is due to their numerous parameters, high storage requirements, and significant off-chip data movement costs [47]. Hence, ReRAM-based PIMs have been proposed as an alternative for accelerating LLMs due to their high-density storage, in-memory processing capability, and energy-efficient computing capability. However, simple LLM tasks such as inferencing requires a significant number of write operations, and ReRAM-based PIMs are still plagued by the write endurance problem. Moreover, LLM inference tasks are usually time-critical applications. Thus, the high write latency of ReRAM cells can potentially limit its adaptability for LLM inferencing.

V. CONCLUSION

Advancement in novel ML algorithms and computing systems design is tightly coupled and advancement in one cannot be achieved without the other. However, the slowing down of Moore's law has impacted the development of new computing platforms, which is detrimental to future developments and applications of ML. In this paper, we have discussed the benefits and challenges associated with three emerging hardware paradigms as enablers for designing energy-efficient and high-performance hardware architectures for accelerating various types of ML applications.

ACKNOWLEDGMENTS

This work was funded in part by grants from the National Science Foundation CCF-1813370 and CCF-2006788 and funding from EU Commission Horizon EU research and innovation program in the framework of PHASTRAC (https://phastrac.eu) with grant no. 101092096.

REFERENCES

[1] C. Schuman, S. Kulkarni and M. Parsa et al., "Opportunities for neuromorphic computing algorithms and applications," *Nat Comput Sci*, vol. 2, 2022.

[2] G. Csaba and W. Porod, "Coupled oscillators for computing: A review and perspective," *Applied Physics Reviews*, vol. 7, 2020.

[3] C. Delacour et al., "Energy-Performance Assessment of Oscillatory Neural Networks Based on VO2 Devices for Future Edge AI Computing.," *IEEE Transactions on Neural Networks and Learning Systems*, 2023.

[4] J. Buhmann, R. Divko and K. Schulten, "Associative memory with high information content.," *Physical Review*, 1989.

[5] M. Abernot et al., "Two-Layered Oscillatory Neural Networks with Analog Feedforward Majority Gate for Image Edge Detection Application.," in *IEEE International Symposium on Circuits and Systems.*, 2023.

[6] C. Delacour, et. al., "A Mixed-Signal Oscillatory Neural Network for Scalable Analog Computations in Phase Domain," *preprint at (hal-03961010)*, 2023.

[7] T. Wang, et. al., "Solving combinatorial optimisation problems using oscillator based Ising machines," *Nat Comput*, vol. 20, 2021.

[8] T. Jackson, S. Pagliarini and L. Pileggi, "An Oscillatory Neural Network with Programmable Resistive Synapses in 28 Nm CMOS," *2018 IEEE International Conference on Rebooting Computing (ICRC)*, pp. 1-7, 2018.

[9] M. Graber, et. al., "A Versatile & Adjustable 400 Node CMOS Oscillator Based Ising Machine to Investigate and Optimize the Internal Computing Principle," *2022 IEEE 35th International System-on-Chip Conference (SOCC)*, 2022.

[10] W. Moy, et. al., "A 1,968-node coupled ring oscillator circuit for combinatorial optimization problem solving," *Nat. Electron.*, vol. 5, 2022.

[11] S. Carapezzi, et. al., "Role of ambient temperature in modulation of behavior of vanadium dioxide volatile memristors and oscillators for neuromorphic applications," *Sci. Rep.*, vol. 12, 2022.

[12] M. Graber, et. al., "A Fast Graph Minor Embedding Heuristic for Oscillator Based Ising Machines," *2022 Austrochip Workshop on Microelectronics (Austrochip)*, 2022.

[13] D. Albertsson, et. al., "Highly reconfigurable oscillator-based Ising Machine through quasiperiodic modulation of coupling strength," *Sci. Rep.*, vol. 13, 2023.

[14] M. Abernot, T. Gil, M. Jimenez, J. Nunez, M. J. Avedillo, B. Linares-Barranco, T. Gonos, T. Hardelon and A. Todri-Sanial, "Digital Implementation of Oscillatory Neural Network for Image Recognition Applications," *Frontiers in Neuroscience*, vol. 15, 2021.

[15] M. Abernot, et al., "Oscillatory Neural Networks for Obstacle Avoidance on Mobile Surveillance Robot E4," *2022 International Joint Conference on Neural Networks (IJCNN)*, pp. 1-8, 2022.

[16] M. Abernot et al., "SIFT-ONN: SIFT Feature Detection Algorithm Employing ONNs for Edge Detection.," in *Proceedings of the 2023 Annual Neuro-Inspired Computational Elements Conference*, 2023.

[17] S. Bahirat and S. Pasricha, "UC-PHOTON: A Novel Hybrid Photonic Network-on-Chip for Multiple Use-Case Applications," in *IEEE ISQED*, 2010.

[18] Y. Xu and S. Pasricha, "Silicon Nanophotonics for Future Multicore Architectures: Opportunities and Challenges," *IEEE Design & Test*, 2014.

[19] F. Sunny, E. Taheri, M. Nikdast and S. Pasricha, "A survey on silicon photonics for deep learning," *ACM JETC*, vol. 17, no. 4, 2022.

[20] S. Pasricha and M. Nikdast, "A Survey of Silicon Photonics for Energy Efficient Manycore Computing," *IEEE Design and Test*, 2020.

[21] A. N. Tait and et al, "Broadcast and Weight: An Integrated Network For Scalable Photonic Spike Processing," *IEEE JLT*, 2014.

[22] S. V. R. Chittamuru, I. Thakkar and S. Pasricha, "LIBRA: Thermal and Process Variation Aware Reliability Management in Photonic Networks-on-Chip," *IEEE TMSCS*, no. Oct-Dec 2018.

[23] S. V. R. Chittamuru, I. Thakkar and S. Pasricha, "PICO: Mitigating Heterodyne Crosstalk Due to Process Variations and Intermodulation Effects in Photonic NoCs," in *IEEE/ACM Design Automation Conference (DAC)*, 2016.

[24] F. Sunny, A. Mirza, M. Nikdast and S. Pasricha, "CrossLight: A Cross-Layer Optimized Silicon Photonic Neural Network Accelerator," in *IEEE ACM Design Automation Conference (DAC)*, 2021.

[25] F. Sunny, A. Mirza, M. Nikdast and S. Pasricha, "ROBIN: A Robust Optical Binary Neural Network Accelerator," *ACM TECS*, Oct 2021.

[26] F. Sunny, M. Nikdast and S. Pasricha, "A Silicon Photonic Accelerator for Convolutional Neural Networks with Heterogeneous Quantization," in *ACM GLSVLSI*, 2022.

[27] F. Sunny, M. Nikdast and S. Pasricha, "SONIC: A Sparse Neural Network Inference Accelerator with Silicon Photonics for Energy-Efficient Deep Learning," in *IEEE/ACM ASPDAC*, 2022.

[28] F. Sunny, M. Nikdast and S. Pasricha, "RecLight: A Recurrent Neural Network Accelerator With Integrated Silicon Photonics," in *IEEE ISVLSI*, 2022.

[29] V. K. Kukkala, S. V. Thiruloga and S. Pasricha, "LATTE: LSTM Self-Attention based Anomaly Detection in Embedded Automotive Platforms," *ACM TECS*, 2021.

[30] S. Afifi, F. Sunny, M. Nikdast and S. Pasricha, "TRON: Transformer Neural Network Acceleration with Non-Coherent Silicon Photonics," in *ACM GLSVLSI*, 2023.

[31] E. Strubell, A. Ganesh and A. McCallum, "Energy and policy considerations for modern deep learning research," in *AAAI 2020 – 34th AAAI Conference on Artificial Intelligence*, 2020.

[32] A. I. Arka, B. K. Joardar, J. R. Doppa, P. P. Pande and K. Chakrabarty, "Performance and Accuracy Tradeoffs for Training Graph Neural Networks on ReRAM-Based Architectures," *IEEE Transactions on Very Large Scale Integration (VLSI) Systems*, vol. 29, no. 10, pp. 1743-1756, 2021.

[33] W. Wen, Y. Zhang and J. Yang, "ReNEW: Enhancing Lifetime for ReRAM Crossbar based Neural Network Accelerators," in *IEEE International Conference on Computer Design (ICCD)*, 2019.

[34] Y. Cai et al., "Long live TIME: Improving lifetime for training-in-memory engines by structured gradient sparsification," in *Proceedings - Design Automation Conference (DAC)*, 2018.

[35] S. Das, J. R. Doppa, P. P. Pande and K. Chakrabarty, "Design-Space Exploration and Optimization of an Energy-Efficient and Reliable 3D Small-world Network-on-Chip," *IEEE Transactions on Computer-Aided Design of Integrated Circuits and Systems, (TCAD)*, vol. 36, 2017.

[36] S. Han, J. Pool, J. Tran and W. Dally, "Learning both weights and connections for efficient neural networks," *Advances in Neural Information Processing Systems (NeurIPS)*, pp. 1135-1143, 2015.

[37] L. Liang, L. Deng, Y. Zeng, X. Hu, Y. Ji, X. Ma, G. Li and Y. Xia, "Crossbar-Aware Neural Network Pruning," *IEEE Access*, 2018.

[38] G. Yuan et. al, "TinyADC: Peripheral Circuit-aware Weight Pruning Framework for Mixed-signal DNN Accelerators," in *DATE*, 2021.

[39] J. Frankle and M. Carbin, "The lottery ticket hypothesis: Finding sparse, trainable neural networks," in *International Conference on Learning Representations (ICLR)*, 2019.

[40] T. Chen et al., "A Unified Lottery Ticket Hypothesis for Graph Neural Networks," in *International Conference on Machine Learning (ICML)*, 2021.

[41] C. Ogbogu, A. I. Arka, B. K. Joardar, J. R. Doppa, H. Li, K. Chakrabarty and P. P. Pande, "Accelerating Large-Scale Graph Neural Network Training on Crossbar Diet," *IEEE Transactions on Computer-Aided Design of Integrated Circuits and Systems*, 2022.

[42] W. Chiang et al., "Cluster-GCN: An efficient algorithm for training deep and large graph convolutional networks," in *Proceedings of the ACM SIGKDD International Conference on Knowledge Discovery and Data Mining*, 2019.

[43] M. Paul, S. Ganguli and G. K. Dziugaite, "Deep Learning on a Data Diet: Finding Important Examples Early in Training," in *Advances in Neural Information Processing Systems 34 (NeurIPS 2021)*, 2021.

[44] H. You, Z. Lu, Z. Zhou, Y. Fu and Y. Lin, "Early-Bird GCNs: Graph-Network Co-Optimization Towards More Efficient GCN Training and Inference via Drawing Early-Bird Lottery Tickets," in *AAAI Conference on Artificial Intelligence*, 2022.

[45] W. Hu, M. Fey, M. Zitnik, Y. Dong, H. Ren, B. Liu, M. Catasta and J. Leskovec, "Open Graph Benchmark: Datasets for Machine Learning on Graphs," in *34th Conference on Neural Information Processing Systems (NeurIPS)*, Vancouver, Canada., 2020.

[46] K. Roy, I. Chakraborty, M. Ali, A. Ankit and A. Agrawal, "In-Memory Computing in Emerging Memory Technologies for Machine Learning: An Overview," in *IEEE Design Automation Conference (DAC)*, 2020.

[47] M. Kang, H. Shin and L.-S. Kim, "A Framework for Accelerating Transformer-Based Language Model on ReRAM-Based Architecture," *IEEE Transactions on Computer-Aided Design of Integrated Circuits and Systems*, vol. 42, no. 9, 2022.

[48] G. Plastiras, M. Terzi, C. Kyrkou and T. Theocharides, "Edge Intelligence: Challenges and Opportunities of Near-Sensor Machine Learning Applications," in *IEEE 29th International Conference on Application-specific Systems, Architectures and Processors (ASAP)*, 2018.

A Comparative Study on Front-Side, Buried and Back-Side Power Rail Topologies in 3nm Technology Node

Sandra Maria Shaji, Lingjun Zhu, Junsik Yoon, Sung Kyu Lim
School of ECE, Georgia Institute of Technology, Atlanta, GA
sshaji3@gatech.edu, limsk@ece.gatech.edu

Abstract—**The standard cells are becoming increasingly smaller due to aggressive device down-scaling, and power rails take up a sizable portion of the available space. Buried Power Rail (BPR) and Back-Side Power (BSP) have been gaining more attention owing to their capacity to reduce the standard cell height from 6-Track in the traditional Front Side Power Rail (FS-PR) to 5-Track and 4-Track, respectively. In this paper, we provide a comprehensive comparison of power rail topologies at the device, standard cell, and full chip design level in terms of Power, Performance and Area (PPA). Our experiments show that nanosheet width scaling for BPR and BSP reduces device gate capacitance by 26% and 40%, respectively, resulting in an improvement of internal power of over 33% and 40%, respectively, at the standard cell level, and total power drop of over 24% and 30%, respectively, at the full chip level. Additionally, the floorplan c an b e s hrunk d own b y 7 % with BPR compared to FSPR, and even further by an additional 17% with BSP. This study also demonstrates the Back-side Power delivery network (BS-PDN) benefits i n I R d rop f or B PR and BSP topologies.**

I. Introduction

Aggressive scaling over the last decade, allowed the three-dimensional fin-structured F ETs (FinFET) t o r eplace t he planar MOSFET structures in and beyond 14 nm technology node [1] up until now at sub-3nm where vertically stacked Nanosheet FETs (NSFET) are gaining attention due to its capacity to mitigate Short Channel effects (SCEs) [2]. However, due to lithography and process limitations, the scaling of device dimension has slowed down and new device architectures like Forksheet devices and Complementary-FET (CFET) are being explored. At the standard cell level, scaling techniques like Contact Over Active Gate (COAG) and Single Diffusion Break (SDB) have been successful in minimizing cell footprints. However, when device size decreases with advanced technology nodes, the standard cell footprint also reduces accordingly with power rails occupying a significant portion of it. Therefore, standard cell layouts that reposition the power rail to create more room will be necessary as cells continue to scale down.

A conventional standard cell with Front-Side Power Rail (FS-PR) in 3 nm technology node has a cell height of 6-Track, with the power rails taking up two of the tracks [3]. With the Buried Power Rail (BPR) technology, only one track need to be reserved to draw connections from the power rail and the cell height can be reduced to 5-Track [3]. The novel technology of Back-Side Contact (BSC) introduced in [4] removes even this single-track requirement for power rails, reducing the cell height to 4-Tracks. Power rails in this technology are moved to backside metal layer underneath the active region and connected through BSC.

There are numerous works studying the scalability and performance of BPR technologies with respect to traditional FS-PR at the standard cell level [5] [3]. However, there isn't a complete study on how the down-scaled BPR cells compare against the FS-PR cells in terms of power, performance, and area in full chip designs. While the study presented in [4] introduces the potential of the BSC technology to further downscale the cells, there is no work presented on its implementation and impact on standard cells and full chip design performance and power.

In this paper, we present a comprehensive study comparing different power rail topologies highlighting their impact on PPA at three levels- device, standard cell, and full chip design. In addition to implementing FS-PR and BPR topologies, we for the first time, present a thorough study on the use of BSC to develop a third power rail topology, Back-Side Power-rail (BSP). As part of this study, we develop Process Design Kits (PDK) and cell libraries for each of the three power rail architectures. We then use them in tandem with appropriate power delivery network design in full chip design simulation to evaluate the PPA and supply voltage drop impact between the different power rail technologies.

II. Standard Cell Design

In this section, we design the layout of a single drive strength inverter cell, INVx1, for the three power rail configurations and discuss the design constraints involved. Fig 1 presents the FS-PR, BPR and BSP layouts for INVx1. The layers used in the layouts are derived from ASAP7 [6] cell library and are scaled to 3 nm technology node. The section view of the three power rail topologies are presented in Fig 2.

In the FS-PR configuration, the power rails are in the M1 metal layer and are 3 × Critical Dimension (CD) wide. The cell height for this configuration is 6-Track, i.e., 144 nm. The nanosheet width for the device is fixed at 32 nm.

979-8-3503-1176-1/23 $31.00 © 2023 IEEE

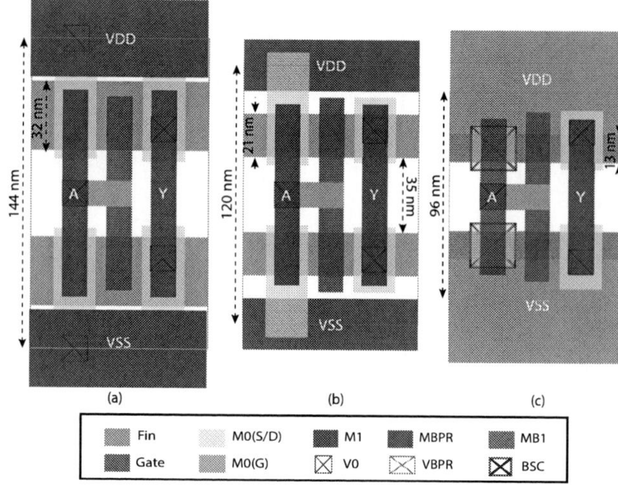

Fig. 1. INVx1 cell layout comparison (a) FS-PR, (b) BPR and (c) BSP power rail configurations.

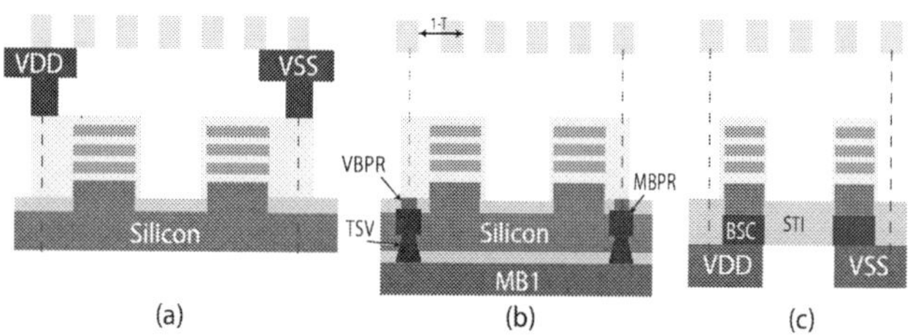

Fig. 2. Standard cell section view comparison FS-PR, BPR and BSP power rail configurations.

In the BPR configuration, the power rails are moved to the buried metal layer, MBPR, and are connected to the Source/Drain (S/D) epi through 12 nm × 18 nm VBPR vias. We assume the following design rules: 25 nm wide MBPR metal, 9.5 nm MBPR-Nanosheet space, and Nanosheet tip-to-tip gap of 35 nm [4]. NS width that complies with these design requirements and fits within the 5-Track cell height is calculated to be 21 nm.

In BSP topology, the power rails are placed in the Backside Metal layer, MB1, and connect to the S/D epi directly from the bottom through BSC. Scaling nanosheet width with cell height could lead to insufficient drive current causing performance to degrade at the chip level, so we scale the nanosheet such that the design rule 35 nm nanosheet tip-tip spacing is met. As only four tracks can be used for interconnects inside a cell, the gate cut for four-track cells must be scaled down to one CD to prevent open circuits on the outermost tracks.

III. PDK Development for Three PR Topologies

To compare the three power rail configurations, we develop PDKs with NSFET devices of nanosheet widths 32 nm, 21 nm and 13 nm for FS-PR, BPR and BSP configurations respectively. We follow the three staged flow in Fig 3 for this purpose.

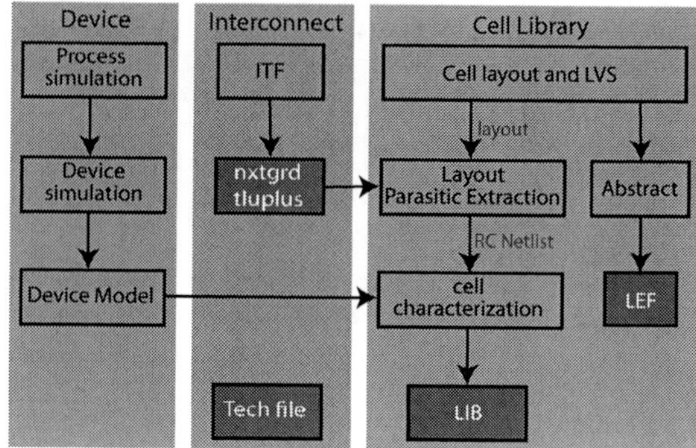

Fig. 3. DTCO flow implemented in this work to develop PDK and cell library for Place and Route in Synopsys EDA tools [5]

A. Device Structure Design

To develop the 3 nm PDK, we start with the NSFET device structure design. We simulate three NSFET devices of nanosheet widths - 32 nm, 21 nm and 13 nm as derived in Section II. The device contact poly pitch is scaled to 42nm from 48nm in the TSMC 5nm finFET device [7]. The gate length and spacer length are scaled to 12 nm and 5 nm, respectively as shown in Table I. The nanosheets in the device design are 10 nm apart and 5 nm thick, respectively. Due to an increase in the access resistance in lower nanosheet layers with number of nanosheets [8], we limit the device design to 3 nanosheets. The S/D epi is made rectangular to avoid epi merging in standard cells. We adopt punch-through stopper (PTS) doping to avoid current leakage through the substrate.

TABLE I
Comparison of Device Physical Parameters of FinFET [7] and NSFET Device.

Physical parameters (nm)	TSMC 5nm	NSFET 3nm
CPP	48	42
Fin/NS pitch	25	NA
Gate length	15	12
spacer length	6	5
# Fin/NS	2	3
Fin/NS thickness	6	5
Fin/NS width	NA	32, 21, 13
Fin/NS stack height	55	55

B. TCAD Simulation and Device Modelling

With Synopsys sdevice platform we perform the steps in the process simulation flow as shown in Fig 4. The nanosheet stack fin is formed by alternating Si and SiGe epitaxy and patterned like a wide fin. This is followed by Shallow Trench Isolation (STI), polygate deposition and patterning, and inner spacer deposition. For S/D epitaxy, we choose Si for n-type FET and $Si_{0.5}Ge_{0.5}$ for p type FET. The S/D epi and PTS doping concentrations for both n-type and p-type NSFETs are 4×10^{20} and 4×10^{18}. The anneal temperature is 1000 °C and anneal time for n-type and p-type MOSFETS are 0.4 s and 0.6 s respectively. The SiGe sacrificial layers between nanosheets are removed after the S/D epitaxy. The polysilicon dummy gate is replaced with High K Metal Gate (HKMG)

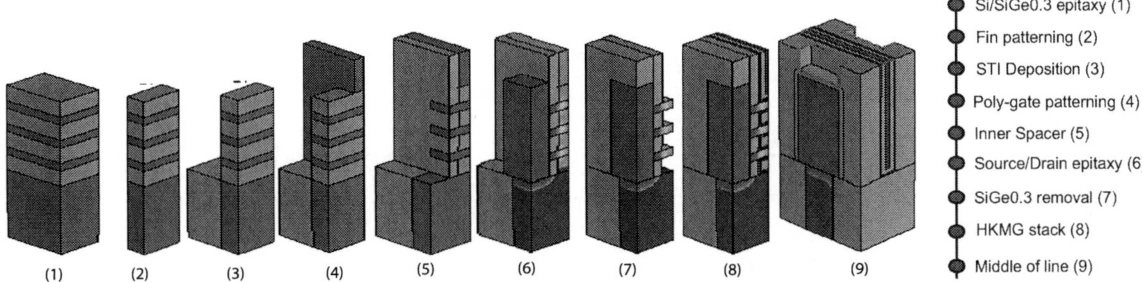

Fig. 4. Steps involved in TCAD process simulation of NSFET device.

stack. Gate, source, and drain contacts are established and the structure is replicated to form the other half.

We perform device simulation to obtain the output and transfer characteristics of the device. The device simulations follow the equations and models used in [5]. For reliable estimation of device performance, the NSFET device is fully calibrated to the IMEC N3 prediction model [9] as shown in Fig 5. Performance deviations with structure and doping profile changes is verified with the sensitivity table in [10]. We simulate the NSFET device to obtain the output characteristics at multiple gate biases as well as the transfer characteristics for the device at drain biases of 0.05V and the operating voltage, 0.7 V. We also perform AC Simulation in TCAD to generate a gate capacitance (C_{gg}) versus gate voltage curve. The compact model card for the device is generated from curve fitting of the device simulation outputs using Berkeley Short channel IGFET model (BSIM) Common Metal Gate (CMG) model with HSPICE.

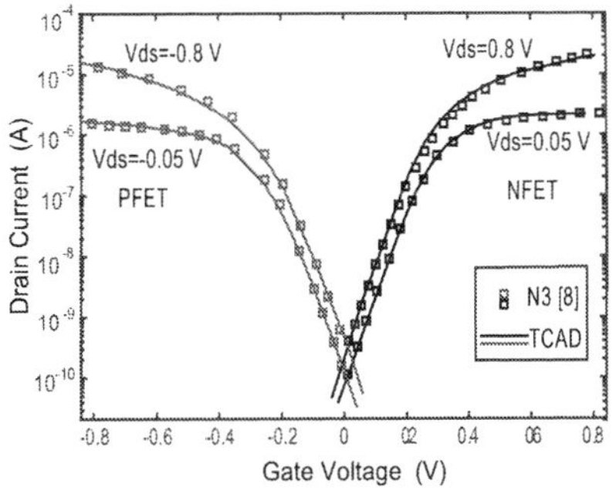

Fig. 5. TCAD calibration to IMEC N3 prediction model [9].

C. Technology and Interconnect Files

The BEOL stack is adopted from ASAP7 [6] and scaled down to 3 nm node as shown in the Table II. The width and resistance of the buried metal layer, MBPR, and the via connecting it to M0 metal, VBPR is derived from [5]. The MBPR power rails in BPR cell are connected to the backside metal layers through wide micro-TSVs of low resistance of 5 Ω. The BSC in BSP cells is assumed to be of Ru with WAC TiN and is of resistance of 74.9 Ω [4] [11]. The backside metals are made wide to provide low resistance path for power delivery.

The BEOL metal dimensions and rules are included in the technology file (.tf). The interconnect file with the parasitics of the metal and dielectric layers was compiled in Synopsys STAR RC to obtain the NXTGRD and TLUPlus files for parasitic extraction of standard cells and full chip designs, respectively.

TABLE II

METAL AND VIA DIMENSIONS AND RESISTANCE USED IN THIS WORK.

	Metal	W (nm)	Resistance (Ω
Back-side	MB2-MB1	61	34
BPR cell layer	MBPR	25	65
Front-Side	M0	20	523
	M1-M3	12	347
	M4,M5	18	101
	M6,M7	24	44
	VIA	**W × L nm^2**	**Resistance (Ω)**
Back-Side	VB1	40 × 40	4.0
BPR cell layer	VBPR	20 × 12	74.6
	μ-TSV	60 × 60	5
BSP cell layer	BSC	20 × 20	74.9
Front-Side	V0-V3	12 × 12	63.5
	V4,V5	18 × 18	19.8
	V6,V7	24 × 24	10.8

D. Standard cell characterization

The "cell library" section of the flow in Fig 3 defines the steps followed to derive the LIB file that captures the electrical behavior and LEF file capturing the pin shape and location. The layouts for 57 standard cells listed in Table III are drawn manually for each of the three power rail configurations. The electrical connectivity is verified in Layout Vs Schematic (LVS) Verification and the LEF is exported from the layout abstract. Parasitic extraction of the layouts with Synopsys STARRC generates the RC Netlist for each of the 57 standard cells and cell characterization on this RC netlist generates the LIB file for the three cell libraries.

IV. STANDARD CELL PPA

The three cell libraries generated have different cell heights and employ devices of different NS widths. The libraries, therefore, vary in cell power and performance. Table IV compares the device metrics in each cell library. Compared to NSFET in FS-PR, on current (I_{on}) reduces by 35% and 50% as effective width (W_{eff}) decreases from 222 nm in the FS-PR device to 156 nm and 108 nm in BPR and

TABLE III
STANDARD CELLS IN THE CELL LIBRARY

Std Cells	(input)x(#parallel transistors)
INV	x1, x2, x3, x4, x5, x6, x7, x8, x10, x12, x14, x16
BUF	x1, x2, x3, x4, x5, x6, x7, x8, x10
NAND, NOR	2x1, 3x1, 4x1, 2x2, 3x2
AND, OR	2x1, 3x1, 4x1
XOR, XNOR	2x1, 3x1, 2x2
AOI, OAI	21x1 , 22x2, 221x1, 222x1, 11x1, 31x1
MUX	2x1
DFF (flipflop	Hx1, HQNx1
DHL (latch)	x1

BSP devices, respectively. The low mobility of holes along the dominant transport at 100 surface orientation reduces I_{on}/W_{eff} in p-type NSFET compared to n-type NSFET. The gate capacitance reduces with nanosheet width owing to the smaller inversion charge and the lower outer fringing bringing down the inversion capacitance and the parasitic capacitance (C_{par}), respectively.

TABLE IV
DEVICE MERIT COMPARISON OF NSFETS IN THE THREE CELL LIBRARIES - 6 TRACK FS-PR, 5 TRACK BPR AND 4 TRACK BSP.

PR	6T FS-PR		5T BPR		4T BSP	
Metric	PFET	NFET	PFET	NFET	PFET	NFET
NS width	32 nm		21 nm		13 nm	
W_{eff}	222 nm		156 nm		108 nm	
I_{on} (mA)	0.14	0.2	0.09	0.13	0.07	0.10
C_{par} (fF)	0.11	0.13	0.08	0.09	0.07	0.08
C_{gg} (fF)	0.15	0.16	0.11	0.11	0.09	0.09

In the standard cell level, a larger device I_{on} indicates a larger cell drive current and hence faster cells. However, wider nanosheet's high gate capacitance leads to high pin capacitance, which can slightly slow down the cell. Table V shows the trend in cell delay degrading and pin capacitance reducing from FS-PR to BPR and BSP. Fig 6 depicts the cell rise delay longer than the cell fall delay due to smaller I_{on} in p-type NSFET than the n-type NSFET causing asymmetry in cell pull-up and pull-down circuits.

TABLE V
STANDARD CELL METRIC COMPARISON OF 6-TRACK, 5-TRACK AND 4-TRACK INVX1.

	6T INVx1	5T INVx1	4T INVx1
Cell height (nm)	144	120	96
Cell width (nm)	84	84	84
Cpin (fF)	0.325	0.249	0.217
Lkg Power (pW)	8649	275	269
Slow case: input slew= 10ps output load=1.44fF			
cell delay (ps)	5.103	8.166	9.434
transition delay (ps)	7.16	10.92	13.33
Int. Power (fJ)	0.067	0.045	0.04
Fast case: input slew= 40ps output load=5.76fF			
Cell delay (ps)	19.5721	31.03	35.9
transition delay (ps)	28.0645	42.2684	51.94
Int Power (fJ)	0.105	0.050	0.044

The cell leakage power improves as we move to BPR and BSP, since the NS width is smaller. While the high pin capacitance increases dynamic power, the decreased cell delay shortens the time when both pull-up and pull-down circuits are active, lowering short circuit current and power. The pin capacitance, however, dominates and we see internal power reducing from FS-PR to BPR to BSP cells.

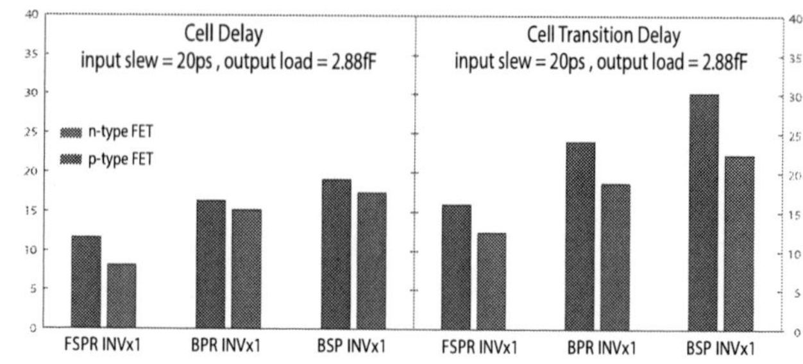

Fig. 6. cell rise and fall delay comparison of 6T FS-PR, 5T BPR and 4T BSP INVx1 cells.

V. FULL CHIP DESIGN

A. Full chip design and simulation setup

For the design-level PPA analysis, we choose two standard cell-only benchmark designs - Elliptical Curve Group (ECG) core and JPEG Encoder core and a CPU design, Arm® Cortex®-A7 64-bit dual-core processor. The physical design implementation steps followed in this work are illustrated in Fig 7. We perform logic synthesis on the three design RTL netlists using each of the three cell libraries in Synopsys Design Compiler. The floorplan and I/O pin for the standard cell-only designs are automatically derived by Synopsys IC Compiler II based on cell density settings optimal for design performance. The CPU design floorplan is derived from the design manual as shown in Fig 9(a). In the power planning step, the P/G mesh is designed and routed for the design. ICC2 performs placement, clock tree synthesis and route optimization on the design.

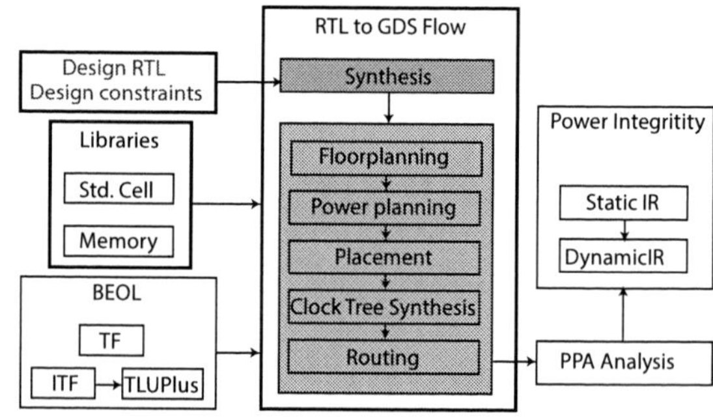

Fig. 7. Design logic synthesis and P&R flow followed in this work.

B. PDN design

The conventional Front-Side Power Delivery Network (FS-PDN) is used in tandem with FS-PR library. To enable a small power source pad pitch, we assume power is delivered to the chip from the off-chip circuit through a Redistribution Layer (RDL) [12] as shown in Fig 8.

For experiments using BPR and BSP PDKs, we employ the Back-Side Power Delivery Network (BS-PDN) scheme that leverages the backside metals under the silicon substrate to route the P/G mesh. In experiments with BPR technology, the P/G pins on the buried metal layer are connected to the backside metals with micro-scale TSVs. The silicon substrate is thinned down to achieve reasonable TSV pitch dimensions. With BSP cells we use BS-PDN architecture since the P/G pins in these cells are on the backside metal layers. The P/G grid for both FS-PDN and BS-PDN are generated automatically with ICC2 using a pattern-based power grid generation flow.

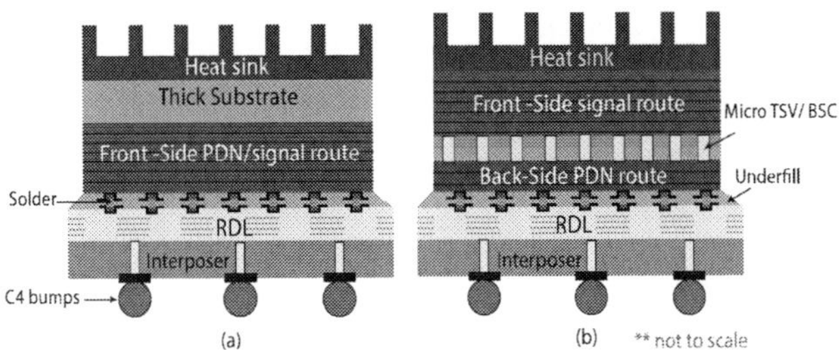

Fig. 8. Power Delivery through RDL in (a) FSPDN and (b) BSPDN

C. Memory macro modeling and implementation

The CPU design, Arm® Cortex®-A7, contains memory macros. Since our work is based on simulations and there are no published 3 nm memory compilers available for use, we build an estimate of the memory macro model by scaling the delay and power tables in memory LIB files and the footprint and area in memory LEF files from 16 nm TSMC technology node to 3 nm node using scaling factors assumed from studying the PPA trend across technology nodes. The memory library so generated has P/G pins on the front side M4 metal layer and is used for FS-PR configuration. For BPR and BSP configurations, the delay-power models are scaled to reflect the trend observed with standard cells in each power rail technology. The memory LEF files for these cells are also altered to reposition the power rail to buried metal and backside metal layers respectively [3]. The memory footprint is scaled down to reflect the cell height reduction in BPR and BSP cells respectively.

D. PPA Analysis

The PPA summary for the three flows - FS-PR 6-Track cell library with FSPDN, BPR 5-Track cell library with BSPDN and BSP 4-Track cell with BSPDN for each of the three chip designs - ECG, JPEG and Cortex A7 is tabulated in Table VI.

The die footprint of the designs reduces with cells. While the BPR cells result in 9% instance cell area saving, with the 4-Track BSP cells, it is over 22% when compared to the traditional FS-PR cells. With smaller footprints enabled by smaller cell heights of cells along with the use of backside metals for power delivery, the wirelength is observed to improve. The benefit in wire capacitance so achieved together with smaller pin capacitance in the shorter cells outlined in Section IV, reduces the design switching power. The improved cell internal power with BPR and BSP translates to an immense reduction of design internal power in all three designs. The total power for all three designs drops by 25% with BPR and over 30% with BSP cells. Hence, scaling enabled by BSC technology presents opportunities for low-power design applications.

The performance metrics in Table VI degrade upto 20% with BPR and further drops with BSP cells. This is mainly due to the longer cell delays and smaller driving currents in these cells as discussed in Section IV. The instance count also increases with shorter cells due to the addition of repeaters to compensate for the long cell delay and output slew of BPR and BSP cells.

The energy efficiency of the three flows is presented in terms of Power Delay Product (PDP). The PDP improves by 9.7% in ECG, 6.2% in JPEG and by 21% in Cortex A7 CPU, with 5-Track BPR technology compared to the flow using FS-PR cells. The PDP of design implemented with BSP library does not mprove further due to the degradation of RC with narrow nanosheets [13].

E. IR analysis

In the advanced 3nm technology node, the front-side metal layer widths are aggressively narrowed resulting in high resistance paths for FSPDN and compromising power integrity. Moving power delivery mesh to the low resistance backside metals with wider wire width reduces the voltage drop in the power grid. Hence, it is argued BSPDN improves the power integrity of the chip [3].

We perform Power integrity analysis on the fully routed design obtained from ICC2 with Ansys RedHawk. Table VI presents the worst static voltage drop for the three designs- ECG, JPEG and Arm® Cortex®-A7 in each of the three experiment flows - FS-PR with FSPDN, BPR with BSPDN and BSP with BSPDN. Due to high cell density in the standard cell-only designs, ECG and JPEG, we see the Voltage drop is well over 10% of the supply voltage 700 mV. When BSPDN is adopted the IR drop reduces to less than 70 mV threshold.

For the FS-PDN in Cortex-A7, we find the static IR-drop hotspot is located in the right side of the floorplan, as shown in Fig. 9 (b). This region has high standard cell density and it is challenging for front-side power delivery due to the high resistivity of the FS metals. For the BS-PDN designs with BPR and BSP, we find the hotspot region shrinks and the worst IR drop reduces by 89% and 87%, respectively, thanks to the less resistive path from the back side. This demonstrates the benefits of BS-PDN in power integrity.

VI. Conclusion

In this work, we simulate and implement NSFETs with three power rail topologies - the convention FS-PR, BPR, and BS-PR enabled by the novel BSC. This paper is the first demonstration of the BSC technology implementation from device to standard cells to full chip design, thoroughly comparing it with other rail topologies at each stage. We find that BPR and BSC enable further scaling of standard cells

TABLE VI
PPA COMPARISON BETWEEN THE THREE FLOWS - FS-PR CELL LIBRARY WITH FSPDN, BPR CELL LIBRARY WITH BSPDN AND BSP CELL WITH BSPDN FOR THE THREE DESIGNS - ECG, JPEG AND CORTEX A7.

PDN	ECG FS-PR FSPDN	ECG BPR BSPDN	ECG BSP BSPDN	JPEG FS-PR FSPDN	JPEG BPR BSPDN	JPEG BSP BSPDN	CORTEX A7 FS-PR FSPDN	CORTEX A7 BPR BSPDN	CORTEX A7 BSP BSPDN
Target Freq. (GHz)		10			8			1.6	
Cell area	3638	3456	2832	13924	12275	10133	181481	163922	138639
#Instance	92 K	113 K	113K	335 K	349 K	366 K	486 K	489 K	502 K
Total WL (μm)	264424	249034	232981	831999	846903	786862	3035500	2604374	2595484
WNS (ps)	4	33.5	51	46.6	87.1	119.6	46.9	63.44	97
Eff. Freq. (GHz)	9.6	7.49	6.62	5.8	4.71	4.08	1.48	1.45	1.38
Total wire cap (pF)	37.5	33.9	32.8	115.7	109.6	106.4	438	352.5	336.7
Total pin cap (pF)	82.7	71.9	65.8	327.8	260.9	239.3	458.6	358.2	323.7
Total cap (pF)	120.2	105.8	98.6	443.5	370.5	345.7	896.6	710.7	700.4
Total Power (W)	304	214	195	888	672	583	262	194	184
Sw Power (mW)	90.3	82.5	74.3	306	294	251	76	61.2	58.9
Int Power (mW)	212	131	121	574	378	331	175	132	123
Lkg Power (mW)	2.1	0.1	0.1	8.45	0.3	0.3	11.6	1.1	0.9
PDP (pJ)	31.6	28.6	29.4	151.8	142.4	142.8	176	138.	132.8
IR (mW)	124.2	33.5	40.6	137.2	32.4	53.9	23.8	2.6 3.4	

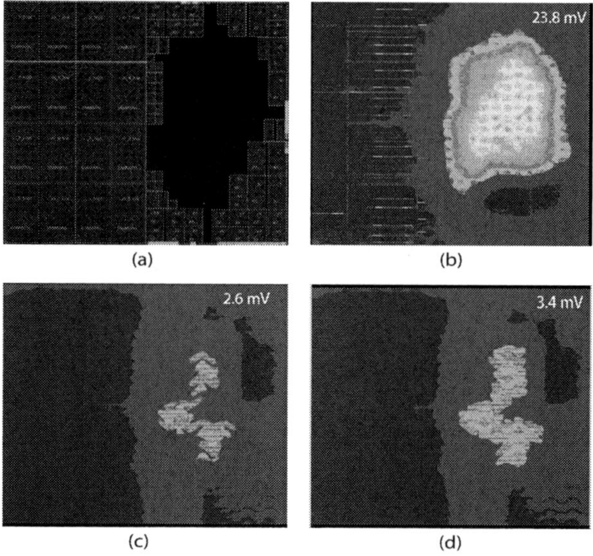

Fig. 9. (a) manually derived floorplan of Cortex A7, Static IR drop map of CortexA7 in flows (b) FS-PR with FSPDN (b) BPR with BSPDN and (c) BSP with BSPDN.

by repositioning the power rails resulting in an average of 7% and 24% floorplan area reduction at the full ship level. In addition to this, the smaller nanosheet width devices used in these technologies offer power-efficient standard cells. The area and power savings along with BSPDN reducing IR drop, we find BSP cells best suited for low-power applications.

ACKNOWLEDGEMENT

This research is partially funded by Samsung Semiconductor, Inc.

REFERENCES

[1] A. Razavieh, P. Zeitzoff, and E. J. Nowak, "Challenges and limitations of cmos scaling for finfet and beyond architectures," *IEEE Transactions on Nanotechnology*, vol. 18, pp. 999–1004, 2019.

[2] G. Bae, D.-I. Bae, M. Kang, S. Hwang *et al.*, "3nm gaa technology featuring multi-bridge-channel fet for low power and high performance applications," in *2018 IEEE International Electron Devices Meeting (IEDM)*, 2018, pp. 28.7.1–28.7.4.

[3] D. Prasad, S. S. Teja Nibhanupudi, S. Das, O. Zografos *et al.*, "Buried power rails and back-side power grids: Arm® cpu power delivery network design beyond 5nm," in *2019 IEEE International Electron Devices Meeting (IEDM)*, 2019, pp. 19.1.1–19.1.4.

[4] S. Song, G. Nallapati, I. Khan, N. Nikfar *et al.*, "System design technology co-optimization for 3d integration at ¡5nm nodes," in *2021 IEEE International Electron Devices Meeting (IEDM)*, 2021, pp. 22.3.1–22.3.4.

[5] J.-S. Yoon, J. Jeong, S. Lee *et al.*, "Performance, Power, and Area of Standard Cells in Sub 3 nm Node Using Buried Power Rail," *IEEE Transactions on Electron Devices*, vol. 69, no. 3, pp. 894–899, 2022.

[6] L. Clark, V. Vashishtha, L. Shifren *et al.*, "ASAP7: A 7-nm finFET predictive process design kit," *Microelectronics Journal*, vol. 53, pp. 105–115, 07 2016.

[7] G. Yeap, S. S. Lin, Y. M. Chen *et al.*, "5nm cmos production technology platform featuring full-fledged euv, and high mobility channel finfets with densest 0.021µm2 sram cells for mobile soc and high performance computing applications," in *2019 IEEE International Electron Devices Meeting (IEDM)*, 2019, pp. 36.7.1–36.7.4.

[8] F. M. Bufler, D. Jang, G. Hellings *et al.*, "Monte Carlo Comparison of n-Type and p-Type Nanosheets With FinFETs: Effect of the Number of Sheets," *IEEE Transactions on Electron Devices*, vol. 67, no. 11, pp. 4701–4704, 2020.

[9] G. Rzepa, M. Karner, O. Baumgartner *et al.*, "Reliability and Variability-Aware DTCO Flow: Demonstration of Projections to N3 FinFET and Nanosheet Technologies," in *2021 IEEE International Reliability Physics Symposium (IRPS)*, 2021, pp. 1–6.

[10] H.-H. Park, W. Choi, M. A. Pourghaderi *et al.*, "NEGF simulations of stacked silicon nanosheet FETs for performance optimization," in *2019 International Conference on Simulation of Semiconductor Processes and Devices (SISPAD)*, 2019, pp. 1–3.

[11] A. Gupta, O. V. Pedreira, G. Arutchelvan, H. Zahedmanesh *et al.*, "Buried power rail integration with finfets for ultimate cmos scaling," *IEEE Transactions on Electron Devices*, vol. 67, no. 12, pp. 5349–5354, 2020.

[12] H. Lu, R. Furuya *et al.*, "Design, modeling, fabrication and characterization of 2–5- μm redistribution layer traces by advanced semiadditive processes on low-cost panel-based glass interposers," *IEEE Transactions on Components, Packaging and Manufacturing Technology*, vol. 6, no. 6, pp. 959–967, 2016.

[13] J.-S. Yoon, J. Jeong *et al.*, "Optimization of nanosheet number and width of multi-stacked nanosheet fets for sub-7-nm node system on chip applications," *Japanese Journal of Applied Physics*, vol. 58, p. SBBA12, 04 2019.

CoolDRAM: An Energy-Efficient and Robust DRAM

Nezam Rohbani
School of Computer Science,
Institute for Research in
Fundamental Sciences (IPM)
Tehran, Iran

Mohammad Arman Soleimani
Department of Computer Engineering,
Sharif University of Technology
Tehran, Iran

Hamid Sarbazi-Azad
School of Computer Science, Institute for
Research in Fundamental Sciences (IPM)
and Department of Computer Engineering,
Sharif University of Technology

***Abstract*—DRAM is the most mature and widely-utilized memory structure as main memory in computing systems. However, energy dissipation and latency of DRAM are two of the most serious limiting factors of this technology. All DRAM main operations are initiated by a Precharge phase, which is time-consuming and power-hungry. This work proposes a novel DRAM cell access scheme that entirely eliminates Precharge phase from DRAM read, write, and refresh operations, with a very slight modification i n c ommodity D RAM s tructure. The proposed DRAM design, called CoolDRAM, operates using a single extra cell row as reference cells. CoolDRAM reduces energy dissipation by about 34% on average, with a negligible area overhead of about 0.4%. The robustness of CoolDRAM against process variation and environmental noises is 61× and 1.78× of the state-of-the-art, respectively, while maintaining the same power consumption and latency.**

***Index Terms*—DRAM, Power Consumption, Precharge, Sense Amplifier, D ata Similarity**

I. Introduction

In a memory hierarchy, DRAM consumes the major portion of energy budget allocated to the memory subsystem [1], [9]. Tiny capacitors, arranged in two-dimensional memory arrays (MATs) are used as storage elements in this memory, which lose their charge over time and during accessing the cells. This characteristic of DRAM has made its cell access complex and energy-hungry [9].

Read, write, and refresh are three major DRAM operations which all of them, more or less, follow the same procedure [6], [17], [22]. Precharge and Activation are the two main micro-operations common to all DRAM operations [1], [9], [16]. These operations need charge and discharge cycles on a large number of bitlines with high parasitic capacitances, which makes them energy-hungry operations.

To reduce power/energy dissipation and latency of DRAM memories, many techniques are presented in the previous works. By utilizing multi-banking feature of DRAMs, some techniques put some cold banks in drowsy state and arrange data to maximize the number of cold banks [4], [6]. Some other techniques propose partitioning memory arrays to activate a smaller number of cells during memory access [7], [12]. DRAM refresh energy reduction is the goal for some other techniques, like selective refreshing and adaptive refresh rate by considering temperature, operating voltage, and application criticality [3], [5], [10], [11], [14]. Some previous works propose optimizations in sensing structures of DRAMs, like overlapping Precharge and Activation by Row-Buffer Decoupling (RBD) [19]. Data encoding to reduce DRAM bus activity [13], [15] and minimizing the number of Activations (row-hit boosting) for data access [20], [21], are the other techniques to reduce DRAM energy consumption. All of the above-mentioned power/energy consumption and delay reduction techniques need Precharge of bitlines to half-VDD (*VDD/2*) before Activating a row of cells. The only technique that eliminates the Precharge phase from DRAM cell access procedure is PF-DRAM [16]. However, this technique imposes significant memory structure modifications and area overhead.

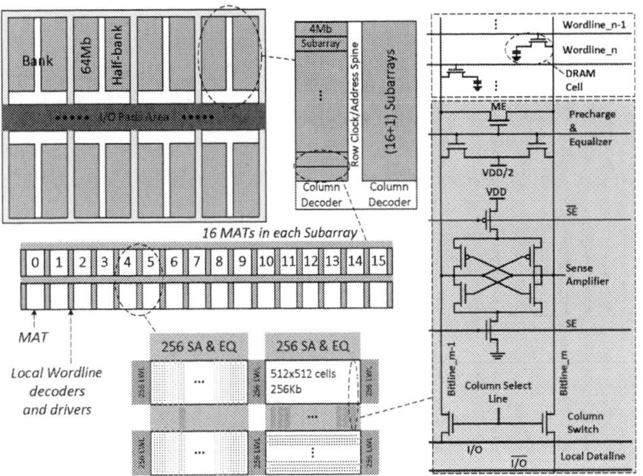

Fig. 1: DRAM chip general structure

This work proposes a DRAM memory design, called CoolDRAM, which not only entirely eliminates the need for the Precharge operation in DRAM during cell access, but also imposes a very slight modification to the commodity DRAM structure. By eliminating the Precharge phase, DRAM energy consumption is reduced by 34% on average. The area overhead of CoolDRAM is less than 0.4% compared with commodity DRAMs. Furthermore, CoolDRAM is 61× and 1.78× more robust against process variation and environmental noises, respectively, compared with the state-of-the-art.

II. Proposed DRAM Design

A. Preliminaries and Motivation

Fig. 1 depicts a DRAM chip structure. A DRAM chip is organized in multiple banks that each bank can serve an operation in parallel with the others. Each bank is composed of two half-banks and each one contains subarrays. Each subarray is composed of multiple (generally 16) MATs, that share adjacent peripherals with each other. In a MAT the cells are arranged in a two-dimensional matrix, composed of rows (wordlines) and columns (bitlines). To access a cell in a MAT, all of the bitlines in the MAT are charged to half-VDD, using precharge and equalizer transistors, in the Precharge phase and then are floated. By Activating a wordline, each cell in the accessed row is connected to the corresponding bitline. Based on the stored charge in the memory cell, *0* or *VDD*, a tiny voltage perturbation (V_S), about ±90 mV, is generated on the corresponding bitline by charge sharing between the cell capacitor and the parasitic capacitor of the floated bitline. This voltage perturbation is detected by differential sense amplifiers connected to the bitlines. To cancel ambient and internal noises on long bitlines for a reliable cell access, differential sensing is mandatory. The sense amplifier compares the voltage of a bitline connected to the target cell with another float bitline in a pair. V_S can be calculated by Eq.(1). C_{BL} and C_{cell} are the parasitic capacitance of a bitline and memory cell, respectively.

$$V_S = \frac{V_{DD}}{2} \cdot \frac{C_{cell}}{C_{BL} + C_{cell}} \tag{1}$$

The most popular sense amplifier in DRAMs is single-ended differential sense amplifier [9]. This sense amplifier is composed of two cross-coupled inverters in a positive feedback to compare bitlines voltage in each pair and detect the accessed cell value. In the Precharge phase, the sense amplifier is disconnected from VDD and GND power rails using *SE* and $\overline{SE}$ (sense amplifier enable) signals (Fig. 1). In this way, the internal nodes of the sense amplifier are precharged to half-VDD together with the bitlines in each pair.

After developing V_S on bitlines in the Activation phase, and consequently, on the sense amplifier internal nodes, the sense amplifier is activated. Since this initial condition on the sense amplifier is unstable, the sense amplifier node with lower voltage drops to *0*, and the other node rises to *VDD*, promptly.

During cell access (read/write/refresh), by activating one row, a large number of cells (e.g. 64 K cells in DDRx) are connected to the bitlines and their value are determined in the bitline pairs sense amplifiers. Because of the large parasitic capacitance of bitlines, a major portion of DRAM power dissipation is consumed for voltage toggling on the bitlines during Precharge and Activation phases [2], [9], [16].

The dissipated energy on a bitline pair during the Precharge phase can be calculated by Eq.(2). In which, Q_{pre} is the charge stored in the bitline during Precharge operation and β is a constant, determined by the width ratio of precharge transistors to equalizer transistor. By connecting bitlines in each pair to each other, through equalizer transistor (*ME* in Fig. 1), sharing the opposite charges on the bitlines aids the precharge transistors (and half-VDD power source) to precharge the bitlines to *VDD/2*, faster and more efficiently [9]. In commodity DRAM, β is about 0.54 [16].

$$E_{pre} = \beta . Q_{pre} . VDD = \beta . C_{BL} . \frac{VDD}{2} . VDD = \beta . C_{BL} . \frac{VDD^2}{2} \tag{2}$$

During the Activation phase, the sense amplifier pulls the voltage of one bitline in each pair to *VDD* and pulls that of the other one to *0*, based on the voltage perturbation on the accessed bitline. The dissipated energy in this phase for commodity DRAM (E_{act}) can be calculated by Eq.(3). In which Q_{act} is the charge drawn from the power source during the Activation phase, that is equal to the electric charge needed to charge C_{BL} plus cell capacitor (C_{cell}) to *VDD*.

$$E_{act_C} = Q_{act} . VDD$$

$$= \begin{cases} (C_{BL} + C_{cell}) . (\frac{VDD}{2} - V_{S_C}) . VDD & , V_{S_C} > 0 \\ = (C_{BL} + C_{cell}) . (\frac{VDD}{2} - \frac{VDD}{2} . \frac{C_{cell}}{C_{BL} + C_{cell}}) . VDD & \\ & \\ C_{BL} . \frac{VDD}{2} . VDD & , V_{S_C} < 0 \end{cases}$$

$$= C_{BL} . \frac{VDD^2}{2} \tag{3}$$

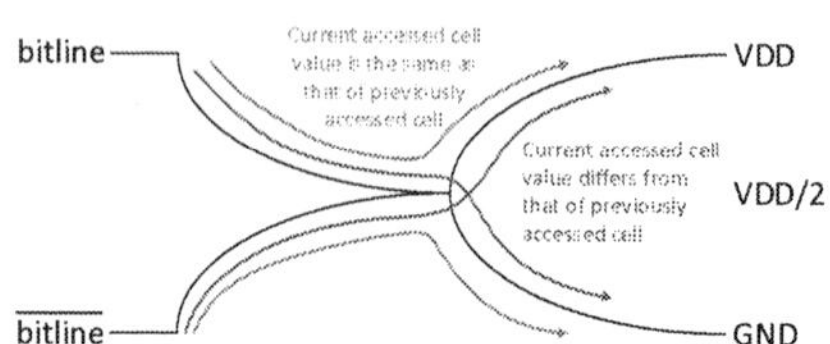

Fig. 2: Bitlines voltage activity in a bitline pair during consecutive Activations

The total dissipated energy for a precharge-activation cycle on a bitline pair, in the case of a row conflict, can be calculated using Eq.(4). This value is constant regardless of the accessed cell value, '0' or '1'.

$$E_{acc_C} = E_{pre} + E_{act_C} = \frac{1}{2}(1 + \beta) . C_{BL} . VDD^2 \tag{4}$$

During the execution of real-world workloads, a very high data similarity (about 90%) exists between the stored data in DRAM main memory [16], [18]. This is because of large number of '0's stored in DRAM memory, narrow-width values, locality of references, data dependency, high-frequently used instructions, and low hamming distance of op-codes. Regarding the very high similarity of stored data in DRAM, it is expected that a large portion of bitlines stabilizes to the same previous value of former Activation. Thus, a useless discharge and charge on each bitline pair (equal to E_{acc_C}) is dissipated with no change on a large number of bitlines states. In other words, during reading the same previous value on a bitline pair, one bitline is discharged to half-VDD (during Precharge phase) and again is charged to the same previous value (in Activation phase), and the other bitline is charged to the half-VDD (during Precharge phase) and is discharged to the *0* (during Activation phase) again. This is while, if the bitlines were not precharged to half-VDD, no voltage swing occurred on bitlines during reading the same value on a bitline pair (reading '0' after '0' or reading '1' after '1'), see blue arrows in Fig. 2.

B. CoolDRAM

Precharging bitlines to half-VDD is a power-hungry and time-consuming operation in DRAM cell access, however, differential sense amplifiers need equal voltage on both bitlines in a pair as an *initial point*. The technique presented in PF-DRAM [16], charges or discharges both bitlines in each pair equally to *VDD* or to *0*, during the Activation phase (does not drive bitlines in a pair to opposite values like commodity DRAMs). PF-DRAM utilizes *VDD* or *0* as an initial point on bitlines in each pair. In this condition, if the initial value on bitlines is equal to the accessed cell value, no voltage swing occurs on that bitline pair and thus, overall power dissipation reduces considerably [16]. However, the main drawback of PF-DRAM is that it utilizes an unbalanced sense amplifier to detect equal voltage on bitlines, which makes it vulnerable to transistors' threshold voltage drift. Furthermore, the DRAM vendors are very reluctant to apply any modification to the DRAM structure, especially the sense amplifier which needs to be very accurately engineered to fit inside the narrow space between DRAM bitlines [9].

In this work we present CoolDRAM which not only entirely eliminates the Precharge phase from the DRAM cell access

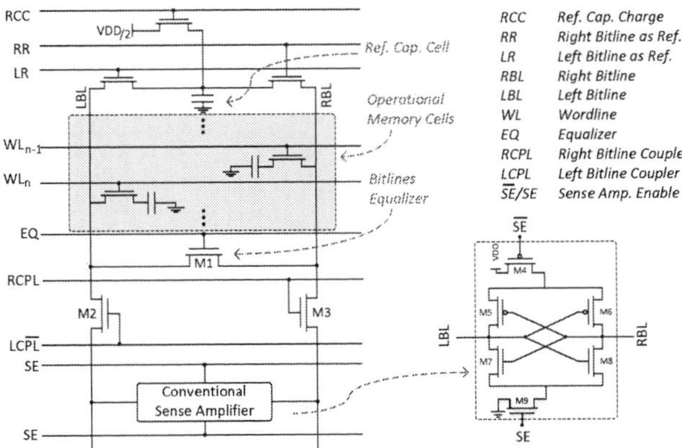

Fig. 3: Proposed CoolDRAM structure

sequence, but also uses exactly the same sense amplifier structure of commodity DRAMs.

Fig. 3 shows the general design of CoolDRAM. CoolDRAM imposes the minimum modifications to the commodity DRAM structure. Its sense amplifier is exactly the same as the sense amplifier utilized in commodity DRAMs and the memory cell array remains untouched. The precharge circuit in CoolDRAM is slightly modified for updating bitlines voltage compared with commodity DRAM. *M1* connects two bitlines in each pair to each other to balance their voltage before Activation phase, just like commodity DRAM memories (*ME* in Fig. 1). *M2* and *M3* are utilized to couple and decouple bitlines from the sense amplifier.

The only added part of CoolDRAM to commodity DRAM is a dual-contact reference cell row to the memory array (shown at top of Fig. 3). The reference cell can be connected to Right Bitline (RBL) or Left Bitline (LBL) in each pair, by activating Right Bitline as Reference (RR) or Left Bitline as Reference (LR) signals, respectively. This cell develops a voltage perturbation on the reference bitline for proper operation of the sense amplifier. Before connecting this cell to a bitline, it is charged to half-VDD by activating Reference Capacitor Charge (RCC) signal, while RR and LR are deactivated. This signal is activated by the activating power source (VPP) [1] of DRAM, 2.5 V in DDR4, to avoid voltage drop on the n-type control transistor. The charge of the reference cell can be performed in parallel with restoring data in accessed cell, thus, no timing overhead is added to the cell access sequence.

In CoolDRAM, during the Activation phase, both bitlines are charged/discharged to *VDD*/*0* and not opposite values. The equal voltage on bitlines during Activation phase eliminates the need for precharging bitlines to half-VDD as the initial point and next cell access can be started right after the previous cell access.

In the first step of cell access, the sense amplifier internal nodes are floated by gating the sense amplifier from the power rails, VDD and GND, by deactivating *M4* and *M9* using $\overline{SE}$ and SE, the same as commodity DRAMs (see Fig. 3. In this phase, *M2* and *M3* are activated using Left Bitline Coupler (LCPL) and Right Bitline Coupler (RCPL) signals. *M1* connects LBL and RBL till right before activating the target wordline, to keep the voltage of bitlines in each pair as close as possible.

By activating the target row, the corresponding memory cell is connected to one of the bitlines in a pair. Connecting the cell to LBL or RBL depends on activating even or odd wordlines. Together with activating the target wordline, one of the RR or LR signals is activated to connect the reference cell to the opposite bitline in that pair.

During cell access in this design, two scenarios are possible: 1- the charges of target cell and floated bitlines are the same and 2- the charges of target cell and floated bitlines differs.

In the first scenario, the voltage of connected bitline to the target cell remains unchanged (*VDD* or *0*). This is while the voltage perturbation amplitude on the bitline connected to the reference cell is equal to the commodity DRAMs ($V_S = \frac{V_{DD}}{2} \cdot \frac{C_{ref\,Cell}}{C_{BL}+C_{ref\,Cell}}$). If the bitlines initial voltage is *VDD*, the voltage of bitline connected to the reference cell drops to $VDD{-}V_S$ and if its initial voltage is *0*, its voltage rises to V_S.

In the second scenario, a voltage perturbation of equal to $2V_S$ is developed on the bitline connected to the target cell ($VDD \cdot \frac{C_{ref\,Cell}}{C_{BL}+C_{ref\,Cell}}$). Thus, the voltage of the bitline corresponding to the target cell will be $VDD-2V_S$ or $2V_S$ at the end of charge sharing phase, while the voltage change on the reference bitline will be $VDD-V_S$ or V_S. In both scenarios and with all of possible bitlines and target cells initial charges, the voltage difference between bitlines is equal to V_S.

After charge sharing, bitlines are decoupled from the sense amplifier by deactivating *M2* and *M3*, and the sense amplifier is activated, by turning *M4* and *M9* on. Based on the initial condition on the sense amplifier internal nodes, the node with higher voltage rises to *VDD* and the other node drops to *0*; the same operation as the Stabilization phase in commodity DRAMs. Decoupling transistors are exploited in the previous works as well, to decrease tRCD and power dissipation during sense amplifier stabilization [19]. By decoupling bitlines with high parasitic capacitance during stabilization of the sense amplifier, the sense amplifier stabilizes faster and more power-efficient [19].

After stabilization of the sense amplifier, *M1* together with one of *M2* or *M3* are activated to update bitlines and restore the accessed cell. If the memory cell connected to LBL is accessed, *M2* and *M1* are activated, and if the cell connected to RBL is accessed, *M3* and *M1* are activated to update bitlines and cell restoration.

All of the control signals are triggered by VPP to prevent voltage drop on the n-type transistors, like the commodity DRAMs. Since a large number of cell accesses lead to reading the same value on bitlines [8], [16], thus, the bitlines voltage swing in a large portion of Activations is limited to charge and discharge of the reference bitline in each pair by V_S. A bitline pair voltage flips only when the stored charge in the target cell differs from the previously accessed cell through that bitline pair.

The column decoder selects the target column(s) right after the stabilization of the sense amplifier to transfer the read value to the global sense amplifier to be sent to the I/O drivers. Bitlines are updated in parallel with the required time for accessing the target column (tCL).

During the restoration of the accessed cell, the reference cell is disconnected from the bitline and is charged/discharged to half-VDD by activating the RCC signal, to prepare the memory array for the next Activation. Thus, right after the previous Activation, the next row Activation can be issued in CoolDRAM with no need to perform a Precharge operation on bitlines. Both open-row and closed-row policies are applicable in CoolDRAM.

The main contributions of CoolDRAM are as follows:

- Very simple modification is applied to DRAM arrays to entirely eliminate Precharge phase from DRAM cell access procedure. In the only previous work [16] that eliminates Precharge phase, the sense amplifier is modified to detect different voltages on bitlines.
- Imbalanced sense amplifier in the previous work [16] is very complex to be tuned in nano-scale technology, with high process variation rate. This is while CoolDRAM uses the same straightforward balanced sense amplifier structure of commodity DRAMs which is more straightforward to fabricate.

III. Experimental Setup

Circuit-level simulations are conducted using Synopsys HSPICE using 14 nm Multi-Gate technology model. DRAM memory MAT size is considered 512×512 cells [9]. Operational DRAM cells and reference cell access transistors' dimensions are considered the smallest in the technology. Coupling transistors and equalizer transistor width-to-length ratio is set at 2-to-1 and the sense amplifier transistors dimensions are considered 10-to-1.

The voltage of activating power source is 2.5 V (the same as commodity DRAMs). The main VDD, used for sense amplifiers, peripheral units, and charging DRAM cells is considered 1.2 V. Half-VDD power source (0.6 V) is generated by the internal voltage regulator of DRAM chips.

IV. Evaluation

A. CoolDRAM Functionality

Bitlines' parasitic capacitance and resistance are accurately modeled by the Lumped-element model by segmenting the bitlines into 3 elements. Regarding the distributed parasitic resistance and capacitance on bitlines, the physical location of accessed cell may generate different V_S on the sense amplifier nodes. The highest V_S is generated by the DRAM cells which are physically located closer to the sense amplifier and the weakest V_S is generated by the furthest cell from the sense amplifier. In our evaluations we considered the worst-case scenario of cell access which is the access to the furthest memory cell from the sense amplifier in the memory array. The reference cell location is fixed, thus the developed signal by connecting the reference cell to the bitline is predictable and experiences fewer fluctuations.

Fig. 4 shows all possible signal activities on the reference cell, accessed cell, bitlines, and sense amplifier during cell access through the LBL (as an example). The reference cell is charged to 0.6 V before Activation. By connecting the reference cell to RBL, by sharing its charge with the RBL parasitic capacitance, its voltage drops or rises to about 1.11 V or 0.09 V, respectively. This voltage perturbation is almost half of the voltage perturbation generated by the target cell charge. See the voltage perturbation on LBL and RBL during the time period between 1 ns and 3 ns.

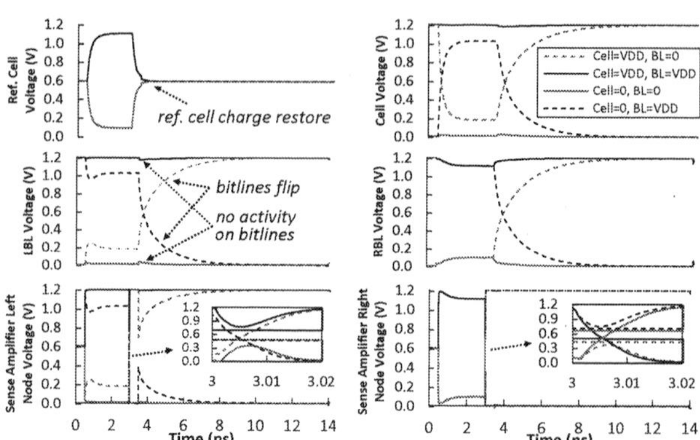

Fig. 4: Voltage waveform of nodes in a bitline pair, showing the functionality of CoolDRAM

Two different conditions can be observed on the LBL after connecting the target cell to this bitline. If the stored charge on the LBL is the same as the stored charge in the cell, black and red solid lines in Fig. 4, no voltage perturbation occurs on LBL. In the other case, if the stored charge in the cell differs from the charge on the LBL, the voltage perturbation is about 180 mV. In both cases, the amplitude of voltage difference between LBL and RBL is about 90 mV.

Sense amplifier internal nodes follow LBL and RBL voltages during the time period 0.5 ns to 3 ns. By decoupling the sense amplifier from the bitlines at 3 ns and activating the sense amplifier, the sense amplifier's node with higher voltage rises to *VDD* and the other node drops to *0*. This is performed in about 20 ps because of the low parasitic capacitance of sense amplifier internal nodes. After stabilization of the sense amplifier, by activating *M2* (the target cell is connected to LBL in this example), together with *M1* (bitlines equalizing transistor), the LBL, RBL, and accessed cell voltages are restored or updated. It is worth mentioning that LBL and RBL voltages flip only if the recently accessed cell value differs from the previously accessed cell value. Otherwise, LBL and connected cell voltages remain untouched and voltage perturbation on the RBL is recovered to *VDD* or *0*. Together with updating bitlines and accessed cell voltage, the reference cell voltage is charged to 0.6 V by triggering the RCC signal.

TABLE I: Simulation parameters

Technology	14 nm Multi-Gate Predictive Technology Model
Operating Voltage	Chip power supply (VDD) = 1.2 V Activating power supply (VPP) = 2.5 V
MAT size	512×512 cell
C_{BL}, C_{cell}	144 fF, 24 fF
Transistors W/L	SA 10/1, M1∼M3 2/1, access transistors 1/1
Process Variation	σ = 2.5%∼15%, $m = V_{th}$, Cap., Res.

B. CoolDRAM Robustness

1) Process Variation: The Gaussian distribution is used to model process variation effect on transistors threshold voltages (V_{th}), as well as cells and bitlines capacitance, and resistance drift. The mean value (m) is the nominal parameter value with a standard deviation of σ (see Table I). In our evaluation, we considered Systematic Process Variation as one of the most important process variation effects in nano-scale chip fabrication. Under this variation, nearby devices experience

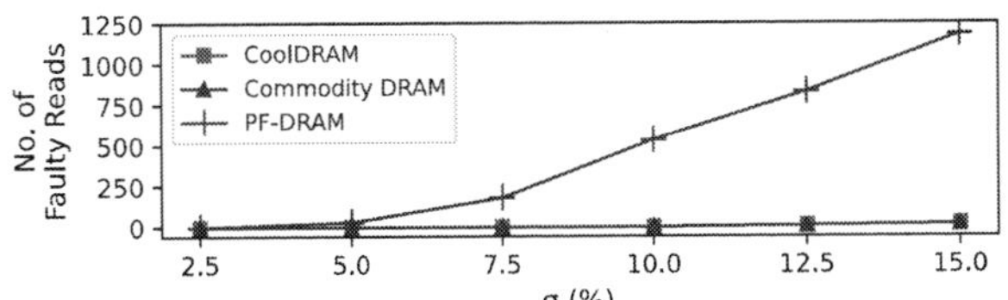

Fig. 5: Number of faulty reads due to process variation over 20 K simulation rounds

the same drift in their nominal parameters. To evaluate this, Monte Carlo simulation is employed.

Fig. 5 shows the number of faulty operations caused by process variation over 20K simulation rounds. The results show that CoolDRAM and commodity DRAM are highly resilient to process variation, with no faulty reads observed even when increasing σ to 10% and only 8 and 16 faults are observed by increasing σ to 12.5% and 15%, respectively. Since PF-DRAM operation is based on an imbalanced sense amplifier, a slight drift in its sense amplifier transistors' threshold voltage may lead to a faulty cell access operation. The study's findings indicate that the fault rate for PF-DRAM significantly increases to 6.1% at σ of 15%, whereas CoolDRAM's fault rate is less than 0.1%.

2) Noise Immunity: There are two types of noise that can impact DRAM cell access: ambient noise and internal noise. Bitlines are particularly vulnerable to ambient noise due to their lengthy structure, which acts like an antenna and absorbs electromagnetic noise during the charge sharing phase. Internal noise in DRAM is dependent on the clearance between the bitlines and wordlines (space between them) and can be categorized into several types, including Wordline Drive Noise, Power-Supply Voltage Bounce, and Bitline-to-Bitline and Bitline-to-Wordline coupling noises.

Ambient Noise: The Monte Carlo simulation with 20K iterations is utilized to detect the number of faulty reads during cell access. Ambient noise on the bitlines is simulated using a current source which simulates a spike noise by Eq.(II). Here, Q and T are the injected charge and duration, and t and Per are simulation time and noise injection moment, respectively.

$$I_{noise} = (\frac{Q}{T})\sqrt{SGN(1+SGN(t-Per))} \times \frac{t-Per}{T} \times \exp(-\frac{t-Per}{T}) \quad (5)$$

Table II presents the number of faulty read operations observed during 20 K simulation rounds in the presence of ambient noise. The results demonstrate that PF-DRAM exhibits the highest fault rate. With an ambient noise charge of 2.5 fC, 3 faulty reads are observed in PF-DRAM. This value increases significantly to 3350 faulty reads with 10 fC ambient noise charge, which is 1.76× and 1.78× of that of commodity DRAM and CoolDRAM, respectively. The slightly higher robustness of CoolDRAM compared with commodity DRAM is because of higher capacitance on the reference bitline by connecting the reference cell to it during cell access.

Fig. 6 illustrates the output voltage of the sense amplifier before it reaches its final stabilization point in the presence of various noise levels on the bitlines during a read operation of a '1' on the accessed cell (2000 samples are presented). In a correct read, the sense amplifier should be stabilized to *VDD*, thus the cases with the sense amplifier output voltage under 0 V lead to a faulty read. As Fig. 6 shows, PF-DRAM sense

TABLE II: Number of faulty read operations at the presence of ambient noise in 20 K simulation rounds

Average Charge	CoolDRAM	PF-DRAM	Commodity DRAM
0C	0	0	0
2.5fC	0	3	0
5fC	40	648	48
7.5fC	694	2060	722
10fC	1873	3350	1902

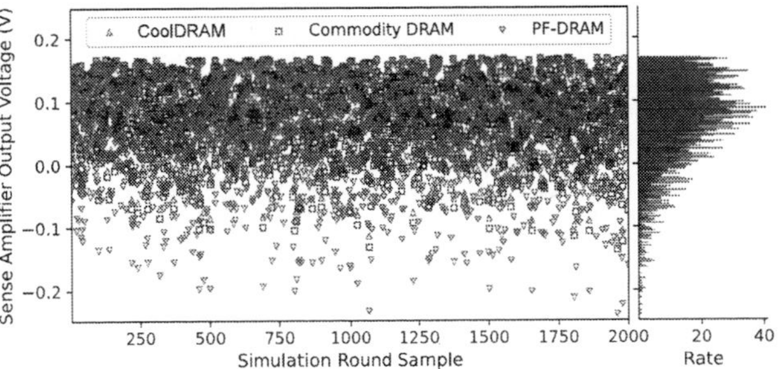

Fig. 6: Noise effect on sense amplifier output voltage (2K Samples are shown)

amplifier output voltage is much more affected by the ambient noise compared with the CoolDRAM, with the variance of 5.8 mV compared with 2.8 mV, respectively.

Internal Noise: Since bitlines in commodity DRAM, with folded structure, swing in opposite directions during the Precharge and Activation phases, the crosstalk effect between even and odd bitlines and wordlines cancels each other out pretty well. However, in open bitline structure DRAMs, Bitline-to-Wordline noise is more significant. In CoolDRAM, bitlines voltage swing reduces considerably in normal system operation, because of data similarity and special data access scheme. Nevertheless, in the worst-case, i.e. a Trojan which arranges and accesses data in DRAM in a way that in any access a swing on all of the bitlines occurs, can increase Bitline-to-Wordline noise. However, since row buffer access and restoration can be performed in parallel in CoolDRAM, by decreasing the steep of signals on the bitlines, this noise source can be controlled.

C. System-Level Evaluation

The energy consumption reduction during CoolDRAM access is dependent on both the row-hit rate and bit-flip probability in consecutive accesses to the same bitline. To determine these parameters, we selected 16 random workloads from the SPEC CPU2017 benchmark suite and used the gem5 full system simulator on an X86 64-bit processor with single-, 2-, 4-, and 16-cores, two levels of 32 KB and 2 MB of caches. Our system-level evaluation considers the main memory as a DDR4-2400 with a capacity of 8 GB and an 8 KB row buffer size, with open-row policy. The Micron power model [2] is used to calculate the energy dissipation of different components of DRAM.

The results show removing Precharge phase from cell access (read/write/refresh) reduces energy consumption by 28%, 29%, 31%, and 34%, on average for single-, 2-, 4-, and 16-cores, respectively. Increasing the number of parallel executing workloads also improves the efficiency of CoolDRAM. Fig. 7 depicts the Activation energy reduction of CoolDRAM compared to commodity DRAM. The required energy for charge-discharge cycles of bitlines, one of the main power-

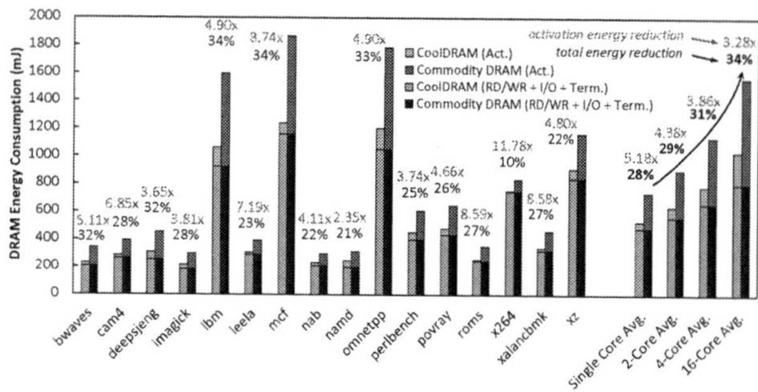

Fig. 7: Comparison of energy consumption between CoolDRAM and commodity DRAM for executing one billion instructions from each workload in the SPEC CPU2017 benchmark suite

hungry operations in DRAM [2], is reduced by an average of 5.18×, 4.38×, 3.86×, and 3.28× for single-, 2-, 4-, and 16-cores, respectively. The energy dissipation reduction for *x264* and *namd* are 11.78× (highest) and 2.35× (lowest), respectively. This difference in reduction is attributed to x264 having lower locality of references to DRAM and a higher bit-flip probability during execution compared to namd. The peripheral energy consumption for Read and write, I/O, and I/O termination power are the same for commodity DRAM and CoolDRAM. As both CoolDRAM and PF-DRAM eliminate the precharge phase, their energy dissipation during SPEC CPU2017 execution is very close, with a difference of less than 1% due to the sense amplifier structure. Hence, the results for PF-DRAM are not presented here.

D. Area Overhead

Sense amplifiers occupy nearly 8% of the entire commodity DRAM memory chip area [16], [19]. The proposed CoolDRAM memory structure keeps sense amplifiers of DRAM untouched. The area of bitlines coupling transistors and bitlines voltage equalizer transistor is exactly the same as the required area for the precharge circuit in commodity DRAMs [9]. The only additional circuit in CoolDRAM, as compared to commodity DRAM, is one row of dual-contact reference cells and one half-VDD charge transistor per bitline pair. The occupied area by the added circuit in CoolDRAM is less than two rows in the memory array. Thus, the area overhead imposed by the added circuit to DRAM MATs with 512×512 cells is less than 0.4%. It is worth mentioning that the two access transistors to the dual-contact cell and the charging transistor are made using the smallest possible DRAM transistors in the technology, the same as DRAM cell access transistors. The area overhead of CoolDRAM is approximately 22× less than that of PF-DRAM (the area overhead of PF-DRAM is about 8.8%).

V. Conclusions

The precharge phase in DRAM cell access is a power-hungry and time-consuming operation. This work proposes a DRAM structure, called CoolDRAM, in which cell access operation is performed without the need for Precharging bitlines. The area overhead of CoolDRAM is less than 0.4%, and its robustness against noise and process variation almost remained untouched compared with commodity DRAM. Meanwhile, CoolDRAM's energy consumption is reduced by an average of 34% compared to commodity DRAM.

References

[1] "Ddr4 sdram," https://www.micron.com/-/media/client/global/documents/products/data-sheet/dram/ddr4/8gb_ddr4_sdram.pdf, 2015.

[2] "Micron technical note: Calculating memory power for ddr4 sdram," https://www.micron.com/-/media/client/global/documents/products/technical-note/dram/tn4007_ddr4_power_calculation.pdf, 2017.

[3] A. Agrawal, A. Ansari, and J. Torrellas, "Mosaic: Exploiting the spatial locality of process variation to reduce refresh energy in on-chip edram modules," in *2014 IEEE 20th International Symposium on High Performance Computer Architecture (HPCA)*. IEEE, 2014, pp. 84–95.

[4] A. M. Amin and Z. A. Chishti, "Rank-aware cache replacement and write buffering to improve dram energy efficiency," in *2010 ACM/IEEE International Symposium on Low-Power Electronics and Design (ISLPED)*. IEEE, 2010, pp. 383–388.

[5] I. Bhati, Z. Chishti, S.-L. Lu, and B. Jacob, "Flexible auto-refresh: Enabling scalable and energy-efficient dram refresh reductions," in *Proceedings of the 42nd Annual International Symposium on Computer Architecture*, 2015, pp. 235–246.

[6] K. K. Chang, A. G. Yağlıkçı, S. Ghose, A. Agrawal, N. Chatterjee, A. Kashyap, D. Lee, M. O'Connor, H. Hassan, and O. Mutlu, "Understanding reduced-voltage operation in modern dram devices: Experimental characterization, analysis, and mechanisms," *Proceedings of the ACM on Measurement and Analysis of Computing Systems*, vol. 1, no. 1, pp. 1–42, 2017.

[7] N. Chatterjee, M. O'Connor, D. Lee, D. R. Johnson, S. W. Keckler, M. Rhu, and W. J. Dally, "Architecting an energy-efficient dram system for gpus," in *2017 IEEE International Symposium on High Performance Computer Architecture (HPCA)*. IEEE, 2017, pp. 73–84.

[8] E. Cooper-Balis and B. Jacob, "Fine-grained activation for power reduction in dram," *IEEE Micro*, vol. 30, no. 3, pp. 34–47, 2010.

[9] K. Itoh, *VLSI memory chip design*. Springer Science & Business Media, 2013, vol. 5.

[10] S. Khan, D. Lee, and O. Mutlu, "Parbor: An efficient system-level technique to detect data-dependent failures in dram," in *2016 46th Annual IEEE/IFIP International Conference on Dependable Systems and Networks (DSN)*. IEEE, 2016, pp. 239–250.

[11] S. Khan, C. Wilkerson, D. Lee, A. R. Alameldeen, and O. Mutlu, "A case for memory content-based detection and mitigation of data-dependent failures in dram," *IEEE Computer Architecture Letters*, vol. 16, 2016.

[12] S.-L. Lu, Y.-C. Lin, and C.-L. Yang, "Improving dram latency with dynamic asymmetric subarray," in *48th Annual IEEE/ACM International Symposium on Microarchitecture (MICRO)*. IEEE, 2015, pp. 255–266.

[13] S. Mittal and J. S. Vetter, "A survey of architectural approaches for data compression in cache and main memory systems," *IEEE Transactions on Parallel and Distributed Systems*, vol. 27, no. 5, pp. 1524–1536, 2015.

[14] M. Patel, J. S. Kim, and O. Mutlu, "The reach profiler (reaper) enabling the mitigation of dram retention failures via profiling at aggressive conditions," *ACM SIGARCH Computer Architecture News*, vol. 45, 2017.

[15] G. Pekhimenko, V. Seshadri, O. Mutlu, M. A. Kozuch, P. B. Gibbons, and T. C. Mowry, "Base-delta-immediate compression: Practical data compression for on-chip caches," in *International Conference on Parallel Architectures and Compilation Techniques (PACT)*. IEEE, 2012.

[16] N. Rohbani, S. Darabi, and H. Sarbazi-Azad, "Pf-dram: a precharge-free dram structure," in *ACM/IEEE 48th Annual International Symposium on Computer Architecture (ISCA)*. IEEE, 2021, pp. 126–138.

[17] N. Rohbani, M. A. Soleimani, and H. Sarbazi-Azad, "Pipf-dram: processing in precharge-free dram," in *Proceedings of the 59th ACM/IEEE Design Automation Conference*, 2022, pp. 1075–1080.

[18] H. Seol, W. Shin, J. Jang, J. Choi, J. Suh, and L.-S. Kim, "Energy efficient data encoding in dram channels exploiting data value similarity," *ACM SIGARCH Computer Architecture News*, vol. 44, no. 3, 2016.

[19] O. Seongil, Y. H. Son, N. S. Kim, and J. H. Ahn, "Row-buffer decoupling: A case for low-latency dram microarchitecture," in *2014 ACM/IEEE 41st International Symposium on Computer Architecture (ISCA)*. IEEE, 2014, pp. 337–348.

[20] S. Srikanth, L. Subramanian, S. Subramoney, T. M. Conte, and H. Wang, "Tackling memory access latency through dram row management," in *Proceedings of the International Symposium on Memory Systems*, 2018.

[21] X. Tao, Q. Zeng, and J.-K. Peir, "Hot row identification of dram memory in a multicore system," in *Proceedings of the International Conference on High Performance Compilation, Computing and Communications*, 2017, pp. 71–75.

[22] A. N. Udipi, N. Muralimanohar, N. Chatterjee, R. Balasubramonian, A. Davis, and N. P. Jouppi, "Rethinking dram design and organization for energy-constrained multi-cores," in *Proceedings of the 37th annual international symposium on Computer architecture*, 2010, pp. 175–186.

Development of Tropical Algebraic Accelerator with Energy Efficient Time-Domain Computing for Combinatorial Optimization and Machine Learning

Qiankai Cao, Xi Chen, Jie Gu
Department of Electrical and Computer Engineering
Northwestern University, Evanston, IL, USA
{qiankaicao2019, xichen2020}@u.northwestern.edu, jgu@northwestern.edu

Abstract— **Tropical algebra solves complex problems with only sum and min/max operations replacing expensive multiplication and addition in linear algebra. Due to the low computing cost, tropical algebra has recently gained significant attention in a broad range of areas such as combinatorial optimization, scheduling, machine learning, etc. In this paper, we propose a generic hardware accelerator architecture for tropical algebra supporting a wide range of applications. Novel time-domain (TD) computing accelerators with special mapping, precision expansion and, unrolling techniques are proposed to further improve hardware efficiency. Test results on various tropical calculations including linear regression, dynamic programming, and neural network are shown to demonstrate an energy saving from 1.5X to 2.1X, latency saving from 2.6X to 5.2X, or an overall energy-delay-product (EDP) improvement from 3.9X-10.5X compared with conventional digital implementation manifesting the promise of the algebraic solution on low power edge devices.**

Keywords—tropical algebra, shortest path problems, dynamic programming, time-domain computing

I. Introduction

Efficient hardware support for complex computing tasks such as combinatorial problems or machine learning algorithms on low power edge devices has become critically important as traditional general-purpose CPUs or microcontroller does not provide sufficient computing power for such tasks. To resolve such challenges, domain-specific accelerators have been rapidly developed. Compared with general-purpose CPU, specialized ASIC accelerators have shown significant power and performance advantage, for instance, using systolic array-based accelerator for massive multiplication and accumulation (MAC) operation in deep learning jobs [1]. However, there are still significant limitations on existing MAC based accelerators including (1) high power and area due to expensive hardware multiplier; (2) lack of flexible data flow or configuration for supporting wide range of applications beyond deep learning, e.g., dynamic programming, regression tasks, etc. As a result, there are still major challenges for efficient hardware solutions for common optimization problems.

To overcome such challenges, two directions of efforts are currently being taken in the community. First, a specific hardware accelerator for different computational tasks is being built. For instance, many "ising" machines were specially developed to improve the efficiency for solving "annealing" or combinatorial problems [2, 3]. Path-finding ASIC chips were used to find optimal cruise path for moving robots [4]. Second, analog and mixed-signal approaches have been explored to improve the efficiency of computation. For instance, analog based compute-in-memory has been rapidly applied to improve MAC efficiency for deep learning operations [5, 6]. Following the viable directions of hardware acceleration, in this work, we explore hardware accelerations on an emerging computing method, i.e., tropical algebra for a broad range of applications including both typical optimization problems and machine learning problems, demonstrating significant enhancement of computing efficiency from the proposed approaches.

Tropical algebra (also referred as "min-plus" or "max-plus" algebra) is an analogue of linear algebra, where conventional addition and multiplication are simply replaced by minimization (maximization) and addition [7]. The interests in tropical algebra were originally motivated by the possibility of dealing with a class of non-linear problems in pure and applied mathematics, operational research, science, and engineering as if they were linear since the underlying structure is a commutative and idempotent semiring. Tropical algebra not only allows succinct transformation from non-linear techniques into linear space, but also enables users to efficiently describe and deal with complex sets, reveal combinatorial aspects of problems and solve a class of challenging problems in a compact format. For example, tropical algebra has been applied to a wide range of problems for operations research and scheduling; discrete event system [8], control [9], integer programming, dynamic programming [10]. More recently, due to its elimination of expensive multiplication operation, it has been popularly studied for machine learning tasks such as linear regression [7] morphological image analysis [11] and deep neural networks (DNNs) [12]. Fig. 1 shows the many applications that exploit the expression power of tropical algebra. For instance, while the classical Euclidean line and parabola are formed using multiplication and addition, in tropical geometry, the tropical line and parabola are shown in piecewise line through addition and min/max. This also turns polynomials into piecewise-linear functions and replaces an algebraic variety by an object from polyhedral geometry, rendering simplified solution space for solving complex problems such as integer programming, dynamic programming [13, 14].

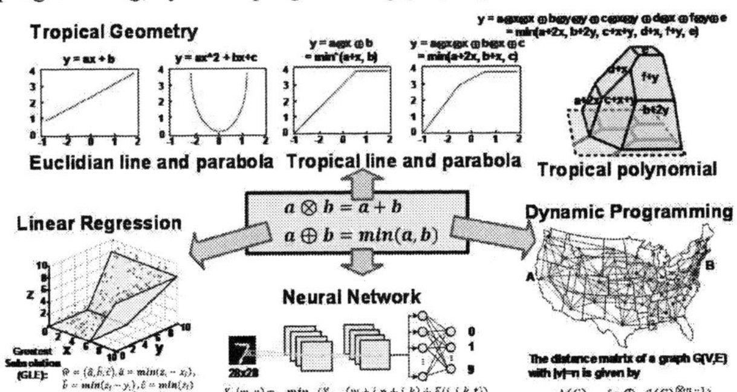

Fig. 1. Overview of tropical algebra and applications.

Interestingly, as will show in this paper, the emerging time-domain computing, which is a variant of analog mixed-signal computing, is highly suitable for realization of tropical algebra [16, 17]. As shown in Fig. 2, time-domain computing encodes information in the format of signal propagation delay or time pulse and performs computing using special time domain logic circuits. TD shows significant advantages in many domain-specific computing tasks. For example, a swarm robot system for reinforcement leaning accelerator was built leading to

979-8-3503-1176-1/23 $31.00 © 2023 IEEE

superior energy consumption compared with digital counterpart [15]. A time-based 2-D wavefront propagation solution for solving single-source shortest path problem was demonstrated [18]. A high throughput dynamic-time-warping engine using special time-domain flip-flop was demonstrated for time series classification showing tremendous power saving compared with digital counterpart [19]. As shown in Fig. 2, an ADD operation is only a few inverters summing up the delay of signals while a MAX and MIN operation can be realized by simple OR and AND gates in time-domain, e.g., late arrived signals are automatically passed by an OR gate for MAX operation while early arrived signals are passed by an AND gate for MIN operation. This succinct signal processing method renders significant saving of the hardware cost and power consumption when excessive ADD, MAX, MIN are being utilized. Similar to other analog mixed-signal computing method, Digital-Time-Converter (DTC) and Time-Digital-Converter (TDC) are being used to convert between digital domain and time domain, which are computational overhead of time-domain computing.

As additions and Min/Max operations are the most basic operations in tropical algebra, time-domain computing is naturally fit for tropical algebra. In this paper, for the first time, a generic hardware accelerator architecture is proposed with special mapping, precision expansion and energy enhancement techniques to enable hardware implementation for solving a wide range of mathematical problems using tropical algebra. The main contributions of this paper are as follows. (1) a generic time-domain accelerator architecture for tropical algebra is proposed using special mapping scheme and reconfigurable signal processing flow for tropical algebra on a diverse set of applications; (2) To minimize the data conversion cost, a special unrolling operation technique is developed for tropical matrix multiplications for efficient time-domain operation; (3) To deal with limited bit precision in time-domain implementation, a noval precision expansion method is developed to separate the original higher precision data into combination of lower precision data while maintaining overall bit accuracy rendering significant higher computing energy efficiency; (4) Our proposed hardware accelerator showed significant advantages in performance and energy efficiency when applied to a diverse range of tropical applications, such as combinatorial problems and machine learning tasks. The results showed a remarkable improvement in energy consumption, ranging from 1.5X to 2.1X, and latency saving, ranging from 2.6X to 5.2X, which resulted in an overall energy-delay-product (EDP) enhancement of 3.9X to 10.5X when compared to a traditional digital implementation. To the best of our knowledge, this is the first generic hardware implementation for tropical algebra.

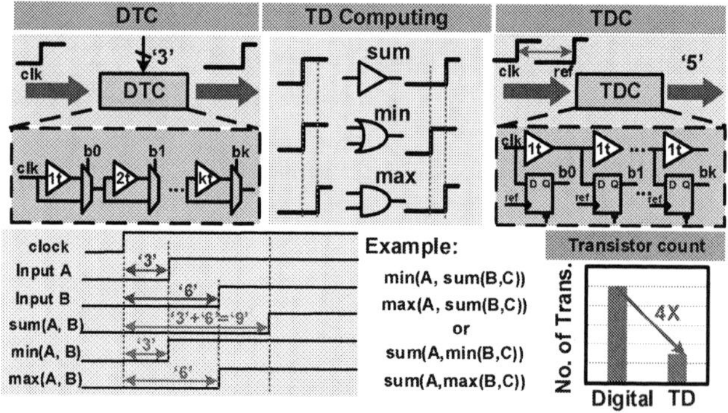

Fig. 2. Overview of TD computing and waveforms of TD computing.

II. Tropcial v.s. Classical Algebra

A. Tropical Semirings

Compared with classical real number ring $(\mathbb{R}, +, \times)$, the min-plus semiring $(\mathbb{R}_{min}, \oplus, \otimes)$ with $\mathbb{R}_{min} = \mathbb{R} \cup \{+\infty\}$ defines tropical multiplication as conventional addition and tropical addition as conventional min operation. With min operation replaced by max operation max-plus semiring is $(\mathbb{R}_{max}, \oplus, \otimes)$ with $\mathbb{R}_{max} = \mathbb{R} \cup \{-\infty\}$.

Tropical multiplication:

$$a \otimes b = a + b \tag{1}$$

Tropical addition:

$$a \oplus b = min\{a, b\} \tag{2}$$

or

$$a \oplus b = max\{a, b\} \tag{3}$$

In formula (1), a tropical multiplied by b is a plus b as classical arithmetic. In formula (2) and (3), a tropical add b is min of a and b for min-plus semiring $(\mathbb{R}_{min}, \oplus, \otimes)$, or max of a and b for max-plus semiring $(\mathbb{R}_{max}, \oplus, \otimes)$. Tropical algebra has found numerous applications in the computing literature particularly in a variety of graph algorithms, such as shortest path, graph matching, alignment, in which matrix multiplication is used to succinctly represent non-linear problems with linear expressions as follow.

Tropical matrix addition :

$$(A \oplus B)_{i,j} = a_{i,j} \oplus b_{i,j} \tag{4}$$

Tropical matrix multiplication :

$$(A \otimes B)_{i,j} = \bigoplus_{k=1}^{n} a_{i,k} \otimes b_{k,j} = \min_{k=1,\dots n}(a_{i,k} + b_{k,j}) \tag{5}$$

In (4), the i-jth result of tropical matrix addition is given by tropical addition of the i-jth elements of matrix A and B. In (5), the i-jth result of tropical matrix multiplication is given by i-th row of matrix A tropical multiplied by j-th column of B.

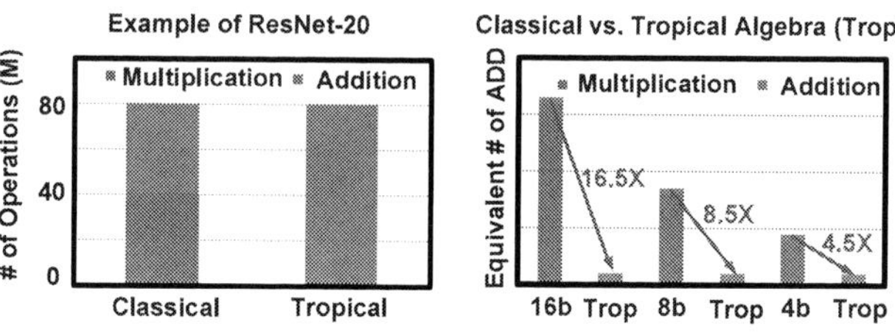

Fig. 3. Comparsion for classical and tropical algebra for example of ResNet-20, computation cost comparison between classical algebra and tropical algebra for 16-bits, 8-bits and 4-bits.

Compared with addition operation, multiplication operation is of much higher computation complexity. The wide-used convolution in deep neural networks involves massive MAC operation. Tropical algebra allows replacing the expensive MAC with low-cost addition and min operation for the convolution neural network as discussed in prior works [12]. As shown in our study in Fig. 3, in the example of ResNet-20, using tropical algebra, a 4.5X to 16.5X reduction of computation cost can be achieved for a tropical CNN operation given a MIN/MAX operation cost similarly as an addition operation.

III. TD Acceleration Technique

A. General Flow

To accelerate tropical algebra for various applications, Fig. 4. shows the proposed overall architecture diagram of the time-domain accelerator which contains a 20 x 20 tropical PE array, TDC for converting time-domain signals back into digital. During operations, each tropical cell is first configured to support different data flow, e.g., diagonal configuration, horizontal configuration, vertical configuration, and orthogonal configuration. For instance, in the diagonal configuration, time-domain signals from horizontal, vertical, and diagonal cells are propagated into Min module for tropical ADD (TADD) operations while inside the tropical MAC, a tropical multiplication (TMUL) is realized by summing delays from two delay lines generated from DTCs based on digital inputs. Similarly, for horizontal, vertical, and orthogonal configuration, time-domain signals travel in corresponding directions. As for bypass function, each tropical cell has the option to either process the time-domain signal or directly bypass the signal to a later stage. The tropical MAC performs tropical multiplication and accumulation through transition of rising edge of signal. As in Fig. 4, the tropical MAC unit consists of two delay chain based digital-time-converters (DTCs) for summation of input1 shown as red arrows and input2 shown as green arrows with various configurations to support MIN/MAX operation in different directions to support applications such as linear regression, dynamic programming, DTW, and neural network as will be discussed in later sections. The "Delay Unit" serves as a register in time domain to synchronize the data with the clock for pipeline operations for our special unrolling operation as will be proposed in later section.

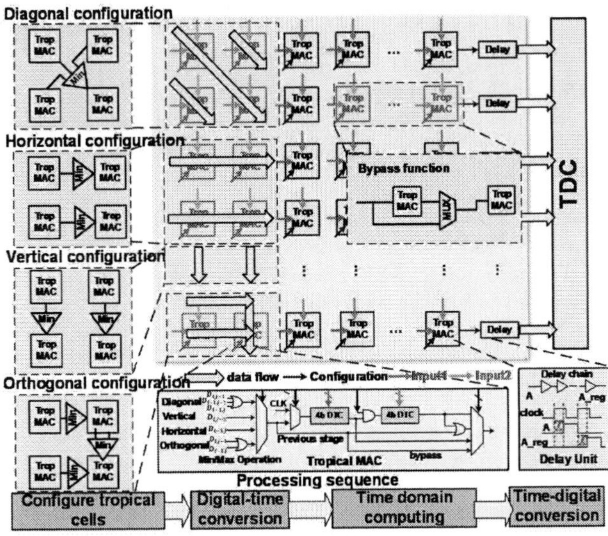

Fig. 4. Architecture diagram, data flow and time domain circuits with TDC for each row.

B. Mapping Method for Tropical Hardware Implementation

The time-domain accelerator that has been proposed can be applied across a variety of scenarios that involve an PE array structure, indicating its potential to effectively address intricate issues in numerous situations. By configuring signal waves in different directions, the tropical PE array can be configured into diagonal, horizontal, vertical, or orthogonal propagations. As illustrated in Fig. 5, in the diagonal setups, once the sum operation (TADD) has been performed, every tropical MAC will select the minimum value from $D_{i-1,j}$, $D_{i,j-1}$, $D_{i-1,j-1}$ and transmit it through all subsequent tropical MACs until the final stage. When in a horizontal arrangement, the tropical MAC located in a particular row will obtain the minimum value from $D_{i,j-1}$. Likewise, in vertical setup, the signals will receive the minimum value from $D_{i-1,j}$ within a column, whereas in an orthogonal operation, the minimum value will be derived from $D_{i-1,j}$, $D_{i,j-1}$. A formula has been provided to enable the mapping of different types of problems onto the array, either through individual configurations or hybrid ones. For instance, when utilizing a diagonal configuration, the formula $D_{i,j} = |A_i - B_j| + min(D_{i-1,j}, D_{i-1,j-1}, D_{i,j-1})$ is employed to generate the initial minimum and secondary summation operations, such as in the case of DTW shown in Fig. 5(a). In contrast, the vertical and horizontal configurations are comparable as they lack any topological linkage with neighboring rows or columns. These configurations are viewed as vector operations, where the formula $D_{i,j} = min(D_{i-1,j}, sum(A_i, B_j))$, $D_{i,j} = min(D_{i,j-1}, sum(A_i, B_j))$ are utilized to facilitate first summation and second minimum applications, such as in CNN and dynamic programming. Within the orthogonal configuration, the first summation and second minimum operations remain intact, albeit with interconnectivity between neighboring rows or columns. As such, the formula $D_{i,j} = min(D_{i-1,j}, D_{i,j-1}, sum(A_i, B_j))$is utilized. When dealing with actual problems, we have the option of hybridizing the configurations, enabling the mapping of any sum-min algebra problem regardless of the sequence of minimum or summation operations.

Furthermore, the diagram shown in Fig. 5(b) illustrates the mapping approach employed to address instances where the problem size is incongruent with the size of tropical PE array. In case where the size of the problem is smaller than PE array, a hybrid configuration comprising of both horizontal and vertical configurations can enable the array to process four groups of the same problem simultaneously. In the bypass function, unused tropical cells can bypass, which helps to better organize problem mapping. In Fig. 5 (b), a split method can be employed to efficiently map over-mapped problems (where the size of the problem exceeds that of the PE array) onto the same hardware. As illustration, a 40 by 40 matrix tropical multiplication task can be partitioned into four distinct sections, each comprising of a 20 by 20 matrix that can be seamlessly mapped onto the hardware.

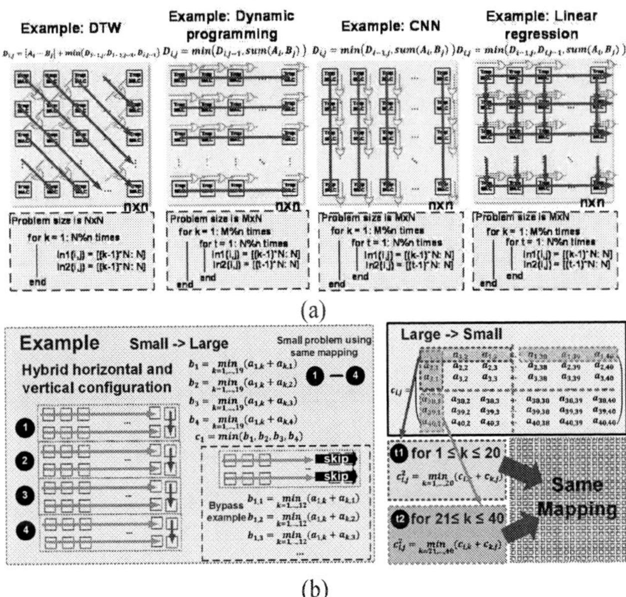

Fig. 5. (a) Mapping method and example with various configurations. (b) Example for small problem to large array mapping and large problem to small array mapping.

C. Precision Expansion Method for Bit-precision Extension

Lemma 1 Precision expansion with recovery residual matrix

Input: Low precision bits matrix $c_{i,j_{norm}} = c_{i,j} \gg k$

residual matrix $res_{i,j} = c_{i,j} - (c_{i,j_{norm}} \ll k)$

Output: Recovered tropical matrix $c_{i,j}^2 = 2^k \times c_{i,j_{norm}}^2 + res_{i,j}^2$

Proof. $c_{i,j_{norm}}^2 = c_{i,j} \otimes c_{i,j} = (\frac{c_{i,j} - res_{i,j}}{2}) \otimes (\frac{c_{i,j} - res_{i,j}}{2})$

$$= \frac{1}{2} min_{k=1,\dots,n}(c_{i,k} - res_{i,k} + c_{k,j} - res_{k,j})$$

$$= \frac{1}{2} min_{k=1,\dots,n}(c_{i,k} + c_{k,j} - res_{i,k} - res_{k,j})$$

Since, $c_{i,j} \le (c_{i,j_{norm}} \ll k)$

$$= \frac{1}{2^k} \min_{k=1,\dots,n} (c_{i,k} + c_{k,j}) - \frac{1}{2^k} \min_{k=1,\dots,n} (res_{i,k} + res_{k,j}) = \frac{1}{2^k}(c_{i,j}^2 - res_{i,j}^2)$$

Similar to other analog computing methods, time-domain computing shows efficient operation in 4~6-bit precision while many machine learning methods require 8-bit precision. To cope with this limitation, a precision expansion method is developed to extend the bit precision with multiple groups of 4-bit operation maintaining higher precision without exponentially increasing the latency of the time-domain operation as in a brute-force method. Precision expansion uses two low precision numbers to represent one high precision number, e.g., two 4-bit numbers to represent one 8-bit number. Because tropical algebra only involves summation and minimum operations, it is feasible to substitute a single round of high-precision numbers with several rounds of low-precision numbers when performing arithmetic calculations. These two numbers can subsequently be summed together to restore the original value. By expanding low precision bits, we replace one 8-bit min-sum operation with two 4-bit min-sum operations. Our proposed time-domain accelerator can process an 8-bit data by splitting it into a 4-bit data and a 4-bit residual matrix using this technique. Lemma1, presented above, provides evidence of the validity of the bit splitting operation using a residual matrix. Fundamentally, in the tropical quantization method, both matrix $c_{i,j}$ and its right shifting version, resulting in the residual matrix $res_{i,j}$, can be mapped to tropical hardware. Upon completing the operation, the original value is restored by adding the residual matrix back. An example of the operation is provided in Fig. 6, where the original matrix is shifted by k bit to obtain a low precision matrix, denoted as $c_{i,j_{norm}}$ and its residual matrix $res_{i,j}$. After tropical matrix multiplication, at the end of the operation, residual matrix $res^2_{i,j}$ is added back to $c^2_{i,j_{norm}}$ to recover its original result following the precision expansion recovery calculation. Compared with a brute-force bit extension with exponential growing cost in time domain, an 8X improvement in energy consumption can be achieved using this method.

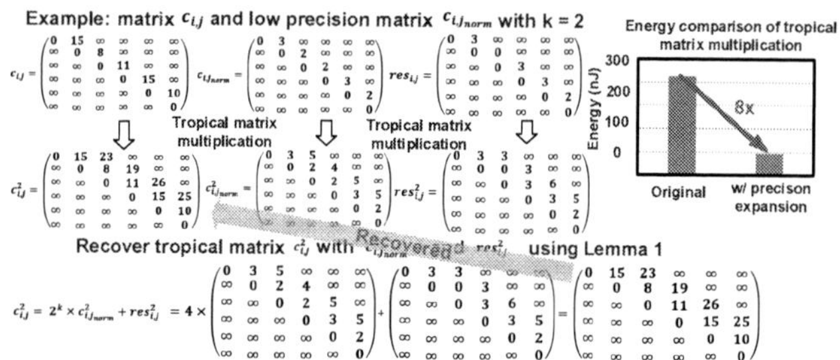

Fig. 6. Example of precision expansion, energy comparison of tropical matrix mutiplication.

D. Unrolling Scheme for Elimiation of Expensive Digital-Time conversion

Similar as other analog computing methods, the conversion from digital to time domain conversion results in significant power consumption as shown in the energy breakdown on the left of fig. 7 (b). Despite the high efficiency of time-domain operation, the conversion between digital and time formats from DTC and TDC accounts for 72% of the energy consumption. Fig. 7(a) shows an unrolling scheme to enhance the efficiency of time-domain tropical acceleration, whereby the time-domain signal is fed back directly as a trigger signal for the subsequent cycles. This method bypasses the TDC operation of the last iteration and the DTC operation at the start of the next iteration since the input data is already in the time domain and contains the necessary information. Fig. 7(b) demonstrates that an additional 26% to 46% energy savings can be attained through the utilization of the unrolling technique leading to a total of 65% energy savings compared with digital solution.

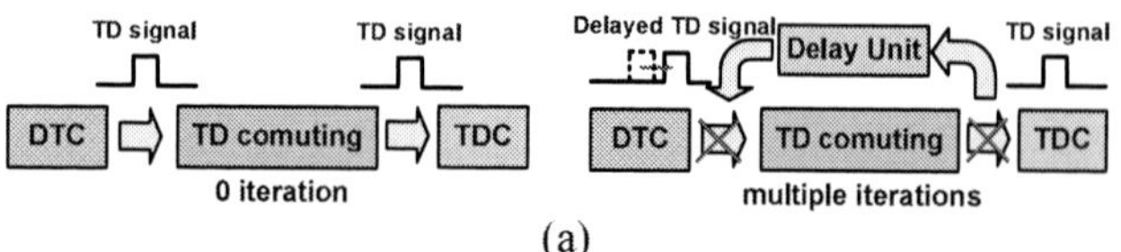

(a)

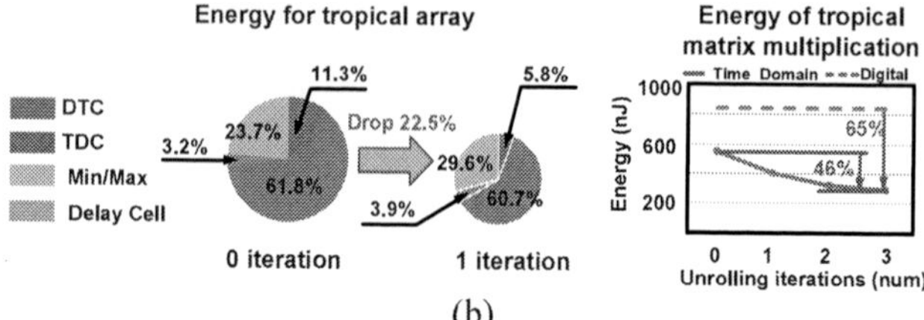

(b)

Fig. 7. (a) Data flow of unrolling scheme, (b) power breakdown and comparison for unrolling technique

IV. Experimental results and Case Study

To demonstrate the proposed tropical accelerator, a design was implemented in a TSMC 65nm technology. Special mapping, precision expansion and unrolling operations are adopted as proposed in Section III. Several important tropical applications are mapped into the hardware accelerator to demonstrate the benefits of the proposed techniques, as described below. Comparisons are made with conventional digital ASIC implementation in terms of energy, latency, and energy delay product (EDP).

A. Linear Regression

As shown in Fig. 8, the problem of tropical linear regression is described as fitting a tropical line $y = max(a + x, b)$ with the parameter $\hat{a}$ given by $min(f_i - x_i)$ and $\hat{b}$ given by $min(f_i)$. To process the calculation using the tropical accelerator, rows 1 through 10 are organized as data $i = 1, \dots, 200$ for $\hat{a}$ and rows 11 through 20 are arranged as data $i = 1, \dots, 200$ for $\hat{b}$. In Fig. 8, the heat map of propagation illustrates the delay of the signal at each tropical MAC cell, akin to the propagation of a wavefront, with each color indicating the arrival time of the time domain signal within the array. As shown in Fig. 8, the time-domain signal output of $\hat{a}$ and $\hat{b}$ propagate horizontally from column 1 to column 19. The deeper blue color indicates a smaller value obtained by computing $min(f_i - x_i), min(f_i)$. Next, in the last column, the time-domain signal travels vertically to calculate the minimum result of 10 rows for $\hat{a}$ and $\hat{b}$ respectively, which are then sent out for TDC operation at row 10 and row 20. Compared to digital solution, time domain exhibits a reduction in energy consumption by a factor of 1.8X and an improvement in latency by 2.6X. The overall energy-delay-product (EDP) of time domain demonstrates a 4.7X advantage over digital counterpart.

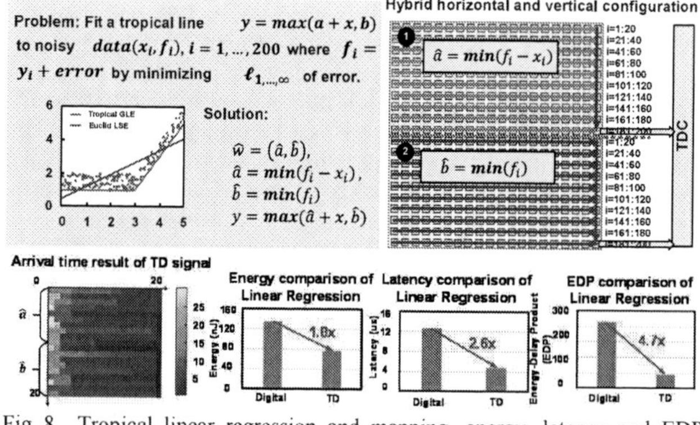

Fig. 8. Tropical linear regression and mapping, energy, latency and EDP results.

B. Dynamic Time Warping

Dynamic time warping (DTW) is one of the most used methods for time-series classification, which involves a series of data points graphed in time order, e.g., voice, stock price and ECG signal. To classify the time series, dynamic time warping calculates the similarity of the time series with variable speed. The computation DTW algorithm is given by the formula below.

$$D_{i,j} = |A_i - B_j| + \min(D_{i-1,j}, D_{i-1,j-1}, D_{i,j-1}) \quad (6)$$

One example of the computation is shown in Fig. 9(a). $D_{5,5}$ is given by adding the absolute value of the difference between A_5 and B_5 to the minimum found among its diagonal, previous row, and previous column. For instance, the result value of $D_{5,5}$ is obtained by computing $|9-8| + min(8,4,3) = 4$. As shown in Fig. 9(a), the tropical MAC is utilized in the DTW application with a diagonal configuration. The time series data is streamed to each column and row for calculations. Fig. 9(a) illustrates the computation process for the dynamic time warping algorithm using a color map. The algorithm calculates the distance between time series A and B and displays the arrival time of the TD signal in different colors. Deeper blue indicates an earlier arrival time, which corresponds to a smaller value in the time domain. The green line, shaped like a valley, depicts the propagating track of the warping path. Along this path, the final similarity result is computed at the corner of the array, and subsequently sent to the TDC for data conversion. Fig. 9(b) presents a comparison of energy, latency and EDP between the digital and time domain implementations. Compared to digital counterpart, the time domain implementation has 1.8X lower energy and 5.2X lower latency. As a result, the overall EDP (energy-delay product) is reduced by a factor of 9.5 in the time domain implementation, leading to significant improvement in both latency and energy consumption.

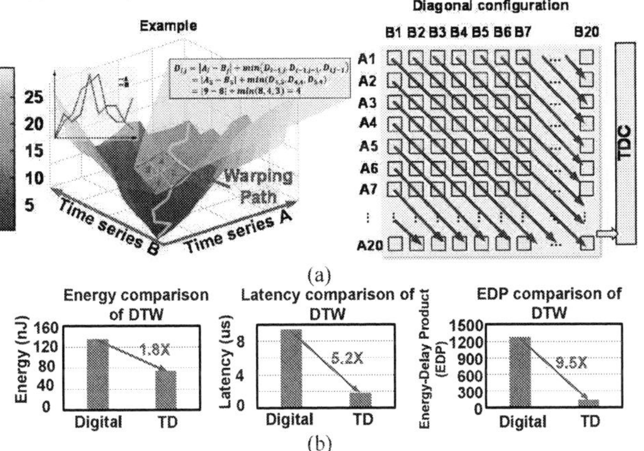

Fig. 9. (a) DTW algorithm and mapping result, (b) energy, latency, and EDP comparison.

C. Dynamic Programming

Dynamic programming is a critical computation task in many optimization problems such as shortest path and graph search. In Fig. 10, an example of a graph with its adjacency matrix is shown. To solve the shortest path problem for this graph, Floyd's algorithm is used, which is represented by the formula $\Delta(G) = (\epsilon \oplus \mathcal{A}(G)^{\otimes n-1})$. Each entry of the matrix $\mathcal{A}(G)$ represents the distance between each node in the graph. And the entry in row i and column j of the matrix $\Delta(G)$ corresponds to the length of the shortest path from node i to node j in G. The proposed hardware is used to implement the tropical matrix multiplication in this work utilizing the unrolling technique. Fig. 10 also illustrates how the arrival time result of time-domain signal depicts the spread of min-sum operation throughout the array.

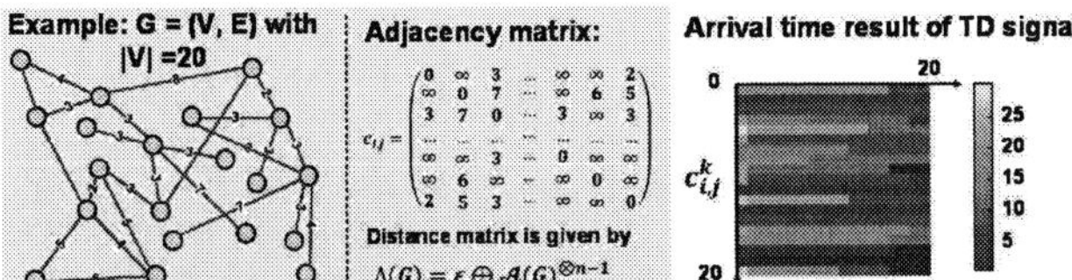

Fig. 10. Shortest path problem, arrival time result.

The mapping detail for dynamic programming is shown in Fig. 11(a). In step one, the tropical array is arranged horizontally and used to perform a min-sum calculation. In step two, the output of step one is temporarily delayed for synchronization with subsequent operations. Finally, in step three, the tropical MAC is set up with an unrolling operation, where the delay result from the previous step is used as a trigger to initiate a second round of min and sum operations on each column of the array. For the array, the input for each row is represented by $c_{i,j}$, while the input for each column alternates between a digital input and a TD signal due to the unrolling operation. This enables the TD signal (represented by $c_{i,j}^k$) to be sent back to each column as a trigger signal. In fig. 11 (b), there is a 1.5X improvement in energy, which increases to 2.1X with the use of unrolling. A 2.6X and 5.2X latency improvement is achieved for TD without and with unrolling technique, respectively. Overall, there is a 3.9X EDP (energy-delay product) improvement, which increases to 10.5X with the use of unrolling.

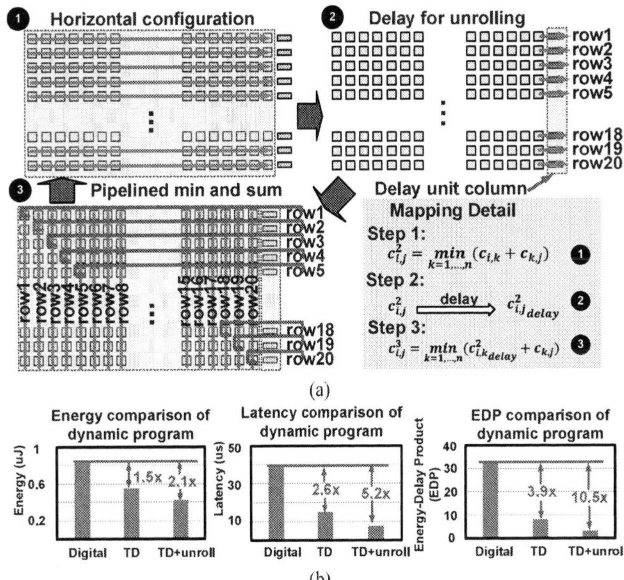

Fig. 11. (a) Mapping detail of dynamic programming, (b) energy, latency and EDP comparison with digital conterpart w/ and w/o unrolling.

D. Neural Network

Convolutional Neural network is widely used in computer vision for image classification, object detection, segmentation and so on. As the network grows larger, computation of multiplication has become the bottleneck for edge device. Recently, there has been exploration of tropical convolution neural networks (also referred to as morphological networks), which replace costly multiplication operations with additional operations. The fundamental calculation is demonstrated in the following formula, which calculates the similarity between the feature and the filter.

$$\min_{k=0,\dots,C_{in}} (X_{n-1}(m+i, n+j, k) + F(i,j,k,t)) \quad (7)$$

(a)

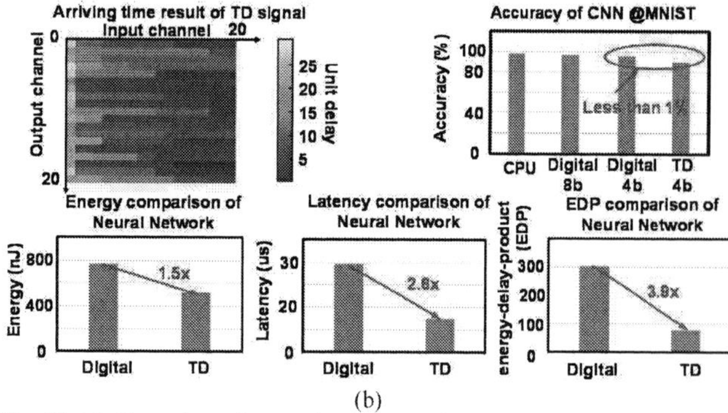

(b)

Fig. 12. (a) Example and mapping of neural network. (b) wavefront and quantizatin result of MNIST; energy, latency and EDP comparison between digital and TD.

In Fig. 12(a), a neural network comprising of two convolutional layers and one fully connected layer is presented. The color map shows the arrival time result of TD signals. The representation of the neural network is shown using split technique in the form of a matrix, where each row indicates distinct output channels, and each column denotes input channels. This technique is utilized for mapping large problems to smaller hardware, as discussed in section III B. To accommodate the large number of input and output channels in the neural network application, the tropical MACs are arranged horizontally. However, as the array size is smaller, only a portion of the data can be processed by the tropical array at each time step. During time step t1, the input channels ranging from 1 to 20 and output channels ranging from 1 to 20 are mapped onto the array for computation. During time step t2, the input channels ranging from 21 to 32 and output channels ranging from 1 to 20 are mapped onto the array with columns 12 to 20 skipped. The tropical array follows a similar process in time step t3 and t4, processing different input and output channels until all calculations for the neural network layers are completed.

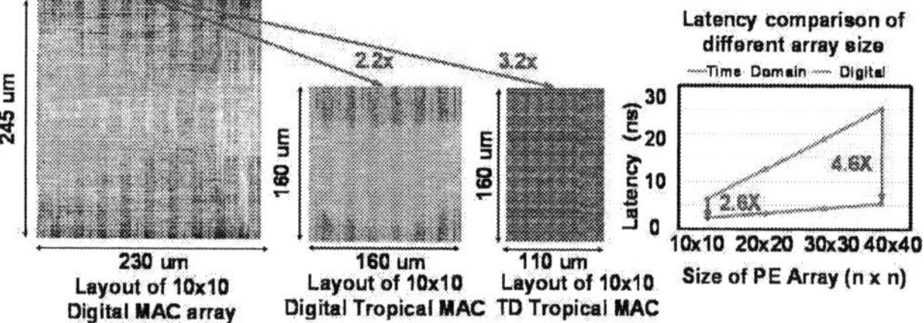

Fig. 13. Area saving and latency benefits using tropical algebra.

The accuracy results of the neural network applied to MNIST dataset are displayed in Fig.12(b). The accuracy drop caused by time domain tropical calculation is less than 1% when compared with the digital 4-bit implementation. Furthermore, the proposed TD design exhibits a 1.5X improvement in energy compared to the digital implementation. The latency of the neural network in TD is 2.6X lower than that of the digital implementation. As a result, the overall EDP is reduced by 3.9X compared to the digital implementation. Fig.13 displays a comparison of the implementation area. The tropical accelerator using standard digital design achieves an area reduction of 2.2X when compared to the digital MAC implementation. Furthermore, for a 10x10 array size, the time-domain tropical accelerator reduces the area by a 3.2X. The figure also illustrates the improvement in latency scaled with MAC array size. Because of the simplicity of MIN/MAX operation, the latency benefit of the time-domain tropical accelerator grows as the array size increases. For instance, in a 10x10 array, the time-domain tropical accelerator has a 2.6X improvement over the digital implementation, whereas in a 40x40 array, the improvement increases to 4.6X due to the benefit of MIN/MAX operations. This highlights the enhanced advantage of the time-domain tropical accelerator in larger scale applications.

V. Conclusion

In this work, a generic hardware accelerator architecture is proposed for tropical algebra. Energy efficient time domain computing circuits are further implemented to accelerate the min/max-sum calculations. For various applications, novel mapping methods are developed to support various problem size matching with the accelerator array size. Tropical precision expansion method is proposed to separate data into lower bit data with residual matrix to help reduce hardware cost, while maintaining data integrity. Special unrolling technique is also proposed in time domain that can reduce data conversion between digital and time domain. Experiments on a 65nm design show that for applications of linear regression, DTW, dynamic programming and neural network, an improvement from 1.5X to 2.1X for energy consumption and 2.6X to 5.2X saving for latency are achieved from the proposed design in comparison with conventional digital implementation. An overall energy-delay-product (EDP) improvement from 3.9X to 10.5X is also observed from our test results.

VI. Acknowledgements

This work was supported in part by NSF under grant number CCF-1846424.

References

[1] N. P. Jouppi et al., "In-datacenter performance analysis of a tensor processing unit," *ACM/IEEE 44th Annual International Symposium on Computer Architecture*, 2017

[2] M. Yamaoka, et al., "A 20k-Spin Ising Chip to Solve Combinatorial Optimization Problems With CMOS Annealing," in *IEEE Journal of Solid-State Circuits*, vol. 51, no. 1, pp. 303-309, Jan. 2016

[3] I. Ahmed, et al., "A Probabilistic Self-Annealing Compute Fabric Based on 560 Hexagonally Coupled Ring Oscillators for Solving Combinatorial Optimization Problems," *2020 IEEE Symposium on VLSI Circuits, 2020*

[4] A. Amravati, et al., "A 55nm time-domain mixed-signal neuromorphic accelerator with stochastic synapses and embedded reinforcement learning for autonomous micro-robots,"*IEEE International Solid - State Circuits Conference - (ISSCC)*, 2018.

[5] Z. Chen, et al., "15.3 A 65nm 3T Dynamic Analog RAM-Based Computing-in-Memory Macro and CNN Accelerator with Retention Enhancement, Adaptive Analog Sparsity and 44TOPS/W System Energy Efficiency," *2021 IEEE International Solid- State Circuits Conference (ISSCC)*, 2021.

[6] S. Xie, et al., "Gain-Cell CIM: Leakage and Bitline Swing Aware 2T1C Gain-Cell eDRAM Compute in Memory Design with Bitline Precharge DACs and Compact Schmitt Trigger ADCs," *IEEE Symposium on VLSI Technology and Circuits*, 2022

[7] P. Maragos, et al., "Tropical Geometry and Machine Learning," in *Proceedings of the IEEE*, vol. 109, no. 5, pp. 728-755, May 2021

[8] F. Bacelli, et.al., Synchronization and Linearity (An algebra for discrete event systems), John Wiley & Sons, 1992.

[9] W. M. McEneaney, Max-Plus Methods for Nonlinear Control and Estimation. Boston, MA: Birkhauser, 2006.

[10] M. Mohri, "Semiring frameworks and algorithms for shortest-distance problems," J. Autom. Lang. Comb., vol. 7, no. 3, pp. 321–350, 2002.

[11] P. Maragos, et al., "Morphological filtering for image enhancement and detection," in Handbook of Image and Video Processing, CA: Academic, 2000.

[12] S. Fan, et al., 2021, February. "An Alternative Practice of Tropical Convolution to Traditional Convolutional Neural Networks." In *The 5th International Conference on Compute and Data Analysis*, 2021.

[13] D. Maclagan and B. Sturmfels, Introduction to tropical geometry. American Mathematical Society, 2021, vol. 161.

[14] Jean-Eric Pin. Tropical Semirings. J. Gunawardena. Idempotency, Cambridge Univ. Press, Cambridge, pp.50-69, 1998, Publ. Newton Inst. 11. ffhal-00113779

[15] N. Cao, et al., "14.1 A 65nm 1.1-to-9.1TOPS/W Hybrid-Digital-Mixed-Signal Computing Platform for Accelerating Model-Based and Model-Free Swarm Robotics," *IEEE International Solid- State Circuits Conference - (ISSCC)*, 2019.

[16] Z. Chen, et al., "Digital Compatible Synthesis, Placement and Implementation of Mixed-Signal Time-Domain Computing," 2019 56th ACM/IEEE Design Automation Conference (DAC), Las Vegas, NV, USA, 2019.

[17] Z. Chen, et al, "A Mixed-Signal Time-Domain Generative Adversarial Network Accelerator with Efficient Subthreshold Time Multiplier and Mixed-Signal On-Chip Training for Low Power Edge Devices," IEEE Symposium on VLSI Circuits, 2020.

[18] L. R. Everson, et al., "2.5 A 40×40 Four-Neighbor Time-Based In-Memory Computing Graph ASIC Chip Featuring Wavefront Expansion and 2D Gradient Control," *IEEE International Solid- State Circuits Conference - (ISSCC)*, 2019.

[19] Z. Chen, et al, "High-Throughput Dynamic Time Warping Accelerator for Time-Series Classification With Pipelined Mixed-Signal Time-Domain Computing," in *IEEE Journal of Solid-State Circuits*, vol. 56, no. 2, pp. 624-635, Feb. 2021.

IMBUE: In-Memory Boolean-to-CUrrent Inference ArchitecturE for Tsetlin Machines

Omar Ghazal*§, Simranjeet Singh†¶§, Tousif Rahman*, Shengqi Yu*, Yujin Zheng*,
Domenico Balsamo*, Sachin Patkar†, Farhad Merchant*, Fei Xia*, Alex Yakovlev*, Rishad Shafik*
*Newcastle University, UK, †Indian Institute of Technology Bombay, India,
¶Forschungszentrum Jülich GmbH, Germany
{simranjeet, patkar}@ee.iitb.ac.in, {O.G.G.Awf2, s.rahman, y.zheng26, s.yu10, domenico.balsamo,
farhad.merchant, fei.xia, alex.yakovlev, rishad.shafik}@newcastle.ac.uk

***Abstract*—In-memory computing for Machine Learning (ML) applications remedies the von Neumann bottlenecks by organizing computation to exploit parallelism and locality. Non-volatile memory devices such as Resistive RAM (ReRAM) offer integrated switching and storage capabilities showing promising performance for ML applications. However, ReRAM devices have design challenges, such as non-linear digital-analog conversion and circuit overheads. This paper proposes an In-Memory Boolean-to-Current Inference Architecture (IMBUE) that uses ReRAM-transistor cells to eliminate the need for such conversions. IMBUE processes Boolean feature inputs expressed as digital voltages and generates parallel current paths based on resistive memory states. The proportional column current is then translated back to the Boolean domain for further digital processing. The IMBUE architecture is inspired by the Tsetlin Machine (TM), an emerging ML algorithm based on intrinsically Boolean logic. The IMBUE architecture demonstrates significant performance improvements over binarized convolutional neural networks and digital TM in-memory implementations, achieving up to a *12.99*x and *5.28*x increase, respectively.**

***Index Terms*—Boolean-to-Current, Tsetlin Machine, ReRAM, In-Memory Computing.**

I. Introduction

Machine Learning (ML) applications are intrinsically data-centric. This creates a compelling case for transitioning from traditional computing systems towards systems utilizing In-Memory Computing (IMC). IMC introduces the advantage of computations being performed within the memory itself [1]. This reduces latency and energy overheads from data movement, the so-called von Neumann bottleneck, and thereby facilitates the better acceleration of ML problems in resource-constrained micro-edge applications [2], [3].

For effective IMC, typically, the memory compute unit is organized to leverage parallelism from the ML algorithm through crossbar-like structures. These crossbars capitalize on performance by exploiting the low power consumption and fast access times of their constituent memory devices [2]–[4].

Emerging analog memory devices such as Resistive RAM (ReRAM) present one promising approach to addressing drawbacks with the continued scaling down of CMOS-based memory devices. ReRAM offers non-volatile storage with high-density crossbars, reduced leakage currents, and memory size overheads [1], [5]. However, one essential challenge with ReRAM devices is the nonlinearity in their operation within memory arrays. These variations can be manifested between devices and between cycles for the same device [3].

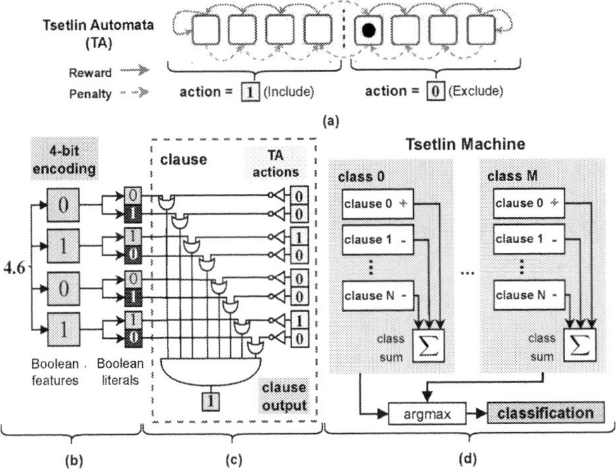

Fig. 1: Diagram showing core components for inference using TMs: (a) the TA, (b) the booleanization of the input space, (c) the computation of a clause output and (d) the TM's architecture for inference.

While metal-oxide based ReRAM crossbars provide the means for both computation and storage through current summations [3], [5], when considering the wider ML inference procedure, there is a significant logic expense required to supplement the ReRAM architectures in the form of high bit precision ADCs and DACs [6] or through increased control logic complexity [6].

This paper explores an alternative approach to tackling the above design challenges through an inherently bitwise inference algorithm called the **Tsetlin Machine (TM)**, see Fig. 1. The TM structure utilizes learning units called Tsetlin Automata (TA) to form logic propositions that relate Boolean inputs to the output for classification tasks without the need for multiply accumulate units. The simplicity of this algorithm is developed into an ReRAM crossbar to propose an In-Memory Boolean-to-Current Inference Architecture (IMBUE). A brief overview of the inference algorithm for the TM is given in the following paragraphs, and a more in-depth expansion of the algorithmic concepts and hardware-oriented nature of the TM can be found in [7]–[9].

Tsetlin Automata: The TAs are the learning units that perform the classification and decision-making tasks in TM (Fig. 1a). Each automaton behaves like a finite state machine

§Equal contribution

979-8-3503-1176-1/23 $31.00 © 2023 IEEE

where half of the states of the TA are dedicated to an action. Reward and penalty stimuli move the automaton between the two actions: include or exclude. During training, rewards and penalties are issued to find optimal state positions for each TA. Once the training is completed, the TA can be viewed as one of the two possible Boolean values '1' and '0', indicating whether to include ('1') or exclude ('0') a particular feature. The input features must also be brought to a '1'/'0' Boolean domain to use this concept for feature selection. In doing so, it introduces two fundamental aspects of the TM: a) the booleanization of the input space, and b) how the logical propositions called clauses can be constructed to relate these Boolean inputs to the TAs.

Booleanization: Fig. 1b shows how a raw value '4.6' can be converted to a 4-bit encoding to form Boolean features. Each Boolean feature is then further broken into Boolean literals which are the features and their complements. Now, these literals may interact with their respective TAs to compute a clause. While Fig. 1b has only one raw value, real TM applications will booleanize all the raw input values to compute a clause output.

Clause Computation: Fig. 1c shows how each Boolean literal can be related to the actions of each dedicated TA through NOT, OR and AND logic to form a 1-bit clause output. The TM contains many clauses, each with its own set of TAs that create different logic propositions from the same Boolean literals. The TM organizes these clauses into an architecture that enables classification.

TM Architecture: For multi-class problems, the TM contains a fixed and equal number of clauses for each class. Within these classes, each clause can also have a polarity. Through the training process, the clauses with positive polarity learn the logic that supports the classification of their class. Clauses with negative polarity learn logic propositions that oppose their class. This is seen through the $+$ and $-$ signs in the clause blocks in Fig. 1d. Each class has an equal number of positive and negative polarity clauses. Clauses with positive polarity have their outputs multiplied by (+1), while negative polarity clauses are multiplied by (-1).

TM Inference: The TM's inference process is outlined by assembling all the above elements. First, the input Boolean literal is seen by all clauses in all the classes of the TM. Each clause will output a 0/1 based on its TA actions. Then the clause outputs will be multiplied according to their polarities and summed for each class. The $argmax$ of these class sums is the predicted classification.

Aside from the simplicity of the clause computation, one key aspect of the TM is the sparsity of TAs with include actions compared to exclude actions once the TM is trained. Upon viewing the clause computation, it is clear to see that only include TAs have an impact on the final clause output. This paper seeks to exploit this behavior through a representation of TA actions in ReRAM as either high and low resistances and the representation of input Boolean literals as Boolean voltages help to realize the Boolean-to-Current cells where the interaction between these literals (V) and these TAs (R) is further simplified into two distinct currents (I) ranges that may

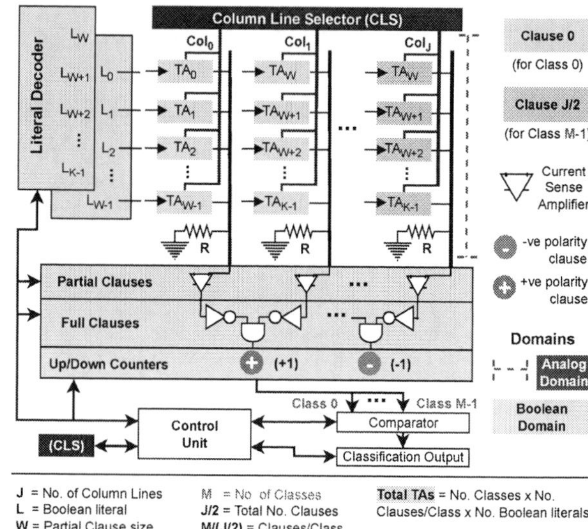

Fig. 2: Proposed TM inference architecture using IMBUE.

be summed according to the principles of Kirchhoff's Current Law (KCL). This forms the basis of the IMBUE Boolean-to-Current mechanism.

This paper proposes IMBUE for an ReRAM crossbar architecture to model the TM inference process. It investigates the non-linearity in current conversion, ReRAM device variations and evaluates the energy efficiency of the full system. In doing so, the following contributions are offered:

- for the first time, integration of 1T1R ReRAM technology into a new ML inference method utilizing Boolean-to-Current paths by following KCL;
- investigation into variation tolerance of the TM under Device-to-Device (D2D), Cycle-to-Cycle (C2C) and CMOS variations; and
- extensive evaluation of the efficiency of IMBUE in terms of power, latency and accuracy compared to it's closest Binary Neural Network and Digital TM state-of-the-art counterparts.

Next, Section II presents IMBUE, Section III presents experimental results and validation for the design, Section IV examines the energy efficiency of the proposed inference system and Section V concludes the paper.

II. Proposed IMBUE Method

The principal design concept of this paper is the ReRAM crossbar architecture for the TM inference presented in Fig. 2. At its core, the crossbar uses IMBUE for Boolean-to-Current conversion to realize clause computation. This section builds the TM inference procedure presented in Fig. 1, first focusing on mapping the TA as a ReRAM cell and creating clause outputs, then building to the full classification process.

A. TA Cell using 1T1R

The design of the TA cell is constructed with one memristor and one PMOS transistor forming the 1T1R structure. Fig. 3a

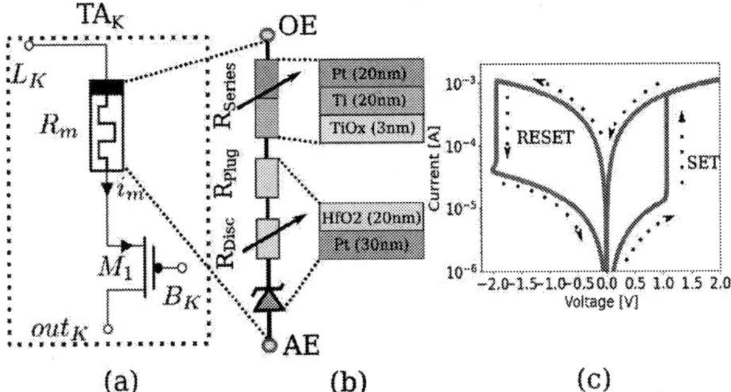

Fig. 3: Equivalent circuit for a TA in (a) represented by 1T1R cell, (b) shows the material stacks of memristor along with the I-V characteristics in (c).

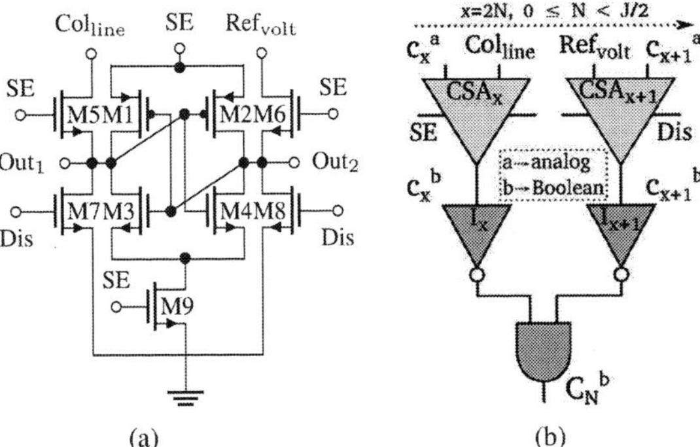

Fig. 4: CSA to convert current to Boolean in (a) and construction of a full clause using partial clauses is given in (b).

shows the architecture this design along with the equivalent stack of the memristor in Fig. 3b. The memristor model used in this study consists of a Pt/Ti/TiO$_x$/HfO$_2$/Pt material stack, which is proven to have good electroforming voltage stability and thermal stability [10]. Fig. 3c shows the I-V characteristics of this memristor device, which has only two states, High Resistive State (HRS) = logic '0' used for TA exclude action and Low Resistive State (LRS) = logic '1' for the TA include action. The Boolean-to-Current mechanism is formed by the Ohm's Law relation when the literals, now seen as input voltages, are dropped across the TA resistances to produce a current. Fig. 2 shows these TAs organized into columns. These columns are partial clauses which constitute only part of the full clause computations. This is done to minimize the effect of sneak currents and device variations. The partial clauses are examined next.

B. Partial Clause Computation

The partial clause evaluates part of the full clause using the Boolean-to-Current mechanism. In Fig. 2 one full clause needs K Boolean literals and respective TAs. This is divided into two partial clauses that now compute $K/2$ Boolean literals with their TAs (seen in yellow). The mapping of this split is done through the literals decoder such that all K Boolean literals interact with their corresponding TAs for every clause. This is done for all clauses for all classes in the TM (the last partial clause of the last class is seen in pink). To perform a particular full clause computation the column line selector activates the two respective partial clause columns.

At the end of each column is an R resistor. This is used to convert the KCL current to a column voltage Col$_{line}$ in Fig. 2. Fig. 4a shows the Current Sense Amplifier (CSA) used to compare this Col$_{line}$ voltage with a Ref$_{volt}$ reference voltage to generate a full-swing rail-to-rail Boolean output. Taking into account the variation of memristor devices, careful design choices must be made to have a larger current range between having at least one include TA in a column and only exclude TAs. Therefore, the proposed CSA utilizes a 65 nm CMOS technology that can function with input voltages close to the ground with minimal variations and low supply voltage (1.2V). From the CSA structure in Fig. 4a, M1-M4 transistors combine to form two cross-coupled inverters of equal size and the input Col$_{line}$ and Ref$_{volt}$ are applied to M5 and M6, respectively. During the partial clause evaluation (reading phase), SE is connected to VDD, allowing the CSA to determine which line has the higher voltage and produces a near Boolean'1' or '0' full swing output to the partial clause. Two additional 1.2 µm NMOS transistors M7 and M8 are used to discharge the two internal sensing nodes Out$_1$ and Out$_2$ to avoid adding a bias to subsequent CSA voltage comparisons. These partial clauses are then combined into full clause outputs.

C. Full Clause Computation

Dividing into partial clauses helps minimize the non-linearity of the 1T1R cell's behavior. As the number of TAs within a clause increases, the variations in current contributions of HRS TA cells (exclude actions) may become large enough to equal an LRS TA cell (include) or affect the CSAs voltage comparison range. In Fig. 4b the inverters and AND gate compute a full clause from the partial clause outputs given by $C_N = c_{2N} \wedge c_{2N+1}$, where $N = 0 \rightarrow J/2$, $c_{2N} = \sum_{i=0}^{W-1} V(TA_i(I)), and\ c_{2N+1} = \sum_{i=W}^{K-1} V(TA_i(I))$. $V(TA_i(I))$ which indicates the Col$_{line}$ voltage sensed by the CSA as a result of the KCL from that respective TAs column.

D. TM System Level Inference

As seen with Fig. 1(d) full clauses must be assigned a polarity before contributing to a class sum. The architecture in Fig. 2 achieves this through up/down counters. The m class sums are then considered through a comparator to produce the final classification. These components are tied together through the control unit.

III. Experimental Results and Validations

This section evaluates the efficiency of IMBUE. We will study the characteristics of the TA cell under writing and reading operations. Later, we will also examine the modularity and scalability of IMBUE using the Boolean-to-Current mechanism and the CSA timing considerations.

A. 1T1R TA Cell Characterization

Initially, the TA cells must be programmed with the appropriate actions found through training. A voltage pulse with a specific duration and amplitude is applied to program the action in TA. Fig. 3c shows the I-V characteristics of a device where

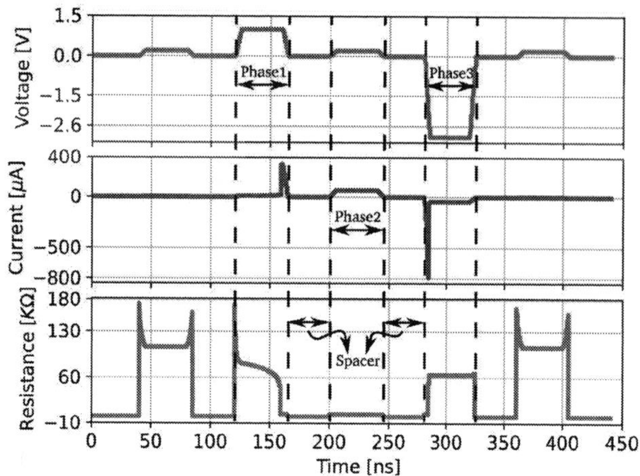

Fig. 5: Programming and computation phases for a single TA. SET (programming to include) and RESET (programming to exclude) paths are marked with arrows. Table I shows the mapping of literal and action values on 1T1R cell in terms of reading voltage, resistance, and current respectively.

a) Programming action value in a TA cell: TA programming involves two different programming voltage values named V_{set} (1V) and V_{reset} (-2.5V). V_{set} changes the action from exclude to include, while V_{reset} changes it from the include to exclude value. A pulse duration of 35 ns is used to store these TA actions. The mechanism for programming any TA cell can be done by activating the column line as the select column and disabling the CSA by setting SE to low. Finally, this voltage is applied to the required TA. V_{set} occurs during phase 1 in Fig. 5 with a spark of rising current at the end of the applied pulse when the current is less than $0.4\ mA$. At the same time, the resistance value gradually decreases until, at the last few nanoseconds, a substantial drop coincides with the current spark reaching the maximum value. Phase 3 describes the reset operation, using V_{reset}. During this phase, the current drops rapidly while the resistance reaches its expected value. To provide a discharge period for the CSA, a Spacer of $0V$ is used between the phases. The programming of TA actions is a one-time process. The programming combinations are described below:

- *Exclude* → *Include*, applying V_{set} to switch to LRS.
- *Include* → *Include*, applying V_{set} to stay in LRS.
- *Exclude* → *Exclude*, applying V_{reset} to stay in HRS.
- *Include* → *Exclude*, applying V_{reset} to switch to HRS.

b) Boolean-to-Current computation: The Boolean literal input voltages are given as logic '1' → $0\ V$ and logic '0' → $0.2\ V$ as seen in Fig. 5. They are then applied to the selected partial clause column. The target TAs' Col_{line} is turned on so that the potential difference across the TA is somewhere between the two programming voltages.According to Ohm's law, the Boolean literal value and TA action interact to produce a proportional current. At the end of each Col_{line}, a voltage divider resistance ($R = 100\ \Omega$) converts the output current to a voltage. This voltage is then sensed using the CSA as shown in the waveform Fig. 6. Table II shows power consumption for all possible TA combinations. The goal is to dominate the *include* with *literal '0'* combination while maintaining the accuracy for less power consumption. This can be done in TM training by reducing the include/exclude ratio as much as possible.

TABLE I: Translate the 1T1R operation into TA concept.

Literal (logic)	Read (V)	Action	Resistance ($K\Omega$)	Current (μA)
'0'	0.2	Include	≈2.5	≈76.07
'0'	0.2	Exclude	≈105.8	≈1.89
'1'	0	Include	≈7.6	$\approx 137e^{-9}$
'1'	0	Exclude	≈33.6	$\approx 9.9e^{-9}$

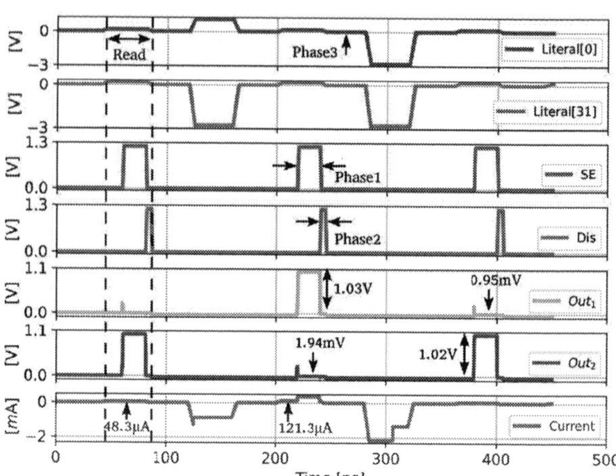

Fig. 6: CSA waveforms in 3 phases.

B. Partial Clausing through CSA Operation

The current created by the partial clause is outlined in three operating phases. The first phase occurs during the Col_{line} reading pulse (35 ns), and SE is held to VDD (20 ns) to enable the latch. The voltage across the column resistor R is sampled and supplied into the CSA's internal nodes Out_1 and Out_2. The higher voltage node is set to a higher output voltage near VDD, whereas the lower voltage node is set near GND. In the second phase, SE is set to GND, while Dis is set to VDD for a (5 ns) spark to discharge the two output nodes to the ground voltage level. The SE and Dis signals are grounded in the final phase to prepare the CSA for the next sensing cycle. The waveforms for the CSA design at 1.2V VDD connected to a Col_{line} of 32 TAs is shown in Fig. 6.

C. Impact of Variations

In this section, we study how manufacturing process variations affect the design choices of architecture. First, we discuss the variation in ReRAM cells followed by variation in CMOS-based components in the architecture, especially CSA.

TABLE II: Power consumption for 1T1R cell.

Operation	TA power consumption (μW)
Program to *Exclude*	54.54
Program to *Include*	215.1
Include × Literal '0'	14.37
Exclude × Literal '0'	$377.2e^{-3}$
Otherwise	≈ 0

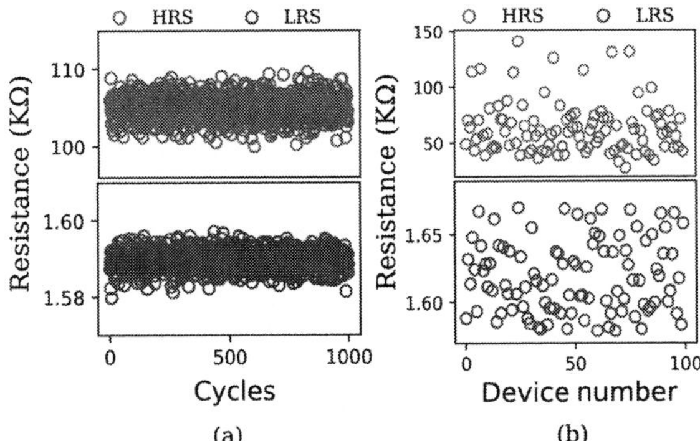

Fig. 7: LRS and HRS distribution for 1000 cycles under C2C variations in (a) D2D variation on 10x10 crossbar in (b).

1) ReRAM cell variations: The D2D and C2C variations can affect the switching behavior of ReRAM devices. The following are the parameters affected during the manufacturing process of ReRAM used in this study. The parameters' ranges used in this study are the same as in [10]:

- Maximum and minimum oxygen vacancy concentration in the disc, ($N_{Disc,max}$ and $N_{Disc,min}$).
- Radius and length of the disc, (r_{disc} and l_{disc}).

a) C2C variation: C2C is analyzed by altering the variable parameters before each SET and RESET cycle. A MATLAB script randomly increases or decreases (with equal probability) each parameter's value from its previous value every cycle. These are then exported as a CSV file. This file contains the varying parameters and their timestamps and is used for the Spectre simulation in this study. Fig. 7a shows the distribution of HRS and LRS for 1000 cycles. The new set of values is applied after each 100 ns period. Within these 100 ns, the device is RESET with 25 ns (20 ns pulse width with 2.5 ns rise and fall time) pulse of -2 V and SET with 2 V pulse with the same duration. The state of the device is read by applying a READ pulse of 0.2 V after every RESET/SET operation. In the next cycle, new values are applied in the simulation. It has been observed that there is a $\pm5\%$ change in HRS and an $\approx1\%$ change in LRS during C2C variations.

b) D2D variation: The experimentally validated Gaussian distribution from [10] is used to generate a random set of values for each device's varying parameters. These variations were then individually applied to the experiment's devices. The distribution of HRS and LRS states in a 10x10 crossbar without a transistor is shown in Fig. 7b. To measure the HRS distributions, all devices are first switched to LRS by applying a 2 V SET pulse with a duration of 150 ns and a rise and fall period of 50 ns. The devices are switched to the HRS state in the next cycle by applying a -2 V RESET pulse. At 0.2 V, the device's resistance status is read. The same method is used to measure LRS distribution. HRS has been found to have a greater impact on variance than LRS. Due to variation, the HRS vary from 31-155 $K\Omega$ with an average of 65.56 $K\Omega$ and LRS from 1.55-1.67 $K\Omega$ with an average of 1.64 $K\Omega$. The ratio of HRS and LRS decides how many TAs can be fitted into a single column to compute partial clauses accurately.

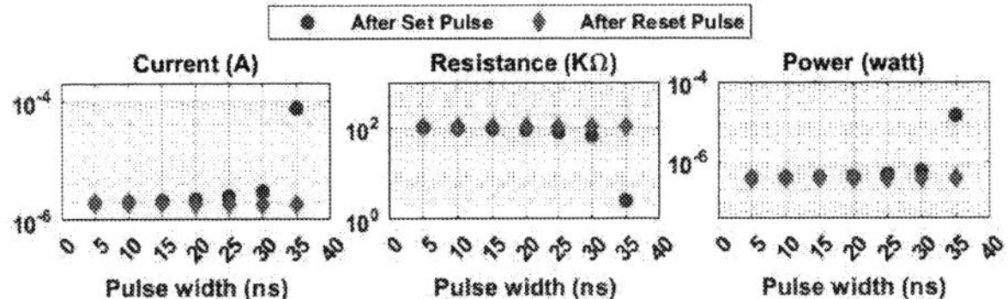

Fig. 8: Current, Resistance and Power variation for different programming pulse duration.

2) Pulse duration study: This study gives the minimum pulse duration required to switch the 1T1R device from LRS to HRS and vice-versa. Fig. 8 shows the current, resistance, and power of a single device on different pulse duration (5 to 100) ns while the voltage of the pulse is constant ($V_{set} = 1$ V and $V_{reset} = -2.5$ V). It is shown in Fig. 8 that the device switched from HRS to LRS at 35 ns pulse width. A pulse width greater than 35 ns will lead to more power consumption and latency of the design.

3) CMOS corner and process variation: Corner analysis and Monte Carlo analysis were performed to evaluate the dependability of the CSA. The worst case scenario of only one include TA in a partial clause column of 32 TAs is used. Each cycle a new set of literals is applied to this column. Out_1 and Out_2 readings are taken from the CSA for each cycle (see Fig. 4a). Table III lists the two output's voltages for the corners analysis and the process variation for 2000 cycles with a small standard distribution.

IV. Comparative Energy Efficiency

The energy efficiency of the proposed IMBUE based architecture is simulated with a Python script using the power consumption values seen in Table II and the timing presented in Fig. 6. The crossbar size was kept as 32 TAs per partial clause column. TMs were trained for datasets including Noisy XOR, MNIST, Kuzushiji-MNIST (K-MNIST), Fashion-MNIST (F-MNIST), and Google Keyword Spotting dataset with 6 keywords (KWS-6), which were booleanized using the method presented in [11]. The TAs from these models were extracted and used to simulate the proposed architecture and compare against a CMOS TM implementation presented in [8]. The results showed that the number of include TAs in each of these TM models significantly impacted the resulting energy efficiency when applied to the proposed IMBUE based architecture. As the complexity of the learning problem increases, the TM is unable to select as many key include actions for the TAs. However, this benefits IMBUE's Boolean-to-Current mechanism greatly as it reduces the contribution of the most power-intensive part of the proposed inference system. Table IV presents the energy/datapoint values obtained for the different datasets using IMBUE and CMOS TM. The results showed that

TABLE III: CSA's corners and process variation analysis.

Operation	CSA output	Corner analysis					Process variations	
		Nominal	FF	SS	SF	FS	Mean	SD
		(mV)						
SET	Out_1	865.7	860.3	867.1	892	819.6	864.86	10.35
SET	Out_2	0.95	3.95	0.32	1.32	0.99	1.06	0.445
RESET	Out_1	28.35	30.06	28.33	28.75	28.38	28.44	0.216
RESET	Out_2	876.4	872.9	874.7	908.1	823.1	875.46	12.33

TABLE IV: Energy estimation and comparison to state-of-the-art TM for different datasets.

Dataset name	Accuracy %	Classes #	Clauses total	TA cells #	Includes #	Includes %	CSAs #	CMOS TM [8]	IMBUE	x energy reduction
								Average energy/datapoint (nJ)		
Noisy XOR	99.2	2	12	576	48	8.3	18	0.0092	**0.02**	**0.36**
MNIST	96.48	10	2000	3136000	18927	0.6	98000	50.01	**13.9**	**3.597**
KWS-6	87.1	6	1800	1357200	7990	0.58	42413	21.64	**5.91**	**3.66**
K-MNIST	88.6	10	5000	7840000	31217	0.39	245000	125.03	**26.47**	**4.722**
F-MNIST	87.67	10	5000	7840000	25742	0.32	245000	125.03	**23.66**	**5.283**

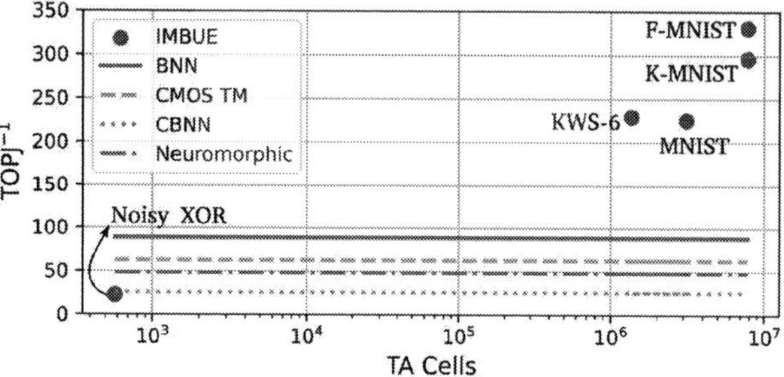

Fig. 9: IMBUE energy efficiency for TM inference applied to various datasets.

IMBUE outperformed CMOS TM in terms of energy efficiency for all datasets except for Noisy XOR which has the highest include/exclude ratio.

The closest comparable ML to TMs are Binary Neural Networks (BNNs) with both reliant on bitwise operations. However to avoid biases in the presence of algorithmic differences, we employed the TopJ^{-1} metric, which measures trillion operations per second per joule of energy consumed. For TMs this can be calculated as the number of TAs used to compute all the clauses divided by the energy per datapoint. Fig. 9 shows how the proposed IMBUE architecture based TM models perform when compared against Neuromorphic [12], BNN [13] and Binarized Convolutional Neural Networks (CBNN) [14]. As seen with Table IV, as the include sparsity increases for more complex TM models, the IMBUE solution offers better performance. Specifically, the TM inference based on IMBUE outperforms CMOS-based TM, BNN, CBNN, and Neuromorphic by up to 5.28x, 3.74x, 12.99x, and 6.87x, respectively, achieving a TopJ^{-1} of 331 for F-MNIST.

V. Conclusion

The IMBUE TM architecture was realized through a ReRAM crossbar and evaluated in terms of performance of the inference procedure and resilience against ReRAM variations. The core component of the TM, the Tsetlin Automaton, was implemented with a 1T1R cell. Each cell is programmed to either LRS or HRS to reflect the include or exclude TA actions respectively. The ReRAM resistances may vary upon reading, however D2D and C2C studies showed sufficient robustness in meeting the CSA margins when reading the stored TA actions. The rationale for IMBUE's Boolean-to-Current mechanism is made evident from the nature of the TM's clause computation; the logic based interaction between Boolean inputs and the TA include/exclude actions in the clause is simplified into current accumulation. To address the issues of sneak currents and non-linearity when converting these currents from the analog domain, the TM clause was broken into partial clauses. These partial clauses help mitigate device variations while still permitting scalability. The power consumption of the TM inference is dominated by two conditions as seen in Table II. Examination of case study datasets exposes the rarity of such conditions. This is further seen through Table IV where increasing model complexity leads to increased sparsity in TM include decisions and therefore offers much better energy efficiency. For the F-MNIST dataset the proposed architecture gives 12.99x better TopJ^{-1} compared with CBNN.

The recent works with TMs have proposed coalesced clause architectures where clauses are shared between classes [15]. Future work aims to explore the associated trade-offs from applying the principles of IMBUE to such an algorithm.

Acknowledgments

The authors would like to gratefully acknowledge the funding support from the UK Northern Accelerator (ref: NACCF 220), Lloyds Registers Foundation (ref: 5thICON-12) and Norwegian Research Council (ref: AIEverywhere project).

References

[1] A. Sebastian *et al.*, "Memory devices and applications for in-memory computing," *Nature Nanotechnology*, vol. 15, no. 7, pp. 529–544, 2020.

[2] R. Wang *et al.*, "Implementing in-situ self-organizing maps with memristor crossbar arrays for data mining and optimization," *Nature Communications*, vol. 13, no. 1, p. 2289, 2022.

[3] Q. Xia *et al.*, "Memristive crossbar arrays for brain-inspired computing," *Nature Materials*, vol. 18, no. 4, pp. 309–323, 2019.

[4] X. Liu *et al.*, "Memristor crossbar architectures for implementing deep neural networks," *Complex & Intelligent Systems*, vol. 8, no. 2, pp. 787–802, 2022.

[5] S. Yu *et al.*, "Current-mode carry-free multiplier design using a memristor-transistor crossbar architecture," in *DATE*, 2020, pp. 638–641.

[6] B. Wu *et al.*, "ReRAM crossbar-based analog computing architecture for naive bayesian engine," in *Proc. ICCD*, 2019, pp. 147–155.

[7] O.-C. Granmo, "The Tsetlin Machine – A Game Theoretic Bandit Driven Approach to Optimal Pattern Recognition with Propositional Logic," 2018. [Online]. Available: https://arxiv.org/abs/1804.01508

[8] A. Wheeldon *et al.*, "Learning automata based energy-efficient AI hardware design for IoT applications," *Philosophical Tran of the Royal Society A*, vol. 378, no. 2182, p. 20190593, 2020.

[9] A. Bakar *et al.*, "Logic-based intelligence for batteryless sensors," ser. HotMobile '22. NY, USA: ACM, 2022, p. 22–28.

[10] C. Bengel *et al.*, "Variability-aware modeling of filamentary oxide-based bipolar resistive switching cells using SPICE level compact models," *IEEE TSCAS I*, vol. 67, no. 12, pp. 4618–4630, 2020.

[11] J. Lei *et al.*, "Low-power audio keyword spotting using tsetlin machines," *MDPI Journal of Low Power Electronics and Applications*, vol. 11, no. 2, 18, 2021.

[12] D. Miyashita *et al.*, "A neuromorphic chip optimized for deep learning and cmos technology with time-domain analog and digital mixed-signal processing," *IEEE JSSC*, vol. 52, no. 10, pp. 2679–2689, 2017.

[13] W. Choi *et al.*, "Content addressable memory based binarized neural network accelerator using time-domain signal processing," in *55th ACM/ESDA/IEEE DAC*, 2018, p. 1–6.

[14] H. Yonekawa *et al.*, "On-chip memory based binarized convolutional deep neural network applying batch normalization free technique on an fpga," in *IEEE IPDPSW*, 2017, pp. 98–105.

[15] S. Glimsdal *et al.*, "Coalesced multi-output tsetlin machines with clause sharing," *ArXiv*, vol. abs/2108.07594, 2021.

iMAT: Energy-Efficient In-Memory Acceleration for Ternary Neural Networks With Sparse Dot Product

Shien Zhu, Shuo Huai, Guochu Xiong, Weichen Liu
School of Computer Science and Engineering, Nanyang Technological University, Singapore
{shien001, shuo001, guochu001}@e.ntu.edu.sg, liu@ntu.edu.sg

***Abstract*—Ternary Neural Networks (TNNs) achieve an excellent trade-off between model size, speed, and accuracy, quantizing weights and activations into ternary values {+1, 0, -1}. The ternary multiplication operations in TNNs equal light-weight bitwise operations, favorably in In-Memory Computing (IMC) platforms. Therefore, many IMC-based TNN accelerators have been proposed. They build dedicated ternary multiplication cells or utilize efficient bitwise operations on IMC architectures. However, existing ternary value accumulation schemes on IMC architectures are inefficient. They extend the sign bit of integer operands or conduct two-round accumulation with specially designed encoding, bringing long latency and extra memory write overhead. Moreover, existing IMC-based TNN accelerators overlook TNNs' sparsity and conduct operations on zero weights, resulting in unnecessary power consumption and latency.**

In this paper, we propose iMAT to accelerate TNNs with operator-, architecture- and layer-level optimizations. First, we propose a single-round Ternary Variable-Bitwidth Accumulation scheme, which efficiently extends the addition result sign bit without extra memory write overhead. Second, we propose an in-memory accelerator with enhanced sensing circuits for the accumulation scheme and a Sparse Dot Product Unit to exploit TNNs' weight sparsity, utilizing zero weights to skip unnecessary operations. Further, we propose Fused Scaling Functions which combine the scaling, activation, normalization, and quantization layers to reduce the hardware complexity without affecting the model accuracy. Simulation results show that compared with dense in-memory TNN accelerators, our iMAT achieves up to 2.7× speedup and 3.7× energy efficiency on ternary ResNet-18.

***Index Terms*—In-Memory Computing, Non-Volatile Memory, Neural Network Compression**

I. Introduction

Deep Neural Networks (DNNs) have shown great success in computer vision, natural language processing, and many other domains. DNNs' model size and computational complexity have significantly increased with their accuracy in the past few years. Therefore, quantization methods are widely applied to reduce DNNs' storage cost and inference latency, e.g. utilizing 8-bit and 4-bit integers (INT8/INT4) instead of 32-bit floating-point values to represent DNN weights and activations [1], [2].

Ternary Neural Networks (TNNs) achieve an excellent trade-off between model size, speed, and accuracy by quantizing the weights and activations into ternary values {+1, 0, -1} [3]. As Table I shows, ternary ResNet-18 achieves 65.7% top-1 accuracy on ImageNet, 3.7-6.8% higher than Ternary-activation Binary-weight Network (TBN) and Binary Neural Network (BNN) [4]. Moreover, TNN can skip null operations with zero values to achieve higher performance, but BNNs

TABLE I
Comparison of Full-Precision and Quantized ResNet-18 with Top-1 Accuracy on ImageNet

Model	Value	Main Ops	Storage Saving	Speedup	Accuracy
FP32 [1]	32-bit	×, +	1×	1×	70.3%
INT8 [1]	8-bit	×, +	4×	4×	69.7%
INT4 [2]	4-bit	×, +	8×	8×	69.4%
TNN [3]	2-bit	Bitwise	16×	16×	65.7%
TBN [4]	2/1-bit	Bitwise	32×	>16×	62.0%
BNN [4]	1-bit	Bitwise	32×	64×	58.9%

using binary values {+1, -1} are dense. Compared with DNNs using INT8 and INT4, 2-bit TNNs have 2-4× smaller model sizes and lower computational complexity. Further, ternary multiplications in TNNs equal 2-bit additions or bitwise operations, favored on In-Memory Computing (IMC) architectures.

IMC is the next-generation Artificial-Intelligence computing platform [5]. IMC architectures compute by using traditional memory arrays and Non-Volatile Memories (NVMs) like Spin-transfer Torque Magnetoresistive Random Access Memory (STT-MRAM) [6], reducing the data movement and bringing high data-level parallelism. As modern IMC designs support addition and Boolean functions, they are widely adopted to accelerate addition-centric networks [7] and Boolean operation-centric TNNs and BNNs [8].

However, it is a challenge to accumulate the ternary multiplication results on IMC architectures efficiently. Traditional methods extend the sign bit of the addition operands to ensure the correct addition result. However, this scheme brings 50% more memory write overhead to 2-bit ternary values on IMC arrays. TiM-DNN [9], TeC-Cell [10], SpinLiM [11] and RTN [12] adopt specially designed encoding schemes for ternary values, leading to a slow two-round accumulation process and extra subtraction operations on partial sums.

More importantly, existing IMC-based accelerators overlook the sparsity nature of TNNs. For example, ResNet-18 released by PyTorch contains 43%-62% zero values among layers, which means we can get ∼2× speedup and energy efficiency by simply skipping operations on zeros. However, TiM-DNN [9] and TeC-Cell [10] conduct dense ternary multiplication when processing TNNs. Similarly, SpinLiM [11] and IMC-CD [13] perform dense bitwise operations to accelerate TNNs. As a result, they suffer from the computational overhead of null operations on zero weights. Moreover, they cannot work as standard memory devices nor traditional IMC accelerators due to the fixed-function ternary processing cells.

In this paper, we propose iMAT as an energy-efficient

sparse in-memory acceleration for TNNs. First, we propose a dedicated Ternary Variable-Bitwidth Accumulation (TVBA) scheme, which reduces the accumulation overhead by extending the sign bit on the addition results rather than operands. Second, we propose a Sparse Dot Product Unit (SDPU) to exploit TNNs' weight sparsity, utilizing zero weights to skip unnecessary operations. Then we design an in-memory TNN accelerator based on standard STT-MRAM arrays with the SDPU and enhanced sensing circuits that support the proposed sign bit calculation. Noticing that TNNs contain scaling and quantization layers besides activation and normalization layers, we further propose Fused Scaling Functions (FSFs) to reduce 50%-87.5% operations in them without affecting the accuracy.

Evaluation results show that the proposed TVBA scheme reduces 36.0% latency and 35.6% energy compared with traditional accumulation methods. The SDPU achieves up to 3.7× speedup and 4.1× energy efficiency on ternary dot product with 75% sparsity than related works. Compared with dense bitwise operation-based TNN accelerators SpinLiM [11] and RTN [12], our proposed iMAT achieves 2.5-2.7× speedup and 3.4-3.7× energy efficiency on ternary ResNet-18.

II. Background and Related Works

In-Memory Computing (IMC) is a memory-centric computer architecture that utilizes memory arrays and crossbars to do computation [6]. The key features of NVM-based IMC include reduced data movement, ultra-high internal memory bandwidth, near-zero leakage power, and ultra-high data-level parallelism. Taking STT-MRAM as an example, Fig. 1 presents the basic idea of IMC. Each STT-MRAM cell contains a Magnetic-Tunnel-Junction (MTJ) which is a variable resistor in a high or low resistance state. We can access two cells simultaneously, as Fig. 1 (a) shows. Based on the equivalent circuit in Fig. 1 (b), we can infer that the sensed voltage V_{sense} has three cases: V_{00}, $V_{01/10}$, and V_{11}. The Sense Amplifier (SA) compares the sensed voltage V_{sense} with reference voltages V_{AND} and V_{OR} to get the AND and OR results between the operands in these two cells, as Fig. 1 (c) and (d) show. For example, the AND result will be "1" only if $V_{sense} = V_{11}$. Further, we can add more logic gates to the SA to realize more complex functions like XOR and Addition.

Table II presents related works on TNN accelerators. These related works either conduct dense bitwise operations or dense ternary multiplications. TiM-DNN [9] proposes a Ternary Processing Cell based on two SRAM cells to conduct 2-bit ternary multiplication in dedicated encoding. TeC-Cell [10] designs a non-volatile Ternary Compute-Enabled memory cell consisting

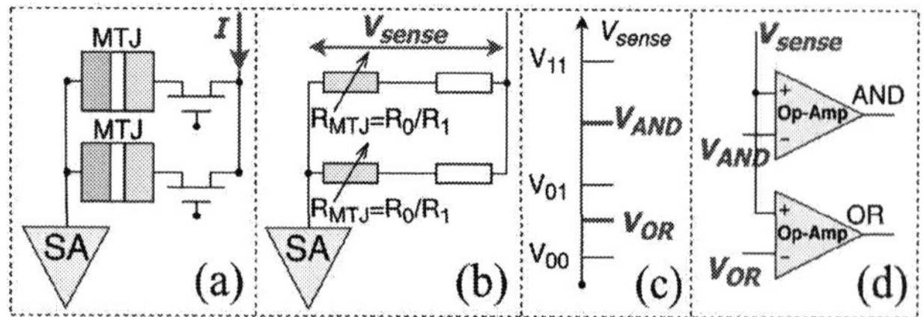

Fig. 1. (a) Two STT-MRAM cells. (b) Equivalent circuit of two STT-MRAM cells. (c) Reference voltages. (d) Voltage comparing inside the SA.

TABLE II
Related works on TNN accelerators

Target Operation	Paper
Dense bitwise operation	SpinLiM [11] RTN [12] IMC-CD [13]
Dense ternary multiplication	TiM-DNN [9] TeC-Cell [10]
Sparse addition	This work

of two ferroelectric and six standard transistors to perform ternary scalar multiplications. SpinLiM [11] realizes the AND-XOR-based ternary bitwise multiplication in a particular encoding utilizing two 2T-2MTJ SOT-MRAM cells. IMC-CD [13] builds ternary bitwise multipliers using two SRAM cells with a particular encoding and realizes vector summation in the charge domain based on capacitors. RTN [12] presents an ASIC accelerator for TNNs with dense XNOR-based logic and a specially designed encoding.

In contrast, this work processes TNNs by utilizing the sparsity to skip null operations on zero weights. Moreover, we provide a novel variable-bitwidth accumulation method for ternary values to reduce the accumulation overhead.

III. Proposed Accelerator

We accelerate TNNs on iMAT at the operator, architecture, and layer levels. First, the Ternary Variable-Bitwidth Accumulation provides an efficient accumulation operator for ternary values. Second, the in-memory architecture with the Sparse Dot Product Unit accelerates ternary General-Matrix-Multiplications (GEMMs), which are dominant operations in convolution and fully-connected layers. Third, the Fused Scaling Functions accelerate ReLU-like activation layers, normalization layers, and other TNN layers.

A. In-Memory Accumulation

The ternary value accumulation is an essential part of ternary convolution and fully-connected layers. Thus, we propose a dedicated TVBA scheme to remove the bit-extension cost of traditional accumulation schemes on signed integers.

1) Ternary Variable-Bitwidth Accumulation (TVBA): Different from traditional accumulation methods that extend the sign bit of operands, we extend the sign bit of addition results. We add N-bit operands and extend the addition result to N+1 bits to guarantee correct accumulation. The extended sign bit is determined following equation (1), which equals the majority of the two most significant SUM bits ($SUM_{bit_{N-1}}$ and SUM_{bit_N}) and the last Carry-out bit $Carry_{bit_N}$. Note that equation (1) is only valid for ternary value accumulation, invalid for standard signed integer accumulation.

$$Sign = \textbf{Majority}(SUM_{bit_{N-1}}, SUM_{bit_N}, Carry_{bit_N}) \quad (1)$$

We provide the truth table of ternary value accumulation in Table III to verify equation (1), which contains adding 2-bit ternary values to get 3-bit results and adding 3-bit values to get 4-bit results. Other longer bitwidth addition in the TVBA also follows equation (1).

Fig. 2 (a) presents a high-level TVBA workflow. First, we add the 2-bit ternary integers and obtain 3-bit immediate results. We add the first two bits as standard integer addition and store the $Carry_{bit_N}$ inside the sensing circuit in steps 1-2,

TABLE III
TERNARY VALUE ADDITION WITH THE EXTENDED SIGN BIT

Bitwidth	a	b	SUM	Carry Out	Extra Sign	a+b
2-bit to 3-bit	01(+1)	01(+1)	10	0	0	010(+2)
		00(0)	01	0	0	001(+1)
		11(-1)	00	1	0	000(0)
	00(0)	00(0)	00	0	0	000(0)
		11(-1)	11	0	1	111(-1)
	11(-1)	11(-1)	10	1	1	110(-2)
3-bit to 4-bit	010(+2)	010(+2)	100	0	0	0100(+4)
		001(+1)	011	0	0	0011(+3)
		000(0)	010	0	0	0010(+2)
		111(-1)	001	1	0	0001(+1)
		110(-2)	000	1	0	0000(0)
	001(+1)	001(+1)	010	0	0	0010(+2)
		000(0)	001	0	0	0001(+1)
		111(-1)	000	1	0	0000(0)
		110(-2)	111	0	1	1111(-1)
	000(0)	000(0)	000	0	0	0000(0)
		111(-1)	111	0	1	1111(-1)
		110(-2)	110	0	1	1110(-2)
	111(-1)	111(-1)	110	1	1	1110(-2)
		110(-2)	101	1	1	1101(-3)
	110(-2)	110(-2)	100	1	1	1100(-4)

then generate the third bit (sign bit) by reading the highest two bits of the addition result and reusing the stored $Carry_{bit_N}$ in step 3. Second, we add the 3-bit integers and get the 4-bit immediate results. Third, we repeat this addition process until we get the accumulation result. To prevent unnecessary addition overhead, we set a bitwidth threshold for immediate results. For example, the highest accumulation bitwidth is 9-bit when accumulating 192 ternary values, which guarantees correct accumulation results in the worst case.

2) Computational Complexity Analysis on Accumulation Schemes: Fig. 3 compares the computational complexity of accumulation by extending the operand sign bits (Ext-Sign), the two-round accumulation using the encoding of TiM-DNN [9] and TeC-Cell [10] (TiM-TeC), and our proposed TVBA based on signed integer encoding. Supposing the vector length is N and n is an iterator variable ($N = 2^n$ in Ext-Sign and TVBA cases, $N = 3 \times 2^{n-1}$ in the TiM-TeC case). Ext-Sign has a large sign bit extension overhead on every addition operand, making the average accumulation overhead for one ternary operand equal to 6-bit addition.

TiM-TeC improves the accumulation speed compared with Ext-Sign greatly but still has performance bottlenecks. As Table IV shows, TiM-DNN [9] and TeC-Cell [10] encodes

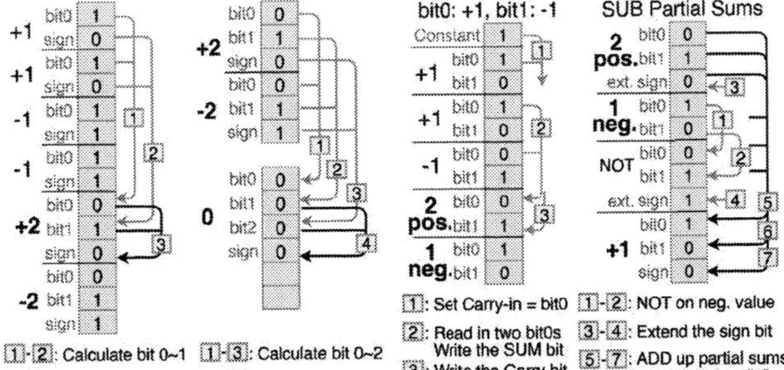

(a) Ternary Variable-Bitwidth Accumulation (b) Accumulation in TiM-TeC Encoding

Fig. 2. Accumulation schemes in iMAT and related works.

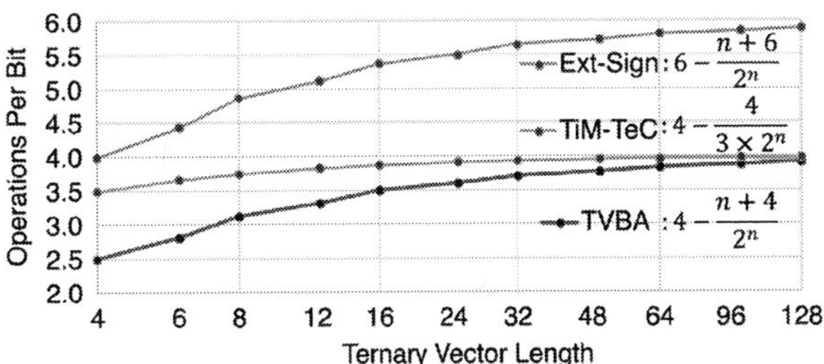

Fig. 3. The computation complexity of ternary accumulation schemes.

the +1 as "10" and the -1 as "01", which means counting the first bits of a ternary vector can get the number of +1 pos and counting the second bits produces the number of -1 neg. The accumulation result of the ternary vector equals $(pos - neg)$, as Fig. 2 (b) shows. Therefore, TiM-TeC encoding results in an inefficient two-round accumulation on the first and second bits with a subtraction on the partial sums.

$$A = N \times (\frac{3}{2} + \frac{4}{4} + \frac{5}{8} + \frac{6}{16} + ... + \frac{n+2}{2^n}) \quad (2)$$

$$= N \times (4 - \frac{n+4}{2^n}) \quad (3)$$

Our TVBA further reduces the accumulation complexity. As Fig. 2 and Table IV show, thanks to the standard signed integer encoding and the one-round accumulation workflow, TVBA accumulates 4 ternary values in 10 steps while TiM-TeC accumulates 3 ternary values in 13 steps. The computational complexity of TVBA is presented in equations (2)-(3). Though our TVBA has the same complexity as TiM-TeC when $n \to \infty$, it has 5.6%-28.6% lower complexity than TiM-TeC style accumulation and 33.4%-37.5% lower complexity than Ext-Sign when $4 \le N \le 512$ (under typical CMA sizes).

B. iMAT Architecture

Our accelerator contains two main components, STT-MRAM-based Computational Memory Arrays (CMAs) and a Digital Processing Unit (DPU), as Fig. 4 (a) shows. The CMAs with sparse dot products conduct ternary GEMM of convolution and fully-connected layers, while the DPU deals with other layers like ReLU and batch normalization. This subsection focuses on the CMA architecture while the next subsection simplifies the DPU by layer-level optimization.

1) Computational Memory Array (CMA): The CMA consists of a memory controller, a row address decoder, a column address decoder, Sense Amplifiers (SAs), and memory cells, as shown in Fig. 4 (b). The SDPU resides inside the memory controller. We load the ternary weights to the SDPU to generate corresponding activation signals, e.g., add up operands in two rows. Subsection 3) will introduce the SDPU details.

Our CMA flexibly supports three work modes: a typical STT-MRAM array with memory Read and Write, a traditional

TABLE IV
THE ENCODING SCHEMES IN RELATED WORKS AND OUR iMAT

Num	RTN [12]	SpinLiM [11]	IMC-CD [13] TiM [9] W	TeC [10] TiM [9] X	**iMAT/ Signed Int**
+1	11	11	10	10	**01**
0	0X	X0	0X	00	**00**
-1	10	01	11	01	**11**

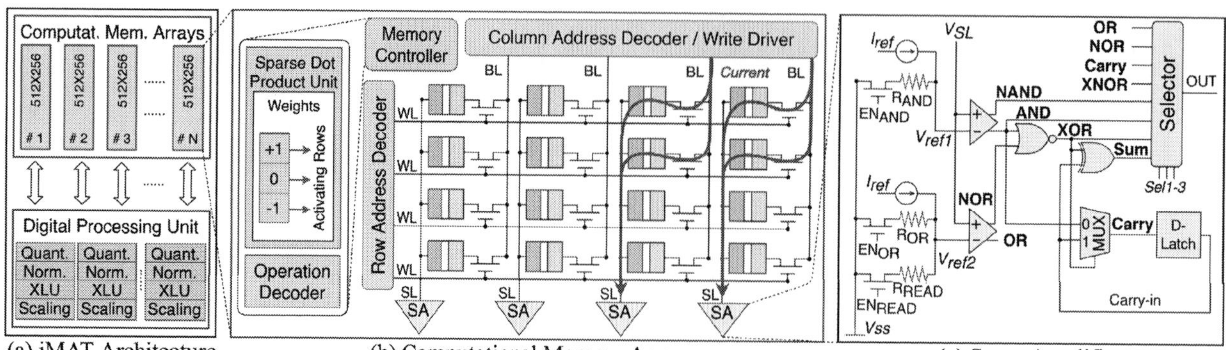

(a) iMAT Architecture (b) Computational Memroy Array (c) Sense Amplifier

Fig. 4. The detailed architecture of our proposed accelerator iMAT.

IMC array with boolean and integer arithmetic operations, and a sparse ternary dot product engine for TNNs. Taking the traditional IMC array mode as an example, the computation workflow of the CMA is as follows. First, the memory controller decodes the instructions from a CPU and sends signals to the memory row and column decoders. Second, the address decoders activate corresponding word-lines (WLs) in memory rows and bit-lines (BLs) in columns. Then reference currents flow over these memory cells to the source-lines (SLs). Last, the SAs compare the sensed voltage from SLs with reference voltages to provide the desired results in the instruction.

2) Enhanced Sense Amplifier (SA): Our CMA supports 10 native functions by configuring the SA, as Table V shows. When conducting TVBA, the SA provides the extended sign bit at the Carry-out port with the majority function. It also supports the subtraction function by one NOT and one addition with the carry-in signal to be "1", as equation (4) shows.

The proposed SA's detailed architecture is presented in Fig. 4 (c). First, the three enable signals EN_{AND}, EN_{OR}, and EN_{READ} control the corresponding circuits to provide the reference voltages for AND, OR, and READ operations. Second, the SA compares the sensed voltage from the source line V_{SL} with reference voltages V_{ref1} and V_{ref2} to obtain the AND, OR, or READ result. Next, other complex functions XOR, SUM, and Carry can be obtained by adding logic gates following equation (5)-(8). Inspired by [7], the Carry is calculated using only one multiplexer (MUX). Finally, the selector selects the calculation result according to Table V.

$$a - b = a \textbf{ ADD } (\textbf{NOT } b),\ Carry_{in,bit0} = 1 \quad (4)$$

$$\textbf{XOR} = AB \textbf{ NOR } \overline{(A+B)} \quad (5)$$

$$\textbf{SUM} = (A \text{ XOR } B) \textbf{ XOR } Carry_{in} \quad (6)$$

$$\textbf{Carry} = \textbf{Majority}(A, B, Carry_{in}) \quad (7)$$

$$= (A \text{ XOR } B) \textbf{ MUX } (AB, Carry_{in}) \quad (8)$$

We optimize the SA by selector port sharing and logic gate sharing. First, the READ and OR share the same selector port because they use the same Operational Amplifier (Op-Amp). So are NOT and NOR. Second, XNOR is calculated by reusing the NOT logic gate inside the MUX.

3) Sparse Dot Product Unit (SDPU): The sparse dot product is the core operation of ternary GEMM. As the weights are {+1, 0, -1}, the ternary multiplication-accumulation operations are equivalent to direct addition and subtraction operations. Therefore, we adopt an addition-based sparse dot product to reduce the computation complexity.

TABLE V
THE CONFIGURATION SIGNALS OF ENHANCED SENSE AMPLIFIER

Function	EN Signals			Selector Signals		
	AND	OR	READ	Sel1	Sel2	Sel3
Read	0	0	1	0	0	0
OR	0	1	0			
NOT	0	0	1	0	0	1
NOR	0	1	0			
Carry	1	1	0	0	1	0
XNOR	1	1	0	0	1	1
NAND	1	0	0	1	0	0
AND	1	0	0	1	0	1
XOR	1	1	0	1	1	0
SUM	1	1	0	1	1	1

The SDPU removes the ternary multiplication overhead and reduces the accumulation overhead with sparsity and TVBA. As Fig. 5 (a) shows, we skip the operations on zero weights by adding the activations corresponding to non-zero weights only. We load the filter weights to SDPU weight registers (W-reg) to generate memory row activation signals. First, we add those ternary activations catering to weight "+1" and store the immediate results in the reserved blank cells. As Fig. 5 (a) shows, the positive partial sums of rows 2, 3, 4, and 5 are stored as ia1, ib1, ..., ix1. Second, we repeat the TVBA on activations related to weight "-1" and get the negative partial summation results ia2, ib2, ..., ix2. Third, a subtraction operation between the partial summation result pairs gets the final dot product results Sa, Sb, ..., Sx. Note that all memory columns of one CMA can compute in parallel.

C. Layer-Level Optimizations

The DPU conducts scaling, activation, normalization, and quantization functions in the TNN computation workflow. These functions need high-precision operators, e.g. 16-bit fixed-point scalar-vector multiplication and addition operations. Therefore, we optimize them at the layer level to improve the DPU efficiency without affecting the TNN accuracy.

1) Computation Workflow in TNNs: Taking the convolution layer as an example, the activations X and weights W in TNNs are quantized into ternary values with scaling factors s_w, s_x and biases b_x, as equation (9)-(11) show. The quantized convolution result is presented in equations (12)-(13), where

$s_c = s_x \cdot s_w$ and $b_c = b_x \cdot s_w \cdot (I * W^t)$. As Fig. 5 (b) shows, the ternary activations are mapped to the memory cells of CMAs in an Img2Col manner. We leave 1/4-1/2 of the memory rows at the bottom to serve as immediate result storage. The ternary filter weights stored in a nearby CMA are loaded to the SDPUs filter by filter. Then we conduct ternary GEMM based on the addition-based sparse dot product to get convolution result Y_c.

$$x^t = \begin{cases} +1, & x > x_p \\ 0, & x_n \leq x \leq x_p \\ -1, & x < x_n \end{cases} \tag{9}$$

$$W = s_w \cdot W^t \tag{10}$$

$$X = s_x \cdot X^t + b_x \tag{11}$$

$$Y = s_x \cdot s_w \cdot (X^t * W^t) + b_x \cdot s_w \cdot (I * W^t) \tag{12}$$

$$= s_c \cdot Y_c + b_c, \quad Y_c = X^t * W^t \tag{13}$$

Then the convolution results are sent to the DPU to perform the scaling in equation (13), activation functions, batch normalization, and ternarization. The frequently used ReLU series activation layers in TNNs like PReLU and ReAct PReLU share a unified form called XLU (X Linear Unit). As equations (14)-(15) show, ReAct PReLU can be generalized to XLU, where $k_1 = 1$, $k_2 = k$, $y_1 = y_0 - x_0$, and $y_2 = y_0 - k \cdot x_0$. Furthermore, the batch normalization (BN) collapses into a linear function during inference [4], as we apply the pre-trained u, γ, σ, and β directly. Thus, we simplify BN in equation (16), where $s_b = \frac{\gamma}{\sqrt{\sigma^2+\varepsilon}}$ and $b_b = \beta - \frac{u}{\sqrt{\sigma^2+\varepsilon}}$.

$$ReActPReLU(x) = \begin{cases} x - x_0 + y_0, & x > x_0 \\ k \cdot (x - x_0) + y_0, & x \leq x_0 \end{cases} \tag{14}$$

$$XLU(x) = \begin{cases} k_1 \cdot x + y_1, & x > x_0 \\ k_2 \cdot x + y_2, & x \leq x_0 \end{cases} \tag{15}$$

$$BN(x) = \gamma \cdot \frac{x - u}{\sqrt{\sigma^2 + \varepsilon}} + \beta = s_b \cdot x + b_b \tag{16}$$

2) Fused Scaling Functions (FSFs): As quantized convolution, ReLU series activation, and normalization are linear transformation functions, we combine equations (13), (15), and (16) to a fused scaling function in equations (17)-(18), where $s_i = s_b \cdot k_i \cdot s_c$ and $b_i = s_b \cdot k_i \cdot b_c + s_b \cdot y_i + b_b$. This fused scaling function (18) with three operations (comparison, multiplication, and addition) is used before residual shortcuts. One quantization is needed after the shortcut.

$$Y = \begin{cases} s_b \cdot (k_1 \cdot (s_c \cdot Y_c + b_c) + y_1) + b_b, & x > \frac{x_0 - b_c}{s_c} \\ s_b \cdot (k_2 \cdot (s_c \cdot Y_c + b_c) + y_2) + b_b, & x \leq \frac{x_0 - b_c}{s_c} \end{cases} \tag{17}$$

$$= \begin{cases} s_1 \cdot Y_c + b_1, & x > x_1 \\ s_2 \cdot Y_c + b_2, & x \leq x_1 \end{cases} \tag{18}$$

$$y^t = \begin{cases} +1, & y_c > (x_p - b_1)/s_1 \\ -1, & y_c < (x_n - b_2)/s_2 \\ 0, & else \end{cases} \tag{19}$$

Furthermore, we can fuse equation (18) with the ternarization function (9) to equation (19). Therefore, the ternarized convolution result Y_c can produce the ternarized activation for the next layer ($Y_i^t = X_{i+1}^t$) by comparing it with the pre-calculated thresholds in equation (19). Compared with the sequential computation with 8 operations and the fused version with 4 operations, our highly optimized function (19) only needs 1 operation between consecutive convolution layers.

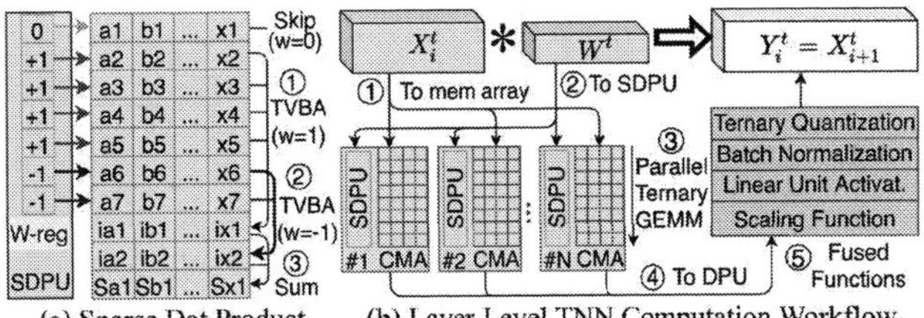

(a) Sparse Dot Product (b) Layer-Level TNN Computation Workflow

Fig. 5. The computation workflow on iMAT.

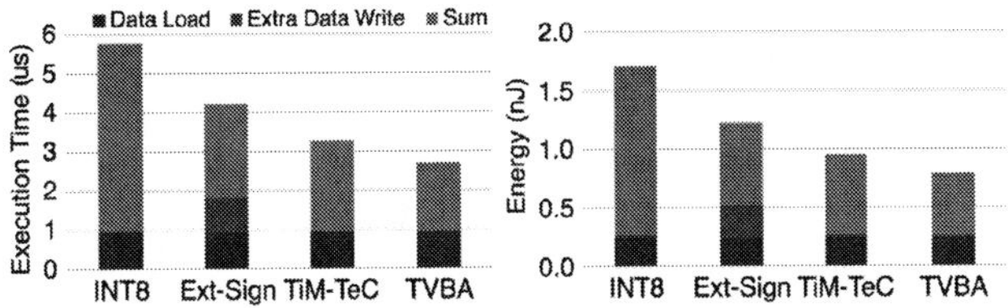

Fig. 6. The performance of different methods on accumulating 128 ternary values. Left: execution time. Right: energy.

IV. Evaluation Results

We implement the Digital Processing Unit with fused scaling functions and the enhanced memory Sense Amplifier of our proposed iMAT in Cadence Virtuoso IC6.1.8 based on the NCSU 45nm FreePDK [14]. We refer to related work STT-CiM [15] for the Op-Amp inside the SA. The power and latency of the DPU and SA are obtained from Virtuoso ADE L using Spretre under 1.2V Vdd, and the STT-MRAM write latency and energy are extracted from [16].

A. Performance of Variable-Bitwidth Accumulation

We compare TVBA with traditional accumulation methods on accumulating 256 ternary vectors on one CMA. Each vector contains 128 2-bit ternary values and all 256 columns of the CMA compute in parallel. The baseline methods are using fixed-bitwidth 8-bit integers to accumulate (INT8), extending the sign bits of operands (Ext-Sign), and using the encoding of TiM-DNN [9] and TeC-Cell [10] (TiM-TeC).

Our TVBA reduces the total accumulation time by 53.4%, 36.0%, and 17.2% and the energy by 53.9%, 35.6%, and 17.1% than using INT8, Ext-Sign, and TiM-TeC, respectively. Fig. 6 provides the execution time and energy breakdown of the accumulation methods. Accumulating using INT8 leads to long summation (Sum) latency and much energy cost. Ext-Sign can reduce the Sum latency by 51.2%, but results in extra data write overhead on extending the sign bits. TiM-TeC has 22.6% lower total latency than Ext-sign thanks to less extra

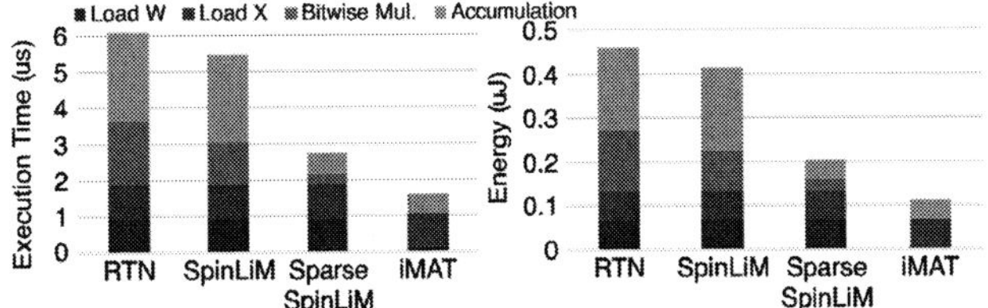

Fig. 7. The dot product performance of iMAT SDPU and related works. Left: execution time. Right: energy.

data write. Our proposed TVBA reduces the Sum latency by 23.6% than TiM-TeC thanks to lower complexity.

B. Performance of Sparse Dot Product

Fig. 7 presents the performance and energy of the ternary dot product of related works and iMAT. We implement the dense bitwise operation-based ternary dot product of RTN [12] and SpinLiM [11] on the same STT-MRAM-based CMA. We benchmark the ternary dot product between 256 activation vectors and one weight vector with 75% sparsity (16 "+1" and 16 "-1"), and the vector length is 128. A Sparse SpinLiM that skips zero operations is provided as a reference to show the effectiveness of sparsity, and it achieves 1.99× speedup and 2.03× energy efficiency than SpinLiM.

Our proposed SDPU achieves 3.45× and 3.83× speedup and 3.71× and 4.13× energy efficiency than SpinLiM and RTN respectively. RTN and SpinLiM store both the weight and activation vectors in the CMA, leading to a large weight loading overhead (Load W). They utilize bitwise operations to conduct the ternary multiplication and then accumulate the ternary multiplication results, leading to a large computation overhead. In contrast, our sparse dot product method converts the ternary multiplication into addition and subtraction operations and fuses the addition and subtraction with the accumulation process to boost performance.

C. Performance on DNN Models

We evaluate the model-level performance based on the official PyTorch ResNet-18 weights quantized using Least-Squared Quantization [4]. Our accelerator iMAT achieves 2.53-2.70× speedup and 3.41-3.72× energy efficiency over related works, as Fig. 8 shows. Our Sparse Dot Product and Ternary Variable-Bitwidth Accumulation contribute to 2.20× speedup and 3.69× energy efficiency over related work RTN. The Fused Scaling Functions (FSFs) further bring 22.73% and 0.81% improvement to the speed and energy efficiency. The ternary ResNet-18 has 43.73%-62.48% layer-level sparsity and the average sparsity is only 51.87%. Our iMAT can achieve higher performance on models with higher sparsity.

V. Conclusion

In this paper, we propose an STT-MRAM-based in-memory accelerator iMAT for TNNs. We propose an efficient Ternary Variable-Bitwidth Accumulation scheme with up to 37.5% lower complexity than traditional accumulation schemes. We propose an in-memory architecture with an enhanced sensing circuit for the accumulation scheme and sparse dot product unit that can skip the null operations on zero weights of convolution and fully-connected layers. We further propose a fused scaling function that removes up to 87.5% operations in scaling, activation, normalization, and quantization layers. Evaluation results show that iMAT achieves 2.5-2.7× speedup and 3.4-3.7× energy efficiency over RTN and SpinLiM on ternary ResNet-18.

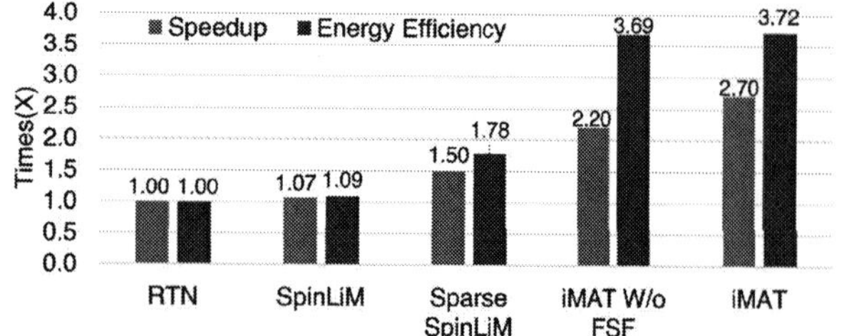

Fig. 8. iMAT's performance on ternary ResNet-18.

Acknowledgment

This work is partially supported by the Ministry of Education, Singapore, under its Academic Research Fund Tier 2 (MOE2019-T2-1-071), and Nanyang Technological University, Singapore, under its NAP (M4082282/04INS000515C130).

References

[1] F. Zhu, R. Gong, F. Yu, X. Liu, Y. Wang, Z. Li, X. Yang, and J. Yan, "Towards unified int8 training for convolutional neural network," in *Proceedings of the IEEE/CVF Conference on Computer Vision and Pattern Recognition*, 2020, pp. 1969–1979.

[2] X. Sun, N. Wang, C.-Y. Chen, J. Ni, A. Agrawal, X. Cui, S. Venkataramani, K. El Maghraoui, V. V. Srinivasan, and K. Gopalakrishnan, "Ultra-low precision 4-bit training of deep neural networks," *Advances in Neural Information Processing Systems*, vol. 33, 2020.

[3] Y. Li, W. Ding, C. Liu, B. Zhang, and G. Guo, "Trq: Ternary neural networks with residual quantization," in *Proceedings of the AAAI Conference on Artificial Intelligence*, 2021.

[4] H. Pouransari, Z. Tu, and O. Tuzel, "Least squares binary quantization of neural networks," in *the IEEE/CVF Conference on Computer Vision and Pattern Recognition Workshops*, 2020, pp. 698–699.

[5] Y. Ma, Y. Du, L. Du, J. Lin, and Z. Wang, "In-memory computing: The next-generation ai computing paradigm," in *Proceedings of the 2020 on Great Lakes Symposium on VLSI*, 2020, pp. 265–270.

[6] A. Sebastian, M. Le Gallo, R. Khaddam-Aljameh, and E. Eleftheriou, "Memory devices and applications for in-memory computing," *Nature nanotechnology*, vol. 15, no. 7, pp. 529–544, 2020.

[7] S. Zhu, S. Li, and W. Liu, "imad: An in-memory accelerator for addernet with efficient 8-bit addition and subtraction operations," in *Proceedings of the Great Lakes Symposium on VLSI*, 2022, p. 65–70.

[8] S. Yu, H. Jiang, S. Huang, X. Peng, and A. Lu, "Compute-in-memory chips for deep learning: Recent trends and prospects," *IEEE Circuits and Systems Magazine*, vol. 21, no. 3, pp. 31–56, 2021.

[9] S. Jain, S. K. Gupta, and A. Raghunathan, "Tim-dnn: Ternary in-memory accelerator for deep neural networks," *IEEE Transactions on Very Large Scale Integration (VLSI) Systems*, vol. 28, no. 7, pp. 1567–1577, 2020.

[10] S. K. Thirumala, S. Jain, S. K. Gupta, and A. Raghunathan, "Ternary compute-enabled memory using ferroelectric transistors for accelerating deep neural networks," in *2020 Design, Automation & Test in Europe Conference & Exhibition (DATE)*, 2020, pp. 31–36.

[11] L. Luo, H. Zhang, J. Bai, Y. Zhang, W. Kang, and W. Zhao, "Spinlim: Spin orbit torque memory for ternary neural networks based on the logic-in-memory architecture," in *2021 Design, Automation & Test in Europe Conference & Exhibition (DATE)*, 2021, pp. 1865–1870.

[12] Y. Li, X. Dong, S. Q. Zhang, H. Bai, Y. Chen, and W. Wang, "Rtn: Reparameterized ternary network," in *Proceedings of the AAAI Conference on Artificial Intelligence*, vol. 34, 2020, pp. 4780–4787.

[13] X. Yang, K. Zhu, X. Tang, M. Wang, M. Zhan, N. Lu, J. P. Kulkarni, D. Z. Pan, Y. Liu, and N. Sun, "An in-memory-computing charge-domain ternary cnn classifier," in *IEEE Custom Integrated Circuits Conference*, 2021, pp. 1–2.

[14] North Carolina State University, "NCSU FreePDK45," 2011. [Online]. Available: https://eda.ncsu.edu/freepdk/freepdk45/

[15] S. Jain, A. Ranjan, K. Roy, and A. Raghunathan, "Computing in memory with spin-transfer torque magnetic ram," *IEEE Transactions on Very Large Scale Integration Systems*, vol. 26, no. 3, pp. 470–483, 2017.

[16] N. Sayed, M. Ebrahimi, R. Bishnoi, and M. B. Tahoori, "Opportunistic write for fast and reliable stt-mram," in *Design, Automation & Test in Europe Conference & Exhibition (DATE)*, 2017, pp. 554–559.

TensorCV: Accelerating Inference-Adjacent Computation Using Tensor Processors

Dongho Ha
Yonsei University
Seoul, Korea
dongho.ha@yonsei.ac.kr

Won Woo Ro
Yonsei University
Seoul, Korea
wro@yonsei.ac.kr

Hung-Wei Tseng
University of California, Riverside
Rivsrside, California, USA
htseng@ucr.edu

Abstract—**The advancements in AI/ML accelerators have made the core AI/ML computation relatively insignificant in application pipelines. For example, inferencing only accounts for 3% of the latency in an image-based ML pipeline with the help of Tensor Cores. The mismatch in performance growth between ML model computation and ML-adjacent computation, the producer and consumer of ML models, will become the bottleneck leading to system inefficiency.**

This paper presents a set of innovative algorithms to allow the entire ML-based computer vision pipelines to leverage AI/ML accelerators. Our proposed algorithms feature matrix-based operations that AI/ML accelerators specialize in. Simply compiler optimizations cannot take full advantage of hardware acceleration without revisiting algorithms.

This paper implements the proposed algorithms as an open-source library, TensorCV, in a system platform with Tensor Cores. TensorCV shows a 6.12× speedup in optimized ML-adjacent functions and saves 81% energy consumption on modern heterogeneous computers. The code is available at https://github.com/escalab/TensorCV.

I. Introduction

The broad spectrum of applications that use camera and video inputs to sense the world make computer vision-related workloads one of the essential categories in artificial intelligence (AI) and machine learning (ML). Recent advancements in AI/ML hardware accelerators, including Google's Tensor Processing Units (TPUs), NVIDIA's Tensor Cores, Apple's Neural Engines, Intel's Gaussian Neural Accelerators, etc., have significantly improved the computation time directly related to the core of AI/ML, namely, inference and training. As a result, inferencing a highly optimized NN model can account for less than 3% of the time in computer vision pipelines. Instead, these non-inference code sections consume the majority of execution time in these applications nowadays.

These code sections adjacent to the core ML inference or training process of the computer vision pipeline typically perform operations that help enhance or extract the most critical part of images to enable more accurate and efficient ML results. Unfortunately, the conventional approach typically relies on CPU code whose performance can only scale with the relatively slow improvement through Moore's Law, but not the rapid growth from emerging, innovative hardware accelerators. The inefficiency of these ML-adjacent stages will lead to under-utilized hardware accelerators and become the performance bottleneck, as these stages cannot feed sufficient inputs to well-optimized ML models.

To address the above problems without increasing hardware costs, a potential solution is leveraging existing hardware accelerators (e.g., Tensor Cores, TPUs) for inference/training-adjacent stages. Using hardware accelerators can bring several benefits. (1) These accelerators' microarchitecture can directly compute on higher dimensional datasets, providing more efficient processing models for inference/training-adjacent stages. (2) Using the same training/inference hardware for adjacent computation can remove unnecessary data movement and transformation overhead. (3) The associative property of tensor algebra enables optimizations crossing the boundaries of stages in the pipeline to further reduce operations. (4) By moving more computation into AI/ML accelerators, the system can free up CPUs/GPUs for more meaningful workloads. (5) The system can reclaim the significantly wasted idle power in AI/ML accelerators [20]. (6) The application can significantly reduce the total energy consumption if the execution time is at the same level, while AI/ML accelerators consume lower power than other computing resources.

However, using AI/ML accelerators for non-training/inference tasks is still challenging for the following reasons. First, existing AI/ML accelerators take domain-specific design approaches and abstract their hardware operations in a domain-specific way. Converting existing general-purpose programming language code to use a domain-specific language interface is non-trivial. Second, and probably the most important, due to the difference in micro-architectures and execution models, existing workloads will need to change their algorithms fundamentally to make use of AI/ML accelerators.

In this paper, we demonstrate the application of AI/ML accelerators in accelerating inference/training-adjacent tasks. We revisited the design of frequently used, performance-critical inference-adjacent functions in modern computer vision (CV) pipelines. We proposed matrix/tensor-based algorithms to allow these functions to enjoy the facilities that AI/ML accelerators provide. Though these algorithms potentially have higher algorithmic complexity than existing solutions, these algorithms can still supersede the performance of optimized implementations on modern general-purpose processors since AI/ML accelerators can execute our underlying operations efficiently. Our implementation, TensorCV, shows our algorithms and implementations can achieve 6.12× and save 81% energy consumption compared to the CPU implementation on a desktop computer using the latest CPU and GPU approaches, while the existing GPU implementation shows 2.98× speedup and 64% energy saving.

In presenting TensorCV, this paper makes the following contributions. (1) It proposes a set of algorithms that enable critical image-processing functions on AI/ML accelerators. AI/ML hardware for the same functions is only possible with these algorithms. (2) It implements and evaluates the proposed algorithms on a real system platform to prove the performance benefits. (3) It provides an open-source implementation aligning with the interface of the most popular computer vision library to impact a broad spectrum of CV applications.

979-8-3503-1176-1/23 $31.00 © 2023 IEEE

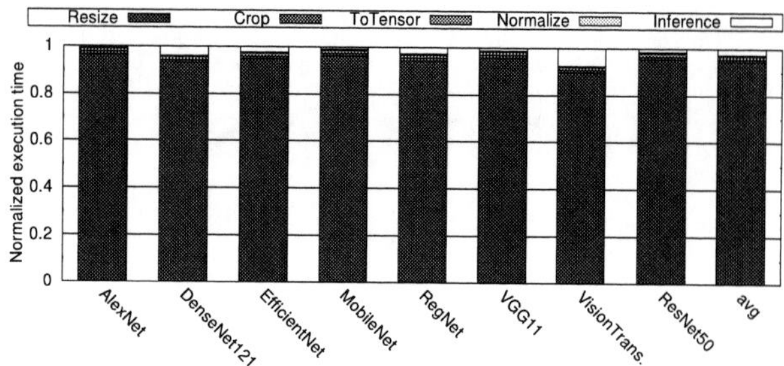

Fig. 1. Latency breakdown when running a MobileNetV3 object detection

II. Background and Motivation

A. AI/ML accelerators

AI/ML accelerators have significant presence in modern computer systems as they are more efficient than conventional GPUs in AI/ML tasks for the following reasons. (1) AI/ML accelerators require fewer operations and cycles in performing the same task. For example, each Tensor Core operation can multiply two tiles of matrices in one cycle. In contrast, using GPUs' vector processing model, a matrix multiplication would require a vector element-wise multiplication and an accumulation operation on each pair of rows and columns. When multiplying two 8K × 8K matrices, tensor cores will take 2^{30} tile MMA operations that require 3.2 million cycles on RTX 3090. Still, conventional CUDA cores will use 2^{39} multiplications and 2^{26} accumulations that require 100 million cycles on the same GPU. (2) As the hardware is specialized for matrix operations, the same circuit area can deliver higher throughput than general-purpose architectures. (3) The design can use smaller circuit areas to reduce power consumption. However, due to AI/ML accelerators' specialization for NN/matrix operations, legacy programs cannot easily take advantage of these accelerators unless their algorithms are presented in matrix algebra.

This paper implements TensorCV algorithms on Tensor Cores for accessibility reasons. Tensor Cores are ubiquitous in NVIDIA's GPU architectures, and NVIDIA made their API available at various level programming frameworks. Conversely, high-performance TPUs are only accessible through Google's cloud services, and Apple's NPUs are only available on their machines without revealing their API to the public. However, as these AI/ML accelerators are essentially matrix processing units, we envision the same algorithm that works efficiently on Tensor Cores would work on other AI/ML accelerators with minimal modifications.

B. Modern CV pipelines and performance bottlenecks

The advantage of avoiding a tremendous amount of manual feature engineering while maintaining high accuracies in classifications and recognitions makes NNs an inevitable component in modern CV applications. In modern AI/ML-assisted CV applications, the application must standardize, shrink and clean up the content before inference because these inference-adjacent operations can reduce both computational operations and memory consumption and improve the accuracy of inference, making NN models more efficient and economically available in applications.

With AI/ML accelerators significantly improve the inference performance, inference-adjacent computation becomes more critical in CV pipelines. Figure 1 shows latency breakdown in running popular image classification applications written in PyTorch and TensorRT libraries. For these CV workloads, the inference-adjacent stages take 4032×3024 images as inputs, resize images to 256×256 (Resize) or crop images to 224×224 (Crop), reshape the tensors (ToTensor), and normalize the images (Normalize). The result shows that inference-adjacent stages account for 97% of overall latency. These applications only spent 3% of time on Tensor Cores for inference.

TABLE I
OpenCV image pre-processing functions

Function Name	Number of GitHub Code	Description
resize()	>1000k	Resize an image or a video frame.
cvtColor()	682k	Convert an image from one color space to another.
rect()	504k	Define a rectangular region of interest in an image.
normalize()	477k	Normalize an image or a matrix.
rotate()	295k	Rotate an image or a matrix by a specified angle.
Canny()	262k	Perform edge detection on an image.
Sobel()	188k	Perform gradient calculation on an image.
dilate()	182k	Perform morphological dilation on an image.
findContours()	169k	Find the contours in an image.
erode()	158k	Perform morphological erosion on an image.

III. Algorithm

As modern AI/ML accelerators cannot work on computation without matrix operations, the most critical task in this paper is revisiting entrenched implementations to promote the use of matrix operations. Therefore, the fundamental idea of this paper is treating each input image as an input matrix $Input$, and our algorithms dynamically create specialized matrices (i.e., matrix kernels) that the algorithm can later perform operations together with $Input$ to achieve equivalent image processing results.

Table I lists the most frequently used image pre-processing functions in Open-CV [2] ranked by their occurrence in public GitHub repositories. Considering the demands of each function in NN applications, this paper targets five functions; `resize`, `cvtColor`, `crop`, `rotate`, and `normalize`. Existing implementations of these functions target conventional CPU/GPU applications and employee scalar or vector computation that AI/ML cannot perform efficiently. The following sections elaborate on the how this paper generates appropriate matrix kernels for the corresponding algorithms.

A. Resize

1) Baseline resize algorithm: Resize is the most frequently used pre-processing function. In AI/ML-assisted CV pipelines, resize function can help the application shrink images to fit the demanding input size of the model. By shrinking input size, resizing helps a training or trained model work with various sizes of input images and, probably the most important, reduce memory consumption and execution time.

In the conventional bi-linear resizing implementation, the code will first compute the relative ratio between input and output matrix sizes, which we call row and column scales ($rowScale$ and $columnScale$). For an input image ($Input$) sized m-by-n, the input matrix size is m-by-$3n$ since each row includes R, G, and B channels. If the target output image size is m'-by-n' the output matrix size is m'-by-$3n'$. Hence, the code computes the row and column scales as follows:

$$rowScale = \frac{m'}{m}, \quad columnScale = \frac{n'}{n} \tag{1}$$

Then, using the scales, the code iterates through every bounding box containing the source pixels to a target pixel and calculates the weighted average as the target pixel value. The following equations show how the baseline code calculates the boundary $(top_i, bot_i, left_j, right_j)$ of bounding boxes in $Input$ and weights for corresponding output pixel (i, j).

$$\begin{aligned} top_i &= \lfloor \frac{i}{rowScale} \rfloor \\ bot_i &= \lceil \frac{i}{rowScale} \rceil \\ left_j &= \lfloor \frac{\lfloor j/3 \rfloor}{columnScalme} \rfloor \\ right_j &= \lceil \frac{\lfloor j/3 \rfloor}{columnScale} \rceil \\ rowWeight_i &= \frac{i}{rowScale} - top_i, \\ colWeight_j &= \frac{\lfloor j/3 \rfloor}{columnScale} - left_j \end{aligned} \tag{2}$$

The finally, the conventional algorithm will calculates each output pixel (i, j) value as follows:

$$\begin{aligned} &output_{i,j} = \\ &(1 - rowWeight_i) \cdot (1 - colWeight_j) \cdot Input_{top_i, left_j} + \\ &(1 - rowWeight_i) \cdot (colWeight_j) \cdot Input_{top_i, right_j} + \\ &(rowWeight_i) \cdot (1 - colWeight_j) \cdot Input_{bot_i, left_j} + \\ &(rowWeight_i) \cdot (colWeight_j) \cdot Input_{bot_i, right_j} \end{aligned} \tag{3}$$

However, the algorithm in OpenCV-CUDA cannot exploit Tensor Cores since the there is no matrix multiplications but only exists element-wise weighted averages.

2) TensorCV resize algorithm: TensorCV transforms the conventional implementation into two matrix multiplications. Considering an m-by-$3n$ input and m'-by-$3n'$ output images, the proposed algorithm creates an m'-by-m matrix as L^{resize} and $3n$-by-$3n'$ matrix as R^{resize}. The content of L^{resize} and R^{resize} is only related to the target image's size; therefore, TensorCV only needs to create them once for every batch. In contrast, the conventional approach not only prevents the code from using Tensor Cores but also recalculates weights for the same pixel position in each channel. The proposed algorithm will fill the content of L^{resize} using the following formula:

$$L^{resize}_{m \times m'} \ni l^{resize}_{i,j} = \begin{cases} 1 - rowWeight_j & \text{if } j = top_i \\ rowWeight_j & \text{if } j = bot_i \\ 0 & \text{else} \end{cases} \tag{4}$$

R^{resize} holds column-related weights, and TensorCV's algorithm fills the R^{resize} using the following formula:

$$\begin{aligned} &R^{resize}_{3n \times 3n'} \ni r^{resize}_{i,j} = \\ &\begin{cases} 1 - colWeight_i & \text{if } \begin{array}{l} \lfloor i/3 \rfloor = left_i \\ \text{and } i \pmod 3 = j \pmod 3 \end{array} \\ colWeight_i & \text{if } \begin{array}{l} \lfloor i/3 \rfloor = right_i \\ \text{and } i \pmod 3 = j \pmod 3 \end{array} \\ 0 & \text{else} \end{cases} \end{aligned} \tag{5}$$

After filling matrices L^{resize} and R^{resize} using Equation 4 and 5 at the beginning of each batch of images, the proposed algorithm can compute the output of each image resizing result as the following.

$$Output_{m' \times 3n'} = L^{resize}_{m' \times m} \cdot Input_{m \times 3n} \cdot R^{resize}_{3n \times 3n'} \tag{6}$$

B. Color space conversion

1) Baseline cvtColor algorithm: As the sensor designs vary, image sources may encode pixels differently. Therefore, CV applications must convert the color spaces between the raw image encoding to the RGB color space that computing devices most frequently use. Taking the most common conversion between YUV color space that most video encoders use to RGB color space as an example, each pixel would require several element-wise multiplications with different coefficients and accumulations of scalar values. [9]. The vectorized version in OpenCV-CUDA already expands the functions into matrix-vector multiplications with each pixel as a 1×3 vector and the coefficients (we call matrix C) as a 3×3 matrix as follows:

$$\begin{aligned} (Y \quad U \quad V) = (R \quad G \quad B) \cdot \begin{pmatrix} 0.299 & -0.147 & 0.615 \\ 0.587 & -0.289 & -0.515 \\ 0.114 & 0.436 & -0.100 \end{pmatrix} \\ (R \quad G \quad B) = (Y \quad U \quad V) \cdot \begin{pmatrix} 1 & 1 & 1 \\ 0 & -0.395 & 2.032 \\ 1.140 & -0.581 & 0 \end{pmatrix} \end{aligned} \tag{7}$$

However, the algorithm in OpenCV-CUDA still cannot fully exploit Tensor Cores since each Tensor Core Unit can work on larger sizes of matrices.

2) TensorCV cvtColor algorithm: The proposed algorithm multiplies the original m-by-$3n$ input matrix with $3n$-by-$3n$ $R^{cvtColor}$, which is similar to an identical matrix. Each element in the $R^{cvtColor}$ is filled by coefficient matrix C in the following formula:

$$\begin{aligned} &R^{cvtColor}_{3n \times 3n} \ni r^{cvtColor}_{i,j} = \\ &\begin{cases} C_{i \pmod 3, j \pmod 3} & \text{if } \lfloor i/3 \rfloor = \lfloor j/3 \rfloor \\ 0 & \text{else} \end{cases} \end{aligned} \tag{8}$$

Still, each three-by-three partial matrix is a coefficient matrix for color space conversion.

$$Output_{m \times 3n} = Input_{m \times 3n} \cdot R^{cvtColor}_{3n \times 3n} \tag{9}$$

C. Cropping

1) Baseline crop algorithm: Cropping (i.e., `rect` in OpenCV) extracts an essential part within an image for the AI/ML model. For a source image with size m-by-n and target image size of m'-by-n' and offset (x, y), the conventional implementation would require at least n' memory operations where each operation copies the length of m' from an offset of x from the beginning of each row. Suppose the cropping operation occurs during the middle stage of the pipeline. In that case, the application typically has to transfer the control to the memory controller and under-utilize the AI/ML hardware.

2) TensorCV crop algorithm: The proposed algorithm keeps the matrix content in matrix units and performs matrix multiplications but requires zero memory operations. Similar to equation 6, the proposed algorithm creates two matrices, $L^{crop}_{m' \times m}$ and $R^{crop}_{3n \times 3n'}$. $L^{crop}_{m' \times m}$ contains an identity matrix sizes m-by-m starting from column y and $R^{crop}_{3n \times 3n'}$ contains an identity matrix sizes $3n$-by-$3n$ starting from row x.

D. Normalize

1) Baseline normalize algorithm: Since the AI/ML models are trained on normalized data, the models must normalize input images before inference. To normalize an image, CV applications calculate the mean ($mean$) and standard deviation ($stddev$) of the input image's pixel values, subtract $mean$ from each pixel value, and divide it by the $stddev$ for each channel. In conventional CV applications, the code calculates mean and standard deviation and conducts normalization as follows:

$$\begin{aligned} mean = \frac{\sum Input_{i,j}}{m \times n}, \quad stddev = \frac{\sqrt{\sum (Input_{i,j} - mean)^2}}{m \times n}, \\ Output_{m \times n} = \frac{Input_{m \times n} - mean}{stddev} \end{aligned} \tag{10}$$

Again, the baseline OpenCV-CUDA algorithm contains no matrix operations for Tensor Cores.

2) TensorCV normalize algorithm: To calculate the mean, TensorCV multiplies 16-by-m matrix $L^{mean}_{16\times m}$ and n-by-16 matrix $R^{mean}_{n\times 16}$ with only ones in the first row and column. We scale the row and column dimensions to $L^{mean}_{16\times m}$ and $R^{mean}_{n\times 16}$ because the current Tensor Core hardware optimizes for 16×8 matrix operations. Since normalization transposes and resizes the matrices, handle all color spaces simultaneously is challenging. Hence, each element in the kernels $L^{mean}_{16\times m}$ and $R^{mean}_{n\times 16}$ are filled in the following formula:

$$\begin{aligned} L^{mean}_{16\times m} \ni l^{norm}_{i,j} &= \begin{cases} 1 & \text{if } i = 0 \\ 0 & \text{else} \end{cases}, \\ R^{mean}_{n\times 16} \ni r^{norm}_{i,j} &= \begin{cases} 1 & \text{if } j = 0 \\ 0 & \text{else} \end{cases} \end{aligned} \tag{11}$$

TensorCV then multiplies matrice $L^{mean}_{16\times m}$ and $R^{mean}_{n\times 16}$ with the $Input$ and results in the sum of all pixel values in the first element of the output matrix as follows:

$$Mean_{16\times 16} = \begin{pmatrix} \sum Input_{i,j} & \cdots \\ \vdots & \ddots \end{pmatrix} = L^{mean}_{16\times m} \cdot Input_{m\times n} \cdot R^{mean}_{n\times 16} \tag{12}$$

Unlike the conventional algorithm calculating the $stddev$ as Equation 10, TensorCV calculates $stddev$ using the square root of the variance of data, allowing matrix multiplications in computing $stddev$.

$$stddev = \sqrt{\frac{\sum Input^2_{i,j}}{m \times n} - mean^2} \tag{13}$$

To compute the sum of the squared values of the input matrix, TensorCV multiplies the $Input$ and transposed one, generating the squared input values in the diagonal elements of the output as the following formula:

$$\begin{aligned} StdDev'_{m\times m} &= Input_{m\times n} \cdot Input^T_{m\times n} \\ &= \begin{pmatrix} \sum Input^2_{0,j} & \cdots & \cdots & \cdots \\ \vdots & \sum Input^2_{1,j} & \cdots & \cdots \\ \vdots & \vdots & \ddots & \vdots \\ \vdots & \vdots & \cdots & \sum Input^2_{m-1,j} \end{pmatrix} \end{aligned} \tag{14}$$

To accumulate the diagonal values with matrix multiplication, the TensorCV utilizes an observation that real matrix values are stored in a single-dimension array, and users can define the length and height of the matrix before computing matrix multiplication. Thus, the algorithm add one to the height of the matrix to align the target values in the first row. Consequently, multiplying the output matrix with the transposed matrix $L^{mean}_{16\times m}$ allows calculating the sum of the squared values of the input matrix as the following formula:

$$\begin{aligned} StdDev_{m\times 16} &= \begin{pmatrix} \sum Input^2_{i,j} & \cdots \\ \vdots & \ddots \end{pmatrix} \\ &= StdDev' \begin{pmatrix} \sum Input^2_{0,j} & \sum Input^2_{1,j} & \cdots & \sum Input^2_{m-1,j} \\ \vdots & \vdots & \vdots & \vdots \\ \vdots & \vdots & \vdots & \vdots \end{pmatrix} \cdot L^{mean\,T}_{16\times m} \end{aligned} \tag{15}$$

After calculating the sum of pixel values and squared pixel values, the proposed algorithm normalizes $Input$ using pair-wise vector operations, again, not matrix operations as the following equation.

$$\begin{aligned} mean &= \frac{Mean_{16\times 16}[0]}{mn} \\ stddev &= \sqrt{\frac{StdDev_{m\times 16}[0]}{mn} - (\frac{Mean_{16\times 16}[0]}{mn})^2} \\ Output_{m\times n} &= \frac{Input_{m\times n} - mean}{stddev} \end{aligned} \tag{16}$$

E. Rotate

1) Baseline rotate algorithm: To acquire accurate inferencing results, the input images' layout must be exact to its original intent [22]. Thus, CV applications must be able to rotate the input matrix. Taking the most common case, the proposed algorithm supports 90, 180, and 270 degrees of counter-clockwise rotation. The existing CV applications rotate an image by allocating corresponding coordinate values to new coordinates. For example, a 90-degree rotation allocates the (x,y) coordinate values to (y,$width - 1 - x$). In that respect, 180- and 270-degree move (x,y) coordinate values to ($width - 1 - x$,$height - 1 - y$) and ($height - 1 - y$,x), respectively.

2) TensorCV rotate algorithm: We observe that rotation can be represented by swapping x and y coordinates and reversing coordinates ($1 - width - x$ and $1 - height - y$). Thus, the TensorCV algorithm swaps the coordinates by transposing the matrix and reverses by multiplying flipped identical matrices $L^{rotate}_{m\times m}$ and $R^{rotate}_{n\times n}$. The matrices $L^{rotate}_{m\times m}$ and $R^{rotate}_{n\times n}$ are filled in the following formula:

$$\begin{aligned} L^{rotate}_{m\times m} \ni l^{rotate}_{i,j} &= \begin{cases} 1 & \text{if } j = m - 1 - i, \\ 0 & \text{else} \end{cases}, \\ R^{rotate}_{n\times n} \ni r^{rotate}_{i,j} &= \begin{cases} 1 & \text{if } j = n - 1 - i, \\ 0 & \text{else} \end{cases} \end{aligned} \tag{17}$$

Like normalize, TensorCV performs the rotation operation on each color channel separately (using a W-H/C format) due to the transpose operation, requiring split and merge operations. The following equations depict how the proposed algorithm computes the 90-, 180-, and 270-degree rotations.

$$\begin{aligned} Output90_{n\times m} &= \left[Input_{m\times n} \cdot R^{rotate}_{n\times n}\right]^T \\ Output180_{m\times n} &= L^{rotate}_{m\times m} \cdot Input_{m\times n} \cdot R^{rotate}_{n\times n} \\ Output270_{n\times m} &= \left[L^{rotate}_{m\times m} \cdot Input_{m\times n}\right]^T \end{aligned} \tag{18}$$

F. Kernel Integration

By transforming algorithms into matrix-based ones, TensorCV can treat this series of operations as a series of matrix operations that the system can fuse these functions into a single one to further reduce matrix operations and memory usage. In TensorCV, we demonstrate the fusion of resize, cropping, color space conversion, and rotation into just two matrix multiplications. Considering a code performs resize, center crop, RGB to YUV color space conversion, and 90-degree rotation, we can represent the process as:

$$\begin{aligned} &\left[L^{crop}_{M''\times M'} \cdot L^{resize}_{M'\times M} \cdot Input_{M\times 3N} \cdot \right. \\ &\left. R^{resize}_{3N\times 3N'} \cdot R^{crop}_{3N'\times 3N''} \cdot R^{cvtColor}_{3N''\times 3N''} \cdot R^{rotate}_{3N''\times 3N''}\right]^T \end{aligned} \tag{19}$$

As mentioned in Section III-A2, the values of two left kernels and four right kernels are only related to the target image's size and function parameters, not input values. Thus, TensorCV simply needs to compute the matrix multiplications between kernels once for every batch.

$$\left[L^{integ}_{M''\times M'} \cdot Input_{M\times 3N} \cdot R^{integ}_{3N\times 3N''}\right]^T \tag{20}$$

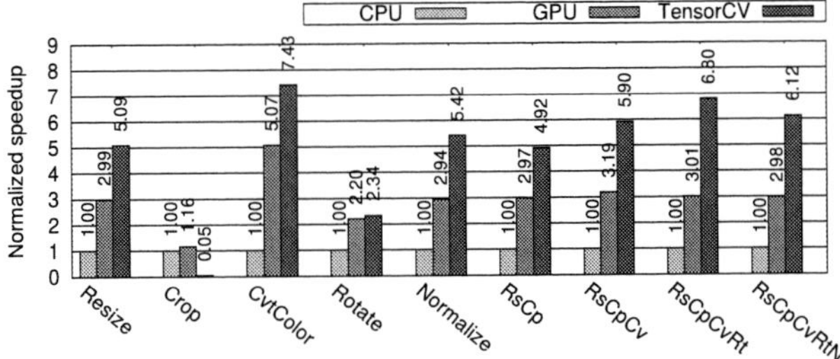

Fig. 2. Performance of TensorCV compared with baseline OpenCV implementation using CPUs and GPUs

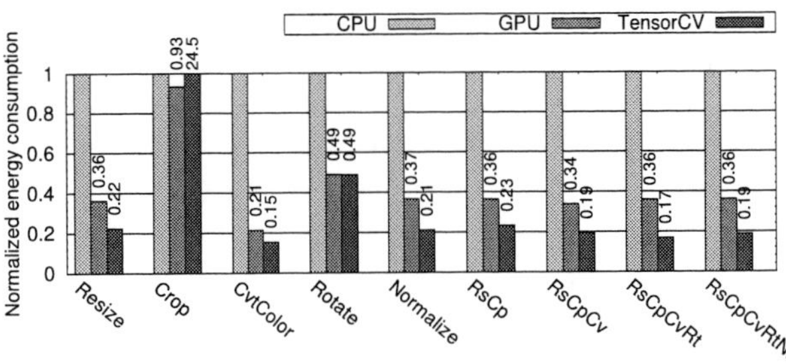

Fig. 3. Energy consumption of TensorCV compared with baseline OpenCV implementation using CPUs and GPUs

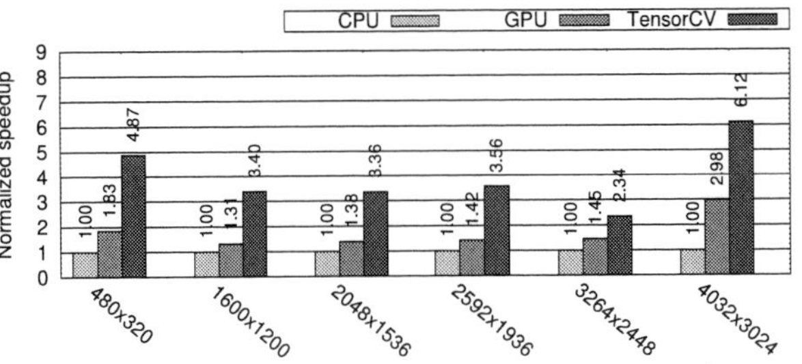

Fig. 4. Performance of RsCpCvRtNm TensorCV function in different input image sizes

One issue is that naively extending $R^{rotate}_{N'' \times N''}$ to $R^{rotate}_{3N'' \times 3N''}$ generates the incorrect output image. TensorCV solves such an issue by transposing the matrix first. Then, the integrated kernels use modified $R^{cvtColor}_{3N'' \times 3N''}$ to convert an RGB format matrix to VUY one. After color space conversion, $R^{rotate}_{3N'' \times 3N''}$ flips VUY to YUV, and the transpose operation generates YUV format output. And finally, transpose the result back to the original format. As such, the algorithm fulfills $R^{cvtColor}_{3N'' \times 3N''}$ and $R^{rotate}_{3N'' \times 3N''}$ as follows:

$$R^{cvtColor\_integ}_{3n \times 3n} \ni r^{cvtColor\_integ}_{i,j} = \begin{cases} r^{cvtColor}_{i,3j+2} & \text{if} \quad \lfloor j/n \rfloor = 0 \\ r^{cvtColor}_{i,3(j-n)+1} & \text{if} \quad \lfloor j/n \rfloor = 1 \\ r^{cvtColor}_{i,3(j-2n)} & \text{if} \quad \lfloor j/n \rfloor = 2 \end{cases},$$
$$L^{rotate\_integ}_{m \times m} \ni l^{rotate\_integ}_{i,j} = \begin{cases} 1 & \text{if} \quad j = m - 1 - i \\ 0 & \text{else} \end{cases},$$
$$R^{rotate\_integ}_{3n \times 3n} \ni r^{rotate\_integ}_{i,j} = \begin{cases} 1 & \text{if} \quad j = 3n - 1 - i \\ 0 & \text{else} \end{cases} \tag{21}$$

IV. Experimental Methodology

We conducted experiments on a machine with an Intel Core i7-12700K processor, 64 GB DDR5 DRAM. The GPU in our experiments is an NVIDIA GeForce RTX 3090 GPU based on Ampere architecture. We implemented TensorCV is implemented using NVIDIA CUDA Toolkit 11.7 and cuBLAS in IEEE 754 half precision. The system runs a Linux 5.15.0 kernel. We ran each function with 100 batches of images, where each batch had 20 samples. The inference-adjacent tasks include resizing the image from 4032×3024 to 256×256, cropping the center of the resized image with 224×224 box, converting the color space RGB to YUV, rotating the image 90-degree counter-clockwise, and normalizing the image values. Note that the applications RsCp, RsCpCv, RsCpCvRt, and RsCpCvRtNm indicate integrated functions of Resize (Rs), Crop (Cp), Color space conversion (Cv), Rotate (Rt), and Normalize (Nm).

V. Results

This section summarizes our evaluation of TensorCV. TensorCV delivered 6.12× speedup in RsCpCvRtNm functions of CV pipelines and saved 81% of the energy.

Figure 2 compares the performance of TensorCV with conventional OpenCV implementations that can only use CPUs or CUDA cores. The baseline of Figure 2 is the CPU-based implementation. TensorCV's algorithm achieves up to 7.43× speedup in the color space conversion function. Note that TensorCV shows a performance drop in the center crop because OpenCV implements the crop with simple memory operations. However, TensorCV also can employ such an approach when it needs to run a single crop, and integrating multiple functions removes the high latency of the crop. Excepting the outlier, center crop, TensorCV achieved 4.67× speedup on average, while OpenCV-CUDA only sped up 3.14× on average.

Another advantage of TensorCV is the ability to optimize across functions and combine several matrix multiplications into fewer ones. We presented four different use cases that combine multiple pre-processing functions. Compared with other implementations, TensorCV achieves 5.93× geometric mean in speedup in integration functions. However, the performance is limited in the conventional GPU implementation as 3.04× speedup, where cross-function optimization is complicated.

Tensor Cores also share the power/energy advantages of other AI/ML accelerators. We use a Watts Up Power Meter to measure the system power. While running these functions using Tensor Cores, the total system power peaked at 203 W. However, the total system power reaches 192 W and 178 W when using CUDA cores and CPU only, respectively. As TensorCV reduces execution time, TensorCV receives huge benefits in energy consumption. Figure 3 compares the energy consumption of TensorCV with its counterparts. TensorCV saves 81% of energy on the RsCpCvRtNm function. In contrast, the existing GPU implementation of OpenCV-CUDA shows only 64% energy saving.

To demonstrate that TensorCV delivers performance advantages regardless of the image size, Figure 4 shows the normalized speedup of TensorCV in a variety of input image sizes. Our evaluation opts for the input image sizes from the default photo sizes of Apple iPhones. TensorCV achieves 4.04× speedup compared to OpenCV implementation on average, while OpenCV-CUDA leads to 1.70× speedup. Moreover, compared to the OpenCV-CUDA implementation, TensorCV shows at least 1.6× higher speedup in all image sizes with 2.36× speedup on average.

VI. Related Work

In addition to aforementioned related work, several other lines of TensorCV-relevant research deserve mention.

Inference-adjacent computation Prior work identified inference-adjacent computation as the bottleneck in many CV applications [3], [10], [13], [19]. However, most prior work focus on scheduling inference-adjacent stages and managing computing resources, not accelerating the inference-adjacent stage itself. Kang et al. [10] propose a runtime engine for efficient resource management and scheduling for ML-adjacent stages. DLBooster [3] offers an FPGA design to selectively offload and compute some critical decoding workloads to provide high performance in inferencing. Tf.Data [13] proposes a framework for building and executing efficient ML-adjacent stages, allowing the user to schedule, compose, and reuse the computations. FastFlow [19] proposes an ML training system offloading some ML-adjacent computations to remote CPUs to mitigate the bottlenecks.

Image processing on GPUs With the growing interest in GPUs as a general-purpose parallel computing platform, there are studies and frameworks exploiting GPUs in image processing algorithms [1], [2], [7], [15]. However, as mentioned in Section II, ML-adjacent stages can be the bottleneck of ML applications, although such parallelized algorithms, because matrix multiplication accelerators can accelerate only AI/ML parts. Although there are a few studies that employ matrix multiplication accelerators for image processing algorithms [6], [17], they focus on accelerating convolution-based algorithms only. In contrast, TensorCV proposes novel algorithms to utilize matrix multiplication accelerators on ML-related image processing algorithms.

Non-AI/ML applications on AI/ML accelerators Prior work demonstrates that exploiting matrix multiplication accelerators in non-AI/ML algorithms by rewriting the algorithms can improve their latency and throughput [4], [5], [8], [11], [12], [14], [16], [18]. They target reduction operation [4], [14], fractal processing [16], database operations [8], stencil computation [12], Fourier transform [5], [11], [18], and general tensor computation [21]. To our best knowledge, TensorCV is the first research exploiting matrix multiplication accelerator on ML-related image processing algorithms.

VII. Conclusion

This paper revisited the algorithms of several most frequently used and under-optimized inference-adjacent functions in CV pipelines. We showed an average of 7.6× compared with state-of-the-art GPU implementations. Furthermore, the energy efficiency is solid – an average of 82% energy saving. As the first work that utilizes Tensor Cores for inference-adjacent computation in CV pipelines, we envision this work would encourage revisits to existing problems and bring more discussions on related topics.

VIII. Acknowledgment

This research was supported by the Super Computer Development Leading Program of the National Research Foundation of Korea(NRF) funded by the Korean government (Ministry of Science and ICT(MSIT)) (2021M3H6A1017683), Intel, and National Science Foundation (NSF) award, CNS-2007124.

References

[1] Yannick Allusse, Patrick Horain, Ankit Agarwal, and Cindula Saipriyadarshan. Gpucv: An opensource gpu-accelerated framework forimage processing and computer vision. In *Proceedings of the 16th ACM International Conference on Multimedia*, MM '08, page 1089–1092, New York, NY, USA, 2008. Association for Computing Machinery.

[2] Gary Bradski. The opencv library. *Dr. Dobb's Journal: Software Tools for the Professional Programmer*, 25(11):120–123, 2000.

[3] Yang Cheng, Dan Li, Zhiyuan Guo, Binyao Jiang, Jinkun Geng, Wei Bai, Jianping Wu, and Yongqiang Xiong. Accelerating end-to-end deep learning workflow with codesign of data preprocessing and scheduling. *IEEE Transactions on Parallel and Distributed Systems*, 32(7):1802–1814, 2021.

[4] Abdul Dakkak, Cheng Li, Jinjun Xiong, Isaac Gelado, and Wen-mei Hwu. Accelerating reduction and scan using tensor core units. In *Proceedings of the ACM International Conference on Supercomputing*, ICS '19, pages 46–57, New York, NY, USA, 2019. Association for Computing Machinery.

[5] Sultan Durrani, Muhammad Saad Chughtai, Mert Hidayetoglu, Rashid Tahir, Abdul Dakkak, Lawrence Rauchwerger, Fareed Zaffar, and Wen-mei Hwu. Accelerating fourier and number theoretic transforms using tensor cores and warp shuffles. In *2021 30th International Conference on Parallel Architectures and Compilation Techniques (PACT)*, pages 345–355, 2021.

[6] Stefan Groth, Jürgen Teich, and Frank Hannig. Efficient application of tensor core units for convolving images. In *Proceedings of the 24th International Workshop on Software and Compilers for Embedded Systems*, SCOPES '21, page 1–6, New York, NY, USA, 2021. Association for Computing Machinery.

[7] Robert Haase, Loic A Royer, Peter Steinbach, Deborah Schmidt, Alexandr Dibrov, Uwe Schmidt, Martin Weigert, Nicola Maghelli, Pavel Tomancak, Florian Jug, et al. Clij: Gpu-accelerated image processing for everyone. *Nature methods*, 17(1):5–6, 2020.

[8] Yu-Ching Hu, Yuliang Li, and Hung-Wei Tseng. Tcudb: Accelerating database with tensor processors. In *Proceedings of the 2022 International Conference on Management of Data*, SIGMOD '22, page 1360–1374, New York, NY, USA, 2022. Association for Computing Machinery.

[9] Jeffrey A. Clark. Python Imaging Library. https://github.com/python-pillow/Pillow/blob/main/src/libImaging/ConvertYCbCr.c, 2021.

[10] Daniel Kang, Ankit Mathur, Teja Veeramacheneni, Peter Bailis, and Matei Zaharia. Jointly optimizing preprocessing and inference for dnn-based visual analytics. *Proc. VLDB Endow.*, 14(2):87–100, oct 2020.

[11] Binrui Li, Shenggan Cheng, and James Lin. tcfft: A fast half-precision fft library for nvidia tensor cores. In *2021 IEEE International Conference on Cluster Computing (CLUSTER)*, pages 1–11, 2021.

[12] Xiaoyan Liu, Yi Liu, Hailong Yang, Jianjin Liao, Mingzhen Li, Zhongzhi Luan, and Depei Qian. Toward accelerated stencil computation by adapting tensor core unit on gpu. In *Proceedings of the 36th ACM International Conference on Supercomputing*, ICS '22, New York, NY, USA, 2022. Association for Computing Machinery.

[13] Derek G. Murray, Jiří Šimša, Ana Klimovic, and Ihor Indyk. Tf.data: A machine learning data processing framework. *Proc. VLDB Endow.*, 14(12):2945–2958, jul 2021.

[14] Cristóbal A. Navarro, Roberto Carrasco, Ricardo J. Barrientos, Javier A. Riquelme, and Raimundo Vega. Gpu tensor cores for fast arithmetic reductions. *IEEE Transactions on Parallel and Distributed Systems*, 32(1):72–84, 2021.

[15] In Kyu Park, Nitin Singhal, Man Hee Lee, Sungdae Cho, and Chris Kim. Design and performance evaluation of image processing algorithms on gpus. *IEEE Transactions on Parallel and Distributed Systems*, 22(1):91–104, 2011.

[16] Felipe A. Quezada, Cristóbal A. Navarro, Nancy Hitschfeld, and Benjamin Bustos. Squeeze: Efficient compact fractals for tensor core gpus. *Future Generation Computer Systems*, 135:10–19, 2022.

[17] Savvas Sioutas, Sander Stuijk, Twan Basten, Lou Somers, and Henk Corporaal. Programming tensor cores from an image processing dsl. SCOPES '20, page 36–41, New York, NY, USA, 2020. Association for Computing Machinery.

[18] Anumeena Sorna, Xiaohe Cheng, Eduardo D'Azevedo, Kwai Won, and Stanimire Tomov. Optimizing the fast fourier transform using mixed precision on tensor core hardware. In *2018 IEEE 25th International Conference on High Performance Computing Workshops (HiPCW)*, pages 3–7, 2018.

[19] Taegeon Um, Byungsoo Oh, Byeongchan Seo, Minhyeok Kweun, Goeun Kim, and Woo-Yeon Lee. Fastflow: Accelerating deep learning model training with smart offloading of input data pipeline. *Proc. VLDB Endow.*, 16(5):1086–1099, mar 2023.

[20] Abenezer Wudenhe and Hung-Wei Tseng. TPUPoint: Automatically Characterizing Hardware Accelerated Data Center Machine Learning Program Behavior. In *2021 IEEE International Symposium on Performance Analysis of Systems and Software*, ISPASS 2021, 2021.

[21] Yunan Zhang, Po-An Tsai, and Hung-Wei Tseng. Simd2: A generalized matrix instruction set for accelerating tensor computation beyond gemm. In *Proceedings of the 49th Annual International Symposium on Computer Architecture*, ISCA '22, page 552–566, New York, NY, USA, 2022. Association for Computing Machinery.

[22] Yue Zhou, Xue Yang, Gefan Zhang, Jiabao Wang, Yanyi Liu, Liping Hou, Xue Jiang, Xingzhao Liu, Junchi Yan, Chengqi Lyu, Wenwei Zhang, and Kai Chen. Mmrotate: A rotated object detection benchmark using pytorch. In *Proceedings of the 30th ACM International Conference on Multimedia*, MM '22, page 7331–7334, New York, NY, USA, 2022. Association for Computing Machinery.

Energy-Harvesting-Aware Adaptive Inference of Deep Neural Networks in Embedded Systems

Gwanjong Park
Sungkyunkwan University
Suwon, Republic of Korea
jesj74@g.skku.edu

Osama Khan
Sungkyunkwan University
Suwon, Republic of Korea
khan980@g.skku.edu

Euiseong Seo
Sungkyunkwan University
Suwon, Republic of Korea
euiseong@skku.edu

Abstract—**In energy harvesting IoT and sensor devices, energy influx is continuously changing and difficult to predict. Recently, the use of deep neural networks (DNNs), which consumes a large amount of energy, has increased in such devices. If a lightweight DNN model is used anticipating low energy influx, it may not achieve satisfactory inference accuracy in the ample energy flow condition. Conversely, using highly accurate sophisticated models may result in frequent inference failures due to energy depletion in situations with low energy influx. In this paper, for energy harvesting embedded systems that periodically perform DNN inference on sensor inputs, we propose an energy-harvesting-aware adaptive inference scheme to maximize inference accuracy while minimizing inference failures due to energy depletion in the long term. The model selector in the proposed scheme, which is a reinforcement learning (RL) agent, selects a DNN model from a model pool in consideration of the energy harvesting state and the accuracy and energy requirement of each model in the model pool. We implemented the proposed scheme on a microcontroller system and evaluated it with six different DNN applications with various types of input data. In the energy harvesting simulation with the real-world solar power traces, our approach, on average across the six workloads, was able to achieve a 65.62% reduction in inference failure rate with only a 6.08% increase in average error rate compared to the base DNN models.**

I. Introduction

Energy harvesting embedded systems collect their operation energy from ambient energy sources. Therefore, compared to traditional battery-backed devices, of which the battery must be regularly replaced or recharged, they require low maintenance costs and are capable of long-term operations in remote areas. For this reason, they are being used in Internet-of-Things (IoT) or sensor devices deployed in environments where power lines are not available.

In many cases, these energy harvesting embedded systems collect information about their surroundings through various sensors and respond to the inputs accordingly. For example, a rural-area surveillance device may transmit a corresponding image to a server or record the occurrence of an event in its storage when a change occurs in the image input obtained from its camera sensor. For processing such ambient sensing data, the use of deep neural networks (DNNs) is getting popular [1]–[3]. However, DNN inference tasks being used in embedded systems consume considerable energy because they are computationally intensive [4].

In an energy harvesting device, the amount of energy influx from harvesting constantly changes, and its energy storage, which is usually a rechargeable battery or a super-capacitor, has limited capacity. When the energy influx is small, the use of a complicated but accurate DNN model will eventually end up in a mission failure from energy depletion. To prevent energy depletion, a simple DNN model can be used instead. However, the use of a simple model leads to the meaningless sacrifice of inference accuracy and the waste of surplus energy when the energy influx is abundant. In particular, in the systems where inference results trigger subsequent actions, such as data communication or additional data processing, incorrect inferences may lead to unnecessary energy consumption, further exacerbating the energy-constrained situations.

Therefore, DNN models to be used for energy harvesting embedded devices should be lightweight enough to avoid depletion of energy while providing a sufficient level of accuracy. However, because of the continuously and unpredictably changing energy influx, it is technically challenging to determine a proper energy budget and thus the structure of a DNN model.

The intermittent computing architecture has been studied for the energy harvesting systems [1], [5]–[10]. It guarantees the progress of long-running tasks, of which execution time may span over multiple power cycles, such as DNN training. However, DNN inference is usually used to provide timely responses to ambient sensing input, and thus performing inference after a certain period is meaningless. Therefore, it is infeasible to apply intermittent computing to inference operations with timeliness requirements.

Early-exit DNNs have been proposed to correspond to varying resource constraints [2], [3]. Depending on the available time or energy, they stop DNN inference midway and return the intermediate result after processing it through a small number of exit layers. However, to apply this approach, an existing DNN model must be modified. Moreover, the DNN, including the early-exit layers, must be trained so that every exit provides a sufficient level of accuracy, which is technically challenging [11]–[13].

This paper proposes an adaptive DNN inference scheme for intelligent energy harvesting embedded systems. Through reinforcement learning (RL), the proposed scheme is trained to take into account the current energy flow state and the

979-8-3503-1176-1/23 $31.00 © 2023 IEEE

characteristics of the models in the model pool to select an appropriate model for an inference operation to maximize the long-term accuracy while minimizing the inference failure due to energy depletion.

We implemented the proposed scheme on a microcontroller board and evaluated it with four DNN models from the MLPerf Tiny benchmark suite [14] and, to ensure a thorough analysis, with two additional time-series classification DNN models [15]. The energy influx was simulated based on the solar irradiance traces [16].

II. Background and Related Work

A. Use of DNNs in Energy Harvesting Embedded Systems

The use of DNNs in embedded systems is getting more and more popular for performing diverse tasks, including anomaly detection, classification, and recognition from audio, image and motion sensing data obtained from diverse sensors [17]. These systems periodically perform inference on the DNN models to the input data, and conduct the follow-up operations based on the inference results.

These embedded systems are often deployed outdoors or in environments where power lines are not available. Therefore, embedded systems that harvest their operation energy from their surroundings have been actively researched [1]–[3].

An energy harvesting system consists of a component that converts an ambient source such as solar, wind, and radio waves into electrical energy, and an energy storage device such as a rechargeable battery or super-capacitor that stores the collected energy. An energy harvesting embedded system starts its operation when the power obtained from energy harvesting is sufficiently stored in the energy storage, and stops when the energy in the storage is exhausted.

If the energy consumption rate of an energy harvesting system exceeds the energy influx, the system will eventually cease to function due to energy depletion. Conversely, if the energy budget for an application is set too low to avoid energy depletion, in cases of abundant energy influx, it may result in a detrimental reduction of service quality and waste of harvested energy due to the limited energy storage capacity.

As DNN structures become larger, they generally exhibit better adaptability to diverse environments and achieve higher inference accuracy. However, as the model size increases, so does the energy consumption required for its inference. Therefore, DNN models running on energy harvesting embedded systems must be neither too large, causing energy depletion, nor too small, compromising the necessary inference accuracy. However, as stated, the continuously changing energy influx and its unpredictability make it difficult to select a DNN model with appropriate inference accuracy and energy consumption for energy harvesting embedded systems [2], [3].

B. Related Work

The intermittent computing approach has been proposed to prevent the continuous restart of programs due to the irregularity of energy influx and to guarantee its completion. It is categorized into three: checkpoint-centric approach [5], [6], use of non-volatile processor architecture [7], [8], and atomic task composition technique [1], [9], [10], in which a task consists of small idempotent atomic code blocks. All of these approaches assume that tasks will be performed across multiple power cycles if they cannot be completed within a single cycle. For this, such tasks must not have an explicit deadline for when they must complete. However, because the event-triggered or periodic inference tasks that we aim to address in this paper have timeliness requirements, it is difficult to apply these intermittent computing approaches.

Adaptive computing approaches complete tasks within given energy and time constraints by adjusting the application service quality depending on the current circumstances. CatNap [9] isolates and reserves energy for time-critical code. If the energy influx is insufficient to run the time-critical code, CatNap reduces energy consumption for job completion by prolonging the operation period and degrading application-specific quality. REHASH [10] modulates the performance of tasks based on a heuristic adaptation scheme to enable higher sensor coverage, completion rates, or throughput of applications. However, these approaches require developers to configure and implement when and how the application quality degradation is performed, respectively.

The use of multi-exit neural network models has been proposed for energy- and time-constrained systems [2], [3]. They create early-exit paths in the intermediate layers of a DNN model and add shallow neural networks to the early-exit paths, performing inference by selecting exits based on the given time or energy budgets. The early-exit architecture enables trade-offs between energy consumption and accuracy. However, because the early-exits share the layers in front of the exits and thus the parameters in the shared layers receive conflicted gradients from different exits during training, training the model for all the exits to produce a sufficient level of accuracy is technically challenging [11]. Additionally, developers have to make a few difficult design decisions, including the number of early-exits, the appropriate layers to add early-exits, and the confidence threshold for selecting each exit [12], [13].

An adaptive DNN model selection scheme was proposed for resource-constrained embedded systems [17]. It selects a model from a model pool depending on the characteristics of the input data and the precision requirement. Our approach shares a commonality with theirs in selecting an appropriate DNN model among those with the same objective. However, unlike theirs that only needs to decide based on the given condition for each inference, the problem we deal with requires adaptation to the continuously and unpredictably changing energy harvesting state, and the current decision has an impact on future outcomes as well.

III. Our Approach

A. Design Overview

As shown in Fig. 1, our approach aims to minimize inference failures caused by energy depletion while maximizing long-term accuracy of DNN inference on embedded systems that harvest energy from ambient sources.

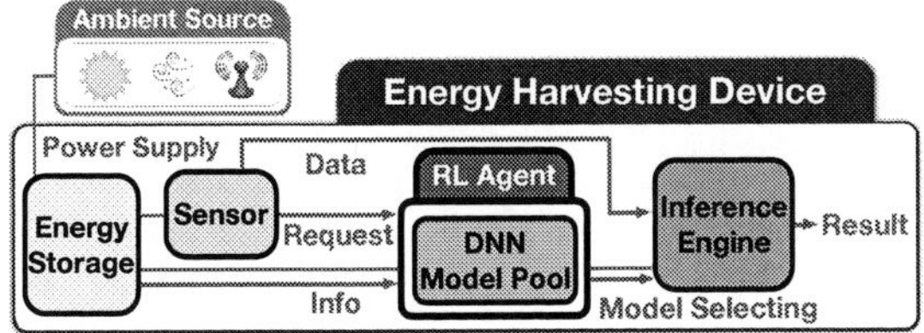

Fig. 1: Design overview of our approach

In the target system, we assume that the capacity of the energy storage is sufficient to perform multiple inference operations. The sensor collects inputs, such as image, motion, or audio data, and the application periodically performs inference on the input data.

If the energy influx from harvesting is smaller than energy consumption for inference operations, the system will eventually exhaust the energy stored in the storage and experience an inference failure. Due to the timeliness requirement of the inference tasks, a missed inference operation will not be performed when the system restarts and the stale data will be discarded. The continuous fluctuations in energy influx further complicate the selection of a DNN model that appropriately balances accuracy and energy requirements.

To minimize inference failures from energy depletion while maximizing inference accuracy, for each inference operation, our proposed approach selects a DNN model from a DNN model pool, which have a few models for the same task but with varying structures. When making the model selection, our approach considers a few factors, such as energy influx state, remaining energy, the accuracy and energy requirement of each model in the model pool. Since it is static and uncontrollable, we do not take into account the energy consumed during the idle state nor the energy used to obtain the input data. The actual model selection is performed by a RL agent.

A DNN model pool may have models that have the same purpose but fundamentally different structures. However, for the ease of development, it is also possible to create a model pool with model variants derived from a base model by applying a varying degree of model compaction techniques. For example, we can derive a few model variants out of a convolutional neural network (CNN) model by adjusting the input image resolution, the size and depth of each convolution layer, the number of layers, the degree of quantization, and so on. When the models in the model pool $\mathcal{M}$ are indexed in descending order of models' energy consumption, $\mathcal{M}$ consisting of n model variants can be defined as $\mathcal{M} = \{m_1, ..., m_n\}$. If the error rate of m_i is defined as $Err(m_i)$, then $Err(m_i) \leq Err(m_j)$ when $i < j$. If m_j violates this condition, it cannot be included in the pool because it has higher energy consumption but lower accuracy than m_i, thus choosing m_i over m_j is always beneficial.

The models are stored in the flash memory, which is inexpensive and consumes low power. Therefore, in contrast to the strictly limited SRAM capacity, the flash memory capacity is sufficiently large enough to accommodate the model pool, which ranges up to only a few hundreds KB in our evaluation.

TABLE I: State Set for RL of Model Selector Agent

Feature	Definition
S1	Current energy remaining in energy storage
S2	Harvested energy during last time step
S3	Harvested energy during last 10 time steps
S4	Difference in harvested energy during last three time steps between now and 10 time steps before

For an inference request, which arrives at a certain time interval, the RL agent selects a model, m, from $\mathcal{M}$ based on system's states. The agent aims to minimize the energy depletion-induced inference failure count and the accumulated inference error rate of inference operations after completing a series of inference operations, which is regarded as an episode in the RL terminology.

B. Design Details of RL-based Adaptive Model Selector

The state transition probability for changes in harvested energy is not defined in most of the cases. Additionally, the remaining energy at a future time is impacted by the energy consumption of the current inference operation. As a result, we can define the model selection problem corresponding to changing energy influx as a Markov decision process (MDP) and train a model selection RL agent using one of the model-free RL approaches [18]. We define the state set, the action set, and the reward for training the RL agent as follows.

Table I described the state set. *S1* represents the amount of energy currently remaining in the energy storage. *S2* and *S3* indicate the short-term and long-term trends of energy influx, respectively. Lastly, *S4* shows the change in the amount of harvested energy for the last three time steps between now and ten-time steps before. It also indicates the volatility of energy influx; a positive value of S4 signifies an increase in recent energy influx, and vice versa.

The available actions for the model selection agent are the models in $\mathcal{M}$. Therefore, the action space is discrete, and as large as the size of the model pool.

We use the error rate of a model as the measure of inference quality obtained from selecting the model for the current inference operation because the error rate is usually smaller than the accuracy of a model, and therefore, more sensitive to change. Suppose that the agent selects a model m_i as an action where $m_i \in \mathcal{M}$. if *S1* is smaller than the energy requirement of m_i, $Egy(m_i)$, the system cannot perform the requested inference and the agent receives a penalty for the inference failure, denoted as p, which should be a relatively large negative number. Otherwise, a positive reward is given based on the ratio of error rate improvement, $IRER(m_i)$, which is defined as Eq.1 where m_n is the smallest model in the pool. $IRER(m)$ denotes the degree of accuracy improvement achieved by choosing model m over model m_n.

$$IRER(m) = \frac{Err(m_n) - Err(m)}{Err(m_n)} \tag{1}$$

The reward is accordingly defined as Eq.2.

TABLE II: Description of Applications Used for Evaluation

Application	Dataset	Model	Size (KB)	Error (%)	MFlops	Energy (mJ)
Keyword Spotting (KS)	Speech Commands [14]	Base model: *DS-CNN* [14]	52.7	8.2	5.3	374
		Removal of 4th layer	41.6	9.5	4.1	291
		Filter count reduction (from 32 to 16)	27.2	11.2	1.6	121
		Layer removal and filter count reduction	21.8	13.0	1.3	96
Anomaly Detection (AD)	ToyADMOS [14]	Base model: *Deep AutoEncoder* [14]	277.0	16.5	0.53	47
		Removal of 2 fully connected layers	242.1	16.9	0.46	42
		Removal of 4 fully connected layers	202.7	17.3	0.39	36
		Removal of 6 fully connected layers	172.3	17.8	0.33	31
Visual Wake Words (VWW)	Visual Wake Words [14]	Base model: *MobileNetV1* [14]	331.8	15.2	15.0	1068
		Removal of 6-7,11-13th layers	67.8	17.2	10.2	741
		Removal of 4-7,10-13th layers	26.5	18.4	7.2	535
		Removal of 3-7,9-13th layers	12.8	23.1	5.6	431
Image Classification (IC)	Cifar10 [14]	Base model: *ResNet* [14]	98.5	14.5	25.0	1558
		6 filters removed from first residual block	47.1	16.4	10.0	653
		8 filters removed from first residual block	34.7	19.5	6.5	443
		10 filters removed from first residual block	24.7	26.4	3.7	268
Gesture Recognition (GR)	UWaveGestureLibraryAll [15]	*MLP* [15] with 200 hidden nodes	275.7	5.6	0.54	45
		MLP with 100 hidden nodes	119.2	6.4	0.23	19
		MLP with 75 hidden nodes	86.4	7.8	0.17	14
		MLP with 50 hidden nodes	56.0	9.0	0.11	9
Latent Variable Detection (LVD)	MoteStrain [15]	*FCN* [15] with 20-32-20 filters	28.7	6.6	3.9	224
		FCN with 16-24-16 filters	19.2	7.1	2.4	137
		FCN with 8-24-8 filters	11.6	7.7	1.2	72
		FCN with 8-8-8 filters	6.6	8.5	0.5	30

$$Reward(m) = \begin{cases} IRER(m), \text{ if } S1 \geq Egy(m) \\ p \text{ where } p < 0, \text{ otherwise} \end{cases} \quad (2)$$

p is a hyperparameter determined by the relative importance of minimizing inference failures to maximizing inference accuracy. The smaller the p value, the smaller models the agent will tend to choose to more proactively avoid inference failures. The p value suitable for a given system, like other hyperparameters, should be explored and determined through the training process of the agent.

The ultimate goal of the agent is to maximize the sum of rewards in an episode. An episode is determined by the mission characteristics of the target system or the energy harvesting environment. For example, in a solar-energy harvesting system, it is desirable for one day to be set as one episode.

To minimize the energy and time overhead, the neural network of the RL agent should be significantly smaller than the DNN models for inference. On the other hand, it should be large enough to adapt to various energy sources and also to the diversity in DNN model characteristics. The proper neural network structure for the agent should be determined together with the p value in consideration of the given energy source and the target DNN models.

IV. Evaluation

A. Evaluation Environment

Our approach was evaluated on an STM32L496ZG microcontroller with an ARM Cortex M4 processor. The clock frequency of the embedded board was set to 50 MHz, and the board was equipped with 320 KB of SRAM and 1 MB of flash memory. The energy consumption of the board was measured with a digital power meter. The energy consumption shown in

TABLE III: Hyperparameters for RL of Model Selection Agent

Hyperparameter	Value
Clip parameter	0.2
Discount factor	0.99
Trajectory buffer size	3072
Learning rate for policy neural network	0.001
Learning rate for value neural network	0.005

the following graphs includes both the energy consumed by the model selection agent and that consumed during the actual inference of the selected DNN model.

Our evaluation workload comprised four benchmark applications from MLPerf Tiny [14] and two time-series classification applications [15]. For each application, we created a model pool containing four models, which are described in detail in Table II. The agent was trained separately for each application using the proximal policy optimization (PPO) [19] implementation and hyperparameter settings of the tinyMan research [20]. Table III presents only the modified hyperparameters from tinyMAN's.

We simulated the energy influx with the solar irradiance traces collected by Oak Ridge National Lab. for six months in 2019 [16]. We trained the agent with 80% of the traces, and evaluated it with the remaining 20%. We assumed that the device has a solar cell size of $14cm^2$ with 10% conversion efficiency and 0.963 tracking factor [21]. The board was assumed to be equipped with a 1.5F capacitor as its energy storage. A single day is set as an episode for RL of the agent. The system was configured to initiate its first inference when the accumulated energy surpasses the energy demand of m_1 for the given application.

As explained in Section III-B, the agent's neural network structure and the p value should be determined depending on the composition of the model pool and the relative importance

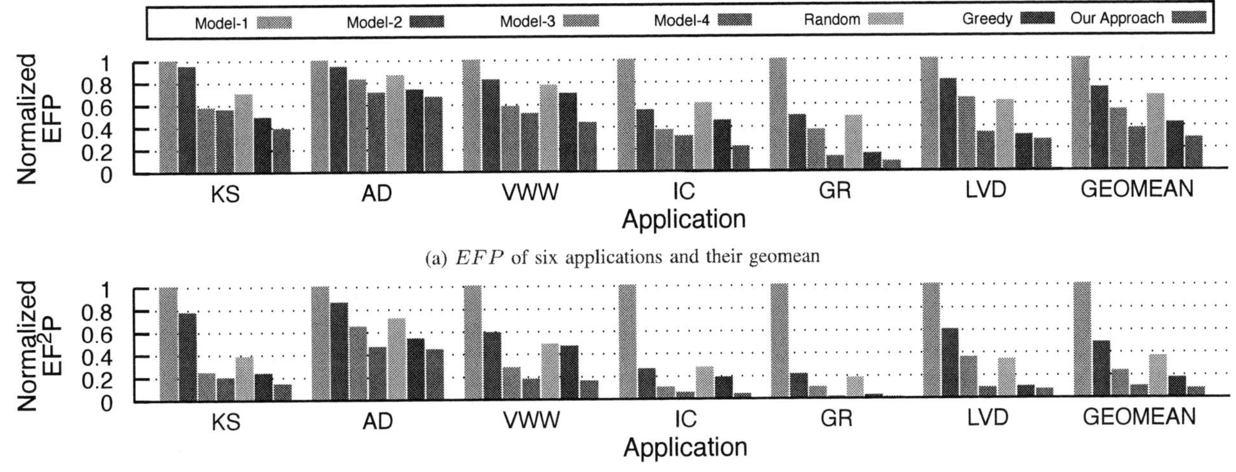

(a) EFP of six applications and their geomean

(b) EF^2P of six applications and their geomean

Fig. 2: EFP and EF^2P of six applications under various inference model selection schemes normalized to *Model-1*

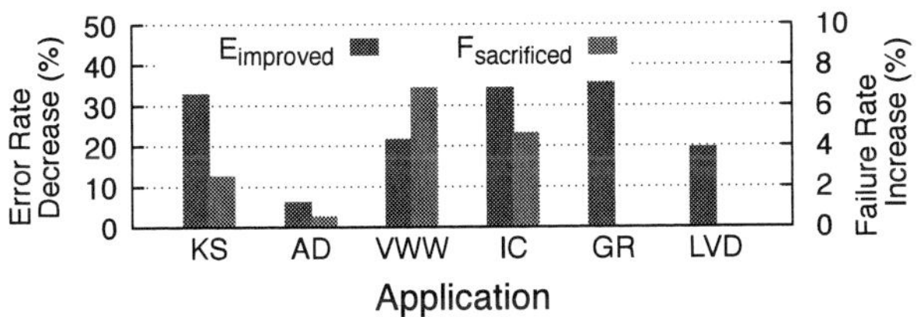

Fig. 3: Decrease in error rate and increase in failure rate under our approach in comparison to the cases of *Model-4*

between accuracy and inference failure. We used two metrics, EFP, and EF^2P, to assess the agent's performance. EFP is the product of the average error rate and the failure rate while EF^2P is the product of the average error rate and the squared failure rate. In comparison to EFP, EF^2P prioritizes avoiding failures over improving long-term accuracy. For both metrics, the smaller, the better.

For each of the six applications, we performed a grid search to find the best combination of the model selection agent's neural network structure and the value of p that minimizes EFP and EF^2P, respectively, and used it for the performance analysis. The explored neural network structures had 2 to 4 hidden layers, and each of them contained 8 to 64 nodes.

The hyperparameter search yielded the following combinations of p and the agent's neural network structure for minimizing EFP: (-3, 16×32) for `KS`, (-3, 16×16) for `AD`, (-7, 16×16) for `VWW`, (-3, 16×32) for `IC`, (-7, 32×32) for `GR`, and (-5, 32×32) for `LVD`. 16×32 denotes a neural network having two layers, each of which has 16 and 32 nodes, respectively. We obtained the same combinations when setting EF^2P as the RL goal.

The energy consumption for the agent's network inference was only 0.092 mJ for the 16×16 network, 0.126 mJ for the 16×32 network, and 0.171 mJ for the 32×32 network. In addition, For the 16×16, 16×32, and 32×32 networks, the agent took memory sizes of 3.57 KB, 4.89 KB, and 7.2 KB, respectively. These sizes are deemed affordable, even when considering the limited size of the SRAM.

B. Performance Analysis

To analyze the effectiveness of the proposed scheme, we measured both EFP and EF^2P under our scheme. For comparison, we also analyzed them under configurations that statically selected one of the models in the model pool, which are denoted as *Model-N* where *N* represents the model index, as well as under two heuristic algorithms, *Greedy* and *Random*. *Random* randomly selects a model from the model pool for each inference while *Greedy* selects the model with the highest accuracy that can be completed using the remaining energy. *Model-1* used the largest model in the model pool and *Model-4* the smallest one.

As shown in Fig. 2, among the static model selection schemes, *Model-1* exhibited the highest EFP in all six applications, while *Model-4* demonstrated the lowest EFP values. The reason for this is that the increase in failure rate by choosing m_1 instead of m_4 outweighed the error rate reduction. For example, in `LVD`'s model pool, m_1 has a 22.4% improved error rate compared to m_4, whereas consumed 646.7% more energy. Consequently, due to this energy consumption difference, *Model-1* exhibited a 290.6% higher failure rate than *Model-4*. On average, *Model-4* showed 63.4% superior results compared to *Model-1* in terms of EFP.

Naturally, *Greedy* outperformed *Random* in terms of EFP across all workloads. However, *Greedy* did not consider the long-term energy consumption, which consequently increased the number of inference failures. This was particularly noticeable in `VWW` and `IC`, where the energy requirement difference between each model was substantial, resulting in a significantly poorer EFP than *Model-4*. Although *Greedy* exhibited lower EFP values compared to *Model-4* in `KS` and `LVD`, EF^2P increased by 18.5% and 10.8%, respectively, due to the increased number of inference failures. The EF^2P values showed similar patterns to EFP across the other workloads.

Our approach outperformed all configurations in terms of EFP across all workloads, and the largest gain was earned for GR. In the case of GR, the energy requirement of m_4 was five times larger than m_1, and the error rate for m_1 was 37.8% better than that of m_4. As a result, our approach had considerable potential to reduce the failure rate and improve the error rate through model selection in response to the circumstance changes. On the other hand, the smallest improvement was seen in AD, where the differences in both energy consumption and error rates among models were the least pronounced. Consequently, regardless of the model chosen, our approach could not achieve significant improvement in the error rate compared to the energy-minimizing selection of m_4.

To better understand the results, we analyzed the increase in error rate and the reduction in failure rate, separately, for the six workloads. As shown in Fig. 3, the degree of error rate improvement was far larger than that of the failure rate increase. Notably, GR and LVD experienced no increase in the number of inference failures due to energy depletion. Conversely, VWW and IC exhibited relatively large increases in energy depletion occurrences because their models have a wide distribution of energy requirements. Consequently, an improper selection of a large model is more likely to result in an inference failure in these applications compared to others.

V. Conclusion

Due to the unpredictably changing energy influx in energy harvesting embedded systems, it is technically challenging but necessary to determine an appropriate DNN model that maximizes accuracy while avoiding energy depletion. Choosing highly accurate DNN models may lead to frequent mission failures due to energy depletion, whereas opting for simpler DNN models may result in inferior accuracy.

We proposed a RL-based adaptive DNN model selection agent that selects an appropriate DNN model among the model variants for an inference operation. The performance evaluation using using six benchmark applications on a commercial microcontroller system showed that our approach can maximize long-term average accuracy while mitigating energy depletion, adapting to fluctuations in energy influx and variations in energy storage states.

Acknowledgements

This research was supported by the Institute of Information and Communications Technology Planning and Evaluation Grant through the Ministry of Science and ICT of Korea, Development of Core Technology for Autonomous Energy-driven Computing System SW in Power-Instable Environment, under Grant 2021-0-00360; and the National Research Foundation of Korea under Grant 2021R1A2C200497612.

References

[1] G. Gobieski, B. Lucia, and N. Beckmann, "Intelligence beyond the edge: Inference on intermittent embedded systems," in *Proceedings of the 24th International Conference on Architectural Support for Programming Languages and Operating Systems*, 2019.

[2] Y. Wu, Z. Wang, Z. Jia, Y. Shi, and J. Hu, "Intermittent inference with nonuniformly compressed multi-exit neural network for energy harvesting powered devices," in *Proceedings of the 57th ACM/IEEE Design Automation Conference*, 2020.

[3] B. Islam and S. Nirjon, "Zygarde: Time-sensitive on-device deep inference and adaptation on intermittently-powered systems," *in Proceedings of the ACM Interactive, Mobile, Wearable and Ubiquitous Technologies*, 2020.

[4] A. Suleiman, Y.-H. Chen, J. Emer, and V. Sze, "Towards closing the energy gap between HOG and CNN features for embedded vision," in *Proceedings of the IEEE International Symposium on Circuits and Systems*, 2017.

[5] K. Maeng and B. Lucia, "Adaptive dynamic checkpointing for safe efficient intermittent computing," in *Proceedings of the 13th USENIX Conference on Operating Systems Design and Implementation*, 2018.

[6] J. Van Der Woude and M. Hicks, "Intermittent computation without hardware support or programmer intervention," in *Proceedings of the 12th USENIX Conference on Operating Systems Design and Implementation*, 2016.

[7] K. Qiu, N. Jao, M. Zhao, C. S. Mishra, G. Gudukbay, S. Jose, J. Sampson, M. T. Kandemir, and V. Narayanan, "ResiRCA: A resilient energy harvesting ReRAM crossbar-based accelerator for intelligent embedded processors," in *Proceedings of the IEEE International Symposium on High Performance Computer Architecture*, 2020.

[8] K. Ma, Y. Zheng, S. Li, K. Swaminathan, X. Li, Y. Liu, J. Sampson, Y. Xie, and V. Narayanan, "Architecture exploration for ambient energy harvesting nonvolatile processors," in *Proceedings of the 21st IEEE International Symposium on High Performance Computer Architecture*, 2015.

[9] K. Maeng and B. Lucia, "Adaptive low-overhead scheduling for periodic and reactive intermittent execution," in *Proceedings of the 41st ACM SIGPLAN Conference on Programming Language Design and Implementation*, 2020.

[10] A. Bakar, A. G. Ross, K. S. Yildirim, and J. Hester, "REHASH: A flexible, developer focused, heuristic adaptation platform for intermittently powered computing," *in Proceedings of the ACM on Interactive, Mobile, Wearable and Ubiquitous Technologies*, vol. 5, no. 3, pp. 1–42, 2021.

[11] Y. Sun, J. Li, and X. Xu, "Meta-GF: Training dynamic-depth neural networks harmoniously," in *Proceedings of the 17th European Conference on Computer Vision 2022*. Springer, 2022.

[12] A. Bakhtiarnia, Q. Zhang, and A. Iosifidis, "Improving the accuracy of early exits in multi-exit architectures via curriculum learning," in *Proceedings of International Joint Conference on Neural Networks*. IEEE, 2021, pp. 1–8.

[13] S. Laskaridis, A. Kouris, and N. D. Lane, "Adaptive inference through early-exit networks: Design, challenges and directions," in *Proceedings of the 5th International Workshop on Embedded and Mobile Deep Learning*, 2021.

[14] C. Banbury, V. J. Reddi, P. Torelli, J. Holleman, N. Jeffries, C. Kiraly, P. Montino, D. Kanter, S. Ahmed, D. Pau, U. Thakker, A. Torrini, P. Warden, J. Cordaro, G. D. Guglielmo, J. Duarte, S. Gibellini, V. Parekh, H. Tran, N. Tran, N. Wenxu, and X. Xuesong, "MLPerf Tiny Benchmark," *in Proceedings of the Neural Information Processing Systems Track on Datasets and Benchmarks*, 2021.

[15] Z. Wang, W. Yan, and T. Oates, "Time series classification from scratch with deep neural networks: A strong baseline," in *Proceedings of the IEEE International Joint Conference on Neural Networks*, 2017.

[16] C. Maxey and A. Andreas, "Oak Ridge National Laboratory (ORNL); Rotating Shadowband Radiometer (RSR); Oak Ridge, Tennessee (Data)," 2007. [Online]. Available: https://www.osti.gov/biblio/1052553

[17] B. Taylor, V. S. Marco, W. Wolff, Y. Elkhatib, and Z. Wang, "Adaptive deep learning model selection on embedded systems," *ACM SIGPLAN Notices*, vol. 53, no. 6, pp. 31–43, 2018.

[18] Y. Li, "Deep reinforcement learning: An overview," *arXiv preprint arXiv:1701.07274*, pp. 10,14, 2017.

[19] J. Schulman, F. Wolski, P. Dhariwal, A. Radford, and O. Klimov, "Proximal policy optimization algorithms," *arXiv preprint arXiv:1707.06347*, 2017.

[20] T. Basaklar, Y. Tuncel, and U. Y. Ogras, "tinyMAN: Lightweight energy manager using reinforcement learning for energy harvesting wearable IoT devices," *arXiv preprint arXiv:2202.09297*, 2022.

[21] Q. Ju and Y. Zhang, "Predictive power management for internet of battery-less things," *IEEE Transactions on Power Electronics*, vol. 33, no. 1, pp. 299–312, 2017.

Precision-aware Latency and Energy Balancing on Multi-Accelerator Platforms for DNN Inference

Matteo Risso*, Alessio Burrello*†, Giuseppe Maria Sarda‡, Luca Benini†,
Enrico Macii*, Massimo Poncino*, Marian Verhelst‡, Daniele Jahier Pagliari*
*Politecnico di Torino, Turin, Italy. †University of Bologna, Bologna, Italy. ‡KU Leuven, Belgium.
Corresponding Email: matteo.risso@polito.it

***Abstract*—The need to execute Deep Neural Networks (DNNs) at low latency and low power at the edge has spurred the development of new heterogeneous Systems-on-Chips (SoCs) encapsulating a diverse set of hardware accelerators. How to optimally map a DNN onto such multi-accelerator systems is an open problem. We propose ODiMO, a hardware-aware tool that performs a fine-grain mapping across different accelerators on-chip, splitting individual layers and executing them in parallel, to reduce inference energy consumption or latency, while taking into account each accelerator's quantization precision to maintain accuracy. Pareto-optimal networks in the accuracy vs. energy or latency space are pursued for three popular dataset/DNN pairs, and deployed on the DIANA heterogeneous ultra-low power edge AI SoC. We show that ODiMO reduces energy/latency by up to 33%/31% with limited accuracy drop (-0.53%/-0.32%) compared to manual heuristic mappings.**

***Index Terms*—Heterogeneous Computing, Edge Computing, Deep Learning Accelerators, Quantization**

I. Introduction

Executing Deep Neural Networks (DNNs) inference at the edge brings several advantages, including lower energy consumption, lower and more predictable response latency, and improved privacy, by eliminating the dependency on a constant Internet connection [1], [2]. However, deploying computationally intensive DNNs on edge devices with tight power envelopes and energy constraints is a daunting task, addressed by current research in two orthogonal ways. On the software side, optimization techniques such as constrained Neural Architecture Search (NAS), pruning, and quantization [2], [3], are applied to DNN models to make them both accurate and resource-efficient. On the hardware side, efficiency is improved through specialization, i.e., by designing increasingly heterogeneous Systems-on-Chip (SoCs), equipped with domain specific accelerators for DNN processing [4]–[7]. In particular, a recent trend goes towards *multi-accelerator* SoCs, in which multiple specialized hardware blocks are either optimized for different DNN operations, or to perform the same operations with different trade-offs in terms of latency, throughput, energy efficiency or accuracy [5]–[7].

How to optimize a DNN model for execution onto these multi-accelerator systems is an open problem. In fact, classic model optimizations are either hardware-independent targeting abstract complexity metrics, or tailored to the scenario in which the entire network runs on a single device (CPU, GPU, etc). While more recent works considered multi-device inference [7]–[14], to our knowledge, they all assumed that all devices could produce equivalently accurate results. This is not always true, with a key counter-example being SoCs including both Digital and Analog In-Memory-Computing (AIMC) accelerators [5], [6], where the latter can be faster and more efficient, but produce approximated results due to very low quantization bit-width used for weights (e.g., binary or ternary), while the former are slower and more energy hungry, but process wider data items at higher numerical precision.

In this work, we propose a novel approach to optimize and map a DNN execution onto such kind of system, which takes into account the quantization supported by different accelerators already at training time. Namely, we leverage a fine-grained, gradient-based, mixed-precision search method [15], [16] to partition each DNN layer onto sub-layers, executed in parallel by the various accelerators using their respective precision. While taking into account the possible accuracy drops due to quantization, our method tries to minimize energy consumption or latency, through appropriate hardware-aware cost models. We name our approach **O**ne-shot **Di**fferentiable **M**apping **O**ptimizer (**ODiMO**).

With experiments on three popular Convolutional Neural Network (CNN) architectures, trained on edge-relevant computer vision benchmarks, we show that our method yields rich Pareto-fronts of mapping solutions in the accuracy versus latency or energy spaces, under different assumptions regarding the accelerators capabilities in a heterogeneous SoC. When deployed on a real-world SoC of this kind, DIANA [6], our optimized models reduce energy/latency by up to 33%/31% with limited accuracy drops (-0.53%/-0.32%) compared to manual mappings based on rules of thumb. Furthermore, we improve accuracy by up to +37% for a 1.12× energy increase compared to a solution that only tries to minimize energy without considering accuracy. Our code is open-sourced at: `https://github.com/eml-eda/odimo`.

II. Background and Related Works

A. Specialized hardware for edge DNN inference

In recent years, specialized architectures for DNN processing at the edge have flourished, with several designs proposed

This work has received funding from the Key Digital Technologies Joint Undertaking (KDT-JU) under grant agreement No 101095947. The JU receives support from the European Union's Horizon Europe research and innovation programme.

both in industry and academia [17]. Many of these modern SoCs contain *multiple* specialized hardware blocks, able to execute the same workload with different trade-offs in terms of latency, throughput, energy consumption, or accuracy. One example is the Jetson AGX Xavier series from NVIDIA, equipped with an 8-cores ARM CPU, a NVIDIA Volta GPU with 512 CUDA cores and two NVIDIA Deep Learning Accelerators (NVDLA). Users can split the workload between the GPU, faster but more energy hungry, and the NVDLAs, slightly slower but more efficient [7].

In the architecture of [5], a control CPU dispatches the workload either to a 590k-cells AIMC accelerator tailored for 1-bit multiply-and-accumulate (MAC) operations, or to a digital Near-Memory Computing (NMC) accelerator, which supports variable precision from 1 to 8bits. In this case, selecting one of the two accelerators results either in more accuracy but higher latency and energy (NMC), or vice versa (AIMC). Similarly, DIANA [6] features a single-core RISC-V CPU as control unit and two DNN accelerators: a 16×16 grid of digital processing elements performing MACs at 8-bit precision, with a 64 kB weight memory, and a 500k-cells AIMC accelerator with ternary weights. The two accelerators share a dedicated 256 kB L1 memory, accessed through Direct Memory Access (DMA).

B. Mixed-Precision Quantization

In parallel to new specialized SoCs, many DNN optimization techniques have been introduced over the years, such as pruning, quantization, and NAS, to design lightweight networks that can fit on edge devices. This section focuses on the main knob explored by ODiMO, i.e., quantization; we refer readers to [2] for details on other techniques.

Quantization improves DNNs' energy-efficiency by reducing the precision of data and operations, e.g., from floating point to low bit-width integer formats (1 to 8-bit) [3]. The default approach is the so-called *fixed-precision* quantization, in which the same bit-width n (usually 8-bit) is used across the model. Recently, *mixed-precision* approaches, that vary n for different parts of a DNN, have been shown to provide additional time, memory and energy savings, especially when native hardware support for sub-byte operations is available [15], [16], [18], [19]. However, finding the optimal assignment of bit-widths to different parts of the network, e.g., to minimize energy under a given accuracy constraint, involves searching a huge space, exponential in the depth of the DNN.

Existing solutions to this problem use techniques inherited from NAS, such as sensitivity-based heuristics [19] or Reinforcement Learning [18]. In particular, a recent approach [15], [16] takes inspiration by Differentiable NAS (DNAS) to speed up the process, optimizing the bit-width assignment *during training*. Multiple copies of each tensor, quantized at different bit-widths, are generated on-the-fly, and linearly combined by means of trainable NAS parameters. The latter are then inserted in a standard DNN training loop, where an appropriately regularized loss function guides the optimization to increase the NAS parameters linked with quantizations that yield a good trade-off between accuracy and inference cost. At the end of training, the bit-widths which have been assigned the largest NAS coefficient are selected for each tensor.

C. DNN mapping on heterogeneous systems

The problem of mapping complex tasks onto a heterogeneous system with accelerators has been studied for a long time. Early works focus on generic workloads (e.g. OpenCL programs) [20], but more recently, the specific case of DNN inference has attracted a lot of attention. The authors of [8] implement a simple form of data parallelism, in which entire inferences are mapped onto a single device, selecting the fastest available between CPUs, GPUs and NPUs at any time. [9] performs a similar mutually-exclusive mapping, but at the level of each DNN layer, using a random forest to predict the latency or energy efficiency of offloading a layer to CPU or to multiple GPUs, based on the tensors geometry. [10] proposes a heuristic for a multi-accelerator system including a GPU (NVIDIA Jetson TX2) and a FPGA (Xilinx Artix7), consisting in offloading all Fully-Connected (FC) layers to the FPGA, and the rest of the DNN to the GPU.

The authors of [7] explore the energy versus latency trade-offs offered by offloading parts of a DNN to the GPU or to the NVDLAs in a NVIDIA Jetson AGX Xavier. Partitioning is done at layer level, and linear programming is used to find the lowest latency mapping under an energy constraint. An alternative mapping scheme for the Xavier is proposed in [11], which explores data parallelism and pipelining among GPU and NVDLAs, focusing only on improving throughput.

In [12], finer-granularity intra-layer partitions are explored, using dynamic programming to optimize DNN training latency on a system composed of multiple Google TPUv2/v3 accelerators, taking into account compute performance and communication overheads. Three partitioning axes are considered (over batches, input channels or output channels). Lastly, other works target DNN mapping problems for networks of distributed devices rather than individual multi-accelerator SoCs, proposing similar data- or model-partitioning schemes [13], [14].

III. One-shot Differentiable Mapping Optimizer

All works discussed in see Sec. II-C assume that mapping part of a DNN to a given accelerator does not affect the final accuracy [7]–[14]. Therefore, they only explore the trade-off between latency and throughput or latency and energy. While reasonable for their targets, this assumption breaks for more extreme-edge-oriented platforms such as [5], [6] in which some of the available accelerators use aggressive quantization. In that case, the mapping choices greatly influence both functional (accuracy) and non functional (e.g., energy) metrics. To our knowledge, no previous work has considered this trade-off in DNN mapping optimizations for multi-accelerator SoCs.

We fill this gap by proposing ODiMO, an optimization method that partitions a DNN execution onto heterogeneous compute domains that include accelerators with different quantization levels and formats, optimizing the trade-off between

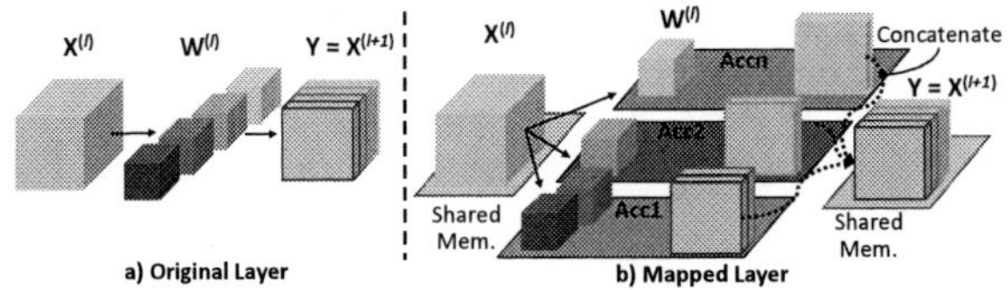

Fig. 1. Mapping strategy for a Convolutional layer.

accuracy and energy or latency. Differently from most conventional mapping strategies, which are coarse-grain (e.g., layer-wise), our tool uses a fine-grain intra-layer partitioning aimed at maximizing the utilization of all accelerators.

A. Mapping optimization strategy

ODiMO considers splitting each Convolutional (Conv) or FC layer in a DNN among N different devices, at the level of *individual output channels/neurons*. That is, all accelerators process the entire layer input and produce a subset of the output activations, as shown in Figure 1 for a Conv layer. For this approach to be effective, the target heterogeneous system must respect two properties: i) different accelerators can have incompatible *weights* quantizations but must have the same *activation* quantization, or at least two similar formats that do not cause a significant difference in terms of accuracy (see Sec. III-B); ii) all accelerators must have access to a shared memory for loading/storing the layer input/partial output [12]. Note that both [5] and [6] respect these constraints. Under these conditions, the problem can be reduced to selecting the best quantization for each channel's weights, where the choice not only influences the overall model accuracy, but also limits the mapping options for that channel to the accelerator(s) that support the selected precision, thus affecting the inference energy/latency costs.

The optimization space is huge: e.g., for just $N = 2$ accelerators and a ResNet18 CNN, there are about 10^{39} possible ways to assign each channel of each layer to one of the two devices. Therefore, ODiMO adopts a DNAS-like optimization method inspired by recent work on fine-grained mixed-precision quantization [16], in which bit-width assignment is performed *during training*, similar to so-called One-shot NAS approaches.

As shown in Fig. 2, for each layer l of a DNN, the weight tensor $W^{(l)}$ is fake-quantized multiple times, simulating the data format supported by *all available accelerators*. Namely, we generate $\hat{W}^{(l)}_{\text{acc}_i}, \forall i \in [1, N]$ different fake-quantized copies of the weights. Each of them is paired with a vector of trainable parameters $\alpha^{(l)}_{\text{acc}_i} \in \mathbb{R}^{C^{(l)}_{\text{out}}}$, where $C^{(l)}_{\text{out}}$ is the number of output channels in the l-th layer. For each channel c in $C^{(l)}_{\text{out}}$ we then compute the *effective weights* as:

$$\hat{W}^{(l)}_c = \sum_{i=1}^{N} \bar{\alpha}^{(l)}_{c,\text{acc}_i} \hat{W}^{(l)}_{c,\text{acc}_i} \quad (1)$$

where $\bar{\alpha}^{(l)}_{c,\text{acc}_i} = \text{sofmax}(\alpha^{(l)}_{c,\text{acc}_i}, \tau)$ and τ is the softmax temperature. The aggregated effective weight tensor $\hat{W}^{(l)}$ for layer l is obtained concatenating the $\hat{W}^{(l)}_c$ tensors of Eq. 1 over the output channel dimension.

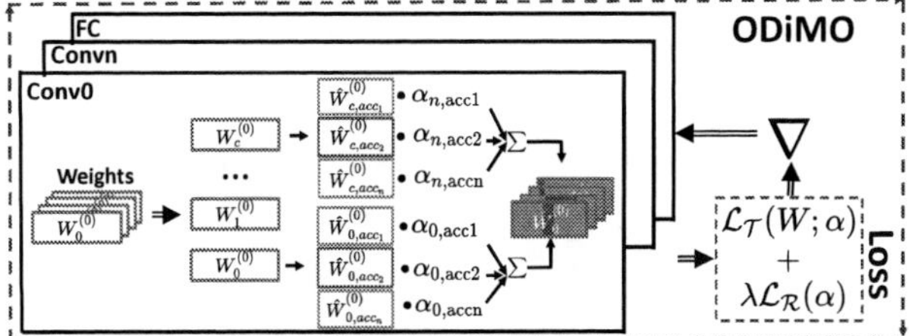

Fig. 2. DNAS-like optimization scheme at training time.

Using $\hat{W}^{(l)}$ in place of $W^{(l)}$ for all layers, ODiMO solves a continuous relaxation of the multi-accelerator mapping problem. In practice, each layer's output becomes a "mix" of what would be produced by all available accelerators, given their quantization formats. The importance of each accelerator in the mix is controlled by $\alpha^{(l)}_{\text{acc}_i}$. We can then train $\alpha^{(l)}_{\text{acc}_i}$ as in DNAS, to learn which mapping provides the best accuracy vs inference cost trade-off for a given channel.

Specifically, the DNN with fake-quantized weights is inserted in a training loop which optimizes:

$$\min_{W,\alpha} \mathcal{L}_\mathcal{T}(W;\alpha) + \lambda \mathcal{L}_\mathcal{R}(\alpha) \quad (2)$$

where $\mathcal{L}_\mathcal{T}$ is the standard task loss, $W = \{W^{(l)}\}, \forall l$ is the set of DNN weights, and $\alpha = \{\alpha^{(l)}_{acc_i}\}, \forall l, i$ is the set of parameters that determine the bit-width assignment for each channel, and consequently its mapping to one of the available accelerators. Lastly, $\mathcal{L}_\mathcal{R}$ is an additional loss term that models the cost of the DNN execution (e.g., energy) as a function of the mapping decisions, and λ is a scalar regularization strength that controls the balance between the two loss terms.

We formulate $\mathcal{L}_\mathcal{R}$ differently when optimizing for energy or latency. For latency, we minimize:

$$\mathcal{L}_\mathcal{R} = \sum_l M^{(l)},\ M^{(l)} = \max(LAT^{(l)}_1, ..., LAT^{(l)}_n) \quad (3)$$

where $LAT^{(l)}_i(\alpha)$ is a differentiable model of the i-th accelerator's latency for layer l, as a function of the channels assigned to it, detailed in Sec. III-C. $M^{(l)}$ is the latency of the entire layer, assuming that the accelerators run in parallel, which except for thermal effects, which are generally negligible for low-power SoCs like DIANA, is the optimal choice for both time and energy reduction, as it minimizes idle consumption. In practice, since we need a fully-differentiable loss term, we substitute the max operation of Eq. 3 with its smooth differentiable approximation. For energy reduction, instead, we use the following model:

$$\mathcal{L}_\mathcal{R} = \sum_l \sum_i P_{act,i} \cdot LAT^{(l)}_i + P_{idle,i} \cdot (M^{(l)} - LAT^{(l)}_i) \quad (4)$$

where $P_{act,i}$ and $P_{idle,i}$ are the average active and idle power consumption of the i-th accelerator.

At the end of training, we *discretize* the mapping. Namely, for each channel, we select the accelerator corresponding to the largest $\alpha^{(l)}_{c,\text{acc}_i}$.

However, the channels assigned to the same hardware are in general not consecutive, which would complicate the merging of partial outputs. Therefore, a layer transformation pass is applied to the DNN before deployment on the target SoC,

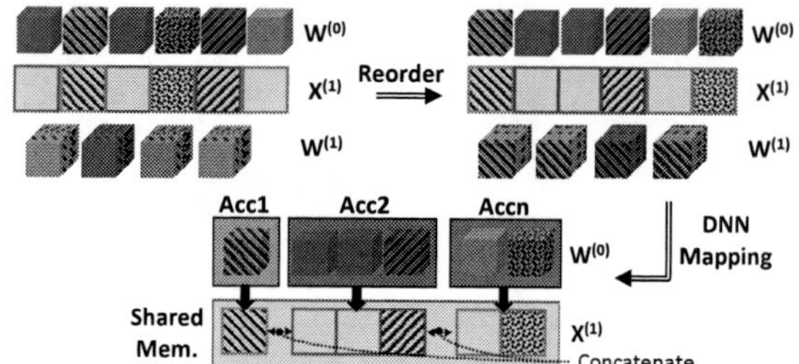

Fig. 3. Final layer re-organization pass to support partitioning.

shown in Fig. 3. Activation channels are represented as side-by-side squares for better visualization, and colors used for weights filters and for activation outlines indicate the assignment to a given accelerator. Black patterns are added to some filters/output slices to clarify the process. The top-left part of the figure shows an example of ODiMO output. On the top-right, the channels in $X^{(1)}$ and the corresponding filters in $W^{(0)}$ are *reordered*, grouping together those that will be dispatched to the same accelerator, to enable a simple concatenation of outputs. To preserve the network functionality, the weights of the next layer $W^{(1)}$ are also reordered across the *input* channels dimensions. After this transformation, the layer is effectively split into N independent sub-layers that can be deployed in parallel onto the N available accelerators, without requiring any data-marshaling overhead to aggregate their outputs (bottom of Fig. 3).

B. Training Details

In this work, we apply ODiMO to the DIANA multi-accelerator SoC of [6], presented in Sec. II-A. This section reports the HW-specific details of our implementation. Note that the general approach is orthogonal to most of these details.

Given a pre-trained floating-point DNN, we first fold Batch Normalization (BN) layers with Conv/FC, since the DIANA accelerators do not implement BN in hardware. Then, we apply fake-quantization following the scheme of [21]:

$$Q(x) = \frac{e^s}{2^{n-1}-1} \cdot \mathrm{round}(2^{n-1} - 1 \cdot \mathrm{clip}(x, -1, 1)) \quad (5)$$

where s is a trainable scale parameter and n is the bit-width. With $n = 2$, Eq. 5 performs ternarization, i.e., the quantization format of DIANA's AIMC accelerator weights, while we use $n = 8$ for the digital accelerator weights. Concerning activations, the AIMC and digital blocks have slightly different formats on 7- and 8-bit respectively. During the optimization phase, we use the worst case of the two (7-bit) as fake-quantization bit-width for layers' inputs/outputs. As long as the DNN is appropriately fine-tuned (see below), we found this approximation not to degrade our results.

The fake-quantized DNN is optimized with the procedure of Fig. 2 until convergence, with an early-stop mechanism. Then, after discretizing the final channel assignment to each accelerator, the model is fine-tuned based on the task loss term $\mathcal{L}_T$ only. In this phase, we use the exact quantization format also for activations, i.e., shared data are stored on 8-bit but the AIMC accelerator D/A and A/D converters are on 7-bit, effectively truncating the LSB of inputs/outputs.

C. Hardware Models

The differentiable latency models plugged in Eq. 3 and 4 are key elements of the proposed method. Latency modeling has been studied extensively in recent NAS literature. A common approach [22] uses a small NN model trained on many profiled layers to predict latency based on the layer geometry. Although this method is compatible with ODiMO, given the predictability of DIANA's AIMC and Digital accelerators execution, we found that using simpler analytical models that account for the respective parallelism and dataflow yields good-enough results while making the optimization faster.

These simplified models neglect non-idealities such as memory stalls, tiling overheads for large activation tensors, and programming overheads. However, comparing them with hardware measurements on a wide set of layer configurations, we verified that they can preserve rank well, i.e., it generally holds that if $LAT^1_{predicted} < LAT^2_{predicted}$, then $LAT^1_{hw} < LAT^2_{hw}$, which makes them usable for mapping decisions. For the AIMC accelerator, our latency model is:

$$LAT^{(l)}_{aimc}(\alpha) = \lceil \frac{C^{(l)}_{in} \times f^{(l)}_x \times f^{(l)}_y}{1152} \rceil \lceil \frac{C^{(l)}_{out}(\alpha)}{512} \rceil \times o^{(l)}_x \times o^{(l)}_y + 2 \times 4 \times C^{(l)}_{in} \times \lceil \frac{C^{(l)}_{out}(\alpha)}{512} \rceil$$

where $C^{(l)}_{in}$, $o^{(l)}_x/o^{(l)}_y$ and $f^{(l)}_x/f^{(l)}_y$ are the layer's input channels, output spatial dimensions and kernel sizes respectively, and the two addends correspond to the cycles of the computation and of the DMA transfer to populate the weights respectively. Note that this model depends on the optimization choices (α) through C_{out}, since ODiMO assign a variable number of output channels to the AIMC accelerator. The digital accelerator model uses the same two terms:

$$LAT^{(n)}_{dig}(\alpha) = \lceil \frac{C^{(l)}_{out}(\alpha)}{16} \rceil \lceil \frac{o^{(l)}_y}{16} \rceil \times C^{(l)}_{in} \times o^{(l)}_x \times f^{(l)}_x \times f^{(l)}_y + C^{(l)}_{in} \times C^{(l)}_{out}(\alpha) \times f^{(l)}_x \times f^{(l)}_y$$

Numeric constants in the two models depend on the sizes of the respective processing element arrays. We do not count activation transfers, because we assume that they are always stored in the shared L_1 scratchpad memory.

IV. Experimental Results

A. Setup

We benchmark ODiMO on three edge-relevant computer vision tasks and DNNs: i) image classification on CIFAR-10, with ResNet20 [23] as reference model; ii) image classification on the 200-classes Tiny-ImageNet [24], with ResNet18 [23]; iii) person detection on Visual Wake Word (VWW), which is based on the MSCOCO 2014 dataset, with a MobileNet-V1 with $0.25\times$ width-multiplier [25]. We pre-train and fine-tune all DNNs using the same epochs and hyper-parameters of the reference papers. ODiMO is written in Python 3.9 and PyTorch v1.11. To deploy our networks on DIANA [6], we adapted the open-source DORY [26] framework.

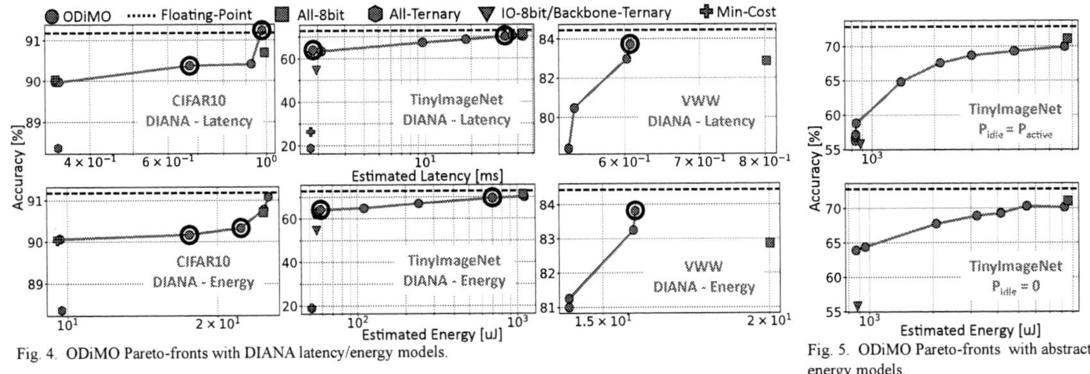

Fig. 4. ODiMO Pareto-fronts with DIANA latency/energy models.

Fig. 5. ODiMO Pareto-fronts with abstract energy models.

We compare ODiMO with several baseline mapping alternatives. Specifically, our baselines are: i) *All-8bit* and *All-Ternary*, the trivial mappings that use only the DIANA digital and AIMC accelerators, respectively; ii) *IO-8bit/Backbone Ternary*, a heuristic solution from [6] that maps the first/last layers to the 8-bit accelerator, and the rest to the AIMC one, based on the rule-of-thumb that aggressively quantizing layers close to the input and the output of the network often causes large accuracy drops; iii) *Min-Cost* an optimized deterministic mapping that uses the same channel-wise partitioning of ODiMO, with the sole goal of minimizing latency or energy without taking accuracy into account. Namely, it statically maps channels of each layer to the AIMC and digital accelerators, before training, to minimize Eq.3 or Eq.4. In case of equivalent solutions, digital channels are maximized since this is expected to improve accuracy.

For the MobileNetV1 on VWW, we only optimize the mapping of pointwise and standard convolutions (and FC layers), since in DIANA, depthwise convolutions can only be executed on the digital accelerator. Further, all baselines that use the AIMC accelerator are not reported for VWW because their training could not converge, resulting in random predictions.

B. Search-Space Exploration

Fig. 4 show the results obtained with ODiMO on the three benchmarks, in the accuracy versus estimated latency (top row) and accuracy versus estimated energy (bottom row) spaces, with latency and energy computed using the DIANA's models described in Sec. III-C. Each ODiMO point is obtained repeating the training procedure of Sec. III-B with a different regularization strength (λ) and using either the energy or latency regularizer. We also report the baselines in green and the floating point DNN accuracy as a horizontal dashed line. In all graphs, *baselines are either dominated or on the Pareto frontier*, demonstrating the effectiveness of our approach. Additionally, ODiMO produces a *rich set of intermediate Pareto-optimal solutions* that could not be obtained otherwise.

With the DIANA cost models, ODiMO can trade-off the estimated latency and accuracy (-32% latency, -0.32% accuracy) w.r.t. the All-8bit baseline on CIFAR-10 (3rd blue dot from the right in the top-row figure). Moreover, energy can be reduced by 29% when accepting a 0.53% accuracy drop (4th point from the right in bottom-row). On TinyImageNet, our tool discovers solutions spanning more than one order of magnitude on the x axis, that can reduce the estimated latency/energy by 15%/35.6% and 77.8%/77.7% for a drop of $<2\%$ and $<5\%$ accuracy w.r.t. the 8bit baseline, respectively. Lastly, on VWW, ODiMO achieves up to 24.3%/20.8% latency/energy reduction while improving accuracy by 0.87%/0.95% w.r.t All-8bit.

Fig. 5 shows the independence of ODiMO from the DIANA SoC specifics. For the sake of space, the figure shows results only on Tiny-ImageNet, and demonstrates how ODiMO is able to find a rich collection of Pareto optimal mappings even with different hardware cost models. Results are obtained considering two abstract models, not related to any specific HW, retaining from DIANA only the presence of two accelerators working with ternary and 8-bit data precision respectively. These models assume that the latency of both accelerators is simply *proportional to the number of operations*, and that the active power of the 8-bit accelerator is 10 times higher than the ternary one ($P_{act,8} = 10 \cdot P_{act,ter}$). Then, for the first model, we assume $P_{idle} = P_{act}$ for both accelerators (no shutdown), while, for the second, we consider $P_{idle} = 0$ (ideal shutdown).

The top graph of Fig. 5 shows ODiMO mappings obtained with the first model. Note that in this corner case, energy and latency minimization coincide, since substituting $P_{idle} = P_{act}$ for all accelerators in Eq. 4 yields Eq. 3, except for a constant. The bottom of Fig. 5, instead, shows the results obtained with $P_{idle,i} = 0$. The two graphs only show accuracies $> 55\%$; the all-ternary and min-cost baselines are not shown, as they reach too low accuracy (see middle graphs in Fig. 4). With these two models, ODiMO reduces energy respectively by 44.2%/51.5% for a drop of $<2\%$ accuracy w.r.t. the 8bit baseline.

C. DIANA Deployment

This section analyzes the results of deploying a subset of the solutions from Fig. 4 on the DIANA SoC, running at a frequency of 260 MHz, substituting modeled with measured latency and energy. For each benchmark, we deploy the All-8bit and Min-Cost baselines and a selection of ODiMO results (highlighted with a black circle in Fig. 4). We select two points from the latency Pareto-front (Large-Lat and Small-Lat) and two from the Energy one (Large-En, Small-En) for all benchmarks except VWW, where given the smaller search space, we deploy a single point from each graph. For all DNNs, we

TABLE I

DEPLOYMENT ON DIANA OF SELECTED SOLUTIONS FROM FIG. 4

	Network	Acc.	lat. [ms]	E. [uJ]	D./A. util.	A. Ch.
Cifar10	All-8bit	90.70	1.55	38.71	100% / 0%	0%
	ODiMO Large - Lat	91.24	1.55	43.20	100% / 21.0%	5.6%
	ODiMO Small - Lat	90.38	1.07	34.43	100% / 44.8%	51.8%
	ODiMO Large - En	90.33	1.05	33.43	100% / 43.1%	50.3%
	ODiMO Small - En	90.17	0.80	25.94	76.2% / 60.0%	72.9%
	Min Cost	90.06	0.47	13.57	9.5% / 93.6%	97.5%
TinyI.	All-8bit	71.29	94.44	2357.3	100% / 0%	0%
	ODiMO Large - Lat	70.16	73.92	2999.8	100% / 8.2%	23.8%
	ODiMO Small - Lat	64.07	4.32	139.2	25% / 87.8%	99.0%
	ODiMO Large - En	69.54	63.55	1648.18	100% / 9.4%	34.2%
	ODiMO Small - En	64.14	5.05	141.25	20% / 84.4%	96.5%
	Min Cost	26.51	4.07	125.96	30% / 89.7%	98%
VWW	All-8bit	82.86	3.05	76.18	100% / 0%	0%
	ODiMO - Lat	83.73	2.80	71.29	100% / 17.8%	39.1%
	ODiMO - En	83.81	2.79	70.74	100% / 17.71%	40%

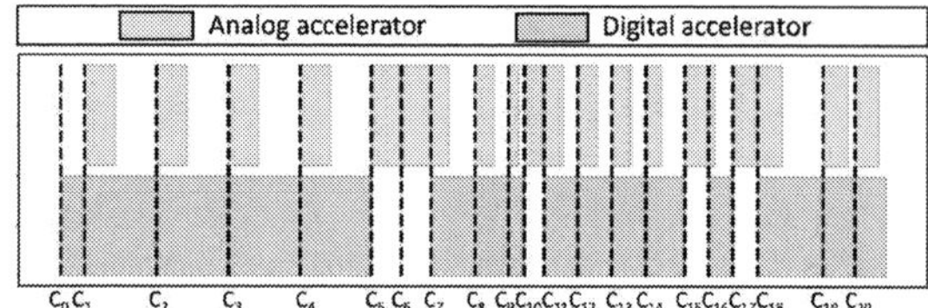

Fig. 6. Utilization of accelerators on convolutional layers of ODiMO-Small-En on CIFAR-10. (C_i = i-th Conv. layer).

report in Table I accuracy, latency, energy consumption, the percentage of time each accelerator is utilized during an end-to-end inference (*D./A. util.*), and the percentage of channels executed on the AIMC accelerator, i.e., the fraction C_{out}^{aimc}/C_{out} for the whole network (*A. Ch.*). On CIFAR-10, ODiMO-Small-En reduces energy by 33% w.r.t All-8bit, for a limited accuracy drop (-0.53%). This result, which is compatible with the 29% reduction estimated by the model (see Sec. IV-B), is achieved offloading a large portion of the channels (72.9% of the total) to the analog accelerator. Further, the digital and AIMC accelerators are active for 76.2% and 60% of the inference time. Fig. 6 shows a breakdown of the utilization of both accelerators throughout an inference with this DNN. For almost the 40% of the time, both accelerators work simultaneously, demonstrating that splitting layers between the two is beneficial to reduce energy consumption, while keeping an almost constant accuracy.

On TinyImageNet, ODiMO-Large-En suffers an accuracy drop compared to All-8bit (-1.75%), but improves the energy by 1.43×, while ODiMO-Small-En achieves 37.63% higher accuracy compared to Min-Cost, at the cost of only 1.12× higher energy consumption. Both solutions exploit the analog accelerator for a large portions of the DNN channels, 34.2% and 96.5%, respectively. It is also worth mentioning that Min-Cost, which offloads only an additional 1.5% of the network to the AIMC accelerator compared to ODiMO-Small-En, fails in reaching a good accuracy; this is because the Min-Cost mapping is built without taking into account accuracy, contrary to our method. Further, notice that ODiMO-Large-Lat effectively improves latency compared to All-8bit (1.27× faster) but at same time fails in reducing the energy consumption (1.27× less efficient) demonstrating the need to optimize energy and latency with two different tailored models, depending on the specific design goals.

On VWW, despite lower benefits, ODiMO-En shows a higher accuracy compared to the All-8bit (+0.95%) solution with 7% lower energy consumption, Pareto dominating it.

V. CONCLUSIONS

We have introduced ODiMO, a tool that partitions a DNN execution at fine grain among multiple accelerators with incompatible quantization formats. To do so, it formulates the problem as a mixed-precision bit-width assignment and uses a DNAS-like approach to optimize the mapping while training the DNN weights. With results on different benchmarks and DNN architectures, we have shown that ODiMO can obtain rich Pareto-fronts in both the accuracy vs energy or latency spaces, and reduce energy by up to 33% with limited accuracy drops compared to a single-accelerator solution. Future work will concentrate on building more accurate hardware models and supporting also activations quantization requiring format conversions whose cost need to be modeled.

REFERENCES

[1] Z. Zhou *et al.*, "Edge Intelligence: Paving the Last Mile of Artificial Intelligence With Edge Computing," *Proc. IEEE*, vol. 107, no. 8, 2019.

[2] V. Sze *et al.*, "Efficient processing of deep neural networks," *Synthesis Lectures on Computer Architecture*, vol. 15, no. 2, pp. 1–341, 2020.

[3] B. Jacob *et al.*, "Quantization and Training of Neural Networks for Efficient Integer-Arithmetic-Only Inference," in *CVPR*. IEEE, 2018.

[4] K. Seshadri *et al.*, "An evaluation of edge tpu accelerators for convolutional neural networks," 2021.

[5] H. Jia *et al.*, "A programmable heterogeneous microprocessor based on bit-scalable in-memory computing," *IEEE J. Solid-State Circ.*, 2020.

[6] K. Ueyoshi *et al.*, "Diana: An end-to-end energy-efficient digital and analog hybrid neural network soc," in *ISSCC*, vol. 65. IEEE, 2022.

[7] I. Dagli *et al.*, "Axonn: Energy-aware execution of neural network inference on multi-accelerator heterogeneous socs," in *DAC*, 2022.

[8] S. Wang *et al.*, "Neural network inference on mobile socs," *IEEE Design & Test*, vol. 37, no. 5, pp. 50–57, 2020.

[9] G. Vasiliadis *et al.*, "The best of many worlds: Scheduling machine learning inference on cpu-gpu integrated architectures," in *IPDPSW*, IEEE, 2022.

[10] Y. Tu *et al.*, "A power efficient neural network implementation on heterogeneous fpga and gpu devices," in *IRI*, IEEE, 2019.

[11] E. Jeong *et al.*, "Deep learning inference parallelization on heterogeneous processors with tensorrt," *IEEE Embed. Syst. Lett.*, vol. 14, 2022.

[12] L. Song *et al.*, "AccPar: Tensor Partitioning for Heterogeneous Deep Learning Accelerators," in *HPCA*, IEEE, 2020.

[13] Y. Kang *et al.*, "Neurosurgeon: Collaborative Intelligence Between the Cloud and Mobile Edge," in *ASPLOS*. ACM, 2017.

[14] D. Jahier Pagliari *et al.*, "CRIME: Input-Dependent Collaborative Inference for Recurrent Neural Networks," *IEEE Trans. Comp.*, 2020.

[15] Z. Cai *et al.*, "Rethinking differentiable search for mixed-precision neural networks," in *CVPR*, IEEE/CVF 2020.

[16] M. Risso *et al.*, "Channel-wise mixed-precision assignment for dnn inference on constrained edge nodes," in *IGSC*, IEEE, 2022.

[17] A. Reuther *et al.*, "AI Accelerator Survey and Trends," in *HPEC*, 2021.

[18] K. Wang *et al.*, "Haq: Hardware-aware automated quantization with mixed precision," in *CVPR*, IEEE/CVF, 2019.

[19] Z. Dong *et al.*, "Hawq-v2: Hessian aware trace-weighted quantization of neural networks," *Adv. Neural Inf. Process. Syst.*, vol. 33, 2020.

[20] K. Moren *et al.*, "Automatic mapping for opencl-programs on cpu/gpu heterogeneous platforms," in *ICCS*, Springer, 2018.

[21] B.-E. Verhoef *et al.*, "Fq-conv: Fully quantized convolution for efficient and accurate inference," *arXiv:1912.09356*, 2019.

[22] H. Cai *et al.*, "Proxylessnas: Direct neural architecture search on target task and hardware," *arXiv:1812.00332*, 2018.

[23] K. He *et al.*, "Deep residual learning for image recognition," in *CVPR*, IEEE, 2016.

[24] Y. Le *et al.*, "Tiny imagenet visual recognition challenge," *CS 231N*, vol. 7, no. 7, p. 3, 2015.

[25] C. Banbury *et al.*, "Mlperf tiny benchmark," *arXiv:2106.07597*, 2021.

[26] A. Burrello *et al.*, "Dory: Automatic end-to-end deployment of real-world dnns on low-cost iot MCUs," *IEEE Trans. Comp.*, 2021.

Florian: Developing a Low-power RISC-V Multicore Processor with a Shared Lightweight FPU

Jina Park†*, Kyuseung Han‡*, Eunjin Choi†, Sukho Lee‡, Jae-Jin Lee‡, Woojoo Lee†#, and Massoud Pedram§
† School of Electrical and Electronics Engineering, Chung-Ang University, Korea
‡ AI SoC Research Division, Electronics and Telecommunications Research Institute (ETRI), Korea
§ Department of Electrical and Computer Engineering, University of Southern California, USA

***Abstract*—As applications running on lightweight RISC-V processors become increasingly diverse and complex, the need for multicore processors supporting floating-point units (FPUs) is riseing, making processor designs using existing open-source RISC-V cores challenging. With the exception of a very few, most open lightweight RISC-V cores are integer cores without FPUs, which greatly reduces the design exploration space, making it impossible to design a processor optimized for each application. For example, most of these applications mainly perform integer operations, but occasionally perform floating-point operations. For them, a multicore processor with FPU per core is overkill and wastes power, which is a critical problem for processors where low-power design is paramount. To address the problem, we propose an external lightweight FPU that can be attached to any RISC-V integer core and a low-power multicore architecture using the designed FPU. For verification, we designed a RISC-V processor that implements all the proposed technologies, prototyped it on an FPGA device, and finally fabricated it as a System-on-Chip. Through experiments, it was confirmed that the proposed technology can cut energy consumption energy by up to 23%.**

I. Introduction

Explosive interest in RISC-V from academia and industry has driven dozens of free and open core releases built on this instruction set architecture [1], [2], [3]. Among them, especially lightweight RISC-V cores are finding increasing use in processors specialized in IoT, wearable, and embedded system applications [4], [5], [6]. These applications are becoming more diverse and complex, and as a result, the number of the applications that require a multicore processor supporting a floating-point unit (FPU) is growing exponentially [7], [8], [9], [10].

Unfortunately, it is very hard to develop a multicore processor supporting FPUs optimized for a target application by utilizing currently available lightweight RISC-V cores. This is due to the fact that most cores are integer cores except for very few cores with FPUs. Even if there is an integer core in the pool that best fits the target application, it is practically impossible to add an FPU to this core. This because the RTL code for that core may not be disclosed, and even if it is, adding FPU to an already completed core is as difficult as designing a new core [11], [12].

This work was partly supported by Institute of Information & communications Technology Planning & Evaluation (IITP) grants funded by the Korea government(MSIT) (No. 2022-0-00957, Distributed on-chip memory-processor model PIM semiconductor technology development for edge applications, and No. 2022-0-00971, Logic Synthesis for NVM-based PIM Computing Architecture).

*Jina Park and Kyuseung Han contributed equally to this work.
#*Corresponding authors*: Woojoo Lee (*space@cau.ac.kr*)

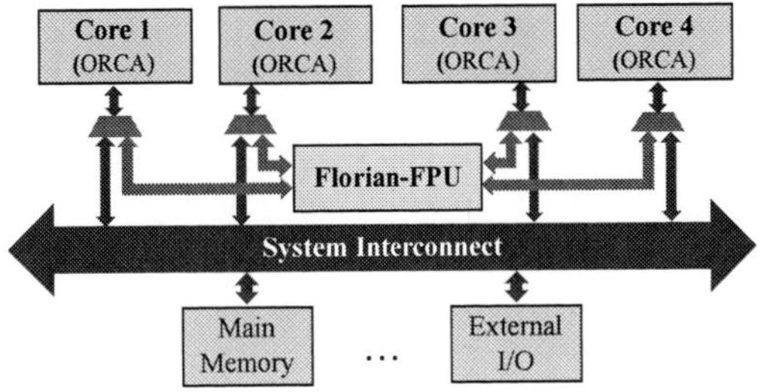

Fig. 1. Concept of the Florian quadcore processor architecture.

As a consequence, there are only a handful of lightweight RISC-V cores supporting a FPU to choose from, namely Rocket [13] and PULP [14], whose inflexibility poses an immediate limit to the development of new multicore processors using them. More specifically, Rocket core is difficult to use for new platform development due to its high platform dependencies. Rocket cores are provided as chip platforms, because of which it is very difficult from an optimization design perspective to create a new processor by extracting only Rocket cores. For example, we compared the performance of a newly created processor by extracting only the Rocket core from the Rocket-Chip platform and a processor made with the ORCA [15] core, which is much lighter than the Rocket core. When each FPGA prototype processor ran a EEMBC Coremark-Pro [16] benchmark program that only performs integer operations, although the Rocket-based processor used about 4 times more FPGA resources than the ORCA-based processor, the execution time was 630 and 372 sec, respectively, it can be seen that serious performance degradation has occurred in the Rocket-based processor.

Since each core has an FPU, there is a limit to designing a low-power multicore processor using PULP. Most IoT, wearables, and embedded system applications that require FPU actually perform integer operations mainly, but only infrequently perform floating-point operations. For these applications, a multicore processor with one FPU per core is over-spec, which wastes power. This is a fatal drawback for lightweight processors, where low-power design is the most important design concern.

In this paper, we present a method to develop the FPU-

979-8-3503-1176-1/23 $31.00 © 2023 IEEE

enabled low-power multicore processors using existing open RISC-V integer cores, which is called *Florian* (*FPU librarian*) method. The proposed Florian includes an external lightweight FPU that can be attached to any RISC-V integer core and a low-power multicore architecture using the designed FPU. Fig. 1 is an example of a quadcore processor architecture that shows the concept of Florian, where four RISC-V integer cores (ORCA cores) share an out-of-core lightweight FPU, called *Florian-FPU*, to handle each floating-point operation. In addition to Florian-FPU, we have designed and implemented an architecture that allows each core to share the FPU efficiently, performed RTL simulation, and completed functional verification through FPGA prototyping. And finally, we have fabricated the processor in Fig. 1 into a chip using the 110 nm process technology and utilized the chip for performance evaluation. Through intensive experiments, we have confirmed that the proposed method can improve energy efficiency by 10.9%, 19.2%, and 23.1% in dualcore, quadcore, and octacore processors, respectively.

II. Florian

The proposed Florian method utilizes an external lightweight FPU, which is deployed independently of the core, so that it is possible to design a processor supporting FPU using existing open cores without changing their structures at all. This method excels especially in the development of multicore processors, enabling energy-optimized designs. This is due to the fact that applications running on recent light-weight processors require floating-point operations, but they account for a small portion of the entire program [17]. Moreover, in the case of AI applications for state-of-the-art edge computing, it is a pragmatic development trend to convert floating-point operations to integer operations as much as possible to reduce processor power consumption [18]. Of course, there are still some floating-point operations left in these applications, which can be covered by a single, lightweight FPU, namely Florian-FPU. Thus, Florian allows processor designers to choose the most customized open core for their applications and design the most energy-efficient processors consuming minimal FPU power.

Fig. 2 shows the structure of the proposed Florian architecture. As shown in the figure, this architecture consists of two main blocks, the aforementioned Florian-FPU, and the Florian-Arbiter. To briefly explain the blocks prior to the detailed description of each, first, Florian-FPU is connected to each core by an independent path without going through the system interconnect. In order to reduce the design difficulty that may arise from the method of devising a new external port on the core for that connection and connecting the entire signal through it, Florian-FPU is designed to easily connect the core using Memory Mapped Input Output (MMIO). In addition, since high-performance communication is not required, the advanced peripheral bus (APB) protocol is simply used. Next, Florian-Arbiter plays a key role in allowing multi-cores to share and use a single FPU. Florian-Arbiter has a FIFO to prevent signal collisions that may occur in communication between cores and the FPU and to sequentially designate the FPU occupancy order of cores for continuous utilization of the FPU. More detailed explanations on the structure and design of Florian-FPU and Florian-Arbiter are described through the following subsections.

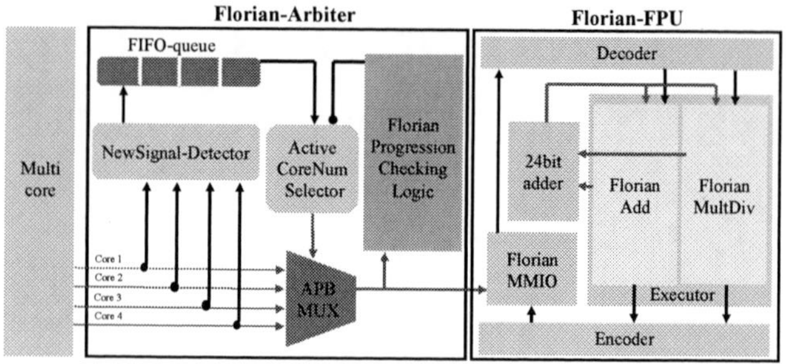

Fig. 2. Block diagram of the proposed Florian architecture.

A. Florian-FPU

The left part of Fig. 3 shows the detailed internal structure of Florian-FPU, consisting of major blocks of MMIO, decoder, encoder, and Executor. Since the Florian MMIO module manages all inputs and outputs of Florian, it can define whether a given signal is an operational command, an operand transfer, or a request for a calculation result. When a signal is defined as an operand, it is converted from a 32-bit IEEE754 [19] format to a 48-bit denormalized format through the decoder, resulting in a structure suitable for floating-point calculations. When the operation is completed through the Executor, the generated denormalized result is properly converted into IEEE754 format through the encoder and then sent back to the core along with the corresponding APB signal.

Florian-FPU is designed to support only addition, subtraction, multiplication and division operations for light weight, and $Executor$ is in charge of these arithmetic parts. The right part of Fig. 3 shows the detailed structure of the Executor architecture, which consists of the *Florian-Add* and *Florian-MultDiv* modules. The modules in Executor receive $24\text{-}bit\_adder\_result$ and $core\_ALU\_result$ signals, the former from a *24-bit adder* located outside the Executor, and the latter from *Concatenator* via the core. In other words, the signals are not generated internally, but are generated externally, which allows us to design a low-power FPU using minimal resources.

To elaborate on the low-power FPU design, for floating-point addition, a 24-bit adder is needed to add two 24-bit mantissa values. Also, since the multiplication and division require enough significant digits during the calculation of the two 24-bit mantissa values, 48-bit multiplier and divider are required to obtain accurate results. Plus, the floating-point multiplication/division requires an 8-bit adder to add or sub the two exponents. When all these calculation modules are built into the FPU, significant power consumption is inevitable. To address this issue, we noted that all the aforementioned calculations are only integer operations, so they can all be performed by the core as well. That is, we devised a method to handle 48-bit multiplication and division using the integer ALU in the core instead of processing it inside the FPU, and ported a 24-bit adder inside the FPU and shared it with $Florian\_MultDiv$ for the 8-bit exponential calculation. As a result, although the execution time inevitably increased slightly

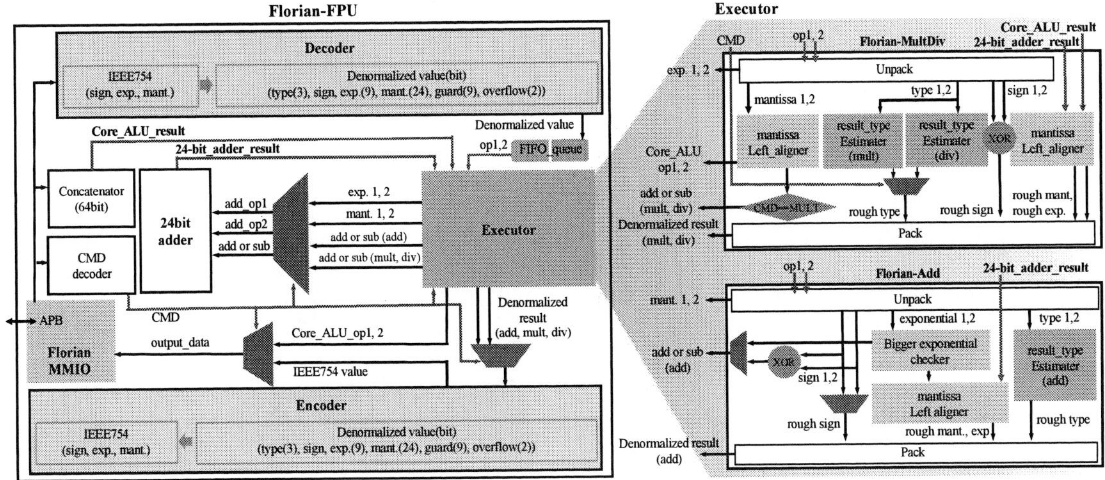

Fig. 3. Structure of the Florian-FPU architecture.

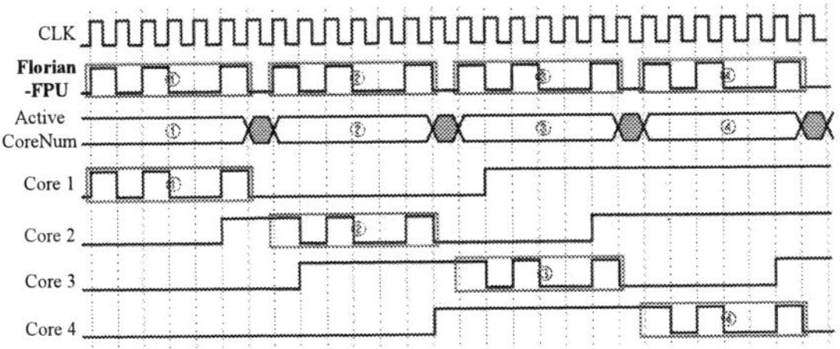

Fig. 4. Waveform of access between Florian-FPU and cores

as additional communication between the core and the FPU was required, energy savings was achieved because the effect of reducing power consumption due to the FPU resource savings was greater than that.

B. Florian-Arbiter

The low power effect of Florian is that multiple cores share the Florian-FPU. To do this, as shown in Fig. 4, only one core must be granted access to the FPU at a time, and the other cores must wait in sequence until the currently executing core completes its task and returns the privilege. This is different from normal bus operation and requires a dedicated arbiter, Florian-Arbiter.

Fig. 5 shows the internal structure of the Florian-Arbiter. The APB signal from each core enters the arbiter, allowing the arbiter to check all signals and select the core to occupy the FPU via the *APB-MUX* and the selected signal $ActiveCoreNum$. $ActiveCoreNum$ represents the index of the current Florian occupied core, defined by two variables: $NextCoreNum$, the index of the highest priority core in the queue, and $is\_florian\_calculating$, the activation signal of the *ActiveCoreNum-Selector*. When *NewSignal-Detector* detects the receipt of a new signal, the core ID of the detected signal is sent to the *FIFO-queue* for updates. Of the list of core IDs waiting in the queue, the one with the highest priority is assigned as $NextCoreNum$ and gets a chance to be assigned as $ActiveCoreNum$.

$NextCoreNum$ is not always immediately assigned to $ActiveCoreNum$. The Florian-FPU needs to interact with

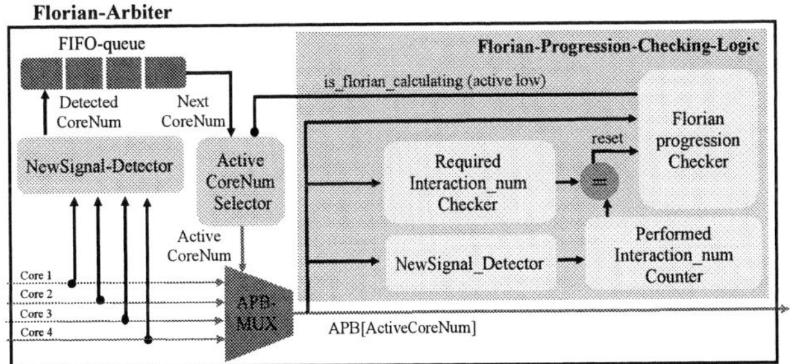

Fig. 5. Structure of the Florian-Arbiter architecture

the core at least three times to complete all floating point operations. So if a signal from another core gets into the FPU before it fills in the required amount of interactivity, it will result in an error. We introduced *Florian-Progression-Checking-Logic* to prevent this error, which serves to check that the currently active cores have communicated sufficiently as needed. After all necessary interactions have been performed, *Florian-Progression-Checking-Logic* can change the register $is\_florian\_calculating$ to 0 to enable *ActiveCoreNum-Selector* and finally update $ActiveCoreNum$ to $NextCoreNum$. Arbitrators designed in this way help each core successfully utilize the Florian-FPU, ensuring that all cores use it fairly.

C. Florian operational process

Fig. 6 shows the operation process of floating-point addition ($Florian\_ADD$) and multiplication ($Florian\_MULT$). First, for $Florian\_ADD$, as soon as the FPU receives the ADD command (cf. 1st line), the next two operands are converted to denormalized form and sent to the *Florian-Add* module in *Executor* for execution.

$Florian\_MULT$ requires a slightly more complex procedure. As aforementioned, multiplication in Florian-FPU requires additional interaction with the core due to integer arithmetic. To obtain the result of a 48-bit multiplication of two given mantissas, we use the MUL and MULHU instructions

```
Function : <Florian_ADD>

1 :      sw      cmd_add,              florian_mmio_input_addr
2 :      sw      operand1,             florian_mmio_input_addr
3 :      sw      operand2,             florian_mmio_input_addr
4 :      lw      result,               florian_mmio_output_addr
```

```
Function : <Florian_MULT>

1 :      sw      cmd_mult,             florian_mmio_input_addr
2 :      sw      operand1,             florian_mmio_input_addr
3 :      sw      operand2,             florian_mmio_input_addr
4 :      lw      multiplicand,         florian_mmio_output_addr
5 :      lw      multiplier,           florian_mmio_output_addr
ALU-1 :  mul     product_lower32bit,   multiplicand, multiplier
ALU-2 :  mulhu   product_upper32bit,   multiplicand, multiplier
6 :      sw      product_lower32bit,   florian_mmio_input_addr
7 :      sw      product_upper32bit,   florian_mmio_input_addr
8 :      lw      result,               florian_mmio_output_addr
```

Fig. 6. Operation process of addition and multiplication of Florian-FPU.

to produce the 64-bit result, then return the value to the FPU so that the multiplication operation in progress can complete. Therefore, a total of 8 interactions between the core and the FPU are required.

In Florian-FPU, floating-point division is performed in a similar way to the multiplication, as are other floating-point operations.

III. Experimental Work

First, to verify the functional correctness of the Florian-based processor, we designed a test processor with the structure shown in Fig. 1 consisting of four ORCA cores and Florian-FPU/Arbiter and prototyped it on a Xilinx Kintex Ultrascale+ FPGA board [20]. We confirmed that the processor operates normally by running applications that include floating-point operations on the prototype processor. In the FPGA prototyping process, the FPGA resource consumption of each major component was compared, which is shown in Table I. As can be seen from the table, the Florian-FPU is lightweight, with only 37.1% of the resources consumed by the single ORCA core and 32.2% of the FPU in Rocket. Moreover, this result is a comparison of only single cores, and if the number of cores in the processor increases further, it is obvious that the difference between the resource consumption of FPU in multicore with FPU per core and multicore using only single Florian FPU will be even greater, which means that the low-power effect of Florian is more pronounced.

Next, we designed and fabricated a prototype processor on a chip. The chip was fabricated in the size of $4500 \times 4500\ \mu m^2$ using the $0.11\ \mu m$ CMOS logic/mixed signal process technology, and the die photo and detailed specifications of the chip are shown in Fig. 7. The power consumption of the ORCA core, Florian-FPU and Arbiter measured from the chip are reported in Table II. All the power values are normalized based on the power value of a single ORCA core. As seen in the table, Florian-FPU and Arbiter consume only 0.3 times and 0.04 times the power of the ORCA Core, respectively, it can be confirmed that the proposed Florian hardware was successfully designed for low power consumption.

We then conducted experiments to evaluate the energy saving of Florian on several multicores with different numbers of cores. To this end, we needed to design a single, dual, quad and

TABLE I
Comparison of FPGA resource consumption.

	Rocket Core	Rocket FPU	ORCA core	Florian-FPU	Florian-Arbiter
LUTs	11711	5456	3230	1546	144
FFs	6653	1775	3056	558	89

Technology:	$0.11\mu m$ CMOS Logic/ Mixed Signal Process
Chip Size:	$4500 \times 4500\ \mu m^2$
Gate Count:	778K GE
Memory	SRAM: 294 kbytes
Supply Voltage	1.2V / 3.3V
Target Freq.	50MHz
Temp. Range	from −40°C to 125 °C
Pads	Signal: 116, Power & ground: 32
Package	168 FBGA

Fig. 7. Die photo & spec. of the Florian prototyping chip.

octa core processor with one Florian-FPU shared, each core has a Florian-FPU, or no FPU, i.e., a total of 11 processors were necessary. For reference, processors without an FPU use soft-float. At this time, manufacturing all processors into chips requires excessive time, effort, and cost, so the execution time and energy consumption of each processor were accurately estimated using the experimental results of the fabricated chip of the quad core processor, execution time of applications in each FPGA prototype processor, and the synthesis results using the $110\ nm$ CMOS technology through Synopsys design compiler. In addition, to evaluate the energy efficiency of the processor with the proposed method depending on the workload of the application's floating point computation, i.e., the lower the floating point workload, the more suitable it is to apply Florian, we combined the benchmark programs of EEMBC Coremark-Pro [16] to create 11 testbench applications where the ratio of floating point computation and integer computation varies from 1:1 to 1:20, respectively. Table III lists the created testbench applications.

Derived from the experiments we performed, Table IV shows the results of running the application on a quad-core processor with no FPU (Case-I), a Florian-FPU on every core (Case-II), and a shared Florian-FPU (Case-III). At this time, the exp10 testbench was used for the application, which was executed simultaneously for all cores. As a result, the execution time of Case-II was reduced by 28% compared to Case-I, indicating that Florian effectively handles floating-point operations. On the other hand, it can be confirmed that the execution time of Case-III is increased than that of Case-II. This is because all cores perform the same application, so all cores perform floating-point operations at the same time,

TABLE II
Power consumption (normalized based on ORCA).

	$P_{dynamic}$	$P_{leakage}$	P_{total}
ORCA	0.96	0.04	1
Florian_FPU	0.299	0.014	0.31
Florian_Arbitor	0.035	0.0016	0.04

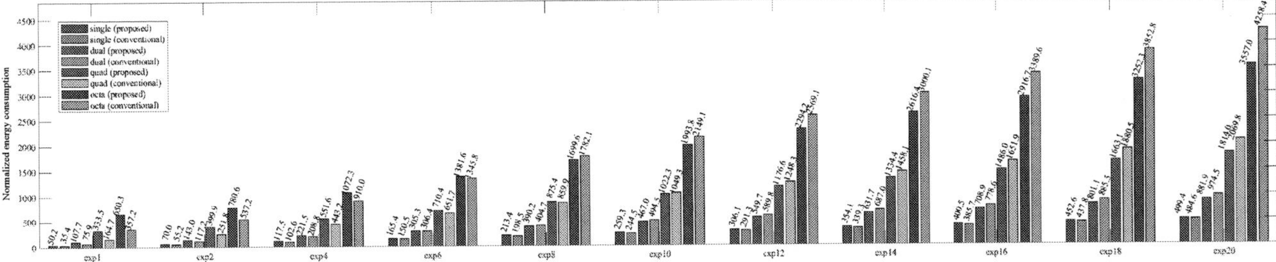

Fig. 8. Energy consumption results of the proposed and conventional single, dual, quad, and octa core processors for 11 applications (each core in the processor runs the same application concurrently).

TABLE III
THE 11 TESTBENCH APPLICATIONS USED IN THE EXPERIMENT.

App.	Workload ratio (FP : Others)	Included EEMBC programs
exp1	1 : 1	linear_alg-mid-100x100-sp, cjpeg-rose7-preset
exp2	1 : 2	linear_alg-mid-100x100-sp, cjpeg-rose7-preset, sha-test
exp4	1 : 4	linear_alg-mid-100x100-sp, sha-test
exp6	1 : 6	linear_alg-mid-100x100-sp, cjpeg-rose7-preset, sha-test
exp8	1 : 8	linear_alg-mid-100x100-sp, cjpeg-rose7-preset, sha-test
exp10	1 : 10	linear_alg-mid-100x100-sp, sha-test
exp12	1 : 12	linear_alg-mid-100x100-sp, sha-test
exp14	1 : 14	linear_alg-mid-100x100-sp, sha-test
exp16	1 : 16	linear_alg-mid-100x100-sp, sha-test
exp18	1 : 18	linear_alg-mid-100x100-sp, sha-test
exp20	1 : 20	linear_alg-mid-100x100-sp, sha-test

TABLE IV
ENERGY CONSUMPTION COMPARISON OF THE QUAD CORE PROCESSORS. POWER VALUES ARE NORMALIZED BASED ON A SINGLE ORCA PROCESSOR.

	Case-I : ORCA × 4	Case-II : (ORCA + Florian) × 4	Case-III : ORCA × 4 + Florian
Power	3.64	5.28	4.10
Exec. Time (sec)	279	199	249
Energy	-	1049	1022

causing a bottleneck in the FPU. However, Case-III can still save 2.57% energy compared to Case-II, because the power consumption can be reduced by sharing the FPU.

Finally, to clearly demonstrate that using the proposed Florian is more beneficial in terms of low-power design than using cores with FPU per core in the multicore development, we report energy comparisons between processors with Florian (denoted as *proposed*) and processors with FPU per core (denoted as *conventional*). For the conventional processors, since the ORCA core does not have an FPU, we assume that each core is equipped with Florian-FPU as a conservative approach. To this end, we manually calculated the application execution time on the conventional processors based on the communication overhead between Florian and Core obtained through RTL simulations.

Fig. 8 shows the energy consumption results for each processor and each application when all cores in the processor simultaneously execute the same application. First, looking at the results according to the change in the number of cores of the processors, the proposed processors can reduce energy consumption by 9.5%, 12.4%, and 16.5% in dual, quad, and octa cores, respectively, which, as expected, increase the

TABLE V
ENERGY SAVING (ES) RESULTS WHEN EACH CORE RUNS THE SAME APPLICATION SIMULTANEOUSLY

App.	$ES_{single}(\%)$	$ES_{dual}(\%)$	$ES_{quad}(\%)$	$ES_{octa}(\%)$
exp1	-41.9	-41.9	-102.5	-82
exp2	-26.8	-22.1	-58.8	-45.3
exp4	-14.4	-6.1	-24.5	-17.8
exp6	-9.84	0.36	-9	-2.66
exp8	-7.46	3.6	-1.8	4.63
exp10	-6.06	5.56	2.57	7.22
exp12	-5.09	6.81	5.74	10.7
exp14	-4.37	8.05	8.48	12.8
exp16	-3.84	8.95	10	14
exp18	-3.38	9.52	11.6	15.6
exp20	-3.06	9.49	12.4	16.5

benefit as more cores share Florian-FPU. For single cores, no gain occurs in all processors, as we conservatively assumed that ORCA core has a built-in Florian-FPU, i.e., the conventional processors is assumed to already have the advantage of introducing the lightweight FPU, and since the FPU is built in, communication overhead is very low compared to external ones. In other words, the benefits of Florian's lightweight FPU are excluded from the results here. Therefore, if it were not for this conservative assumption, it is possible to expect that Florian could have made some gains even in a single core, and it is clear that the gains in dual, quad, and octa cores would have increased further.

Looking at Table V, which reports the change in energy saving (ES) factor of Florian as the floating-point operation workload varies by application, it can be seen that the smaller the proportion of floating-point operations, the greater ES. In the case of exp20, in contrast to the maximum gain in dual, quad, and octa cores, in exp1, losses occur at -41.9%, -102.5%, and -82.0%, respectively. This is because, as the cores share the FPU, if many cores request the use of the FPU at the same time, a bottleneck occurs, inducing delays. Since all of the lightweight applications targeted in this paper have a very small proportion of floating-point arithmetic, it is obvious that there must be benefits through Florian, but if not, it would be better to design a processor with an FPU for each core.

Fig. 8 and Table V are the results of all cores in the processor performing floating point operations at the same time, which may be too far from the actual behavior of the multicore. In other words, it is very likely that the timing required for floating-point operation is different for each core in multicore in general. To reflect this, we modified the test-

TABLE VI
ENERGY CONSUMPTION AND SAVINGS RESULTS OF THE PROPOSED AND CONVENTIONAL PROCESSORS FOR 11 TESTBENCH APPLICATIONS (APPLICATIONS ARE SET TO DIFFERENTIATE THE TIMING OF PERFORMING FLOATING-POINT OPERATIONS PER CORE).

App.	Dual core			Quad core			Octa core		
	$E_{proposed}$	$E_{conv.}$	ES (%)	$E_{proposed}$	$E_{conv.}$	ES (%)	$E_{proposed}$	$E_{conv.}$	ES (%)
exp1	99.5	71.7	-38.7	295.7	156.2	-89.3	620.6	345.9	-79.4
exp2	128.5	111.8	-14.9	330	245.6	-34.3	725.8	525	-38.2
exp4	211.5	208.3	-1.5	424.7	442.8	4.1	914.8	913.2	-0.2
exp6	294.2	304.9	3.5	574.2	652.6	12	1199.7	1346.2	10.9
exp8	377.3	401.6	6	732.9	855.8	14.4	1439.2	1759.7	18.2
exp10	456.1	493.5	7.6	883.8	1051.2	15.9	1734.4	2156.4	19.6
exp12	538.3	589.2	8.6	1042.3	1253.4	16.8	2020.1	2576.1	21.6
exp14	619.4	683.7	9.4	1197.5	1455.9	17.7	2344.4	3012.4	22.2
exp16	699	776.2	9.9	1350.2	1653.4	18.3	2638.8	3408.3	22.6
exp18	789.8	882.4	10.5	1526.5	1879.5	18.8	2968.4	3847.8	22.9
exp20	869.1	975	10.9	1679.2	2079.5	19.2	3286.3	4271.2	23.1

bench applications so that the timing of performing floating-point operations is different for each core, and conducted experiments based on this, and the results are reported in Table VI. The results of this table confirm that Florian's energy saving effect has increased compared to those of the previous Table V because the bottleneck of FPU has been resolved. In the previous result, the gains of the quad and octa core occur from exp10 and exp8, respectively, whereas in this result, they occur from exp4 and exp6, respectively. In other words, in the case of the quad core, it can be seen that Florian has gain unless the floating-point operation is too frequent as the proportion of floating-point operation and other operations is 1:2 or less. Finally, the maximum ES appears when exp20 is performed on the octa core, and the result reaches 23.1%.

IV. CONCLUSION

The open lightweight RISC-V cores are rapidly increasing in use in processors specialized in IoT, wearable, and embedded system applications. As these applications become more diverse and complex, more and more applications require floating-point operations along with the need for multicore processors, which is a major limitation in designing processors using the existing open cores. Because most lightweight RISC-V cores are integer cores without FPU, the design space is greatly narrowed to very few lightweight cores that provide FPU together, making it impossible to design a processor optimized for each application. To tackle this problem, we proposed Florian that includes an external lightweight FPU that can be attached to any RISC-V integer core and a low-power multi-core architecture using the designed FPU. To verify the effectiveness and performance of the proposed Florian, we designed RISC-V processors that implement all the proposed technologies, made prototypes with FPGAs, and finally fabricated a quad core processor chip using $0.11\mu m$ CMOS technology. We conducted intensive experiments on single, dual, quad, and octa core processors with 11 different testbenches, and confirmed that the proposed Florian achieves energy savings of up to 23.1%.

REFERENCES

[1] K. Han *et al.*, "Tip: A temperature effect inversion-aware ultra-low power system-on-chip platform," in *2019 IEEE/ACM Int'l Symposium on Low Power Electronics and Design (ISLPED)*, 2019, pp. 1–6.

[2] C. Heinz, Y. Lavan, J. Hofmann, and A. Koch, "A catalog and in-hardware evaluation of open-source drop-in compatible RISC-V softcore processors," in *Int'l Conf. on ReConFigurable Computing and FPGAs (ReConFig)*, 2019, pp. 1–8.

[3] H. Jang *et al.*, "Developing a multicore platform utilizing open RISC-V cores," *IEEE Access*, vol. 9, pp. 120 010–120 023, 2021.

[4] N. Bruschi *et al.*, "GVSoC: A highly configurable, fast and accurate full-platform simulator for RISC-V based IoT processors," in *IEEE Int'l Conf. on Computer Design (ICCD)*, 2021, pp. 409–416.

[5] K. Han *et al.*, "Developing TEI-aware ultralow-power SoC platforms for IoT end nodes," *IEEE Internet of Things Journal*, vol. 8, no. 6, pp. 4642–4656, 2021.

[6] J. Park *et al.*, "Developing an ultra-low power RISC-V processor for anomaly detection," in *Design, Automation & Test in Europe Conf. & Exhibition (DATE)*, 2023, pp. 1–2.

[7] J. Hormigo and J. Villalba, "Hub floating point for improving fpga implementations of dsp applications," *IEEE Trans. on Circuits and Systems II: Express Briefs*, vol. 64, no. 3, pp. 319–323, 2017.

[8] W. Lee *et al.*, "K-means clustering-specific lightweight RISC-V processor," in *International SoC Design Conference (ISOCC)*, 2021, pp. 391–392.

[9] M. Franceschi, A. Nannarelli, and M. Valle, "Tunable floating-point for artificial neural networks," in *IEEE Int'l Conf. on Electronics, Circuits and Systems (ICECS)*, 2018, pp. 289–292.

[10] Y. Tortorella *et al.*, "RedMulE: A compact FP16 matrix-multiplication accelerator for adaptive deep learning on RISC-V-based ultra-low-power SoCs," in *Design, Automation & Test in Europe Conf. & Exhibition (DATE)*, 2022, pp. 1099–1102.

[11] S. Mach, F. Schuiki, F. Zaruba, and L. Benini, "FPnew: An open-source multiformat floating-point unit architecture for energy-proportional transprecision computing," *IEEE Trans. on Very Large Scale Integration (VLSI) Systems*, vol. 29, no. 4, pp. 774–787, 2021.

[12] Z. Lei, F. Cai, J. Zhou, and Z. Guo, "A floating-point unit architecture based on SweRV EH1 core," in *IEEE Int'l Conf. on Anti-counterfeiting, Security, and Identification (ASID)*, 2022, pp. 1–5.

[13] SiFIVE, https://github.com/chipsalliance/rocket-chip, accessed 19 March. 2022.

[14] PULP, https://pulp-platform.org/, accessed 19 March. 2022.

[15] Vectorblox, https://github.com/riscveval/orca-1, accessed 19 March. 2022.

[16] EEMBC, https://www.eembc.org/coremark-pro/, accessed 19 March. 2022.

[17] A. Pullini *et al.*, "Mr.Wolf: An energy-precision scalable parallel ultra low power SoC for IoT edge processing," *IEEE Journal of Solid-State Circuits*, vol. 54, no. 7, pp. 1970–1981, 2019.

[18] T. Iwashita, K. Suzuki, and T. Fukaya, "An integer arithmetic-based sparse linear solver using a gmres method and iterative refinement," in *2020 IEEE/ACM 11th Workshop on Latest Advances in Scalable Algorithms for Large-Scale Systems (ScalA)*, 2020, pp. 1–8.

[19] IEEE, "Iso/iec/ieee int'l standard - floating-point arithmetic," *ISO/IEC 60559:2020(E) IEEE Std 754-2019*, pp. 1–86, 2020.

[20] Xilinx, https://www.xilinx.com/products/silicon-devices/fpga/kintex-ultrascale-plus.html, accessed 19 March. 2022.

Energy Efficient Real-Time Scheduling on Heterogeneous Architectures with Self-Suspension

Wenwen Xu*[1], Zheyu Zhang*[1], Yuankai Xu[1], Jing Li[2], Yehan Ma[1], Yier Jin[3], Christopher D. Gill[4], Xuan Zhang[4], An Zou[1]
[1]Shanghai Jiao Tong University, [2]New Jersey Institute of Technology,
[3]University of Science and Technology of China, [4]Washington University in St. Louis

***Abstract*—It is witnessed that heterogeneous architectures, such as GPUs, TPUs, and FPGAs, have made complex algorithms practical in the last decade. Despite multiple efforts to study the scheduling of these parallel and complex tasks on heterogeneous architectures, the power and energy consumption of the platforms have yet to be well managed under real-time task deadlines. To establish high schedulability in heterogeneous architectures, many scheduling strategies and models, such as multi-segment self-suspension (MSSS), have been proposed by pioneer researchers. However, directly applying this model to heterogeneous architectures with multiple CPUs and many processing elements (PEs) suffers aggravated power consumption due to the pessimism in the scheduling algorithm and the tolerance margin in the worst-case execution time (WCET) model. Therefore, this paper presents an energy-efficient real-time scheduling approach called *EESchedule*, which works on heterogeneous architectures with guaranteed schedulability and improved power efficiency. In *EESchedule*, we build a general task execution model for the general heterogeneous architectures integrating multiple CPUs and many PEs. Then, an energy-efficient real-time scheduling strategy is introduced. Next, the response time and corresponding schedulability analysis are presented for *EESchedule*. Finally, extensive experiments on heterogeneous NVIDIA Jetson TX2 embedded systems and GPU servers with the Intel i9-10900x CPU and RTX 3080 GPU demonstrate that the *EESchedule* could achieve the same schedulability with 16.8%-40.7% and 39.0%-48.2% reduced power and energy consumption in comparison with state-of-the-art scheduling algorithms.**

I. Introduction

Modern computing systems for embedded or cloud applications embrace architectural heterogeneity to support intensive computation in emerging applications such as artificial intelligence [1]. Heterogeneous computing platforms, like CPU-GPU servers [2], Xilinx UltraScale [3], and TI Keystone II [4], integrate CPUs and parallel processing elements (PEs), such as GPU Streaming Multiprocessors or FPGA IP cores, into a single architecture. For performance and energy efficiency, a task's serial and lightweight computation segments are usually allocated to the CPUs, while parallel and computation-intensive segments are good candidates for offloading to the PEs. The intensive computing on the PEs leads to heavy power and energy consumption in such heterogeneous computing platforms [5]. Fig. 1 shows the rated power consumption

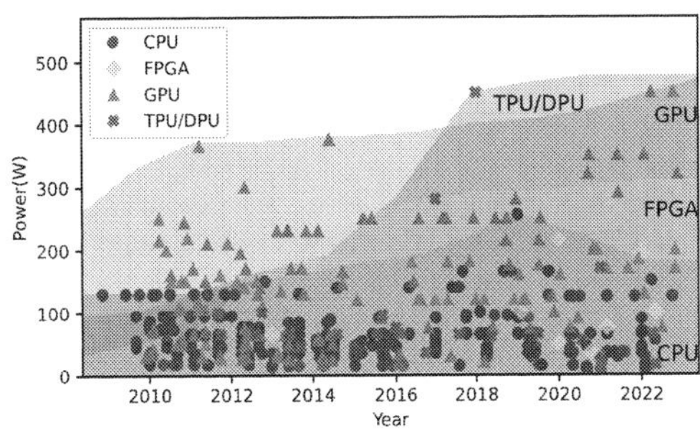

Fig. 1. The power consumption of CPU and heterogeneous processors.

of mainstream homogeneous CPUs and heterogeneous PEs, such as the GPU, FPGA, and TPU/DPU, in recent years. The representative heterogeneous PEs have dramatically increased their rated power to 450 watts, four to five times that of CPUs.

The interleaved execution pattern between CPU and PEs exacerbates inter-dependency and competition among parallel tasks [6]. Diving into the performance of each scheduling instance, there remains a considerable margin to tolerate pessimism in both the scheduling algorithm and the (only occasionally realized) worst-case execution time (WCET). Such pessimism further aggravates the power and energy consumption on heterogeneous computing platforms.

In heterogeneous systems, reducing the frequency and corresponding voltage of PEs at runtime can effectively reduce energy consumption at the cost of longer computation time, which may lead to tasks failing to meet deadlines [7]. To address this issue, we propose *EESchedule*, an energy-efficient scheduling strategy and response time analysis for heterogeneous architectures with multiple CPUs and PEs. Extensive experiments on representative embedded and server-scale CPU-GPU heterogeneous systems show that *EESchedule* can reduce power and energy consumption by 16.8%-40.7% and 39.0%-48.2% without loss of schedulability.

The main contributions of this paper are three-fold.

- Starting with modeling general heterogeneous computing architectures, an energy-efficient **schedule** strategy (*EESchedule*) is presented, which significantly reduces power and energy consumption.
- An end-to-end response time analysis is performed. It demonstrates that the proposed *EESchedule* approach does not compromise system schedulability under the MSSS model on heterogeneous computing platforms.

*Authors contributed equally to this research. An Zou is the corresponding author. This research project is supported by NSFC 62202287, NSFC 62103268, and Shanghai Chenguang Program 21CGA11.

- Extensive experiments on both NVIDIA Jetson TX2 boards and CPU-GPU servers, with varying numbers of tasks and segments, show that EESchedule can reduce energy consumption while guaranteeing schedulability.

As we target general heterogeneous architectures, *EESchedule* can be directly applied to off-the-shelf heterogeneous computing systems with guaranteed deadlines and improved power and energy efficiency.

II. Background and Related Work

A. Scheduling on Heterogeneous Architectures

We examine architectures comprising CPUs and PEs. Generally, in a heterogeneous system, the CPU handles serial tasks such as I/O and serial communication, while PEs execute intensive parallel computations, like large matrix multiplications. Real-time tasks on heterogeneous computing platforms exhibit an interleaved execution pattern (i.e., switching execution between CPUs and PEs). Consequently, scheduling may suffer from both the margin in WCET and inherent defects in the scheduling algorithm. The MSSS model and its analysis can achieve superior schedulability on heterogeneous computing systems among existing scheduling and response time analyses. In MSSS, each task τ_i comprises M_i computational segments separated by $M_i - 1$ suspension segments. Each task τ_i can be expressed as the tuple

$$\tau_i = ((C_i^0, S_i^0, C_i^1, S_i^1, \ldots, S_i^{M_i-2}, C_i^{M_i-1}), D_i, T_i). \quad (1)$$

where D_i and T_i indicate the deadline and release period of task τ_i. C_i^j and S_i^j denote the execution time of the $(j+1)$th computation and suspension segment. Because of the uncertainty in executing the computation and suspension segment, their execution time C_i^j and S_i^j are often considered as a value between their lower bounds and upper bounds $[\check{C}_i^j, \hat{C}_i^j]$ and $[\check{S}_i^j, \hat{S}_i^j]$. The workload function [8] of the h-th computation segment in the task τ_i is defined as follows

$$W_i^h(t) = \sum_{j=h}^{l} \check{C}_i^{j \bmod M_i} + \min\left(\check{C}_i^{(l+1) \bmod M_i}, t - \sum_{j=h}^{l} \left(\check{C}_i^{j \bmod M_i} + S_i(j)\right)\right), \quad (2)$$

where l is the maximum integer satisfying that

$$\sum_{j=h}^{l} \left(\check{C}_i^{j \bmod M_i} + S_i(j)\right) \le t,$$

and $S_i(j)$ is the minimum inter-arrival time between execution segments C_i^j and C_i^{j+1}, which is defined by

$$S_i(j) = \begin{cases} \check{S}_i^{j \bmod M_i} & \text{if } j \bmod M_i \neq (M_i - 1) \\ T_i - D_i & \text{else if } j = M_i - 1 \\ T_i - \sum_{j=0}^{M_i-1} \check{C}_i^j - \sum_{j=0}^{M_i-2} \check{S}_i^j & \text{otherwise.} \end{cases}$$

Heuristically, the workload function $W_i(t)$ represents the maximum amount of computational work given the time interval t, from task τ_i. Once we have the workload function $W_i(t)$ defined, the response time of each computation segment R_i^j can be calculated as the smallest value t that satisfies the recurrence:

$$t \ge C_k + \sum_{\tau_i \in hp(k)} W_i(t), \quad (3)$$

where $hp(k)$ denotes the tasks with higher priority than τ_k. Therefore, the response time R_k of task τ_k is calculated as the smallest value t that satisfies the recurrence:

$$t \ge C_k + \hat{S}_k + \sum_{\tau_i \in hp(k)} W_i(t). \quad (4)$$

Although the MSSS model achieves improved schedulability, a redundant computation margin must be reserved to tolerate the task's WCET and other inherent pessimism.

B. Related Work

As the mainstream computing architecture for heavy computation applications, heterogeneous architectures face trade-offs between computing performance and power and energy efficiency. Popular scheduling models [9], [10] for heterogeneous architectures include but are not limited to the partitioning-based MSSS model [11]–[13], Priority Ceiling Protocol (PCP) [14], and Directed Acyclic Graph (DAG) [15]. Santriaji et al. [16] presented MERLOT, a hardware-based resource manager for GPUs that enforces software-specified timing guarantees for tasks with minimal energy consumption. On the theory side, Yang et al. [17] introduce an approximation scheme, Wang et al. [18] present a mathematical optimization, and Mei et al. [7] propose a heuristic scheduling algorithm to balance energy efficiency and real-time performance. Most energy-efficient real-time scheduling for heterogeneous architectures is based on either offline optimization or heuristic approach.

III. System Model and Scheduling

A. System Model and Notations

In this paper, we consider a heterogeneous architecture with N_{CPU} CPUs and N_{PE} PEs. The heterogeneous architecture executes a set of n independent parallel real-time tasks $\tau = \{\tau_0, \tau_1, ..., \tau_{n-1}\}$. Each task τ_i is composed of M_i CPU segments with $M_i - 1$ PE segments that separate them. In this way, the task can be represented with the MSSS model

$$\tau_i = ((CL_i^0, PL_i^0, CL_i^1, \ldots, PL_i^{M_i-2}, CL_i^{M_i-1}), D_i, T_i), \quad (5)$$

where CL_i^j and PL_i^j are the length of the $(j+1)$th CPU and PE segments in task τ_i respectively.

B. Scheduling Strategy:

Standard Scheduling. In scheduling tasks on heterogeneous architectures, the MSSS model integrates spatial partitioning for PEs and temporal access to CPUs.

For spatial partitioning, the N_{PE} PEs are partitioned into n groups, where group i has N_{PE_i} PEs dedicated to the i-th task. The partitioning and response time analysis follow that of the federated scheduling [19]. From a theoretical analysis perspective, the spatial partitioning significantly reduces the pessimism in the analysis and thus achieves better theoretical schedulability. Without PE assignments, a huge but infrequently released task would block a small but frequently released task and cause a deadline miss. Meanwhile, it may also cause the CPU to idle because each task has a segmented execution pattern. Compared to running a PE segment with entire PEs, running with allocated SMs will make the GPU

979-8-3503-1176-1/23 $31.00 © 2023 IEEE

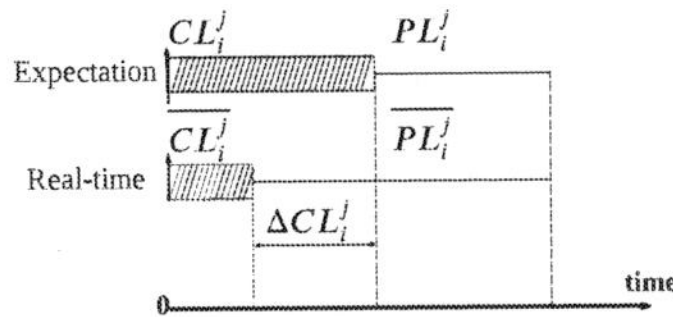

Fig. 2. The frequency update strategy

segment longer. As a result, most of the time, the assigned PEs are running PE segments, and there is only a small amount of time when the PEs are idle, especially for the computation-intensive tasks.

For temporal access, the CPU segments access N_{CPU} CPUs following a preemptive fixed-priority manner. Therefore, the end-to-end real-time scheduling strategy coordinates a grid search on PEs spatial partitioning and CPU temporal access. Following this scheduling strategy, the execution time CE_i^j and PE_i^j of each CPU and PE segment CL_i^j and PL_i^j are obtained as

$$CE_i^j = CL_i^j, PE_i^j = \frac{PL_i^j}{N_{PE_i}}. \tag{6}$$

To account for the unstable and unpredictable execution times of the tasks, the execution time of computation and suspension segment CE_i^j and PE_i^j has upper bound $\hat{CE}_i^j$ and $\hat{PE}_i^j$ and lower bound $\check{CE}_i^j$ and $\check{PE}_i^j$. Finally, the schedulability test will pass when a schedulable case is found by the response time analysis in Section II-A.

Energy-Efficient Scheduling. The key idea of *EESchedule* is to dynamically adjust the computing configurations, i.e., the frequency and voltage of the PEs, based on the task progress at runtime. For task τ_i with M_i CPU segments and $M_i - 1$ PE segments, there are $M_i - 1$ potential adjustments of the processor frequency and voltage located before each PE segment. The expected response time $R_{CL_i^j}$ of each computational segment CL_i^j is calculated offline in the schedulability test. When the task finishes one CPU segment, the computation frequency (and corresponding voltage) of the next PE segment PL_i^j is updated according to the difference between the expected response time $R_{CL_i^j}$ and actual response time $\overline{R}_{CL_i^j}$. Since the actual response time $\overline{R}_{CL_i^j}$ can be much shorter than the expected $R_{CL_i^j}$, we assign the saved response time $\Delta R_{CL_i^j} = R_{CL_i^j} - \overline{R}_{CL_i^j}$ to the following PE segments PL_i^j, so that the updated frequency of PEs allocated to task τ_i can be reduced to

$$f_i = \frac{f_{max} PL_i^j}{PL_i^j + N_{PE_i} \Delta R_{CL_i^j}}, \tag{7}$$

where f_i is the desired frequency (and corresponding voltage) for task τ_i and f_{max} is the default highest processor frequency.

As depicted in Fig. 2, the time for PL_i^j is extended to $PL_i^j + \Delta R_{CL_i^j}$. Since a PE task execution time is inversely proportional to its PE frequency, the frequency can be reduced by a factor of $\frac{PL_i^j}{PL_i^j + N_{PE_i} \Delta R_{CL_i^j}}$.

After finishing the last computation segment $CL_i^{M_i - 1}$, the processor frequency will be set to the lowest value until being set back to the default highest value at the beginning of the next task period. Since each task has allocated PEs, the frequency f_i (and corresponding voltage) should only affect the PEs assigned to task τ_i. If a cluster of cores shares the same frequency and voltage, *EESchedule* conservatively chooses the highest frequency in the cluster to ensure schedulability, as follows.

$$\overline{f} = max(\overline{f_0}, \overline{f_1}, \overline{f_2} \ldots) \tag{8}$$

We will prove that *EESchedule* will not affect the guaranteed schedulability in the next section.

C. Case Study on EESchedule

Given a task set with three parallel tasks noted in Eq. (9) running on a heterogeneous architecture where all the PEs share the same frequency and voltage. Tasks are listed in order of priority from highest to lowest. We assume each task is allocated one PE.

$$\begin{aligned} \tau_0 &= \big((10, 15, 5, 10, 5, 10, 5), 200, 220\big) \\ \tau_1 &= \big((5, 15, 10, 10, 10), 360, 380\big) \\ \tau_2 &= \big((5, 15, 5, 30, 5), 360, 400\big) \end{aligned} \tag{9}$$

The tasks τ_0, τ_1, τ_2 follow the response time analysis in Eq. (3), and thus, the worst-case response time of a segment can be obtained offline:

$$\begin{aligned} &R_{CL_0^0} = 10, R_{CL_0^1} = 5, R_{CL_0^2} = 5; R_{CL_0^3} = 5; \\ &R_{CL_1^0} = 15, R_{CL_1^1} = 20, R_{CL_1^2} = 20; \\ &R_{CL_2^0} = 30, R_{CL_2^1} = 30, R_{CL_2^2} = 30. \end{aligned} \tag{10}$$

Assume at time t_0, the processor frequency (and corresponding voltage) f_i from each task is $f_0 = \frac{10}{13} f_{max}$, $f_1 = \frac{10}{17} f_{max}$, $f_2 = \frac{10}{19} f_{max}$. Therefore, the PEs sharing a same voltage and frequency adopt the highest frequency $f = max(f_0, f_1, f_2) = \frac{10}{13} f_{max}$. From time t_0, each task's actual releasing and finishing time is shown in Fig. 3. According to the task execution pattern, the segment CL_2^1 in task 2 is released at time t_0 and gets executed until segment CL_0^2 is released at time 1. The segment CL_0^2 has worst-case response time 5, but actual execution time 3 because of the pessimism in the estimation of computational task length and response time analysis. Although segment CL_1^1 in task 1 is released at time 2, it cannot preempt CL_0^2 and is executed at time 4. Since the segment CL_0^2 finishes and the next PE segment starts at time 4, the processor frequency for task 0 is adjusted to $f_0 = \frac{f_{max} PL_0^2}{PL_0^2 + \Delta R_{CL_0^2}} = \frac{5}{6} f_{max}$ following Eq. (7). The frequency for the entire processor is obtained by $f = max(f_0, f_1, f_2) = \frac{5}{6} f_{max}$ according to Eq. (8). Similarly, after the CPU segments CL_1^1, CL_0^3, and CL_2^1 finish at time 14, 19, 21, the frequency of entire processor is kept at or adjusted to $\frac{5}{6} f_{max}$, $\frac{15}{23} f_{max}$, and $\frac{10}{13} f_{max}$ accordingly.

IV. System Schedulability

This section demonstrates that the proposed energy-efficient scheduling strategy will not invalidate the schedulability of the MSSS model either in the response time analysis or in the run-time scheduling on real systems.

A. Response Time Analysis

In the response time analysis, we will first prove a shorter actual response time (i.e., the earlier finishing of the CPU segment and starting of the PE segment) does not reduce the

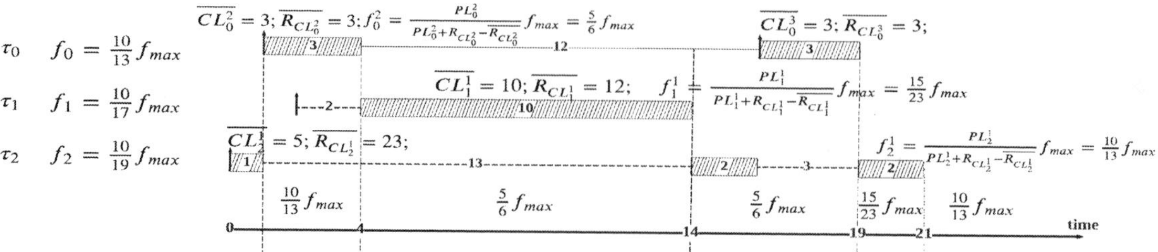

Fig. 3. The actual task execution pattern in the case study.

schedulability. Then, we will present a proof that reducing the computing frequency and voltage of PEs accordingly does not cause any schedulability loss.

Definition IV.1. *Given a CPU segment CL_i^j, its actual response time $\overline{R}_{CL_i^j}$ would be smaller than the expected $R_{CL_i^j}$ from the MSSS model. This can be caused by both or either one of the reasons. (1) The mismatch between the actual execution time $\overline{CE_i^j}$ in this round of execution and the WCET estimate CE_i^j. (2) The inherent pessimism in the response time analysis of MSSS models.*

Lemma IV.1. *For any task τ_i and its CPU segment CL_i^j, a smaller actual response time $\overline{R}_{CL_i^j}$ and earlier starting of next PE segment PL_i^j will not change the workload function $W_i^j(t)$ of task τ_i.*

Proof. The workload function upper bounds the amount of execution that the jobs of task τ_i can perform in the time interval of duration t. It considers the worst-case execution and release scenarios, including early computation (i.e., CPU) and suspension (i.e., PE) completions, as proved in [20]. □

Lemma IV.2. *For any task τ_i, a shorter actual response time $\overline{R}_{CL_i^j}$ will not invalidate the response time analysis of itself and other tasks.*

Proof. The response time of task τ_i is only related to the properties of τ_i, and has nothing to do with the actual response time of itself and other tasks. □

Thus far, we have proved that finishing the CPU segment and starting PE segment earlier do not affect schedulability.

Lemma IV.3. *For the $(j+1)$th segment of any task τ_i, a shorter actual response time $\overline{R}_{CL_i^j}$ by $\Delta R_{CL_i^j}$ and adding a $\Delta R_{PL_i^j}$ to the following PE segment PL_i^j, which satisfies $\Delta R_{PL_i^j} \leq \Delta R_{CL_i^j}$, will not increase the workload $W_i^j(t)$.*

Proof. This can be proved either from the definition of the workload function or the derivation of the workload function. The extended response time of the PE segment PL_i^j indicates a longer execution time of PE segment PL_i^j. By definition, the workload function describes the maximum amount of CPU workload within the time interval t. The extension on PE segment will not add to the CPU workload. Meanwhile, from the derivation of the workload function in Eq. (2), the workload function is always inversely scaled to the suspension (i.e., PE) lengths. Thus, an extended response time of the PE segment PL_i^j does not increase task τ_i's workload function. □

Theorem IV.4. *For any task τ_i and its $(j+1)$th CPU segment with a smaller actual response time $\overline{R}_{CL_i^j}$ (by $\Delta R_{CL_i^j}$) adding $\Delta R_{PL_i^j}$ to the following PE segment PL_i^j, which satisfies $\Delta R_{PL_i^j} \leq \Delta R_{CL_i^j}$, will not increase the response time R of any tasks in the task set.*

Proof. For task τ_i, the expected response time analysis is not impacted by its actual response time, as proven in Lemma IV.2. For task τ_k, which has a lower priority than task τ_i, its response time is calculated with the workload function of τ_i, which is not increased with the shorter actual response time of the CPU segment and extended response time of the following PE segment as proven in Lemma IV.3. Therefore, its response time is not increased. □

Corollary IV.4.1. *Adding $\Delta R_{PL_i^j}$ to the PE segment PL_i^j can be implemented by reducing the PE processor frequency and corresponding voltage to $f_i = \frac{f_{max} PL_i^j}{PL_i^j + N_{PE_i} \Delta R_{PL_i^j}}$.*

Proof. This comes directly from the property of parallel computing processors. The computing frequency indicates the operation speed of digital circuits inside the processor. Since the PE segment PL_i^j is executed in parallel on the N_{PE_i} allocated PE cores with a computing frequency f_i, the response time is $R_{PL_i^j} = \frac{PL_i^j}{N_{PE_i}} * \frac{f_{max}}{f_i}$, which gives $f_i = \frac{f_{max} PL_i^j}{PL_i^j + N_{PE_i} \Delta R_{PL_i^j}}$. □

B. Optimal Priority Assignment

In the proposed approach, the federated scheduling prevents blocking and competition in accessing PEs, and the fixed-priority scheduling supports task preemption in CPUs. Therefore, a vital advantage of the proposed EESchedule is that many essential properties of classic fixed-priority scheduling remain true. Meanwhile, in the analysis, we do not introduce any new constraints that conflict with the classic Audsley's Optimal Priority Assignment (AOPA) assumptions [21].

C. Computational Complexity

1) Offline Schedulability Test: The proposed EESchedule and response time analysis includes grid search on PEs spatial partitioning and fixed-priority CPU scheduling with priority assignments, which has a complexity of

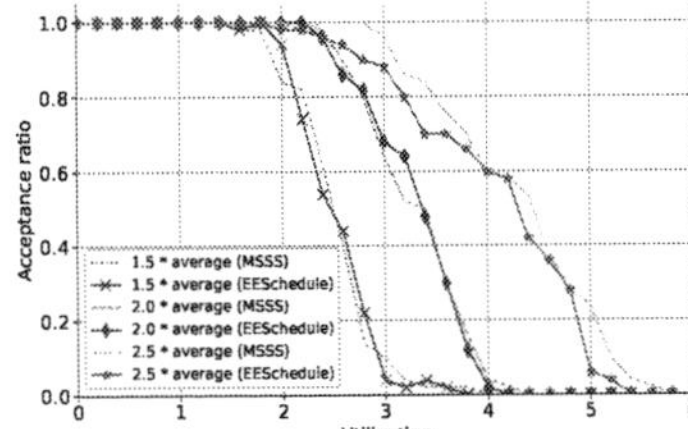

Fig. 4. Schedulability under different WCET model.

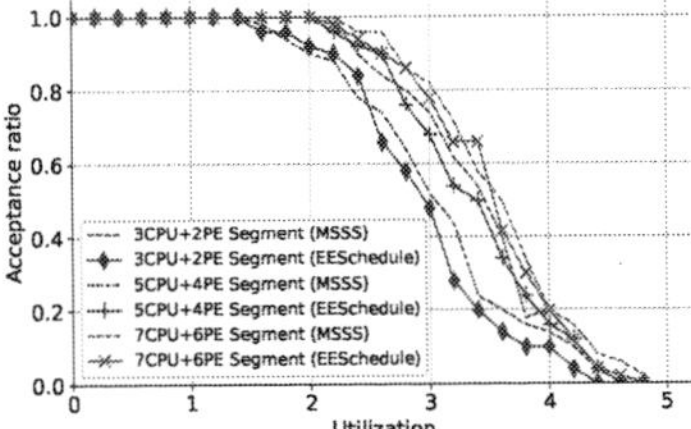

Fig. 5. Schedulability under segment numbers.

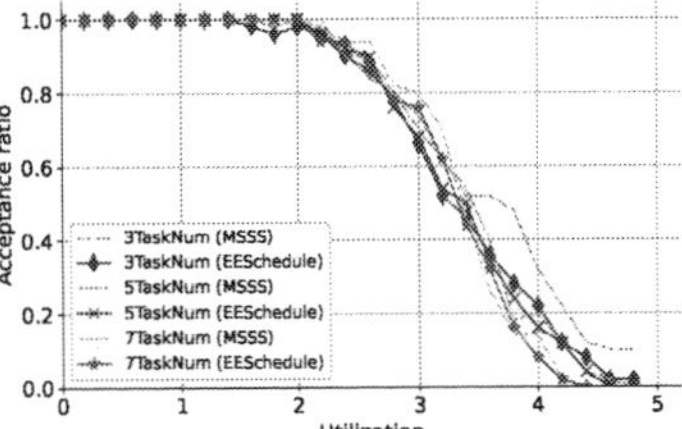

Fig. 6. Schedulability under task numbers.

$min(O({N_{PE}}^n), O(n^{N_{PE}}))$. The priority assignment has a complexity of $O(n^2)$. The analysis of fixed-priority tasks on multi-core CPUs has a complexity of $O(M_i^2)$. Therefore, the time complexity of the entire scheduling strategy is

$$min(O({N_{PE}}^n n^2 M_i^2), O(n^{N_{PE}+2} M_i^2)). \quad (11)$$

2) Online Scheduler: At run-time, the tasks are scheduled with static fixed-priority by the real-time scheduler. Compared with the Linux default Completely Fair Scheduler (CFS), with a complexity $O(logn)$ at run-time, the static fixed-priority scheduling has a complexity of $O(nM_i)$ at run-time. The scaling of processor voltage and frequency can be implemented either at the application level through a system call (syscall) or at the Linux kernel with a kernel module. It takes about 1ms to 5ms for each voltage and frequency change. The extra time and power in changing the voltage and frequency are incorporated in evaluations.

V. System Evaluation

A. System Implementation

The proposed scheduling approach is evaluated on ① an NVIDIA Jetson TX2 CPU-GPU heterogeneous computing platform with ARM Cortex®-A57 CPU and Pascal™ GPUs; and ② a GPU server with an Intel i9-10900x CPU and a RTX 3080 GPU. The run-time power and energy of Jetson TX2 are measured with a Tektronix MDO32 oscilloscope with TCP2020 current probe, and the power and energy of GPU server are measured with the IT9121 power analyzer. We implement persistent threads to support the partitioning of PEs. Each PE segment performs a given amount of floating-point operations on a vector determined by the segment length.

B. System Schedulability

To compare the schedulability for different approaches, we generated tasksets for each utilization level, with the task configurations in Table I. The acceptance ratio was the number of schedulable tasksets over the total number of tasksets. For a complete comparison, we use the methods in previous work [11] to randomly generate the CPU and PE segment lengths, uniformly distributed within their ranges.

TABLE I
PARAMETERS FOR UNIFIED TASK GENERATION

Parameters	Value
Number of tasks N in taskset	3,5,7
Task type	periodic tasks
Worst-case execution time (WCET)	1.5/2/2.5×average time
Number of CPU/PE segments in each τ	3/2;5/4;7/6
Number of releases in each experiment	100
CPU segment length (ms)	1-50
Heterogeneous segment length (ms)	1-50;50-100;100-200
Voltage and frequency scaling time	3ms
Task period and deadline	T_i/D_i
Priority assignment	rate monotonic

We evaluate system schedulability under WCET models, different segment numbers in each task, and different parallel task numbers. Fig. 4 presents the measured system schedulability on Jetson TX2 with different WCET models (i.e., 1.5, 2.0, 2.5 × average execution time). Fig. 5 shows the system schedulability under different numbers of segments (i.e., 3, 5, 7 CPU segments and 2, 4, 6 GPU segments). Meanwhile, the system schedulability with different numbers of parallel tasks is shown in Fig. 6. In the above experiments, the proposed *EESchedule* approach achieves schedulability close to that of the standard MSSS model: as it takes about 1ms to 5ms for each voltage and frequency change, there is, however, a slight variation in schedulability under the proposed *EESchedule* approach. Similar results are also reported on the GPU servers.

C. Power and Energy Saving

Finally, to test the energy consumption when the computing platform is loaded with enough workloads, the processor frequencies and computing energy consumption are measured at the highest utilization rates with a 100% acceptance ratio. Previous work [22] reports that the execution time in Jetson TX could be 3× of average execution time in the worst scenario. To conservatively evaluate the power benefits brought by *EESchedule*, the WCET is set to only 2× of average execution time in our experiments. Fig. 7 and Fig. 8 present the frequency distribution of the Jetson TX2 GPU processors at runtime for different numbers of parallel tasks and segments. In all the tested cases, the ratio of the highest frequency only takes less than 8%, and the median frequency is also around 800MHz. The reduced processor frequency indicates a significant saving at runtime.

Meanwhile, Table II summarizes the average measured the power consumption of running the parallel tasks. We separate the power consumption by the background applications, 2.90W. Compared with the MSSS, *EESchedule* achieves 16.8% - 40.7% energy savings. A close examination of the energy saving performance shows that a more significant percentage of energy saving will be obtained when the parallel tasks are lightweight. Similar experiments are also performed on the GPU servers, and the power consumption of running the parallel tasks is presented in Table II after removing the

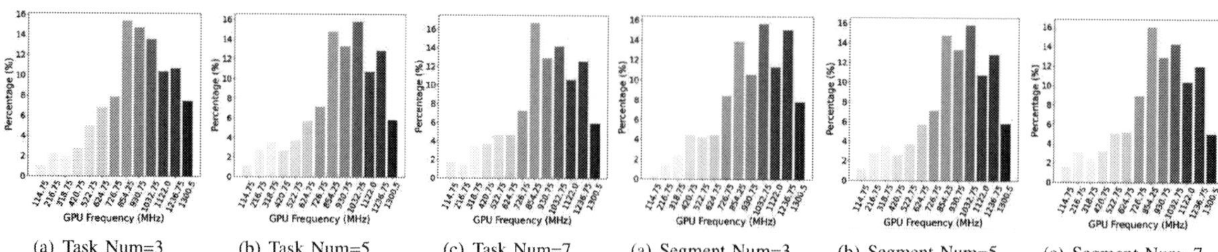

(a) Task Num=3 (b) Task Num=5 (c) Task Num=7

Fig. 7. Processor frequency under different task numbers.

(a) Segment Num=3 (b) Segment Num=5 (c) Segment Num=7

Fig. 8. Processor frequency under different segment numbers.

TABLE II
MEASURED RUN-TIME POWER CONSUMPTION IN JETSON TX2 AND GPU SERVERS

Configuration	MSSS (Jetson TX2)	*EESchedule* (Jetson TX2)	Saving (Jetson TX2)	MSSS (GPU Server)	*EESchedule* (GPU Server)	Saving (GPU Server)
$N = 3$; 5CPU + 4PE Seg.	0.790W	0.585W	26.0%	50.043W	30.532W	39.0%
$N = 5$; 5CPU + 4PE Seg.	1.078W	0.897W	16.8%	49.968W	27.353W	45.3%
$N = 7$; 5CPU + 4PE Seg.	1.074W	0.870W	19.0%	55.533W	28.789W	48.2%
$N = 5$; 3CPU + 2PE Seg.	0.703W	0.417W	40.7%	52.341W	28.645W	45.3%
$N = 5$; 7CPU + 6PE Seg.	0.847W	0.689W	18.7%	50.394W	27.618W	45.2%

background power of 110W. As reported by the results, about 39.0% - 48.2% energy savings are achieved by *EESchedule*. Compared with Jetson TX2, more energy is saved in GPU servers as the heterogeneous PEs (i.e., the GTX 3080 GPU) have more dominant power consumption in the servers. If we compare the energy saving in each scenario, slightly more energy saving will be achieved with more segment numbers and tasks. The above experiments validate that the proposed *EESchedule* can be applied to the general heterogeneous architectures from lightweight heterogeneous embedded systems to power-hungry heterogeneous servers.

VI. CONCLUSION

This paper proposes an energy-efficient real-time scheduling approach called EESchedule that targets heterogeneous architectures. EESchedule preserves the schedulability of state-of-the-art scheduling strategies, while reducing energy consumption. Experiments on a real CPU-GPU system demonstrate that the proposed EESchedule achieves significant energy savings without sacrificing schedulability. Since EESchedule is developed for general heterogeneous architectures, it can be applied to off-the-shelf heterogeneous computing platforms, such as CPU-GPU and CPU-FPGA systems.

REFERENCES

[1] Jeff Anderson, Armin Mehrabian, Jiaxin Peng, and Tarek A El-Ghazawi. Extreme heterogeneity in deep learning architectures., 2019.

[2] Jack Choquette and Wish Gandhi. Nvidia a100 gpu: Performance & innovation for gpu computing. In *IEEE Hot Chips 32 Symposium*, 2020.

[3] Steve Leibson and Nick Mehta. Xilinx ultrascale: The next-generation architecture for your next-generation architecture. *Xilinx White*, 2013.

[4] Benjamin Schwaller, Barath Ramesh, and Alan D George. Investigating ti keystone ii and quad-core arm cortex-a53 architectures for on-board space processing. In *High Performance Extreme Computing*. IEEE, 2017.

[5] Yuki Abe, Hiroshi Sasaki, Martin Peres, Koji Inoue, Kazuaki Murakami, and Shinpei Kato. Power and performance analysis of gpu-accelerated systems. In *Workshop on Power-Aware Computing and Systems*, 2012.

[6] Ronald B Brightwell. Resource management challenges in the era of extreme heterogeneity. Technical report, Sandia National Lab., 2018.

[7] Xinxin Mei, Xiaowen Chu, Hai Liu, Yiu-Wing Leung, and Zongpeng Li. Energy efficient real-time task scheduling on cpu-gpu hybrid clusters. In *IEEE INFOCOM 2017-IEEE Conference on Computer Communications*.

[8] Lea Schönberger, Wen-Hung Huang, Georg Von Der Brüggen, Kuan-Hsun Chen, and Jian-Jia Chen. Schedulability analysis and priority assignment for segmented self-suspending tasks. In *2018 IEEE 24th International Conference on Embedded and Real-Time Computing Systems and Applications (RTCSA)*, pages 157–167. IEEE, 2018.

[9] Jian-Jia Chen and et al. Many suspensions, many problems: a review of self-suspending tasks in real-time systems. *Real-Time Systems*, 2019.

[10] Jinghao Sun, Jing Li, Zhishan Guo, An Zou, Xuan Zhang, Kunal Agrawal, and Sanjoy Baruah. Real-time scheduling upon a host-centric acceleration architecture with data offloading. In *2020 IEEE Real-Time and Embedded Technology and Applications Symposium*. IEEE, 2020.

[11] Sujan Kumar Saha, Yecheng Xiang, and Hyoseung Kim. Stgm: Spatio-temporal gpu management for real-time tasks. In *Conf. on Embedded and Real-Time Computing Systems and Applications*. IEEE, 2019.

[12] Yuankai Xu, Tiancheng He, Ruiqi Sun, Yehan Ma, Yier Jin, and An Zou. Shape: Scheduling of fixed-priority tasks on heterogeneous architectures with multiple cpus and many pes. In *Proceedings of the 41st IEEE/ACM International Conference on Computer-Aided Design*, pages 1–9, 2022.

[13] An Zou, Jing Li, Christopher D Gill, and Xuan Zhang. Rtgpu: Real-time gpu scheduling of hard deadline parallel tasks with fine-grain utilization. *IEEE Transactions on Parallel and Distributed Systems*, 2023.

[14] Pratyush Patel and et al. Analytical enhancements and practical insights for mpcp with self-suspensions. In *Real-Time and Embedded Technology and Applications Symposium*. IEEE, 2018.

[15] Zahaf Houssam-Eddine, Nicola Capodieci, Roberto Cavicchioli, Giuseppe Lipari, and Marko Bertogna. The hpc-dag task model for heterogeneous real-time systems. *Transactions on Computers*, 2020.

[16] Muhammad Husni Santriaji and Henry Hoffmann. Merlot: Architectural support for energy-efficient real-time processing in gpus. In *Real-Time and Embedded Technology and Applications Symposium*. IEEE, 2018.

[17] Chuan-Yue Yang, Jian-Jia Chen, Tei-Wei Kuo, and Lothar Thiele. An approximation scheme for energy-efficient scheduling of real-time tasks in heterogeneous multiprocessor systems. In *DATE*. IEEE, 2009.

[18] Yidi Wang, Mohsen Karimi, Yecheng Xiang, and Hyoseung Kim. Balancing energy efficiency and real-time performance in gpu scheduling. In *2021 Real-Time Systems Symposium*, pages 110–122. IEEE, 2021.

[19] Jing Li, Jian-Jia Chen, Kunal Agrawal, Chenyang Lu, Christopher D Gill, and Abusayeed Saifullah. Analysis of federated and global scheduling for parallel real-time tasks. In *ECRTS*, 2014.

[20] Wen-Hung Huang and Jian-Jia Chen. Schedulability and priority assignment for multi-segment self-suspending real-time tasks under fixed-priority scheduling. In *Technical report*. TU Dortmund, 2015.

[21] Neil C Audsley. *Optimal priority assignment and feasibility of static priority tasks with arbitrary start times*. Citeseer, 1991.

[22] Waqar Ali and Heechul Yun. Protecting real-time gpu kernels on integrated cpu-gpu soc platforms. *arXiv preprint arXiv:1712.08738*.

CARMA: Context-Aware Runtime Reconfiguration for Energy-Efficient Sensor Fusion

Yifan Zhang[§], Arnav Vaibhav Malawade[§], Xiaofang Zhang, Yuhui Li, DongHwan Seong,
Mohammad Abdullah Al Faruque, Sitao Huang
Department of Electrical Engineering and Computer Science, University of California, Irvine
{yifanz58, malawada, xiaofaz7, yuhuil10, dseong1, alfaruqu, sitaoh}@uci.edu

Abstract—**Autonomous systems (AS) are systems that can adapt and change their behaviors in response to unanticipated events and include systems such as aerial drones, autonomous vehicles, and ground/aquatic robots. AS require a wide array of sensors, deep learning models, and powerful hardware platforms to perceive the environment and safely operate in real-time. However, in many contexts, some sensing modalities negatively impact perception while increasing the system's overall energy consumption. Since AS are often energy-constrained edge devices, energy-efficient sensor fusion methods have been proposed. However, existing methods either fail to adapt to changing scenario conditions or to optimize system-wide energy efficiency. We propose CARMA, a context-aware sensor fusion approach that uses context to dynamically reconfigure the computation flow on a field-programmable gate array (FPGA) at runtime. By clock gating unused sensors and model sub-components, CARMA significantly reduces the energy used by a multi-sensory object detector without compromising performance. We use a deep learning processor unit (DPU) based reconfiguration approach to minimize the latency of model reconfiguration. We evaluate multiple context identification strategies, propose a novel system-wide energy-performance joint optimization, and evaluate scenario-specific perception performance. Across challenging real-world sensing contexts, CARMA outperforms state-of-the-art methods with up to 1.3× speedup and 73% lower energy consumption.**

I. Introduction

Autonomous systems (AS) radically improve productivity, logistics, and safety by enabling systems such as aerial drones, ground and aquatic robots, and consumer autonomous vehicles (AVs) to operate without direct human control. These applications require closely coupled perception and state estimation algorithms to navigate complex and unpredictable real-world scenarios in real time. Advanced deep learning models and multiple heterogeneous sensors (cameras, radars, and LiDARs) are necessary for perception across different weather and lighting conditions. However, the increasing complexity of modern AS comes with rising energy costs [1], which can be fatal for energy-constrained AS. The thermal design power of modern AS System-on-Chips (SoCs) can exceed 800 W [2], and the combined sensing, computation, and thermal loads can reduce operating range by over 11.5% [3].

Since the perception system is a major energy consumer in AS [1], [4], several efficient sensor fusion methods have been proposed. However, these methods use static architectures (*e.g.*, early or late fusion) that can fail in complex visual contexts where one or more sensors may be compromised [5]. To address these limitations, context-aware dynamic architectures for sensor fusion have been proposed [5], [6], where the model adapts to changing environmental conditions to enable robust and energy-efficient perception across diverse sensing conditions. Still, existing methods only focus on reducing algorithmic energy usage and ignore large energy consumers, such as the sensors and the hardware computation platforms.

In summary, the key challenges include: (i) effectively perceiving complex and adverse driving scenarios; (ii) reducing the energy consumption of the complete perception system, including sensors, hardware, and algorithms; and (iii) adapting the system configuration to different contexts, improving energy efficiency without compromising performance.

To overcome these challenges, we propose CARMA, a context-aware dynamic sensor fusion approach that uses *runtime model reconfiguration* to adapt its architecture on an FPGA. CARMA uses deep learning processing unit (DPU) [7] on FPGA for efficient, low-latency runtime reconfiguration. CARMA implements a tunable energy-performance optimization over the whole system, including sensors, model architecture, and hardware platform, to maximize energy savings without compromising performance. To our knowledge, this is the first work to propose energy-efficient sensor fusion via context-aware runtime model reconfiguration on FPGAs.

Our major contributions can be summarized as follows:

1) We propose CARMA, an approach for dynamically reconfiguring a complete sensor fusion system for object detection at runtime using contextual information. CARMA uses DPUs on FPGA to enable runtime model reconfiguration with negligible model switching latency.
2) We propose a method for intermittently performing context identification to enable intelligent sensor and submodel clock gating to maximize energy efficiency.
3) We use tunable joint optimization between perception performance and energy consumption to maximize energy efficiency while minimizing perception impacts.
4) We show that CARMA significantly reduces system-wide energy usage compared to state-of-the-art sensor fusion methods and achieves equivalent or better object detection performance across diverse autonomous driving scenarios with up to $1.3\times$ inference speedup and 73% lower energy consumption.

[§]Equal contribution

979-8-3503-1176-1/23 $31.00

II. Related Works

A. Adaptive Computing Systems on FPGA

Self-adaptive systems can modify their runtime behavior according to changing environments and system goals. [8] presents a dynamically reconfigurable convolutional neural network (CNN) accelerator optimized for throughput. In [9], an FPGA reconfigures at runtime to use a lower power design when the battery level decreases. However, its reconfiguration latency is proportional to the bitstream size, which limits it from applying to large components. The DPUs enable users to reconfigure CNN models at runtime with minimal latency overhead. [10] explored a DPU-based energy-efficient hardware accelerator. However, it does not optimize energy efficiency system-wide or handle complex environments.

B. Energy-Performance Optimization

Several works have explored methods on energy-performance trade-off of deep learning algorithms at runtime targeting single-modality image classification task [11], [12], [13]. Recent works have extended these optimizations to multi-sensor fusion for perception [14]. [6] proposes a dynamic-width sensor fusion model that aims to select lower energy submodels while maintaining performance. Although this approach incorporates multimodality, it only optimizes the object detection model parameters and omits *system-wide* energy optimizations. In contrast, we propose using runtime model reconfiguration on a heterogeneous FPGA-driven computing platform to maximize the energy saved by dynamic model selection while applying system-wide energy optimizations to reduce energy usage further.

C. Intermittent Sensing and Control in Autonomous Systems

Due to the energy constraints of many AS, several methods for intermittent sensing and control have been proposed to reduce energy consumption without compromising performance [15], [16]. [17] proposes using an intermittent control strategy for autonomous driving to emulate human-like control behavior. Like these works, CARMA targets energy efficiency by intermittently reconfiguring the model architecture and the set of active sensors to match the current environment context.

III. Methodology

A. Problem Formulation

1) Object Detection Model: AS use object detection to avoid collisions, predict motion, and enable safe path planning. The goal of the object detector ϕ is to use the sensor measurements $\mathbf{X}$ to accurately identify the objects $\mathbf{Y}$ in the environment:

$$\mathbf{Y} = \phi(\mathbf{X}), \text{ where } \mathbf{Y} = \{\mathbf{Y}^i_{class}, \mathbf{Y}^i_{reg}\}_{i=1...d} \quad (1)$$

where $\mathbf{Y}^i_{class}, \mathbf{Y}^i_{reg}$ denote the class and bounding box, respectively, of object i. Extending this framework to multi-sensor perception, early fusion across s sensors can be modeled as:

$$\hat{\mathbf{Y}} = \phi(\psi(\mathbf{X}_1, \mathbf{X}_2, \ldots, \mathbf{X}_s)), \quad (2)$$

where ψ is the function for fusing the sensor data before the object detector processes it, and $\hat{Y}$ represents the object predictions. Similarly, late fusion across s sensors can be modeled as fusing the outputs of an ensemble of independent object detectors:

$$\hat{\mathbf{Y}}_1, \hat{\mathbf{Y}}_2, \ldots, \hat{\mathbf{Y}}_s = \phi_1(\mathbf{X}_1), \phi_2(\mathbf{X}_2), \ldots, \phi_s(\mathbf{X}_s) \quad (3)$$

$$\hat{\mathbf{Y}} = \phi_f(\hat{\mathbf{Y}}_1, \hat{\mathbf{Y}}_2, \ldots, \hat{\mathbf{Y}}_s), \quad (4)$$

where $(\phi 1, \phi_2, ..., \phi_s)$ represent independent object detectors, and ϕ_f represents the late fusion function for combining their outputs. Our proposed approach uses context to identify the best combination of early and late fusion to improve the accuracy of the resultant predictions across driving contexts. As such, the object detection model becomes:

$$\hat{\mathbf{Y}} = \phi_f(\phi_1(\mathbf{X}_1), \phi_2(\mathbf{X}_2), \ldots, \phi_3(\psi(\mathbf{X}_2, \mathbf{X}_s))) \quad (5)$$

Where ϕ_1 and ϕ_2 represent single-sensor object detectors, ϕ_3 is a multi-sensor object detector using early fusion, and ϕ is the late fusion function for fusing the detectors' outputs to obtain $\hat{Y}$. Section III-B2 describes how CARMA identifies context and selects the appropriate model configuration.

2) Energy Model: We model the energy usage of the complete AV driving system E_{sys} as the total energy consumed by the sensors E_s and the execution of the algorithm E_a on the hardware platform.

$$E_{sys} = E_s + E_a \quad (6)$$

We omit factors such as drivetrain energy usage and battery lifetime as these factors have been studied in existing work [18], [19] and can be used in conjunction with our approach. Typical AS contain some combination of static sensors (*e.g.*, cameras, ultrasonic sensors, front-facing radar) and rotating sensors (*e.g.*, spinning top-mounted LiDAR). The energy consumption per sensor $s \in S$ can be computed from the measurement power $P^{meas.}_s$, measurement frequency f_s, and, for spinning sensors, the motor power P^{motor}_s, as follows:

$$E_s = (P^{meas.}_s + P^{motor}_s) * 1/f_s \quad (7)$$

To reduce the energy consumption of the complete system, we clock gate sensors unused in the current visual context. The LiDAR and radar sensors in our testbed, discussed in Section IV-A, are top-mounted spinning sensors, while the cameras are fixed sensors without motors. Since the LiDAR and radar have inertia and require several seconds to start and stop rotating, we assume that we only clock gate the measurement components while keeping the motor spinning so they can be quickly re-enabled to ensure safety. The total power consumption of the Navtech CTS350-X radar is 24 W [20], while the Velodyne HDL-32E LiDAR uses 12 W [21] and the ZED camera uses 1.9 W [22]. The Navtech CTS350-X needs 2.4 W to spin the motor, so $P^{meas.}_{radar} = 21.6$ W. Using comparable LiDAR motor models for the Velodyne HDL-32E, we estimate $P^{meas.}_{LiDAR} = 9.6$ W.

Since our object detection model is reconfigurable, the algorithm energy consumption E_a can be computed as:

$$E_a(\phi, \mathbf{X}) = P_a(\phi, \mathbf{X}) * t(\phi, \mathbf{X}), \quad (8)$$

where $t(\phi, \mathbf{X})$ represents the processing latency in seconds and $P_a(\phi, \mathbf{X})$ represents the power consumption in Watts of processing input $\mathbf{X}$ through the current model configuration ϕ on the hardware platform. We measured the power and latency

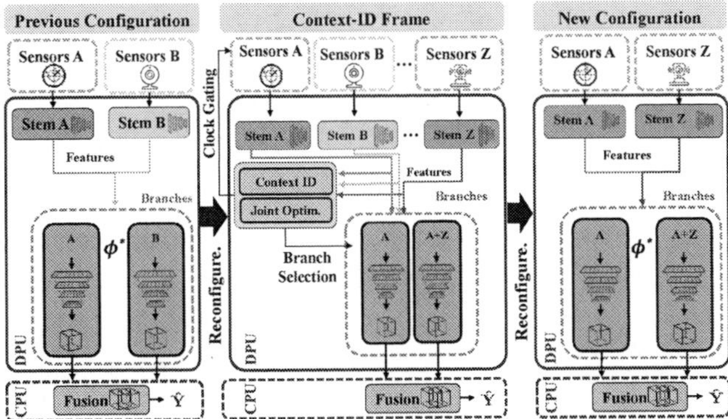

Fig. 1: CARMA System Architecture and Reconfiguration Workflow

of each model configuration on our hardware platform, the Xilinx Kria KV260 FPGA, to compute E_a offline for use in our multi-objective optimization.

3) Multi-Objective Optimization: We implement a tunable joint optimization between system-wide energy consumption and model performance to enable our approach to minimize energy without compromising performance. Since there is typically a trade-off between these two objectives, we use a λ_E term to allow model designers to specify the preference for energy efficiency over performance depending on the application of the system. Given that we know the expected prediction performance L of configuration ϕ for an input $\mathbf{X}$, denoted as $L(\phi, \mathbf{X})$, and the expected system-wide energy consumption of that configuration $E_{sys}(\phi, \mathbf{X})$, our optimization can be formulated as:

$$L_{opt}(\phi, \mathbf{X}) = L(\phi, \mathbf{X}) * (1 - \lambda_E) + E_{sys}(\phi, \mathbf{X}) * \lambda_E \quad (9)$$

$$\phi^*(\mathbf{X}) = \arg\min_{\phi \in \Phi}(L_{opt}(\phi, \mathbf{X})), \quad (10)$$

where $\phi^*(\mathbf{X})$ represents the model configuration that best minimizes the joint optimization loss L_{opt} for input $\mathbf{X}$ for the given λ_E. [6] used a similar optimization to select which branches to execute, with all other system components remaining fixed. However, our proposed approach includes clock gating of unused sensors and stems, drastically increasing the potential energy savings and enabling system-wide optimization.

B. System Architecture

CARMA's architecture is shown in Fig. 1. CARMA consists of a runtime reconfigurable multi-branch sensor fusion model for object detection. Section III-D elaborates on our runtime reconfiguration approach on hardware, while the following text describes our sensor fusion model. The model consists of four key components, (i) feature extraction, (ii) context identification, (iii) submodel selection, and (iv) output fusion. First, multi-modal sensor data is processed by modality-specific *Stem* models to extract an initial set of features for each sensor. These features are then used by the *Gate* model to identify the current visual context. This context is used to select the set of submodels (*Branches*) to execute that optimizes performance and energy efficiency. Each active branch outputs a set of object detections collected and fused by the *Fusion Block* to produce a final set of refined detections.

1) Stem and Branches: We utilize the single shot multibox detector (SSD) [23] for object detection, known for its superior speed and performance compared to Faster R-CNN [24]. SSD employs a single-pass CNN to perform region proposal and object detection, eliminating the need for a separate Region Proposal Network. With a smaller model size and fewer intermediate feature maps, SSD requires fewer hardware resources and has lower memory bandwidth, making it faster to execute on FPGAs. Our proposed architecture incorporates SSD's ResNet-18 backbone, using the first six layers as modality-specific preprocessors (*stem*) and the remaining 23 layers as branches. We implement single-sensor branches for four inputs (two cameras, one LiDAR, and one radar) and three early-fusion branches that take multiple sensors as input: dual camera, LiDAR and radar, and dual camera with LiDAR. These branches include a single merge convolution layer to combine the sensors across the channel dimension before continuing with processing.

2) Context Identification and Gating: To identify the current visual context and perform branch selection, we propose three variants of context-identification, or *gate*, models. The *knowledge* gate uses fixed domain-knowledge rules to select submodels using external contextual information (e.g., weather, time of day, road type). The rules encode domain knowledge on the sensor modalities least likely to be degraded by current environmental factors such as rain, snow or fog. The *deep* gate uses a 3-layer CNN to infer the current context from the stem output features and directly output the set of branches it infers will perform best in the current visual context. Here, context refers to an abstract visual state estimate generated within the CNN's hidden layers, while the gate output indicates which branches to execute. The *attention* gate is the same as the *deep* gate with the addition of a self-attention layer. Given the set of all possible model configurations Φ, the objective of the gate is to estimate the performance L of each configuration ϕ for the current set of input features F:

$$L(\Phi, \mathbf{F}) = \pi(\phi, \mathbf{F}), \forall \phi \in \Phi \quad (11)$$

$$\rho(L(\Phi, \mathbf{F}), \gamma) = \{\phi \in \Phi \text{ s.t. } L(\phi, \mathbf{F}) \leq L(\phi', \mathbf{F}) + \gamma\} \quad (12)$$

$$\Phi^* = \rho(L(\Phi, \mathbf{F}), \gamma), \quad (13)$$

where π represents the gating model and ρ represents a function for identifying the set Φ^* of top performing configurations with an estimated error within γ of the best configuration ϕ'.

3) Fusion Block: The fusion block in CARMA combines object detections from active branches to produce more accurate final bounding box predictions. We employ *weighted boxes fusion* [25], which averages proposed boxes based on confidence scores. In CARMA, the fusion block runs on the CPU due to its complex program logic, which is better supported on the CPU than the DPU. It can also utilize idle CPU resources during DPU inference.

C. Hardware Design Choices

CARMA is adaptive to various platforms. Still, safety-critical real-time tasks require careful hardware design choices.

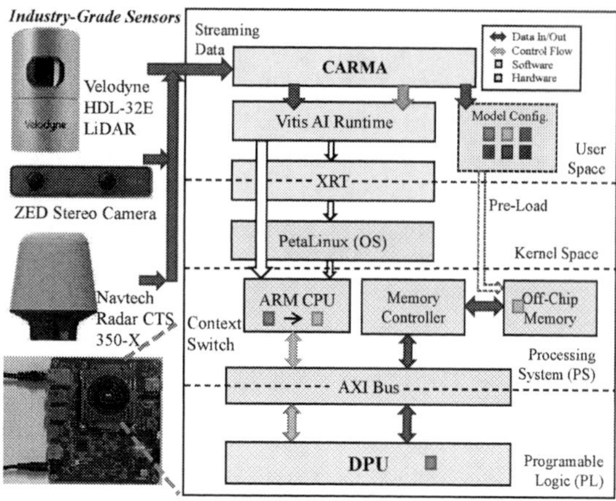

Fig. 2: CARMA System Stack and Experimental Testbed

1) High Throughput: In autonomous systems (AS), real-time data processing with low latency is crucial for safe and efficient vehicle operation. A minimum rate of 10 frames per second (FPS) [1] is typically required to enable accurate control in dynamic environments. The DPU offers user-configurable parameters for optimizing performance, including pixel, input channel, and output channel parallelism. For different branch configurations, the computation workload can vary from 9 to 27 GOP (10^9 operations). With tailored parallelism settings of 8, 16, and 16, respectively, the DPU achieves a theoretical speed of 1228.8 GOP/s at a clock frequency of 300 MHz ($2\times8\times16\times16 = 4096$ operations per cycle), to maintain safe FPS and cover possible tail latency. Additionally, our on-board profiling results show that with 2000 MB/s memory bandwidth, we can guarantee 700 GOP/s average throughput when memory-bounded. (For a 12 GOP workload model with 33 MB estimated memory access).

2) Fast Context Switch Interval: CARMA changes branch configurations (*context switch*) at runtime. Fast context switch intervals are necessary to handle various tasks and events that may occur during vehicle operation. CARMA uses Vitis AI Runtime to load the instruction files into the DPU for inference and switch the context by changing the calling threads corresponding to different configurations. Loading of model instruction files and inference are performed simultaneously, reducing the context switch time to the time of the thread switch (less than 1ms), while traditional FPGA runtime reconfiguration waits until the new bitstream is fully deployed on-board. Since each model file is <25 MB and our system has 4 GB on-board memory, our system can store all model configurations in DDR memory.

D. Hardware Execution Model

Fig. 2 illustrates our hardware execution model. CARMA runs in the application layer on PetaLinux and controls our complete sensor fusion system. It uses Vitis AI Runtime, a set of high-level APIs, to interact with the DPU. Xilinx Runtime (XRT) provides a set of low-level APIs that connect the User Space and Kernel Space and control the hardware. The CPU serves as the hardware host control node and controls the DPU, services interrupts, and coordinates data transfers. The processing system (PS) connects to the DPU via the Advanced eXtensible Interface (AXI) bus for transferring data and control signals. When initializing the system, the compiled models for all sensor-fusion configurations are loaded into the off-chip memory, waiting to be called. At runtime, the DPU fetches compiled instructions from off-chip memory to control the operation of its computing engine.

E. Runtime Workflow and Intermittent Context Identification

Several works have demonstrated safe and effective intermittent perception and control approaches, as discussed in Section II-C. These approaches are intuitive since real-world visual contexts often remain the same for several seconds, especially in the case of broad visual contexts like *rainy weather* or *night driving*. We propose using *intermittent* context-identification to enable broader energy optimizations such as clock-gating unused sensors and stem models for brief periods before re-enabling them to identify the current context. CARMA can directly integrate with existing methods for safe intermittent perception since they use similar strategies, such as clock gating, to control sensing frequency.

To reduce the overhead of context identification and switching, we propose the Context-ID Frame design, shown in Fig. 1. In sensor fusion mode, we only execute the stems and branches needed for a particular model configuration, minimizing energy consumption. In Context-ID mode, we reconfigure the DPU to the Context-ID Frame to select the next model configuration. The following two algorithms describe the workflow of our proposed approach. Alg. 1 shows the

Algorithm 1: Runtime Sensor Fusion Algorithm

Input: t, ϕ^*, $active\_sensors$, T_c
Output: Object Detections ($\hat{\mathbf{Y}}$)
Initialize feature vector $\mathbf{F}$ and branch output vector $\hat{\mathbf{Y}}^*$
for *s in active_sensors* **do**
    $X_s \leftarrow s(t)$ // data input
    $\mathbf{F}[s] \leftarrow stem_s(X_s)$ // extract features
for *branch in* ϕ^* **do**
    $\hat{\mathbf{Y}}^*[branch] \leftarrow branch(\mathbf{F}^*)$ // pass subset of $\mathbf{F}$
$\hat{\mathbf{Y}} \leftarrow fusion\_block(\hat{\mathbf{Y}}^*)$ // fuse detections
if $t/T_c = 0$ **then**
    $\phi^*, active\_sensors \leftarrow \mathbf{Algorithm2}(t+1)$

typical operation of CARMA. For each time step t, data is retrieved from the active set of sensors and processed by the current branch configuration ϕ^* to produce the output detections $\hat{\mathbf{Y}}$. T_c represents the context re-identification interval; when $t/T_c = 0$, execution transfers to Alg. 2 for the next time step $t+1$. Here, T_c can be dynamically configured by an intermittent algorithm, such as those from Section II-C. In Alg. 2, all sensors and stems are activated, and the sensor features F are passed to the gate module π to estimate the loss of each branch configuration. The lowest loss branches are selected by ρ as described in Equation 13. Then, this set Φ^* is passed to the joint optimization to identify the optimal configuration ϕ^*. The outputs of the active branches are fused to produce $\hat{\mathbf{Y}}$. After this step, we clock gate the unused sensors, switch to the new model configuration ϕ^*, and continue executing Alg. 1 with the new ϕ^* and $active\_sensors$ at the next time step.

Algorithm 2: Context ID and Reconfigure Algorithm

Input: t, λ_E, Φ, γ, $E_{sys}(\Phi)$, $all\_sensors$
Output: Object Detections ($\hat{\mathbf{Y}}$), ϕ^*, $active\_sensors$

Initialize feature vec. $\mathbf{F}$ and output vec. $\hat{\mathbf{Y}}^*$
for s *in all_sensors* **do**
  $X_s \leftarrow s(t)$ // data input
  $\mathbf{F}[s] \leftarrow stem_s(X_s)$ // extract features
$L(\Phi) \leftarrow \pi(\mathbf{F}, \Phi)$ // estimate model losses
$\Phi^* \leftarrow \rho(L(\Phi), \gamma)$ // select candidates
for ϕ *in* Φ^* **do**
  $L_{joint}(\phi) \leftarrow (1 - \lambda_E) * L(\phi) + \lambda_E * E_{sys}(\phi)$
$\phi^* \leftarrow \arg\min_{\forall \phi \in \Phi^*}(L_{joint}(\phi))$ // joint opt.
$load\_branches(\phi^*)$ // reconfiguration
for *branch in* ϕ^* **do**
  $\hat{\mathbf{Y}}^*[branch] \leftarrow branch(\mathbf{F}^*)$ // pass subset of $\mathbf{F}$
$\hat{\mathbf{Y}} \leftarrow fusion\_block(\hat{\mathbf{Y}}^*)$ // fuse detections
Initialize empty set $active\_sensors$
for s *in all_sensors* **do**
  if ϕ^* *requires* s **then**
    $active\_sensors \leftarrow active\_sensors \cup \{s\}$
  else
    $clock\_gate(s)$ // clock gate sensors
    $disable\_stem(stem_s)$ // reconfiguration

IV. Experiments

A. Experimental Setup

CARMA can be applied to any multi-sensor AS to enable energy-efficient perception. In our experiments, we evaluate CARMA on a popular AS use case: autonomous driving for AVs. Our hardware testbed is shown on the left side of Fig. 2. We use the Xilinx Kria KV260 FPGA as our computing platform. Due to its portability and compatibility, our design could feasibly be implemented on Xilinx automotive-grade FPGAs in a similar manner. Each model is trained on the RADIATE dataset [26], which contains three hours of high-resolution radar, LiDAR, and stereo camera data across challenging perception contexts. We compare against Faster R-CNN object detectors for single sensor inputs, early and late multi-sensor fusion, and the state-of-the-art method, EcoFusion [6]. To measure the object detection performance of each model, we use the object detection loss function from [27], which combines bounding box loss with classification loss. The object detection metrics we present are for a Faster R-CNN variant of our model trained using the same hyperparameters as [6] for fairer comparison with EcoFusion [6]. However, we verified experimentally that the SSD-based model achieves 50% lower average loss and consumes 15% less energy than the Faster R-CNN version. We used built-in functions in the host code and system commands to measure the end-to-end latency and power consumption of different configurations.

B. Performance on FPGA

We compare the object detection performance and energy consumption of different fusion techniques in Table I. Across different gating and λ_E configurations, CARMA achieves lower average energy usage and loss than almost every early fusion, late fusion, and single sensor model. The only exceptions were the camera-only configurations, which had higher losses than our method but lower energy usage due to the efficiency of the camera sensors. Notably, with an equivalent model loss, CARMA ($\lambda_E = 0, deep$) achieves a **41.3%** reduction in energy compared to EcoFusion ($\lambda_E = 0, attn$). With a higher $\lambda_E = 0.01$ for both models, CARMA achieves **73.7%** lower energy usage with only a 3.2% higher loss than EcoFusion. EcoFusion's inability to account for sensor energy or apply sensor and model clock gating leads to higher average energy consumption, putting it on par with high-energy early fusion and late fusion variants. CARMA also exhibits faster speeds, achieving **6%-33%** speed-up compared to EcoFusion, with lower model latencies for higher λ_E values. The results highlight trade-offs among sensing modalities, with radar branches consuming more energy but providing reliability in camera failure contexts, as supported by lower loss in the late fusion model.

Fusion Type	Configuration	Avg. Loss	Energy (J)	Latency (ms)
None	Radar (R)	2.858	6.73	14.2
	LiDAR (L)	4.682	3.73	14.2
	Camera (C)	1.680	1.81	14.2
Early	$R + L$	2.784	9.16	17.1
	$C_L + C_R$	1.203	2.31	17.1
	$L + C_L + C_R$	3.476	3.73	19.7
Late	$R + L + C_L + C_R$	0.967	10.48	42.6
EcoFusion [6]	$\lambda_E = 0, attn$	0.915	10.41	54.0
	$\lambda_E = 0.01, attn$	0.924	10.36	48.0
	$\lambda_E = 0.1, attn$	1.147	10.18	27.7
CARMA (Ours)	$\lambda_E = 0, attn$	0.915	7.35	51.9
	$\lambda_E = 0, deep$	0.915	6.12	51.2
	$\lambda_E = 0.0001, attn$	0.920	6.68	50.2
	$\lambda_E = 0.001, deep$	0.944	3.31	42.6
	$\lambda_E = 0.001, attn$	**0.959**	**3.23**	**38.5**
	$\lambda_E = 0.01, deep$	**0.954**	**2.73**	**36.1**

TABLE I: Performance and energy comparison between different fusion methods (T_c = 30 samples)

Fig. 3 illustrates the trade-off between system-wide energy consumption and model performance for each gate module at different values of λ_E. Both *deep* and *attn* gates present a clear trade-off between performance and energy efficiency as λ_E increases. However, the large flat region along the right side of both Pareto frontiers illustrates how system-wide energy can be reduced significantly with a minimal performance impact. The results for *loss-based* gating indicate the performance of an optimal gate module and serve as a theoretical upper bound, since it uses the posteriori ground truth loss to select branch. The *knowledge* gate is ineffective in minimizing either objective. Overall, the *deep* and *attn* gate reduce energy consumption by over **55%** while maintaining an average loss within 5% of the $\lambda_E = 0$ models.

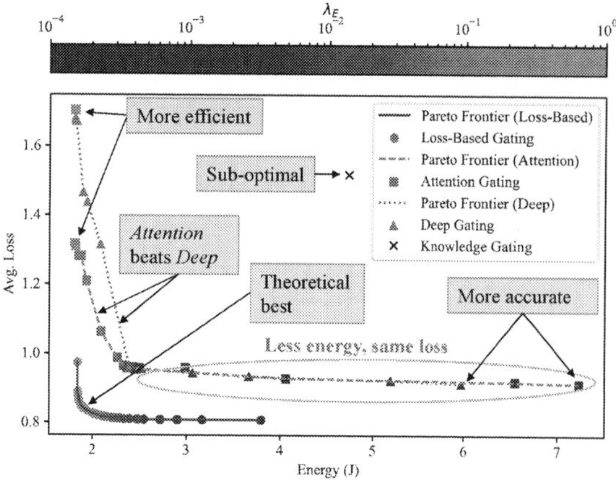

Fig. 3: System-wide energy consumption vs. object detection loss of different gate modules for varying values of λ_E.

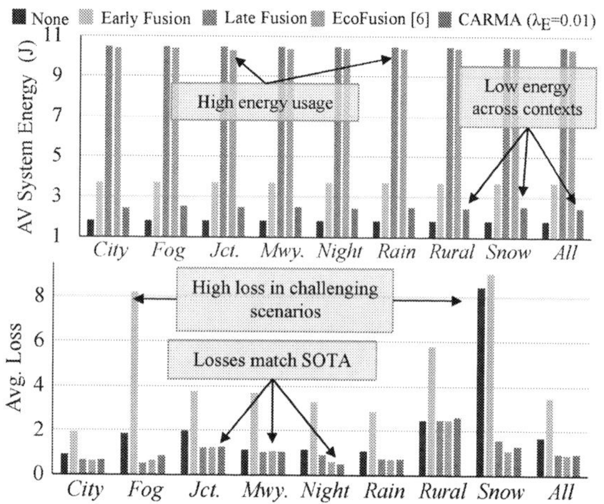

Fig. 4: Scenario-specific energy usage and object detection loss for: No Fusion (C_R), Early Fusion ($L + C_L + C_R$), Late Fusion ($R + L + C_L + C_R$), EcoFusion ($\lambda_E = 0, attn$), and CARMA ($\lambda_E = 0.01, attn$).

C. Scenario-Specific Performance

Fig. 4 shows how different driving scenarios affect the energy consumption and performance of different fusion methods. The results show that CARMA can reduce energy consumption below that of early fusion, late fusion, and EcoFusion across all scenarios. Interestingly, our model minimizes energy consumption in the *Snow* scenario by selecting camera branches only throughout the context (C_L, C_R, and $C_L + C_R$). Early fusion is especially weak in the *Fog*, *Rural*, and *Snow* contexts, likely due to its susceptibility to sensor noise. Late fusion, EcoFusion, and CARMA are robust across all scenarios, with *Rural* being the most challenging.

V. Conclusion

In this work, we proposed a context-aware sensor fusion approach that uses context to reconfigure the perception model on an FPGA at runtime dynamically. CARMA is capable of switching model computation paths with negligible latency while intermittent context identification, system-wide energy-performance optimization, and sensor clock gating maximize energy savings without compromising performance. Overall, CARMA achieves up to 1.3× speedup and reduces energy consumption by over 73% over leading static and dynamic sensor fusion techniques across complex driving contexts.

Acknowledgment

This work was partially supported by the NSF under award CCF-2140154 and hardware donations from AMD-Xilinx University Program.

References

[1] S.-C. Lin, Y. Zhang *et al.*, "The architectural implications of autonomous driving: Constraints and acceleration," in *ASPLOS 2018*.
[2] S. Abuelsamid, "NVIDIA Cranks Up And Turns Down Its Drive AGX Orin Computers," *Forbes*, Jun 2020.
[3] X. He, H. Kim *et al.*, "Energy consumption simulation for connected and automated vehicles: Eco-driving benefits versus automation loads," *SAE Int. J. of Connected and Autonomous Vehicles*, vol. 6, 2022.
[4] A. Malawade *et al.*, "Sage: A split-architecture methodology for efficient end-to-end autonomous vehicle control," *ACM TECS*, vol. 20, 2021.
[5] A. V. Malawade, T. Mortlock, and M. A. Al Faruque, "HydraFusion: Context-aware selective sensor fusion for robust and efficient autonomous vehicle perception," in *ICCPS '22*. IEEE, 2022, pp. 68–79.
[6] A. V. Malawade *et al.*, "EcoFusion: Energy-aware adaptive sensor fusion for efficient autonomous vehicle perception," in *DAC '22*.
[7] Xilinx, "Vitis AI User Guide (UG1414)."
[8] H. Irmak *et al.*, "Increasing Flexibility of FPGA-based CNN Accelerators with Dynamic Partial Reconfiguration," in *FPL '21*.
[9] E. Youssef, H. A. Elsemary *et al.*, "Energy adaptive convolution neural network using dynamic partial reconfiguration," in *MWSCAS 2020*.
[10] A. S. Hussein *et al.*, "Implementation of a DPU-based intelligent thermal imaging hardware accelerator on FPGA," *Electronics*, 2022.
[11] R. T. Mullapudi, W. R. Mark *et al.*, "HydraNets: Specialized dynamic architectures for efficient inference," in *CVPR '18*, 2018, pp. 8080–8089.
[12] Z. Takhirov, J. Wang *et al.*, "Energy-efficient adaptive classifier design for mobile systems," in *ISLPED '16*, 2016, p. 52–57.
[13] C. Hao, X. Zhang *et al.*, "FPGA/DNN Co-Design: An Efficient Design Methodology for IoT Intelligence on the Edge," in *Proceedings of the 56th Annual Design Automation Conference 2019*, ser. DAC '19, 2019.
[14] D. Balemans *et al.*, "Resource efficient sensor fusion by knowledge-based network pruning," *Internet of Things*, vol. 11, p. 100231, 2020.
[15] V. Gokhale *et al.*, "Feel: fast, energy-efficient localization for autonomous indoor vehicles," in *ICC '21*. IEEE, 2021, pp. 1–6.
[16] C. Huang, S. Xu *et al.*, "Opportunistic intermittent control with safety guarantees for autonomous systems," in *DAC '20*. IEEE, 2020.
[17] R. Dash *et al.*, "Intermittent control in autonomous vehicle steering control and lane keeping," in *5th International Conference of The Robotics Society*, 2021.
[18] K. Vatanparvar, S. Faezi *et al.*, "Extended range electric vehicle with driving behavior estimation in energy management," *IEEE transactions on Smart Grid*, vol. 10, no. 3, pp. 2959–2968, 2018.
[19] D. Baek, Y. Chen *et al.*, "Battery-aware energy model of drone delivery tasks," in *ISLPED '18*, 2018.
[20] N. Radar, "Navtech CTS Series," May 2021. [Online]. Available: https://navtechradar.com/clearway-technical-specifications/compact-sensors
[21] V. Lidar, "Velodyne HDL-32e Datasheet," May 2021. [Online]. Available: https://velodynelidar.com/products/hdl-32e/
[22] Stereolabs, "ZED Camera and SDK Overview." [Online]. Available: https://cdn.stereolabs.com/assets/datasheets/zed-camera-datasheet.pdf
[23] W. Liu, D. Anguelov *et al.*, "SSD: Single shot multibox detector," in *European conference on computer vision*. Springer, 2016, pp. 21–37.
[24] S. Ren, K. He *et al.*, "Faster R-CNN: Towards real-time object detection with region proposal networks," *NIPS 2015*.
[25] R. Solovyev, W. Wang, and T. Gabruseva, "Weighted boxes fusion: Ensembling boxes from different object detection models," *Image and Vision Computing*, vol. 107, p. 104117, 2021.
[26] M. Sheeny, E. De Pellegrin *et al.*, "RADIATE: A radar dataset for automotive perception," *arXiv preprint arXiv:2010.09076*, 2020.
[27] R. Girshick, "Fast R-CNN," in *CVPR '15*, 2015, pp. 1440–1448.

Uncertainty-aware Online Learning for Dynamic Power Management in Large Manycore systems

Gaurav Narang[1], Raid Ayoub[2], Michael Kishinevsky[2], Janardhan Rao Doppa[1], Partha Pratim Pande[1]
[1]School of EECS, Washington State University, Pullman, WA, USA
[2]Intel Corporation, Hillsboro, OR, USA
{gaurav.narang, jana.doppa, pande}@wsu.edu, {michael.kishinevsky, raid.ayoub}@intel.com

Abstract— **Large-scale manycore System-on-Chips (SoCs) need to satisfy the conflicting objectives of maximizing performance and minimizing energy consumption for dynamically changing applications. In this paper, we consider the problem of dynamic power management (DPM) in large manycore SoCs for unseen applications at runtime. We employ a machine learning (ML) based DPM policy, which selects the voltage/frequency (V/F) levels for different cluster of cores as a function of the application features such as core computation, inter-core traffic etc. We propose a novel uncertainty-aware *online* learning framework to learn the DPM policy, which can adapt to unseen applications at runtime. It relies on two key ideas. First, an entropy-based uncertainty measure is used to distinguish between seen and unseen system states. Second, we employ conformal prediction to compute uncertain V/F sets for unseen system states. We perform bounded-search over the uncertain V/F configurations using power/performance models to identify the best V/F configurations to minimize the energy-delay product (EDP) and create supervised examples for online learning. Our experiments on 64-core system show that the EDP is reduced by up to 50% and 60% when compared to existing online-imitation learning and reinforcement learning methods, respectively.**

Keywords—uncertainty quantification, power management, machine learning, conformal prediction, large manycore systems

I. Introduction

With many companies taking the pledge to be carbon negative by 2030, emphasis on green computing is dominating large-scale manycore systems design [1]. Datacenters employing large-scale manycore chips constitute 1.8% of the annual total energy consumption [1]. The carbon output of computing - from the smallest chip to the largest data center - must be dramatically moderated. In this respect, Voltage-Frequency Island (VFI) is a well-established design paradigm to create scalable and energy-efficient manycore chips [2]. It works on the premise that each core's computation and communication patterns vary during the execution of the application and similar cores and associated routers/links should be clustered together. VFI control is a dynamic power management (DPM) strategy for manycore chips. Voltage/Frequency (V/F) knobs of the VFIs can be dynamically tuned to reduce the energy while maintaining the application's quality of service (QoS) [3].

To leverage an application's varying energy demands over time, machine learning (ML) methods have been used to create DPM policies as a function of features of the system state [4]. Since all the application workloads executed at runtime may not be known at the design-time, DPM policies created *offline* at the design time may perform sub-optimally when used for unseen applications. Therefore, there is a strong need for *online* learning approaches that adapt to new applications. Prior online learning methods for DPM make different trade-offs between sample-efficiency (number of samples needed to learn an accurate policy) vs. execution-efficiency (overhead to perform each online learning step). Reinforcement Learning (RL) methods update the policy using weak supervision in the form of a reward function [5]. RL has high execution-efficiency, but poor sample-efficiency, which means the DPM policy can be sub-optimal for a long time resulting in missed opportunity for energy savings. Imitation Learning (IL) methods update the policy using strong supervision from power/performance models [6]. IL has relatively high sample efficiency, but poor execution efficiency due to the need to search over large number of V/F configurations to identify the best configuration to minimize energy. Importantly, both RL and IL methods perform online learning at each decision epoch without any reasoning mechanism to distinguish unseen system states where online learning is needed from seen system states at design time.

In this paper, we propose a novel framework referred to as uncertainty-aware online learning (UaOL) to overcome the drawbacks of prior work by achieving a significantly improved trade-off between sample-efficiency and execution-efficiency to learn accurate DPM policies for new applications. UaOL based DPM policy utilizes workload-aware features of a system state (input) such as instructions per cycle (IPC) and inter-VFI traffic, to predict the power management decision (output) for each VFI at each decision epoch. UaOL improves both sample-efficiency and execution-efficiency by formally reasoning about the uncertainty of the DPM policy at a given system state and performs online learning only over unseen system states. It computes an entropy-based uncertainty measure using the probability distribution over candidate V/F outputs from the current DPM policy. A threshold over this uncertainty measure is used to identify unseen system states.

UaOL further improves execution-efficiency of online learning by significantly reducing the search space of candidate V/F configurations without losing the optimal V/F configurations, which minimizes energy-delay product (EDP). We adapt the principles behind conformal prediction (CP) [7] to obtain uncertain V/F sets (a subset of V/F levels) using the ML-based DPM policy for each VFI for unseen system states. CP provides formal guarantees for a user-specified coverage: optimal V/F is contained in the uncertain V/F set with a high probability (e.g., 90%). Additionally, CP is adaptive and will reflect the difficulty of making DPM decisions at a given system state. We perform bounded search over the V/F configurations

This work was supported in part by the National Science Foundation grants CNS-1955353, OAC-1910213, and in part by the Semiconductor Research Corporation's AI Hardware program task 3014.001.

979-8-3503-1176-1/23 $31.00 © 2023 IEEE

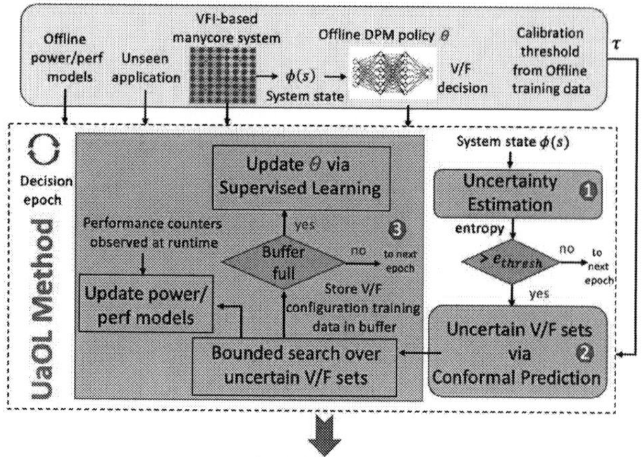

Fig.1: High-level overview of the UaOL framework and three key steps.

from uncertain V/F sets for different VFIs using power/performance models to identify the V/F configuration with minimum EDP, which serves as a supervised training example for online learning. Fig.1 shows the high-level overview of the UaOL framework.

Contributions. The key contribution of this paper is the development and evaluation of UaOL framework for online update of DPM policies for manycore systems on unseen applications at runtime. Specific contributions include:

- Formal reasoning method to identify unseen system states to perform online learning based on the entropy of probability distribution over V/F levels from ML-based DPM policy.
- Conformal prediction-based method to construct uncertain V/F sets for unseen system states to reduce the search space of candidate V/F configurations for improving the execution-efficiency of online learning.
- Experimental results on a representative manycore system demonstrate that online DPM policies learnt from our novel UaOL framework reduce the EDP by up to 50% and 60% when compared to online-IL and RL methods respectively.

II. Related Work

VFI based DPM strategies for manycore chips are designed using a set of known applications at the design-time [8]. Since there is a significant growth in the number and type of applications (graph, ML, cloud etc.), DPM policies need to adapt at runtime to optimally configure the V/F levels of the VFIs. Recent works have used reinforcement learning (RL) methods such as deep Q-learning to solve the VFI control problem [9]. It employs a simple reward function to trade-off power and performance. However, due to its exploratory form of learning, RL requires lot of samples to converge to near-optimal policy. Imitation learning (IL) has addressed the challenges of RL and demonstrated its superiority over RL for VFI-based power management in large-scale manycore systems [10]. IL relies on strong supervision from an expert policy to improve sample-efficiency and accuracy. Prior work has constructed expert policies in offline setting to successfully apply IL for DPM in mobile SoCs and manycore systems [8] [3]. To adapt DPM policies to new applications at runtime for mobile platforms, online-IL method performs a bounded search over candidate configurations guided by power-performance models to identify the best configuration in each decision epoch [6]. These new workload-aware training examples serve as online queries from an expert policy, which are used to update DPM policy using online learning at runtime. However, this online-IL method incurs huge timing and energy overhead to perform the search over candidate configurations. Unlike existing online methods, UaOL provides reasoning mechanism to distinguish unseen system states (where online learning is needed) from seen system states (no benefit in learning).

To summarize, we improve state-of-the-art online learning of DPM policy for manycore platforms using the proposed UaOL framework. It is independent of the specific manycore architecture and type of unseen application. Moreover, it generates high quality DPM policy with significantly less EDP over existing online learning methods.

III. Problem Setup

We consider a manycore system with C cores (e.g., systems with 64 cores) divided into m VFIs where each VFI can operate at $(f_1, f_2, \cdots, f_k)$ V/F levels. We are given a set of application workloads to create an offline DPM policy to effectively bootstrap the initial online learning process. Our goal is to create high-quality DPM policy using online learning to adapt to unseen application APP to optimize EDP. In this work, we consider DPM policies represented as functions of the system state with parameters $\Theta \in R^b$ (e.g., multi-layer perceptron (MLP)). The online DPM policy π takes into account the current system state $\Phi(s)$ (e.g., performance counters and workload features) and produces a decision vector $\pi(\Phi(s), \Theta) = (d_1, d_2, \cdots, d_m)$, where each decision variable allocates the V/F for a single VFI and Θ stands for parameters. Given a manycore architecture, our goal is to learn the DPM policy parameters Θ to adapt to unseen application workload APP at runtime such that EDP is minimized without loss in performance. The system state is represented by the workload features such as each VFI's average and peak inter-VFI communication (or traffic), VFI's average and peak IPC, and VFI's V/F level for previous epoch, where each epoch constitutes of 10k instructions as an example. These features capture the average computation and communication patterns of the VFI, variance of the computation and communication patterns within the VFI, and use the contextual knowledge of the previous V/F prediction.

To adapt to the unseen application workload, we perform a bounded search guided by power-performance models for the optimized decision vector $(d_1, d_2, \cdots, d_m)$ corresponding to the features' values of the system state $\Phi(s)$ for the unseen workload. Next, we update the parameters of DPM policy in an online manner using such supervised training examples: $\Phi(s)$ is input and $(d_1, d_2, \cdots, d_m)$ is the output. UaOL employs CP to generate uncertainty sets $(PS_1, PS_2, \cdots, PS_m)$ for each VFI, where PS_m is a subset of k V/F levels such that the best V/F is present with high probability.

IV. Uncertainty-aware Online Learning Framework

In this section, we first provide a high-level overview of the proposed UaOL framework to create online dynamic power management policies. Subsequently, we describe the details of the key elements of the framework.

Algorithm 1. UaOL for uncertainty-aware online learning

Input: *ARCH* = target manycore system with C cores,
APP = application workload at runtime,
$\Phi(s)$= features of the system state s
$\pi(\Phi(s), \Theta)$= neural network based DPM policy with parameters Θ,
Output: parameters of the DPM policy optimized for *APP*

Initialize: b_0 = empty buffer; Θ_0=parameters of the DPM policy created offline; $D_{offline}$= input-output training data from the offline policy $\pi(\Phi(s), \Theta_0)$ and $t = 0$
Repeat:
  Execute current DPM policy and compute entropy (Equation 1)
  Update online power/performance models for *APP*
  If entropy $< e_{thresh}$, *(no online learning)*
  $D^* = f_k$ i.e., the V/F level with max probability from the distribution of $(f_1, f_2, \cdots, f_k)$ V/F levels.
  break; Go to next epoch
  else: *(perform online learning)*
  Generate uncertain V/F sets PS_m using conformal prediction (section IV B)
  Search for optimal V/F D^* in uncertain V/F sets PS_m by scoring V/F configurations with minimum EDP using power/perf models (section IV C)
  Store training example in a buffer
  $$b_{t+1} = b_t \cup \{\Phi(s)_t, D^*\}$$
  If buffer is full; reset the buffer.
  Supervised learning (SL) on training examples to update the parameters of DPM policy to Θ_{t+1}
  $t = t + 1$
Until end of *APP* execution
return parameters of the DPM policy optimized for *APP*

Overview of UaOL framework. We initialize the DPM policy using known applications at the design-time. This policy can be created offline using any existing method including RL and IL. Since the DPM policy is a mapping from system states $\Phi(s)$ to V/F levels for each VFI of the manycore platform, this knowledge can be useful with varying degree even for unseen applications seen at runtime. In each decision epoch at runtime, UaOL performs the following sequence of algorithmic steps. UaOL measures the uncertainty of the DPM policy to predict V/F level for a given system state. If the uncertainty measure is less than a pre-defined threshold, then UaOL determines that this is a previously seen system state and won't benefit from online learning. If UaOL determines this is an unseen system state, it performs online learning. It first computes uncertain V/F sets for each VFI using the CP method. CP ensures that the best V/F for each VFI is present in the corresponding uncertain V/F set (coverage) with a high probability. The size of uncertain V/F sets vary depending on the degree of difficulty in predicting V/F levels of unseen system states. UaOL restricts the search space of candidate V/F configurations to uncertain V/F sets and selects the best V/F configuration that minimizes EDP using the power-performance models. The newly created training examples of system state features (input) and best V/F configuration (output) is added to a buffer. If the buffer is full, UaOL updates the parameters of the DPM policy using the aggregated training data. Algorithm 1 shows the pseudocode of UaOL framework.

A. Uncertainty of DPM Policy via Entropy

We estimate the uncertainty of the DPM policy for system states of unseen application's features for two reasons. First, we want to search for the optimal V/F candidate configuration only if the policy is uncertain about $\Phi(s)$, i.e., has not seen the application's state features before. Intuitively, the distance between $\Phi(s)$ and features of the states seen in the past is large (aka out-of-distribution inputs). Second, we want to reduce the number of online policy updates since it adds penalty to the overall performance. The DPM policy generates a probability distribution over the k candidate V/F levels (via soft-max scores of MLP classifier). This categorical distribution is utilized to compute the entropy for m^{th} VFI using equation (1) at each decision epoch, where $\mathbb{P}(d_m = f_k \mid \Phi(s), \Theta)$ represents the probability of predicting k^{th} V/F level. We use the terminology of V/F levels throughout the paper to refer to classification labels in the ML literature. We choose entropy since it considers all V/F levels' probabilities to measure uncertainty, which is essential to automate UaOL for all unseen applications [11]. The higher the entropy value, more uncertain the DPM policy is about the application features in the ith decision epoch.

$$en_{i,m} = \sum_{j=1}^{k} \mathbb{P}(d_m = f_j | \Phi(s), \Theta) \ln \mathbb{P}(d_m = f_j | \Phi(s), \Theta) \qquad (1)$$

An entropy threshold (e_{thresh}) over the uncertainty estimate is employed to decide whether the policy is uncertain or not about a given system state based on the features.

B. Uncertain V/F Sets for VFIs via Conformal Prediction

Searching and evaluating all possible V/F configurations to find the one with least EDP for the unseen system state is prohibitive due to a very high number of configurations (k^m possible configurations) in every decision epoch. Therefore, UaOL utilizes CP [12] to generate adaptive size uncertainty sets over V/F levels to reduce the search space of V/F candidate configurations without loss in accuracy.

Overview of Conformal Prediction. CP is a framework for uncertainty quantification that provides formal guarantee for a user-specified coverage: optimal output is contained in the uncertainty set (subset of V/F levels) with a high probability $\alpha = 90\%$. Additionally, uncertainty sets from CP are adaptive and will reflect the difficulty of making predictions on unseen inputs: size of the uncertainty set will be large for difficult inputs and small for easy inputs. There are two key steps in CP. First, in the prediction step, we use a trained model (e.g., deep neural network) to compute *conformity scores,* which measure similarity between calibration examples and a testing input. Second, in the calibration step, we use the conformity scores on a set of calibration examples to find a threshold to construct uncertainty sets, which meets the coverage constraint (e.g., $\alpha = 90\%$). The key element of CP is a conformity scoring function $V(x, y)$, measures similarity between labeled examples, which is used to compare a given testing input to the calibration set. A typical method based on split CP has a threshold parameter τ to compute uncertainty sets for a given testing input and a machine learning classifier. A small set of calibration examples (different from training data) are used to select the threshold τ for achieving the given coverage α empirically on the calibration data. In classification tasks, we select the τ as $(1 - \alpha)$-quantile of conformity scores $V(x, y^*)$ on the calibration set, where y^* is the correct output for input x. The uncertainty set for a new testing input x is given by $\{y: V(x, y) \le \tau\}$.

Applying CP for Uncertain V/F Sets. In our problem setting each input x corresponds to features of the system state $\Phi(s)$, output comes from the set of k candidate V/F levels

$(f_1, f_2, \cdots, f_k)$, and uncertain V/F set is a subset of V/F levels for each VFI. To apply split CP method as explained above, we first use an existing ML method such as RL or IL to create a MLP classifier offline using the applications known at the design time. We use this trained MLP classifier to label each system state with the corresponding V/F level for each known application. This aggregate data is divided into training and calibration set. The set of calibration examples is used to configure CP method for a user-specified coverage α by selecting the appropriate threshold τ. We employ the recently proposed conformity scoring function based on ordered probabilities using the soft-max scores from the MLP classifier [7]. For a given input $\Phi(s)$, we get the sorted probabilities for all V/F levels using the trained MLP classifier $\pi(\Phi(s), \Theta)$ and the conformity score for $\Phi(s)$ and a candidate V/F level pair is obtained by summing all the probabilities up to the given V/F level.

One key challenge in applying CP to UaOL framework is the imbalanced nature of data, i.e., training and calibration data from offline applications have different proportions of examples for different V/F levels. In other words, there are more examples for some V/F levels (majority classes) and less examples for other V/F levels (minority classes). A naïve training and calibration process would lead to biased predictions and uncertain V/F sets to favor majority V/F levels over minority V/F levels. We overcome this challenge by performing oversampling of data from minority V/F levels to create a balanced data [13]. This transformed balanced data is used for both training and calibration purposes to produce V/F uncertainty sets at runtime for unseen system states.

C. Creation of Supervised Training Examples

Once we identify the uncertainty sets over V/F levels for VFIs at the uncertain system states, we perform bounded search over the candidate V/F configurations from these V/F uncertainty sets (only in those epochs where uncertainty exceeds entropy threshold) to select the best V/F configuration, which minimizes the EDP. To guide and perform this search, we use power/performance models [14] which are parametric functions of the performance counters (e.g., IPC, L2 cache miss, number of branch-mispredictions etc.) and are trained using a non-linear neural network regressor [15]. We evaluate and score the V/F configurations from the V/F uncertainty set for m VFIs $(PS_1, PS_2, \cdots, PS_m)$ to find the optimized configuration with minimum EDP. One challenge is that power/performance models are not known for the unseen application in advance. These models are learned and updated in each decision epoch, adapting to unseen application workloads. We employ performance counters of the unseen application's system state to update the parameters of power/performance models. Mean absolute percentage error (MAPE) loss of the neural network regressor remains within acceptable 3 % for power/performance models. We store each supervised training example in the form of features of the uncertain system state $\Phi(s)$ as input and best V/F configuration (one V/F for each VFI) minimizing EDP as output in a buffer. When the buffer is full, we perform online supervised learning to update the parameters of the DPM policy $\pi(\Phi(s), \Theta)$.

In summary, we improve power/performance models with new data from the unseen applications. The improved power/performance models create higher-quality supervised training examples for improving the DPM policy. The improvement in DPM policy would further help in better estimation of uncertainty and uncertain V/F sets for future system states of unseen applications.

V. Experiments and Results

A. Experimental Setup

Manycore platform. We employ GEM5 [16], a full-system simulator, to obtain detailed processor and network-level information. In all the experiments, we consider a system with 64 X86 cores running Linux within the GEM5 platform in full-system mode, noting that UaOL principles are applicable to higher core count as well. The MOESI_CMP_directory memory setup [16] is used with private 64-KB L1 instruction and data caches, and a shared 8-MB (128 KB distributed per core) L2 cache. We consider wireless NoC (Network-on-Chip) as the communication backbone for our manycore architecture [17]. We consider nominal range of operation in the 28-nm technology node. The adopted VFI DPM strategy uses following eight discrete V/F levels that maintain linear relationship (in Volts/GHz): 1.0/3.0, 0.95/2.75, 0.9/2.5, 0.85/2.23, 0.8/1.94, 0.75/1.64, 0.7/1.33, 0.65/1.02.

Benchmarks. Three SPLASH-2 [18] benchmarks (***set1***): FFT, LU, and WATER; and two PARSEC [19] benchmarks (***set2***): CANNEAL, FLUIDANIMATE (FLUID) are considered for experimental evaluation. These benchmarks represent varying compute- and memory-intensive workloads for manycore platforms. The performance counters generated by GEM5 simulations are given as input to McPAT [20] to determine the power values. At any time, one of the two sets is reserved for offline learning phase and other set is reserved for online learning. For example, for online evaluation of FFT benchmark (***set1***), offline DPM policy is learnt from applications in ***set2***. SPLASH-2 and PARSEC principally represent different types of workloads [19]. Hence, evaluation on ***set1*** while learning the policy parameters from ***set2*** and vice-versa is suitable for demonstrating the runtime adaptivity of UaOL.

VFI system. We consider four VFI clusters while imposing a minimum VFI cluster size of four cores. By using the k-means algorithm, we cluster the cores to minimize each VFI's intra-cluster variation in the time-varying computation and traffic statistics for a given set of known applications [17]. We learn the optimal DPM policy for the unseen application on the given VFI configurations learnt from a set of known applications. Please note that analysis of VFI clustering methods lies beyond the scope of this paper.

Decision space for DPM policies. The DPM decision space is defined by the number of VFIs and their respective V/F values. As we have four VFIs and eight V/F pairs, there are 4096 possible DPM decisions for each system state.

DPM policy representation. One function is used for each VFI to predict V/F values at each decision epoch using the following input features: each VFI's average and peak traffic, average and max IPC, and the previous epoch V/F level [8]. The MLP configuration used to represent each of the four VFI controllers are as follows: one input layer with the ReLU activation and an output layer with the softmax activation. The number of output layer neurons is equal to 8 i.e., number of discrete V/F levels.

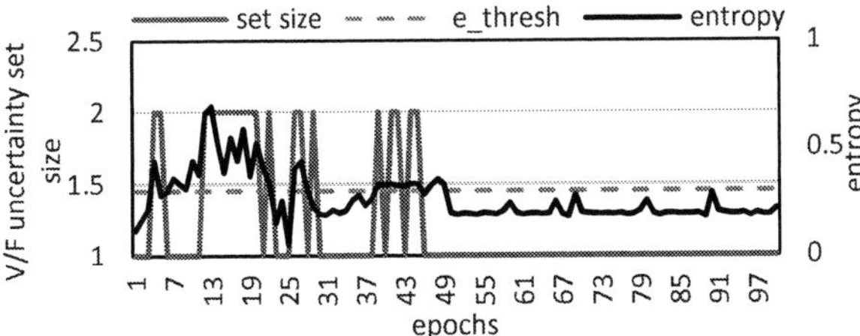

Fig.2: UaOL performance (entropy vs uncertainty set size) for FFT.

B. UaOL and Baseline Online DPM Methods

UaOL. Coverage and entropy threshold are the two critical parameters for UaOL. Coverage $\alpha = 0.1$ (i.e., 90% target coverage) is employed to configure CP for generating adaptive V/F uncertainty sets. Entropy threshold e_{thresh} is learnt from offline application set and is set to 0.3 to incur minimum policy updates while satisfying coverage α constraint. Power and performance models adapt to unseen application at runtime and are updated in each epoch by supervised learning (using non-linear neural network regressor with one hidden layer of size 10) on aggregated training examples [15].

Reinforcement learning. We use the state-of-the-art RL method, namely, proximal policy optimization (PPO) in our experiments. PPO is shown to achieve high accuracy for the learned policy [21]. We employ the same online policy representation as UaOL for actor-critic networks for PPO. Reward function $R(s,a)$ for given state s and action a, for two objectives (energy $E(s,a)$ and performance $T(s,a)$) is shown in equation (2). The scalarization parameter λ is varied in the range: [1,10,100,1000] to achieve the desired trade-off for each unseen application workload at runtime.

$$R(s,a) = E(s,a) + \lambda \cdot T(s,a) \quad (2)$$

Online-Imitation learning. Power/performance models are updated in each decision epoch with the same non-linear regression model as in UaOL. Supervised training examples are constructed using resource bounded search to find optimized V/F configuration in the local neighborhood of the predicted V/F decision by the policy. Online-IL performs learning in every decision epoch using training data from this search [6]. Online DPM policy updates using supervised learning employs the same policy function and parameters as UaOL.

C. Evaluation of UaOL's Performance

In this section, we evaluate UaOL's performance in terms of adaptivity of CP. As shown in Fig.2, UaOL computes entropy in each epoch to measure policy's uncertainty about system state. Policy is uncertain (entropy > e_{thresh}) about system states in the initial epochs of unseen benchmark and generates uncertainty sets of size 2 (as opposed to size 1 in subsequent epochs) to maintain target coverage α. This demonstrates that UaOL generates high quality (i.e., set size is 2 out of possible 8 V/F levels) and adaptive V/F uncertainty sets.

D. Comparative Performance Evaluation

Since EDP is a metric that captures both energy and execution time (performance) in one parameter, we use it as the relevant measure to evaluate the quality of DPM policies from UaOL, RL, and online-IL methods. EDP of the online policies

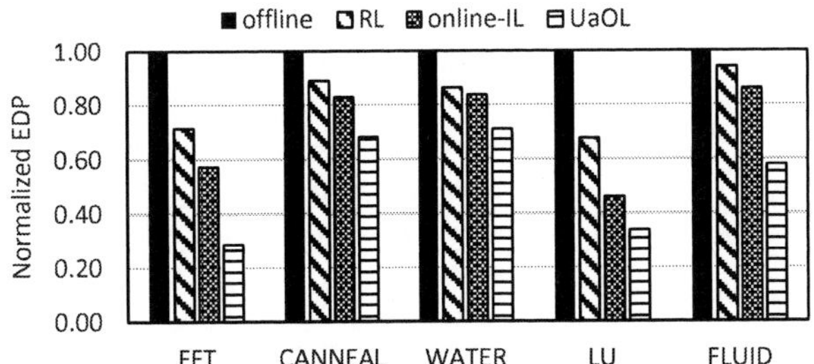

Fig.3: Comparison of EDP savings of various DPM policies on wireless NoC enabled manycore system for various benchmarks.

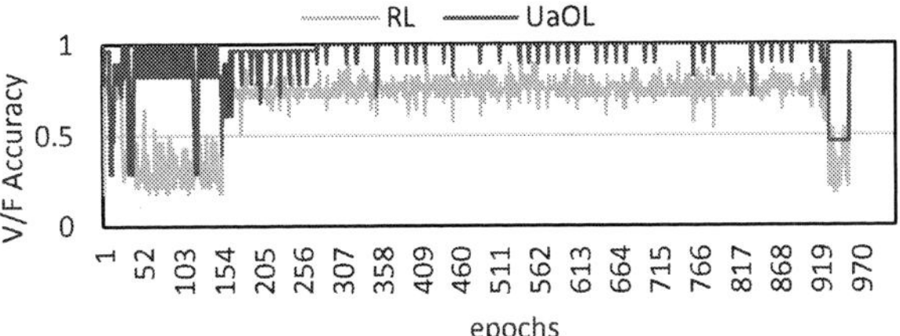

Fig.4: V/F prediction accuracy for RL and UaOL policies for CANNEAL

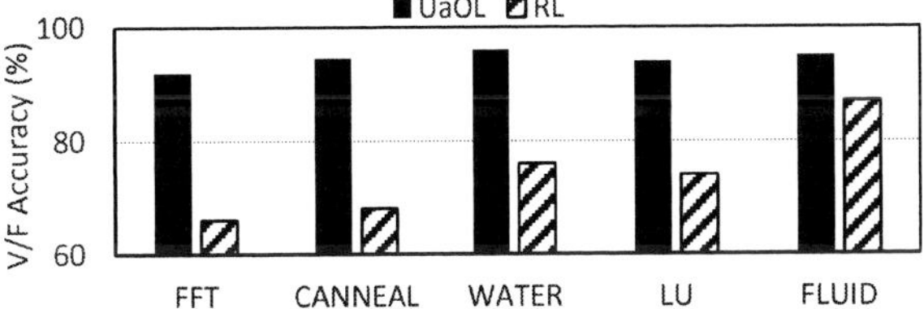

Fig.5: Accuracy comparison of UaOL and RL based online policies.

learnt with UaOL, RL, and online-IL are normalized with respect to offline policies. First, we demonstrate the need for online learning. As shown in Fig.3, DPM policies learnt offline and executing unseen applications have highest EDP (worst policies for unseen applications). Online DPM policies learnt by UaOL reduces EDP by 71% when compared to an offline policy for the FFT benchmark, thereby showing the need of online learning. UaOL policies are sample-efficient when compared to RL and adapt better to unseen applications at runtime. Therefore, RL policies are sub-optimal in terms of EDP savings and UaOL reduces EDP by up to 60% compared to RL. Since online-IL has high execution overhead due to V/F search and larger number of policy updates, UaOL reduces EDP by up to 50% compared to online-IL. All these results demonstrate that online DPM policies from UaOL method adapt better to unseen applications and result in least EDP across all benchmarks.

E. Accuracy Comparison with RL

We evaluate online DPM policies learnt by RL method against proposed UaOL framework in terms of sample-efficiency. For this evaluation, we compare the accuracy of predicted V/F levels of both RL and UaOL with respect to the best offline policy for the application at hand using equation (3), where f_π, f_{ref} is V/F level predicted by online and reference offline policy respectively.

$$Accuracy = 100 \times \left(1 - \frac{|f_\pi - f_{ref}|}{k-1}\right) \quad (3)$$

Fig.4 shows that the online policy learnt from UaOL achieves better accuracy than RL throughout the application's

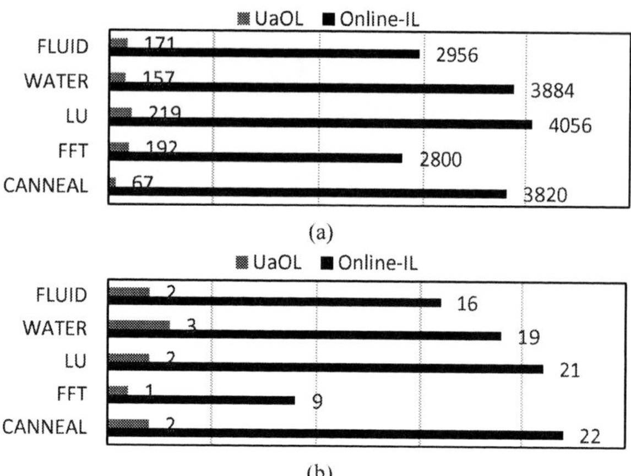

Fig.6: Comparison of (a) total number of V/F search steps and (b) number of online policy updates over applications' lifetime.

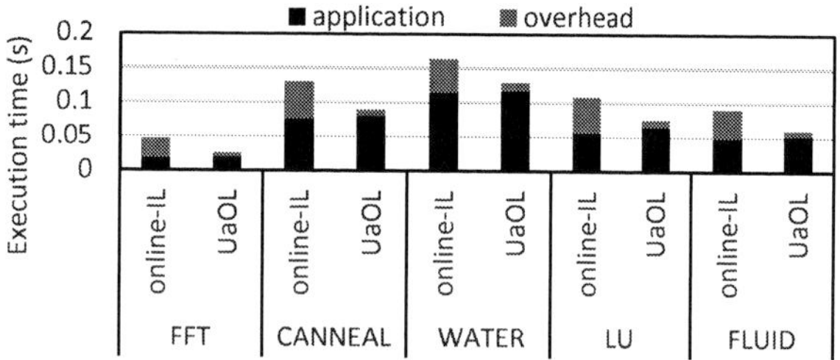

Fig.7: Timing overhead comparison for online-IL vs UaOL

lifetime for the CANNEAL benchmark as an example. We show that UaOL policy is sample-efficient as it takes 100 epochs fewer than RL to converge to 80% accuracy. We see a similar trend for rest of the benchmarks. In Fig.5, we show accuracy (averaged over N decision epochs) for all the benchmarks. For example, RL V/F accuracy drops up to 66 % compared to 92% for UaOL policy for FFT.

F. Overhead Considerations

UaOL performs online learning (including optimal V/F configuration search, power/performance model update and policy update) in only 0.57% of the decision epoch time. We compare the performance (execution time) overhead of online learning with online-IL and UaOL methods. In Fig.6, we show that the total number of V/F search steps for finding optimized V/F configuration in CANNEAL's lifetime using online-IL method is 57 × higher compared to UaOL counterpart. Similarly, number of epochs where online policy is updated in online-IL are up to 10× higher compared to UaOL method. This leads to higher execution time in online-IL as shown in Fig.7.This highlights the significance of uncertainty-aware search and online learning for DPM policy updates.

CONCLUSION

Due to exponential growth in new types of applications, designing online dynamic power management (DPM) policies for unseen applications at runtime is critical. The proposed framework UaOL performs uncertainty-aware, low-overhead online learning using conformal prediction (CP) to generate high quality DPM policies. We show for a 64 core manycore system that online DPM policies learnt using UaOL reduce energy-delay product (EDP) by up to 50% and 60% compared to baseline online-Imitation learning (IL) and reinforcement learning (RL) respectively. We further show that UaOL is sample efficient (contrast to RL), i.e., takes up to 3 × less samples to converge to near-optimal policy and reduces the performance overheads (contrast to online-IL) by up to 57× less V/F search steps and 10× fewer online policy updates.

REFERENCES

[1] C. Zhiwei et al., "Towards a systematic survey for carbon neutral data centers," in *IEEE Communications Surveys & Tutorials*, 2022.

[2] U. Ogras et al., "Design and management of voltage-frequency island partitioned networks-on-chip," *TVLSI*, vol. 17, no. 3, pp. 330-341, 2009.

[3] M. Rapp et al., "NPU-Accelerated Imitation Learning for Thermal- and QoS-Aware Optimization of Heterogeneous Multi-Cores,," in *DATE*, 2022.

[4] S. Pagani et al., "Machine learning for power, energy, and thermal management on multicore processors: A survey.," *TCAD*, vol. 39.1, pp. 101-116, 2018.

[5] L. chen et al., "Improve the Stability and Robustness of Power Management through Model-free Deep Reinforcement Learning," in *DATE*, 2022.

[6] S. K. Mandal et al, "An energy-aware online learning framework for resource management in heterogeneous platforms," *TODAES*, vol. 25.3, pp. 1-26, 2020.

[7] A. Anastasios N. et al, "A gentle introduction to conformal prediction and distribution-free uncertainty quantification," in *arXiv preprint arXiv:2107.07511*, 2021.

[8] R. G. Kim et al., "Imitation learning for dynamic VFI control in large-scale manycore systems," *TVLSI*, vol. 25, no. 9, pp. 2458-2471, 2017.

[9] V. Mnih et al., "Human-level Control Through Deep Reinforcement Learning," pp. 529-533, 2015.

[10] S. K. Mandal et al., "Dynamic Resource Management of Heterogeneous Mobile Platforms via Imitation Learning," *TVLSI*, vol. 27, no. 12, pp. 2842-2854, 2019.

[11] W. Keze et al., "Cost-effective active learning for deep image classification," *TCSVT*, vol. 27, no. 12, pp. 2591-2600, 2016.

[12] A. Anastasios et al., "Uncertainty sets for image classifiers using conformal prediction," in *arXiv preprint arXiv:2009.14193*, 2020.

[13] G. Lemaître et al., "Imbalanced-learn: A python toolbox to tackle the curse of imbalanced datasets in machine learning," *JMLR*, vol. 18, no. 1, pp. 559-563, 2017.

[14] B. Su et al., "PPEP: Online Performance, Power, and Energy Prediction Framework and DVFS Space Exploration," in *MICRO*, 2014.

[15] P. Fabian et al., "Scikit-learn: Machine learning in Python.," *JMLR*, vol. 12, pp. 2825-2830, 2011.

[16] N. Binkert et al., "The gem5 simulator," *ACM SIGARCH Computer Architecture News*, vol. 39, no. 2, pp. 1-7, 2011.

[17] R. G. Kim et al., "Wireless NoC and dynamic VFI codesign: Energy efficiency without performance penalty.," *TVLSI*, vol. 24.7, pp. 2488-2501., 2016.

[18] S. C. Woo et al., The SPLASH-2 programs: Characterization and methodological considerations, ISCA, 1995.

[19] C. Bienia et al., "Parsec vs. splash-2: A quantitative comparison of two multithreaded benchmark suites on chip-multiprocessors," in *IISWC*, 2008.

[20] S. Li et al., "McPAT: An integrated power, area, and timing modeling framework for multicore and manycore architectures," in *MICRO*, 2009.

[21] J. Schulman et al, "Proximal Policy Optimization Algorithms," *CoRR*, vol. abs/1707.06347, 2017.

A Multicore GNN Training Accelerator

Sudipta Mondal, Ramprasath S., Ziqing Zeng, Kishor Kunal, and Sachin S. Sapatnekar
University of Minnesota, Minneapolis, MN, USA

Abstract—**Graph neural networks (GNN) are vital for analytics on real-world problems with graph models. This work develops a multicore GNN training accelerator and develops multicore-specific optimizations for superior performance. It uses enhanced multicore-specific dynamic caching to circumvent the costs of irregular DRAM access patterns of graph-structured data. A novel feature vector segmentation approach is used to maximize on-chip data reuse with high on-chip computation per memory access, reducing data access latency, using a machine learning model for optimal performance. The work presents a major advance over prior FPGA/ASIC GNN accelerators by handling significantly larger datasets (with up to 8.6M vertices) on a variety of GNN models. On average, training speedup of 17× and energy efficiency improvement of 322× is achieved over DGL on a GPU; a speedup of 14× with 268× lower energy is shown over GPU-based GNNAdvisor; and 11× and 24× speedups are obtained over ASIC-based Rubik and FPGA-based GraphACT.**

I. Introduction

In recent years, graph neural networks (GNNs) have achieved unprecedented success on many real-life problems (recommender systems, IC design, embedded sensing, e-commerce, etc.). To accelerate the GNN inference several works have been proposed (GNNIE [1], HyGCN [2], AWB-GCN [3], GNNerator [4], BlockGNN [5], DyGNN [6], BoostGCN [7], etc.) for small- to medium-scale graph workloads. However, a well-trained model is a prerequisite for efficient inference.

Energy-efficient and scalable acceleration of GNN training is an open problem that involves several major challenges:
(i) *High computation and communication costs*: GNN training is more compute-intensive than inference, especially with backpropagation, and incurs high access time and energy costs for communication between memory and on-chip buffers;
(ii) *Scalability for large graph sizes*: Graph sizes in real-world datasets have grown exponentially in recent years [8], necessitating multiple accelerator engines to work together;
(iii) *Load balancing during computation*: High and variable input feature vector sparsity, high adjacency matrix sparsity, and power-law distributions of vertex degrees, result in irregular and random memory accesses during GNN computations, with low utilization of processing elements [1]–[3].
(iv) *Versatility*: A GNN training accelerator must be able to accommodate a wide range of GNN architectures.

GPU-based solutions are energy-inefficient. GNNAdvisor [9], a single-GPU solution is limited to small-to-medium-sized graphs. Multi-GPUs platforms can handle large graphs: RoC [10] uses dynamic techniques for graph partitioning and memory management; NeuGraph [11] employs 2-D graph partitioning and inter-GPU vertex-chunk swapping (with increased communication overhead); PaGraph [12] replicates boundary vertices to reduce communication among partitions, but faces scalability issues due to replica synchronization.

Several FPGA- and ASIC-based accelerators with better energy efficiency have been proposed. Among FPGA-based approaches, GCoD [13] implements algorithm-accelerator co-design, but requires large on-chip buffers due to scatter-based aggregation and incurs high preprocessing overhead for sparsification and polarization; GraphACT [14] proposes a CPU+FPGA platform, with graph sampling and loss gradient calculation offloaded to the CPU, and forward- and back-propagation handled in the FPGA. Among ASIC-based approaches, Rubik [15], uses a hierarchical array of processing elements; GNNear [16] uses an ASIC-based central acceleration engine for some computations and offloads others to near-memory processing engines that reside in the buffer chips of DIMMs. As single-core structures, these methods are not scalable for larger graphs; they largely neglect input feature vector sparsity and power-law degree distribution problems.

Any single-core solution has limited scalability. We accelerate large GNN training by moving past the limitation of single cores and using an *array* of processing cores for training, offering substantial speedup and energy-efficiency improvements. We target much larger graphs than previous ASIC/FPGA training accelerators (we show results on datasets with up to 8.6M vertices in Section VI). We believe this is the first multicore GNN training accelerator to support a wide range of GNNs; the only other multicore accelerator [17] known to us handles inference only and not training. The existing multicore inference accelerators can not handle backpropagation efficiently due to: (i) massive computation/communication overhead for the calculation/propagation of error gradients. (ii) large gradient synchronization overhead. (iii) lack of support for various special functions, e.g., log and softmax.

For the core, we choose the GNNIE inference accelerator [1] over other candidates [2]–[7] as it can handle sparsity in input vertex feature vectors and adjacency matrix, support a wide range of GNN topologies (e.g., GCN, GraphSAGE, GAT, GINConv), and shows speedup and efficiency advantages over other methods. However, simply arraying a set of GNNIE cores leads to performance bottlenecks due to: (i) suboptimality in GNNIE's caching scheme in a multicore scenario; (ii) lack of multicore-specific optimizations that consider both DRAM accesses and inter-core communication. We develop novel techniques to address these challenges and develop methods that are *scalable* for training large graphs. Degree-Quant [18] proposes integer-based GNN training and we leverage this in our implementation. Our contributions are:

- A novel *feature vector segmentation* scheme that reduces memory accesses, and a random-forest-based machine learning (ML) model for optimal segmentation.
- *Multicore-specific graph-specific caching* with reduced random DRAM accesses and limited on-chip communication.
- *Demonstrated gains in scalability, speedup, and energy*

979-8-3503-1176-1/23 $31.00

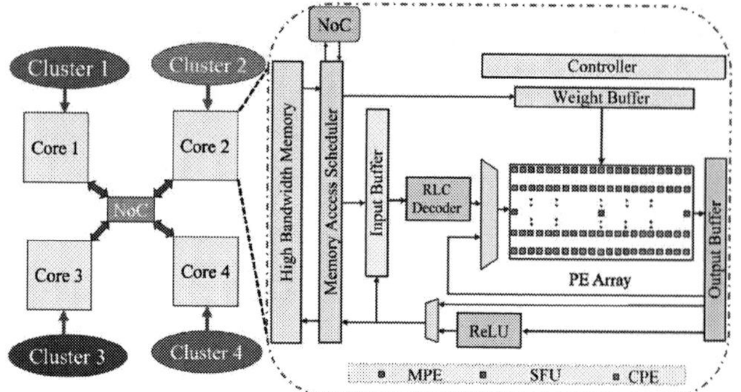

Fig. 1. Block diagram of the proposed architecture (core architecture in inset) with 4 cores; our evaluation considers accelerators with up to 36 cores.

efficiency over prior GPU/FPGA/ASIC solutions across multiple GNN topologies.

II. GNN Training Steps

GNN training involves a *forward pass* similar to inference, and a *backward pass* that feeds gradients back to update weights.

Forward Pass Computations. The forward pass has two steps [1], [2], [16]: ***(a) Weighting*** in layer l multiplies feature vector $\mathbf{h}_i^l$ (dimension F^l) of each vertex i by a weight matrix, W^l (dimension $F^{l-1} \times F^l$). ***(b) Aggregation*** for vertex i combines (sum/max/mean/pool) the weighted feature vectors in a set $\mathcal{N}_i$. For GCN/GAT/GINConv, $\mathcal{N}_i$ is the neighbors $N(i)$ of i; for GraphSAGE, $\mathcal{N}_i$ randomly samples $N(i)$.

Backward Pass Computations. The output node features of the forward pass are compared against the ground truth to compute the loss function. Then, starting from the last layer, the gradients of the loss with respect to the feature vectors and weights are calculated, and weight updates are performed at each layer using the chain rule until the input layer is reached. Backward pass computations consist of Weighting and Aggregation steps similar to the forward pass, and MAC operations for loss computations and gradient updates.

III. Multicore Architecture and Computations

Architecture. Our GNN training engine, shown in Fig. 1, has multiple cores connected by a network-on-chip (NoC).

GNNIE core [1]. A GNNIE core (inset of Fig. 1) consists of an $M \times N$ array of computational PEs (CPEs) for ALU computations; merge PEs (MPEs) within the CPE array that aggregate partial results in their CPE column during Weighting; and special function units (SFUs) for nonlinear functions. Three on-chip buffers cache the input, weight, and output data. The controller for each core orchestrates operations in the PE array (workload reordering for CPEs, sending partial results to MPEs). The memory access scheduler from [19] is modified to handle memory requests from both DRAM and NoC.

Partitioning. For a multicore training engine with m GNNIE cores, the input graph is partitioned into m clusters, and each cluster is the workload for one core. *Intra-cluster edges* are connections between vertices ("*intra-cluster vertices*") within a cluster and can be processed entirely within a core; *inter-cluster edges* connect vertices in the cluster to vertices in another cluster ("*inter-cluster vertices*"). We preprocess the graph with METIS [20] to create clusters that (a) are balanced, i.e., have roughly equal numbers of intra-cluster vertices, (b) have a minimal number of inter-cluster edges.

Weighting. Weighting is performed separately on the feature vectors of each vertex, and can be carried out independently in each core, with no inter-vertex/inter-core communication. The matrix-vector multiplication computations in this step are very structured, but input vector sparsity variations can lead to load imbalance during this computation. These issues are tackled using the workload imbalance strategies using GNNIE's flexible MAC architecture and load redistribution [1].

Aggregation. Aggregation consolidates data from the neighbors of each vertex, and may involve *intra-cluster and inter-cluster* edges. For most GNNs (GCN/GINConv/GraphSAGE), this involves summation, but GATs require nonlinear computation of attention coefficients. Aggregation for each vertex in a cluster is performed on its own core, with no synchronization. Operations on inter-cluster vertices, fetched via NoC from the buffers of other cores, are read-only because there are no data dependencies between operations in the same layer.

Ultra-high sparsity of the adjacency matrix and power-law behavior incur numerous irregular and random memory accesses even on one core; this is exacerbated for large graphs on multiple cores by heavy and irregular *communication between cores* due to long feature vectors, limited NoC bandwidth, and small on-chip buffers. We will address novel methods for overcoming these bottlenecks in Sections IV and V.

Dynamic caching. Since the data for each cluster is too large for the cache (input buffer), GNNIE uses a dynamic caching scheme to fetch vertex data from the DRAM. It processes a subgraph of the cluster, called the **computational subgraph**, which is the subset of intra-cluster and inter-cluster vertices of a core currently in the cache, and edges between these vertices.

For our multicore training engine, the intra-cluster vertices of a core and their edge data are stored in CSR format in the DRAM. For intra-cluster vertices, this data for a computational subgraph is fetched into the input buffer from DRAM (*off-chip communication*), and for inter-cluster vertices, the data is fetched from the input buffers of other cores (which are responsible for DRAM fetches) via the NoC (*on-chip communication*). Within each core, the CPEs process the computational subgraph using efficient load-balancing techniques [1].

IV. Dynamic Cache Replacement Policy

We first review the hardware-centric graph-specific caching technique in [1] for the GNNIE inference engine. To maximize cache data reuse, the number of unprocessed edges, e_v, for each vertex v is tracked during Aggregation. Since nodes with larger e_v are involved in a larger number of future computations, they are more likely to be reused; hence they are prioritized for retention in the cache. Specifically, a node is replaced in the cache if $e_v \leq \gamma$, where γ is a threshold.

This strategy is used to promote cache reuse and minimize DRAM fetches. The graph undergoes lightweight preprocessing to store the vertices in descending order of degree in DRAM (initially, $e_v = d_v$, the degree of vertex v). The cache is initially populated with the first set of DRAM blocks with the highest degrees. An iteration is completed when all edges of the computational subgraph in the cache are processed. At this time, the set of cache blocks that meet the replacement criterion are evicted and the next sequential set of DRAM blocks (the DRAM is stored in degree order) is brought into cache. Multiple iterations are needed until all edges are traversed. By construction, DRAM fetches exclusively use sequential blocks, avoiding expensive random access. All random accesses are limited to the on-chip SRAM cache, which is much more inexpensive than random DRAM access.

Multicore-specific Graph-specific Caching. The direct application of the graph-specific caching scheme of [1] to large graphs in the multicore scenario results in bottlenecks related to *stagnation* (described next), and requirements for increased retention of inter-cluster vertices whose data must be sent over the NoC to other cores. We alter the scheme, using new methods that use dynamic thresholds to prevent stagnation.

In the multicore scenario, the preprocessing step stores each cluster of the graph (instead of the entire graph) in degree order. The retention requirements for intra-cluster and inter-cluster vertices are different. Due to the min-cut objective function of clustering, intra-cluster vertices in a core tend to have higher connectivity within the cluster, while inter-cluster vertices are connected to fewer vertices. Using the same γ threshold for both types of vertices would disadvantage inter-cluster vertices, which might then require frequent fetches across the NoC, with high latency and cost overheads. Therefore, separate thresholds γ_{intra} and γ_{inter} are required for retaining intra-cluster and inter-cluster vertices, respectively.

As the distribution of vertex degrees varies across clusters, the values of these γ parameters must be cluster-specific, and using a uniform value of these γ variables for all clusters is inefficient. For each core, we set γ_{intra} (γ_{inter}) to a certain percentile value, κ_{intra} (κ_{inter}) of the degree distribution of intra-cluster (inter-cluster) vertices of the cluster assigned to the core. Empirically, we find that setting κ_{intra} and κ_{inter} to the 50th percentile of the intra-cluster and inter-cluster vertex degree distribution, respectively, is a good choice. Our choice of γ parameters based on the above criterion also makes the approach generalized (i.e., not tuned for a particular dataset).

We track the unprocessed intra-cluster (inter-cluster) edges of an intra-cluster (inter-cluster) vertex through a simple decrement operation, i.e., whenever an intra-cluster (inter-cluster) edge of an intra-cluster (inter-cluster) vertex is processed in the CPE array, the controller decrements the intra-cluster (inter-cluster) edge count of the vertex by 1. Then, based on γ_{intra} (γ_{inter}) cache replacement is performed. Fetch operations for the next set of intra-cluster and inter-cluster vertices via off-chip and on-chip communication for the next subgraph are overlapped with the computation in the CPE array.

Dynamic Thresholds for Preventing Stagnation. Intra-cluster (inter-cluster) stagnation occurs when the number of cached intra-cluster vertices that meet the eviction criterion based on γ_{intra} (γ_{inter}) is small, as the changes in the computational subgraph across iterations are minor. This results in low computation and low PE utilization per iteration.

We define the metric $e_{intra}[i]$ ($e_{inter}[i]$) as the ratio of the number of intra-cluster (inter-cluster) edges processed up to iteration i, to the total number of intra-cluster (inter-cluster) edges of the cluster associated with the core. After a *detection interval* of every I iterations, we detect stagnation as:

$$e_{intra}[i] \leq (1+\delta)e_{intra}[i-I], \quad e_{inter}[i] \leq (1+\delta)e_{inter}[i-I]$$

where δ is a user-defined threshold. If this is satisfied, we

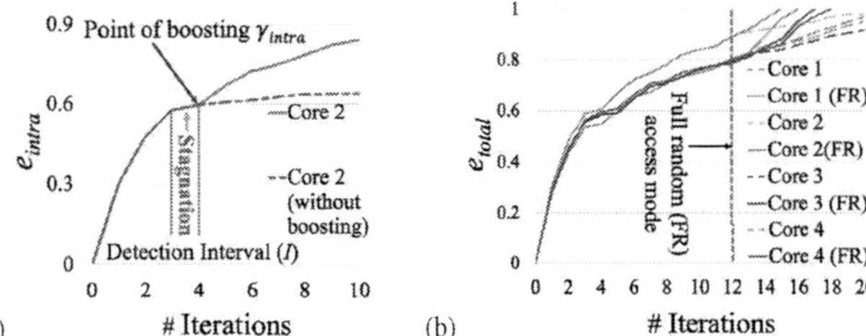

Fig. 2. (a) Boosting γ_{intra} to break intra-cluster stagnation on Core 2. (b) Invoking full random access after most edges are processed on all cores.

boost the relevant γ to κ_{boost}^{th} percentile of the vertex degree distribution. After one iteration with the boosted value evicts numerous vertices and overcomes stagnation by changing the computational subgraph, we revert to the original γ.

We tune the parameter values over a range of datasets. Varying $\kappa_{boost} \in [70, 95]$, the optimal value was found to be $\kappa_{boost} = 90$; varying $I \in [1, 15]$, an optimum was found at $I = 5$; varying $\delta \in [0.01, 0.1]$ yielded the best value of $\delta = 0.05$. For the amazon0601 dataset on a 4-core system, Fig. 2(a) shows the change in e_{intra} with each iteration and shows regions of stagnation that is found after the detection interval of I. At this point, γ_{intra} is boosted, and as shown in Fig. 2(a) this increases the rate of progress of e_{intra} (the dotted line shows the slower trajectory of e_{intra} without boosting).

As the Aggregation computation nears completion, when a large fraction of edges has been processed, it becomes increasingly difficult to find unprocessed edges in the computational subgraph. To detect this, we monitor e_{total}, the ratio of the number of intra-cluster/inter-cluster edges processed up to iteration i, to the total number of edges in the cluster. When e_{total} exceeds a threshold, we move to full random access: the cache now has random access to the DRAM to complete Aggregation. This is shown in Fig. 2(b), where the original trajectory (dotted lines), is accelerated to faster completion (solid lines). At this stage, the number of random DRAM accesses is relatively small and the benefit of faster convergence outweighs the cost of slower random DRAM accesses during this final phase. We find the e_{total} threshold values of 0.8 to be optimal over a range of datasets.

V. Scaling on Large GNNs

A. *Bottlenecks of Scaling on Large GNNs*

While graph-specific caching significantly improves latency and power/energy due to increased data reuse, the benefits of this approach face bottlenecks due to *fundamental limitations* in the traditional structure of GNN computations. Since node feature vectors for each node can be long and the input buffer size is small, the computational subgraph in each iteration constitutes a very small fraction of the total number of vertices in each cluster. Thus, only a small fraction of edges can be processed in each iteration, leading to high rates of cache replacement and slow convergence. Switching to full random access mode can overcome this issue, with significant costs due to the high energy of random DRAM access. The problem becomes more acute for large GNNs that require more cores: with more cores, more inter-cluster vertices are sent over NoC, leading to higher injection rates and/or larger packet sizes. This increases NoC latency, worsening performance.

The key to overcoming this problem is to increase the size of the computational subgraph during Aggregation, subject to the cache size. We achieve this by proposing *feature vector segmentation*, splitting a vertex feature vector into multiple segments, processing one segment at a time. We show that the choice of segment size involves balancing off-chip and on-chip communication latency as we seek to efficiently overlap computation with communication for high performance.

B. Feature Vector Segmentation

During Aggregation, there is no dependency between operations in different elements of the feature vector of a vertex. Therefore, Aggregation over the neighborhood for each feature vector segment can be carried out independently. We develop the concept of feature vector segmentation under a fixed cache size, illustrated in Fig. 3. The conventional approach at left ("Full") uses full feature vectors of length F. For a cache size of C, the number of vertex feature vectors that can fit in the cache is roughly $n_F = C/F$, and this limits the size of the computational subgraph. We can increase the subgraph size by using a *subset* of the entire feature vector. If we split the feature vector into two segments ("2-segments," middle), we can fit a subgraph of $2n_F$ vertices in the cache. For the j-segments case (right), where each segment length is $q = \lceil F/j \rceil$, we increase the size of the computational subgraph by a factor of j relative to the "Full" case.

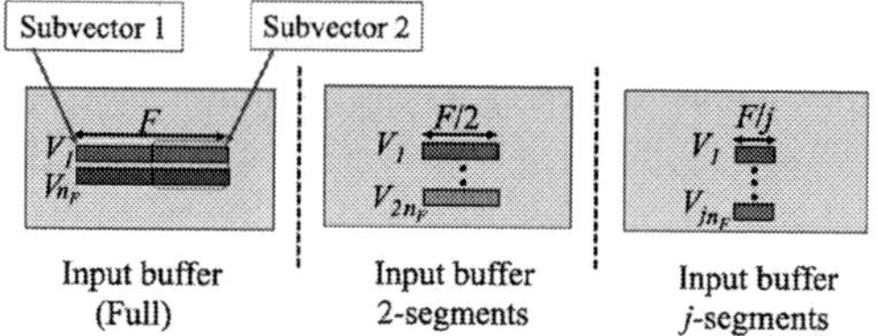

Fig. 3. Feature vector segmentation.

Using j segments, Aggregation operations are performed on one feature vector segment at a time, over all nodes, using the graph-specific caching method of Section IV. For larger j, the computational graph in each iteration is larger, and more edges are available for Aggregation, so that CPEs in each core are kept busy. However, as j increases, more vertices fit into each core and have more neighbors in other cores. Hence, traffic in the NoC also increases as more vertices are sent to other cores, increasing the injection rate and slowing communication.

A few prior approaches have used concepts similar to segmentation, but have significant limitations: our solution gains efficiency by exploiting segmentation in harmony with other schemes that reduce cache access latencies, including graph-specific caching and on-chip fetches from other cores using the NoC. P^3 [21] uses a superficially similar method that segments the feature vector into $\lceil F/c \rceil$ segments, where c is the number of GPU cores. This accommodates the entire graph into each core, but their power-hungry GPU-based solution requires a much larger cache than our power-efficient ASIC solution. BoostGCN [7] and GNNerator [4] implement 2-D graph partitioning and use feature vector segmentation to increase the number of vertices in each partition so that the frequency of DRAM communication decreases. However, as shown in [1], the lower bound of DRAM access for 2-D partitioning is always higher than the graph-specific caching of GNNIE. In addition, while BoostGCN and GNNerator rely solely on DRAM communication to fetch neighboring vertices, we fetch neighbors from DRAM (off-chip communication) or from the cache of other machines (on-chip communication); since DRAM accesses are more costly than on-chip communication, our approach is more efficient.

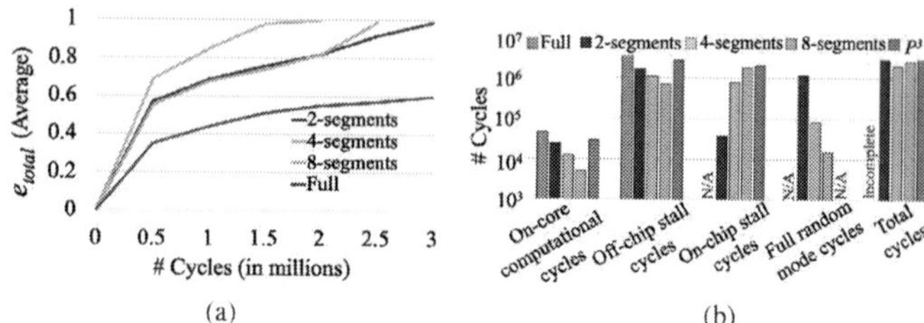

Fig. 4. Performance analysis of feature vector segmentation: (a) e_{total} (Average) vs. Execution Cycles (b) Aggregation cycle comparison.

Performance Analysis. Fig. 4 shows the results of combining feature vector segmentation with dynamic graph-specific caching (Section IV), showing the number of cycles required for Aggregation in the first GCN layer for the amazon0601 dataset (403,394 vertices/3,387,388 edges) on a 4-core system. The results are shown for $j = 1$ (Full), 2, 4, and 8.

Fig. 4(a) shows the progress in processing graph edges as the execution progresses, showing the fraction e_{total} of all edges that are processed (averaged over all cores). For the Full case, e_{total} rises very slowly and does not reach the threshold of 0.8 required to transition to full random access mode. The 2-segments approach progresses faster, and the 4-segments approach is still faster; however, increasing to 8-segments slows down the progress of e_{total}. We will understand this trend based on Fig. 4(b), which shows the total number of cycles for the computation, and the components of this total.

Fig. 4(b) shows improvement in *on-core computation* cycles from the Full to the segmented cases with larger j. Compared to the Full approach, the 2-, 4-, and 8-segment cases reduce *off-chip stall cycles* by 60%, 74%, and 83%, respectively: increasing j reduces DRAM access frequency as the computational subgraph becomes larger. While for the Full case the computational subgraph, on average, contains just 8% of the vertices in a cluster; for 2, 4, and 8 segments, the fraction rises to 16%, 32%, and 64%, respectively. The number of *full random mode cycles* reduces as j increases, because the computation subgraph grows larger as j increases and fewer edges are unprocessed at the switch to full random mode.

Under segmentation, the NoC injection rate increases with higher j due to the increased size of the computational subgraph on each core, which results in the transmittal of multiple feature vector segments of inter-cluster vertices over the NoC. However, since individual segments are smaller, the message size is reduced. This tradeoff between the increased injection rate and the reduced message size implies there is an optimal j for which the on-chip stall cycles are minimized. We develop an ML model to optimize j. The impact of stalls can be further mitigated by overlapping on-chip and off-chip communication during each iteration. In particular, no on-chip stall cycles are required for the Full approach, because on-chip communication per iteration is so low (due to a small number of cached vertices) that it can be completely overlapped.

Fig. 4(b) also shows that the *on-chip stall cycles* increase from 2- to 4- to 8-segment cases. Hence, the total cycle count required to complete the computation has its minimum for the 4-segments case. This explains the trend in Fig. 4(a).

ML Model for Optimizing j. To optimize the number of segments j, we trained a machine learning (ML) model using a random forest (RF) regressor. This trained ML model is then used on unseen graphs. Input parameters include graph attributes (vertex and edge counts, a power-law metric capturing the fraction of edges adjacent on the top 10 percentile of high-degree vertices, and the number of cores). The total number of samples corresponds to 144 synthetic graphs, ranging from 100K–10M vertices and 200k–100M edges, and with power-law metric from 0.27–0.95. A train/test split of 80/20 was used. The RF regressor used 100 estimators, and the model achieved 95% training and test accuracy based on R^2 score.

On real datasets, the model prediction was close to the results of a much more costly enumeration of all $j \in [2, 16]$. For example, the optimal j predicted by the RF model is 5 for amazon0505 and amazon0601 datasets, close to optimal enumerated value of 4 ; the prediction for com-amazon dataset is $j = 4$, which matches the optimum from enumeration.

Comparison with P^3. Fig. 4(b) also show results for P^3 [21], which switches between a segmented method (model parallelism, across feature vector segments) to the Full method (data parallelism, across cores) after the first layer, incurring a large communication overhead due to a burst of communication at that stage; by its very nature, this step provides no opportunities for overlapping off-chip and on-chip communication. Hence, it incurs high on-chip stalls: 89%, 63%, and 7% higher on-chip stall cycles compared to 2-, 4-, and 8-segments, respectively, in our implementation of their method.

PE Utilization. The segmented approach also leads to higher PE utilization compared to full-length feature vectors. The larger computational subgraph size allows more edges to be processed per iteration, increasing the computational intensity of data fetched from DRAM and keeping GNNIE PEs busy. For the first layer of Aggregation of amazon0601 for GCN, the average PE utilization for Full, 2-segments, 4-segments, and 8-segments are 67%, 86%, 100%, and 100%, respectively.

VI. Evaluation

Hardware/Simulation Setup. Each core is implemented in Verilog, synthesized with Synopsys DC in a 12nm standard VT library, placed and routed using Innovus, and verified via RTL simulations. The area, energy, and latency of on-chip buffers are estimated using CACTI 6.5 [22]. Post-P&R metrics for each core are: 4.97mm^2, 0.93W, 934 MHz. The controller has 0.26 mm^2 area and 0.1W power. For the NoC, latency and throughput were analyzed using BookSim2 [23], and power and area using Orion3.0 [24]. The NoC power overhead ranges between 2.9%–6.3% of the total chip power. An in-house simulator computes the execution cycles for our accelerator, with Ramulator [25] modeling off-chip HBM access (256 GB/s, 3.97pJ/bit [26]).

Configuration of the Multicore Accelerator.

Individual GNNIE cores Configuration per core is as follows: *Buffer sizes*: Output: 1MB; Weight: 128KB; Input: 512KB *CPE array with flexible MACs*: 16×16 array; 4 MACs (rows 1–8), 5 MACs (rows 9–12), 6 MACs (rows 13–16).

NoC Buffer size: 128 KB, 4 links per router, 50GB/s BW/link.

Table I: Type A datasets (DD: D&D, TW: TWITTER-Partial, YT: Yeast, SW: SW-620H, OV: OVCAR-8H)

Dataset	Vertices	Edges	(FL, m, j)
DD	335K	1.7M	(89, 2, 4)
TW	581K	1.4M	(1323, 4, 2)
YT	1.7M	3.6M	(74, 16, 2)
SW	1.9M	3.9M	(66, 16, 2)
OV	1.9M	3.9M	(66, 16, 2)

Table II: Type B datasets (SB: soc-BlogCatalog, CA: com-amazon, A-05: amazon0505, A-06: amazon0601, EN: enwiki, A-8M: amazon8M)

Dataset	Vertices	Edges	(FL, m, j)
SB	89K	2.1M	(128, 1, 2)
CA	335K	1.9M	(96, 2, 4)
A-05	410K	4.9M	(96, 4, 4)
A-06	403K	3.4M	(96, 4, 4)
EN	3.6M	276.1M	(300, 16, 16)
A-8M	8.6M	231.6M	(96, 36, 16)

Number of GNNIE cores The number of cores for a dataset is based on the ratio, ϑ, of vertices per computational subgraph (i.e., the full-length vertex features that can fit in cache) to the vertices assigned per core. Empirically, we determined that its optimal range is $0.03 \leq \vartheta \leq 0.15$. Using this, we find the number of cores m (see Tables I and II) for the optimal ϑ that optimizes speedup gain vs. area/power overhead.

We analyze the change in speedup when the number of cores is altered from the optimal m. For the A-06 dataset, $m = 4$; for 2, 16, and 36 cores, the speedup changes by 0.43×, 3.1×, and 7.29×, respectively. In each case, the speedup change is sublinear, indicating that $m = 4$ is optimal.

Benchmark GNN Datasets and Models. We evaluate the performance of our platform using Type A and Type B benchmark graph datasets from Table I and II, respectively. Type A datasets consist of multiple small graphs with no inter-graph edges, while Type B datasets are large monolithic graphs with a high amount of structural irregularity, i.e., higher adjacency matrix sparsity and power-law behavior. Table I and II also provide the input feature length (FL), number of cores (m), and feature vector segments (j) used for each dataset.

We evaluate the accelerator for training four GNN models: GCN, GINConv, GAT, and GraphSAGE. All GNNs have one hidden layer, except GINConv which has five; for GCN, GINConv, and GraphSAGE each hidden layer has 16, 64, and 256 channels, respectively. The GAT hidden layer uses eight 16-dimensional attention heads. All speedup and energy numbers include preprocessing times, including runtime for graph partitioning, degree-based vertex reordering, workload reordering, and neighborhood sampling time (performed on Intel Xeon Gold@2.60GHz CPU) for GraphSAGE. The preprocessing overhead over 500 epochs for amazon0601 is 18%.

Performance comparison with DGL. We compare all GNNs against Deep Graph Library (DGL) [27] on a V100 Tesla GPU with V100S-PCIe@1.25GHz, 32GB HBM2 ("*DGL+Tesla V100*"). The training latency for speedup comparison are averaged over 500 epochs. As shown in Fig. 5(a) and (b), the average speedup of our approach against DGL+Tesla V100 for GCN, GINConv, GAT, GraphSAGE ranges from 8.9×–46.6× across Type A datasets and 3.3×–15.5× for Type B.

The speedup comes from several of our optimizations: (i) Feature vertex segmentation improves scalability for large GNNs. (ii) Dynamic cache replacement mitigates irregular random memory accesses and on-chip communication overhead. (iii) Distributed computation across multiple batches ensures weight reuse. The speedup is particularly high for GINConv: unlike DGL, we use dimension-aware stage reordering (DASR) [1], [3], which requires fewer computations. To determine their impact, we removed these optimizations successively on A-06. Without segmentation, the computation did not complete (as in Fig. 4). With optimal segmentation,

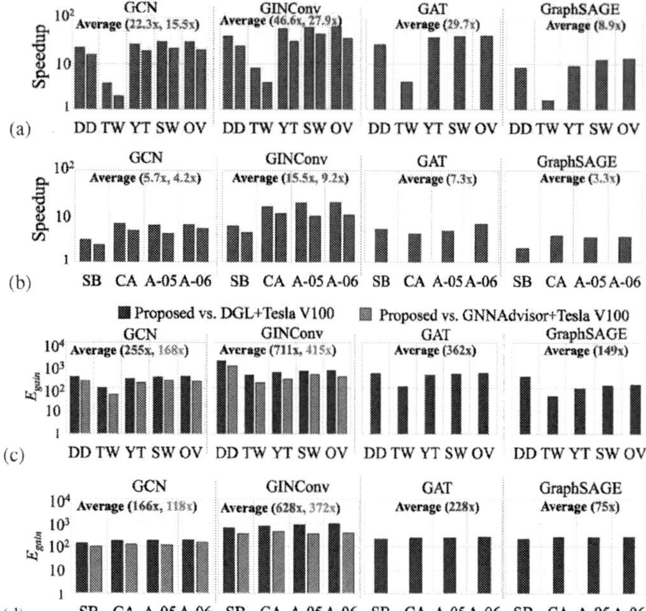

Fig. 5. Speedup and energy efficiency vs. DGL+Tesla V100 and GNNAdvisor+Tesla V100: (a), (c): Type A datasets (b), (d): Type B datasets.

removing dynamic cache replacement increases runtime by 34%; also removing weight reuse raises the penalty to 43%.

GraphSAGE shows lower speedup than other models due to: (i) inclusion of preprocessing time for neighborhood sampling on our platform, but not on DGL+Tesla V100. (ii) mitigation of power-law behavior in real-world graphs by sampling. Type A datasets have higher speedups than Type B datasets due to the lack of on-chip communication overheads. Larger datasets (e.g., OV, A-06) show higher speedups than smaller datasets (e.g., DD, SB) for both Type A and B, indicating scalability.

Comparison with GPU-based accelerators. Speedup: GNNAdvisor implements only GCN and GINConv. For the same configurations for these GNNs, Fig. 5(a) and (b) shows that relative to GNNAdvisor, we achieve 15.5×–27.9× speedup for Type A and 4.2×–9.2× for Type B datasets.

NeuGraph uses 2-D graph partitioning to process large graphs using one NVIDIA Tesla P100 GPU. We achieve 12.2× and 16.9× speedup for GCN on EN and A-8M, respectively, over NeuGraph. The corresponding speedups over GNNAdvisor are 3.1× and 6.8×, respectively.

Energy: Fig. 5(c) and (d), illustrate the energy efficiency comparison with Tesla V100, reporting E_{gain}, the ratio of the energy required by the GPU to the energy of our approach. Compared DGL+Tesla V100, our average E_{gain} ranges from 149×–711× over Type A datasets and 75×–628× over Type B. Against GNNAdvisor+Tesla V100, E_{gain} ranges from 168×–415× and 118×–372×, respectively.

Comparison with FPGA-/ASIC-based accelerators. Our approach achieves an average speedup of 11× and 24× over Rubik and GraphACT, respectively; neither reports absolute power numbers. Our speedup over Rubik is due to its inefficient reuse of cache data which incurs high on-chip and off-chip communication costs, and over GraphACT since it does not consider the power-law behavior of real-world graphs and makes no explicit efforts to address the random off-chip memory accesses. In comparison with GNNear, we achieve 17× average speedup over DGL+Tesla V100, but the speedup of GNNear is only 2.5×. Unlike our approach, the graph partitioner of GNNear is oblivious to community structure in real-world graphs, resulting in high communication costs due to the high number of cut edges between the partitions. GCoD handles only small graphs (up to 233K vertices, as against 8.6M vertices for our approach), and uses a whopping 180W of power even for these graphs, which can be handled by our approach on a single core using < 1W.

VII. Conclusion

Our multicore GNN training accelerator has GPU-like scalability and accelerator-like efficiency for large GNNs, leveraging novel feature vector segmentation and dynamic caching schemes for scalability and to mitigate communication costs.

References

[1] S. Mondal *et al.*, "A Unified Engine for Accelerating GNN Weighting/Aggregation Operations, with Efficient Load Balancing and Graph-Specific Caching," *IEEE T. Comput. Aid. D.*, 2022.
[2] M. Yan *et al.*, "HyGCN: A GCN Accelerator with Hybrid Architecture," in *HPCA*, 2020.
[3] T. Geng *et al.*, "AWB-GCN: A Graph Convolutional Network Accelerator with Runtime Workload Rebalancing," in *MICRO*, 2020.
[4] J. Stevens *et al.*, "GNNerator: A Hardware/Software Framework for Accelerating Graph Neural Networks," in *DAC*, 2021.
[5] Z. Zhou *et al.*, "BlockGNN: Towards Efficient GNN Acceleration Using Block-Circulant Weight Matrices," in *DAC*, 2021.
[6] C. Chen *et al.*, "DyGNN: Algorithm and Architecture Support of Dynamic Pruning for Graph Neural Networks," in *DAC*, 2021.
[7] B. Zhang *et al.*, "BoostGCN: A Framework for Optimizing GCN Inference on FPGA," in *FCCM*, 2021.
[8] W. Hu *et al.*, "Open Graph Benchmark: Datasets for Machine Learning on Graphs," in *NeurIPS*, 2020.
[9] Y. Wang *et al.*, "GNNAdvisor: An Adaptive and Efficient Runtime System for GNN Acceleration on GPUs," in *OSDI*, 2021.
[10] Z. Jia *et al.*, "Improving the Accuracy, Scalability, and Performance of Graph Neural Networks with Roc," in *MLSys*, 2020.
[11] L.Ma *et al.*, "NeuGraph: Parallel Deep Neural Network Computation on Large Graphs," in *USENIX ATC*, 2019.
[12] Z. Lin *et al.*, "PaGraph: Scaling GNN Training on Large Graphs via Computation-Aware Caching," in *SoCC*, 2020.
[13] H. You *et al.*, "GCoD: Graph Convolutional Network Acceleration via Dedicated Algorithm and Accelerator Co-Design," in *HPCA*, 2022.
[14] H. Zeng *et al.*, "GraphACT: Accelerating GCN Training on CPU-FPGA Heterogeneous Platforms," in *FPGA*, 2020.
[15] X. Chen *et al.*, "Rubik: A Hierarchical Architecture for Efficient Graph Neural Network Training," *IEEE T. Comput. Aid. D.*, 2021.
[16] Z. Zhou *et al.*, "GNNear: Accelerating Full-Batch Training of Graph Neural Networks with Near-Memory Processing," in *PACT*, 2021.
[17] G. Sun *et al.*, "Multi-Node Acceleration for Large-Scale GCNs," *IEEE Transactions on Computers*, 2022.
[18] S. Tailor *et al.*, "Degree-Quant: Quantization-Aware Training for Graph Neural Networks," in *ICLR*, 2021.
[19] S. Mondal *et al.*, "GNNIE: GNN Inference Engine with Load-Balancing and Graph-Specific Caching," in *DAC*, 2022.
[20] G. Karypis *et al.*, "A Fast and High Quality Multilevel Scheme for Partitioning Irregular Graphs," *SIAM J. Sci. Comput.*, 1998.
[21] S. Gandhi *et al.*, "P^3: Distributed Deep Graph Learning at Scale," in *OSDI*, 2021.
[22] "CACTI 6.5." https://github.com/Chun-Feng/CACTI-6.5.
[23] J. Nan *et al.*, "A Detailed and Flexible Cycle-Accurate Network-on-Chip Simulator," in *ISPASS*, 2013.
[24] A. B. Kahng *et al.*, "ORION3.0: A Comprehensive NoC Router Estimation Tool," *IEEE Embedded Sys. Lett.*, 2015.
[25] Y. Kim *et al.*, "Ramulator: A Fast and Extensible DRAM Simulator," *IEEE Comp. Arch. Lett.*, vol. 15, no. 1, 2015.
[26] M. O'Connor *et al.*, "Fine-Grained DRAM: Energy-Efficient DRAM for Extreme Bandwidth Systems," in *ISCA*, 2017.
[27] M. Wang *et al.*, "Deep Graph Library: Towards Efficient and Scalable Deep Learning on Graphs," in *ICLR*, 2019. https://github.com/dmlc/dgl/.

Joint Optimization of Cache Management and Graph Reordering for GCN Acceleration

Kyeong-Jun Lee[1], ByungJun Kim[2], Han-Gyeol Mun[1], Seunghyun Moon[1], and Jae-Yoon Sim[1,2]
Pohang University of Science and Technology, Pohang, Republic of Korea.
[1]Department of Convergence IT Engineering [2]Department of Electrical Engineering
(leekj9444, kbj1213, oxigen2, thearth, jysim)@postech.ac.kr

***Abstract*—Graph Convolutional Networks (GCNs) have demonstrated their efficacy in various real-world applications such as social networks and recommendation systems. Accelerating GCNs presents unique challenges due to their large number of nodes, sparse and heavily skewed connections. The reordering of the adjacency matrix has been the main strategy to effectively reduce the amount of re-access. Existing techniques of the reordering are categorized into i) degree-based sorting to identify high-degree nodes so that their data could be stored in the cache and ii) graph partitioning to maximally reuse the clustered data. However, as connections among the nodes vary significantly, processing various GCNs with a single strategy would cause performance degradation. This paper presents a software/hardware co-optimized platform for processing of general GCNs. We propose a hybrid scheme in the graph reordering that combines a sorting and a clustering in an adaptively optimized two-way partitioning. The two-way partitioning enables an efficient allocation of the on-chip cache memory space to reduce off-chip memory access by 4-to-12%. The implemented accelerator in 28nm demonstrates full functionalities with improved energy-efficiency by 2.2-to-3.7× compared to the previous GCN accelerators.**

***Index Terms*—cache, graph convolutional network, graph reordering, hardware accelerator, machine learning.**

I. Introduction

Inspired by the convolutional neural network (CNN), the graph convolutional network (GCN) has been actively researched to handle various real-world problems that can be interpreted as graph problems [1], [2]. The GCN is uniquely featured in connections among nodes that makes the processing of GCN be very different from that of CNN. Fig. 1a visualizes the statistics of non-zeros (NZs) in the adjacency matrix (A) of Pubmed. As a NZ in A represents a connection in a graph, the statistics of NZs of A are extremely sparse (0.028% for Pubmed) and the majority of connections are resulted from only a small number of influencing (high-degree) nodes (Fig. 1b). For example, top-20% high-degree nodes (HDNs) account for 60% NZs of the total, while remaining 80% are low-degree nodes (LDNs) only accounting for 40% NZs.

The extreme sparsity in a large A causes a low data reuse and requires multiple re-access of data stored in off-chip memory. Therefore, locality of data is much more important in GCN than that of CNN [3]. Previous works on the GCN accelerators [4]–[6] assumed enough internal memory in FPGA and applied loop optimization [7] for enhancing the locality of data. Algorithm/hardware co-optimization [8], [9] has been investigated to reduce the off-chip memory access, but with little improvement on large graphs [9]. Considering that processing of GCN is matrix multiplication, reordering of the adjacency matrix was proposed to maximally reuse the data stored in the cache. The reordering schemes have been applied in a form of degree-based sorting [10] or local clustering [11]. The degree-based sorting rearranges the A in the order of degrees so that data for the HDNs can be stored in the cache (Fig. 1b). On the other hand, the local clustering rearranges the A to form dense sub-blocks with a predefined size and stores the data for the sub-blocks in the cache (Fig. 1c).

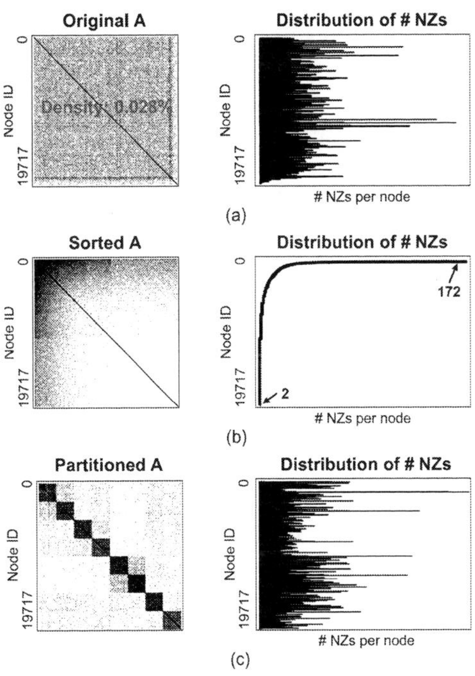

Fig. 1. Visualized statistics of #NZs in the adjacency matrix (A) of Pubmed; (a) original, (b) sorted, and (c) partitioned.

Though both the degree-based sorting and the graph partitioning (or clustering) increase overall data reuse, the improvement suffers from a high sensitivity to NZ statistics. Since the sorting uses the cache to hold data for the HDNs to raise the hit rate, it degrades data reuse for the LDNs. On the other hand, the partitioning focuses on locality in given sub-blocks and has demonstrated that it generally outperforms the

TABLE I
MATRIX DIMENSION AND DENSITY OF 6 GCN DATASETS.

	Matrix dimension					Matrix density						# of multiplications	
	N	F_0	K_0	F_1	K_1	d_A	d_{X0}	d_{X1}	d_W	d_{AX}	d_{XW}	(A×X)×W	A×(X×W)
Cora	2,708	1,433	16	16	7	0.1809%	1.27%	78%	100%	100%	100%	63M	1.33M
Citeseer	3,327	3,703	16	16	6	0.1781%	0.85%	89%	100%	100%	100%	198M	2.39M
Pubmed	19,717	500	16	16	3	0.0279%	10%	78%	100%	100%	100%	166M	18.6M
Flickr	89,250	500	128	128	128	0.0124%	46%	77%	100%	100%	100%	7.5G	4G
Reddit	232,965	602	128	128	128	0.0432%	100%	64%	100%	100%	100%	37.8G	26.4G
Yelp	716,847	300	128	128	128	0.0027%	100%	77%	100%	100%	100%	44.8G	40.1G

sorting on average [12]. However, the partitioning can destroy the inherent locality of the HDNs by distributing their NZs. Therefore, applying the partitioning to a graph having clear distinction between HDNs and LDNs would cause serious performance degradation. It leads to the need of developing a new reordering scheme which optimally combines the sorting and the partitioning.

There have been a few efforts on ASIC for GCN [12]–[14]. [13] employed loop optimizations [7] with an adaptive tiling. [12] proposed an architecture for the partitioning [11]. [14] showed an implementation with a chip fabrication. However, they degrade performance on larger GCN while showing good performance on smaller models. Since the size and network characteristics can change, processing various GCNs with a single strategy would cause failure in keeping optimal performance in the cache management.

The contributions of this paper are summarized as follows:

- We propose a hybrid graph reordering that combines a sorting and a clustering with a two-way partitioning optimized for cache utilization.
- We propose a design platform to find the best combination of the sorting and the clustering that maximizes the cache hit rate.
- We implemented a real accelerator chip and evaluated performance on various datasets of Cora, Citeseer, Pubmed, Flickr, Reddit and Yelp.

II. PRELIMINARY

A. Graph Convolutional Network

Fig. 2 describes how GCN is processed with matrix multiplications. The GCN consists of two main phases: aggregation and combination. These phases are iteratively applied to propagate information through the network and generate node embeddings. The aggregation phase is responsible for collecting information from the neighboring nodes in the graph. The combination phase involves updating the node features based on the aggregated information. The matrices are N×N adjacency matrix (A), N×F feature matrix (X) and F×K weight matrix (W). N is the number of nodes, F is the number of input features and K is the number of output features. Each parameter is described in Table I. The benchmark graphs (Cora, Citeseer, Pubmed, Flickr, Reddit and Yelp) were sourced from [1] and [2]. Matrix A has a density of d_A, matrix X has a density of d_X, and matrix W has a density of d_W. The densities of matrices AX and XW are denoted as d_{AX} and d_{XW}, respectively. The Lth Gconv layer receives A, $X_{(L)}$, and $W_{(L)}$. Matrix A is normalized according to the degree of each node during offline preprocessing, and this normalized adjacency matrix is commonly used across all layers. The output feature matrix $X_{(L+1)}$ is calculated by Eq. (1).

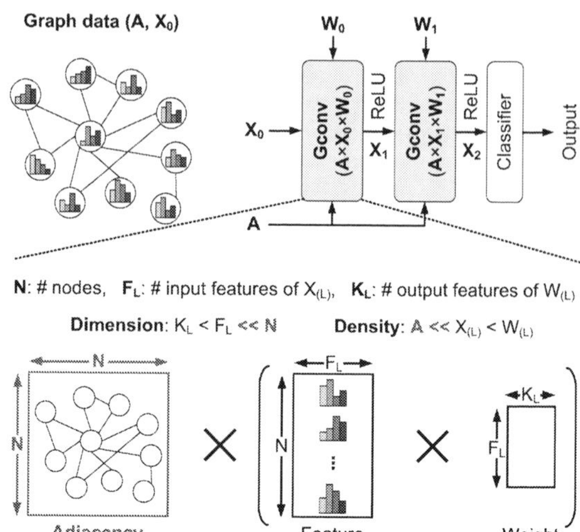

Fig. 2. GCN and its matrix representation.

$$X_{(L+1)} = ReLU(A \times X_{(L)} \times W_{(L)}) \quad (1)$$

B. Acceleration of GCN

During training, the aggregation step with A×X needs to be processed first, and the combination step with AX×W is followed. On the other hand, during inference, X×W should be processed first and followed by A×XW [5]. It is because this computation order reduces the total number of multiply-and accumulate (MAC) operations as shown in Table I. Since this work is for the acceleration of inference task, the latter order is taken. Matrix X has balanced sparsity or is dense, while matrices W and XW have dense NZs. So, the matrix multiplications for X×W can be well supported by existing machine learning acceleration hardware [3].

However, the multiplication with A (called aggregation), cannot be efficiently supported without consideration of graph characteristics. Processing for the aggregation causes a huge amount of irregular off-chip memory access.

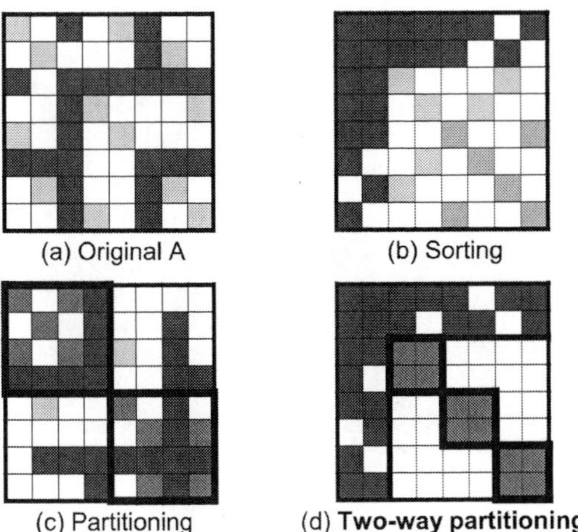

Fig. 3. Effect of different reordering schemes when they are applied to the same adjacency matrix A. Top-2 HDNs are denoted with red color. Thick boxes indicate clustered blocks.

Algorithm 1 The Proposed Two-way Partitioning

Input: Adjacency matrix $\mathbf{A}$

Output: Two-way partitioned of adj. matrix $\mathbf{A}_{TP}$

1: Calculate node degree $\mathbf{D}_i$ for each node

2: Split high/low degree indicies $\mathbf{idx}_{high}$, $\mathbf{idx}_{low}$ from $\mathbf{D}_i$

3: $\mathbf{A}_{low}$ = $\mathbf{A}[\mathbf{idx}_{low}, \mathbf{idx}_{low}]$

4: $\mathbf{idx}_{low}^{new}$ = Partitioning($\mathbf{A}_{low}$)

5: idx^{new} = Concatenate($\mathbf{idx}_{high}$, $\mathbf{idx}_{low}^{new}$)

6: $\mathbf{A}_{TP}$ = $\mathbf{A}[\mathbf{idx}^{new}, \mathbf{idx}^{new}]$

III. Methodology

A. The Proposed Two-Way Partitioning

Fig. 3 illustrates the impact of different reordering schemes (sorting, partitioning, and the proposed two-way partitioning) when applied to a given adjacency matrix A. As shown in Fig. 3b, HDNs are likely to have global interactions over the whole graph. So, if graph partitioning is applied to the graph, the NZs of the HDNs also appear in the non-clustered region, mitigating the local clustering (Fig. 3c). These non-clustered NZs can constitute significant portions depending on graph characteristics, leading to irregular access patterns from off-chip memory.

As described in Algorithm 1, the proposed reordering firstly applies a global sorting to the A and divides it into two regions by permutating the node indicies according to the node degree, *i.e.*, one with HDNs and the other with LDNs (*line 1∼2*). Then, the partitioning is applied only to the region of the LDNs using open source partitioning algorithm [11] (*line 3∼4*). Combining the two regions completes the proposed reordering (*line 5∼6*). As shown in Fig. 3d, this two-way partitioning can achieve better clustering of NZs. Note that selection of the threshold for HDNs and LDNs is a variable and gives an extra degree-or-freedom in clustering. According to setting the threshold, the two-way partitioning can also cover from the sorting (Fig. 3b) to the partitioning (Fig. 3c).

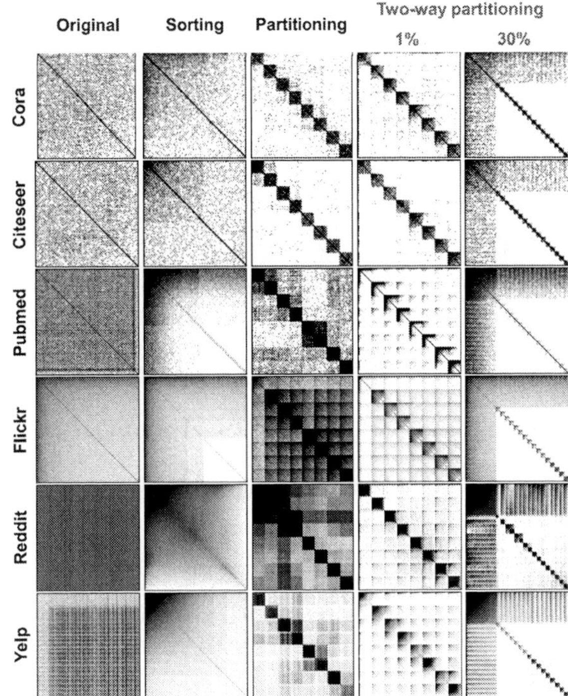

Fig. 4. Visualized results of the reordering schemes on six benchmark datasets.

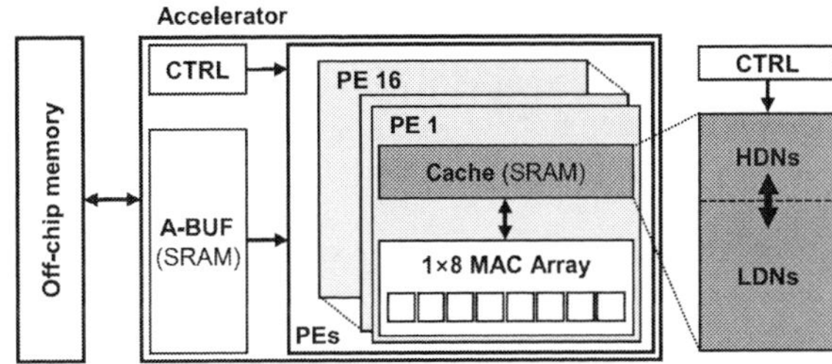

Fig. 5. High-level architecture for the proposed accelerator.

Fig. 4 visualizes results of the reordering schemes on various graphs. Two cases of top-1% and top-30% HDN thresholds are also shown for the proposed two-way partitioning. The two-way partitioning clearly shows improvements compared to the vanilla partitioning. The more HDNs are separated from LDNs, the smaller the proportion of NZs in the non-clustered area. As the result depends on the HDN threshold, this work also proposes an algorithm to find the optimum HDN threshold that maximizes the improvement for a given capacity of cache in Section III-D.

B. Hardware Architecture

Fig. 5 shows the proposed accelerator which consists of 16 processing elements (PEs), a host controller (CTRL), a 20 KB SRAM (A-BUF) to store A. The A-BUF assigns 8 KB for values and 12 KB for indices which are stored in coordinate (COO) format. Each PE includes 8 multiply-accumulator (MAC) units and 16 KB SRAM for cache. A MAC performs

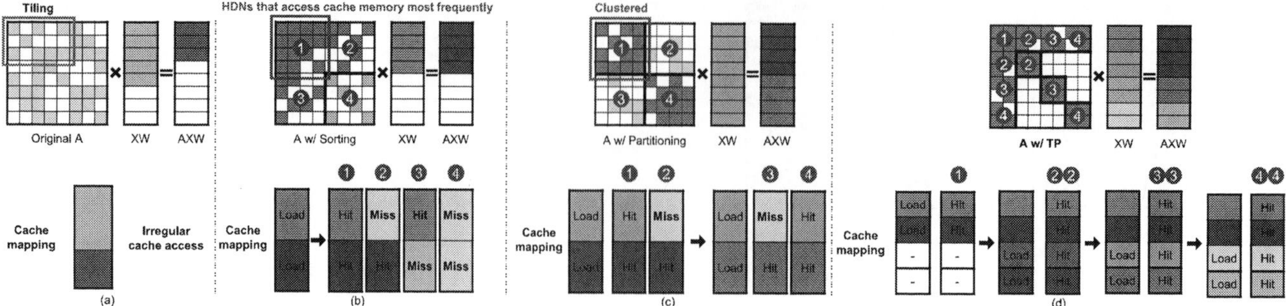

Fig. 6. Cache mappings with 4 different graph reordering; (a) original, (b) sorting, (c) partitioning and (d) two-way partitioning.

a 4-stage pipelined multiplication and accumulation using 16-bit brain floating point format. It produces 4 outputs per 4 clock cycles. The A-BUF stores NZs of A in the aggregation phase and stores X in the combination phase. The cache stores XW (space for input) and AXW (space for output) during the aggregation phase, and stores W (space for input) and XW (space for output) during the combination phase.

For aggregation phase, the cache of each PE stores both HDNs and LDNs. It is managed by the CTRL to support the proposed two-way partitioning algorithm. The CTRL manages the indices of HDN and LDN nodes stored in A-BUF, ensuring that HDNs and the corresponding XW and AXW values for computation remain in the on-chip memory for continued access. The proposed accelerator employs an output stationary dataflow. When processing A×(XW)=AXW, the partial results of AXW are written to the cache after the multiplication of A and XW in the cache has been completed.

C. Cache Management

Fig. 6 describes 4 different strategies of reordering and their optimal cache managements. When the original matrix A is processed without reordering, the loop optimization (Fig. 6a) has been generally used [5], [6], [13]. When processing with a sorted A, the space for the HDNs resides in the cache (Fig. 6b). With a partitioning, the data for the clustered nodes are fetched into the cache (Fig. 6c). [12] compared these 3 methods and demonstrated that the partitioning-based approach generally outperforms the others. As shown in Fig. 6d, the proposed two-way partitioning further improves the clustering of A. By using small cache space for HDNs, the remaining room in the cache can be assigned for clustered nodes.

D. Hardware-Aware Search for Optimal Threshold

The two-way partitioning ideally works when the memory space for the cache is large enough to hold both the HDNs and the clustered nodes. However, as more space is allocated for the HDNs, there will be less space remained for the clustered nodes (Fig. 7a). Since a partitioning with smaller blocks generally degrades the efficiency of cache use, there would be an optimal HDN threshold according to the A and the on-chip cache size. Fig. 7b graphically illustrates the search algorithm, where C represents the total capacity of cache. C_{HDN} and C_P are parts assigned for HDNs and clustered LDNs, respectively.

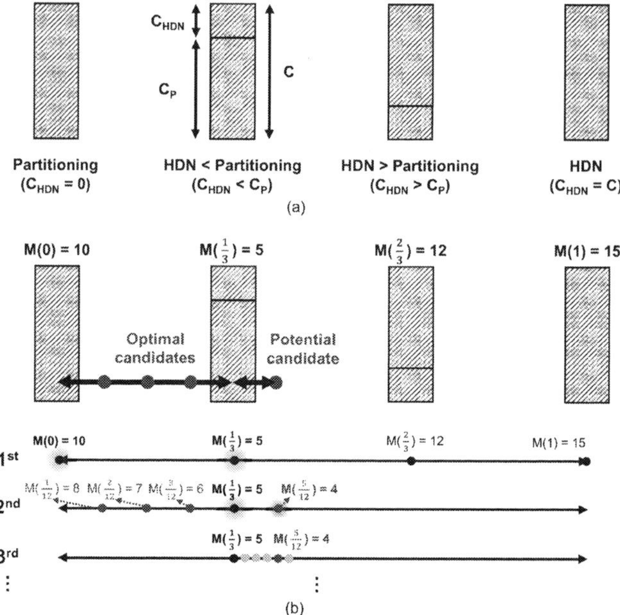

Fig. 7. (a) Cache mapping as a function of c and (b) search algorithm to find the optimal HDN threshold.

The goal is to find the best C_{HDN}, $\widehat{C}_{HDN}$, for given C, represented by Eq. (2), where M_C(c) is the amount of memory access when c = C_{HDN}/C.

$$\frac{\widehat{C}_{HDN}}{C} = \underset{C_{HDN}}{argmin}\ M_C(\frac{C_{HDN}}{C}) \tag{2}$$

It is inefficient to scan all the cases with a fine resolution. We propose a 4-pillar search that can be performed with 4 cores in parallel. The 4 pillars equally divide the search space into 3 regions with c of 0, 1/3, 2/3 and 1. The search begins with estimations of the total memory access with given c's. The score for each region is the sum of the total memory access with the two boundary c's of the corresponding region. The region with the lowest score becomes the inner section whereas a neighboring region with the next lowest score becomes the outer section. Then, the 2 regions in the inner section and 1 region in the outer section become 3 regions for the next step. Recursively repeating the sequence leads to a

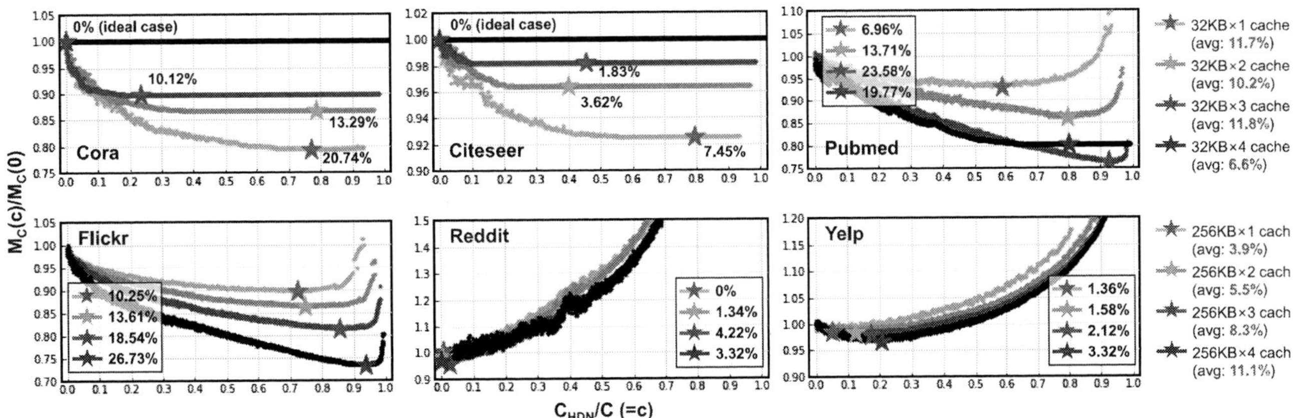

Fig. 8. Normalized amount of off-chip memory access of XW for different datasets as HDN threshold varies.

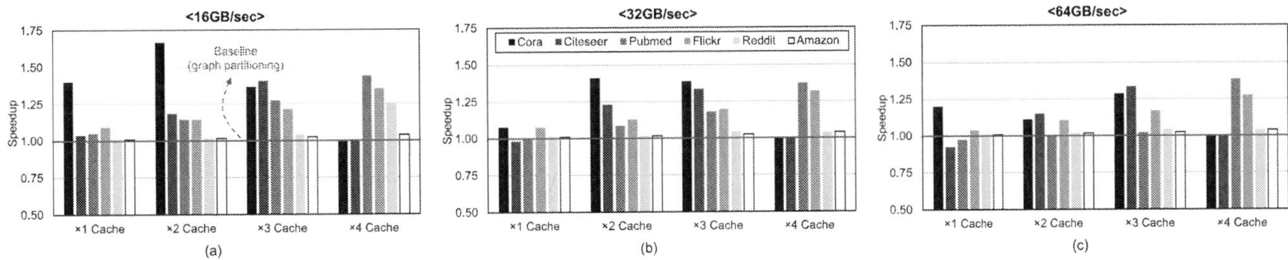

Fig. 9. Speedup of the proposed two-way partitioning method over graph partitioning during the aggregation phase for varying memory bandwidth: (a) 16 GB/sec, (b) 32 GB/sec, and (c) 64 GB/sec.

global optimum for c. The complexity of the search algorithm is logarithmic with respect to C. For given graph data, the two-way partitioning should be processed offline as a one-time operation.

IV. Evaluation

A. Two-Way Partitioning

1) Off-chip Memory Traffic: Fig. 8 shows the total off-chip memory access of XW on the proposed accelerator architecture (Fig. 5) as the HDN threshold varies. The horizontal and vertical axes represent c and the normalized amount of the off-chip memory access (= $M_C(c)/M_C(0)$), respectively. The optimal ratios are denoted by colored stars. The evaluations were conducted for different cases (1×, 2×, 4×, 8×) of the cache memory. Since the cache of 256 KB in our chip is able to store the whole required data for the smaller networks of Cora, Citeseer and Pubmed, the evaluations for these networks were performed with a 32KB cache that was used in [14]. We used open source partitioning algorithm [11] with the size of the sub-block fitted to the cache size. The two-way partitioning generally reduces off-chip memory accesses compared to graph partitioning, and the effect increases with larger on-chip memory capacity. While the improvement is relatively modest for datasets such as Reddit or Yelp, due to their significantly larger number of nodes, the two-way partitioning method consistently mitigates off-chip memory access, even within given memory capacity constraints. It enables an efficient allocation of the on-chip cache memory space to reduce off-chip memory access by 6.6-to-11.7% for small graphs (Cora, Citeseer and Pubmed) and 3.9-to-11.1% large graphs (Flickr, Reddit and Yelp).

2) Speedup: Fig. 9 illustrates the performance improvement of the proposed two-way partitioning method over graph partitioning when varying memory bandwidth (16, 32, and 64 GB/sec). The same hardware configuration as in Fig. 8 was used, and the two-way partitioning method selected the optimal point for each case. Each bar in the graph represents the speed normalized to that of the corresponding graph partitioning. The two-way partitioning method achieves an average speedup of 1.14× compared to graph partitioning, with the maximum speedup reaching 1.67×. The improvement can be attributed to the reduction in off-chip memory accesses and increased on-chip data reuse facilitated by the two-way partitioning. Additionally, as described in Fig. 6, the two-way partitioning method requires less memory bandwidth for each step (*e.g.,* smaller partitioning size), which helps to hide the latency caused by off-chip memory during computation. In some cases with small datasets, the speedup is worse as memory bandwidth increases. This is because the partitioning size is too small to utilize the available memory bandwidth, even though it reduces off-chip memory access. On the other hand, the two-way partitioning consistently delivers improved performance on larger datasets.

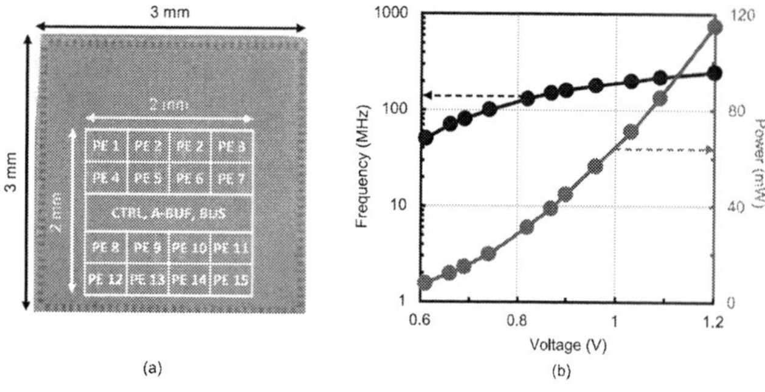

Fig. 10. (a) Die photo and (b) measured operating frequency and power consumption.

TABLE II
PERFORMANCE COMPARISON

		HPCA'21 [13]	HPCA'23 [12]	TCAS-II'22 [14]	This work
Implementation		ASIC [a)]	ASIC [a)]	ASIC [b)]	ASIC [b)]
Technology (nm)		40	65	28	28
Optimization		Loop optimization	Graph partitioning	Loop optimization	Two-way partitioning
Frequency (MHz)		1000	1000	6.25-15.6, 100-250	50-250
Core area (mm^2)		6.51	5.785	3.24	4
Supply voltage (V)		N/A	N/A	0.77-1.18	0.61-1.2
Bit precision		64b FP	64b FP	16b INT	16b BFP
# of multipliers		16	N/A	256	128
Power (mW)		365	16	5.5-35.0	7.5-115.2
Energy Efficiency (Graph/J)	**Cora**	1 [c)]	1.7 [d)]	58k-143k	151k-523k
	Citeseer	1 [c)]	2.1 [d)]	38k-92k	96k-330k
	Pubmed	1 [c)]	1.4 [d)]	5k-13k	11k-39k
	Flickr	1 [c)]	1.4 [d)]	-	40-137
	Reddit	1 [c)]	0.8 [d)]	-	9-30
	Yelp	1 [c)]	2.8 [d)]	-	7-23

[a)] Synthesis only. [b)] Chip fabrication.
[c)] and [d)] are reported as the normalized ratio in [12].

B. Evaluations on Chip

The proposed accelerator was implemented with a real chip fabrication in an active area of 4mm^2 using 28nm LP CMOS technology (Fig. 10a). Fig. 10b summarizes operating frequency and power consumption as the supply voltage varies. The chip operated in a supply range of 0.61-to-1.2V at 50-to-250MHz and showed full functionalities on all the tested six benchmark datasets. We tested the GCN models [1] for smaller datasets (Cora, Citeseer, Pubmed) and [2] for larger datasets (Flickr, Reddit, Yelp), using model configurations described in Table I. Table II presents a comparison between our results and the ASIC accelerators previously reported. When comparing the loop optimization based accelerator [13] and the graph partitioning based accelerator [12], the latter improves the energy efficiency by 1.7-to-2.1×. The proposed scheme shows 2.2-to-3.7× improvement in the energy efficiency over [14] with a reduction of off-chip memory access for small graphs (Cora, Citeseer and Pubmed).

V. CONCLUSION

The reordering technique has been the strategy in the acceleration of GCN since it is the most effective way of raising the cache hit rate. Though the sorting and the partitioning have been extensively investigated in the literature, the combination of them has been barely considered. The main contribution of this work is proposing a way of combining the proven two approaches to enjoy their benefits while compensating their drawbacks. A searching scheme for finding the global optimal value for the combination was also presented. The proposed algorithm co-optimized acceleration platform was verified through simulations and a real chip fabrication using 28nm LP CMOS technology. The measured chip shows full functionalities with reduction of off-chip access by 4-to-12% and energy-efficiency by 2.2-to-3.7× compared to the previously reported GCN accelerators.

ACKNOWLEDGMENT

This work was supported in part by the National Research Foundation funded by the Korea Ministry of Science and ICT under Grant 2019R1A5A1027055.

REFERENCES

[1] B. Jiang *et al.*, "Semi-supervised learning with graph learning-convolutional networks," in *2019 IEEE/CVF Conference on Computer Vision and Pattern Recognition (CVPR)*, 2019, pp. 11 305–11 312.

[2] H. Zeng *et al.*, "GraphSAINT: Graph sampling based inductive learning method," in *International Conference on Learning Representations*, 2020.

[3] J.-S. Park *et al.*, "9.5 a 6k-mac feature-map-sparsity-aware neural processing unit in 5nm flagship mobile soc," in *2021 IEEE International Solid- State Circuits Conference (ISSCC)*, vol. 64, 2021, pp. 152–154.

[4] M. Yan *et al.*, "Hygcn: A gcn accelerator with hybrid architecture," in *2020 IEEE International Symposium on High Performance Computer Architecture (HPCA)*, 2020, pp. 15–29.

[5] T. Geng *et al.*, "Awb-gcn: A graph convolutional network accelerator with runtime workload rebalancing," in *2020 53rd Annual IEEE/ACM International Symposium on Microarchitecture (MICRO)*, 2020, pp. 922–936.

[6] Y. Zhang *et al.*, "G-cos: Gnn-accelerator co-search towards both better accuracy and efficiency," in *2021 IEEE/ACM International Conference On Computer Aided Design (ICCAD)*, 2021, pp. 1–9.

[7] J. R. Allen and K. Kennedy, "Automatic loop interchange," in *Proceedings of the 1984 SIGPLAN Symposium on Compiler Construction*, ser. SIGPLAN '84. New York, NY, USA: Association for Computing Machinery, 1984, p. 233–246.

[8] H. You *et al.*, "Gcod: Graph convolutional network acceleration via dedicated algorithm and accelerator co-design," in *2022 IEEE International Symposium on High-Performance Computer Architecture (HPCA)*, 2022, pp. 460–474.

[9] T. Geng *et al.*, "I-gcn: A graph convolutional network accelerator with runtime locality enhancement through islandization," *MICRO-54: 54th Annual IEEE/ACM International Symposium on Microarchitecture*, 2021.

[10] P. Faldu *et al.*, "A closer look at lightweight graph reordering," in *2019 IEEE International Symposium on Workload Characterization (IISWC)*, 2019, pp. 1–13.

[11] G. Karypis *et al.*, "A fast and high quality multilevel scheme for partitioning irregular graphs," *SIAM Journal on Scientific Computing*, vol. 20, no. 1, pp. 359–392, 1998.

[12] M. Kang *et al.*, "Grow: A row-stationary sparse-dense gemm accelerator for memory-efficient graph convolutional neural networks," 2022, arXiv:2203.00158.

[13] J. Li *et al.*, "Gcnax: A flexible and energy-efficient accelerator for graph convolutional neural networks," in *2021 IEEE International Symposium on High-Performance Computer Architecture (HPCA)*, 2021, pp. 775–788.

[14] K.-J. Lee *et al.*, "A 384g output non-zeros/j graph convolutional neural network accelerator," *IEEE Transactions on Circuits and Systems II: Express Briefs*, pp. 1–1, 2022.

ITA: An Energy-Efficient Attention and Softmax Accelerator for Quantized Transformers

Gamze Islamoglu*, Moritz Scherer*, Gianna Paulin*, Tim Fischer*,
Victor J.B. Jung*, Angelo Garofalo*†, Luca Benini*†
*ETH Zürich, Switzerland , †University of Bologna, Italy
*{gislamoglu,scheremo,pauling,fischeti,jungvi,lbenini}@iis.ee.ethz.ch, †angelo.garofalo@unibo.it

***Abstract*—Transformer networks have emerged as the state-of-the-art approach for natural language processing tasks and are gaining popularity in other domains such as computer vision and audio processing. However, the efficient hardware acceleration of transformer models poses new challenges due to their high arithmetic intensities, large memory requirements, and complex dataflow dependencies. In this work, we propose ITA, a novel accelerator architecture for transformers and related models that targets efficient inference on embedded systems by exploiting 8-bit quantization and an innovative softmax implementation that operates exclusively on integer values. By computing on-the-fly in streaming mode, our softmax implementation minimizes data movement and energy consumption. ITA achieves competitive energy efficiency with respect to state-of-the-art transformer accelerators with 16.9 TOPS/W, while outperforming them in area efficiency with 5.93 TOPS/mm^2 in 22 nm fully-depleted silicon-on-insulator technology at 0.8 V.**

***Index Terms*—neural network accelerators, transformers, attention, softmax**

I. Introduction

The transformer is a deep learning architecture introduced in 2017 [1], which has revolutionized natural language processing tasks by achieving superior accuracy with respect to recurrent neural networks (RNNs) at comparable compute and memory requirements. Recently, transformers have been adopted across multiple modalities, including text [2], [3], image [4], audio [5], and video [6]. The ubiquity of the transformer model highlights its general-purpose capabilities [7] and stresses the need for efficient hardware acceleration.

While most transformer models require gigabytes of memory for their parameters, and billions of operations for each inference, recent research has proven that smaller transformers have applications that suit low-power embedded systems [8]. Besides architectural optimization, research into the compression of transformers has shown that 8-bit quantized models perform on par with their floating-point equivalents [9], [10].

A key component of transformers is the attention mechanism which generates a square matrix of order input length, resulting in a superlinear number of operations and memory size [1]. This computation- and memory-intensive nature of the attention severely impacts the energy cost of deploying transformers on embedded systems, requiring specialized hardware to improve performance and energy efficiency.

A peculiar challenge with transformers is the softmax operation which is applied over the rows of the attention matrix and becomes a bottleneck in low-precision architectures due to its nonlinear and non-element-wise nature. The nonlinearity of softmax restricts performing it on quantized values while the utilization of floating-point units incurs significant area and power costs. Furthermore, the non-element-wise nature of the softmax operation necessitates multiple passes through the attention matrix's row vectors, resulting in substantial data movement and power consumption within the system.

In this work, we present ITA, Integer Transformer Accelerator, an architecture targeting low-power embedded applications. To maximize ITA's energy efficiency, we focus on minimizing data movement throughout the execution cycle of the attention mechanism. In contrast to throughput-oriented accelerator designs, which typically employ systolic arrays, ITA implements its processing elements with wide dot-product units, allowing us to maximize the depth of adder trees, thereby further increasing efficiency.

To overcome the complex dataflow requirements of standard softmax, we present a novel approach that allows performing softmax on 8-bit integer quantized values directly in a streaming data fashion. Our approach also enables a weight stationary dataflow by decoupling denominator summation and division in softmax. The streaming softmax operation and weight stationary flow, in turn, minimize data movement in the system and power consumption of ITA.

Our contributions can be summarized as follows:

- We present ITA, a hardware accelerator utilizing the parallelism of attention mechanism and 8-bit integer quantization to improve performance and energy efficiency. To minimize data movement and power consumption, ITA adopts weight stationary dataflow over output stationary.
- We propose an energy- and area-efficient softmax implementation that fully operates in integer arithmetic with a footprint of only 3.3 % over the total area of ITA and a mean absolute error of 0.46 % compared to its floating-point implementation. The streaming operation further saves energy by reducing data movement.
- We evaluate our architecture in GlobalFoundaries' 22FDX fully-depleted silicon-on-insulator (FD-SOI) technology and achieve an energy efficiency of 16.9 TOPS/W and area efficiency of 5.93 TOPS/mm^2 at 0.8 V, performing similarly to the state-of-the-art in energy efficiency, despite being implemented in a much less aggressive technology, and 2× better in area efficiency.

979-8-3503-1176-1/23 $31.00 © 2023 IEEE

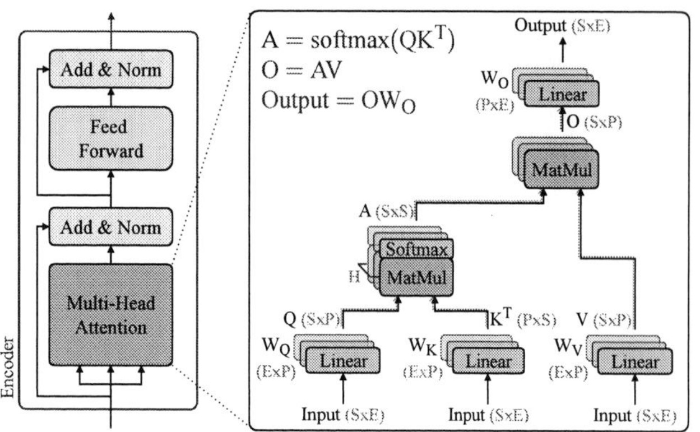

Fig. 1. Transformer encoder and multi-head attention. S: sequence length, E: embedding size, P: projection space, H: number of heads.

II. Preliminaries and Related Work

In this section, we describe the operations in transformer-based networks, focusing on the softmax operation since it is a critical computation in transformers and creates a significant bottleneck in acceleration.

A. Transformers

A transformer network consists of multiple encoder and/or decoder stages, each containing an attention block, and a task-specific final layer. Figure 1 shows a transformer encoder and multi-head attention. In decoders, the inputs are slightly modified but the attention mechanism remains the same.

Multi-head attention is the main building block of transformers. In attention, three linear transformations are applied to inputs of size $S \times E$, where S is the sequence length and E is the embedding size, to generate Query (Q), Key (K), and Value (V) matrices. Q, K, and V are of size $S \times P$, where P is the projection space. Then, matrix multiplication is performed between Q and K^T, and softmax is applied to obtain probabilities. The resulting $S \times S$ attention matrix (A) can be considered as probabilities showing the relationship between Queries and Keys. A is then multiplied by V, which weights the input tokens according to their relevance. By performing these operations in parallel with multiple sets of Query, Key, and Value matrices, multiple heads of attention are obtained. The outputs of these heads are then concatenated and linearly transformed to produce the final output of the attention, which has the same size as the input ($S \times E$).

B. Softmax

Softmax is a key operation in transformers and encountered in every attention layer in the computation of matrix A. It is applied row-wise to the attention matrix to normalize it to probabilities. The softmax function is an $R^n \rightarrow R^n$ function, defined as follows for a vector $\boldsymbol{x}$ of length n:

$$\text{softmax}(\boldsymbol{x})_i = \frac{e^{x_i - \max(\boldsymbol{x})}}{\sum_{j=1}^{n} e^{x_j - \max(\boldsymbol{x})}} \tag{1}$$

and it produces a new vector of length n whose elements sum to 1. Softmax presents two challenges: nonlinearity and non-element-wise operation. The nonlinearity means that we cannot perform softmax on the quantized values directly because $\text{softmax}(\varepsilon x_q) \neq \varepsilon \cdot \text{softmax}(x_q)$ given the quantized value x_q and scaling factor ε. In some accelerators, the input of softmax is first dequantized, softmax is calculated, and then output is quantized again [10], [11]. However, this approach is not hardware-friendly as it involves floating point units. I-BERT [9] proposes a method to approximate softmax using second-order polynomials, eliminating the need for dequantization entirely. However, it operates at a higher precision of 32-bit, as opposed to the 8-bit quantization used in the rest of the network, and requires 32-bit multipliers and dividers.

Furthermore, softmax is not an element-wise operation and requires both a maximum search and a summation over a row of the attention matrix. This results in multiple passes over the row and multiple reads from memory, leading to high data movement and power consumption. Therefore, transformer accelerators usually compute the attention matrix row by row and accumulate the summation over the row. After completing one row, the division is performed to obtain probabilities [12], [13]. However, this method is not feasible for weight stationary accelerators as the attention matrix is not produced row by row. ITA overcomes this issue and minimizes memory traffic by calculating a tight softmax approximation for 8-bit integers in three steps over multiple rows as explained in section IV.

C. Related Work

Accelerating inference of transformer networks is an active area of research, with most accelerators focusing on the attention layer and using integer data formats like our approach. Some architectures exploit the sparsity of the attention matrix, such as OPTIMUS [14] which uses a sparse matrix format and redundant computation skipping in decoding. SpAtten [15] proposes token and head pruning and progressive quantization to reduce memory accesses and computations using a special engine to rank token and head importance scores. ELSA [16] utilizes an approximate self-attention algorithm to filter irrelevant query and key pairs and only performs exact computation for relevant pairs that are selected by hash and norm computation units. Similarly, Wang et al. [12] propose a big-exact-small-approximate processing unit to save power and a bidirectional asymptotic speculation unit to skip redundant computations. However, the sparsity of transformers is limited to the attention matrix and depends on the network itself, and supporting sparsity in these accelerators comes with a cost in the area, such as additional top-k engine in SpAtten and hash and norm computation units in ELSA. Therefore, ITA does not utilize sparsity of attention to achieve higher area efficiency.

SpAtten and ELSA perform softmax in floating point by dequantizing before and quantizing after the softmax. However, this approach requires additional floating point units that are not utilized during the majority of computation, making it less preferable than integer equivalents due to larger area occupancy. Keller et al. [13] use the Softermax algorithm [17], which uses fixed-point arithmetic and replaces base e with 2 to simplify the hardware. In this paper, we present an alternative

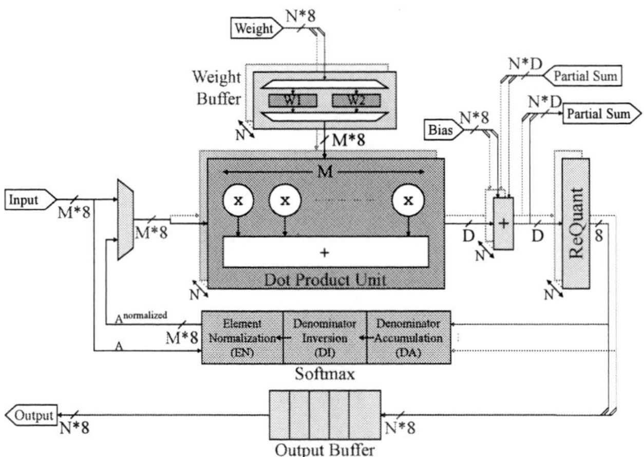

Fig. 2. Architecture of ITA with 8-bit inputs and weights. The softmax block is detailed in Figure 4.

approach to compute softmax in integer with minimal area overhead in hardware, without approximating the softmax with base 2. While Wang et al. [12] and OPTIMUS [14] also compute softmax without conversion to floating point, they do not provide information about the implementation details and errors introduced by their softmax implementation.

III. Architecture

The architecture of our transformer accelerator is shown in Figure 2, targeting 8-bit integer quantized matrices. The accelerator is parametric: it includes N processing engines (PEs), each computing the dot product between two vectors of M elements, and works on tiles of size $M \times M$. Each PE uses 8-bit weights and activations, producing dot product results with higher precision of D-bit. N, M, and D are configured at design time. The adders after PEs accumulate partial sums. Once outputs are fully accumulated, 8-bit biases are added to outputs, which are then converted back to 8-bit format by requantization modules (ReQuant in Figure 2).

The softmax module computes the softmax of the attention matrix A and works in two passes. In the first pass, when it takes the elements of A from the matrix multiplication $Q \times K^T$, it finds the maximum and accumulates the denominator of softmax. In the second pass, when the attention matrix is supplied as input for the $A \times V$ computation, the softmax module normalizes them to probabilities before entering PEs. To achieve high throughput and low power consumption, we propose a novel and hardware-friendly softmax implementation, which is detailed in section IV. The explained clipping operation is performed by the requantization module and the clipping threshold is obtained from quantization-aware training that incorporates our softmax implementation.

Finally, the output FIFO buffers the results temporarily to prevent stalling the accelerator in case the output cannot be written to the memory immediately.

ITA follows a weight stationary approach to reduce the bandwidth and energy requirements. Weights are reused M times and stored in a double-buffered weight buffer, where $W1$ and $W2$ have a capacity of M bytes. Double buffering allows the accelerator to fetch weights for the next computation while simultaneously performing the current computation. This reduces the bandwidth requirement for the weight interface from NM to N bytes per cycle. While the weight stationary approach of ITA only requires a bandwidth of $8(M+3N)+2ND$ bits per cycle (M bytes read for input, N bytes read for weight and bias, N bytes write for output, and ND bits read and ND bits write for partial sums), output stationary approaches typically require substantially more at $8(NM+3N)+2ND$ bits per cycle (NM bytes read for weight, N bytes read for input and bias, N bytes write for output, and ND bits read and ND bits write for partial sums). As the number of processing elements in the accelerator increases, ITA can sustain higher utilization compared to an output stationary flow, with fewer data movements leading to lower power consumption. However, the downside of this approach is the size of the weight buffer ($2NM$ bytes). An output stationary accelerator can double-buffer inputs with a buffer size of $2M$ bytes without buffering weights since they are updated every cycle. We prefer the former because memory bandwidth is often the bottleneck, especially for accelerators, since only a small portion of network parameters can be stored locally and they have to access higher levels of memory continuously. Another difficulty of weight stationary flow is the row dependency of softmax, as explained in subsection II-B. In section IV, we discuss our proposed method to handle this dependency.

The workload mapping and schedule of ITA are summarized in Figure 3. The accelerator operates on tiles of size $M \times M$ and iterates over dimension L to achieve output stationarity in the outer loop. Within each tile, ITA employs a weight stationary regime and shares inputs among N PEs, achieving spatial input reuse. Each PE operates on vectors of size M in the innermost loop and computes the dot product of input and weight vectors. If M is not an integer multiple of matrix dimensions, inputs/weights are padded with zeros.

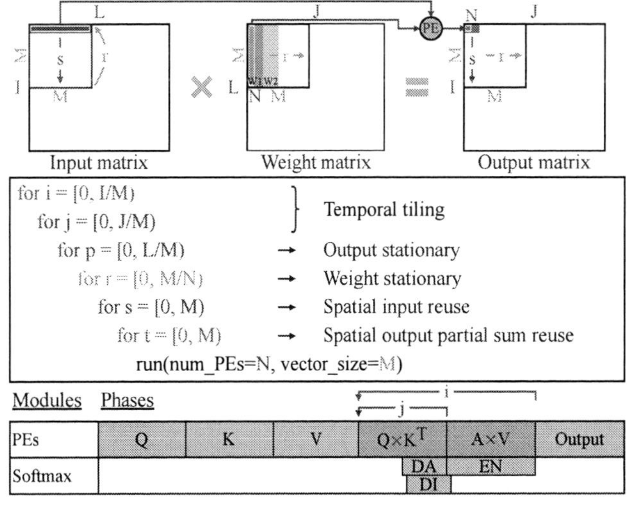

Fig. 3. Workload mapping and computation phases.

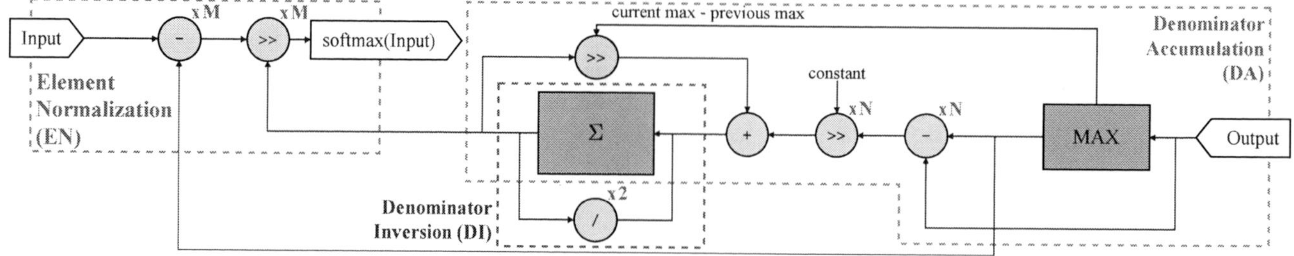

Fig. 4. Softmax implementation. Buffers are shown in blue.

ITA computes linear layers sequentially but fuses $Q \times K^T$ and $A \times V$ in iterations of i. In each final iteration of a $Q \times K^T$ tile, the softmax module accumulates denominators partially (*DA*). Once a row of $M \times J$ section of the attention matrix is completed, the softmax module inverts the denominator of the row (*DI*) and stores the inverted denominator. Then, M rows of $A \times V$ are computed while normalizing elements of A in the softmax module (*EN*). At the start of the next iteration of i, the softmax module is reset and the same steps are repeated until all iterations are completed.

IV. Softmax

We propose a novel hardware-friendly implementation of softmax, shown in Figure 4, with the following features:

- The softmax is computed on the quantized values directly.
- To prevent underflow, both the nominator and denominator are scaled with an integer value. Therefore, the accumulation and inversion of the denominator are performed in 15-bit and 16-bit integer formats, respectively.
- We add minimal memory overhead to store the maximum and sum values. Both maximum and sum buffers contain M elements, equal to the number of rows of a tile.
- We do not use any exponentiation modules and multipliers which are costly in terms of area and power.
- Softmax is computed on-the-fly and does not add any latency to the computation as shown in Figure 3.
- By computing softmax on streaming data, we avoid fetching the same vector multiple times, reducing the data movement and power consumption.

Our main observation is that above a certain value of the scaling factor, softmax quantizes to zero for all inputs except for the maximum of the input. This means that the range of the inputs can be clipped to the inputs that will end up with a softmax greater than 0, and the scaling factor can be tuned accordingly in training time as shown in Figure 5. Secondly, we can hide the factor $\log_2 e$ in the scaling factor ε and change the base to 2 to simplify the hardware, as follows:

$$e^x = e^{\varepsilon x_q} = (2^{\log_2 e})^{\varepsilon x_q} = 2^{((\log_2 e)\varepsilon) x_q} = 2^{\varepsilon' x_q} \quad (2)$$

where x_q is the quantized value ($x = \varepsilon x_q$) and $\varepsilon' = (\log_2 e)\varepsilon$.

The maximum meaningful scaling factor, computed based on the range of inputs with non-zero quantized softmax, is $\varepsilon = B/(2^B \log_2 e)$, where B is the number of bits used in quantized representation (equals 8 in our case). Using this scaling factor, ε' becomes:

$$\varepsilon' = (\log_2 e)\varepsilon = (\log_2 e)B/(2^B \log_2 e) = B/2^B \quad (3)$$

and softmax can be written as follows:

$$\begin{aligned} \mathrm{softmax}(\boldsymbol{x})_i &= \frac{2^{\varepsilon'(x_{qi} - \max(\boldsymbol{x_q}))}}{\sum_{j=1}^{n} 2^{\varepsilon'(x_{qj} - \max(\boldsymbol{x_q}))}} \\ &= \frac{2^{\frac{B}{2^B}(x_{qi} - \max(\boldsymbol{x_q}))}}{\sum_{j=1}^{n} 2^{\frac{B}{2^B}(x_{qj} - \max(\boldsymbol{x_q}))}} \\ &= \frac{2^{(x_{qi} - \max(\boldsymbol{x_q})) \gg (B - \log_2 B)}}{\sum_{j=1}^{n} 2^{(x_{qj} - \max(\boldsymbol{x_q})) \gg (B - \log_2 B)}} \end{aligned} \quad (4)$$

Using the above formula, the softmax module is implemented as depicted in Figure 4 and softmax is computed in three steps as shown in Figure 3. In *Denominator Accumulation (DA)*, we find the maximum of the first computed part of a row and store it in the MAX buffer. Then, we subtract the maximum from all the elements, accumulate the sum, and store it in the Σ buffer. When we get the next parts of the row, we compare the previous maximum that is stored in the MAX buffer with the current maximum. If the current maximum is greater, we update the maximum. The difference between the two maximums is used to update the accumulated sum in Σ and added up with the accumulation over the current part of the row. These operations are repeated over M rows of the attention matrix and the maximum and accumulated sum are stored in the respective buffers for each row. Once the denominator of the softmax is accumulated for a row in *DA*, the inverse of the denominator is computed in *Denominator Inversion (DI)* using serial dividers and stored in the Σ buffer. Since *DI* is overlapped with *DA*, we have plenty of time to compute the inverse of the denominator. Therefore, only two serial dividers suffice to compute the inverse without causing

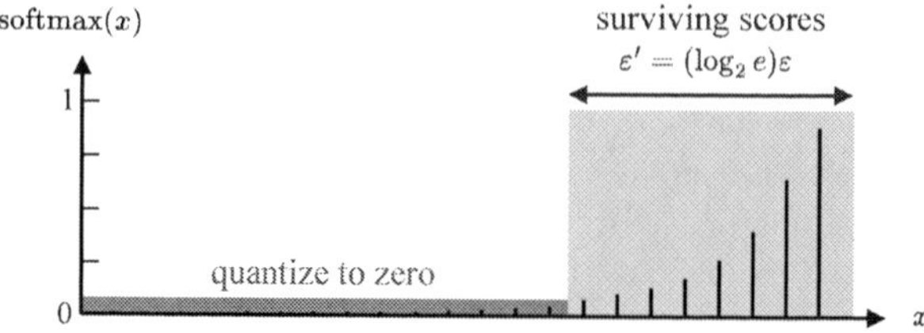

Fig. 5. Effect of softmax and quantization on attention probabilities.

any stalls. After obtaining the inverse of the denominator ($\Sigma_{inverse}$), we compute the softmax by shifting it as follows in *Element Normalization (EN)*:

$$\text{softmax}(\boldsymbol{x})_i = \Sigma_{inverse} \gg ((\max(\boldsymbol{x_q}) - x_{qi}) \gg (B - \log_2 B)) \quad (5)$$

As $B = 8$ is a constant in the architecture, a programmable shifter is not required for shifting by $B - \log_2 B$. Here, $B - \log_2 B$ evaluates to $8 - \log_2 8 = 5$, and we can simply take the most significant 3 bits of $(\max(\boldsymbol{x_q}) - x_{qi})$ to perform the shift.

V. Evaluation

A. Physical Implementation and Measurements

We evaluate ITA with 16 processing engines consisting of 64 multiply-accumulate (MAC) units ($N = 16$ and $M = 64$) and D is selected 24-bit to allow up to 256-element dot products, enough for the targeted compact models [4]. The memory buffers for weights and for storing the maximum and sum values in the softmax module are made of latch-based memories and clock-gated.

ITA is implemented in GlobalFoundries' 22FDX FD-SOI technology and targets an operating frequency of 500 MHz in worst-case conditions (SS/0.72 V/125 °C). Synopsys' Fusion Compiler 2022.03 is employed for both synthesis and implementation of the accelerator. The power consumption of ITA is estimated using Synopsys' PrimeTime 2022.03, which takes into account the switching activities obtained from a post-layout gate-level simulation using a synthetic benchmark at the operating frequency of 500 MHz. The power consumption is estimated under typical conditions (TT/0.80 V/25 °C).

B. Experimental Results

The total area occupied by ITA is 0.173 mm^2. The area breakdown of ITA is presented in Figure 6. The PEs take 58.1 % of the total area, while the weight buffer occupies 19.6 %. Others include the remaining components of ITA's datapath (6.3 %), control circuitry (2.3 %), and output buffer (1.1 %). The hardware-friendly softmax solution implemented in this work proves to be very area efficient, with only 3.3 % area contribution, corresponding to 28.7 KGE.

The entire accelerator consumes a total power of 60.5 mW over the execution of attention. Figure 6 shows the power breakdown of ITA. The majority of power is consumed in PEs with 59.5 %. Clock tree and I/O registers (22.9 %) also lead to significant power consumption due to their high toggling rate. Others consist of remaining datapath elements of ITA (6.7 %), weight buffer (1.7 %) and output buffer (0.7 %). The softmax module only consumes 1.4 % of the power. Although the weight buffer of ITA takes a significant portion of the area, its power consumption is less than 2 % due to clock-gating.

C. Softmax

To assess the accuracy of our softmax implementation, we compare the Mean Absolute Error (MAE) of our implementation with the 32-bit integer-only softmax from I-BERT [9]. We

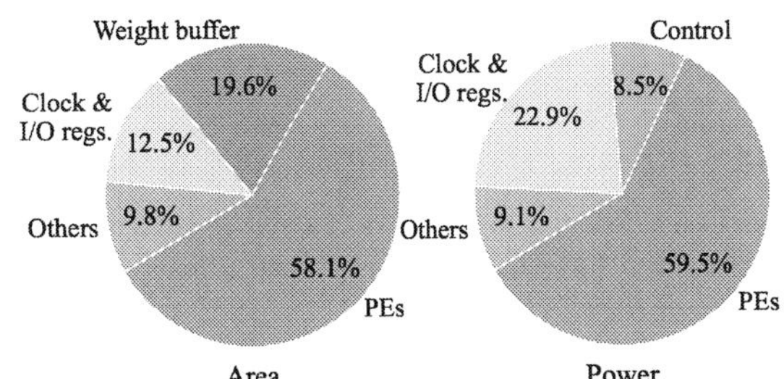

Fig. 6. Area and power breakdown of ITA.

use the activation of the Compact Transformer [18] as input in order to simulate the data distribution of a real transformer. Our implementation achieves an MAE of 4.6e−3, meaning that the average distance to the floating point value is 0.46 %. The MAE of I-BERT's softmax is 0.35 %, the slightly lower MAE is explained by the difference in input precision (32-bit for I-BERT vs 8-bit for ours). Compared to the I-BERT implementation which uses 32-bit multipliers and dividers, our approach operates at a lower precision and features a much simpler datapath, resulting in better latency and energy consumption.

D. Performance Evaluation

We compare ITA with a software baseline executed on MemPool, consisting of 256 32-bit RISC-V cores with single instruction, multiple data (SIMD) support [19]. We use a highly optimized kernel for matrix multiplications and the I-BERT algorithm for softmax. Compared to MemPool, ITA achieves 6× speedup and 45× energy efficiency in attention computation.

E. Comparison to State-of-the-Art

We present a comparison of ITA to state-of-the-art transformer accelerators in Table I. To have a fair comparison, we evaluate ITA as a standalone accelerator and integrate it into a system with 64 KiB static random-access memory (SRAM). The latter we call ITA System.

ITA achieves an energy efficiency of 16.9 TOPS/W standalone, and 8.46 TOPS/W integrated into the system, which is superior to all other accelerators except for Keller et al. [13]. If we hypothetically scale down the voltage to 0.46V, using V_{dd}^2 scaling, ITA would be 1.3× more efficient, and the system would be only 1.5× less efficient than [13], despite being implemented in much less advanced technology. Wang et al. [12] report higher efficiency in 12-bit, but only at lower voltage and with the assumption of 90 % sparsity. Furthermore, this sparsity exploitation reduces the area efficiency, which is much lower than ITA not only because of higher precision but also because of additional speculation and out-of-order execution logic.

ITA outperforms all other accelerators in terms of area efficiency, except for Keller et al.'s accelerator [13], which uses a 5 nm technology. To provide a technology-independent metric, we present the area efficiency in terms of gate-equivalent (GE) as well and ITA surpasses all accelerators both as a standalone accelerator and at the system level.

TABLE I
COMPARISON OF THE PROPOSED ARCHITECTURE TO STATE-OF-THE-ART TRANSFORMER ACCELERATORS.

	OPTIMUS [14]	SpAtten [15]	ELSA [16]	Wang et al. [12]	Keller et al. [13]		This work	
					INT4	INT8	ITA	ITA System
Technology [nm]	28	40	40	28	5	5	22	22
Area [mm^2]	5.2	18.71	1.26/2.15[1]	6.82	0.153	0.153	0.173	0.407
Voltage [V]	1	1.1	1.1	0.56-1.1	0.46-1.05	0.46-1.05	0.8	0.8
Frequency [MHz]	200	1000	1000	50-510	152-1760	152-1760	500	500
Data formats	INT16	INT8-16/FP	INT8/FP16	INT12	INT4	INT8	INT8	INT8
Number of MAC units	1024	1024	528	512	1024	512	1024	1024
On-chip memory [KB]	1420	392	4.5/148[1]	336	141	141	2.24	67.8
Power [mW]	731.8	2600	969.4/1494.2[1]	12.06-272.8	-	-	60.5	121
Throughput [TOPS]	0.5	1.61	1.09	0.52-4.07[2]	3.6[4]	1.8[4]	1.02	1.02
Energy efficiency [TOPS/W]	0.68	0.62	1.12/0.73[1]	1.91-27.56[3]	**95.6**[5]	**39.1**[5]	16.9	8.46
Area efficiency [TOPS/mm^2]	0.096	0.086	0.865/0.507[1]	0.0765-0.597[3]	**23.3**[4]	**11.7**[4]	5.93	2.52
Area efficiency*[TOPS/MGE]	0.0310	0.0566	0.569/0.333[1]	0.0247-0.192[3]	0.242[4]	0.121[4]	**1.18**	**0.500**

* Gate-equivalents (GE) of other technologies are scaled based on the GE of 22 nm technology.
[1] with external memory modules (on-chip SRAM).
[2] measured at 1.1 V, 510 MHz. The highest throughput is measured for $A \times V$ assuming 90 % near-zero and zero probabilities.
[3] the lowest value measured at 1.1 V, 510 MHz for linear layers that compute Q, K, V and the highest value measured at 0.56 V, 50 MHz for $A \times V$ assuming 90 % near-zero and zero probabilities.
[4] at 1.05 V.
[5] at 0.46 V.

VI. Conclusion

We presented ITA, a hardware accelerator for quantized transformers that exploits parallelism of attention mechanism and 8-bit integer quantization to achieve efficient inference on embedded systems. Our architecture features a novel and hardware-friendly softmax implementation that operates directly on quantized values and facilitates weight stationary dataflow, reducing power consumption. ITA is evaluated on an advanced 22 nm technology, achieving energy efficiency of 16.9 TOPS/W and area efficiency of 5.93 TOPS/mm^2.

Acknowledgment

This work is supported in part by the NeuroSoC project funded under Horizon Europe Grant Agreement n° 101070634.

References

[1] A. Vaswani, N. Shazeer, N. Parmar, J. Uszkoreit, L. Jones, A. N. Gomez, L. Kaiser, and I. Polosukhin, "Attention is all you need," 2017. [Online]. Available: http://arxiv.org/abs/1706.03762

[2] J. Devlin, M.-W. Chang, K. Lee, and K. Toutanova, "BERT: Pre-training of deep bidirectional transformers for language understanding," 2018. [Online]. Available: http://arxiv.org/abs/1810.04805

[3] A. Radford, J. Wu, R. Child, D. Luan, D. Amodei, and I. Sutskever, "Language models are unsupervised multitask learners," 2019.

[4] A. Dosovitskiy, L. Beyer, A. Kolesnikov, D. Weissenborn, X. Zhai, T. Unterthiner, M. Dehghani, M. Minderer, G. Heigold, S. Gelly, J. Uszkoreit, and N. Houlsby, "An image is worth 16x16 words: Transformers for image recognition at scale," 2020. [Online]. Available: https://arxiv.org/abs/2010.11929

[5] P. Verma and J. Berger, "Audio transformers: Transformer architectures for large scale audio understanding. Adieu convolutions," 2021. [Online]. Available: https://arxiv.org/abs/2105.00335

[6] G. Bertasius, H. Wang, and L. Torresani, "Is space-time attention all you need for video understanding?" 2021. [Online]. Available: https://arxiv.org/abs/2102.05095

[7] N. Benaich and I. Hogarth, "State of AI report 2022," 2022. [Online]. Available: https://www.stateof.ai/2022-report-launch.html

[8] A. Burrello, M. Scherer, M. Zanghieri, F. Conti, and L. Benini, "A microcontroller is all you need: Enabling transformer execution on low-power IoT endnodes," in *2021 IEEE International Conference on Omni-Layer Intelligent Systems (COINS)*, 2021, pp. 1–6.

[9] S. Kim, A. Gholami, Z. Yao, M. W. Mahoney, and K. Keutzer, "I-BERT: Integer-only BERT quantization," 2021. [Online]. Available: https://arxiv.org/abs/2101.01321

[10] O. Zafrir, G. Boudoukh, P. Izsak, and M. Wasserblat, "Q8BERT: Quantized 8bit BERT," in *2019 Fifth Workshop on Energy Efficient Machine Learning and Cognitive Computing - NeurIPS Edition (EMC2-NIPS)*, 2019, pp. 36–39.

[11] A. Bhandare, V. Sripathi, D. Karkada, V. Menon, S. Choi, K. Datta, and V. Saletore, "Efficient 8-bit quantization of transformer neural machine language translation model," 2019. [Online]. Available: http://arxiv.org/abs/1906.00532

[12] Y. Wang, Y. Qin, D. Deng, J. Wei, Y. Zhou, Y. Fan, T. Chen, H. Sun, L. Liu, S. Wei, and S. Yin, "A 28nm 27.5TOPS/W approximate-computing-based transformer processor with asymptotic sparsity speculating and out-of-order computing," in *2022 IEEE International Solid-State Circuits Conference (ISSCC)*, vol. 65, 2022, pp. 1–3.

[13] B. Keller, R. Venkatesan, S. Dai, S. G. Tell, B. Zimmer, W. J. Dally, C. Thomas Gray, and B. Khailany, "A 17-95.6 TOPS/W deep learning inference accelerator with per-vector scaled 4-bit quantization for transformers in 5nm," in *2022 IEEE Symposium on VLSI Technology and Circuits*, 2022, pp. 16–17.

[14] J. Park, H. Yoon, D. Ahn, J. Choi, and J.-J. Kim, "OPTIMUS: Optimized matrix multiplication structure for transformer neural network accelerator," in *MLSys*, 2020.

[15] H. Wang, Z. Zhang, and S. Han, "SpAtten: Efficient sparse attention architecture with cascade token and head pruning," 2021, pp. 97–110. [Online]. Available: https://arxiv.org/abs/2012.09852

[16] T. J. Ham, Y. Lee, S. H. Seo, S. Kim, H. Choi, S. J. Jung, and J. W. Lee, "ELSA: Hardware-software co-design for efficient, lightweight self-attention mechanism in neural networks," in *2021 ACM/IEEE 48th Annual International Symposium on Computer Architecture (ISCA)*, 2021, pp. 692–705.

[17] J. R. Stevens, R. Venkatesan, S. Dai, B. Khailany, and A. Raghunathan, "Softermax: Hardware/software co-design of an efficient softmax for transformers," in *2021 58th ACM/IEEE Design Automation Conference (DAC)*, 2021, pp. 469–474.

[18] A. Hassani, S. Walton, N. Shah, A. Abuduweili, J. Li, and H. Shi, "Escaping the big data paradigm with compact transformers," 2021. [Online]. Available: https://arxiv.org/abs/2104.05704

[19] M. Cavalcante, S. Riedel, A. Pullini, and L. Benini, "MemPool: A shared-L1 memory many-core cluster with a low-latency interconnect," in *2021 Design, Automation, and Test in Europe Conference and Exhibition (DATE)*, 2021, pp. 701–706.

Energy-Efficient RISC-V-Based Vector Processor for Cache-Aware Structurally-Pruned Transformers

Jung Gyu Min, Dongyun Kam, Younghoon Byun, Gunho Park, and Youngjoo Lee
Department of Electrical Engineering, POSTECH, Pohang, Korea
{mjg1104, rkaehddbs, byh1321, gunho1123, youngjoo.lee}@postech.ac.kr

Abstract—**Based on recent RISC-V designs, we present in this paper a low-power vector processor architecture for efficiently deploying vision transformer (ViT) models. To fairly measure the processing efficiency of different processor designs with instruction/data cache memories, we first develop the evaluation framework based on numerous design tools for jointly considering the algorithm, architecture, and circuit performances together, numerically revealing that the previous CSR-based data compression cannot accelerate pruned transformer models at all due to under-utilization of the vector-extended processing units. We then introduce a series of algorithm-hardware co-optimization approaches to greatly minimize cache misses by applying 1) the accuracy-preserved structured ViT pruning, 2) the vertical-CSR (vCSR) data storing format, and 3) vCSR-aware custom memory-accessing instructions. Experimental results show that the proposed optimization schemes eventually improve the processing efficiency of pruned transformers in resource-limited computing platforms, e.g., achieving 11 times lower energy consumption for handling the 0.7-pruned ViT model.**

Index Terms—**Embedded vector processor, machine learning, RISC-V, transformer compression**

I. Introduction

The RISC-V instruction-set architecture (ISA) for embedded devices has continuously evolved to increase CPU performance and the recent extension even supports vector arithmetic operations for utilizing data-intensive workloads such as machine learning (ML) models [1]. Due to the increased number of parameters as well as the computational complexity, however, the advanced transformer models achieving state-of-the-art performance cannot be deployed in the contemporary RISC-V-based computing platforms in practice, requiring additional accelerator associated with high-bandwidth on-/off-chip memories [2]. For realizing memory-reduced ML operations for resource-limited devices, the pruning-based model compression gains popularity to reduce the overall model size without degrading accuracy [3]; however, due to the serialized decompress operations for a specific compressed data format, e.g., compressed sparse row (CSR) form, the system-level efficiency for operating recent pruned-transformers in general-purposed platforms cannot be improved as expected [4].

This research was supported by the Super Computer Development Leading Program of the National Research Foundation of Korea(NRF) funded by the Korean government (Ministry of Science and ICT(MSIT)) (No. : 2020M3H6A1085498), the MSIT, Korea, under the Information Technology Research Center support program (IITP-2021-2020-0-01461) and under the Grand Information Technology Research Center support program (IITP-2022-2020-0-01612) supervised by the IITP (Institute for Information & Communications Technology Planning & Evaluation).

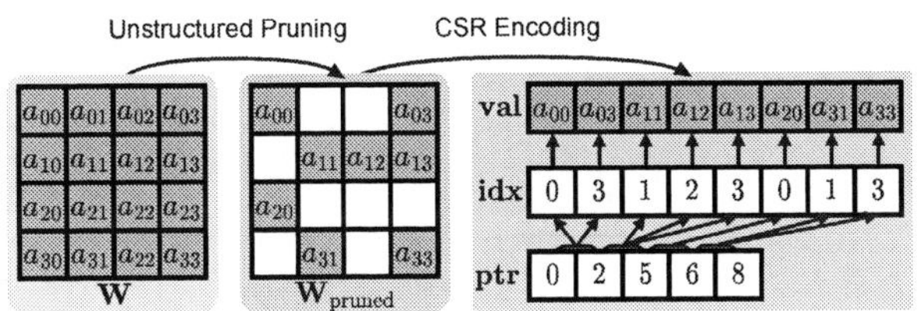

Fig. 1. CSR format for a randomly pruned matrix [7].

In this paper, we present novel algorithm-hardware co-optimization approaches to efficiently operate the pruned-transformer inference at the RISC-V-based processor platform. The hardware evaluation framework is developed to accurately measure the vision transformer (ViT) [5] efficiency with various design configurations by calculating the processing cycles from Gem5 simulator [6] and by estimating the power consumption from Synopsys Design Compiler, reporting the hardware-level pruned-ViT performance with the conventional CSR format on the vector-extended RISC-V architecture for the first time. To enhance the overall processing efficiency, we then present an advanced structured pruning method that forces the survived weight patterns to form the same-size vectors, maintaining the accuracy of the dense-ViT model. Adopting the conventional horizontal-CSR (hCSR) data encoding, accessing structured patterns from the data cache can be simplified, improving the hardware performance by reducing the number of instructions. Considering the cache miss ratio, we then propose the vertically-pruned vector patterns with the vertical-CSR (vCSR) format, greatly reducing the memory-accessing cycles while achieving attractive recognition accuracy. In addition, the vCSR-aware memory-accessing instructions are newly defined as custom instructions, further reducing the number of activated instructions for saving energy consumption. As a result, the fully-optimized vector processor reduces the energy dissipation by 11.0 times for operating the 0.7-pruned ViT model compared to the baseline processing.

II. Background

A. Sparse Matrix Computations on RISC-V Vector Extension

The RISC-V ISA extensions include vector operations performing multiple data elements simultaneously, utilizing the

979-8-3503-1176-1/23 $31.00

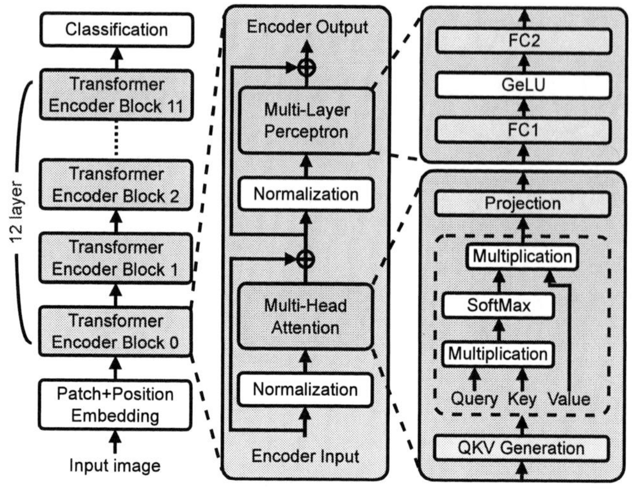

Fig. 2. Layer structure of vision transformer (ViT) [5].

TABLE I
RATIO OF PROCESSING TIME FOR DENSE VIT PROCESSING

	QKV Generation	Projection	FC1	FC2	Others
Ratio	0.22	0.08	0.29	0.36	0.05

data-level parallelism by realizing the vector processing units on the baseline scalar core [8]. Note that the vector instructions in the RISC-V ISA extensions consider resource-limited environments, i.e., introducing only simple element-wise parallel operations on dense vectors. Considering the vector computations from sparse matrices for activating pruned-transformer models on lightweight devices, however, the current vector extension of RISC-V ISA cannot enhance overall processing efficiency. More precisely, the conventional sparse matrices are in general encoded by the well-known CSR forms to effectively reduce the memory footprints according to the sparsity level, with the detailed format employing three components (**val**, **idx**, **ptr**) shown in Fig. 1 [7]. However, computing CSR-encoded matrices requires additional serialized steps to regenerate dense vectors whose elements are only nonzero values in the sparse matrices. High-performance GPU platforms may support the dedicated computing kernels for CSR-encoded data, greatly accelerating the sparse matrix computations by minimizing these format-oriented overheads [9]. Since decompressing CSR-encoded data should be performed on the RISC-V scalar core with a series of element-wise memory-accessing instructions, on the other hand, the sparse matrix computations require even more processing time than the dense version of the same matrix. Moreover, cache misses rapidly increase especially for small-size cache memories of resource-limited computing platforms when handling the randomly-sparse matrices with CSR-encoded forms [10], requiring extra processes on accessing the lower-level memories.

B. Transformer Pruning for Model Compression

Figure 2 illustrates the general structure of a vision transformer, consisting of four weight matrix multiplications: QKV generation, projection, and two fully-connected blocks (FC1 and FC2). Due to the numerous multiply-and-accumulate (MAC) operations for performing these weight matrix multiplications, the deployment of parallel MAC operators for accelerating the overall processing speed has gained wide acceptance while simply performing the remaining operations (including non-linear functions) in a serialized manner. [4]. When operating the baseline ViT model [11] on a scalar processor realizing RV32IMF ISA with the L1 cache of 2KB, as shown in Table I, the four weight matrix operations occupy more than 95% of overall inference time. Therefore, pruning-based transformer compressing becomes essential to reduce the number of weights and construct sparse matrix computations while maintaining recognition accuracy. For example, [12] reports that weight pruning on ViT model can remove more than 50% of total weights with only a 0.29% accuracy drop. The memory requirement is also relaxed by adopting a sparse-matrix compression method like the CSR format.

Although the number of MAC operations and memory footprint can be reduced according to the pruning ratio, achieving the desired speed-up is challenging, even with the vector-extended ISA, due to the time-consuming memory access sequences at the scalar core constructing a dense vector with only surviving weights [13]. This results in a mere 2.10x speed-up from the 0.8-pruned ViT model on the RISC-V vector extension core with 2KB cache memories, which is far less than expected by the pruning ratio. Therefore, efficient acceleration of pruned-transformer models requires considering both the hardware-aware transformer pruning method with efficient data compression and the pruning-aware vector processor architecture with practical resources.

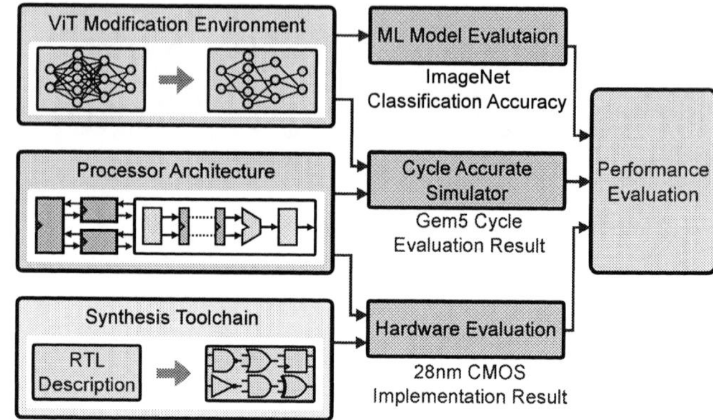

Fig. 3. Proposed performance evaluation framework.

III. PROPOSED PERFORMANCE EVALUATION FRAMEWORK

For accurate comparisons of different vector-extended RISC-V architectures running end-to-end transformer inference operations, in this work, we newly introduce advanced performance evaluation frameworks by combining various existing development environments. As shown in Fig. 3, the proposed framework consists of four parts: 1) the ML model evaluation part, 2) the cycle evaluation part, 3) the hardware evaluation part, and 4) the efficiency evaluation part

TABLE II
ACCURACY RESULTS OF VIT MODELS WITH UNSTRUCTURED PRUNING

Computing Method	Pruning ratio	Accuracy (%)	Cycle ($\times 10^9$)		
			Ideal	Additional	Total
Scalar	0	81.80	580.08	-	580.08
Vector	0	81.80	130.43	-	130.43
Vector*	0.5	81.34	65.21	99.08	164.30
	0.6	81.28	52.71	58.50	111.21
	0.7	81.16	39.13	47.48	86.61
	0.8	80.50	26.08	35.90	61.99

* Encoded by the conventional CSR format [7]

TABLE III
EVALUATION RESULTS OF PRUNED PROJECTION LAYER

Encoding Method	Vector length	Accuracy (%)	Cycle ($\times 10^6$)		
			Ideal	Additional	Total
CSR*	2	81.16	630.44	143.67	774.11
	4		295.92	316.51	612.42
	8		168.55	356.82	525.36
hCSR**	2	81.01	630.44	0.54	630.98
	4	80.74	295.92	3.00	298.92
	8	79.48	168.55	11.58	180.12

* Unstructured pruning
** Structured pruning with given vector length

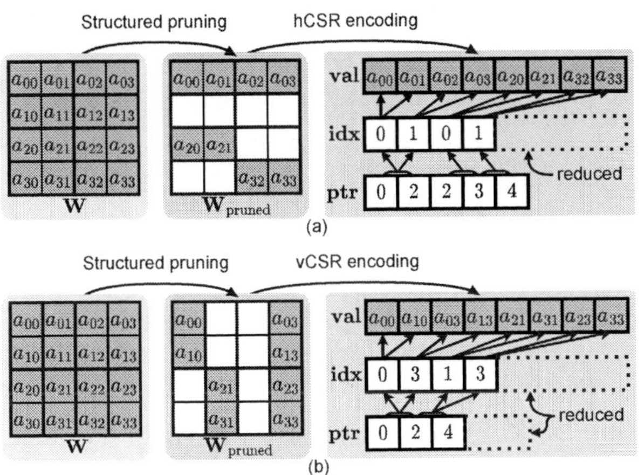

Fig. 4. (a) Previous hCSR format for horizontally-structured pruning patterns [15], (b) and proposed vCSR format for vertically-structured sparse matrices.

for generating the final processing efficiency of end-to-end transformer operations by collecting all the evaluation results. More specifically, ML model evaluation is performed using Pytorch environments to analyze the recognition accuracy on ImageNet benchmark [14] for various pruning approaches applied to the DeiT-based ViT model [11]. With the architectural description, our framework includes Gem5 simulator for the cycle evaluation, fairly investigating the advantages of different RISC-V configurations, e.g., vector operators and cache memory structures. Accepting RTL descriptions, we finally adopt Synopsys Design Compiler for the gate-level hardware evaluation, calculating energy consumption as well as the area complexity in 28nm CMOS technology.

Utilizing dedicated tools for each evaluation step, in contrast to the previous works independently focusing on the limited parts [1] [16], the proposed framework offers a wide range of performance evaluations from the recognition accuracy to the end-to-end processing efficiency, which can allow finding the system-level optimization approaches considering algorithm, architecture, and implementation all together. For example, as reported in Table II, the processing latency of various pruned-ViT models can be examined across different RISC-V-based processor architectures. Considering the embedded devices, small-size cache memories are assumed in this case study, i.e., 2KB L1 cache and 64KB L2 cache. With the well-known magnitude-based pruning method resulting in random sparsity [3], the pruning ratio has been set from 0.5 to 0.8 by considering the significant compression performance [9] and acceptable accuracy drops [12], where the conventional CSR method in Fig. 1 is applied to encode the pruned-ViT models. Note that the processing latency can be significantly reduced due to the acceleration of operations and the reduction of L2 cache access by introducing new vector instructions operated by additional 4-lane vector arithmetic operators. Due to the serialized decompressing operations of CSR-encoded sparse matrices, however, despite accommodating a significant amount of accuracy drop, only the aggressively pruned-ViT models can enjoy additional performance improvements, which are far from the ideal latency only for the non-zero computations as depicted in Table II. To further improve pruned-ViT processing on the vector-extended RISC-V computing platform, in this paper, we newly introduce several algorithm-hardware co-optimization schemes for minimizing the unwanted CSR-oriented processing overheads effectively.

IV. PROPOSED OPTIMIZATION APPROACHES

A. Cache-Aware Structured Pruning with vCSR Encoding

In order to reduce additional overheads for handling CSR-encoded data, grouped CSR format can be applied to represent the sparse matrices [15]. Figure 4(a) depicts how the previous horizontal CSR (hCSR) format reduces the number of **idx** values. More precisely, unlike the original CSR method, a single index of hCSR format represents multiple elements in a row-wise direction, decreasing memory accesses per element.

To efficiently utilize the hCSR method, it is required to make the structured pruning patterns to maximally utilize the data-level parallelism, i.e., setting the same vector length at the pruning phase and operator architecture. Introducing larger vectors for massive-parallel operations generally increases the processing efficiency by relaxing the CSR-oriented overheads; however, structured pruning with large vectors may significantly sacrifice the accuracy [17]. Therefore, selecting an ideal vector length becomes vital for optimizing performance while preserving acceptable accuracy. Changing the vector length, using the proposed evaluation framework, Table III reports the cycle consumed on the projection layer in a single block of the 0.7-pruned ViT model for each processing option, numerically showing the gradual performance improvement according to the vector length. Considering the accuracy results, we select

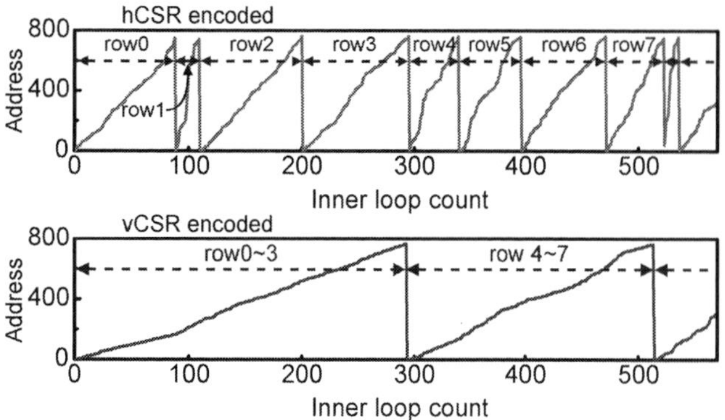

Fig. 5. Address distributions for loading elements from pruned-ViT models using different encoding schemes.

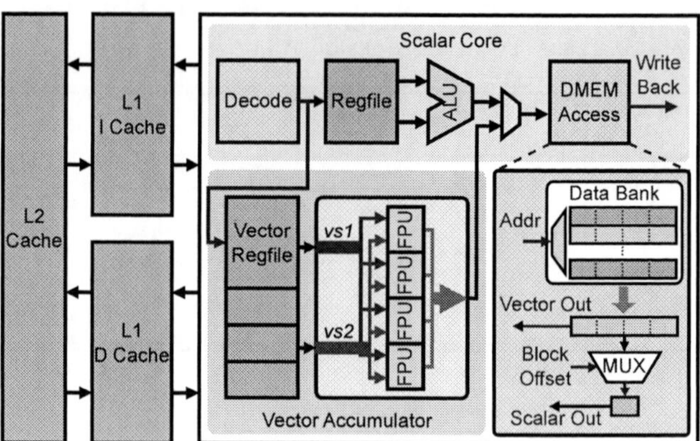

Fig. 6. Block diagram of the proposed vector processor.

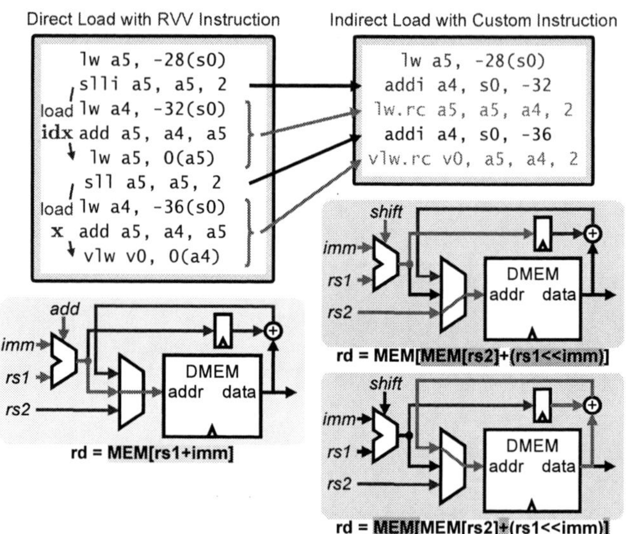

Fig. 7. Performance improvements by the custom memory instructions.

the optimal vector length to be four in this work, allowing only a 1.1% accuracy drop compared with the unpruned version.

The processing latency can be further optimized by carefully managing the cache misses with small-sized cache memories in the resource-limited computing platforms [18]. The processing steps for CSR-based data are generally composed of two loops for computing a sparse matrix, where the inner loop finds the non-zero elements in a row and the outer loop counts the number of processed rows. The previous hCSR scheme in fact accelerates the inner loop processing by grouping the row-wise elements; therefore, the accessing address of the input vector **x** increases rapidly for fetching multiple elements inside of a row, resulting in numerous cache misses in resource-limited environments. In this work, we propose column-wise structured pruning, where the survived patterns are based on vertical vectors. Then, as shown in Fig. 4(b), the vCSR format is newly introduced to group non-zero weights in a vertical direction, still reducing the number of **idx** values similar to the previous hCSR method. In contrast to the horizontal weight accesses in the previous work, as depicted in Fig. 5, the new vCSR approach greatly improves the spatial locality of data accessing patterns as it accelerates the outer loops by handling multiple rows simultaneously. Combined with structured pruning with vertical vectors, therefore, we can reduce cache misses by adopting the vCSR format, leading to an additional latency reduction for pruned-ViT processing.

B. Vector-Extended RISC-V Design with Custom Instructions

To fairly estimate the energy consumption of end-to-end pruned-ViT processing in our evaluation framework, we designed the vector processor architecture in 28nm CMOS technology, which is dedicated to the proposed algorithm-level optimization schemes. The proposed processor essentially implements the recent RVV 1.0 ISA [8] for realizing the vector-level arithmetic operations, where the overall architecture is illustrated in Fig. 6. The vector arithmetic units have four processing lanes to efficiently support the data-level parallelism according to the proposed algorithm-level optimization, i.e., structured ViT pruning with the new vCSR format. Note that we also implemented SRAM-based L1 cache memories for instruction and data accesses, and the SRAM-based L2 cache is realized as a unified last-level cache. The memory-accessing unit is also carefully designed to fetch multiple non-zero weights described by CSR-based formats, where the address calculation steps are performed in the scalar core.

In addition to the baseline RVV 1.0 ISA, we newly define custom instructions to reduce the energy consumption from instruction cache accesses. More precisely, as depicted in Fig. 7, the custom instructions are related to the memory-accessing operations with indirect accessing mode. As exemplified in the figure, the proposed indirect load instruction eliminates the previous address calculation steps for finding the data address of CSR-encoded non-zero weights. The dedicated address calculator is also developed to prevent increasing the critical delay of the baseline RVV 1.0 processor, i.e., we simply re-utilize the existing data path and resources to support the indirect load operations. Note that the custom instructions preserve data memory accesses where instruction memory accesses can be greatly reduced as depicted in Fig. 7. To

TABLE IV
END-TO-END VIT EVALUATION RESULTS OF DIFFERENT PROCESSING OPTIONS

Method			Scalar	Vector	CSR	PRO1	PRO2	PRO3	CSR	PRO1	PRO2	PRO3	CSR	PRO1	PRO2	PRO3
Pruning ratio			0		0.5				0.6				0.7			
Top-1 Accuracy (%)			81.80		81.34	81.23	81.20		81.28	81.04	81.05		81.16	80.74	80.72	
L1 D Cache Size	1KB	Cycle ($\times 10^9$)	638.8	174.3	184.5	79.6	77.4	76.9	128.6	65.7	63.6	63.2	102.2	52.1	50.1	49.7
		Energy (J)	14.1	5.97	6.42	2.71	2.52	2.32	4.20	2.13	2.03	1.86	3.27	1.65	1.55	1.42
	2KB	Cycle ($\times 10^9$)	580.1	130.4	164.4	69.8	64.1	63.6	110.2	58.3	53.1	52.7	86.6	46.5	42.2	41.9
		Energy (J)	13.8	4.80	6.22	2.58	2.24	2.07	3.92	2.05	1.81	1.67	3.02	1.59	1.39	1.26
	4KB	Cycle ($\times 10^9$)	408.8	116.2	150.0	58.8	57.1	56.6	98.2	49.3	47.3	47.0	76.2	39.8	37.6	37.3
		Energy (J)	11.8	4.35	5.81	2.29	2.24	2.07	3.92	1.94	1.81	1.67	2.96	1.52	1.38	1.27
	8KB	Cycle ($\times 10^9$)	366.9	110.0	141.0	53.9	53.4	52.9	89.9	44.9	44.3	43.9	70.0	35.9	35.2	34.9
		Energy (J)	14.2	5.36	7.46	2.75	2.53	2.33	4.27	2.11	2.04	1.87	3.22	1.63	1.55	1.42
	16KB	Cycle ($\times 10^9$)	354.6	108.6	137.8	51.9	51.9	51.4	86.3	43.2	43.0	42.6	67.3	34.6	34.3	33.9
		Energy (J)	20.6	6.94	9.88	3.57	3.27	3.02	5.45	2.70	2.64	2.42	4.12	2.08	2.00	1.83

handle the same vCSR-encoded pruned-ViT model, as a result, we can expect additional energy reduction by inactivating the L1 instruction cache in Fig. 6. Moreover, the reduced number of instructions also reduces cache misses, further saving the energy consumption caused by L2-memory accessing cases.

V. EXPERIMENTAL RESULTS

Using the proposed evaluation framework to demonstrate the impacts of the proposed algorithm-hardware co-optimization approaches, we realized end-to-end ViT operations on various vector architectures in 28nm CMOS technology, equally running at the operating speed of 500 MHz with 2KB L1 I cache and 64KB L2 cache. Varying the cache memory size, Table IV summarizes the experimental results of different model/hardware configurations regarding recognition accuracy, processing latency, and energy consumption. For simplicity, we define three proposed configurations denoted as PRO1, PRO2, and PRO3, gradually adopting the structure ViT pruning with a vector length of four, the vCSR format, and the cache-aware custom instructions, respectively. To preserve model accuracy close to the baseline unstructured pruning, as detailed in Table III, we set the vector length as four to realize PRO1 method with structured sparsity. Note that the PRO2 design with vertical length-4 vectors provides similar accuracy to the horizontal grouping used for the PRO1 option, achieving a recognition accuracy of 80.72% for the 0.7-pruned model. Compared with the unpruned ViT operations, the latency reduction by pruned-ViT models encoded by the conventional CSR method is generally limited due to the large portion of additional processing cycles for the serialized decoding operations as detailed in Table II. In contrast, the proposed optimization schemes successfully solve the previous limitations by applying the hardware-aware structured pruning methods and the algorithm-aware hardware designs. Targeting the 0.7-pruned ViT model with 2KB cache memories, the processing speed is improved by 1.86 times by utilizing PRO1 option compared with the previous unstructured pruning method. Subsequently, we reduce the latency by 10.2% with PRO2 method, eventually reducing processing cycles by 3.11 times,

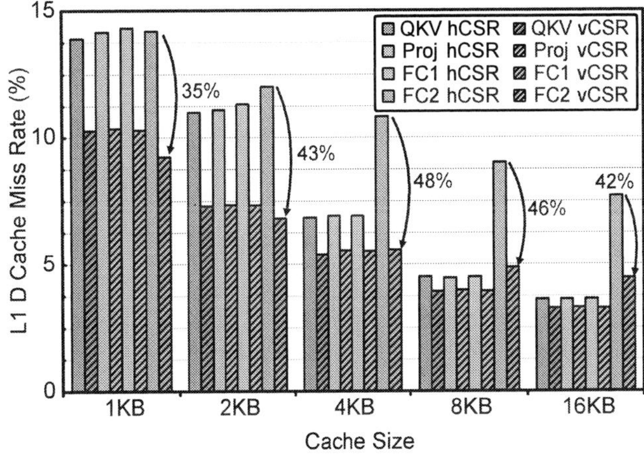

Fig. 8. Cache miss ratios of ViT linear layers with structured pruning patterns encoded by hCSR and vCSR formats.

compared with the unpruned ViT model running at the RVV 1.0 architecture [8]. Note that the PRO3 version minimizes energy consumption by reducing instruction memory accesses without increasing the overall processing cycles, saving total energy consumption of end-to-end pruned-ViT processing by 11.0 times from the scalar core, as detailed in Table IV.

Note that the performance improvements from the proposed optimization schemes are basically from the reduced cache miss ratios by increasing the locality of data accessing patterns, allowing to fully enjoy the advantages from pruned-ViT models even with the small-size L1 cache memories. In order to provide detailed analyses, Fig. 8 shows how the proposed PRO2 design reduces the miss ratio of L1 data cache compared with the PRO1 architecture. As the vCSR-based PRO2 method reduces the range of data access patterns by accelerating outer-loop operations rather than the inner counterparts, which is described in Fig. 5, it is clear that we can greatly lower the miss cases, especially for the small cache memories. For large L1 cache memories, moreover, we can still offer a sufficient amount of miss-rate reductions for FC2 calculations whose

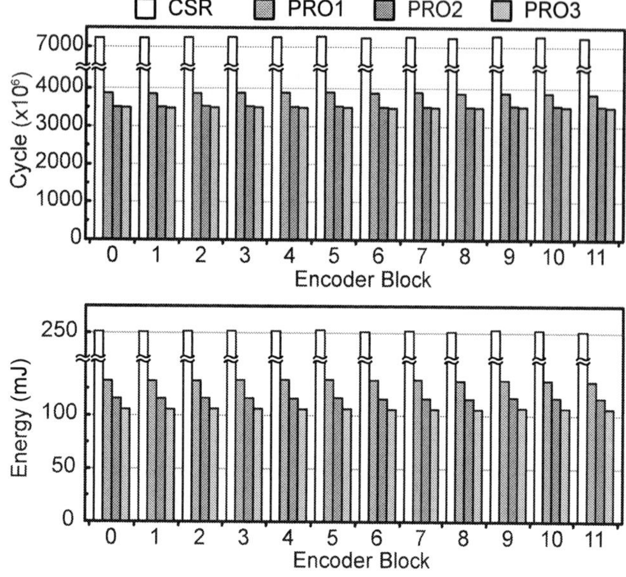

Fig. 9. Block-wise evaluation of processing cycles and energy consumption for operating 0.7-pruned ViT model.

corresponding matrix is a very long rectangle, maximizing the locality advantages from the vCSR format. Note that a large cache obviously reduces the number of cycles by increasing the overall hardware complexity; therefore, the reasonable cache size is determined to be 2KB by considering the processing latency as well as the energy consumption.

Fig. 9 finally illustrates how the proposed optimizations gradually improve the processing cycles for each encoder block of the 0.7-pruned ViT model, where the cache memory size is set to 2KB. It is noticeable that the proposed optimization schemes always improve the performance regardless of the encoder block, showing the generality of our approaches focusing on the memory-accessing patterns. Therefore, the proposed work successfully offers advanced algorithm-hardware co-optimization schemes for developing the efficient vector processor design dedicated to the pruned ML processing for the recent transformer models.

VI. Conclusion

We have presented in this paper novel algorithm-hardware co-optimization schemes for pruned-ViT models on resource-limited vector processor architecture. The advanced evaluation framework is developed by carefully connecting the existing design environments, catching the essential insights for optimizing the end-to-end ViT processing. To minimize the extra overheads for constructing the vector operands from CSR-based representations, we introduce vector-level structured pruning with the vCSR format, greatly improving the data-accessing locality. Subsequently, custom instructions are newly designed for vCSR-encoded patterns, further saving energy consumption. For the resource-limited platforms with 2KB L1 cache, the proposed work achieves 2x latency and 2.4x energy reductions compared to the previous pruning-based design.

References

[1] A. Garofalo, M. Rusci, F. Conti, D. Rossi, and L. Benini, "Pulp-nn: A computing library for quantized neural network inference at the edge on risc-v based parallel ultra low power clusters," in *2019 26th IEEE International Conference on Electronics, Circuits and Systems (ICECS)*. IEEE, 2019, pp. 33–36.

[2] H. Genc, S. Kim, A. Amid, A. Haj-Ali, V. Iyer, P. Prakash, J. Zhao, D. Grubb, H. Liew, H. Mao, A. Ou, C. Schmidt, S. Steffl, J. Wright, I. Stoica, J. Ragan-Kelley, K. Asanovic, B. Nikolic, and Y. S. Shao, "Gemmini: Enabling systematic deep-learning architecture evaluation via full-stack integration," in *2021 58th ACM/IEEE Design Automation Conference (DAC)*, 2021, pp. 769–774.

[3] S. Han, H. Mao, and W. J. Dally, "Deep compression: Compressing deep neural networks with pruning, trained quantization and huffman coding," *arXiv preprint arXiv:1510.00149*, 2015.

[4] J. Park, H. Yoon, D. Ahn, J. Choi, and J.-J. Kim, "OPTIMUS: OPTImized matrix multiplication structure for transformer neural network accelerator," *Proceedings of Machine Learning and Systems*, vol. 2, pp. 363–378, 2020.

[5] A. Dosovitskiy, L. Beyer, A. Kolesnikov, D. Weissenborn, X. Zhai, T. Unterthiner, M. Dehghani, M. Minderer, G. Heigold, S. Gelly *et al.*, "An image is worth 16x16 words: Transformers for image recognition at scale," *arXiv preprint arXiv:2010.11929*, 2020.

[6] N. Binkert, B. Beckmann, G. Black, S. K. Reinhardt, A. Saidi, A. Basu, J. Hestness, D. R. Hower, T. Krishna, S. Sardashti *et al.*, "The gem5 simulator," *ACM SIGARCH computer architecture news*, vol. 39, no. 2, pp. 1–7, 2011.

[7] A. Buluç, J. T. Fineman, M. Frigo, J. R. Gilbert, and C. E. Leiserson, "Parallel sparse matrix-vector and matrix-transpose-vector multiplication using compressed sparse blocks," in *Proceedings of the twenty-first annual symposium on Parallelism in algorithms and architectures*, 2009, pp. 233–244.

[8] K. Asanovic *et al.*, "RISC-V Vector Extension 1.0," [Online] Available :https://github.com/riscv/riscv-v-spec/releases/tag/v1.0, 2021.

[9] J. L. Greathouse and M. Daga, "Efficient sparse matrix-vector multiplication on gpus using the csr storage format," in *SC'14: Proceedings of the International Conference for High Performance Computing, Networking, Storage and Analysis*. IEEE, 2014, pp. 769–780.

[10] X. Feng, H. Jin, R. Zheng, K. Hu, J. Zeng, and Z. Shao, "Optimization of sparse matrix-vector multiplication with variant csr on gpus," in *2011 IEEE 17th International Conference on Parallel and Distributed Systems*. IEEE, 2011, pp. 165–172.

[11] H. Touvron, M. Cord, M. Douze, F. Massa, A. Sablayrolles, and H. Jégou, "Training data-efficient image transformers & distillation through attention," in *International conference on machine learning*. PMLR, 2021, pp. 10 347–10 357.

[12] T. Chen, Y. Cheng, Z. Gan, L. Yuan, L. Zhang, and Z. Wang, "Chasing sparsity in vision transformers: An end-to-end exploration," *Advances in Neural Information Processing Systems*, vol. 34, pp. 19 974–19 988, 2021.

[13] W. Wen, C. Wu, Y. Wang, Y. Chen, and H. Li, "Learning structured sparsity in deep neural networks," *Advances in neural information processing systems*, vol. 29, 2016.

[14] O. Russakovsky, J. Deng, H. Su, J. Krause, S. Satheesh, S. Ma, Z. Huang, A. Karpathy, A. Khosla, M. Bernstein *et al.*, "Imagenet large scale visual recognition challenge," *International journal of computer vision*, vol. 115, pp. 211–252, 2015.

[15] J. Yu, A. Lukefahr, D. Palframan, G. Dasika, R. Das, and S. Mahlke, "Scalpel: Customizing dnn pruning to the underlying hardware parallelism," in *Proceedings of the 44th Annual International Symposium on Computer Architecture*, 2017, pp. 548–560.

[16] F. Yu, K. Huang, M. Wang, Y. Cheng, W. Chu, and L. Cui, "Width & depth pruning for vision transformers," in *Proceedings of the AAAI Conference on Artificial Intelligence*, vol. 36, no. 3, 2022, pp. 3143–3151.

[17] S. Moon, Y. Byun, J. Park, S. Lee, and Y. Lee, "Memory-reduced network stacking for edge-level cnn architecture with structured weight pruning," *IEEE Journal on Emerging and Selected Topics in Circuits and Systems*, vol. 9, no. 4, pp. 735–746, 2019.

[18] S. Teerapittayanon, B. McDanel, and H.-T. Kung, "Branchynet: Fast inference via early exiting from deep neural networks," in *2016 23rd International Conference on Pattern Recognition (ICPR)*. IEEE, 2016, pp. 2464–2469.

Machine Learning Driven Synthesis of Clock Gating

Doyeon Won, Soomin Kim, and Taewhan Kim
Dept. of Electrical and Computer Engineering, Seoul National University, Seoul, Korea
Email : {dywon, smkim, tkim}@snucad.snu.ac.kr,

***Abstract*—One of the key issues in the synthesis of clock gating is how the flip-flops with similar activity patterns in the target design are identified and grouped, so that all flip-flops in each group should be clock-gated in a way to make a full effectiveness in power saving. As yet, due to the excessive runtime and explosive memory usage demand, the conventional grouping methods have relied on flip-flops' toggling probability or toggling pattern of 'short' length, which clearly results in the power saving far off that of the optimal grouping. In this work, we overcome this limitation by proposing a machine learning (ML) based flip-flop grouping for clock gating. Precisely, we devise (1) a convolutional autoencoder (CAE) model to produce a 'short' embedding vector corresponding to the 'very long' input activity pattern of every flip-flop, (2) a convolutional neural network (CNN) based ranker model to predict the degree of flip-flop activity similarity between two input embedding vectors, and (3) a CNN-based model to produce an embedding vector that combines two input embedding vectors. Then, we propose an ML based clock gating synthesis algorithm, which is able to reduce the total dynamic power on circuits by 6.3% further on average over that by the conventional state-of-the-art clock gating with no timing violation by the gated logic delay as well as the satisfaction of physical proximity constraint on flip-flops for clock gating.**

***Index Terms*—clock gating, machine learning, filp-flop, low-power, timing constraints**

I. Introduction

As a way of reducing dynamic power, clock gating is generally recognized as one of the most effective techniques. Clock gating saves power by deactivating the clock signal delivered to a certain group of flip-flops if it detects in advance that the outputs of all flip-flops in the group will not toggle in the next clock cycle.

Fig. 1(b) shows the synthesized logic structure by clock gating for the circuit in Fig. 1(a), in which three flip-flops f_1, f_2, and f_3 are grouped to be synchronized by the signal *gated_CLK*. The grey boxes shown in Fig. 1(b) represent the clock gating logic, which can be divided into three parts: XORs, an OR-tree, and an ICG (integrated clock gating) cell. Boolean expression of the three XOR gates in Fig. 1(b) is

$$g_i = D_i \oplus Q_i \tag{1}$$

where D_i and Q_i are the input and output signals of flip-flop f_i, i = 1, 2, 3. The XOR gate attached to each flip-flop detects if the flip-flop will toggle in the next clock cycle or not. That is, g_i = 0 implies that the next clock signal can be deactivated. Then, Boolean expression of clock signal gating condition G in Fig. 1(b) is

$$G = g_1 + g_2 + g_3, \tag{2}$$

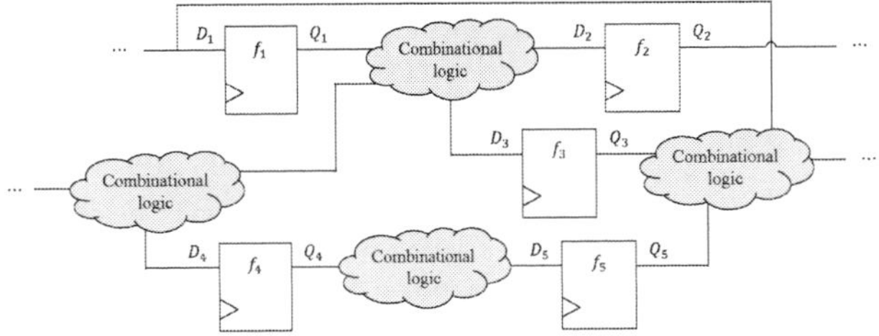

(a) A logic circuit with no clock gating

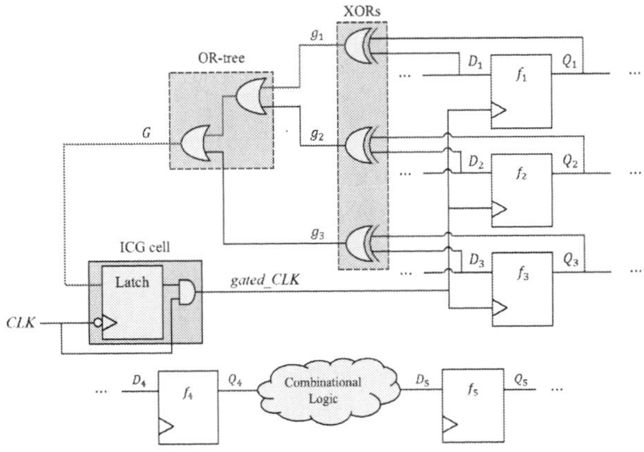

(b) Logic structure transformed from (a) by clock gating on f_1, f_2, and f_3.

Fig. 1. An example of clock gating to sequential logic. The red lines indicate the timing path on the clock gating logic.

which forms an OR-tree. Finally, the latch in the ICG cell is used to remove potential glitches from the gating logic function G.

The most impacting factor on power saving in clock gating is the way of grouping flip-flops in the circuit. Specifically, the essential items to be considered in flip-flop grouping are that (*item 1*) the flip-flops to be grouped should be physically placed as close as possible, (*item 2*) the increase of clock signal delay caused by XORs, OR-tree, and ICG should not violate the circuit timing constraint, and (*item 3*) the input signal activity sequences of all flip-flops to be grouped should not only exhibit similar patterns but also stay very often in the untoggling state. Among the three items, items 1 and 2 are the constraints to be satisfied for flip-flop grouping while item 3 directly affects the effectiveness of power saving. Most of the prior clock gating methods have focused on items 1 and 3 with no or little consideration on item 2.

However, due to the excessive runtime and explosive memory usage demand for manipulating similarity checking and toggling state counting operations over the extremely *long* flip-flops' original input signal activity sequences, the conventional flip-flop grouping methods to deal with item 3 have resorted to flip-flops' toggling probability (e.g., [1]–[4]) or toggling pattern of *short* length (e.g., [5]–[7]), which clearly does not provide an accurate prediction on the actual power saving. In this work, we overcome this fundamental limitation by proposing a machine learning (ML) based flip-flop grouping for clock gating. The contributions of this work are summarized as:

1. We devise a set of ML models not only to fully reflect the flip-flops' original input signal activity sequences but also to efficiently manipulate the similarity checking and toggling state counting operations for flip-flop grouping.
2. We propose an ML based clock gating synthesis algorithm which is able to effectively and efficiently group flip-flops, maximizing the total amount of dynamic power saving while not violating the timing constraint caused by gating logic delay as well as satisfying the physical proximity constraint for grouping flip-flops.

II. Motivation

(1) **Impractical/inefficient manipulation of flip-flops' toggling sequences**: Assume that a logic circuit has K flip-flops f_0, f_1, $\cdots$, f_{K-1} that are in the consideration of clock gating, together with a set, S, of their output toggling (0 indicating untoggling, 1 indicating toggling) sequences S_0, S_1, $\cdots$, S_{K-1}, each of which has length $L = |S_i|$, extracted from the logic simulation. It should be noted that sequence S_i, $i = 0, \cdots, K-1$, is equivalent to the clock signal activity (0 for disabling, 1 for enabling) sequence for clock gating. Most of the conventional flip-flop grouping methods have installed the following two steps as a core engine:

- *Step 1*: In order to find a pair of flip-flops that lead to the most power-saving grouping, 0/1 comparison operations for checking pair-wise toggling activity similarity and counting pair-wise untoggling are performed, taking $O(K^2 \cdot L^2)$ time where $K \ll L$.
- *Step 2*: Once a pair of flip-flops, say f_i and f_j, to be grouped for clock gating are selected, the next step is to generate a new clock signal activity sequence, say $S_{i,j}$, from the sequences S_i and S_j, updating S by replacing S_i and S_j with $S_{i,j}$, taking $O(L)$ time.

Then, a repeated application of the core engine produces group(s) of flip-flops for clock gating. Here, the most time-consuming process is the 0/1 comparison in Step 1 due to the long value of L, which is usually in between 0.1M and 1M clock cycles. Consequently, shortening S_i (i.e., reducing $|S_i|$ ($= L$)) is critical to reduce the runtime as well as memory usage. However, simply shortening S_i may lose the prediction accuracy on the clock signal activity of the resulting clock gating.

For example, for the initial clock signal activity sequences S_1, S_2, and S_3 with L = 12, as shown in Fig. 2(a), it is better to group f_1 and f_3 for clock gating than to group f_1 and f_2 since the latter uses two more clock enabled cycles than the former among the L clock cycles. On the other hand, if we use only the non-shaded left-half part on S_1, S_2 and S_3 for selecting flip-flops for grouping, as indicated in Fig. 2(b), f_1 and f_2 shall be chosen rather than f_1 and f_3, resulting in two more clock enabled clock gating than that in Fig. 2(a). Thus, the quadratic run time reduction by using half of L sacrifices prediction accuracy on clock gating. In this work, *we overcome this fundamental limitation by inventing a new approach* by encoding the activity sequences to embedding vectors of length much shorter than that of L to be practically feasible in processing Step 1 while deriving a new embedding vector that closely reflects the intended (i.e., accurate) process in Step 2.

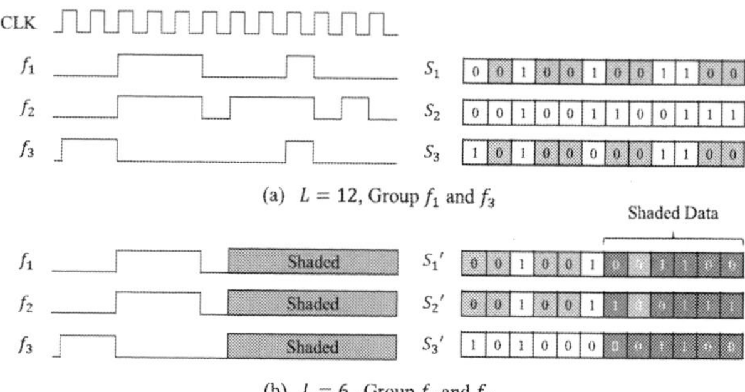

Fig. 2. (a) Grouping flip-flops using activity sequences of L = 12, which performs 36 (= 12 × 3) 0/1 comparisons, synthesizing a clock gating that **disables 7 clock cycles** among 12, i.e., predicting untoggling probability of 7/12 = 0.58, indicated by the green boxes. (b) Grouping flip-flops using sequences of L = 6 (i.e., left-half), which performs 18 (= 6 × 3) 0/1 comparisons, synthesizing a clock gating that **disables 5 clock cycles** among 12, i.e., predicting untoggling probability of 5/12 = 0.42.

(2) **Inaccurate prediction on timing overhead**: The clock gating signal delay caused by the clock gating function should not violate the timing constraints imposed on the circuit. The delay increase by clock gating, as indicated by the red lines in Fig. 1(b), is the sum of the delay of an XOR gate and the delay of OR-tree. Clock gating signal delay induced by grouping flip-flops under no consideration of timing may cause a setup time violation on the latch in ICG cell. For example, as shown by the red lines in Fig. 3(a), clock gating signal G induced by grouping f_1 and f_2 has arrival time of 1.0 (= $max\{a(D_1), a(D_2)\}$ + D_{XOR} + D_{OR} = 0.7 + 0.2 + 0.1), which exceeds the latch's setup time limit of 0.9 (= T_{clk} - t_{setup}), violating timing constraint.[1] On the other hand, the clock gating signal G induced by grouping f_1 and f_3 has arrival time of 0.8 (= 0.5 + 0.2 + 0.1), which is below the latch's setup time limit, satisfying the timing constraint. Thus, *a delicate and accurate clock gating delay evaluation is necessary in the course of grouping flip-flops to avoid any timing violation.* (It should be noted that a number of prior works have controlled the clock gating delay by downsizing the Boolean logic of clock gating functions in the input circuits

[1] T_{clk}, t_{setup}, $a(D_i)$, D_{XOR}, and D_{OR} are the clock period, the setup time of latch in ICG, the arrival time of data input D_i to flip-flop f_i, an XOR-gate delay, and an OR-gate delay, respectively.

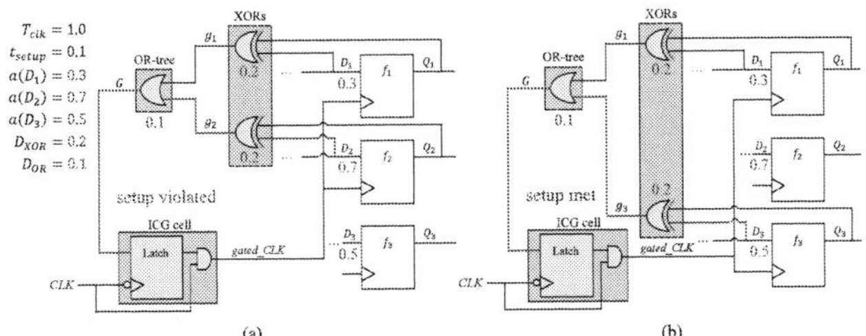
Fig. 3. Illustration of the impact of clock gating delay on circuit timing. The longest paths to G are shown by red lines. (a) Violating latch's setup time constraint: $(T_{clk} - t_{setup}) - (max\{a(D_1), a(D_2)\} + D_{XOR} + D_{OR}) =$ (1.0 - 0.1) - (0.7 + 0.2 + 0.1) = -0.1. (b) Meeting latch's setup time constraint: (1.0 - 0.1) - (0.5 + 0.2 + 0.1) = 0.1.

at the cost of less power saving by clock gating (e.g., [8], [9]). However, this technique is not applicable to the logic circuits where the exact Boolean logic is not available or more importantly, designers do not want to change the already optimized logic structure.)

III. The Proposed ML Based Clock Gating

A. Overall Flow

Fig. 4 depicts our synthesis framework of clock gating. For an input circuit $\mathcal{C}$, together with the setup time slacks on the individual flip-flops $f_0, f_1, \cdots, f_{K-1}$, their physical location, the physical proximity constraint δ, and a set $\mathcal{S}$ of their output toggling sequences, it applies, as a preprocessing step, Encoding-net model in Sec. III-C to convert the toggling sequence $S_i \in \mathcal{S}$ of flip-flop f_i into an embedding vector $V_i \in \mathcal{V}$ such that $|V_i| \ll |S_i|$, and generates a set, $\mathcal{R}$, of embedding vector pairs such that $\mathcal{R} = \{(V_i, V_j) \mid i, j \in \{0, \cdots, K-1\}\}$. Then, the framework *iteratively* performs the following two steps using $\mathcal{R}$ until there is no more gain in power saving $P(\cdot)$ in *Eq.6* or there is no more pair of embedding vectors to group the corresponding flip-flops with no violation of timing/proximity constraints: (Step 1) selecting a pair (V_i, V_j) of embedding vectors by applying Similarity-net model in Sec. III-C that satisfy proximity constraint δ and timing feasibility in Sec. III-D as well as have the largest power saving according to the cost formulation in Sec. III-B, and (Step 2) generating a new embedding vector, $V_{i,j}$, by applying Grouping-net model in Sec. III-C to V_i and V_j, and updating $\mathcal{R}$ by replacing V_i and V_j with $V_{i,j}$. We repeat this process as long as there is a profiable (i.e., saving power) pair of embedding vectors in $\mathcal{R}$.

B. Cost Formulation for Flip-flop Grouping

Let $F_{i,j} = F_i \cup F_j = \{f_{i_1}, f_{i_2}, \cdots\} \cup \{f_{j_1}, f_{j_2}, \cdots\}$ be the set of flip-flops selected to group for clock gating and $\hat{\rho}(F_{i,j})$ be an estimate of *toggling probability of all flip-flops in* $F_{i,j}$. Then, we can estimate the power ($P_f(F_{i,j})$) consumed by all flip-flops in $F_{i,j}$ and the clock gating logic power ($P_{cg}(F_{i,j})$) consumed by grouping the flop-flops with a single clock gating signal:

$$P_f(F_{i,j}) = P_f \cdot \hat{\rho}(F_{i,j}), \tag{3}$$

$$P_{cg}(F_{i,j}) = K \cdot P_{XOR} + (K-1) \cdot P_{OR} + P_{ICG} \tag{4}$$

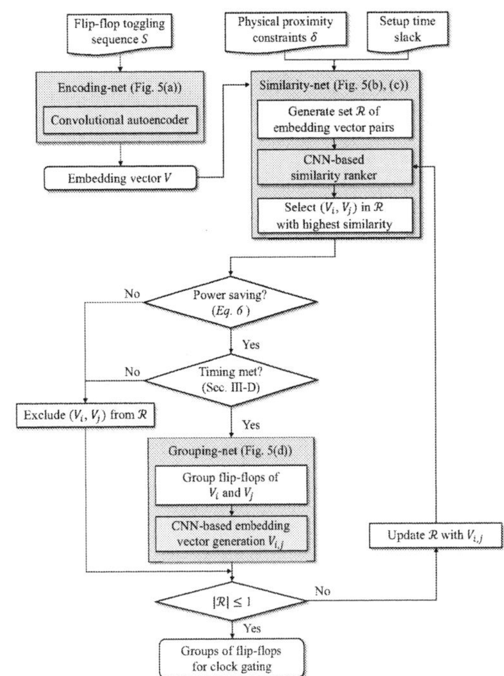

Fig. 4. The synthesis flow of clock gating. The flip-flop grouping is iteratively performed based on the cost formulation on power saving, the proximity constraint, and the outcome of timing feasibility checking.

where P_f represents the power consumed by a flip-flop when toggling, $K = |F_{i,j}|$, and P_{XOR}, P_{OR}, and P_{ICG} are the averaged power consumptions of XOR, OR, and ICG cells whose values are extracted through power simulation.
Thus, the power saving by grouping all flip-flops in F_i and F_j can be expressed as:

$$\triangledown P(F_{i,j}) = (P(F_i) + P(F_j)) - (P_f(F_{i,j}) + P_{cg}(F_{i,j})) \tag{5}$$

where the sum of $P(F_i)$ and $P(F_j)$ represents the amount of power consumption of all flip-flops in F_i and F_j and two clock gating logics produced by separately clock gating the flip-flops in F_i and the flip-flops in F_j.
We can recursively formulate $P(F_i)$ and $P(F_j)$ as we express $P(F_{i,j})$ in *Eq.5* and the basis is the case when $|F_i| = 1$ or $|F_j| = 1$, which exactly corresponds to $P_f \cdot \hat{\rho}(F_i)$ or $P_f \cdot \hat{\rho}(F_j)$ in *Eq.3*. Then, the sufficient and necessary conditions for grouping all flip-flops satisfying proximity constraint in two sets F_i and F_j by a single clock gating signal satisfying timing constraint is

$$\triangledown P(F_{i,j}) > 0. \tag{6}$$

C. Neural Network Models for Flip-flop Grouping

Our framework of grouping flip-flops for clock gating is built up by three neural network models called Embedding-net, Similarity-net, and Grouping-net. Fig. 5 shows the overall structure of the framework.

- Embedding-net: We implement Embedding-net with a *convolutional autoencoder* (CAE) model [10] to convert the

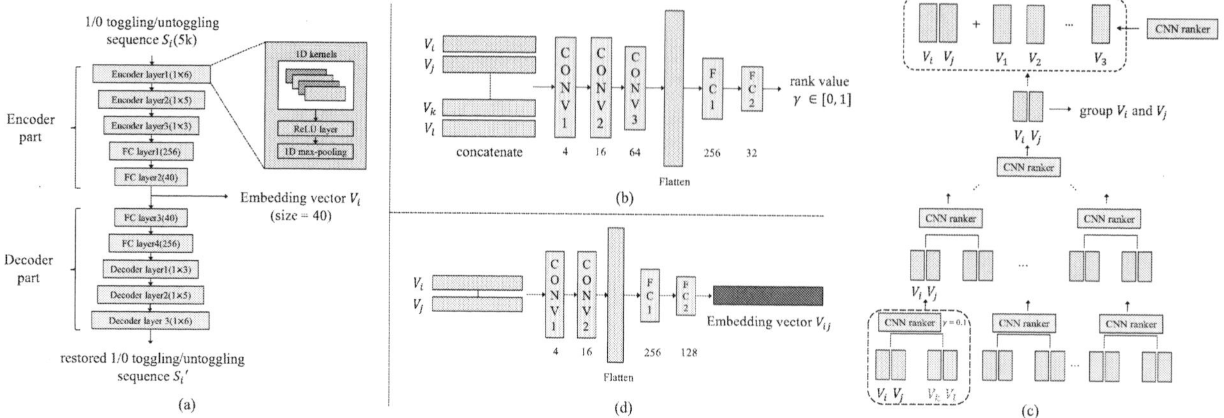

Fig. 5. The structure of our ML based flip-flop grouping for clock gating. (a) A convolutional autoencoder which produces an embedding vector corresponding to the original input toggling/untoggling sequence of a flip-flop. (b, c) A CNN model based ranker that answers, for two input pairs of flip-flops or groups of flip-flops, which pair is more profitable than the other in terms of $\nabla P(\cdot)$ in *Eq.6* if the flip-flops in each pair were grouped for clock gating. (d) A CNN model which produces a new embedding vector corresponding to the bitwise-ORed sequence of all input toggling/untoggling sequences of the flip-flops corresponding to the two input embedding vectors.

original input toggling/untoggling sequence of each flip-flop to an embedding vector of short length. Each of the encoder and decoder consists of three convolutional layers and each layer is then composed of 1D-convolutional sub-layer, ReLU sub-layer, and 1D max-pooling sub-layer, as shown in Fig. 5(a). The input and output on the model are long toggling/untoggling sequences of the same size in the form of 1/0 vector. We applied unsupervised learning to the full structure of the CAE model with the objective of restoring the 1/0 sequence of input vector as much as possible. Note that the 1D kernels used in our model are able to extract the local toggling behavior, which will be an important feature in assessing the toggling/untoggling sequence similarity among flip-flops. Once model training is done, we use the encoder part in the model to produce the short embedding vector from the original long input toggling/untoggling sequence on the model. We can conduct the inferencing as well as training on Embedding-net before performing the iterative flip-flop grouping for clock gating.

• Similarity-net: By intuition, we can use a simple CNN model with 2D kernels to predict the similarity, as a value in [0, 1], between two input embedding vectors, say V_i and V_j. However, since we are interested in selecting a pair of flip-flops or groups of flip-flops which has the closest similarity,[2] in terms of the most profitable in power saving, rather than computing an 'absolute' value of similarity, we adopt a *CNN-based ranker* model in [11]. Our model receives two pairs of embedding vectors, say (V_i, V_j) and (V_k, V_l), and outputs a rank value, denoted by γ, in between 0 and 1 where as γ is close to 0, the similarity between V_i and V_j is higher than that of V_k and V_l. Likewise, as γ is close to 1, the similarity between V_k and V_l is higher than that of V_i and V_j, as illustrated in Fig. 5(b). Once training on CNN ranker model is done, we can use CNN ranker model to find the best pair of flip-flops or groups of flip-flops for grouping by ranking their embedding vector pairs, as illustrated in Fig. 5(c). We iterate this rank-based flip-flop grouping until no power saving is expected.

• Grouping-net: We use a CNN-based model called Grouping-net to perform the flip-flop grouping, as illustrated in Fig. 5(d). It takes two embedding vectors as input and produces a new embedding vector. During training, the model aims to minimize the difference between the original bitwise-ORed 1/0 toggling/untoggling sequences of the input vectors and the 1/0 toggling/untoggling sequence of the output vector.

D. Timing Feasibility Checking by Clock Gating

With the synthesis of clock gating logic, we are required to check two timing constraints. Fig. 6(a) shows two types of delay path (red and blue) incurred by clock gating.

1. *Latch's setup timing*: As mentioned in Sec. II and shown in Fig. 6(b), the constraint to satisfy is

$$T_{clk} - t_{setup} \geq D_{XOR} + D_{OR-tree} \quad (7)$$

where $D_{OR-tree}$ is the arrival time of the output signal of the OR-tree in the clock gating logic.
Let $a(D_0), a(D_1), \cdots, a(D_{K-1})$ be the arrival times of data inputs D_0, D_1, $\cdots$, D_{K-1} to the flip-flops for clock gating. Then, for an initial set $\mathcal{A} = \{a(D_0), a(D_1), \cdots, a(D_{K-1})\}$, we construct a timing-minimal OR-tree recursively as follows: (step 1) pick two inputs, D_i and D_j, of the smallest (i.e., earliest arrival) values in $\mathcal{A}$, (step 2) allocate a two-input OR gate with D_i and D_j as input, (step 3) set $a(D_{i,j})$, which is the arrival time of the output signal of the OR gate produced in step 2, to $max\{a(D_i), a(D_j)\} + D_{OR}$, and (step 4) replace $a(D_i)$ and $a(D_j)$ in $\mathcal{A}$ with $a(D_{i,j})$. We repeat steps 1 $\sim$ 4 until $|\mathcal{A}| = 1$. Then, $D_{OR-tree}$ becomes the

[2]We quantify the degree of similarity between V_i and V_j by predicting *the portion of untoggling state* on the flip-flips or groups of flip-flops corresponding to V_i and V_j, if they were grouped for clock gating.

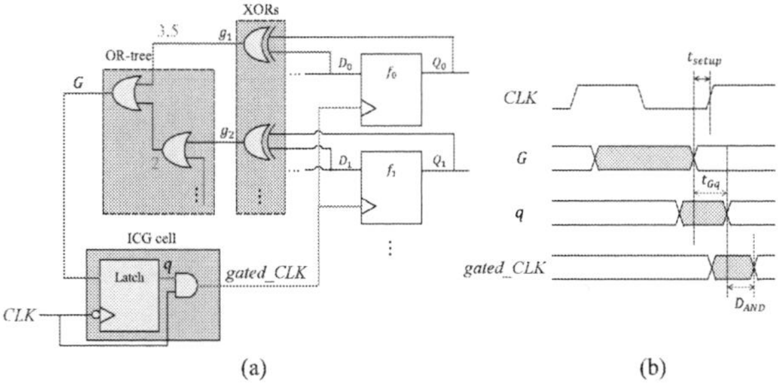

Fig. 6. Checking the delay paths incurred by clock gating logic. (a) Latch's setup timing (red line) and flip-flop's setup timing (blue line). (b) Timing diagram of signals in clock gating logic.

value in $\mathcal{A}$. Thus, the arrival time of G, which is an input to the latch in ICG cell, is D_{XOR} + $D_{OR-tree}$.

2. *Flip-flop's setup timing*: Let $slk_0, slk_1, \cdots slk_{K-1}$ be the setup time slacks of the flip-flops with data inputs D_0, D_1, $\cdots$, D_{K-1}, respectively. Then, the delayed clock signal to all flip-flops should satisfy the flip-flops setup time constraint:

$$max\{D_{XOR} + D_{OR-tree} + t_{Gq} - T_{clk}, 0\} + D_{AND} \leq min\{slk_0, slk_1 \cdots, slk_{K-1}\} \quad (8)$$

where t_{Gq} and D_{AND} represent the latch's input-to-q delay and two-input AND-gate delay in the ICG cell, respectively.

IV. Experimental Results

A. Experimental Setup

To evaluate the performance of our proposed ML based clock gating synthesis framework, we implemented the IWLS 2005 benchmark circuits in [12]. We used Synopsys *Design Compiler* for logic synthesis and *IC Compiler* for physical implementation by using Synopsys 28nm cell library and physical design kit (PDK). We set the clock frequency in the range of 500MHz to 1GHz for the testcases to identify the conditions under which timing violations occur and verify our timing-aware clock gating framework. Clock gating logic is implemented physically with a self-gating method using XOR gates, OR-tree, and ICG cell as illustrated in Fig. 1(b). When grouping flip-flops, we constrain that the maximum number of flip-flops to be included in the same group should not exceed 10% of the total number of flip-flops in the circuit. For physical proximity constraint, it is set to the value of 20%$\sim$ 70% of the averaged distance of all flip-flop pairs. We synthesized each circuit with utilization of 80%. We conduct RTL simulation to obtain flip-flop activity sequences with a duration interval ranging from 300K to 1M clock cycles for each flip-flop.

B. Assessment of Our ML based Clock Gating

• **Comparing clock gating performance:** Table I summarizes clock gating results implemented by commercial (Synopsys) tool [13], which is a baseline for comparison, and results of our ML-based clock gating framework, which consists of the three models described in Sec.III-C. In particular, the variable that can be adjusted for clock gating in the commercial tool is maximum bitwidth, which is set to the maximum group size we used in the experiment. The evaluation metrics include the clock gating (CG) ratio, power consumption of flip-flops, total power consumption, worst negative slack (WNS), and area overhead of the clock gating logics (i.e., XORs, OR-Tree, and ICG cell). Our ML-based clock gating framework results in reduced power consumption of flip-flops (P_{ff}) with no degradation on timing. Although the area overhead increases slightly due to the added combinational logic, the overall power is significantly reduced due to the clock gating. Compared to the commercial tool based clock gating, our ML-based clock gating framework reduces P_{ff} by 9.4% and P_{tot} by 6.3%. Runtime refers to the duration of time elapsed from the completion of the embedding process to the completion of flip-flop grouping by Similarity-net and Grouping-net. Fig 7 shows the distribution of flip-flops for circuit SPI synthesized by Synopsys clock gating and ours. It shows that ours groups 34.9% of flip-flops while the commercal one groups 27.0%.

• **Machine learning model performance:** Table II summarizes the performance evaluation of our ML models. These models are essential components of our ML-based clock gating framework, and they are trained to accomplish specific objectives. For Encoding-net, we evaluate the performance of the CAE model by how well model restores the original flip-flops' activity sequence. The goal of Encoding-net is to extract features from the input activity sequence and use them to reconstruct the original sequence. Since input sequence consists of 0 or 1 values, we assume the restored value is properly restored if the difference between the restored value and original value is less than 0.3 (= ρ in Table II). The performance of Encoding-net is measured in terms of its reconstruction accuracy. For Similarity-net, the objective is to prioritize grouping of flip-flops. The CNN ranker is trained to predict the similarity between two activity patterns and rank them accordingly. The performance of Similarity-net is evaluated by its accuracy on a general classification problem. Finally, Grouping-net goal is to group flip-flops with similar activity patterns into a single group. The minimization cost in training is the difference between the bitwise-ORed 1/0 toggling/untoggling sequences corresponding to the input vectors and the 1/0 toggling/untoggling sequence corresponding to the output vector. The performance of Grouping-net is measured by the difference between the newly created embedding vector of grouped flip-flops and the embedding vector produced by Encoding-net using the union of input activity sequences. Overall, the performance scores of all three models are high enough. For each embedding vector value, the ratio at which the value does not differ by more than the threshold was over 76%, which demonstrates that our proposed models can accurately capture the activity patterns of flip-flops and group them based on their similarity, which is critical for achieving optimal clock gating performance.

TABLE I
COMPARISON OF CLOCK GATING IMPLEMENTATIONS IN TERMS OF CLOCK GATING RATIO OF FLIP-FLOPS, POWER SAVING, CRITICAL PATH TIMING (ns), AND AREA OVERHEAD (um^2) PRODUCED BY COMMERCIAL (SYNOPSYS) TOOL, AND OUR ML BASED FLIP-FLOP GROUPING FOR CLOCK GATING. P_{tot} INDCATES THE SUM OF POWER CONSUMPTIONS BY FLIP-FLOPS P_{ff}, CLOCK NETWORK P_{clk}, AND COMBINATIONAL LOGIC P_{comb}.

Circuit	# of FFs	State-of-the-art commercial tool [13]					Our ML based clock gating					
		CG ratio (%)	Power (mW)		Timing	Area	CG ratio	Power (mW)		Timing	Area	runtime
			P_{ff}	P_{tot}	WNS (ns)	$Cell_{tot}$		P_{ff} ($Red.$)	P_{tot} ($Red.$)	WNS (ns)	$Cell_{tot}$	
SPI	229	27.0%	0.46	0.78	-0.10	5172.1	34.9%	0.45 (2.2%)	0.78 (0.0%)	-0.07	5371.5	35m
WB_DMA	611	32.6%	1.10	1.67	-0.11	10203.9	49.7%	1.02 (7.2%)	1.65 (1.2%)	-0.09	10367.2	1h
AES_CORE	530	25.1%	1.16	5.41	-0.36	32526.1	23.2%	1.02 (12.1%)	4.21 (22.1%)	-0.08	31554.7	7h 26m
WB_CONMAX	818	35.2%	1.62	6.80	-0.06	54140.6	25.3%	1.49 (8.0%)	6.22 (8.5%)	-0.00	52499.7	1h 15m
MEM_CTRL	1120	28.5%	2.76	4.34	-0.18	20663.4	41.7%	2.43 (11.9%)	4.07 (6.2%)	-0.15	20651.9	3h 45m
AC97_CTRL	2229	13.9%	4.94	6.70	-0.06	28557.1	31.4%	4.18 (15.3%)	6.71 (-0.1%)	-0.11	33611.5	3h 50m
Avg. reduction								**9.4%**	**6.3%**	met		

TABLE II
MACHINE LEARNING MODEL TRAINING AND TEST PERFORMANCE USED IN OUR CLOCK GATING SYNTHESIS FRAMEWORK

Model	Scoring function	Score	
		Training set	Test set
Encoding-net	$\lvert S_i[n] - S'_i[n] \rvert < \rho$	98.6%	96.2%
Similarity-net	$Accuracy = \frac{TP+TN}{TP+TN+FP+FN}$	91.2%	76.2%
Grouping-net	$\lvert V_{i,j}[n] - V'_{i,j}[n] \rvert < \rho$	84.8%	84.2%

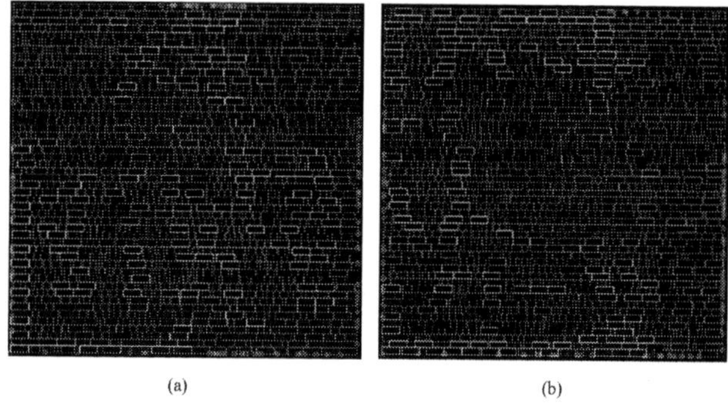

Fig. 7. Distribution of flip-flops for circuit SPI by (a) a commercial tool and (b) our ML based clock gating. The flip-flops involved in clock gating are colored in red, occupying 27.0% in (a) and 34.9% in (b).

V. CONCLUSION

This paper proposed an ML based clock gating approach to overcome the inherent limitation of manipulating the extremely long activity patterns on flip-flops, which is an essential task for making full effectiveness on clock gating. Precisely, we devised (1) a convolutional autoencoder model to produce a short embedding vector corresponding to the activity pattern of every flip-flop, (2) a CNN-based ranker model to predict the degree of flip-flop activity similarity between two input embedding vectors, and (3) a CNN-based model to produce an embedding vector that combines two input embedding vectors. Through experiments, it was confirmed that our proposed approach was practically very effective in saving power by clock gating while not violating the timing by the gated logic delay as well as satisfying physical proximity constraint on flip-flops for clock gating.

ACKNOWLEDGEMENT

This work was supported in part by Samsung Electronics Company, Ltd. under IO201216-08205-01, IO230102-04421-01, and IO221227-04376-01, in part by Samsung Advanced Institute of Technology (SAIT) under IO230223-05124-01, in part by the National Research Foundation of Korea (NRF) grant funded by the Korea Government (MEST) under 2021-R1A2C2008864, in part by the Institute of Information and communications Technology Planning and Evaluation (IITP) grant funded by Korea government (MSIT) under 2021-0-00754, Software Systems for AI Semiconductor Design, and in part by the BK21 Four Program of the Education and Research Program for Future ICT Pioneers, Seoul National University. The EDA tool was supported by the IC Design Education.

REFERENCES

[1] A. Bonanno, A. Bocca, A. Macii, E. Macii, and M. Poncino, "Data-driven clock gating for digital filters," in *International Workshop on Power and Timing Modeling, Optimization and Simulation*, 2009.
[2] S. Wimer and I. Koren, "Design flow for flip-flop grouping in data-driven clock gating," *TVLSI*, 2013.
[3] D. Gluzer and S. Wimer, "Probability-driven multibit flip-flop integration with clock gating," *TVLSI*, 2016.
[4] L. Li, K. Choi, and H. Nan, "Activity-driven fine-grained clock gating and run time power gating integration," *TVLSI*, 2012.
[5] Q. Tong and K. Choi, "Activity correlation-based clustering clock-gating technique for digital filters," *International Journal of Electronics*, 2017.
[6] C.-H. Lin, S.-H. Huang, J.-H. Jian, and X.-J. Chen, "New activity-driven clock tree design methodology for low power clock gating," in *IEEE International Symposium on Next Generation Electronics*, 2017.
[7] A. Ranganayakulu and K. Satyaprasad, "Subword partition based data driven clock gating scheme for low power vlsi design," *International journal of computer applications*, 2014.
[8] E. Arbel, C. Eisner, and O. Rokhlenko, "Resurrecting infeasible clock-gating functions,," in *DAC*, 2009.
[9] Y.-M. Kuo, S.-H. Weng, and S.-C. Chang, "A novel sequential circuit optimization with clock gating logic,," in *ICCAD*, 2008.
[10] J. Xie, R. Girshick, and A. Farhadi, "Unsupervised deep embedding for clustering analysis," in *ICML*, 2016.
[11] T.-C. Lee, C.-Y. Yang, and Y.-L. Li, "itplace: machine learning-based delay-aware transistor placement for standard cell synthesis," in *ICCAD*, 2020.
[12] C. Albrecht. IWLS 2005 benchmarks. [Online]. Available: :http://www.iwls.org/iwls2005/benchmarks.html
[13] "Design compiler - user guide," in *https://picture.iczhiku.com/resource/eetop/wYKfeQTQHSRITVVC.pdf*. Synopsys, Inc., 2019.

Automatic Generation of Structured Macros using Standard Cells – Application to CIM

Christian Lanius, Jie Lou, Johnson Loh, Tobias Gemmeke
RWTH Aachen University
Aachen, Germany
{lanius,lou,loh,gemmeke}@ids.rwth-aachen.de

***Abstract*—Regularity can be exploited to efficiently describe, place and route logic blocks with a repetitive structure. We present a design flow to automatically generate regular, standard-cell based designs, which can be seamlessly integrated into a traditional digital flow in commercial EDA software. The generated arrays can be seamlessly integrated into a "sea of gates", with no guard-rings or keep-out areas.**

The flow takes a description of a regular design as an input and generates netlist, placement, constraints, routing, initial parasitics estimates and timing information. We show that, in example designs, the run-time of EDA tooling is up to 2.5x faster, reduces the critical path by 47%, reduces the metal utilization by 45% and achieves a utilization of 93%.

***Index Terms*—design automation, compute-in-memory, regular designs, structured data path, standard cells**

I. Introduction

A lot of modern, data-driven applications rely on hardware with massive parallelism on the level of atomic function calls. In the case of machine learning, these are for example Multiply-and-Accumulate (MAC) operations or associative data retrieval mapped to a ternary content addressable memory (TCAM). To feed the massively parallel computations, high bandwidth connectivity of stored information is required, thus leading to designs integrating memory and compute elements - the so-called 'Compute-in-Memory' (CIM) architectures.

Relying on the high density of classic 6T-cell-based SRAM memories, CIM can be realized as analog computations by precisly steering word-line pulse shapes and current or voltage levels on bit-lines.

However, computations in the analog domain often suffer from limited accuracy and the expensive analog-to-digital conversion. Thus, digital computations using standard cells introduce a trade-off: Larger area and more power consumption in the memory array on the one side. One the other side, it enables localized co-integration of storage and complex computation as any modulation of the wordline or analog-digital conversion on the bitline becomes superfluous.

In CIM architectures, the arrays implementing the functionalities typically dominate area and energy, thus a careful design of the atomic building block is paramount for a well performing implementation. To fine-tune the performance, manual selection of standard cells and their placement can be used. At the same time, some peripheral logic to control the design is required, which is not as critical in terms of performance and can thus be synthesized easily and placed directly around the array together with all the other unstructured logic. Besides performance benefits, the custom realization of the compute elements and the array also has the potential to significantly reduce the design space and complexity of the netlist during synthesis, drastically reducing the run-time of the tools.

The concept of exploiting regularity has been investigated by different groups in the past but with different goals: Either focusing on coarse-grain placement of lower density [1] or the pervasive application of custom designed basic cells [2]. Another direction is macro compilers realizing very limited logic functionality such as SRAM generation, to be used as a black-box in the digital design [3]. The flow presented in this work lies in-between: Exact placement and routing of the array are created outside the standard tool chain, but a proper presentation within the electronic design automation (EDA) tools down to the cell level is preserved providing full transparency - a key aspect considering co-integration with other parts of the digital design including all steps of verification and timing closure (extraction, AOCV, IR-drop). In any case, the database of the structured macro should be clean by itself considering all requirements imposed by the design rules or aspects suggested by the design for manufacturability (DFM), which have massively grown compared to the earlier works in this field. Examples are coloring aware placement for double patterning of lower metal layers, edge constraints and area optimal insertion of diffusion breaks.

What is more, overly stringent regularity requirements in the flow lead to inflexible and inefficient design exploration: Buffers on high fan-out nets, logical functions with many inputs, tap-cell insertion, equivalent but mirrored arrays as well as specific aspect ratios. The following flow includes appropriate knobs for describing such irregularities and resolving them automatically. This reduces repetition in the description and provides a generic definition of the array.

In this work, we describe a design flow to exploit the advantages of regular designs. The burden on the designer is kept as small as possible, while providing direct control of all relevant design aspects as beneficial to the realization at hand. To summarize, our contributions are:

- Application specific textural design description for regu-

This work was partially funded by the Federal Ministry of Education and Research (BMBF, Germany) in the project NEUROTEC II (project number 16ME0399).

lar designs, including the specification of specific irregularities.
- Flow for automatic generation of netlist, placement and routing information, parasitics and timing information.
- Integration of the regular design into a commercial EDA flow with forward annotation of physical information in synthesis and RTL simulation.
- Validation in silicon of different designs created by the design flow in an advanced 22nm FDSOI node.

II. Related Work

Today, commercial EDA software for the synthesis and implementation of digital designs is extensively used: Vendors like Cadence (Genus/Innovus) and Synopsys (DC/ICC2) provide software packages to handle all parts of the digital flow. While these tools are customizable using TCL and provide some capability of structuring the placement [4], the inner workings of their placement and routing algorithms are proprietary and can not be adapted to use the additional a priori knowledge of the designer about fine-grained regularity.

OpenTimer [5] is an open-source non-linear dealy model (NLDM) based timing engine. Its integration into other software is straight forward. Due to its fast performance and multiprocessing capability, it is well suited to analyze even larger designs during the exploration.

FPGAs also feature a regular structure and there has been recent work to build academic, open-source FPGAs [6]. Similar to the flow proposed in this paper, technology mapped netlists are created and duplicated. However, in previous work, connectivity is described explicitly and placement is done by creating hard macros, which are then placed according to a simple grid. The flow presented in [6] is more involved than the simple use of standard-cells, requiring larger amount of customization and is not as tightly coupled.

On the other end of generators, projects like [7] focus on providing the ability to describe complex systems on a high level: They produce RTL level HDL for entire SoCs which can then be mapped to the target technology. Projects like [8], [9] share with the proposed flow in that they use high level languages like Scala and Python, but they instead focus on providing the user a new way to describe their functionality on an RTL level.

Single bit-cells (e.g. 6T cells) used in SRAM macros are around 10x smaller than standard-cell memory. However, for small capacities, the peripherals introduce such an overhead, that standard-cell memory are in total smaller than traditional SRAM macros [10].

The break-even point between the two approaches moves towards larger capacities with modern FinFET technologies, as the SRAM cells can not reach the scaling targets of standard-cells [11]. Thus the approach proposed in this paper benefits in modern technologies from a shrinking area gap between SRAM macros and standard-cell memory.

III. Proposed Methodology

In the following, we will describe the key aspects of the proposed flow. The flow is implemented as a standalone tool, which takes as input technology information (abstract view of standard cells, verilog descriptions and timing information as NLDM tables) and a parameterized json description of the circuit to be realized.

The outputs generated are

- A comprehensive verilog netlist of the array
- Timing information for simulation, taking into consideration routing parasitics
- A constraint file to provide information about timing exceptions,
- Files to specify placement and routing for the implementation tool
- A script to forward annotate the expected parasitics to the synthesis tool
- A human readable report with key information, such as area and utilization

The proposed flow traverses the design from the top element and recursively generates building blocks, described in Section III-B while connectivity is derived according to the rules discussed in Section III-C.

To allow for a more generic description of the designs, the textual representation can be parameterized. Inside the description, values can be replaced with variable names or expressions. Capacity and word-size are important parameters for a lot of memory-like designs. To annotate the design intent assertions, logic expressions which have to evaluate to true, can be specified. After variable substitution and evaluation, these assertions are checked, thus making sure that the variables stay in the range expected by the designer.

This parameter approach can be extended to make the entire design even more flexible: By specifying a template folder, all files in the folder will be parsed and the parameters will be replaced by their value. This allows us to generate arbitrary files for use in a variety of settings. The most common use-case for this is the generation of setup scripts exporting the values as environment variables or the automatic generation of headers and array instantiation templates.

A. Basic Element

We call the basic building block of the array a MULTI-CELL: This is the smallest logical unit in our system. It is made up of a number of standard cells which are connected within the multi-cell. The functionality of such an element is not divisible into smaller meaningful units. As an example, consider a standard-cell based SRAM bit-cell as proposed in [10] (cf. Fig. 1).

The corresponding multi-cell description in json is also shown. First the standard cells making up the multi-cell are defined, followed by the internal connectivity of the cell: The first cell's output Z (index 0) is connected to the input A of the second one and the other way around, forming the latch. Finally, the output of the latch is connected to the input of the AOI gate.

While these connections are described, most of them will be inferred using the names and the properties associated with the ports of the multi-cell, which are defined next, as

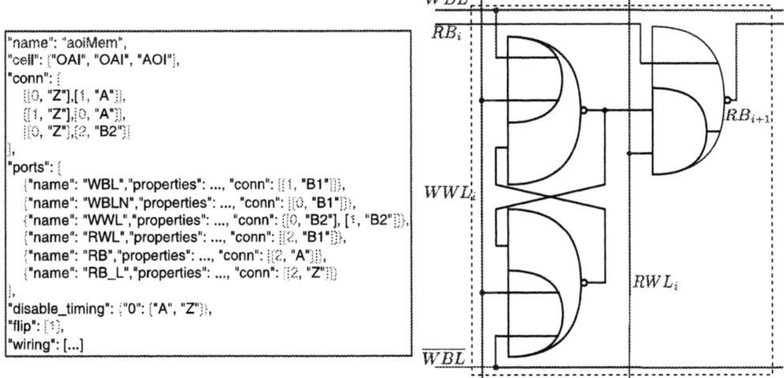

"name": "aoiMem",
"cell": ["OAI", "OAI", "AOI"],
"conn": [
[[0, "Z"],[1, "A"]],
[[1, "Z"],[0, "A"]],
[[0, "Z"],[2, "B2"]]
],
"ports": [
{"name": "WBL","properties": ..., "conn": [[1, "B1"]]},
{"name": "WBLN","properties": ..., "conn": [[0, "B1"]]},
{"name": "WWL","properties": ..., "conn": [[0, "B2"], [1, "B2"]]},
{"name": "RWL","properties": ..., "conn": [[2, "B1"]]},
{"name": "RB","properties": ..., "conn": [[2, "A"]]},
{"name": "RB_L","properties": ..., "conn": [[2, "Z"]]}
],
"disable_timing": {"0": {"A", "Z"}},
"flip": [1],
"wiring": [...]
WBL_i
$\overline{RB_i}$
RB_{i+1}
WWL_i
RWL_i
$\overline{WBL_i}$

Fig. 1: Example for a multi-cell and its textual representation.

well as the connections between the multi-cell's ports and the standard cells. As a combinational loop is formed to act as a latch, we mark one of the connections as "disable_timing" to remove the loop. Further implementation details, such as flipping of the gates to improve routing, can be included as well. In the example, the second OAI gate is flipped. Once all of the basic multi-cells have been defined, the array can be assembled, according to patterns specified in the iteration rules, as described in the next section.

B. Cell Repetition

A recurring pattern in digital CIM architectures are a repetitive two dimensional area surrounded by row- and column peripherals, such as gating logic. Thus, the basic repetition scheme should map well to this.

Array constructs are assembled out of multi-cells or other arrays. An array is defined by an up-to 3x3 template, i.e. up to 9 different multi-cells or arrays, and a repetition count in horizontal and vertical direction. An example for an array consisting of the cells A to I being replicated 4x2 times is shown in Fig. 2 (top left).

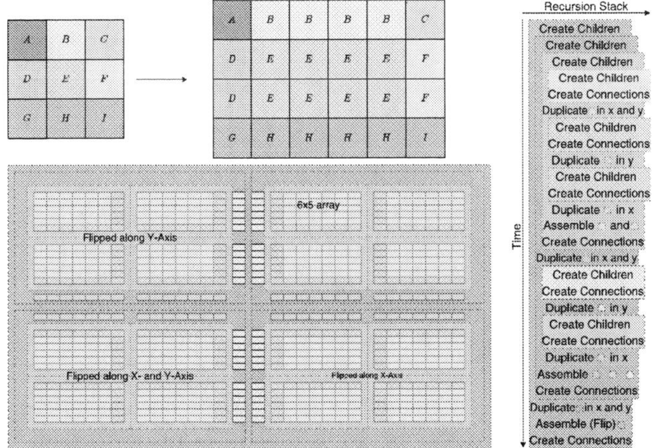

Fig. 2: Visualization of the replication scheme for an array (top left). The order of generation, and the corresponding layout are visualized for a multi-level array.

For the sake of regularity in the description, the logical layout can differ from the physically implemented layout: A design which is symmetric around an axis can be more concisely described as the same sub-design, where one part of it is flipped around the axis. In other cases, the logical layout might not feasible to implement because it is very narrow in one direction, such as the design presented in [12].

We thus introduce the concept of reshaping: An array can be flipped around its axes and its aspect ratio can be changed, resulting in a deviation of logical layout and physical layout. The connectivity of the design is not impacted by the reshaping, only the physical arrangement. The array shown in Fig. 2 has the violet sub-arrays being flipped and Fig. 3 shows how a very long narrow design is reshaped into a more rectangular shape.

Fig. 2 on the right shows how the proposed flow generated the array on the left: Starting from the red top level array, it recursively traverses down to the multi-cell level, while children are generated on the respective levels. Once all children of an array are generated and interconnected, the recursion returns. Repetitions correspond to direct copies, thus the run-time of the flow is roughly linear with the number of arrays.

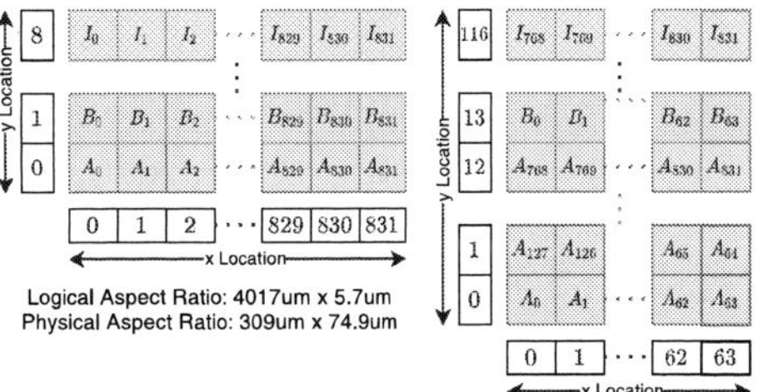

Fig. 3: Reshaping allows a change in aspect ratio

C. Connectivity Inference

After the cells in an array have been replicated, a connectivity step is performed: According to the steps described in the following, logical connections between the multi-cells are created. These connections either correspond directly to a electrical connection, or to a distributed logical reduction function such as OR-reduce. Also, arithmetic reductions such as 'popcount' [13] can be supported in the future.

The generation of the connections is controlled by the name of the ports of the multi-cells and the properties assigned to them:

- A **direction** which can be horizontal, vertical, or both
- A **level** describing on which hierarchy level the connection is created
- Special types of nets to describe more specific connection methods: **Local** connections of cells to their neighbors, signals with multiple drivers (for **tristate** buffers) and **tree-like** reduction operators

Algorithms such as TCAM are often operating word parallel. To further speed-up the look-up, the comparison can happen bit-wise in parallel [14]. In that case, the result of each comparison has to be true and the word is found if the logical AND over all of them (word-wise) is true. Similarly, standard-cell memory use distributed mux schemes [10], [15], where muxing is implemented as masking the data with a 1-hot encoded wordline, followed by an OR over all words. Thus, logic operations over bit-slices or a word are desired. As the design should be flexible in its size, the gates for these operations are inserted by the tool. Based on the fan-in, a properly sized tree of inverting gates is generated and optimally (based on wire-length) placed in the array.

If the logic function is applied horizontally, the logic gates are inserted in-line with the multi-cells they belong to. If applied vertically on a continuous part of the array, the logic function is implemented adjacent to it, similar to the work done in [15], minimizing routing congestion and making alignment straight forward.

D. Routing

The controlled placement of the cells leads to high densities, where routing the signals can become an issue for commercial EDA tools. Additionally, in certain applications, such as time-domain computing (discussed in Section IV-D), matching and predictable wiring capacitances are important. Therefore, the proposed flow provides alternate means in lieu of random routing solutions created in the standard EDA flow.

We define routing as a set of possible routing solutions for each pin in all multi-cells. We call one such solution a routing stencil, and they are part of the textual description of the standard-cell macro (omitted in the above example for brevity). During the routing, a stencil for each pin attached to the net is selected, such that a valid routing solution is obtained. The trivial straight connection between non adjacent standard-cells is automatically inserted. This is necessary as the distance between the cells depends on the parameterization. The routing stencil is selected based on the local environment of the cell, such as row and signal direction, and compatible routing stencils of the other cells connected to the same net.

One possible routing solution is shown in Fig. 4a: The first four rows show a template for each of the standard cells with the fifth row being the final routing solution for the multi-cell.

To bring the routing into the EDA tool flow, the stencils are converted into a TCL script which uses the proprietary commands of the corresponding commercial EDA tool.

E. Finishing Steps

At the end of the generation, a design consisting of uniquified standard cells and nets is available. If required, tap-cells are inserted at this point to drive the wells. Given some maximum distance between columns of tap-cells, they are evenly distributed in the design. Due to the advanced technology, additional edge constraints have to be satisfied: The placement is legalized at this step by inserting appropriate diffusion breaks where necessary or flipping cells if allowed.

Afterwards, the final location of the cells can be determined and output files are generated. If a routing solution is available, we use the actual wire length to estimate the parasitic loading of the interconnects. Otherwise we use a rectilinear Steiner tree to estimate the wire length. To perform timing checks, we extend OpenTimer with a python API using PyBind11 [16].

Fig. 4b shows a simplified digital design flow. We generate supporting files for all necessary steps to convey design intent of the generated design.

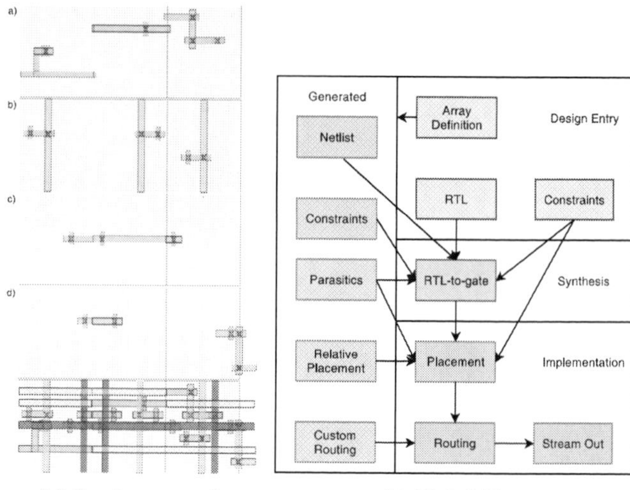

(a) Routing example (b) Digital Flow

Fig. 4: Left: Example routing for the memory cell shown in Fig. 1. a) - d) shows the different parts of the routing. Last row is a) - d) plus automatic connections (unfilled) and other signals (dark gray). Right: Simplified digital flow. Orange files are generated by the proposed flow.

IV. Design Case Studies

In the following section, we will discuss selected applications that benefit from such highly regular structure. Each of the discussed applications will exploit some aspect of the regular structure, but often in a different manner.

A. Design Space Exploration

One major benefit is the parameterizability of the flow and the fast turn-around time because of the incorporated delay and parasitics estimation: Instead of having to run, for example an FIR filter, through the design flow, which can take significant time, an initial design space exploration can happen entirely based only on the described tool's outputs. We show-case this capability by performing a design space exploration of FIR filters designs.

Insertion of registers in an FIR design increases the area and they take certain amount of energy. At the same time they reduce activity in the design, thus decreasing the energy consumption and decrease the the timing path length. An example for an experiment supported by the proposed flow is shown in Fig. 5: Parameterized by the number of additions/cycle, FIR filter layouts are generated and evaluated. The resulting design is competitive with an implementation placed by a standard EDA tool in most metrics while reducing the metal utilization in the upper layers by 90% as more signals are purely local.

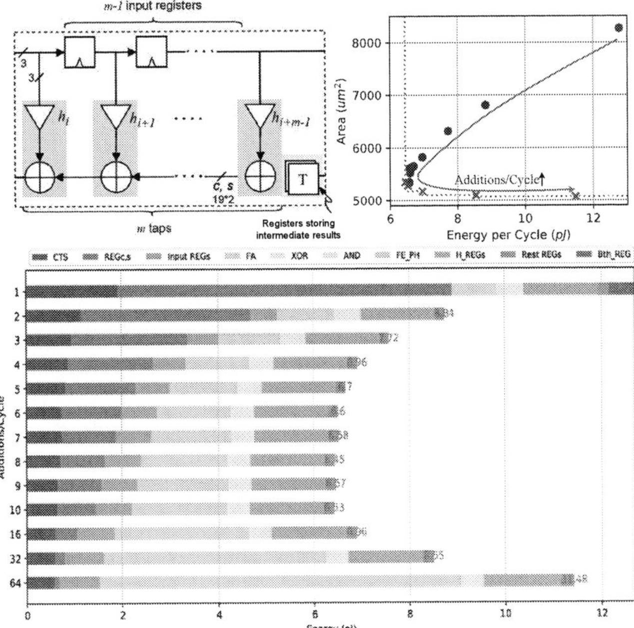

Fig. 5: Design space exploration of an FIR filter for the number of additions/cycle.

B. Design Density & Routability

Standard-cell memories target capacities between synthesized register files and SRAM macros. They are suitable for scaling the supply voltage in a wider range compared to SRAM at the cost of larger area/bit. Additionally, the peripheral overhead for 6T-SRAM leads to traditional SRAM actually being larger than a standard-cell memory [10]. At the same time, once generated, interspersed memory macros do not increase the run-time of the implementation.

We have implemented a banked standard-cell memory and implemented it using our structured place&route surrounded by control logic synthesized with a standard design flow. Table I shows the key metrics for different designs, tested in three different configurations: Fully automatic (i.e. using a traditional digital flow), regular placement and additionally with mostly predefined routing as sketched in Fig. 4a.

The area for the different runs was fixed. The regularly placed standard-cell memory was placed and routed successfully in the provided core area, while the commercial EDA tool was not able to find a valid routing solution, thus generating the large number of DRC. As the area constraint is not as tight for the other designs, the EDA tool is able to place and route the designs without DRC violations. Fig. 6 shows images of the placement and routing for the different approaches. Even visually an improvement is apparent as orange shapes (metal 6) have almost disappeared.

C. Compute-In-Memory Architectures

Because of the use of standard cells, not only memory can be implemented with the proposed flow, but also CIM designs, where logic is co-integrated with the memory elements. Common functionality to implement are multiply-accumulate

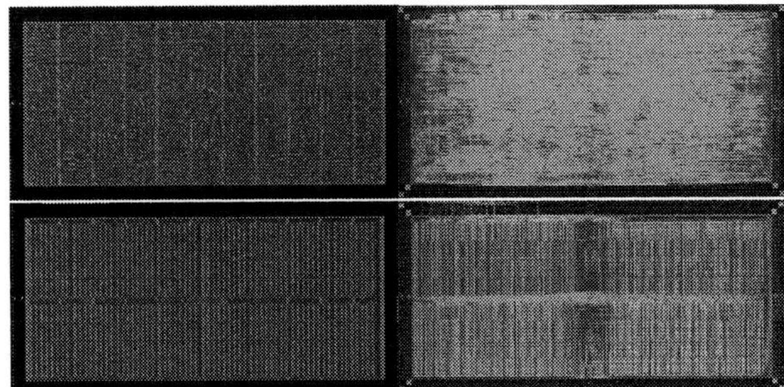

Fig. 6: Difference between traditional digital p&r (top) and our solution implemented with the proposed flow. Left is placement, right is routing.

for neural network and cryptography applications [17]–[19], (T)CAM and basic logic operations on multiple words. The design presented in [14] shows a TCAM implementation using standard-cells, highlighting the benefits of the proposed approach. The authors in [20] implement a design with 16 CIM arrays, where each array can perform multiple functions, mainly sorting and searching.

TABLE I: Results of different case studies.

Mode / Metric	Standard Digital Flow	Regular Placement	+Regular Routing
Design	**Standard Cell Memory**		
Total Run-time [mm:ss]	98:13	80:04	39:15
Number of DRC	210	93483	0
Critical Path [ps]	5440	2939	2924
Metal Utilization	1.22	0.78	0.67
C_{tot}[pF]	83.00	46.30	43.96
Design	**Multi-Task CIM Array [20]**		
Total Run-time [mm:ss]	201:40	202:45	191:02
Number of DRC	0	0	0
Critical Path [ps]	12472	14287	14139
Metal Utilization	0.90	0.83	0.73
C_{tot}[pF]	282.1	258.24	233.4
Design	**TDC CIM Array [12]**		
Total Run-time [mm:ss]	211:22	209:35	208:55
Number of DRC	11	23	0
Metal Utilization	0.41	0.48	0.48
C_{tot}[pF]	45.5	58.7	58.18
Design	**TCAM Array [14]**		
Total Run-time [mm:ss]	10:35	7:32	8:07
Number of DRC	0	0	0
Critical Path [ps]	389	385	383
Metal Utilization	0.74	0.69	0.66
C_{tot}[pF]	19.12	17.62	17.04

D. Time-Domain Computing

An emerging field at the intersection of the analog and digital domain is time domain computing [21]: Instead of encoding values as binary values or voltages/currents, the value can be encoded as a delay between two edges or a pulse width. Accumulation in time domain to simple concatenation of appropriately set delay stages, leading to a very efficient binary MAC operation, which is interesting for low-power neural network inference. Such approach relies on highly accurate delays as any variation will increase the uncertainty and potential inaccuracy of the computation. Using the presented flow, a physical implementation of a binary TDC was realized [12] that reduces the delay variation by 79% by employing regular placement and an additional 28% by forcing a regular routing pattern, as shown in Fig. 7.

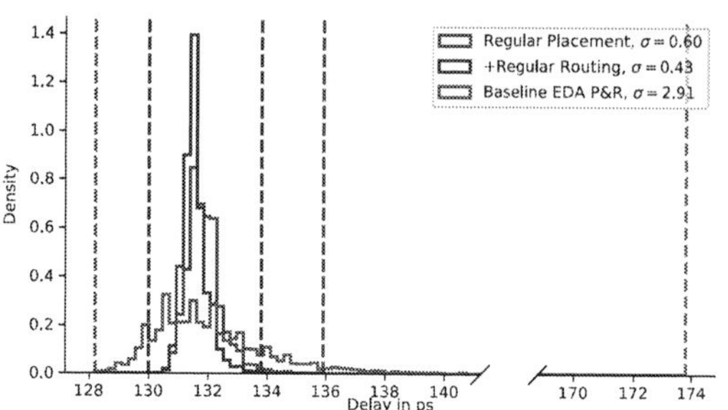

Fig. 7: Stepwise reduction in variance from standard P&R to regular placement and regular routing. The dashed lines correspond to the minimum and maximum delay of each version respectively.

E. Custom Cells

In the domain of time domain computing, a variable delay cell with an analog tuning input has been designed to fit into a typical digital flow [22]. It uses a current starved inverter design with long channel devices to minimize variability. In general, long channel devices can not be placed inside the regular structure of digital logic due to DRC violations. By exploiting our knowledge of their more complex placement requirements, we are able to integrate them seamlessly in a core device array. The resulting poly pitch is shown in Fig. 8. Note the aligned long channel devices, which break the regular poly pitch but are integrated into the digital area with no DRC errors.

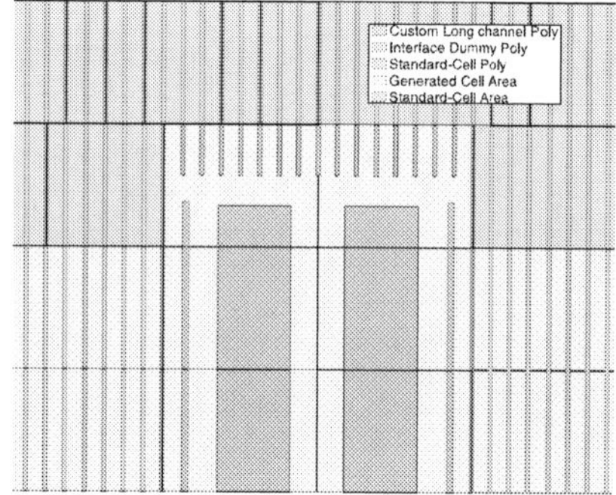

Fig. 8: The proposed flow can generate layouts with non-CPP poly pitch tightly integrated with standard-cells.

V. Conclusion

In this paper, we have shown how regular structures of standard cells can be derived from a textual description with supporting files being automatically generated. The information can be used throughout the traditional digital flow. The generated regular structures have a large number of applications, making the designs denser, faster, routable (despite high utilization) and provide opportunities to reduce parasitics as well as systematic delay variations. With the proposed flow, utilizations of upwards of 90% can be placed and routed, runtime of the implementation tool is up to 2.5x faster, the critical path is 2x shorter and the metal utilization can reduced by 45%. Different designs, from simple standard-cell memory to CIM architectures for TCAM, sorting and time domain computing have been implemented using the proposed flow.

References

[1] S. Bae, H.-O. Kim *et al.*, "Coarse-Grained Structural Placement for a Synthesized Parallel Multiplier," in *Proceedings of the 2015 Symposium on International Symposium on Physical Design*, ser. ISPD '15. New York, NY, USA: ACM, 2015, p. 17–24.

[2] M. Gansen, F. Richter *et al.*, "A datapath generator for full-custom macros of iterative logic arrays," in *Proceedings IEEE International Conference on Application-Specific Systems, Architectures and Processors*, 1997, pp. 438–447.

[3] M. R. Guthaus, J. E. Stine *et al.*, "OpenRAM: An open-source memory compiler," in *2016 IEEE/ACM ICCAD*, 2016, pp. 1–6.

[4] C. Raibaut, E. Ponsot, and L. Bouetel, "Method and apparatus for logic layout with controlled placement of structured cells," aug 2002, patent No. US20020188919A1, Filed June 7th., 2001, Issued Dec. 12th., 2002.

[5] T.-W. Huang and M. D. F. Wong, "OpenTimer: A High-Performance Timing Analysis Tool," in *2015 IEEE/ACM ICCAD*, 2015, p. 895–902.

[6] X. Tang, E. Giacomin *et al.*, "OpenFPGA: An Opensource Framework Enabling Rapid Prototyping of Customizable FPGAs," in *29th International Conference on Field Programmable Logic and Applications (FPL)*, 2019, pp. 367–374.

[7] A. Amid, D. Biancolin *et al.*, "Chipyard: Integrated Design, Simulation, and Implementation Framework for Custom SoCs," *IEEE Micro*, vol. 40, no. 4, pp. 10–21. 2020.

[8] J. Bachrach, H. Vo *et al.*, "Chisel: Constructing hardware in a Scala embedded language," in *DAC Design Automation Conference*, 2012.

[9] J. Decaluwe, "MyHDL: A Python-Based Hardware Description Language," *Linux J.*, no. 127, p. 5, 2004.

[10] X. Fan, J. Stuijt *et al.*, "Synthesizable Memory Arrays Based on Logic Gates for Subthreshold Operation in IoT," *IEEE TCAS I: Regular Papers*, vol. 66, no. 3, pp. 941–954, 2019.

[11] A. Lu, X. Peng *et al.*, "NeuroSim Simulator for Compute-in-Memory Hardware Accelerator: Validation and Benchmark," *Frontiers in Artificial Intelligence*, vol. 4, 2021.

[12] J. Lou, C. Lanius, and T. Gemmeke, "All-Digital Time-Domain Compute-in-Memory Engine for Binary Neural Networks with 1.05 POPS/W Energy Efficiency," in *IEEE ESSCIRC*, 2022.

[13] T. Stadtmann, C. Latotzke, and T. Gemmeke, "From Quantitative Analysis to Synthesis of Efficient Binary Neural Networks," in *2020 IEEE ICMLA*, 2020, pp. 93–100.

[14] X. Fan, N. Meyer, and T. Gemmeke, "Compiling All-Digital Embedded Content Addressable Memories on Chip for Edge Application," *IEEE TCAD*, 2021.

[15] A. Teman, D. Rossi *et al.*, "Controlled placement of standard cell memory arrays for high density and low power in 28nm FD-SOI," in *The 20th ASP-DAC*, 2015.

[16] W. Jakob, J. Rhinelander, and D. Moldovan, "pybind11 – Seamless operability between C++11 and Python," 2017.

[17] H. Fujiwara, H. Mori *et al.*, "A 5-nm 254-TOPS/W 221-TOPS/mm2 Fully-Digital Computing-in-Memory Macro Supporting Wide-Range Dynamic-Voltage-Frequency Scaling and Simultaneous MAC and Write Operations," in *2022 IEEE ISSCC*, vol. 65, 2022, pp. 1–3.

[18] Y.-D. Chih, P.-H. Lee *et al.*, "An 89TOPS/W and 16.3TOPS/mm2 All-Digital SRAM-Based Full-Precision Compute-In Memory Macro in 22nm for Machine-Learning Edge Applications," in *2021 IEEE ISSCC*, vol. 64, 2021, pp. 252–254.

[19] S. Zhang and T. Gemmeke, "A 22-nm 1,287-MOPS/W Structured Data-Path Array for Binary Ring-LWE PQC," in *IEEE ESSCIRC*, 2023.

[20] C. Lanius and T. Gemmeke, "Multi-Function CIM Array for Genome Alignment Applications built with Fully Digital Flow," in *2022 IEEE NorCAS*, 2022.

[21] Z. Chen, H. Zhou, and J. Gu, "Digital Compatible Synthesis, Placement and Implementation of Mixed-Signal Time-Domain Computing," in *56th ACM/IEEE Design Automation Conference (DAC)*, 2019, pp. 1–6.

[22] J. Lou, F. Freye *et al.*, "Scalable Time-Domain Compute-in-Memory BNN Engine with 2.06 POPS/W Energy Efficiency for Edge-AI Devices," in *Proceedings of GLSVLSI 2023*. New York, NY, USA: Association for Computing Machinery, 2023, p. 665–670. [Online]. Available: https://doi.org/10.1145/3583781.3590220

Multi-objective optimization for Floating Point Mix-Precision Tuning

Zeqing Li
Tsinghua University
Department of Computer Science
Email: zq-li17@mails.tsinghua.edu.cn

Yongwei Wu
Tsinghua University
Department of Computer Science
Email: wuyw@tsinghua.edu.cn

Youhui Zhang
Tsinghua University
Department of Computer Science
Zhongguancun Laboratory
Beijing 100094, China
Email: zyh02@tsinghua.edu.cn

***Abstract*—This paper proposes a multi-objective optimization method for mixed-precision computation. Unlike previous studies that often take mantissa length reduction as the only optimization target, our work models the actual performance and power consumption of mixed precision programs on the corresponding hardware platforms, and based on this model searches for the pareto optimal set of all precision configurations.**

Experiments show that this tool can obtain performance improvements of $15\% - 71\%$ on floating-point benchmarks while satisfying accuracy requirements. Compared to some typical counterpart-work, an average 21% improvement can be obtained in SIMD scenarios.

I. Introduction

In high-performance computing tasks, floating-point computation is a primary means of real-value computation. As computation scales increase, the efficient employment of floating-point computation has an increasingly significant impact on overall performance [1]. While higher precision can improve output accuracy, it can also increase program runtime, energy consumption, and memory-access pressure.

However, not all programs require high-precision operations throughout to guarantee output accuracy. Ideally, applications should use as little high-precision as possible to gain performance and power advantages while maintaining output precision. Developing mixed-precision codes manually is challenging, as it requires a deep understanding of the numerical behavior of the algorithm, as well as details of floating-point rounding errors, particularly for large HPC programs.

Fortunately, many studies aim to automate the development of mixed-precision versions of a given program [2]–[5]. However, all existing studies have some limitations.

Firstly, they only use the number of variable bits as the feedback signal for optimization, and do not precisely model program performance. This approach is inconsistent with reality, mainly because SIMD instructions are widely used as an effective speedup in floating-point programs, and the acceleration effect of SIMD depends on information such as the degree of vectorization and specific hardware parameters. Thus, signals such as the number of variable bits can only provide rough trends. Additionally, the additional overhead generated at the hardware level is not considered, as some reduced precision variables and their operations will not necessarily improve performance.

Secondly, existing work lacks modeling of power consumption, which should also be a key optimization goal in some scenarios.

In this paper, we propose a method to model accuracy, performance, and power consumption by sampling on the corresponding hardware platform using automatic differentiation (AD) and Gaussian processes (GP), respectively. A multi-objective Bayesian optimization algorithm then explores the search space to find the optimal precision configuration with a guaranteed precision threshold. In the vast majority of our experiments, the tool chain yields better precision configurations compared to existing work.

This work was supported in part by National Natural Science Foundation of China (NSFC) under Grant No. 62250006, 62072266 and 62202254, in part by Beijing National Research Center for Information Science and Technology (BNR2022RC01003), in part by Tsinghua University Initiative Scientific Research Program, and in part by Suzhou-Tsinghua Innovation Leadership Program.

II. Related Work

The field of approximate computation has a range of well-established work, and we refer the reader to a detailed survey of the entire domain [6], [7], presenting here only a small part of the work most relevant to that described in this paper.

One direction of work in mixed precision is to analyze floating-point codes to find unstable numerical computations. Some work [8], [9] give tight bounds on the rounding error based on symbolic Taylor expansions, but the precision strategy obtained is too conservative. Precimonious [3], also dedicated to numerical analysis, is the first approximate computation tool that can be applied to large programs. Nhut [10] analyzes the variables of large programs in parallel for further efficiency, and can adjust the floating-point by any number of precision bits. ADAPT [11], in order to study the effect of variable precision on the output error in a more refined way, analyzes it with automatic differentiation to obtain a more radical precision configurations. We differ from these techniques in that we go beyond static analysis of the floating-point code and model the program performance power consumption by taking into account the hardware influences.

979-8-3503-1176-1/23 $31.00 © 2023 IEEE

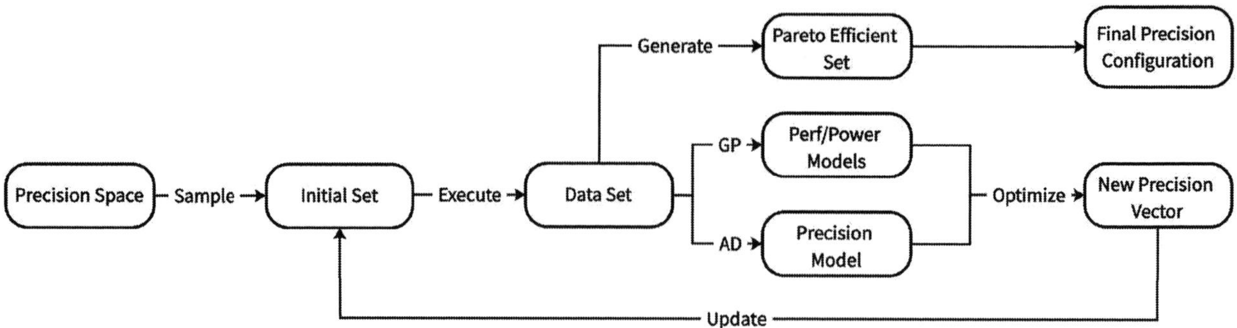

Fig. 1. Flowchart for searching mixed precision configurations. (a) Sample: A series of initial precision configurations are obtained from the precision space by a sampling algorithm. (b) Execute: These precision configurations are executed on the simulator or hardware platform to obtain the performance and power consumption corresponding to each configuration. (c) GP/AD: Modeling with Gaussian processes and automatic differentiation and using them as surrogate functions in the optimization process. (d) Optimize: Build acquisition function and use Bayesian optimization to search for better precision configurations. (e) Update: Add the new configuration to the initial set and consider it as a complete iteration after executing process b again. (f) Generate: After several iterations, a pareto efficient set is generated from the data set, and a proper precision configuration is selected according to the specific constraints.

III. Problem Formulation

Definition 1 (Program Input Precision). *The given program has n floating-point variables $X = \{x_1, x_2, ..., x_n\}$. The precision of each variable is inscribed by a two-tuple (exponent and mantissa), and in this paper we only consider the mantissa, such that the program precision is represented by the precision vector $P = \{p_1, p_2, ..., p_n\}$ and p_i represents the mantissa of the variable x_i.*

Definition 2 (Output Error). *Error of the program's output for a given precision configuration relative to the initial precision output.*

Definition 3 (Performance). *Performance is defined as the acceleration ratio achieved by the program in a mixed precision configuration.*

Definition 4 (Power). *Power consumption is defined as the sum of the power consumption of each module.*

In the present problem, there is a large correlation between output error, performance, and power consumption, and they are often not optimized simultaneously. Therefore, we introduce Pareto optimality to measure the advantages and disadvantages of the precision configuration.

Definition 5 (Pareto Efficient Set). *Given distinct $\boldsymbol{y}_i \in \mathbb{R}^L$ for $i = 1, \ldots, n$, we write $\boldsymbol{y}_j \succeq \boldsymbol{y}_k$ when $y_{j,l} \geq y_{k,l}$ for each $l = 1, \ldots, L$, and say " $\boldsymbol{y}_j$dominates $\boldsymbol{y}_k$ ". For the set of distinct points $\mathcal{Y} = \{\boldsymbol{y}_1, \cdots, \boldsymbol{y}_n\}$, the subset of Pareto efficient points, $\mathcal{P}(\mathcal{Y}) \subseteq \mathcal{Y}$, is defined as $\mathcal{P}(\mathcal{Y}) = \{\boldsymbol{y}_i \in \mathcal{Y} : \boldsymbol{y}_j \nsucceq \boldsymbol{y}_i \forall \boldsymbol{y}_j \in \mathcal{Y} \backslash \{\boldsymbol{y}_i\}\}$. In other words, the Pareto efficient set is the set of non-dominated points, and is always non empty.*

Problem 1 (Floating Point Precision Tuning). *Given a precision space $\mathcal{D}$, each precision configuration corresponds to a vector p in the space $\mathcal{D}$. Output error, performance and power consumption are the elements y in the output space Y. Precision Tuning requires finding the set of vectors $P = \{p | f(p) \in Y, y' \nsucceq f(p), \forall p \in \mathcal{D}, \forall y' \in Y\}$.*

In this article, we aim to explore the Pareto efficient set consisting of output error, performance, and power consumption under different precision configurations, and to choose the optimal configuration according to the actual output precision requirements.

IV. Methodology

A. Overview

Figure 1 illustrates the full process of the mixed precision searching algorithm. First, we sample in the precision space and optimize the sampling process using active learning and domain knowledge (IV-B). Then, it is tested in real hardware or clock-driven processor simulator to obtain the power-consumption and performance under different configurations. After obtaining the dataset, we model the accuracy using automatic differentiation (IV-C) and model the performance and power consumption with Gaussian processes (IV-D). Taking the above models as surrogate functions, we apply a multi-objective Bayesian optimization algorithm (IV-E) to search for a better precision vector and add it to the dataset. After several iterations, we find the Pareto effective set from the dataset and select the optimal precision configuration based on actual constraints.

B. Sampling

Considering the impact of sampling time and data set size on modeling speed, we aim to utilize a small but information-rich data set. Although a naive uniform sampling method exists, it is not ideal for capturing representative samples in our problem.

To address this challenge, we leverage transductive experimental design (TED), a technique that has shown significant promise in architecture design [12]. As outlined in Algorithm

1, TED constructs a distance matrix K between newly sampled vectors and unsampled spaces using a suitable distance function. We then maximize the trace of this matrix to obtain more representative sampling points [12].

To further enhance sampling performance, we incorporate domain knowledge and manually prune the sampling space. Specifically, we specify original sampling points and apply the TED algorithm to neighborhoods of these points. Our experiments demonstrate that incorporating domain knowledge can lead to a 13% reduction in the number of required samples on average (as shown in Algorithm 1).

Algorithm 1 Sampling algorithm

Input: D is the unsampled space, α is the normalization factor, N is the amount of sampled data, and C is the set of artificially introduced sampling points.
Output: S is the sampled data set with $|S| = N$.

$S \leftarrow \emptyset$
$D_i \leftarrow$ neighborhood of $c_i \in C$
for x in D_i **do**
  $S_i \leftarrow TED(x, \alpha, \lfloor N/|C| \rfloor)$
  $S \leftarrow S \cup S_i$
end for

Procedure $TED(D, \alpha, N)$
Input: K is the metric matrix.
$s \leftarrow \emptyset$
for $i = 1$ to N **do**
  $s_* \leftarrow argmaxTr[K_{Ds}(K_{ss} + \alpha I)^{-1} K_{sD}]$
  $s \leftarrow s \cup s_*$, $D \leftarrow D \backslash s_*$
  $K \leftarrow K - K_{Ds_*}(K_{s_* s_*} + \alpha I)^{-1} K_{s_* D}$
end for
return s
EndProcedure

C. Output Error Estimation Model

We construct our error model by approximating the program with a first-order Taylor series, where the output error is assumed to be linear in the rounding error. Specifically, we consider the input of the program, denoted by x, and the output, denoted by y, which together can be viewed as a function f that satisfies $y = f(x)$. Assuming a small perturbation Δx in the input caused by a change in input precision, we can estimate the output error of the program using the following equation: $\Delta y \approx \left|f'(a)^T \overline{\Delta x}\right|$.

To measure the error of a specific precision, we use the absolute error due to rounding in variable x, which is less than $|x|\epsilon$, where $\epsilon = 2^{-p}$ and p represents the number of bits in the mantissa for that precision [11]. We define a metric $\mathcal{E}$ to capture the sensitivity of any input or intermediate variable to rounding errors:

$$\mathcal{E}_x = |f'(x) \times x| \tag{1}$$

D. Performance and Power Models

Research on building reliable prediction models quickly for a given initial training set has been a popular topic in recent years. Gaussian process (GP) models have gained significant attention due to their non-parametric and robust nature, which have contributed to their success in traditional optimization problems. In architecture design, GP has been successfully applied in several studies. For instance, in the design of 64-bit prefix adders, GP was used to efficiently and effectively explore the parameter space of EDA tools [13]. Similarly, BOOM-Explorer [14] uses GP to characterize the design space in microarchitecture design. Experiments show that this method can accurately fit processor performance and further derive optimal microarchitecture parameters.

In light of these successes, we aim to employ GP to construct performance and power consumption models. We assume that we have feature vectors $\boldsymbol{X}$ that correspond to a set of power or performance values $\boldsymbol{y}$. GP provides a prior over the value function f as $f(\boldsymbol{x}) \sim \mathcal{GP}(\mu, k_{\boldsymbol{\theta}})$, where μ represents the mean value and the kernel function k is parameterized by $\boldsymbol{\theta}$. Using this prior, we can construct Gaussian distributions for any collection of value functions f, as specified by the following equation:

$$\boldsymbol{f} = [f(\boldsymbol{x}_1), f(\boldsymbol{x}_2), \ldots f(\boldsymbol{x}_n)]^T \sim \mathcal{N}(\boldsymbol{\mu}, \mathrm{K}_{\boldsymbol{XX}|\boldsymbol{\theta}}) \tag{2}$$

where $\mathrm{K}_{XX|\boldsymbol{\theta}}$ is the intra-covariance matrix among all feature vectors and $\left[\mathrm{K}_{\boldsymbol{XX}|\boldsymbol{\theta}}\right]_{ij} = k_{\boldsymbol{\theta}}(\boldsymbol{x}_i, \boldsymbol{x}_j)$. Thus, given a newly sampled feature vector $\boldsymbol{x}_*$, the predictive joint distribution f_* that depends on $\boldsymbol{y}$ can be calculated according to Equation:

$$f_* \mid \boldsymbol{y} \sim \mathcal{N}\left(\begin{bmatrix} \boldsymbol{\mu} \\ \mu_* \end{bmatrix}, \begin{bmatrix} \mathrm{K}_{\boldsymbol{XX}|\boldsymbol{\theta}} + \sigma_e^2 \boldsymbol{I} & \mathrm{K}_{\boldsymbol{X}\boldsymbol{x}_*|\boldsymbol{\theta}} \\ \mathrm{K}_{\boldsymbol{x}_*\boldsymbol{X}|\boldsymbol{\theta}} & k_{\boldsymbol{x}_*\boldsymbol{x}_*|\boldsymbol{\theta}} \end{bmatrix}\right) \tag{3}$$

In general, the performance of GP depends on the expressiveness of the kernel function $k_{\boldsymbol{\theta}}$. The common kernel functions are linear kernels, polynomial kernels, and radial kernels, etc. DNN is used in BOOM-Explorer [14] to find the optimal gaussian kernel. However, in the context of this particular problem, our testing revealed that although DNN-based methods have the potential to attain higher prediction accuracy, they are not necessarily capable of generating optimal precision configurations. This can be attributed to the restricted range of variable precision options, which are limited to 16-32-64 bits. Upon careful consideration of factors such as time overhead, training data size, and prediction accuracy, we have decided to adopt the ARD Matern 5/2 kernel, which is defined as

$$\begin{gathered} k_{M52}(\boldsymbol{x}, \boldsymbol{x}') = \theta_0^2 \left(1 + \sqrt{5r^2} + \tfrac{5}{3}r^2\right) \exp\left(-\sqrt{5r^2}\right) \\ r^2 = \sum_{d=1}^{D} (x_d - x_d')^2 / \theta_d^2 \end{gathered} \tag{4}$$

E. Multi-objective Optimization

Next we need to use Bayesian optimization to obtain the Pareto set based on the above model. Suppose the sampled precision configurations are $X = \{x_1, x_2, ..., x_t\}$ and the model to evaluate them is $y_i = f(x_i)$. Our goal is to find a new precision configuration x_{t+1} such that $reward = f(x_{t+1}) - f(x_*)$

979-8-3503-1176-1/23 $31.00 © 2023 IEEE

is maximized. This is very simple with a single objective, but in this problem a better precision configuration should not only have higher output accuracy, but also lower power consumption and runtime. There is a clear correlation between these multiple optimization objectives before, so how to generalize the single-objective optimization framework to the multi-objective case is the main problem we face.

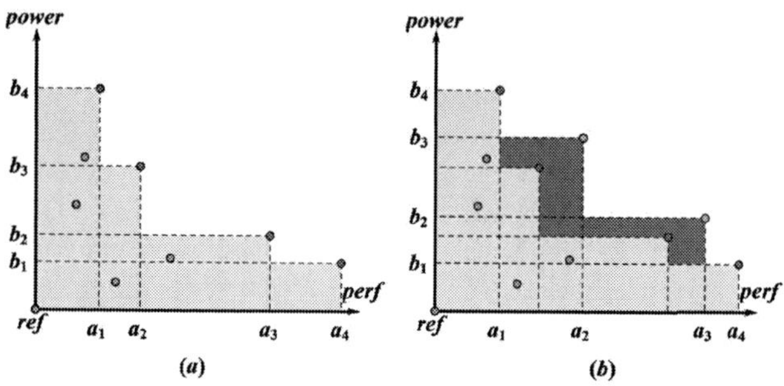

Fig. 2. Two-dimensional schematic of the Pareto hypervolume when only performance and precision are considered. (a) The four red points represent the currently obtained pareto-optimal set, and the shaded area is their PV relative to the origin. The green points included in it are the suboptimal precision configuration, dominated by the red points. (b) The EIPV is the expectation of the area of the dark region, and the blue points are candidates for a more optimal configuration. Algorithm 2 selects new sampling points by maximizing EIPV.

To solve this problem, we introduce the concept of Pareto hypervolume (PV) [15], [16]. Given a set of distinct points $\mathcal{Y} = \{\boldsymbol{y}_1, \ldots, \boldsymbol{y}_n\}$, we have defined its Pareto efficient subset, $\mathcal{P}(\mathcal{Y})$. Define a reference point, $v_{\text{ref}} \in \mathbb{R}^L$, which is dominated by each element of $\mathcal{P}(\mathcal{Y})$ i.e. $\boldsymbol{u} \succeq \boldsymbol{v}_{\text{ref}}$ for each $\boldsymbol{u} \in \mathcal{P}(\mathcal{Y})$. The Pareto hypervolume of $\mathcal{P}(\mathcal{Y})$ with respect to v_{ref} is

$$\operatorname{Vol}_{\boldsymbol{v}_{\text{ref}}} \mathcal{P}(\mathcal{Y})) = \int_{\mathbb{R}^L} \mathbb{I}\left[\boldsymbol{y} \succeq \boldsymbol{v}_{\text{ref}}\right] \left[1 - \prod_{\boldsymbol{u} \in \mathcal{P}(\mathcal{Y})} \mathbb{I}[\boldsymbol{u} \nsucceq \boldsymbol{y}]\right] dy \tag{5}$$

where $\mathbb{I}(.)$ is the indicator function, which outputs 1 if its argument is true and 0 otherwise.

In this way, we convert the problem of simultaneously optimizing the three objectives into optimizing the super volume of these three variables relative to the reference point. By definition a larger PV indicates a better configuration, so the optimal configuration point should maximize the total increment of PV, i.e., maximize the following expression

$$r_T = \operatorname{Vol}_{\boldsymbol{v}_{\text{ref}}} \left(\mathcal{P}\left(\tilde{\mathcal{Y}}_T\right)\right) - \operatorname{Vol}_{\boldsymbol{v}_{\text{ref}}} \left(\mathcal{P}\left(\mathcal{Y}^*\right)\right) \tag{6}$$

where $\mathcal{Y}^*$ is the true Pareto frontier and $\tilde{\mathcal{Y}}_T$ is the suggested Pareto frontier after T evaluations of each of the objectives.

Unfortunately, the variable proves to be computationally infeasible when T is sufficiently large. In this problem, we use the expected improvement in Pareto hypervolume introduced by [17]. In other words, our goal is to maximize the area of the dark shadows in Figure 2.

Algorithm 2 Mixed-precision Search Algorithm

Input: D is the unsampled space, α is the normalization factor, N is the amount of sampled data, and T is the maximal iteration number.

Output: Final Precision Configuration P

$X_0 \leftarrow sampling(D, \alpha, N)$
Use gem5 to obtain corresbonding clock cycles and power
$L \leftarrow X_0$
$U \leftarrow D \backslash L$
for $x = 1 \leftarrow T$ **do**
  Establish and train GP on L
  $\boldsymbol{x}_* \leftarrow \arg\max_{\boldsymbol{x} \in \mathrm{U}} \operatorname{EIPV}(\boldsymbol{x} \mid \mathrm{U})$
  Update : L, U
end for
Construct Pareto-optimal set X from L
Select precision configuration P based on constraints
return P

V. Evaluation

In the experimental section, we searched for the optimal configurations of different scalar and vector programs for a given accuracy constraint and compared them with previous work. In order to get power consumption conveniently, we used the cycle-driven CPU simulator, Gem5, as a test platform for most of our experiments. Further, we also did real machine tests in x86 environment to verify the correctness of the method.

A. Scalar Program

We compared our search results with those of ADAPT [11]. Given our output results produced precisely in-place precision vectors, we mapped the mantissa back to half-precision, single-precision, and double-precision for comparison. To obtain the program's precision configuration, we initially utilized ADAPT's test programs with equivalent input files and error metrics. As ADAPT solely supports 32-64 bit mixing, we ensured a fair comparison by setting the lowest precision to single precision in the first set of experiments. Subsequently, we extended the search logic of ADAPT to support half-precision search.

Figure 3 presents the experimental results, with each group representing a distinct test case on the horizontal coordinates. The two left columns in each group represent the data obtained by running ADAPT, while the vertical coordinates represent the speedup ratio obtained by each program relative to the full-precision version. Unlike the sampling operation, we conducted five runs of each program and calculated the average to obtain the exact running time.

Our experiments revealed that our tool could provide better precision configurations in the vast majority of cases in scalar program tests. Our tool yielded an average speedup ratio of 32.7%, which is better than ADAPT's 30.5% for $32 - 64$ bit configurations. Our tool also obtained an average speedup ratio of 42.5% when the precision options expanded, which is more advantageous than the existing work's 36%.

This improvement resulted from two primary reasons: firstly, a decrease in the number of bits of variables did not necessarily equate to faster speed. In some mixed-precision versions of programs, the compiler implicitly added data type conversions, increasing overhead at the hardware level, diminishing or offsetting the performance benefits of lower precision. Type conversion operations could account for $10-20\%$ of the total time in a given program, as observed in [18]. Secondly, greedy search methods risked being restricted to local optima, which became more pronounced as the search space increased, i.e., when the number of optional precision configurations per variable increased.

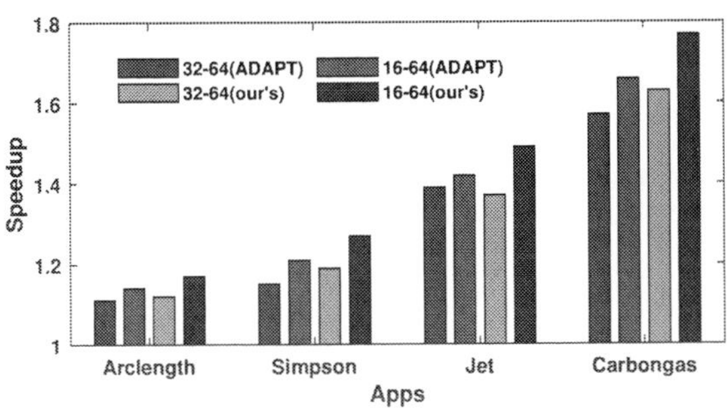

Fig. 3. Acceleration ratios obtained for each program with different precision configurations in different precision ranges

B. SIMD Program

In order to further test the performance improvement of the program by mixing precision in a vector environment, three representative programs were selected for testing, namely HPCCG, SVM and GMRES. HPCCG is an application in the mantevo benchmarking suite, and the core computational algorithm is the conjugate gradient method. SVM is the prediction stage of support vector machines, and is a representative algorithm in the field of machine learning. GMRES is the generalized minimum residual method for iteratively solving systems of linear equations and is the core algorithm for many scientific computing problems.

To maximize the impact of vectorization on the performance of the program, we mark the code segments that can be vectorized and vectorize them manually. We also keep the scalar version of the program for ADAPT analysis and accuracy model construction. To further exploit the performance and power benefits of vectorization, here we set floating-point types for each variable, i.e., half-precision, single-precision, and double-precision.

Experiments show that our tool can provide better precision configurations in a SIMD environment than ADAPT, obtaining an average performance improvement of 21%.

In the case of solving systems of linear equations (GMRES), our advantage is particularly clear. Specifically, we found that excessive reduction in the precision of the variables leads to a decrease in the convergence speed. Compared to the original precision, GMRES requires more iterations to reach convergence, and in extreme cases, the number of iterations can even increase by a factor of 2-3. Therefore, a proper search strategy should not only reduce the variable precision, but also keep the number of additional iterations within a suitable range. If the focus is only on the number of variable bits, the extra number of iterations tends to weaken the performance gain from mixing precision. Using performance rather than the number of variable bits as the optimization goal in our work can largely avoid this problem.

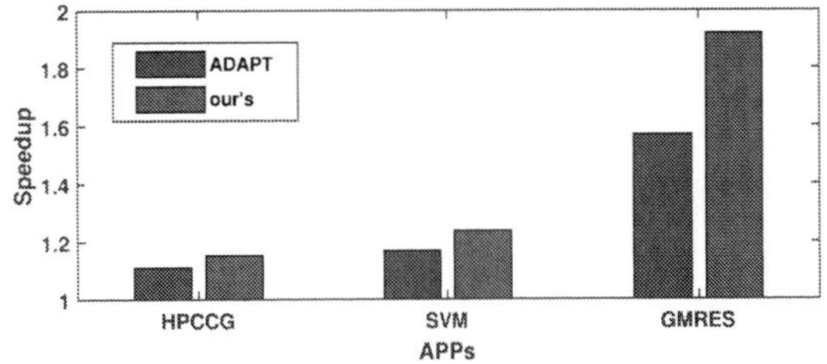

Fig. 4. The acceleration ratios obtained by each test program in the vector environment. Where blue represents the precision configuration obtained by ADAPT and orange represents the precision configuration obtained by our tool.

Regarding power consumption, our program achieves an average power reduction of 17% in a vector environment through mixed precision. However, similar to performance, low-precision operations may be affected by certain operations, such as type conversion, that may negate the advantages of mixed precision. Nevertheless, after vectorization, there is a reduction of $5-7\%$ in power consumption, primarily due to the significant reduction in floating-point operations and memory accesses (up to 40% in SVM) enabled by SIMD.

To validate our approach on real hardware, we limited the precision levels to 32 and 64 bits. For scalar programs, we achieved the same precision configuration as the existing tool for arclength and Simpson, and observed a 5% and 6.2% performance gain, respectively, due to better precision configurations in the other two programs. For vector programs, we obtained different precision configurations and an average additional performance gain of 21%, consistent with the simulator tests. Notably, we found that using the precision configuration obtained from the simulator as an initial point and fine-tuning it by hardware sampling could further optimize the search overhead. This insight suggests that future work can employ a two-step approach to first coarse-grain the search on the fast model and then fine-tune it on the accurate model.

C. Tuning Overhead

While our tool offers more accurate precision configurations than existing tools such as ADAPT [11] and Precimonious [3], it requires sampling, leading to a larger search overhead. The search time can range from tens of seconds to tens of minutes, with a significant portion of the time devoted to obtaining performance and power consumption through a clock-driven processor simulator. On average, our tool requires 10-20 iterations to achieve the final precision configuration, depending on the program and the initial training set distribution.

TABLE I
ERROR THRESHOLDS, OUTPUT ERRORS, AND ESTIMATION ERRORS FOR VARIOUS APPLICATIONS

Apps	Error Threshold	Output Error	Estimate Error
Arclength	1e-12	2.3e-13	2.5e-13
Simpson	1e-12	4.8e-14	4.3e-14
Jet	1e-13	2.2e-13	8.9e-14
Carbongas	1e-10	6.5e-11	3.7e-11
HPCCG	1e-9	2.3e-10	4.3e-10
SVM	1e-10	7.2e-12	3.1e-12
GERES	1e-9	5.9e-11	1.2e-10

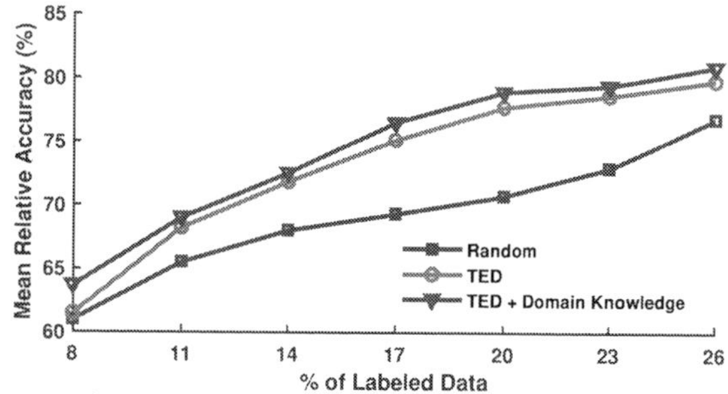

Fig. 5. The performance of MRA varies with changes in the number of samples under different sampling strategies.

To reduce the search time, we introduce domain knowledge and active learning ideas in the initial sampling process. This reduces the search time by an average of 31%. We show in Figure 5 that using the transductive experimental design algorithm individually can improve the prediction accuracy compared to random sampling. Furthermore, a better initial set with the introduction of domain knowledge can further reduce the number of samples needed, optimizing the search time.

However, there is still room for optimization of the search process, such as parallelizing fragments of the search process as done in [10], and sampling with faster simulators or building hardware models to predict program performance more quickly. These are left for future work.

VI. Conclusion

This paper presents an algorithm for solving mixed precision configurations of programs. The algorithm is modeled by sampling and uses multi-objective Bayesian optimization to search for the precision configuration with optimal performance and power consumption. Its novelty is that it uses information about the program's running on hardware to guide the search, implicitly including the impact of operations such as data type conversion and SIMD on program performance. Experiments show that it yields precision configurations that are better than those given by existing work in most cases. For programs with a higher degree of vectorization, the performance advantage we achieve will be even more apparent.

References

[1] Ganesh Gopalakrishnan, Paul D. Hovland, Costin Iancu, Sriram Krishnamoorthy, Ignacio Laguna, Richard A. Lethin, Koushik Sen, Stephen F. Siegel, and Armando Solar-Lezama. Report of the hpc correctness summit, jan 25–26, 2017, washington, dc, 2017.

[2] Michael O Lam, Jeffrey K Hollingsworth, Bronis R de Supinski, and Matthew P LeGendre. Automatically adapting programs for mixed-precision floating-point computation. In *Proceedings of the 27th international ACM conference on International conference on supercomputing*, pages 369–378, 2013.

[3] Cindy Rubio-González, Cuong Nguyen, Hong Diep Nguyen, James Demmel, William Kahan, Koushik Sen, David H Bailey, Costin Iancu, and David Hough. Precimonious: Tuning assistant for floating-point precision. In *SC'13: Proceedings of the International Conference on High Performance Computing, Networking, Storage and Analysis*, pages 1–12. IEEE, 2013.

[4] Alexey Solovyev, Marek S Baranowski, Ian Briggs, Charles Jacobsen, Zvonimir Rakamarić, and Ganesh Gopalakrishnan. Rigorous estimation of floating-point round-off errors with symbolic taylor expansions. *ACM Transactions on Programming Languages and Systems (TOPLAS)*, 41(1):1–39, 2018.

[5] Eva Darulova, Einar Horn, and Saksham Sharma. Sound mixed-precision optimization with rewriting. In *2018 ACM/IEEE 9th International Conference on Cyber-Physical Systems (ICCPS)*, pages 208–219. IEEE, 2018.

[6] Qiang Xu, Todd Mytkowicz, and Nam Sung Kim. Approximate computing: A survey. *IEEE Design Test*, 33(1):8–22, 2016.

[7] Gennaro Rodrigues, Fernanda Lima Kastensmidt, and Alberto Bosio. Survey on approximate computing and its intrinsic fault tolerance. *Electronics*, 9(4), 2020.

[8] Thierry Braconnier and Philippe Langlois. From rounding error estimation to automatic correction with automatic differentiation. In *Automatic differentiation of algorithms*, pages 351–357. Springer, 2002.

[9] Alexey Solovyev, Marek S Baranowski, Ian Briggs, Charles Jacobsen, Zvonimir Rakamarić, and Ganesh Gopalakrishnan. Rigorous estimation of floating-point round-off errors with symbolic taylor expansions. *ACM Transactions on Programming Languages and Systems (TOPLAS)*, 41(1):1–39, 2018.

[10] Nhut-Minh Ho, Elavarasi Manogaran, Weng-Fai Wong, and Asha Anoosheh. Efficient floating point precision tuning for approximate computing. In *2017 22nd Asia and South Pacific Design Automation Conference (ASP-DAC)*, pages 63–68. IEEE, 2017.

[11] Harshitha Menon, Michael O Lam, Daniel Osei-Kuffuor, Markus Schordan, Scott Lloyd, Kathryn Mohror, and Jeffrey Hittinger. Adapt: Algorithmic differentiation applied to floating-point precision tuning. In *SC18: International Conference for High Performance Computing, Networking, Storage and Analysis*, pages 614–626. IEEE, 2018.

[12] Kai Yu, Jinbo Bi, and Volker Tresp. Active learning via transductive experimental design. In *Proceedings of the 23rd international conference on Machine learning*, pages 1081–1088, 2006.

[13] Yuzhe Ma, Ziyang Yu, and Bei Yu. Cad tool design space exploration via bayesian optimization. In *2019 ACM/IEEE 1st Workshop on Machine Learning for CAD (MLCAD)*, pages 1–6, 2019.

[14] Chen Bai, Qi Sun, Jianwang Zhai, Yuzhe Ma, Bei Yu, and Martin DF Wong. Boom-explorer: Risc-v boom microarchitecture design space exploration framework. In *2021 IEEE/ACM International Conference On Computer Aided Design (ICCAD)*, pages 1–9. IEEE, 2021.

[15] Samuel Daulton, Maximilian Balandat, and Eytan Bakshy. Differentiable expected hypervolume improvement for parallel multi-objective bayesian optimization. *Advances in Neural Information Processing Systems*, 33:9851–9864, 2020.

[16] Amar Shah and Zoubin Ghahramani. Pareto frontier learning with expensive correlated objectives. In *International conference on machine learning*, pages 1919–1927. PMLR, 2016.

[17] Michael Emmerich. *Single-and multi-objective evolutionary design optimization assisted by gaussian random field metamodels*. PhD thesis, Dortmund, Univ., Diss., 2005, 2005.

[18] Giuseppe Tagliavini, Stefan Mach, Davide Rossi, Andrea Marongiu, and Luca Benini. A transprecision floating-point platform for ultra-low power computing. In *2018 Design, Automation & Test in Europe Conference & Exhibition (DATE)*, pages 1051–1056. IEEE, 2018.

REFROM: Responsive, Energy-efficient Frame Rendering for Mobile Devices

Tsung-Yen Hsu*, Yi-Shen Chen†, Yun-Chih Chen*, Yuan-Hao Chang‡, Tei-Wei Kuo*§¶
*Department of Computer Science and Information Engineering, National Taiwan University
†Department of Electronic and Computer Engineering, National Taiwan University of Science and Technology
‡Institute of Information Science, Academia Sinica
§Mohamed bin Zayed University of Artificial Intelligence
¶High Performance and Scientific Computing Center, National Taiwan University
htsungyen@gmail.com, billy195375@gmail.com, f07922039@ntu.edu.tw, johnson@iis.sinica.edu.tw, ktw@csie.ntu.edu.tw

Abstract—**The increasing demand for high-quality graphics on mobile devices necessitates a high frame rate for display refresh. However, current process scheduling and memory management policies fail to consider the computation demands of frame rendering because they are optimized for saving energy and resource utilization. This leads to unresponsive displays for mobile users due to rendering delays. Accurately estimating computation demands is challenging for the mobile operating system, particularly under memory pressure, without display-specific semantics from user space. Moreover, the complexity of frame rendering makes it infeasible to schedule them with real-time policies. To address these issues, we propose a new framework called REFROM that utilizes a history-based frame time estimator to analyze frame time samples from UI threads and predict the computation requirements of upcoming frames. Experimental results demonstrate that REFROM reduces the number of delayed frames by up to 40% and improves up to 4% energy efficiency compared to the existing approaches.**

I. Introduction

The growing demand for immersive visual experiences on mobile devices has driven the development of graphics-intensive applications, such as animation-rich web pages and movie-like mobile games. However, achieving such experiences requires a high-resolution display with an intensive frame rate. While most smartphones require a minimum frame rate of 60Hz, recent flagship models such as Samsung Galaxy S21 and Google Pixel 6 Pro go even further with screens that can reach up to 120Hz.

However, ensuring power efficiency is equally important for smartphone manufacturers since mobile devices run on batteries [1], [2]. To achieve this, manufacturers typically equip devices with heterogeneous processors composed of performant big cores and energy-efficient little cores. In addition, devices are often designed to be memory-restricted [3], but users may run and switch among multiple applications, causing memory demands to exceed physical capacity. To free up memory, Android compresses infrequently accessed memory pages and decompresses them if they are re-accessed. However, the compression process is CPU-intensive and may interfere with the timely execution of the display render threads, resulting in noticeable flickers, or "janks."

A jank occurs when an application's graphics rendering falls behind the required frame rate, causing the system to skip frames, leading to stuttering that significantly impacts the user experience. According to a recent study covering 15 phone models, up to 18% of the rendered frames appear as janks, and of these, 39% can be attributed to the render thread's inability to acquire sufficient CPU time due to competition with other system services [4].

To prevent jank, some systems have attempted to secure render threads with a real-time scheduling policy, but this approach often degrades performance and power consumption [5]. Instead, the current practice is to use a fairness- and energy-driven scheduler, such as the Energy Aware Scheduler (EAS) in Android, to estimate thread computation demands and dispatch them to the CPU with the lowest possible energy consumption.

Nonetheless, accurately estimating the computation requirements of render threads, particularly when the system is under memory pressure, is difficult due to its complex nature. Android's current estimator fails to capture the thread dependency in the rendering pipeline, resulting in missed deadlines due to inaccurate estimation. To address this, we develop a new framework on Android, called REFROM (**R**esponsive, **E**nergy-efficient **F**rame **R**endering f**O**r **M**obile devices), that feeds frame time samples from UI threads into a history-based frame time estimator to predict the computation requirements of upcoming frames. This userspace-assisted estimation framework improves the scheduler's task placement, reserving sufficient CPU resources for the render thread while avoiding unnecessary energy consumption.

The experimental results indicate that REFROM reduces janks by up to 40% compared to the existing Android scheduling policy. Moreover, REFROM improves up to 4% and 10.8% energy efficiency than the baseline and performance-oriented approach of pinning all render threads to performant cores, respectively.

The rest of the paper is organized as follows. §II illustrates the complexity of Android's display rendering pipeline and why existing strategies fail to recognize such subtlety. §III presents REFROM, our userspace-assisted solution to achieve energy-efficient, responsive frame rendering. §IV shows RE-

979-8-3503-1176-1/23 $31.00 © 2023 IEEE

FROM's effectiveness through extensive experiments. §V introduces the related work. §VI concludes this work.

II. Background and Motivation

A. Android's Display Rendering

Today's mobile devices provide users with stunning visual experiences through high-refresh-rate displays. In Android, this experience is made possible by a complex pipeline that involves both user- and kernel-space threads. The pipeline begins with the display controller in the kernel periodically prompting the user application to produce a new frame by sending a VSYNC signal to the Android module, *Choreographer*. Then, the *Choreographer* enacts the application's user-interface thread (UI thread) to create a tree of drawing commands and passes them to the render thread, which also runs in user space. The render thread then requests a memory buffer for the new frame from the *SurfaceFlinger* kernel thread and instructs the GPU to perform the rendering. Finally, *SurfaceFlinger* takes the GPU-generated frame, glues together all visual components, and presents a new frame to the screen. It is important to note that the entire frame rendering process must be completed within 16.6 ms on display with a frame rate of 60Hz. Failure to meet this deadline will result in noticeable flickers by janks, resulting in Quality of Service (QoS) problems. On devices with higher frame rates, the deadline is even more strict.

However, assigning the display pipeline to a real-time scheduling policy for timely execution is not an adequate solution. While Android assigns the latency-critical kernel threads in the display pipeline to the *SCHED_FIFO* real-time scheduling policy, applying the same policy to application threads running in user space is risky because they are developed by third parties and can monopolize system resources. Moreover, using a real-time scheduling policy for render threads cannot guarantee timely execution for two reasons. Firstly, Android's current real-time scheduler is based on thread priority and does not consider the heterogeneous computing capability of different CPU cores. Thus, high-priority render threads may be dispatched to little cores. Secondly, render threads can *self-suspend* when they try to allocate critical system resources, making them unsuitable for scheduling by a real-time policy[1]. For example, the UI thread may sleep while accessing a compressed memory page that has yet to be swapped in, and the render thread may also sleep while waiting for the *SurfaceFlinger* to allocate memory. Therefore, application threads in the display pipeline, even though timing-critical, must be scheduled by the Energy Aware Scheduler (EAS), an extension of the fairness-driven default Completely Fair Scheduler (CFS), to ensure system stability.

B. Delayed Frame Rendering under Scarce Memory

Mobile users usually multi-task heavily among multiple applications, leaving a tremendous memory footprint that often exceeds the physical capacity. Android uses the *ZRAM* kernel module to free up memory, which runs a kernel thread called

[1]Chen et al. [6] have discussed the inadequacy of self-suspending task in real-time systems

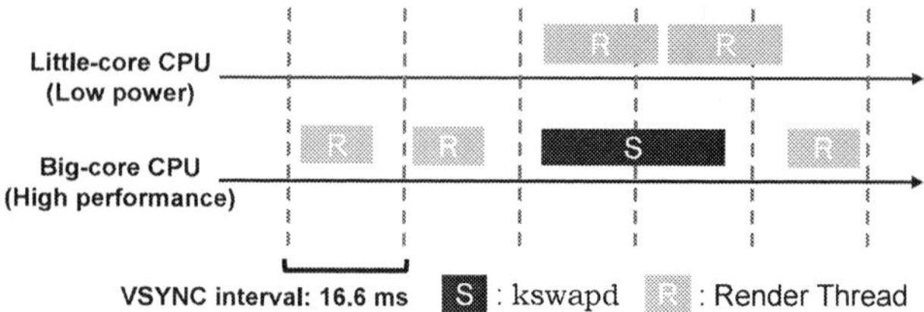

Fig. 1: `kswapd`'s interference on render threads

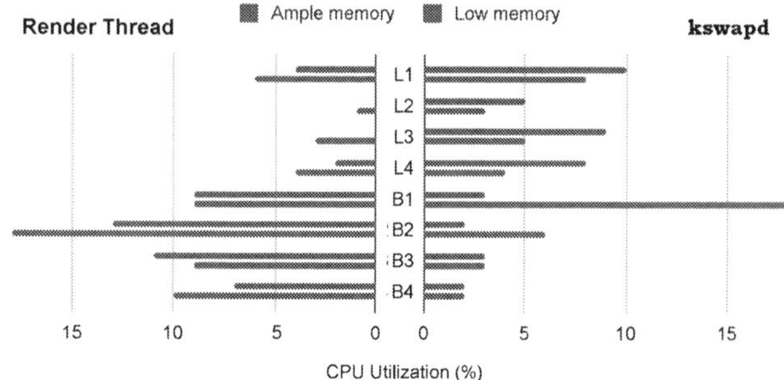

Fig. 2: Process placement distribution under CFS-based EAS

`kswapd` to compress and swap out infrequently accessed memory pages. When the memory pages are re-accessed, Android decompresses and swaps them back in. However, `kswapd` competes with the application threads in the display pipeline for CPU resources. This competition becomes especially problematic for heterogeneous processors that rely on EAS to dispatch processes to the most energy-efficient cores. This is because EAS estimates CPU demand on a per-thread basis and does not consider the execution dependency in the display pipeline. As a result, sometimes `kswapd` overrides the render threads of applications and runs on the performant cores, causing rendering delays or janks. Fig. 1 depicts this scenario.

TABLE I: Distribution of render threads and `kswapd`

Thread type	CPU type	Ample memory	Low memory
Render threads	Little-core	13%	23%
	Big-core	87%	77%
`kswapd`	Little-core	76%	41%
	Big-core	24%	59%

Fig. 2 illustrates the distribution of CPU utilization among the Qualcomm Snapdragon 855's four big cores (B1 to B4), and four little cores (L1 to L4) for both render threads and `kswapd`. Table I presents the percentage of time these two types of threads operate on the little or big cores. When memory is overprovisioned (i.e., ample memory), `kswapd` compresses less and is primarily scheduled on the little cores, freeing up the big cores for render threads to use. Table I shows that `kswapd` is scheduled on little cores for 76% of the time, while render threads are scheduled on big cores for 87% of the time. However, when the free memory drops below 5% of the total capacity due to increased multitasking (i.e., low memory), `kswapd`'s compression operations significantly consume more

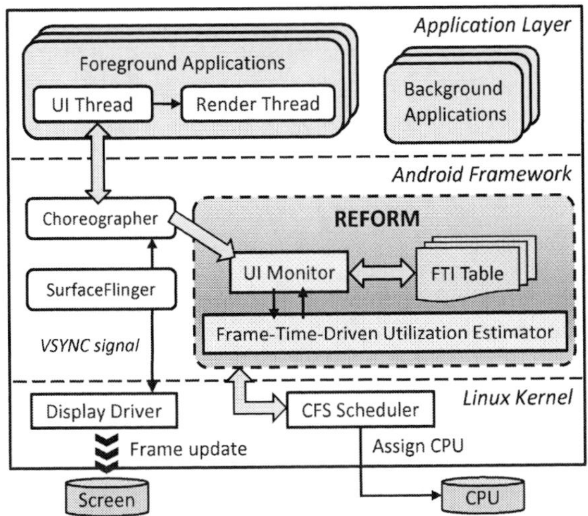

Fig. 3: System architecture of the proposed REFROM

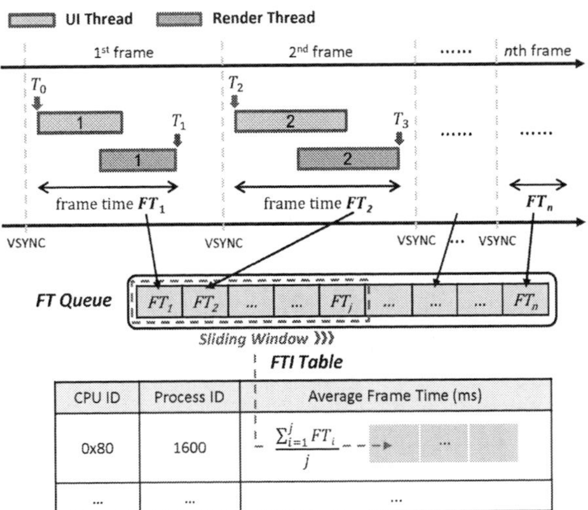

Fig. 4: The implementation of the proposed UI monitor

CPU resources. This pushes the render thread onto the little cores, resulting in 23% of the time being dispatched to the little cores, compared to 13% when there has ample memory.

C. Motivation: Cross-layer Frame Time Estimation

One way to prevent "jank" is to pin all application threads to the big-core CPUs. However, such an approach is energy-inefficient because the application threads may not always use all the CPU resources available. Moreover, this approach may cause potential starvation of `kswapd`, resulting in Android's low memory killer daemon killing other applications to free up memory.

The memory-restricted execution environment and complex composition of the display pipeline make it challenging for Android to maintain both energy efficiency and smooth rendering with the existing schedulers. Our goal is to overcome these limitations by developing a cross-layer solution that can precisely capture the fluctuating CPU demands of the display pipeline and communicate the timing requirements to the CFS scheduler. By doing so, we aim to enable timely rendering of frames without unnecessary energy consumption.

III. REFROM: Responsive, Energy-efficient Frame Rendering for Mobile Devices

A. Overview

We introduce REFROM, a lightweight module integrated into the Android framework that aims to enhance rendering performance while minimizing energy consumption. The architecture of REFORM is depicted in Fig. 3. REFROM consists of the UI monitor (discussed in §III-B) and the Frame-Time Driven Utilization Estimator (FTD-UE) (discussed in §III-C). By utilizing a short window of past CPU demands of rendered frames, REFROM predicts the CPU demands of rendering the upcoming frame. The CFS scheduler can then leverage REFROM's prediction to make better scheduling decisions.

B. UI Monitor

Fig. 4 illustrates how the UI monitor is implemented to track the rendering performance of foreground applications with the help of Android's Choreographer module. The Choreographer module acts as a mediator responsible for monitoring UI threads and forwarding the VSYNC signal from the display driver to application threads. As a mediator, the Choreographer is capable of recording the start and end timestamps of rendering a new frame, as well as detecting skipped frames (i.e., "janks").

To estimate upcoming workloads, the UI monitor maintains a lookup table called the Frame-Time Information (FTI) table. Each application has an associated FTI entry created during its startup. Whenever a new frame is produced, the Choreographer invokes the UI monitor to collect the combined execution time of the UI and render thread from each application. The UI monitor appends the execution time, along with the CPU IDs of the UI thread, to the Frame-Time (FT) queue. Once several frame information in the FT queue has accumulated, the UI monitor calculates the running average of the frame time within the sliding window and appends it to the FTI entry indexed by the corresponding CPU ID. The sliding window continues to move and calculate the next average frame time.

C. Frame-Time-Driven Utilization Estimator

We present the Frame-Time-Driven Utilization Estimator (FTD-UE) algorithm to predict the frame time of upcoming frames and evaluate the required task utilization to ensure optimal UI rendering performance. Task utilization is a metric used to measure the amount of CPU capacity required by a task when running on a heterogeneous CPU with cores of varying processing frequencies. The term "CPU capacity" refers to the computing throughput of a core operating at a specific frequency. For example, in a heterogeneous processing architecture that uses two types of processors (i.e., big.LITTLE processor), a performant core would have greater CPU capacity than an energy-efficient core.

Algorithm 1: Frame-Time-Driven Utilization Estimator (FTD-UE)

Input: Previous CPU ID C_{ID}
Input: Process ID P_{ID}

1 $list\_ft = FTItable(C_{ID}, P_{ID})$;
2 $forecast\_ft = WeightedAverage(list\_ft)$;
3 $\Delta U = CalTaskUtil(forecast\_ft, C_{ID})$;
4 **if** $\Delta U >= 0$ **then**
5 $\quad UCmin = UCmin + \Delta U$;
6 **else**
7 $\quad UCmax = UCmax + \Delta U$;
8 $AssignCPU(UCmin, UCmax)$;

Android's current CFS scheduler employs the Per-Entity Load Tracking (PELT) algorithm to estimate the potential utilization of upcoming tasks. However, PELT only accounts for the utilization of individual tasks and does not consider the combined utilization of an entire task set. As tasks in a task set can be dependent, and their computational demands can influence one another, separately evaluating the utilization of each task can lead to underestimation. This may mislead the CFS scheduler to allocate the victim task set to a core with insufficient CPU capacity, resulting in premature preemption and prolonged execution times. To address this, REFROM records the combined execution time of the UI and render thread received from the Choreographer, thus reflecting their execution dependency.

Algorithm 1 outlines the specifics of FTD-UE. FTD-UE is executed whenever the CFS scheduler chooses a process from the ready queue and runs it on a given CPU. Let us now go through FTD-UE, step by step.

Step 1: Retrieve a window of past frame time (Line 1): An operating system tracks essential metadata for each process, such as its state, in the Process Control Block (PCB). In Linux, the system records the ID of the last CPU core on which a process ran in the PCB. FTD-UE uses the last CPU core ID and the process ID to search the FTI table. If no entry is found, the process is not related to frame-rendering, and its task utilization can be calculated using the default estimator, PELT. However, if an entry is found, an array of average frame times previously calculated by the sliding window is retrieved, and the corresponding FTI entry is deleted.

Step 2: Predict the upcoming frame time (Line 2): Given the array of average frame times, we calculate their weighted average and use it as a prediction of the upcoming frame time using Eq. (1):

$$\bar{x} = \frac{\sum_{i=1}^{n} w_i^k x_i}{\sum_{i=1}^{n} w_i^k}, w_i = i \tag{1}$$

$\bar{x}$ represents the frame time to be predicted, and n is the sample size. The weight of the recorded average frame time x_i is denoted by w_i, with x_n being the most recent record and x_1 being the least recent one. The exponential coefficient k determines the significance of recent records. Recent records have a greater impact on the predicted frame time than distant records. For example, when $n = 4$ and $k = 1$, the weights of x_1 to x_4 increase linearly: 1, 2, 3, 4. When $k = 2$, the weights of x_1 to x_4 increase exponentially: 1, 2^2, 3^2, 4^2. Alternatively, one could predict the frame time by taking the mean of the recorded average frame times or simply the most recent one. We have evaluated all of these possibilities and found that setting $k = 2$ provides the most accurate prediction.

Step 3: Calculate the required task utilization (Line 3): Having the predicted frame time, FTD-UE evaluates the required task utilization with Eq. (2) and Eq. (3) to guarantee that the upcoming frame can be rendered before the deadline (the arrival of the next VSYNC signal).

$$\Delta perf = \bar{x} - T_{vsync} \tag{2}$$

In Eq. (2), $\bar{x}$ is the prediction made by Eq. (1). T_{vsync} is the period of the display synchronization (e.g., 16.7 ms for a 60Hz display). $\Delta perf$ means the CPU demand shortage that should be compensated.

$$\Delta U = \frac{\Delta perf}{T_{vsync}} \times \frac{Cap(C_{ID})}{g} \tag{3}$$

$Cap(C_{ID})$ refers to the CPU capacity of the previous CPU on which the frame rendering process ran. The parameter g is configurable and determines the aggressiveness of the utilization adjustment. A small value of g may result in overestimated utilization that exceeds the actual CPU capacity. In this paper, we set g to 10.

If the value of $\Delta perf$ is positive, this indicates a deficit in the amount of CPU utilization provisioned in the previous cycle. As a result, we suggest to the CFS scheduler that it should select a CPU core with higher capacity for rendering the upcoming thread. Conversely, if $\Delta perf$ is negative, this indicates a surplus amount of CPU utilization provisioned in the previous cycle. In this case, we suggest to the CFS scheduler that it should select a CPU core with lower CPU capacity to avoid unnecessary energy consumption.

Step 4: Select a proper CPU core to run on: The Linux kernel employs the *Utilclamp* mechanism to aid the CFS scheduler in determining which CPU core to dispatch a process with a specific utilization demand. Utilclamp exposes two parameters for each process: *UCmin* and *UCmax*. *UCmin* represents the minimum CPU utilization the scheduler must ensure for the process. A process can boost itself by specifying a higher *UCmin*, which prompts the scheduler to dispatch it to a more powerful CPU core. Conversely, *UCmax* represents the maximum CPU utilization the process may require. A process can reduce its CPU demand, thereby saving energy, by specifying a lower *UCmax*, which prompts the scheduler to dispatch it to a more energy-efficient CPU core. Utilclamp helps Linux meet a process's CPU demands while consuming the least energy possible. This, in turn, allows Linux to reduce processing frequency using the *SchedUtil* module when there is excess utilization.

UCmin and *UCmax* are stored in a process's PCB. FTD-UE reads the previous *UCmin* and *UCmax* and adjusts them (from Line 4 to Line 7) to reflect the CPU utilization demands of rendering the upcoming frames. This adjustment assists the CFS scheduler in selecting a CPU core with the requested utilization (at Line 8).

IV. Performance Evaluation

A. Experimental Setup and Evaluation Metrics

TABLE II: Specifications and Benchmarks

Specifications of Pixel 4	
Processor	Qualcomm Snapdragon 855
Memory	6GB LPDDR4x
Storage	64GB UFS 2.1
Screen	90Hz P-OLED 1080x2280 pixels
Network	WiFi IEEE 802.11 a/b/g/n/ac
Battery	2800 mAh
OS	Android 11.0.0, Linux Kernel 4.14
zRAM	2GB (default)
Attributes of benchmarks	
Application	Twitter, Facebook
Footprint	450MB
Rendered Frames	5000

To evaluate the performance of REFROM, we conducted a series of experiments on popular mobile applications using the Android smartphone Pixel 4. We also implemented Legacy and All-On-Big schemes for comparison. The Legacy scheme represents Linux's task placement policy based on CFS and EAS, while All-On-Big always places all UI rendering tasks on the big-core CPUs to obtain the best rendering performance. Pixel 4 is equipped with four 1.7GHz little-core CPUs and four 2.4GHz big-core CPUs, allowing for 17 different CPU frequencies. Detailed specifications of Pixel 4 are listed in Table II.

We evaluated the rendering performance and energy efficiency of Pixel 4 while scrolling through Twitter and Facebook, which are popular social network apps. Twitter comprises abundant images and hypertext content, while Facebook has more diverse multimedia, such as various lengths of video. We selected screen scrolling as our benchmark scenario, as it is the most common usage scenario for Android mobile devices. Users usually spend about 60% of their time on screen scrolling [7]

Our preliminary experiments showed that QoS problems occur frequently when the speed of screen scrolling is higher than 12.5 pixels/ms. To ensure a fair comparison, we limit the speed of screen scrolling to 12.5 pixels/ms. We implemented REFROM by dynamically adjusting the CPU frequency via the CPU frequency governor equipped in the Linux kernel. We used *Dumpsys* to reveal the service information of Android devices, which provided us with information on graphics rendering and energy consumption."

B. Experimental Results

1) Rendering Performance and QoS: To evaluate rendering performance and QoS on mobile devices, we measured the number of big janks and FPS variance during the execution of representative benchmarks. If a frame takes longer than 100 ms to render, it is considered a big jank, which can significantly impact the user experience. FPS variance describes the dispersion of FPS and reflects the stability of frame rendering. We measured the number of big janks and FPS variance while exploring Twitter and Facebook with screen scrolling, and also in scenarios where many applications were simultaneously running in the background, which we refer to as "Twitter+Upload" and "Facebook+Upload."

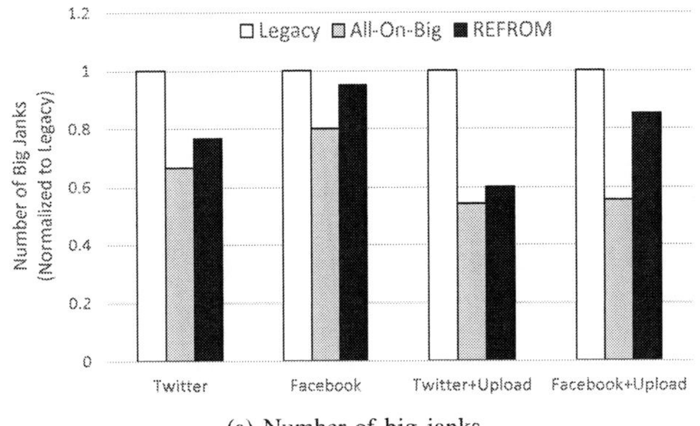

(a) Number of big janks

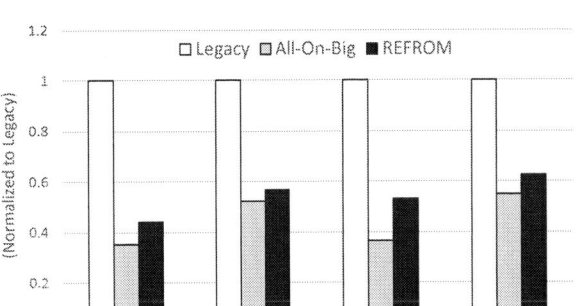

(b) FPS variance

Fig. 5: Evaluation of rendering performance

Our proposed REFROM outperforms Legacy in all evaluated cases, reducing big janks by 23.3% to 40% for Twitter and 5% to 14.8% for Facebook (as shown in Fig. 5a), and alleviating 37.4% to 55.9% FPS variance compared to the Legacy scheme (as shown in Fig. 5b). Based on our observations, Twitter consumes more CPU utilization than Facebook, making our REFROM more effective in improving QoS for Twitter. This implies that REFROM can aggressively refine task utilization to optimize rendering performance and QoS for applications, particularly when the system workload is heavy and varies significantly. In contrast, Legacy fails to prevent big janks because it is not aware of UI tasks' QoS requirements. Although the All-On-Big scheme induces less big janks, it is not energy-efficient for battery-powered mobile devices.

2) Energy Efficiency: Fig. 6 illustrates the energy efficiency evaluation results obtained while scrolling through Twitter and Facebook on mobile devices. The energy efficiency metric is based on the energy consumed for each rendered frame during the execution of representative benchmarks. REFROM demonstrates superior energy efficiency compared to the Legacy and All-On-Big schemes. It provides an improvement of up to 4% and 10.8% in energy efficiency compared to the Legacy and All-On-Big schemes, respectively. The experimental results clearly show that REFROM optimizes rendering performance to ensure QoS with higher energy efficiency than Legacy and All-On-Big schemes.

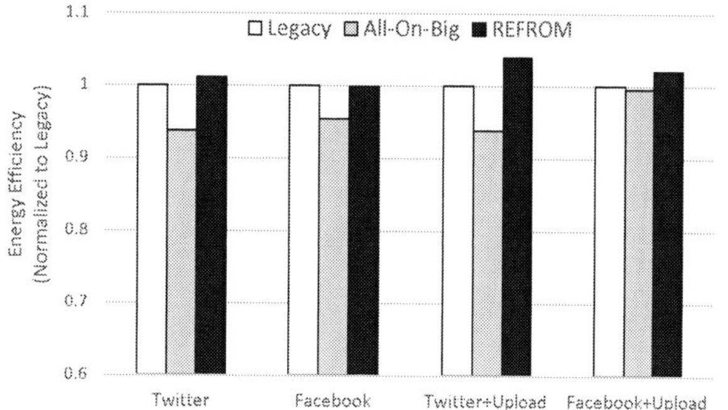

Fig. 6: Evaluation of energy efficiency

V. Related Work

Mobile devices have long faced challenges with poor responsiveness due to delays in graphics rendering. Numerous studies have suggested various solutions to reduce the computational burden of frame rendering by avoiding redundant rendering of similar frames. For example, Lee et al. [8] and Chang et al. [9] have proposed skipping the rendering of similar frames in social media and multimedia-rich apps, respectively. Lin et al. [4] have proposed a dynamic layout trimming approach to remove duplicate frames, while Ahn et al. [10] have proposed reusing previously rendered frames during fast screen scrolling to reduce pressure on the graphics rendering pipeline. While previous approaches have focused on reducing rendering, REFROM aims to ensure that the necessary amount of computation is guaranteed under heavy rendering demands while reducing unnecessary computation under light rendering demands. REFROM can be integrated with the frame-deduplication solutions mentioned earlier to further enhance user experience and energy efficiency.

Another critical challenge faced by mobile devices is delivering a seamless user experience under a limited memory budget. To address this, Android compresses cold memory pages using ZRAM to enable smooth app switching. However, we have observed that this can lead to competition for CPU resources, causing display flickers. Previous studies by Wang et al. [11], Wu et al. [12], and Son et al.[13] have suggested extensions to ZRAM, huge page management, and adaptive pre-paging strategies to mitigate memory pressure and reduce application switch delays. While these studies focus on memory management, our work concentrates on CPU scheduling and can be integrated with these earlier efforts.

VI. Conclusion

In this study, we investigate the QoS issues arising from the increasing demand for high-quality frame rendering on modern mobile devices. Unlike existing process scheduling and memory management policies that underestimate the computational needs of frame rendering, we propose REFROM to provide responsive displays. REFROM uses a history-based frame time estimator to analyze frame time samples from UI threads and estimate the computation requirements of upcoming frames. Our experimental results demonstrate that REFROM reduces the number of delayed frames by up to 40% and improves energy efficiency by up to 4% compared to existing approaches.

Acknowledgments

This work was supported in part by National Science and Technology Council under grant nos. 111-2221-E-002-152-MY3, 108-2221-E-002-062-MY3, 111-2223-E-001-001, 111-2923-E-002-014-MY3, 111-2221-E-001-013-MY3, and 112-2927-I-001-508 and Academia Sinica under grant nos. AS-IA-111-M01 and AS-GCS-110-08.

References

[1] Y.-S. Chen, C.-F. Wu, Y.-H. Chang, and T.-W. Kuo, "A write-friendly arithmetic coding scheme for achieving energy-efficient non-volatile memory systems," in *Proceedings of the 26th Asia and South Pacific Design Automation Conference*, 2021, pp. 633–638.

[2] Y.-S. Chen, Y.-H. Chang, and T.-W. Kuo, "Drift-tolerant coding to enhance the energy efficiency of multi-level-cell phase-change memory," in *Proceedings of the ACM/IEEE International Symposium on Low Power Electronics and Design*, 2022, pp. 1–6.

[3] M.-C. Yang, Y.-H. Chang, C.-W. Tsao, and P.-C. Huang, "New era: New efficient reliability-aware wear leveling for endurance enhancement of flash storage devices," in *Proceedings of the 50th Annual Design Automation Conference*, 2013, pp. 1–6.

[4] H. Lin, C. Liu, Z. Li, F. Qian, M. Li, P. Xiong, and Y. Liu, "Aging or glitching? what leads to poor android responsiveness and what can we do about it?" *IEEE Transactions on Mobile Computing*, 2023.

[5] Y.-S. Chen, C.-F. Wu, Y.-H. Chang, and T.-W. Kuo, "Energy efficiency enhancement of scm-based systems: Write-friendly coding," *IEEE Transactions on Computer-Aided Design of Integrated Circuits and Systems*, 2022.

[6] J.-J. Chen, G. Nelissen, W.-H. Huang, M. Yang, B. Brandenburg, K. Bletsas, C. Liu, P. Richard, F. Ridouard, N. Audsley *et al.*, "Many suspensions, many problems: a review of self-suspending tasks in real-time systems," *Real-Time Systems*, vol. 55, no. 1, pp. 144–207, 2019.

[7] J. Yu, H. Han, H. Zhu, Y. Chen, J. Yang, Y. Zhu, G. Xue, and M. Li, "Sensing human-screen interaction for energy-efficient frame rate adaptation on smartphones," *IEEE Transactions on Mobile Computing*, vol. 14, no. 8, pp. 1698–1711, 2014.

[8] G. Lee, S. Lee, G. Kim, Y. Choi, R. Ha, and H. Cha, "Improving energy efficiency of android devices by preventing redundant frame generation," *IEEE transactions on mobile computing*, vol. 18, no. 4, pp. 871–884, 2018.

[9] Y.-C. Chang, W.-M. Chen, P.-C. Hsiu, Y.-Y. Lin, and T.-W. Kuo, "Lsim: Ultra lightweight similarity measurement for mobile graphics applications," in *Proceedings of the 56th Annual Design Automation Conference 2019*, 2019, pp. 1–6.

[10] W.-H. Ahn, C.-K. Hong, K.-M. Han, S.-H. Choi, J.-W. Oh, and S.-H. Lim, "Scrolling-aware rendering to reduce frame rates on smartphones," *Electronics*, vol. 10, no. 17, p. 2177, 2021.

[11] Y.-X. Wang, C.-H. Tsai, and L.-P. Chang, "Killing processes or killing flash? escaping from the dilemma using lightweight, compression-aware swap for mobile devices," *ACM Transactions on Embedded Computing Systems (TECS)*, vol. 20, no. 5s, pp. 1–24, 2021.

[12] C.-F. Wu, Y.-H. Chang, M.-C. Yang, and T.-W. Kuo, "When storage response time catches up with overall context switch overhead, what is next?" *IEEE Transactions on Computer-Aided Design of Integrated Circuits and Systems*, vol. 39, no. 11, pp. 4266–4277, 2020.

[13] S. Son, S. Y. Lee, Y. Jin, J. Bae, J. Jeong, T. J. Ham, J. W. Lee, and H. Yoon, "Asap: Fast mobile application switch via adaptive prepaging," in *USENIX Annual Technical Conference*, 2021, pp. 365–380.

DCIM-3DRec: A 3D Reconstruction Accelerator with Digital Computing-in-Memory and Octree-Based Scheduler

Yiqi Jing[1], Yiyang Sun[1], Xiao Wang[1], Wentao Zhao[1], Meng Wu[1], Fengyun Yan[2], Yufei Ma[1], Le Ye[1,3] and Tianyu Jia[1,*]

[1] School of Integrated Circuits, Peking University, Beijing, China
[2] Nano Core Chip Electronic Technology Co., Ltd., Hangzhou, China
[3] Advanced Institute of Information Technology of Peking University, Hangzhou, China

***Abstract*—Learning-based 3D reconstruction has evolved rapidly with promising quality, while it requires high-performance hardware for interactive applications. In this work, a reconstruction accelerator called *DCIM-3DRec* is presented which leverages *digital computing-in-memory (DCIM)* design to facilitate learning-based reconstruction deployment on real-time and low-power edge platforms. The *DCIM-3DRec* is designed with the following features: a *reconfigurable DCIM* macro array for high data reuse and macro utilization, and an *Octree-based subdivision scheduler* for efficient management of 3D space prediction. The *DCIM-3DRec* accelerator is implemented and evaluated in TSMC 55 nm technology, with a DCIM macro efficiency of 19.4 TOPS/W at INT8. Overall, the *DCIM-3DRec* accelerator achieves 23× performance gain and four orders of magnitude energy efficiency improvement compared to a Nvidia RTX3090 GPU.**

I. Introduction

3D reconstruction is a key technique to support human-machine interactive use cases, such as medical imaging, autonomous machine, AR/VR, etc. The goal of 3D reconstruction is to infer objects' 3D geometry and structure from input source data, such as 2D RGB, 2.5D RGB-D images [1], and 3D point clouds [2]. Conventionally, 3D objects or scenes are reconstructed by approaches like multi-view stereo [1] or structure from motion, which depends on multi-view images. During the reconstruction, the surfaces of a 3D object are identified based on ray tracing operations for the space occupancy or signed distance function (SDF). The object is then represented via different data structures, such as voxel, point cloud, or mesh, and rendered through ray-casting or rasterization algorithms, such as the Marching Cubes [2]. However, the conventional representations have their shortcomings: voxel representation is memory-consuming; point cloud requires irregular memory access and lacks topological knowledge; mesh only performs well for specific scenes and lacks flexibility.

Recently, deep-learning-based reconstruction became a promising high-quality alternative approach and has drawn significant industry attention in recent years, such as Meta [2], Nvidia [3], and Tesla [4]. Compared with conventional 3D representation methods, implicit representation, where the object surface is represented indirectly as a contour plane, is widely used in learning-based reconstruction [8]. As the contour plane is continuous, 3D output mesh can be generated through sampling and interpolation at any given resolution with the local occupancy or SDF results predicted by the neural network. Currently, learning-based reconstruction with implicit representation has been explored using various neural network models such as auto-encoder, variational autoencoder (VAE) [2-6], or Transformer [7], and has shown high-quality 3D reconstruction compared with the conventional approaches.

*Corresponding Email: tianyuj@pku.edu.cn

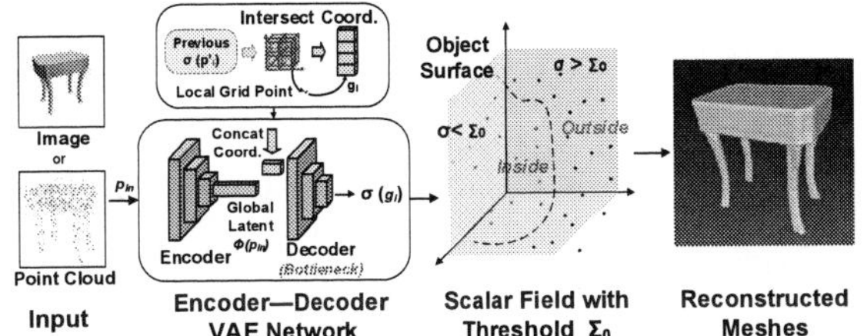

Fig. 1. Illustration of learning-based implicit 3D reconstruction.

Fig. 1 illustrates the computation flow of a typical VAE-based learning 3D reconstruction with implicit representation. The input 2D images or point clouds p_{in} are first fed into the *Encoder* network to extract spatial features as a global latent code $\Phi_L(p_{in})$. Meanwhile, a set of local grid point g_i is generated by the uniform sampling of 3D space. The position of local grid points is combined with global latent code and sent into Decoder to predict occupancy probability $\sigma(g_i)$ of grid point [3][4] or signed distance function (SDF) to object surface [2] as following

$$\sigma(g_i) = f(g_i, \Phi_L(p_{in})), \quad (1)$$

$$SDF(g_i) = f(g_i, \Phi_L(p_{in})), \quad (2)$$

where the function f represents the neural network model of the *Decoder* network. The predicted occupancy σ of all the grid points forms a 3D scalar field, representing the object surface implicitly as an isosurface of a certain threshold Σ_0. Based on the learned implicit surface, the object can be rendered through the Marching Cubes [2]. Other emerging learning models have also been explored for 3D reconstruction, such as the recent Transformer-based model [7]. However, VAE is still the mainstream model in this area, thus we focus on the most widely used VAE-based models in this paper.

Although learning-based 3D reconstruction has demonstrated promising quality, it has high requirements for computing hardware. Our experiments on a RTX3090 GPU show that 0.3s to 12s is consumed to reconstruct a single object from the *ShapeNet* dataset, which is not tolerable for real-time applications. Our analysis indicates that latency is highly related to the complexity of the reconstruction object, such as surface evenness or volume, etc. It is also interesting to observe that the *Decoder* network consumes more than 80% of the overall latency and is the computation bottleneck, due to repetitive inference predictions for each local grid point inside the 3D space. The high reuse rate of the *Decoder* network in fact fits well with the computing-in-memory (CIM) technique, which maintains model weight stationary. Hence this observation motivates us to design a CIM-based accelerator for learning-based 3D reconstruction.

979-8-3503-1176-1/23 $31.00 © 2023 IEEE

To accelerate vision applications, a few hardware accelerators have been previously developed. For example, Intel demonstrated a ray-casting accelerator with near-memory search, while it only supports single ray-casting operation in scene reconstruction [9]. Stereo image accelerators have also been designed for different 3D applications, such as hand-gesture recognition [10], navigation [11], or single stereo matching [12]. In recent years, there is increasing interest in designing point cloud accelerators [13-14], since random point cloud processing is challenging for conventional memory management. However, there is limited acceleration for the 3D reconstruction task. In fact, 3D reconstruction is computation-heavy and requires specific hardware acceleration for real-time use scenarios, e.g. the 3D reconstruction consumes the longest latency in the perception pipeline of the ILLIXR AR/VR testbed [15].

To facilitate deploying learning-based 3D reconstruction, we demonstrate a digital computing-in-memory (DCIM) based accelerator *DCIM-3DRec*. The *DCIM-3DRec* is designed by leveraging the in-memory computing at the circuit level, and Octree-based subdivision scheduling at the architecture level. The contributions of this work are summarized as follows. First, the learning-based 3D reconstruction algorithms are analyzed in depth for the computation requirements and bottlenecks. Our observations show that learned-based 3D reconstruction is an interesting workload and suits well for CIM technique. Second, based on the analysis, a DCIM-based accelerator is designed with a reconfigurable DCIM macro array and Octree-based scheduling. Finally, the accelerator is implemented and evaluated using TSMC 55 nm technology, with an average of 23× and up to 28× performance gain and significant energy saving compared to a Nvidia RTX3090 GPU and Jetson NX.

II. Learning-Based 3D Reconstruction

A. Model Architecture

Learning-based reconstruction with implicit representation is rapidly developing and has shown promising performance. We focus on the mainstream VAE-based 3D reconstruction models due to their leading reconstruction quality, e.g. VAE-based model [5] outperform the Transformer model [7] by 16% in term of reconstruction metric *Intersection-over-Union (IoU)*. In this work, we take a deep dive into two representative models, i.e. Meta's *DeepSDF* [2] and *Convolutional Occupancy Network (ConvOnet)* [5], which are state-of-the-art implicit 3D reconstruction models based on mainstream encoder-decoder. As shown in Fig. 2, *ConvOnet* infers the 3D surfaces using fully-connected layers and a global latent code encoding. The latent code of each grid point $\Phi_L(g_i)$ is acquired through bilinear interpolation between feature planes and grid points g_i and then fed into an ResNet Decoder to predict the occupancy probability $\sigma(g_{in})$ of the grid point. A grid voxel will be marked as an *active voxel* if it contains both occupied (σ >0.5) and vacant (σ<0.5) grid points. All the *active voxels* will be subdivided into 8 finer spaces each, and the decoder network steps are repeated until the desired resolution is reached. *DeepSDF* utilized a variant of the VAE model, i.e. auto-decoder, which only maintains the decoder network and uses a *maximum-a-posterior (MAP)* optimization to generate the latent code. Although *DeepSDF* eliminates the computation of the encoder network, the newly introduced MAP optimizer consumes even longer computation latency based on our experiments.

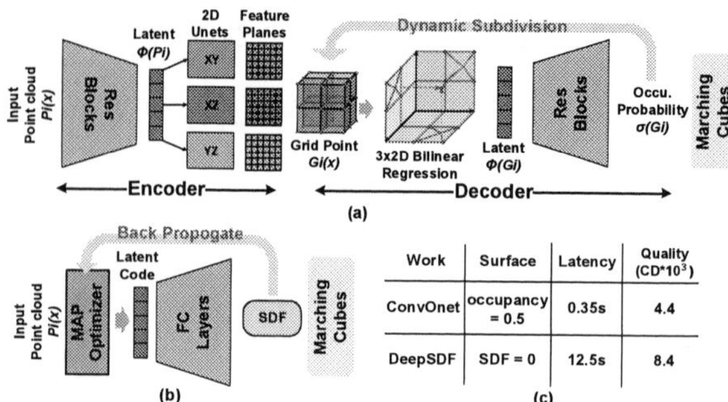

Work	Surface	Latency	Quality (CD*10³)
ConvOnet	occupancy = 0.5	0.35s	4.4
DeepSDF	SDF = 0	12.5s	8.4

Fig. 2. Computations in the (a) Conv. Occupancy Network [5] and (b) DeepSDF [2], and (c) the performance comparison.

B. Computation Analysis

To evaluate the hardware requirements of the learning-based 3D reconstruction models, we deploy these models onto a Nvidia RTX3090. Fig. 2(c) shows the latency and reconstruction quality of both *ConvOnet* and *DeepSDF* on the same *ShapeNet* dataset. It is notable to observe *ConvOnet* performs much better than *DeepSDF* in terms of performance and quality, e.g. 35× faster. One key reason is the dynamic intersection scheme in *ConvOnet* significantly reduces the grid point number for *Decoder* inferences. As for the reconstruction quality, *DeepSDF*'s global encoding fails to notice the logistics within the structure of objects. Although *ConvOnet* shows better quality, it still consumes an average of 0.35s latency to reconstruct one 3D object on such a high-performance GPU, which is not tolerable for many real-time applications.

We further break down the runtime latency of *ConvOnet* to analyze the performance bottleneck. As shown in Fig. 3, the computation latency is composed of three main parts, i.e. *Encoder*, *Decoder*, and Marching cubes for mesh generation. The *Encoder* and Marching cubes consume constant latency, as their operations only relate to the target reconstruction resolution and input size. However, the *Decoder* has the largest average latency and it varies greatly across different shape classes. These observations provide two insights: *Decoder network is the computation bottleneck*, and *Decoder latency is related to the target reconstruction category.* We further find out the *Decoder* latency is strongly related to the iterations of space *subdivision*, which generates local grid points as the *Decoder* inference input.

When reconstructing an object, the voxel space around the object surface will be identified as *active voxels*, which will be further subdivided into finer-grained sub-voxels. This process is the space *subdivision*. The local grid points of these sub-voxels will be combined with the global latent code as the *Decoder* input to predict the occupancy of the sub-voxel space. For a challenging object with uneven or angled surfaces, the number of active voxels caused by *subdivision* significantly increases leading to more iterations of *Decoder* inference. As the curve of subdivision number shown in Fig. 3, the loudspeaker class, which usually has an obtuse-angled front face, is more likely to have more space subdivisions, leading to longer latency. The subdivision number is also related to the overall volume of the target object, e.g., a small object normally has fewer active voxels leading to a shorter latency. For example, the rifle category, which has the smallest size, ends up with the least number of divisions and the shortest latency.

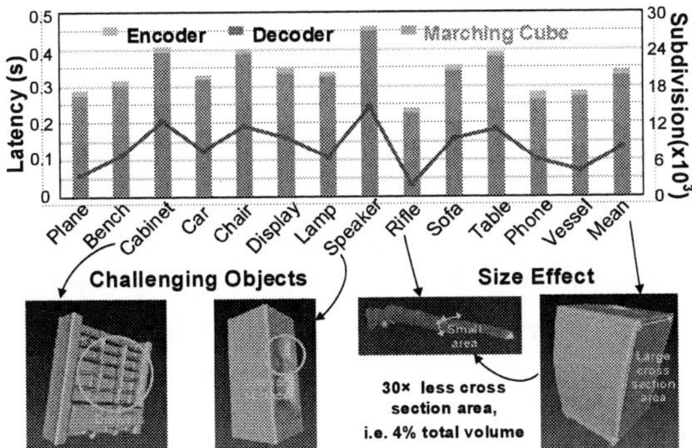

Fig. 3. Latency of each object category for *ConvOnet* on GPU.

Based on the above analysis, we find the *Decoder* network is the main computation bottleneck and requires repetitious model inferences for each voxel subdivision. This motivates us to leverage the computing-in-memory technique, which has the advantage of maintaining model weights in memory to eliminate the data movement cost. Therefore, we present a digital computing-in-memory based 3D reconstruction accelerator *DCIM-3DRec* to significantly improve energy efficiency. Meanwhile, it is also observed that the generated local grid points during *subdivision* consumes a large amount of storage, i.e. more than 1MB, due to the randomness of the object surface voxels. To effectively manage the voxel *subdivision* and the grid point storage, the accelerator is designed with an Octree-based voxel subdivision storage and scheduling. In this work, we focus on the acceleration of learning models and ignore the marching cube, since it is the conventional algorithm that has been well explored.

III. Architecture of DCIM-3DRec

A. Overall Architecture

In this work, a 3D reconstruction accelerator *DCIM-3DRec* is developed with DCIM, as shown in Fig. 4. We chose DCIM as it has better scalability for high-bit precision computation than analog CIM. The accelerator contains two main computing engines, i.e. stereoscopic processing engine (SPE) and plane processing engine (PPE). These two heterogeneous engines enable simultaneous plane and spatial feature processing to support optimal hardware mapping. There is also a global memory buffer for data storage and a top-level controller for computation flow management.

The SPE unit consists of two DCIM macro arrays to support the major computation, i.e. *Encoder* and *Decoder* networks. Since these two models are residual networks with high repetitious inferences, especially the *Decoder*, DCIM suits well for this computation scenario by highly reusing stored weights. The DCIM macro array supports reconfigurable data flows by the local scheduler, which contains configurable DCIM interconnect datapath. An Octree-based subdivision scheduler is designed to manage the voxel subdivision and *Decoder* computation flow. The Octree-based scheduler first stores the local grid points g_i generated during the voxel subdivisions using the Octree data structure, which significantly reduces the memory requirement. Then it schedules the grid points of *active voxels*, which are combined with the global latent code from the Global Feature Unit via bilinear interpolation, as the *Decoder* inference inputs. An index-based data storage is adopted in the Global Feature Unit to efficiently store the 3D global latent code $\sigma(p_{in})$.

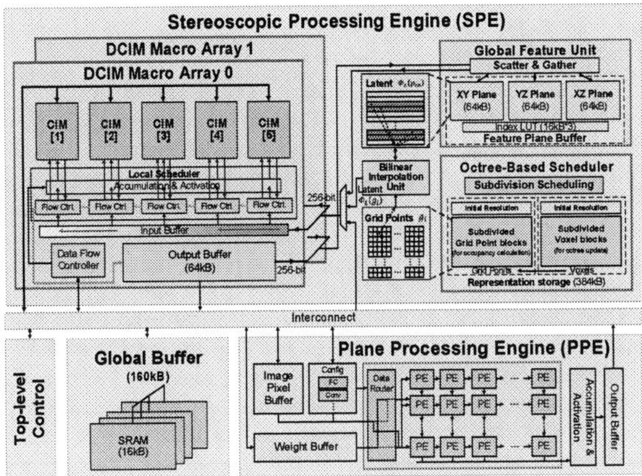

Fig. 4. Overall architecture of the *DCIM-3DRec* accelerator.

The PPE engine, which contains a 24×32 MAC array, is deployed to support convolutions and transposed convolutions in 2D-Unet during feature plane generation and the input/output fully-connected layer. The heterogeneous SPE+PPE architecture enables more flexible model deployment and optimal model mapping for balanced computation throughput of 3D reconstruction models, e.g. PPE runs in parallel with DCIM arrays during *Decoder* inference.

B. Reconfigurable CIM Macro Array

To enable optimal dataflow for *Encoder* and *Decoder*, the size of the DCIM macro array is chosen as 1×5 and the datapath is designed in a reconfigurable fashion. The computation flow is explained using three successive time steps in Fig. 5. For *Encoder* computation, the input feature is processed in a pipeline fashion between time steps. Each CIM macro array computes one Resnet block of an input point. Once the statistics of all points in each network layer are completed, the active word lines of the macro array are changed to update weights and start the computation of the next layer. For the *Decoder* network, all weights are stored in subarray and the successive Resnet blocks of one input feature are orderly processed in different rows. The reconfigurability of the DCIM macro array improves resource utilization and minimizes weight storage requirement, i.e. about 62.5% reduction compared with a scheme with static datapath.

We designed DCIM macro using INT8 data precision with negligible reconstruction quality degradation, i.e. reconstruction metric *IoU*=0.855 at FP32 versus *IoU*=0.853 at INT8. The DCIM macro size is chosen as 256×256 based on the storage and computing ratio in the 3D reconstruction models. Since the *Decoder* network has a high reuse rate and tends to store all weights in CIM to improve computing throughput, more resources are tilted toward computing density. The balance of the power consumption and latency of bit-line and word-line are also taken into consideration for optimal overall performance. Depending on the area proportion and arrangement of storage and computing units, DCIM macro size is chosen as 256×256.

As shown at the bottom of Fig. 5, each DCIM macro is made up of 32 identical columns. The SRAM word lines and input activations are shared across all columns, and each column contains 256×8 bits SRAM cell. Each column consists of one SRAM read/write module, one global adder tree, and 32 processing units (PUs) with one subarray and one shift and adder. The subarray contains an 8×8-bit cell and a local computing unit

(LCU) composed of 64 NOR gates. The ratio of the SRAM bit cell and NOR gate is selected as 1:1.

The DCIM macro has a memory mode to realize normal read and write operations. During the DCIM computing mode, 32×8-bit data selected in different subarrays are read out simultaneously to reduce the read energy. Furthermore, these data are directly sent to LCUs for 1b multiplication and products are sent to shift and adder for accumulation, which prevents energy waste of data movement. The prior scheme that sends input feature serially in single bit [16] would degrade the performance, thus we use LCU to receive multi-bit width input simultaneously. The 8×8 NOR gate array implements 64 1-bit multiplication and generates eight 8b products at a time. The products are summed by the following shift and adder, and the partial sums are accumulated by the global adder tree to obtain a final 21-bit MAC result as the output of a CIM column.

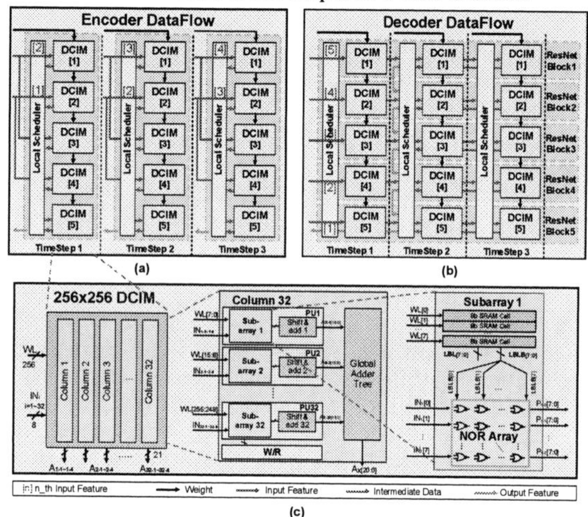

Fig. 5. Computation dataflow for (a) Encoder and (b) Decoder, and (c) the design details of DCIM macro.

IV. Octree-Based Subdivision Storage and Scheduler

A. Octree-Based Management for Local Grid Points

Around the object boundary, voxels that contain both occupied σ>0.5 and vacant σ<0.5 grid points are called *active voxels*, as shown in Fig. 6(a). To reconstruct a 3D object, the occupancy probability of the local grid points in the space will be predicted via the *Decoder* network. The identified *active voxels* will be further subsequently subdivided into sub-voxels to characterize object edges with higher resolution. The depth of sub-voxel division is defined by reconstruction resolution, i.e. 3 by default.

As mentioned earlier, a large number of grid points are generated during the voxel *subdivision*, requiring >1MB of memory to store the generated grid points. Meanwhile, the coordinates of active voxels and their sub-voxels are random and vary greatly across different object categories. To efficiently manage the storage dynamic voxel subdivision, we developed an Octree-based subdivision storage and scheduler, which stores the predicted voxels and local grid points based on an *Octree* and guides the following space subdivision. *Octree* is a data storage structure that has a prominent advantage for 3D data management [17]. Since the voxels scheduled for the Decoder inference are very sparse in the real 3D scenes, an Octree-based subdivision can significantly reduce storage requirements.

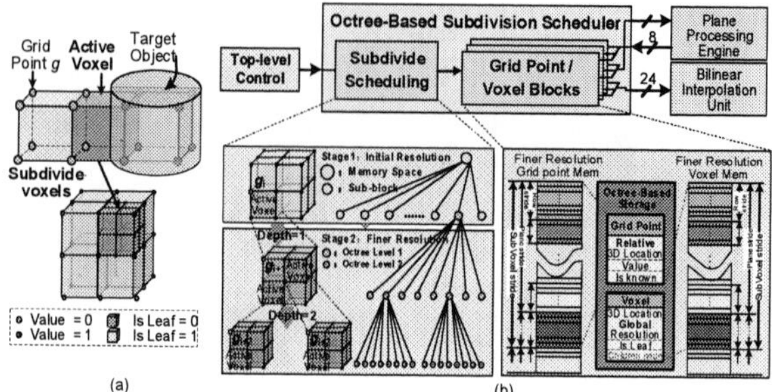

Fig. 6. (a) Active voxels around the object boundary, (b) Octree-based subdivision scheduler and its data storage structure.

The Octree-based subdivision scheduler is designed in SPE to manage the voxel subdivide scheduling as well, as shown in Fig. 6(b). A subdivide scheduling module is used to manage the voxel subdivisions based on the identified/non-identified grid points inside the Octree. The active voxel is determined based on the occupancy of its contained grid point g_i and will be subdivided to generate new g_{i+1} in a finer resolution depth. The scheduler also contains a group of grid point and voxel storage blocks, in which two data fields are stored for each identified *active voxel*: voxel and grid point g_i. The voxel file contains its *3D location*, *resolution*, whether it's an active voxel *(is Leaf)*, and the information of its *children nodes*. The grid point field contains its *3D location*, the occupancy σ of g_i *(value)*, and whether the occupancy is calculated *(is known)*. The location of the voxel is defined by its grid point at the upper left corner, which is closely related to the spatial location.

During the process of object reconstruction, the memory space is first pre-segmented into sub-blocks and the occupancy of the grid points in the sub-blocks is predicted to identify *active voxels* in the initial resolution. Since only voxels divided in the initial resolution could be potentially subdivided again, only voxels and grid points in active sub-blocks are stored and processed for more details at a finer resolution. Active voxels are subdivided into eight finer voxels with the relative position indexed by address.

B. Index-Based Feature Storage for Global Features

To reconstruct a 3D object, the global 3D feature, i.e. latent codes $\Phi_L(p_{in})$, which need to be combined with local grid points, is generated by the *Encoder* network and continuously accessed during the *Decoder* inference. To facilitate the storage and access of the global latent code, an index-based storage and access module is deployed in the SPE engine. As shown in Fig. 7, it mainly consists of a Feature Plane Buffer (FPB), and a Scatter and Gather Unit (SGU). This module is designed to aggregate all input 3D coordinates $p_{in}(x, y, z)$ and map them onto three feature planes, i.e. *XY* plane, *YZ* plane and *XZ* plane, as latent code.

In the learning-based reconstruction, the coordinates of all input points can be normalized to (0,1) and mapped onto three unique indexes, i.e. one integer for each plane, e.g. *Index_xy(p)*. Thus the 3D features can be linked by index values to achieve more efficient data management. During the *Encoder* inference, the ResNet block generates the latent codes, i.e. 32×8b, for all input points. The SGU receives the latent code $\Phi_L(p_{in})$ and its corresponding indexes to execute max pooling using *scatter_max* and *gather* operations. All the points with the same index are grouped, following a max latent code search to get the *scatter_max* result. The *gather* function sums up the values that have the same index from all three planes. This scatter and gather process is repeated for each ResNet block, except the last ResNet block which does not need max pooling. The output of the last ResNet block goes through an FC layer and a *scatter_mean* function to generate the feature planes, which are stored in FPB.

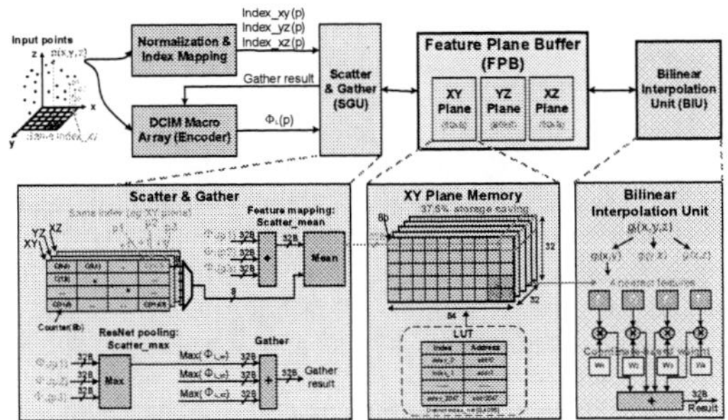

Fig. 7. Index-based storage and access for global features.

The FPB unit is designed with independent *XY*, *YZ* and *XZ* plane memory, with each containing 32×64×32 bytes (64kB) and a 16kB look-up table (LUT). When an input point with an index value is to be stored in the FPB, it will first access the LUT, which stores each index value and its address in the corresponding feature plane memory. When a feature in FPB is to be read, the unit fetches its address in LUT based on the index value.

There is also a bilinear interpolation unit, which accesses the FPB data and performs bilinear interpolating of given grid point g_i among its 4 nearest feature latent code $\Phi_L(p_{in})$ during *Decoder* inference. For a given grid point $g_i(x, y, z)$, it is first projected into 3 planes and gets 3 index values respectively. BIU finds the 4 nearest features based on the index values and calculates four coordinate-based weights using the projected coordinates.

$$W_1 = k(y_1 - y)(x_1 - x) \tag{3}$$

$$W_2 = k(y_1 - y)(x_2 - x) \tag{4}$$

$$W_3 = k(y_2 - y)(x_1 - x) \tag{5}$$

$$W_4 = k(y_2 - y)(x_2 - x) \tag{6}$$

where $k = 64^2$, (x,y) is the 2D projected coordinate of given grid point g_i, and (x_1, y_1), (x_1, y_2), (x_2, y_1), (x_2, y_2) are coordinates of the 4 nearest features $\Phi_L(p_{in})$ which can be calculated by their index values. Assuming the corresponding 4 nearest features are F_1, F_2, F_3, F_4, the interpolating result $F(g_i)$ can be derived by:

$$F(g_i) = W_1F_1 + W_2F_2 + W_3F_3 + W_4F_4 \tag{7}$$

V. Evaluation

A. Implementation and the DCIM Design

We implemented the *DCIM-3DRec* accelerator using TSMC 55nm process, as shown in Fig. 8. The *DCIM-3DRec* accelerator is developed using RTL and synthesized at 200 MHz with an area of 7.1× 3.4mm^2. The accelerator is physically implemented in a hierarchical fashion, where the custom DCIM column/macro designs are duplicated at the top hierarchy. In total, ten DCIM macros and 936kB SRAM are deployed. All the SRAMs are generated from the commercial SRAM compiler. The accelerator area is dominated by two DCIM macro arrays, i.e. 67%, which is after our memory optimizations via Octree-based subdivision scheduler and index-based global feature storage. The energy, performance, and area results are reported based on post-P&R netlist at the typical corner, i.e. 1.2V at 25°C.

The digital CIM macro is implemented using a custom bit-cell design and synthesizable adder and logic. The area of each DCIM macro is 2.1 mm×0.65 mm, which contains 32 columns and 32×32 subarrays. The power of DCIM macro is simulated after post-layout for computing mode with 50% bit-level weights sparsity. Each macro consumes 21.1 mW at 200 MHz at the nominal voltage of 1.2 V, in which the subarrays and adder trees contribute about 30% and 70%, respectively. The DCIM macro achieves 410 GOPS in INT8 precision for both the input features and weights, i.e. energy efficiency of 19.4 TOPS/W. Compared with advanced CIM macros of either digital-CIM [16] or analog-CIM [18], our DCIM macro achieves comparable energy efficiency with custom optimizations, as shown in Table 1.

TABLE I. Comparison Between Out DCIM and Other Work

	ISSCC'20 [18]	ISSCC'21 [16]	This work
Technology	28nm	22nm	55nm
CIM Type	Analog	Digital	Digital
Area	not provided	0.202 mm^2	1.365 mm^2
Throughput*	30.48 GOPS	917 GOPS	410 GOPS
Efficiency*	16.63 TOPS/W	24.7 TOPS/W	19.4 TOPS/W

* 8b weight, 8b input

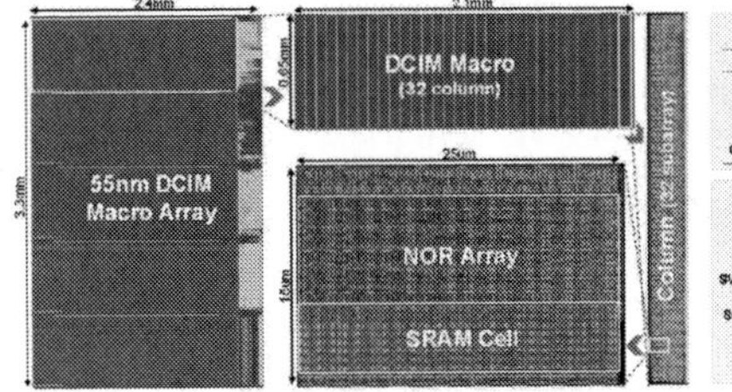

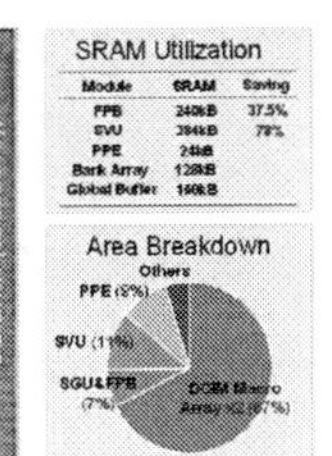

Fig. 8. 55nm layout of DCIM in *DCIM-3DRec* accelerator.

B. DCIM-3DRec Accelerator Evaluations

We compared our accelerator with both a high-performance Nvidia RTX3090 GPU and an edge Nvidia Jetson NX platform. *DCIM-3DRec* supports INT8 operation to achieve a reconstruction quality *IoU* of 0.853 for the *ShapeNet* dataset, compared to an *IoU* of 0.855 using floating-point operations on GPU. For a single object reconstruction, a 2048 input 3D point cloud data p_{in} is initially sent to the input buffer. With the help of our memory optimizations, all intermediate data is stored on-chip leading to no other off-chip IO access.

We first evaluate performance improvements of accelerator design. As shown in Fig. 9, a RTX 3090 GPU consumes 0.2s on average to finish the encoder-decoder flow, while the Jetson NX platform is about 18× slower, i.e. around 4s. In comparison, *DCIM-3DRec* achieves an average of 23× improvement across different 3D reconstruction categories compared to RTX 3090 GPU. Our accelerator performs the best for the *Plane* and *Rifle* categories with a 28× speedup, due to the smooth shapes of these objects requiring the least *dynamic division.* We also show the runtime breakdown for the reconstruction. The bottleneck of the *Decoder* network is alleviated from 94% to 78% due to the *Decoder* computation being pipelined in the accelerator.

We further elaborate on the energy consumption based on the post layout. *DCIM-3DRec* consumes about 334 mW at 200 MHz, in which the two DCIM macro array contributes about 45%. For a single object, *DCIM-3DRec* consumes about 2.51 mJ within 7.5ms. As a comparison, RTX 3090 and Jetson NX are estimated to consume about 30 J/object and 40 J/object, which is 4 orders of magnitude worse energy efficiency. Since our *DCIM-3DRec* is implemented using mature 55 nm, the performance and efficiency can be further improved using more advanced technologies.

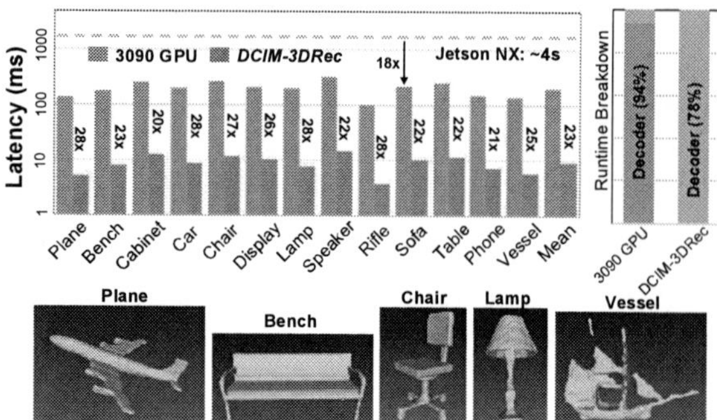

Fig. 9. Performance comparison between DCIM-3DRec and 3090 GPU, runtime breakdown, and reconstructed 3D objects.

We also evaluate the storage savings from the Octree-based storage for local grid points and the index-based storage for global features. In terms of local grid points storage, we compared two other alternative storage approaches with our Octree-based storage: *Coordinate-based* approach that only stores selected grid points and their corresponding location based on the subdivision, and *Address-based* approach stores all grid points in order and accesses the grid points according to the memory address. The *coordinate-based* stores the least number of grid points while requiring extra incompressible location information to access the points. The *address-based* simplifies the content information and the accessing process while requiring a large number of extra points to ensure proper address indexing. Following the observation that most categories of objects in the *ShapeNet* dataset have 3000 to 12000 subdivision times for *Decoder* inference, the *Octree-based* storage in our scheduler has 3.5× to 3.8× storage saving compared to *coordinate-based* storage and 5.1× to 23.1× saving compared to *address-based* storage in the majority of reconstruction cases, As shown in Fig. 10(a). In terms of global feature storage, we also compared our index-based feature storage with a baseline scheme with continuous feature storage. Since the index values of any two addresses are not identical, with a 64×64 feature plane resolution, our index-based storage scheme with LUT saves 60% and 2.2× when the number of input points p_{in} is 2048 and 1024 due to the sparsity of the indexed features, as shown in Fig. 10(b).

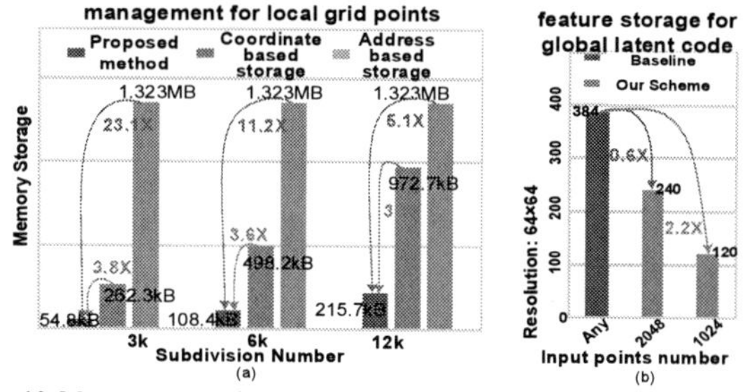

Fig. 10. Memory storage benefits for (a) local grid points and (b) global latent feature storage.

To the best of our knowledge, *DCIM-3DRec* is the first learning-based 3D reconstruction accelerator that leverages the DCIM technique to alleviate computation bottlenecks. The prior 3D vision accelerators [9-13] only focus on object detection or recognition. With our accelerator, the learning for reconstructing a single 3D object can be finished at an average of 7.5ms, which can satisfy many real-time interactive applications, such as AR/VR [15]. The low power consumption of *DCIM-3DRec* also facilitates the deployment of 3D reconstruction at the edge.

VI. Conclusion

The learning-based implicit representation is a promising method for 3D reconstruction. In this paper, a digital-CIM based accelerator *DCIM-3DRec* is presented for the 3D reconstruction task. Based on our analysis, it is observed that the learning-based reconstruction is computation bottlenecked the repetitious *Decoder* network inference. Due to the high model reuse rate, we naturally utilize the DCIM technique to improve 3D object reconstruction performance and efficiency. *DCIM-3DRec* is designed with the features of reconfigurable CIM macro array, Octree-based subdivision storage and scheduling, and index-based global feature storage. Based on evaluations on a 55nm technology, the *DCIM-3DRec* achieves 23× performance gain and 4 orders of magnitude energy efficiency improvement.

Acknowlegement

This work was supported in part by National Key R&D Program of China No. 2022YFB4400600, NSFC Grant No. 92164301 and 62225401, and the 111 Project.

References

[1] S. Seitz, et al., "A comparison and evaluation of multi-view stereo reconstruction algorithms", *CVPR*, 2006.
[2] J. Park, et al., "Deepsdf: Learning continuous signed distance functions for shape representation", *CVPR*, 2019.
[3] T. Takikawa, et al., "Neural geometric level of detail: real-time rendering with implicit 3D shapes", *CVPR*, 2021.
[4] A. Elluswamy, et al., "Occupancy networks", CVPR Workshop on Autonomous Driving (WAD), 2022.
[5] S. Peng, et al., "Convolutional occupancy networks", *ECCV*, 2020.
[6] J. Tang, et al., "SA-ConvONet: sign-agnostic optimization of convolutional occupancy networks", *ICCV*, 2021.
[7] D. Wang, et al., "Multi-view 3D reconstruction with Transformers", *ICCV*, 2021.
[8] B. Mildenhall, et al., "NeRF: representing scenes as neural radiance fields for view synthesis", *ECCV*, 2020.
[9] M. Kar, et al., "A ray-casting accelerator in 10nm CMOS for efficient 3D scene reconstruction in edge robotics and augmented reality applications," *IEEE Symp. on VLSI Circuits (VLSI)*, 2020.
[10] S. Choi, et al., "A 9.02mW CNN-stereo-based real-time 3D hand-gesture recognition processor for smart mobile devices", *ISSCC*, 2018.
[11] Z. Li, et al., "A 1920×1080 30-frames/s 2.3 TOPS/W stereo-depth processor for energy-efficient autonomous navigation of micro aerial vehicles", *IEEE Journal of Solid-State Circuits (JSSC)*, 2018.
[12] Z. Yue, et al., "MC-CIM: a reconfigurable computation-in-memory for efficient stereo matching cost computation", *DAC*, 2022.
[13] D. Im, et al., "DSPU: A 281.6mW real-time depth signal processing unit for deep learning-based dense RGB-D data acquisition with depth fusion and 3D bounding box extraction in mobile platforms", *ISSCC*, 2022.
[14] Y. Feng, et al., "Crescent: taming memory irregularities for accelerating deep point cloud analytics", *ISCA*, 2022.
[15] M. Huzaifa, et al., "ILLIXR: enabling end-to-end extended reality research", IEEE International Symp. on Workload Characterization (IISWC), 2021.
[16] Y. Chih et al., "An 89TOPS/W and 16.3TOPS/mm2 all-digital SRAM-based full precision compute-in-memory macro in 22nm for machine-learning edge applications", *ISSCC*, 2021.
[17] A. Hornung, et al., "Octomap: An efficient probabilistic 3d mapping framework based on octrees", Autonomous Robots, 2013.
[18] X. Si, et al. "A 28nm 64Kb 6T SRAM Computing-in-Memory Macro with 8b MAC Operation for AI Edge Chips", *ISSCC*, 2020

Processing-in-Memory Using Optically-Addressed Phase Change Memory

Guowei Yang*, Cansu Demirkiran*, Zeynep Ece Kizilates*, Carlos A. Ríos Ocampo†, Ayse K. Coskun*, Ajay Joshi*
*Boston University, †University of Maryland, College Park
{guoweiy, cansu, zecek, acoskun, joshi}@bu.edu, riosc@umd.edu

***Abstract*—Today's Deep Neural Network (DNN) inference systems contain hundreds of billions of parameters, resulting in significant latency and energy overheads during inference due to frequent data transfers between compute and memory units. Processing-in-Memory (PiM) has emerged as a viable solution to tackle this problem by avoiding the expensive data movement. PiM approaches based on electrical devices suffer from throughput and energy efficiency issues. In contrast, Optically-addressed Phase Change Memory (OPCM) operates with light and achieves much higher throughput and energy efficiency compared to its electrical counterparts.**

This paper introduces a system-level design that takes the OPCM programming overhead into consideration, and identifies that the programming cost dominates the DNN inference on OPCM-based PiM architectures. We explore the design space of this system and identify the most energy-efficient OPCM array size and batch size. We propose a novel thresholding and reordering technique on the weight blocks to further reduce the programming overhead. Combining these optimizations, our approach achieves up to 65.2× higher throughput than existing photonic accelerators for practical DNN workloads.

***Index Terms*—optical computing, phase change memory, processing-in-memory, deep neural networks**

I. Introduction

Deep Neural Networks (DNNs) are commonly used today for a variety of tasks such as image classification [1]–[3] and natural language processing [4]. The size of the DNNs has grown over the years, and current state-of-the-art DNNs contain hundreds of billions of parameters [5]. Moving DNN parameters as well as the DNN activation data between memory and compute causes large time and energy overheads. Moreover, the gap between computation and memory speed is continuously increasing, exacerbating the problem.

Processing-in-Memory (PiM) has emerged as a viable approach to mitigate this data movement cost. Existing PiM architectures incorporate various types of devices: DRAM [6], [7], ReRAM [8], [9], Spin-Transfer Torque RAM (STT-RAM) [10], and Phase Change Memory (PCM) [11]–[15]. DRAM-based PiM designs [6], [7] perform computation across the DRAM cells, thus, avoiding expensive data transfers between DRAM and compute. However, DRAM requires frequent refreshing to maintain the stored data, increasing power consumption. Both ReRAM [8], [9] and STT-RAM [10] are non-volatile memory. They do not consume power to maintain data storage, and offer higher throughput and consume lower power than DRAM-based devices. However, ReRAM suffers from process variation challenges and endurance issues, and STT-RAM has low storage density.

This research was partially funded by the NSF CCF 2131127 grant.

PCM-based PiM consumes lower power and provides higher throughput compared to PiM based on other technologies [12]. PCM is also a non-volatile memory that does not consume power to retain the stored data. Moreover, the multi-level capability of a PCM cell achieves higher throughput compared to a 1 bit/cell storage in DRAM; e.g., a PCM cell can store up to 6 bits/cell [16] and perform analog computation on multi-bit data [12]. We can access and perform computations in PCM cells using electrical or optical signals. Electrically-addressed PCMs (EPCMs) [11] offer higher storage density but have lower throughput [17] than Optically-addressed PCMs (OPCMs). OPCM-based PiM systems achieve high compute density (up to 162 TOPS/mm^2 [12]), making OPCM a promising solution for next-generation computing systems.

Recent works [12]–[15] have explored the idea of PiM using OPCM. These works, however, focus on small OPCM arrays such as 4×4 [13]. Moreover, these works evaluate their designs using small DNNs (such as four 2×2 kernels [12]), which can easily fit in small OPCM arrays. So, in those works, one needs to perform the expensive OPCM programming step only once, and so they do not consider programming cost (programming latency and energy are $2-3$ and $4-5$ orders of magnitude higher than compute latency and energy, respectively) in DNN inference. In contrast, state-of-the-art DNN models contain hundreds of billions of parameters and, therefore, cannot fit in a single OPCM array. In a practical scenario, we need to periodically program the OPCM array, and the cost of this reprogramming should be considered in the overall inference cost. In fact, the reprogramming latency and energy can easily dominate the total latency and energy and become bottlenecks. Therefore, to make OPCM-based PiM practical, we need to reduce OPCM's programming cost.

To this end, in this paper we make the following contributions:

- We provide a full system-level design of an OPCM-based PiM architecture that explicitly accounts for programming of the OPCM cells for performing DNN inference. To the best of our knowledge, we are the first to identify that the programming overhead dominates the DNN inference time and energy on OPCM, and the first to present a solution for this issue.
- We investigate the impact of OPCM array size and batch size on latency and energy efficiency in OPCM-based PiM and identify the optimal OPCM array size and inference batch size that achieve the highest energy efficiency in DNN inference, considering the programming cost.

979-8-3503-1176-1/23 $31.00

- We present a novel thresholding and reordering technique to reduce the OPCM programming overhead further. Our method applies thresholding to limit the number of OPCM cells that we should reprogram. It also reorders the programming of the matrix blocks to the OPCM array in a way that minimizes the number of OPCM cells we need to reprogram.

We evaluate the proposed OPCM-based PiM system using practical DNN workloads, including VGG-11 [1], AlexNet [2], ResNet-50 [3], and BERT-Large [4]. Our solution achieves 4918× higher IPS and 41.3% worse IPS/W than Eyeriss v2 [18], and 4.5× higher IPS and 1.2× higher IPS/W than TPU v3 [19]. Compared to photonic accelerators ADEPT [20] and Albireo-C [21], our solution has lower IPS/W. However, our system still achieves 2.3× and 65.2× higher IPS than ADEPT and Albireo-C, respectively.

II. OPCM Background

The basic building block of the OPCM design is an optical waveguide with embedded $Ge_2Sb_2Te_5$ (GST), as shown in Figure 1. The ratio between the amorphous and crystalline areas of the GST cell determines the device transmittance, which is encoded into multiple bits. Both the amorphous and the crystalline states are non-volatile. Therefore, the data is stored permanently, allowing its use as an optical multi-bit memory [22]. Experimental demonstrations have achieved up to 64 deterministic states in a GST cell, which allows storing up to 6 bits/cell [16]. Given these properties and developments, GST cells have been used as a platform for optical in-memory computing [23] by realizing scalar-scalar multiplications that map a multiplicand to the GST transmittance (i.e., stored in memory) and the multiplier to the amplitude of an input pulse.

To change the phase state of the GST cell and, thus, to modulate the transmission of the waveguide, electrical or optical pulses create the transition-triggering heat stimuli. Electrical switching uses a variety of waveguide-embedded microheaters to electro-thermally induce the reversible phase transitions. Researchers have demonstrated switching energies as low as 5.55 nJ and 860.71 nJ to amorphize and crystallize, respectively, using graphene microheaters [24]. Optical switching, on the other hand, uses the optical absorption of GST to trigger the amorphization or crystallization using pulses inside the waveguide. This method consumes less energy at the GST cell; 460 pJ for amorphization and 140 pJ for crystallization [22]. However, in a large architecture, the optical pulses must be routed through a switching network, which imposes the need for energy-hungry phase-shifters. Alternatively, dedicated couplers can be used to reach each GST cell, an optical solution that requires a large footprint and complicated experimental setups [12]. Even though electrical programming consumes more energy per switching event, the scalability, compatibility with microelectronics, and form factor make it a more suitable solution for setting the optical transmission of the GST array. Therefore, for optimized DNN inference operations, we assume electrical programming for efficiently writing the GST array weights and optical addressing for high-throughput General Matrix Multiply (GEMM) operations [13], [24].

III. Related Work

Moving data between memory and compute units causes significant overhead, and PiM is a potential solution to this problem.

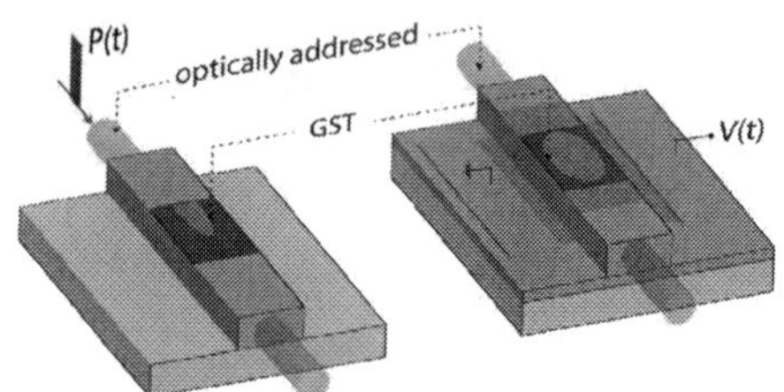

Fig. 1. Optical waveguide with embedded GST. The transmitted light is correlated to the phase state of GST, which can be controlled optically using pump pulses or electrically using fast electro-thermal heating in waveguide-integrated microheaters.

By performing computation inside the memory cells, PiM reduces the data movement overhead and improves the throughput and energy efficiency. PiM architectures based on DRAMs [6], [7] are promising, but they suffer from high latency due to charging/discharging of the capacitance [6], which also results in higher energy consumption. The latency of a single operation in DRAM-based PiM architectures can take almost 28 ns and requires at least 3 nJ of energy because of the activation, read/write, and precharge processes [7]. Furthermore, the technology scaling for DRAM is slowing down. Considering the increase in the amount of data to be processed, this scalability issue becomes more critical [25].

ReRAM-based PiM architectures are being actively explored, and their nonvolatile nature makes ReRAMs a favorable candidate for implementing PiM designs [8], [9]. STT-RAM [10] is also a non-volatile memory that shows great potential to accelerate DNN inference. Both ReRAM and STT-RAM achieve higher throughput and energy efficiency compared to DRAM-based PiM system. However, ReRAM suffers from process variation challenges and endurance issues, and STT-RAM has a low storage density. They are also fundamentally limited by the energy-throughput tradeoff.

PCM is another type of device that has gained attention as a suitable candidate for PiM. EPCM devices [11] are accessed with electrical signals, while OPCM devices are accessed with optical signals. When programming the PCM cells, existing designs use either electrical switching [15], or optical switching [12], [14].

In contrast to DRAM, both OPCM and EPCM are fully passive and do not need power to retain data. Their multi-level capability provides higher throughput, and their analog computing approach consumes less power than digital computing [15]. We argue that OPCM is a better choice to accelerate DNN inference. Compared to EPCM whose frequency is limited by energy, OPCM can operate at high frequency (up to 25 GHz) to provide extremely high throughput. Thus, in this paper, we focus on OPCM-based PiM architectures.

IV. OPCM-Based PiM Design and Operation

In this section, we first describe the architecture of our proposed OPCM-based PiM system and the dataflow for mapping DNN inference to that system. Then, we discuss the microarchitecture of the OPCM array and how it performs GEMM operations. Finally, we present the thresholding and reordering technique for the DNN weights to reduce the PCM programming overhead during inference.

A. Full System Architecture and Dataflow

1) System Architecture: Figure 2 shows our proposed 2.5D-integrated OPCM-based PiM system that consists of the host processor chiplet, DRAM chiplet, OPCM chiplet containing multiple crossbar arrays for GEMM operations, the CMOS chiplet for electrical-optical and optical-electrical (E-O-E) conversions and non-GEMM operations (referred to as "E-O-E + non-GEMM" chiplet henceforth), and the laser source chiplet. The non-GEMM operations include inter-block accumulation, non-linear activation functions, pooling, and batch normalization. In addition, the E-O-E + non-GEMM chiplet contains SRAM arrays for storing inter-block accumulation outputs and buffering the data received from/sent to the DRAM. The host processor chiplet is connected to the "E-O-E + non-GEMM" chiplet using electrical links embedded in the interposer. The DRAM chiplet and the laser source chiplet are connected to the "E-O-E + non-GEMM" chiplet using optical links embedded in the interposer. The laser source is also connected to the DRAM chiplet using optical links embedded in the interposer. Prior to performing the DNN inference operations, it is more efficient to program the weights in the OPCM array electrically than optically; however, during DNN inference, it is more efficient to perform the GEMM operations with OPCM than EPCM, which uses optical links [13], [24]. So the "E-O-E + non-GEMM" chiplet connects to the OPCM array through both electrical and optical links embedded inside the interposer.

2) Dataflow of DNN Inference: The host processor chiplet executes the DNN-based application. When executing the DNN inference part of the application, the processor uses Application Programming Interfaces (APIs) to transfer the control to the control logic in the "E-O-E + non-GEMM" chiplet. Given that an OPCM cell requires an area of $30 \times 30\,\mu m^2$ [12], and DNNs contain hundreds of billions of parameters, it is not practical to build an OPCM array that is large enough to fit an entire DNN at one time. For example, VGG-11 has 133 million parameters and requires an OPCM array with an area of 239,400 mm^2, which is impractical. To perform DNN inference on the OPCM chiplet, we process the DNN layer by layer. For each layer, we apply the standard blocking technique [12], which breaks down the large matrix into smaller blocks and processes one block at a time. We have multiple OPCM crossbar arrays in the OPCM chiplet that run in parallel. For every OPCM array, the control logic first reads a weight block of a layer from DRAM using optical links, saves it in the SRAM buffer in the "E-O-E + non-GEMM" chiplet, and programs it into the OPCM array using electrical links. The control logic then reads the inputs to this block from DRAM and feeds them to the OPCM array via optical links. Note that the inputs get converted from the electrical domain to the optical domain in the DRAM and are then routed directly to the OPCM array. The OPCM array performs GEMM operation in the optical domain. The output of the OPCM array is routed to the "E-O-E + non-GEMM" chiplet, where the data is converted into the electrical domain and fed to the digital logic performing the non-GEMM operations, including inter-block accumulation and non-linear operations. When the OPCM array is performing operations on one block, the control logic loads the next block to be processed into the SRAM buffer in a pipelined fashion. The above operations are repeated for all blocks in every layer. Once the operations for a layer finish, the results are transferred back to DRAM, and these results serve as the input for the following layer. After all the layers of the DNN are executed, the inference result is written back into the DRAM and is accessible to the host processor. The traditional communication between the host processor chiplet and the DRAM chiplet is through the "E-O-E + non-GEMM" chiplet.

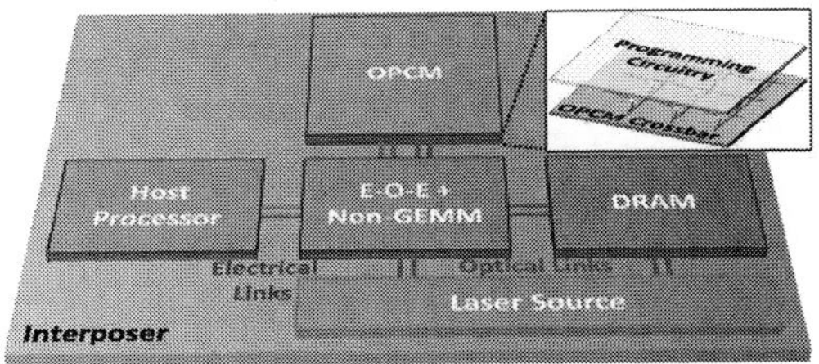

Fig. 2. Architecture of an OPCM-based computing system.

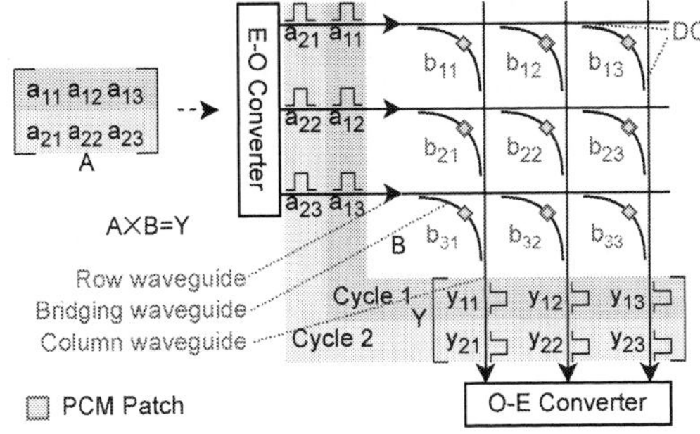

Fig. 3. GEMM example of 2×3 input matrix A multiplied with a 3×3 weight matrix B on OPCM array.

B. OPCM Array Architecture

The OPCM chiplet consists of multiple OPCM crossbar arrays for GEMM computation and the programming circuitry for setting the state of each OPCM cell. The OPCM crossbar array (see Figure 3) is based on the Photonic Tensor Core [12], which consists of row waveguides, column waveguides, bridging waveguides, and Directional Couplers (DCs). We deposit the phase change material (e.g., GST) on top of every bridging waveguide. The DCs connect the horizontal and vertical waveguides through the bridging waveguide. The split ratio of the DC is carefully designed such that the horizontal DCs split the light from a row evenly into every column, and the vertical DCs combine the attenuated light received from every row in a single column [12].

C. GEMM operations in OPCM

1) GEMM Operation: We follow the standard scheme to perform GEMM operations in OPCM [12]. Consider two matrices, matrix A that is $P \times M$ and matrix B that is $M \times N$. To multiply A and B, assume we have an OPCM array of M rows and N columns. We map matrix B to that OPCM array. To perform the matrix-matrix multiplication, we split the matrix A into P $1 \times M$ vectors and then perform P matrix-vector multiplications (MVMs) in the OPCM array. Below we describe the matrix multiplication with a concrete example.

To understand the matrix-matrix multiplication process, assume A is a 2×3 matrix and B is a 3×3 matrix. So $M = 3$, $N = 3$, $P = 2$ (see Figure 3). Assume we have a 3×3 OPCM array. Elements of the matrix B are electrically programmed into the PCM cells of the OPCM array. Then three elements of the first row of matrix A (i.e., a_{11}, a_{12}, and a_{13}) are converted into the optical domain (each element is mapped to a unique wavelength), and then these three optical signals are routed into the three rows of the OPCM array. Each optical signal is split equally across the three columns and routed into the bridging waveguides. As the optical signal passes through the bridging waveguide, it is attenuated by the PCM material on the bridging waveguide and then coupled into the column waveguide. This attenuation process performs the multiplication operations. At the output of the first column, we get three products in three different wavelengths– $a_{11} \times b_{11}$, $a_{12} \times b_{21}$ and $a_{13} \times b_{31}$. These three products are accumulated using a photodetector, i.e., $y_{11} = a_{11}b_{11} + a_{12}b_{21} + a_{13}b_{31}$ to give us the first element of the first row of the product matrix. Similarly, the second column and the third column of the OPCM array give us the second element and third element of the first row of the product matrix. Then we similarly send the second row of matrix A through the OPCM array to get the second row of the product matrix.

2) Handling negative weights and inputs: DNN weights are usually signed; however, the intensity of light waves and the loss in PCM cells are always non-negative. To support negative weights, we need to decompose the weight matrix B into positive and negative component matrices B^p and B^n, where $b_{ij}^p = \max(0, b_{ij})$ and $b_{ij}^n = \max(0, -b_{ij})$. The final result of matrix multiplication then becomes $A \times B = A \times B^p - A \times B^n$ [14], where the subtraction happens in the electrical domain. With two 6-bit PCM cells [16] representing the positive and negative components, we effectively achieve 7-bit quantization. To support negative inputs, we need to offset the input and include an extra column of precomputed weight references [13].

3) Mapping DNN to OPCM: DNNs include a variety of layers, including convolution and fully-connected layers. The operations in these layers can be mapped to GEMM operations. We use a weight-stationary approach for performing these GEMM operations. The blocking approach requires us to program the weights within a block onto the OPCM array, perform multiplication using that block, then repeat this process for another block. Unfortunately, for large DNNs, the cumulative latency of programming all the blocks into the OPCM array is on the order of seconds, and cumulative energy is on the order of joules. This is because we need to re-program the OPCM array frequently due to its small capacity compared to large DNN weights. For example, a 64×64 array is programmed at least 30,000 times during one inference of the four example DNNs. The programming time and energy are $2-3$ orders of magnitude and $4-5$ orders of magnitude larger than the time and energy for performing a single inference (more details in Section V-B, see Figure 4). Effectively, this reprogramming overhead cancels OPCM's performance and energy advantages for DNN inference. Therefore, we need to reduce the programming cost of OPCM.

D. OPCM Programming Cost Reduction

1) Choosing Batch Size and Array Size: One way to reduce the programming cost is to use large batch sizes during inference. After we program a block of a matrix into the OPCM array, the same block can operate on all the inputs in a batch. This way, we can amortize the time and energy overhead of programming the OPCM array across the large batch. However, the intermediate memory required to perform the accumulations between blocks increases proportionally with batch size, which poses an upper bound for the batch size.

For example, Figure 5 (a) shows the IPS/W for VGG-11 with various batch sizes and array sizes. It is clear that in terms of performance per unit power, array size $= 64$ yields the highest IPS/W, and the larger the batch size, the better the IPS/W.

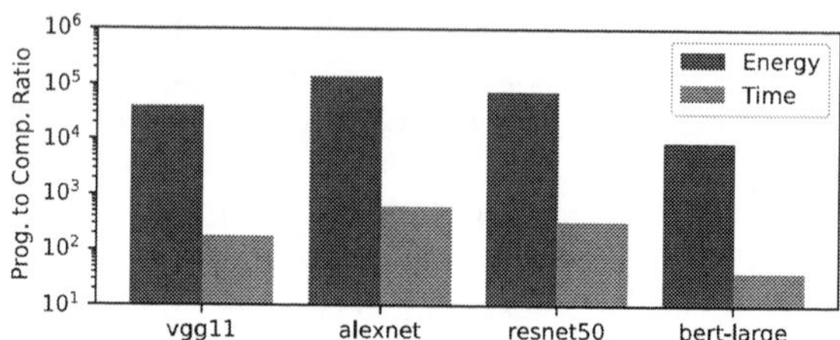

Fig. 4. Ratio of programming cost to computation cost for one inference, with 16 arrays each of size 64×64. The time and energy spent on programming the OPCM cells are $2-3$ and $4-5$ orders of magnitude higher than that on computation only, showing that programming cost dominates the inference performance and energy.

Figure 5 (b) shows the SRAM required for inter-block accumulation for different batch sizes and OPCM array size combinations. SRAM requirement increases with batch size and OPCM array size. For an array size $= 64$, the SRAM requirement reaches 392 MB at batch size $= 4{,}096$, taking up about 600 mm^2 on GF22FDX technology node. Further increasing the batch size would increase the IPS/W, but the SRAM area will become prohibitively large. Therefore, we need to choose the batch size and array size for each DNN to maximize energy efficiency under an area constraint.

2) Block reordering to minimize state changes: We propose a block reordering technique to reduce the energy required for programming the OPCM array. This technique exploits the observation that there exists a similarity between the elements across blocks. Essentially, during inference, when programming a new block into the OPCM array, not all OPCM cells need to be programmed. For example, suppose the difference between the value of an element at a specific position in the new block and the value of an element at that same position in the current block is less than a certain threshold, we don't need to program the OPCM cell corresponding to that element, i.e., we do not need to overwrite it. We can reuse the old value as DNNs are tolerant to minor weight variations.

Moreover, when performing DNN inference, the blocks within a layer can be mapped to the OPCM array in arbitrary order without degrading inference accuracy. This provides a chance to further reduce the programming cost as we can follow a block processing order that minimizes the programming latency and energy for a given layer. Figure 6 shows a toy example where the matrix is divided into four blocks. We construct an undirected graph where each vertex represents a block. Suppose block 1 is currently in the OPCM array, and we now program block 2 into that array; the cost of this programming, i.e., the number of cells to be overwritten, is denoted on the edge between block 1 and block 2. There are 24 possible orders in which the four blocks can be programmed into the OPCM array. Out of the 24 possible orders, the $3 \rightarrow 1 \rightarrow 4 \rightarrow 2$ order has the least latency and energy. We can use this order to minimize the programming cost. Note that one needs to determine the order of using blocks only once, and that can be done offline. The overhead of the control logic for processing the blocks in non-sequential order is minimal.

In today's DNNs, for a typical 64×64 OPCM array, we have up to 50,000 blocks in a layer. The number of possible block processing orders is up to $50{,}000! = 10^{200{,}000}$. Performing an exhaustive search to determine the order of processing the blocks is impractical, even if the process is done offline. In fact, finding the minimum cost of traversing all the vertices in a graph is a well-studied problem called

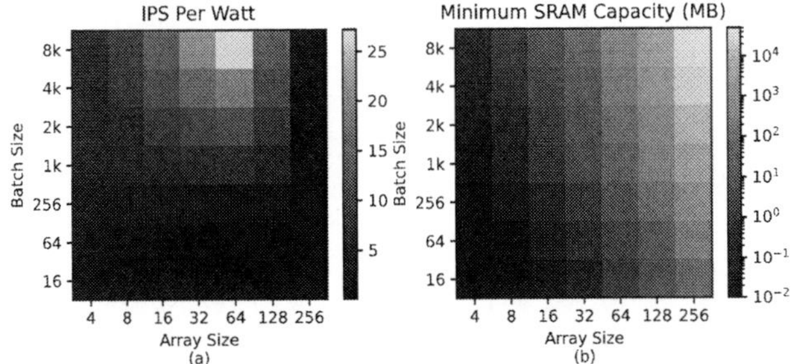

Fig. 5. (a) IPS/W and (b) Minimum SRAM capacity (for inter-block accumulation) for VGG-11. Plots for AlexNet, ResNet-50 and BERT-Large look similar.

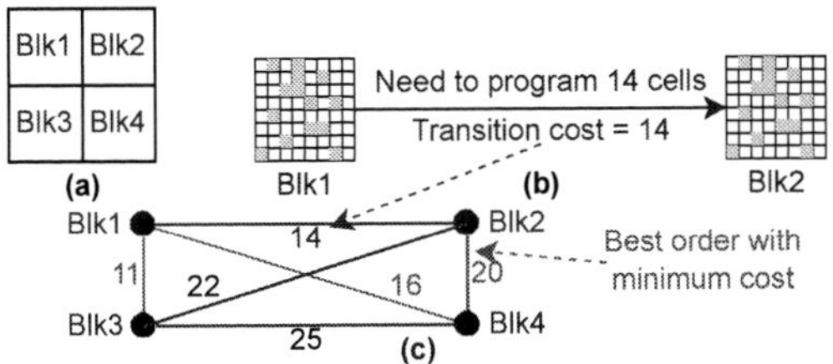

Fig. 6. Reordering the weight blocks to reduce programming cost. (a) Split the weight matrix into blocks of the same size as the OPCM array. (b) Compute the transition cost between each pair of blocks. (c) Construct an undirected graph whose vertices are weight blocks and edges are the transition cost between the two blocks. Solve Traveling Salesman Problem to find the order that minimizes cost.

Traveling Salesman Problem (TSP) [26], and there are various tools to solve it. We use Google OR-Tools [27], a widely used library for combinatorial optimization, to solve our TSP problem.

V. Evaluation

A. Methodology

In this section, we evaluate the OPCM-based PiM in terms of its performance and energy efficiency for performing DNN inference. We limit the total number of OPCM crossbar cells to 256×256, which is divided into multiple OPCM arrays. The OPCM array uses electrical signals to program the PCM cells and optical signals to perform GEMM operation. During GEMM operation, the power consumption is primarily in the laser source and the E-O-E conversion unit. The required laser power depends on the optical losses experienced by the optical signals as it passes through the OPCM array. We assume the losses of the GST cell, waveguide crossing, and DC are 0.6 dB, 0.0028 dB, and 0.01 dB, respectively, and the combined quantum efficiency of the laser and photodetector is assumed to be 10% [12]. The OPCM array is a passive device with passive micro-ring resonators (MRRs) [28] and does not consume any thermal tuning power. The OPCM arrays perform MVM operations at 25 GHz. We assume the programming energy of each cell to be the average of amorphizing (5.55 nJ) and crystallizing (860.71 nJ), and it takes 400 ns to program the entire array [24]. Each crossbar cell occupies an area of 30×30 μm^2 [12], and the diameter of MRRs is 10 μm.

For the "E-O-E + non-GEMM" chiplet, we synthesize the SRAM and electrical non-linear units in GF22FDX technology node to get the power and area estimations. The ADC (which also performs the O-E conversion) power and E-O power are 194 mW/channel and 1 pJ/bit, respectively, at 25 GHz [12]. DRAM accesses are assumed to be 20 pJ/bit [29].

We choose the following four popular DNNs as workloads: VGG-11 [1], AlexNet [2], ResNet-50 [3], and BERT-Large [4]. The first three DNNs run image classification on the Imagenet dataset, and BERT-Large runs a question-answering task on the SQuAD 1.1 dataset. All models are pre-trained with FP32 and quantized to 7-bit precision (combining two 6-bit cells, as discussed in Section IV-C2). We use Google OR-Tools [27] to solve the reordering problem, apply thresholding to weights with in-house scripts, and then use PyTorch to test the inference accuracy after thresholding and reordering.

B. Results

1) OPCM Programming Cost: We first focus on the OPCM array's programming overhead during a DNN inference. We perform inference using four DNNs with the OPCM array size of 64×64 and observe the time and energy spent on programming versus computation. As shown in Figure 4, the time and energy spent on programming the OPCM cells for one inference operation are $2-3$ orders of magnitude and $4-5$ orders of magnitude higher than those for computing in the OPCM cells. This is because the capacity of the OPCM array is at least $30{,}000\times$ smaller than the weights in the four DNNs, so we must reprogram it frequently.

2) Choosing Batch Size and Array Size: Using an extremely large batch size is a straightforward way to amortize away the programming cost per inference. From Figure 5 (a), it is evident that array size $= 64$ yields the best energy efficiency, while the larger the batch size, the better. However, the batch size is limited by the SRAM capacity. Figure 5 (b) shows the minimum SRAM requirement. As discussed earlier in Section IV-D1, with array size $= 64$ and batch size $= 4{,}096$, the minimum SRAM required reaches 392 MB, which is a feasible design. We perform similar analyses for all DNNs evaluated and find that array size $= 64$ and batch size $= 4{,}096$ is the practical and most energy-efficient configuration.

3) Using Thresholding and Reordering: Next, we evaluate the effectiveness of the thresholding and reordering technique. Figure 7 shows the programming cost reduction for different thresholds. For each threshold value, we find the mapping order with the maximum programming cost saving. With only reordering (i.e., threshold $= 0$), we observe a $27.8\% - 29.7\%$ reduction in the programming cost across the four DNNs. As we increase the threshold, the savings increase. We observe up to $62.2\% - 77.6\%$ reduction in programming energy for threshold $= 16$.

The thresholding approach can, however, introduce errors in the weights and might cause accuracy degradation. Figure 8 shows the accuracy metrics for various thresholds. "f" means the accuracy of the float-point model, and "0" means only quantization and no thresholding applied. The figure shows that all DNNs can tolerate small thresholds without having a significant accuracy drop. With less than 5% accuracy loss, the thresholding and reordering approach can achieve 42.9%, 46.5%, 47.4%, and 45.2% programming cost reduction for AlexNet, VGG-11, ResNet-50, and BERT-Large, respectively.

4) Comparison with Related Work: In Table I, we compare our OPCM-based PiM solution against the results reported by previous works, including Eyeriss v2 [18], TPU v3 [19], ADEPT [20], and Albireo-C [21]. Our solution achieves $4.5\times$ higher IPS and $1.2\times$ higher IPS/W than TPU v3, and $4{,}918\times$ higher IPS but 41.3% worse IPS/W than Eyeriss v2. Compared with photonic accelerators, ADEPT, and Albireo-C, our solution has lower IPS/W. However,

TABLE I

PERFORMANCE AND ENERGY EFFICIENCY OF OPCM-BASED PIM ARCHITECTURE

	This work				Eyeriss v2 [18]	TPU v3 [19]	ADEPT [20]	Albireo-C [21]
Configuration	Array size 64×64, 16 arrays, 25 GHz, batch size 4096, 7-bit quantization				ASIC	ASIC	Photonic	Photonic
Model	VGG-11	BERT	ResNet-50	AlexNet	AlexNet	ResNet-50	AlexNet	AlexNet
Threshold*	6	7	4	5				
IPS	91,493	10,162	148,166	501,629	102	32,716	217,201	7,692
IPS/W	26.05	0.76	21.55	102.64	174.8	18.18	7476.78	344.17

* Thresholds are chosen to achieve maximum programming cost savings with less than 5% accuracy loss.

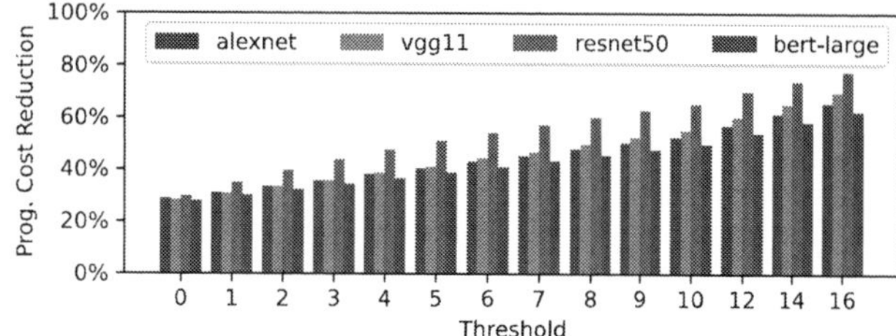

Fig. 7. Reduction in programming energy after thresholding and reordering for 64×64-sized blocks. A threshold of "0" means we quantize the model to 7-bit precision and do not use thresholding. With reordering alone and no thresholding (i.e., threshold $= 0$), we observe a $27.8\% - 29.7\%$ reduction in the programming cost. Using a larger threshold leads to even more reduction. The thresholding and reordering technique can reduce the programming cost by $42.9\% - 47.4\%$ with less than 5% accuracy loss (See Figure 8).

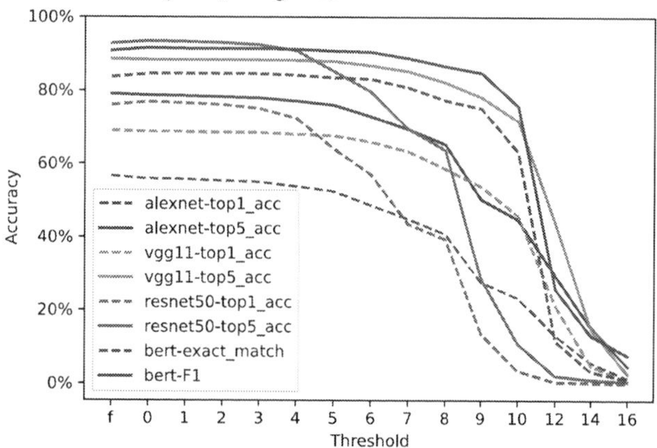

Fig. 8. Accuracy after thresholding and reordering for 64×64 blocks. "f" on the X-axis corresponds to the original FP32 model, and a threshold of "0" means we quantize the model to 7-bit precision and do not apply thresholding. All these four DNNs can tolerate a threshold of at least 4 without a significant accuracy drop.

it still achieves $2.3\times$ and $65.2\times$ higher throughput than ADEPT, and Albireo-C, respectively.

VI. CONCLUSION

The performance of DNN inference is limited by the data movement cost. To tackle this problem, we propose a complete OPCM-based PiM system architecture, which combines the OPCM array with photonic links. In the OPCM-based PiM system, the OPCM programming cost dominates. We propose to use three techniques - strategically choosing the batch size, reordering the blocks during matrix multiplication, and using thresholding when updating the OPCM array, to amortize the programming cost. Using our approach, we achieve 42.9%, 46.5%, 47.4%, and 45.2% programming energy reduction for AlexNet, VGG-11, ResNet-50, and BERT-Large, respectively.

REFERENCES

[1] K. Simonyan and A. Zisserman, "Very deep convolutional networks for large-scale image recognition," *ICLR*, 2014.

[2] A. Krizhevsky *et al.*, "Imagenet classification with deep convolutional neural networks," in *NIPS*, 2012.

[3] K. He *et al.*, "Deep residual learning for image recognition," in *CVPR*, 2016.

[4] J. Devlin *et al.*, "Bert: Pre-training of deep bidirectional transformers for language understanding," *NAACL HLT 2019*, pp. 4171–4186, 10 2018.

[5] P. Villalobos *et al.*, "Machine learning model sizes and the parameter gap," *arXiv*, 7 2022. [Online]. Available: http://arxiv.org/abs/2207.02852

[6] X. Xin *et al.*, "Roc: Dram-based processing with reduced operation cycles," *DAC*, 2019.

[7] D. Quan *et al.*, "Lacc: Exploiting lookup table-based fast and accurate vector multiplication in dram-based cnn accelerator," in *DAC*, 2019.

[8] P. Chi *et al.*, "Prime:a novel processing-in-memory architecture for neural network computation in reram-based main memory," in *ISCA*, 2016.

[9] A. Shafiee *et al.*, "Isaac: A convolutional neural network accelerator with in-situ analog arithmetic in crossbars," in *ISCA*, 2016.

[10] S. Jain *et al.*, "Computing in memory with spin-transfer torque magnetic ram," *TVLSI*, 2018.

[11] B. C. Lee *et al.*, "Architecting phase change memory as a scalable dram alternative," in *ISCA*, 2009.

[12] J. Feldmann *et al.*, "Parallel convolutional processing using an integrated photonic tensor core," *Nature*, 2021.

[13] F. Brückerhoff-Plückelmann *et al.*, "Broadband photonic tensor core with integrated ultra-low crosstalk wavelength multiplexers," *Nanophotonics*, 2022.

[14] P. Guo *et al.*, "Starlight: a photonic neural network accelerator featuring a hybrid mode-wavelength division multiplexing and photonic nonvolatile memory," *Optics Express*, p. 37051, 2022.

[15] H. Zhu *et al.*, "Elight: Towards efficient and aging-resilient photonic in-memory neurocomputing," *TCAD*, 2022.

[16] C. Wu *et al.*, "Programmable phase-change metasurfaces on waveguides for multimode photonic convolutional neural network," *Nat. Commun.*, 2021.

[17] A. Narayan *et al.*, "Architecting optically controlled phase change memory," *ACM Trans. Archit. Code Optim.*, vol. 19, pp. 1–26, 12 2022.

[18] Y. H. Chen *et al.*, "Eyeriss v2: A flexible accelerator for emerging deep neural networks on mobile devices," *IEEE J. Emerg. Sel. Topics Power Electron.*, vol. 9, pp. 292–308, 7 2018.

[19] N. P. Jouppi *et al.*, "A domain-specific supercomputer for training deep neural networks," *Commun. ACM*, vol. 63, no. 7, p. 67–78, jun 2020.

[20] C. Demirkiran *et al.*, "An electro-photonic system for accelerating deep neural networks," *arXiv preprint arXiv:2109.01126*, 2021.

[21] K. Shiflett *et al.*, "Albireo: Energy-efficient acceleration of convolutional neural networks via silicon photonics," in *2021 ISCA*, 2021.

[22] C. Ríos *et al.*, "Integrated all-photonic non-volatile multi-level memory," *Nature Photonics*, 2015.

[23] C. Ríos *et al.*, "In-memory computing on a photonic platform," *Science advances*, vol. 5, no. 2, p. eaau5759, 2019.

[24] Z. Fang *et al.*, "Ultra-low-energy programmable non-volatile silicon photonics based on phase-change materials with graphene heaters," *Nature Nanotechnology*, vol. 17, no. 8, pp. 842–848, 2022.

[25] S. K. Kim and M. Popovici, "Future of dynamic random-access memory as main memory," *MRS Bulletin*, 2018.

[26] M. Jünger *et al.*, "Chapter 4 the traveling salesman problem," in *Network Models*, ser. Handbooks in Operations Research and Management Science. Elsevier, 1995, vol. 7, pp. 225–330.

[27] L. Perron and V. Furnon, "Or-tools," Google. [Online]. Available: https://developers.google.com/optimization/

[28] A. A. Nikitin *et al.*, "Optical bistable soi micro-ring resonators for memory applications," *Optics Communications*, vol. 511, p. 127929, 5 2022.

[29] M. Horowitz, "Computing's energy problem (and what we can do about it)," *IEEE ISSCC*, vol. 57, pp. 10–14, 2014.

LAXOR: A Bit-Accurate BNN Accelerator with Latch-XOR Logic for Local Computing

Dongrui Li[1, 2*], Tomomasa Yamasaki[1*], Aarthy Mani[2], Anh Tuan Do[2], Niangjun Chen[1], Bo Wang[1]
[1]Singapore University of Technology and Design
[2]Institute of Microelectronics, Agency for Science, Technology & Research (A*STAR), Singapore

***Abstract*—Binary Neural Network (BNN) accelerators are attractive solutions for Artificial Internet-of-Things (AIoT) applications thanks to the compact models and low computational cost while maintaining satisfactory classification performance. Various analog/mix-signal compute-in-memory macros have been proposed to boost the energy efficiency of binary convolution tasks. However, this approach incurs inaccurate computation due to its sensitivity to temperature, noise, and process variations. In this work, we present a full-digital BNN architecture that leverages a novel Latch-XOR logic array for local bitwise multiplication, suppressing massive data movement and achieving 4.2× lower energy per operation compared to the decoupled standard cell approach. An optimized population count circuitry is also proposed for data accumulation, which obtains 1.37× Energy-Delay-Area saving compared to Binary-Adder-Tree-based implementation. To enable seamless hardware-software co-optimization, we have developed an in-house simulator for design space exploration as well as flexible mapping with various network topologies and kernel sizes. Our experiment shows the Latch-XOR-based architecture in 28nm CMOS technology achieves an enhanced energy efficiency of 2315 $TOPS/W$, 3.4× higher compared to the state-of-the-art synthesized digital architecture. This manifests that the proposed accelerator is highly suited for AIoT applications.**

I. Introduction

Domain-specific hardware accelerators for Deep Neural Networks (DNNs) have become prevailing due to their optimized architectures, data flow, and circuits to enhance critical design metrics. Binary Neural Networks (BNNs), known as single-bit-precision DNNs, have demonstrated satisfactory accuracy in a variety of workloads [1], [2]. The weights and activations of BNNs are binarized to either '+1' or '−1' resulting in tiny model size and lightweight convolution implemented by XNOR/XOR and population count operations [1]. Consequently, its hardware implementation requires a much smaller memory capacity and dissipates substantially less power compared to multi-bit precision, real-valued neural networks [3], [4].

Contemporary on-chip BNN acceleration falls into two major categories: the analog/mixed-signal Compute-In-Memory (CIM) circuitry and the full-digital approach. In CIM-based architectures [5]–[7], parallel read-out of multiple memory rows allows for one-shot weighted sum accumulation (i.e., convolution), leading to considerably improved energy efficiency.

This research is supported by the Ministry of Education, Singapore (MOE-T2EP50122-0024), National Research Foundation, Singapore (NRF-CRP23-2019-0003), and SUTD (SRT3IS20162, SKI 2021_02_06). *Both authors contributed equally to this work.

However, the analog/mixed-signal approach is susceptible to temperature, noise, and process variation, causing potential inaccurate computation and thus accuracy degradation [4]. To avoid this, pure-digital BNN accelerators were proposed [4], [8]. Nevertheless, these works usually separate local weight storage from XNOR/XOR operation, resulting in intensive data movement, which deteriorates energy efficiency.

Another challenge in neural network accelerator design is how to effectively and accurately predict the hardware performance at the application level, under a wide variety of datasets and design parameters of the architecture. Moreover, designers must capture the trade-offs between performance and energy consumption early to reduce the rounds for hardware optimization. To achieve this, various simulators for CNN accelerators have been proposed [9]–[11]. Nevertheless, these works rarely model BNN topologies and their bitwise multiplication (i.e., XNOR or XOR). HAWIS [12] provides a generator for BNN models and estimates their energy consumption. However, the architecture primitive is based upon a ReRAM model and thus not very useful for synthesized, bit-accurate digital implementation.

In view of the above, we propose LAXOR, a BNN accelerator with a cycle-accurate simulator for hardware-software co-optimization. The essence of LAXOR lies in a novel local computing paradigm that fuses the weight storage (i.e., latch) and the compute unit (i.e., XOR gate) in a single logic to minimize data movement, achieving 4.2× lower energy consumption. We have validated our architecture with a variety of workloads, showing an accuracy of 65.21% for CIFAR-100 and 85.25% for CIFAR-10 with a total energy per classification of 103.74 μJ and 3.82 μJ, respectively in 28nm CMOS technology. In particular, LAXOR consums 1.8× lower energy compared to the state-of-the-art digital implementation [3] with the CIFAR-10 workload. Assisted with the optimized population count circuits, the proposed accelerator attains an energy efficiency of 2315 $TOPS/W$, 3.4× $\sim$ 37.8× higher compared to the advanced BNN accelerator architectures [13], [3], [14], respectively.

II. Architectural Overview

A. Many-core Architecture

To enable modular hardware implementation and network mapping, we propose a many-core architecture with compact XOR arrays and energy-efficient population count (i.e., pop-count) logic for local computing, whose detailed implementation will be discussed in Section III. As shown in Fig.1(a), the architecture consists of 4 Processing Engine (PE) clusters, a global controller and configuration unit, an accumulation unit to calculate total sums for activation layers or Fully-Connected (FC) layers, and a comparison block to determine the final inference result based on the maximum value. Specifically, the PE clusters account for most of the compute workload, the power consumption as well as the silicon area. Each cluster consists of 64 PEs where every single PE incorporates a Latch-XOR array (i.e., 1024 cells) for bitwise multiplications, a popcount unit that performs kernel-level accumulation, and a 9-bit bias unit for kernel mapping compensation or activation bias. The output of the popcount logic is connected to a local activation unit for the binarization, generating output feature maps. The three components are closely integrated into a single PE to minimize data movement. We also employ an input buffer (i.e. IB_{Ci} in Fig.1) to load weight or input activation before loading them to the PEs in a time-multiplexing manner. The datapath supports various kernel sizes (e.g., 3×3, 4×4, etc.), and their mapping methodology will be illustrated hereafter.

B. Dataflow Mapping

As most BNN models adopt small kernel dimensions, ranging from 1×1 to 11×11 [15], we enable the proposed LAXOR

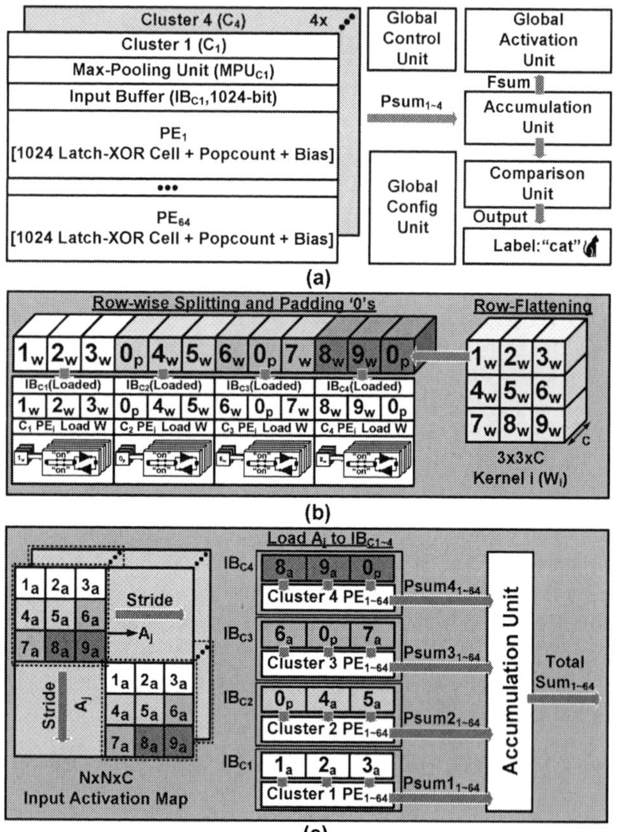

Fig. 1. Block diagram of the (a) many-core architecture (b) Dataflow mapping scheme for odd-sized weight and (c) Dataflow mapping scheme for convolution operation.

architecture with a flexible dataflow mapping scheme to support them correspondingly.

Our mapping strategy is to leverage the parallelism of the 4 clusters. For a 3D odd-sized kernel W_i (e.g., $(2N+1)^2 \times C$, where C is the number of channels and N is a non-negative integer), we can rewrite the equation into $(4N^2+4N+4) \times C - 3 \times C$ while 3 groups of '0' are padded, each with a size of C. By doing so, we transform it into a $(4N^2+4N+4) \times C$ shape before equally splitting it into four segments and mapping them onto the corresponding PEs in the clusters.

Specifically, by taking a 3D $3 \times 3 \times C$ kernel as an example, we partition the transformed $12 \times C$ kernel into four cluster windows evenly. As the size of each partition does not exceed the input buffer partition size, we then map the kernel data onto the 4 input buffers (i.e., $IB_{C1} \sim IB_{C4}$), each corresponding to an input feature pixel in the window (Fig. 1(b)). As shown in Fig. 1(c), after all the kernels have been loaded into the PEs, the input activation patch A_j will also be transformed, following the same way as W_i. The input buffer is reused to load the input activation and broadcast them to all the PEs within one cluster. Concurrently, the bias is loaded to each cluster to compensate for the transformation loss (i.e., pad $3 \times C$ with '0' to Psums). Thereafter, the PEs are enabled to perform convolution operations and generate 4 groups of 64 independent partial sums in parallel (i.e., Psum1 $\sim$ Psum4) from the clusters of each 64 PEs. At last, the accumulation unit sums up all the partial sums to produce 64 total sums (i.e., Fsums). On the other hand, for a 3D even-sized kernel W_i (e.g., $4 \times 4 \times C$), it can be partitioned into four segments and mapped onto four clusters directly.

III. Local Computing in PE

A. Compact 10-Transistor Latch-XOR Computing Cell

Convolution is the most energy-exhausting operation in neural network computation [16]. A convolution consists of multiple weight-input multiplications, followed by an accumulation. In BNN, due to its single-bit precision, we can implement the bitwise multiplication efficiently with an inverted-XOR gate. Besides, the accumulation can be realized by popcount logic which counts the number of '0' of the XOR output. Existing BNN accelerators such as [3], [4] implement this binary convolution with separate latches/flip-flops and XOR/XNOR gates, which can be seen in Fig. 2(a). However, this approach results in significant energy overhead due to frequent data movement between the local memory unit and the compute unit.

To improve the overall energy efficiency of the accelerator, we propose a tightly coupled 10T Latch-XOR cell (Fig. 2(b)) in which the computation is in-situ with the data storage for local computing. It comprises a transmission gate, a cross-coupled latch, a two-transistor (i.e., M1, M2) switching path, and an inverter. During the weight loading, the transmission gate is turned on to write the weight to the latch by asserting XOR_EN to high and sending the value to the cell via the IN signal. After all the weights have been programmed to the array, the transmission gates are turned off to isolate IN and the storage

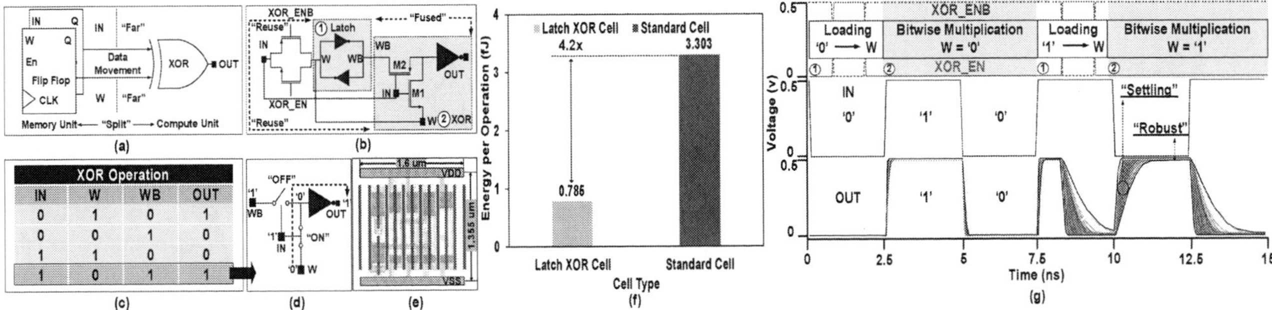

Fig. 2. (a) Standard cell-based separate implementation of weight storage and XOR logic. (b) Proposed 10T Latch XOR cell design. (c) The truth table of XOR operation. (d) Operating principle of (*Weight* ⊕ *Input*) logic using 2 transistors and 1 inverter. (e) The layout of the proposed 10T cell. (f) Energy gain of the proposed 10T cell with respect to the design in (a). (g) A waveform sample of 10,000-point Monte-Carlo simulation of the proposed 10-T cell at 0.5 V.

node of the latch. During inference, input data is sent to the cell to perform the XOR operation with two transistors M1, M2, and an inverter, as shown in Fig. 2(b). Note that the IN signal is reused to deliver both weight and data to the cell in a time-multiplexing manner for area-saving purposes.

The truth table (Fig. 2(c)) lists all the input patterns and the corresponding output values for an XOR operation. Specifically, when the weight value is logic '0' (i.e., W is '0' and WB is '1') and the input activation (IN) is logic '1', the NMOS transistor M1 is turned on whereas the PMOS transistor M2 is off. Consequently, the weight passes through M1 and gets inverted so that a logic '1' is generated at the output terminal (OUT) as illustrated in Fig. 2(d). Fig. 2(e) captures the layout of our 10T custom latch-XOR cell, occupying only $1.6\mu m \times 1.355\mu m$ in a 28nm CMOS technology, which is 5× smaller than the overall area of a standard XOR gate with a latch cell using the same technology. As shown in Fig. 2(f), our proposed 10T latch XOR cell only consumes 0.785 fJ energy per operation, 4.2× lower compared to the decoupled standard cell approach. We also verified the reliability of the proposed cell through 10,000 Monte-Carlo simulations with 3-sigma variations on top of the TT corner. Fig. 2(g) plots the timing diagram of the XOR operations as the truth table lists. The output inverter improves the driving capability of the output when a weak '0' passes, enabling the cell to be fully functional at 0.5V, 200MHz against process variations.

B. Popcount Unit PCL Design

Three design strategies have been investigated for the popcount unit implementation, which are Binary Adder Tree (BAT), Look-Up Table (LUT), and Parallel-counter-Carry-Look-ahead (PCL) circuitry. They are illustrated in Fig. 3(a), (b) and (c), respectively. BAT is a common technique for the addition of multiple input operands through a $\lceil N/2 \rceil$ layer reduction tree hierarchy, where N is the number of the operands. However, the computation depends on the height of the tree, leading to a long latency if the hierarchy is deep. LUT can store the numbers of '0's in the 8-bit data patterns without the need to count the bits. For instance, it outputs 8 if the input data pattern is '00000000'. When the data pattern is longer than 8-bit, the logic will compare the first 8-bit section followed by the next and repeat it till the last section is processed. An accumulator sums up the outputs from all the sections before transforming the counting result into a total sum. However, the LUT approach incurs more power and delay compared to the BAT counterpart, causing a higher Energy-Delay-Area (EDA) product.

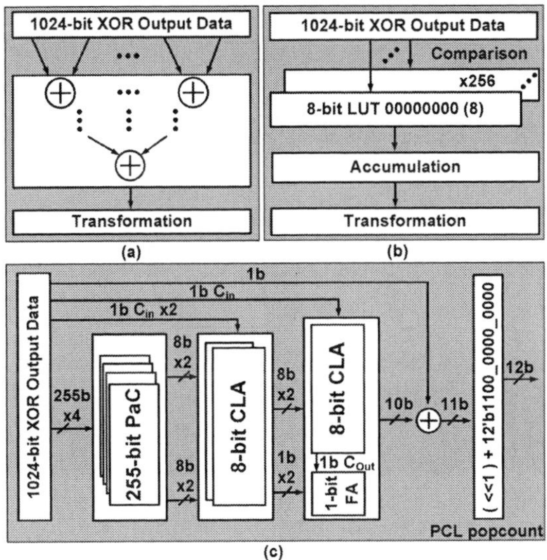

Fig. 3. Block diagram of (a) Binary adder tree (b) Popcount logic using LUT approach and (c) Proposed popcount design using PCL unit.

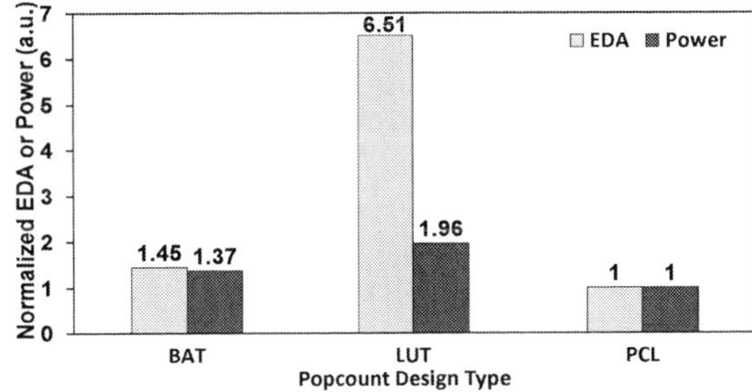

Fig. 4. Popcount design analysis of normalized EDA product and power values of proposed PCL unit vs. conventional BAT and LUT design at 0.9V.

As Fig. 3(c) illustrates, the PCL unit is a hybrid combinational circuitry of the parallel counter (PaC) and Carry

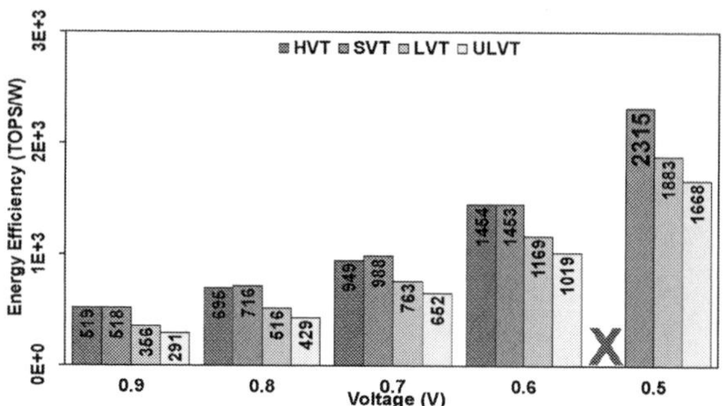

Fig. 5. Energy efficiency of the PE evaluated by sweeping the voltage from 0.9 V down to the minimum voltage required for successful operations at 200 MHz. Note that HVT devices fail at this operating condition.

Look-Ahead adder (CLA) to accomplish the counting and transforming functions. Specifically, the PaC works as a large parallel counter to compress the output of the 1020-bit XOR operation into four 8-bit data so that the two pairs of data packets can be sent to the two CLAs for addition. Thereafter, the results are processed by the third CLA with 1-bit carry-in, generating 10-bit data before being added with the last 1-bit data of the XOR output. Finally, the 11-bit data undergoes a shift-and-add operation to be transformed into the exact 12-bit result by left-shifting one bit and adding with 1100_0000_0000, which is the 2's complement of 1024 in binary.

To achieve higher energy efficiency, we compare the three techniques for the 1024-bit popcount operation on Energy-Delay-Area (EDA) product and power consumption. Fig. 4 reveals that our proposed PCL scheme outperforms the counterparts by 6.51× and 1.45× on EDA, respectively. For power consumption, it can reduce power by a factor of 1.96 and 1.37 compared to them, respectively.

To optimize the energy efficiency of the PE for binary convolution, we have investigated the implementation of the Latch-XOR array using four distinct threshold voltage (i.e., V_{th}) devices and varied the supply voltage from the nominal voltage (i.e., 0.9 V) to the minimum voltage that supports a frequency of 200MHz. As Fig. 5 shows, the Standard V_{th} (SVT) array is fully functional at 0.5 V with the specified frequency and achieves the highest energy efficiency of 2315 $TOPS/W$ among all the device types. Note that the High V_{th} (HVT) macro fails at the operating corner. Consequently, we implement the array with SVT devices for optimal energy efficiency.

IV. Design Analysis with LAXOR Simulator

We design a Python-based simulator for the proposed LAXOR accelerator to (i) map and verify the functionality of a BNN model onto the proposed architecture and (ii) generate application-specific, cycle-accurate results (e.g., latency, energy, utilization, etc.) for design analysis. Apart from that, the simulator supports a variety of BNN models and architectural parameters, ensuring a framework to adapt to a wide range of workloads and hardware configurations for scalability. Specifically, the simulator can emulate all the computation and dataflow in convolution (CONV), max-pooling (MP), and FC layers.

Fig. 6 depicts the overview of the LAXOR simulator, which consists of a front-end tool, Areca, and a back-end tool, Bits-Island. Areca interfaces with the pre-trained model and user configurations before generating the data stream in a format tailored to the accelerator. Bits-Island replicates the LAXOR architecture, maps the data stream onto different PEs, and simulates the functionality layer by layer. Eventually, the toolchain reports the mapping results, layer output, and other critical design metrics. To ensure accurate estimation, latency and energy per atomic operation (e.g., 2-input XOR, 1-bit weight loading) are provided after post-layout simulation as in [17].

Previous DNN energy models [9], [10] emphasize layer-wise dynamic energy but rarely take leakage energy into account. As leakage energy could dominate in low voltage operations, we rectify this and incorporate the leakage energy in our model. As Equation (1) shows, the total energy consumption E_{total} consists of the computation energy E_{cp}, the data movement energy E_{dm}, and the leakage energy E_L. The computations are decomposed into different types of atomic operations so that E_{cp} can be calculated as a sum of the contributions from all the operation categories (e.g., XOR, popcount, 4-input OR, etc). Similarly, the evaluation of data movement energy E_{dm} is based upon the contributions from all dataflow, each is considered as the product of the data amount and its unit movement energy. Total leakage energy E_L is obtained by multiplying total leakage power with the latency of the entire application. This allows us to accurately capture the realistic energy consumption of LAXOR when simulating different networks, datasets and eventually to obtain the optimal design choice. It is worth noting that the simulator runs more than 8×10^6 faster than the Cadence Spectre tool when simulating a 1024b binary convolution, which significantly improves the efficiency of the architectural exploration.

$$E_{total} = \sum_{l \in Layers} (E^l_{cp} + E^l_{dm}) + E_L \quad (1)$$

LAXOR simulator supports a variety of BNN models and architectural parameters. Users can perform a design analysis to profile the impact of these parameters at the architecture level. For instance, we can configure the number of PEs, the

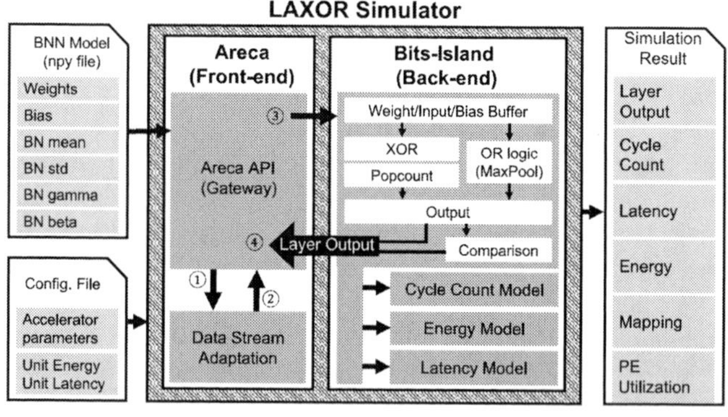

Fig. 6. High-level overview of the LAXOR Simulator.

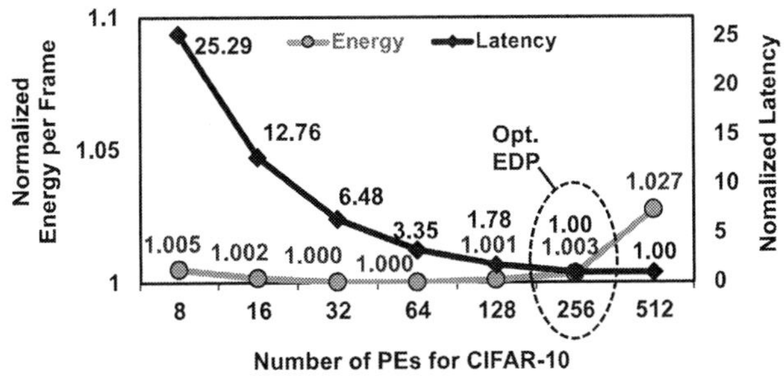

Fig. 7. Design space exploration regarding the number of PEs with 1024 Latch-XOR cells per one PE and 1024-bit buffer with respect to [3].

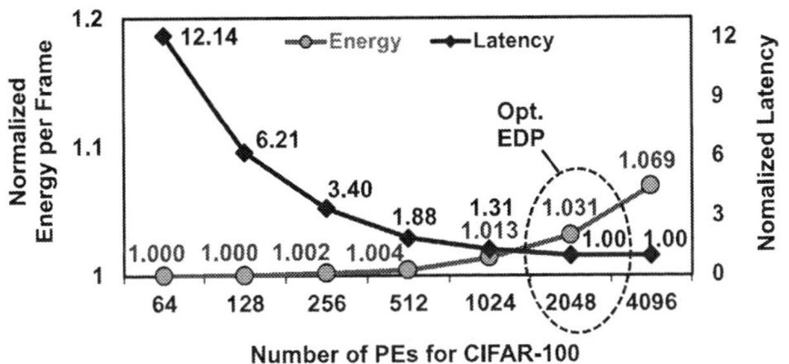

Fig. 8. Design space exploration regarding the number of PEs with 1024 Latch-XOR cells per one PE and 1024-bit buffer with respect to CIFAR-100.

number of Latch-XOR per PE, buffer size, etc. to investigate the performance and energy of the architecture with a given benchmark. Hereafter, we utilize CIFAR-10 as our driving application to demonstrate the design analysis process. The model consists of 8 CONV (256 filters, 2×2 kernel), 2 MP (2×2 kernel), and 1 FC layer. Firstly, we decide the number of PE to deploy in the architecture. We start with 1024 Latch-XOR cells per PE as this size can adequately fit one (2×2×256) kernel in the model. Fig. 7 shows the normalized energy per inference and the latency with different numbers of PEs. With 256 PEs, the mapped application can achieve the lowest Energy-Delay Product (EDP). This is because the 256 PEs configuration provides the highest concurrency to compute all the filters while avoiding leakage overhead from idle ones compared to using 512 PEs. The simulator can search for the optimal configuration subject to network topologies and workloads. Fig. 8 exhibits the normalized energy and latency when running a CIFAR-100 task. The optimal EDP is obtained with 2048 PEs where each PE is configured with 1024 10T cells. Regardless of network models, we adopt a 1024-bit global buffer in our architecture and divide the PEs into 4 clusters due to layout considerations. Our source code for LAXOR simulator is available at https://github.com/tomomasayamasaki/LAXOR.

V. Experimental Results

We have implemented various BNN workloads on the LAXOR architecture with the assistance of the in-house developed simulator. The accelerator buffers a 9-bit bias for each filter and accommodates 256 4-input OR logic for 2×2 max pooling operation. Each parameter is set according to the optimal design analysis of the CIFAR-10 workload which is our driving application (see Section IV). As Table I shows, we utilize a four-layer FC network with 1024 neurons per layer for the MNIST inference task and achieve 98.31% accuracy. For the fashion-MNIST workload, the BNN model has more FC layers and can achieve an accuracy of 87.58%. We also implement two BNN models for CIFAR-10 workloads and CIFAR-100 workloads, respectively. The BNN model for CIFAR-10 is from [3] and the BNN topology for CIFAR-100 is inspired by ResNet-18. Our experiment shows that the former can predict the result with an accuracy of 85.25% while the latter can achieve 65.21% and 86.89% for top-1 and top-5 accuracy, respectively.

After being mapped onto the LAXOR accelerator, the MNIST-MLP workload is estimated to consume a total energy of 0.042 μJ per classification with 0.4 μs latency while the Fashion-MNIST workload can consume an energy of 0.069 μJ per classification with 0.7 μs latency. The CIFAR-10 benchmark consumes 3.82 μJ per inference in 0.02 *ms* while the CIFAR-100 task dissipates 103.74 μJ per inference in 0.56 *ms*. Particularly, LAXOR has a 1.8× lower energy consumption compared to the estimated hand-designed digital implementation in [3] when running the CIFAR-10 workload with the same model. Table I lists the simulation output from the tool including accuracy, energy consumption, latency and PE utilization of various workloads. It is worth noting that the energy consumption does not necessarily increase with the model size as different computation types such as CONV or FC also plays an important role. The PE utilization ratios show that the PE clusters in the LAXOR accelerator are efficiently mapped thanks to our mapping scheme.

TABLE I
The Simulated Result of LAXOR

Dataset	MNIST	Fashion MNIST	CIFAR-10	CIFAR-100
Accuracy (%)	98.31	87.58	85.25	65.21 (top-1) 86.89 (top-5)
No. of CONV	0	0	8	15
No. of FC	4	6	1	1
No. of MP	0	0	2	1
Model size (*MB*)	0.79	1.29	0.51	7.45
Dynamic Energy (*uJ*/classification)	0.024	0.040	3.73	101.50
Leakage Energy (*uJ*/classification)	0.018	0.029	0.09	2.24
Total Energy (*uJ*/classification)	0.042	0.069	3.82	103.74
Cycle count	887	1469	4628	112787
Latency (*ms*)	0.004	0.007	0.02	0.56
PE utilization (%)	92.61	95.42	89.45	96.6

Table. II compares the proposed BNN accelerator with state-of-the-art implementations. LAXOR achieves an energy efficiency of 2315 $TOPS/W$ with respect to Multiply-Accumulate (MAC) operation, outperforming all the synthesized digital BNN accelerators in the table, i.e., [3], [13], [14]. In particular, the energy efficiency of LAXOR is 3.4× higher compared to [13] which was implemented with a more advanced technology node. As a mixed-signal design, [19] is energy efficient but

TABLE II
BENCHMARKING LAXOR WITH STATE-OF-THE-ART BNN ACCELERATORS

	ISSCC'22 [18]	ISSCC'18 [19]	CICC'18 [20]	ISCAS'18 [13]	TCAD'18 [14]	JSSC'19 [3]	**This Work**
CMOS Technology	28nm	28nm	28nm	22nm	65nm	28nm	**28nm**
Design Type	Compute-In-Memory	Mixed-Signal	Digital	Digital	Digital	Digital	**Digital**
Result Type	Silicon	Silicon	Silicon	Synthesis	Synthesis	Synthesis	**Synthesis**
VDD (V)	0.5-1.1	0.6-0.8	0.66-0.9	0.65	0.6, 1.2	0.6, 0.8	**0.5-0.9**
Bit Width	1	1	1	1	1	1	**1**
Frequency (MHz)	280M	1.5M-10M	1.5-48M	-	480M	10M	**200M**
Core Area (mm^2)	0.033 (Macro Area)	4.6	1.4	0.46	1.9	-	**2.73**
Performance ($TOPS$)	9.175-20.032 (0.9-1.1V)	0.072-0.478	0.09-2.8	91.12	1.5	0.478	**104.8**
Compute Density ($TOPS/mm^2$)	-	0.0157-0.1039	0.064-2	198.1	0.79	-	**38.388**
MAC Energy Efficiency ($TOPS/W$)	625 @0.5V* (100% toggle rate)	532-772 (0.8-0.6V)	145-230 (0.9-0.66V)	672.6 @0.65V	61.2 @0.6V	299 @0.8V	**2315 @0.5V (100% toggle rate)**
Bit Accurate	No	No	Yes	Yes	Yes	Yes	**Yes**

Note: * Estimated energy efficiency from the graph of the paper [18].

its computation precision can be limited by process, voltage and temperature variations as explained in Section I. CIM design in [18] shows an impressive energy efficiency in silicon with a 25% input toggle rate. However, the metric downgrades significantly when increasing the toggle rate to 100%, which is much lower than ours.

Due to approximate computing, its inference accuracy degrades compared to the exact arithmetic approach even after being compensated with additional approximate-aware training. On the contrary, LAXOR ensures bit-accurate, deterministic computation, maximally free from accuracy loss induced by hardware non-linearity and non-ideality.

VI. CONCLUSION

This paper introduces LAXOR, a flexible BNN accelerator with a novel local computing paradigm. Thanks to the proposed Latch-XOR logic and the architecture-circuit co-optimization, LAXOR achieves an energy efficiency of 2315 $TOPS/W$ at 0.5V, 200MHz for binary convolution operation, outperforming the digital state-of-the-art by 3.4×. Apart from that, we present a Python-based simulator that enables fast design analysis for optimal design points and flexible mapping for a variety of network layers and kernel sizes. With the hardware-software co-design methodology, we demonstrate a full-digital, bit-accurate BNN accelerator that is highly suited for AIoT applications.

REFERENCES

[1] I. Hubara *et al.*, "Binarized neural networks," *Advances in neural information processing systems*, vol. 29, 2016.

[2] M. Courbariaux *et al.*, "Binaryconnect: Training deep neural networks with binary weights during propagations," 2015.

[3] D. Bankman *et al.*, "An always-on 3.8 μJ/86% cifar-10 mixed-signal binary cnn processor with all memory on chip in 28-nm cmos," *IEEE Journal of Solid-State Circuits*, vol. 54, no. 1, pp. 158–172, 2019.

[4] P. C. Knag *et al.*, "A 617-TOPS/W all-digital binary neural network accelerator in 10-nm finfet cmos," *IEEE Journal of Solid-State Circuits*, vol. 56, no. 4, pp. 1082–1092, 2021.

[5] S. Yin *et al.*, "XNOR-SRAM: In-memory computing sram macro for binary/ternary deep neural networks," *IEEE Journal of Solid-State Circuits*, vol. 55, no. 6, pp. 1733–1743, 2020.

[6] W.-S. Khwa *et al.*, "A 65nm 4kb algorithm-dependent computing-in-memory sram unit-macro with 2.3ns and 55.8TOPS/W fully parallel product-sum operation for binary dnn edge processors," in *2018 IEEE International Solid-State Circuits Conference - (ISSCC)*, pp. 496–498, 2018.

[7] S. Xie *et al.*, "16.2 edram-cim: Compute-in-memory design with reconfigurable embedded-dynamic-memory array realizing adaptive data converters and charge-domain computing," in *2021 IEEE International Solid-State Circuits Conference (ISSCC)*, vol. 64, pp. 248–250, 2021.

[8] J. Lee *et al.*, "Unpu: An energy-efficient deep neural network accelerator with fully variable weight bit precision," *IEEE Journal of Solid-State Circuits*, vol. 54, no. 1, pp. 173–185, 2019.

[9] T.-J. Yang *et al.*, "A method to estimate the energy consumption of deep neural networks," in *2017 51st Asilomar Conference on Signals, Systems, and Computers*, pp. 1916–1920, 2017.

[10] Y. Zhao *et al.*, "Dnn-chip predictor: An analytical performance predictor for dnn accelerators with various dataflows and hardware architectures," in *ICASSP 2020 - 2020 IEEE International Conference on Acoustics, Speech and Signal Processing (ICASSP)*, pp. 1593–1597, 2020.

[11] A. Samajdar *et al.*, "Scale-sim: Systolic cnn accelerator simulator," in *https://arxiv.org/abs/1811.02883*, arXiv, 2018.

[12] Q. Tang *et al.*, "Hawis: Hardware-aware automated width search for accurate, energy-efficient and robust binary neural network on reram dot-product engine," in *2022 27th Asia and South Pacific Design Automation Conference (ASP-DAC)*, pp. 226–231, 2022.

[13] M. Rusci *et al.*, "Design automation for binarized neural networks: A quantum leap opportunity?," in *2018 IEEE International Symposium on Circuits and Systems (ISCAS)*, pp. 1–5, 2018.

[14] R. Andri *et al.*, "YodaNN: An architecture for ultralow power binary-weight cnn acceleration," *IEEE Transactions on Computer-Aided Design of Integrated Circuits and Systems*, vol. 37, no. 1, pp. 48–60, 2018.

[15] T. Simons *et al.*, "A review of binarized neural networks," *Electronics*, vol. 8, no. 6, 2019.

[16] P. Judd *et al.*, "Stripes: Bit-serial deep neural network computing," in *2016 49th Annual IEEE/ACM International Symposium on Microarchitecture (MICRO)*, pp. 1–12, 2016.

[17] B. Wang *et al.*, "Shenjing: A low power reconfigurable neuromorphic accelerator with partial-sum and spike networks-on-chip," in *2020 Design, Automation Test in Europe Conference Exhibition (DATE)*, pp. 240–245, 2020.

[18] D. Wang *et al.*, "Dimc: 2219TOPS/W 2569F2/b digital in-memory computing macro in 28nm based on approximate arithmetic hardware," in *2022 IEEE International Solid-State Circuits Conference (ISSCC)*, vol. 65, pp. 266–268, 2022.

[19] D. Bankman *et al.*, "An always-on 3.8 μJ/86% cifar-10 mixed-signal binary cnn processor with all memory on chip in 28nm cmos," in *2018 IEEE International Solid-State Circuits Conference - (ISSCC)*, pp. 222–224, 2018.

[20] B. Moons *et al.*, "Binareye: An always-on energy-accuracy-scalable binary cnn processor with all memory on chip in 28nm cmos," in *2018 IEEE Custom Integrated Circuits Conference (CICC)*, pp. 1–4, 2018.

AR-PIM: An Adaptive-Range Processing-in-Memory Architecture

Teyuh Chou[1], Fernando Garcia-Redondo[2,3], Paul Whatmough[2,4], and Zhengya Zhang[1]

[1]University of Michigan, Ann Arbor, [2]Arm Research, [3]Now with IMEC, [4]Now with Qualcomm AI Research

Abstract—**The crossbar-based processing-in-memory (PIM) architecture has garnered considerable attention for its potential in achieving high energy efficiency for deep neural networks (DNNs). The PIM hardware's accuracy depends heavily on the design and resolution of the analog-to-digital converters (ADCs). Regrettably, high-resolution ADCs tend to be costly and often dominate the overall energy and area of the PIM designs. We propose adaptive-range PIM (AR-PIM) architecture that enables the use of lower-resolution ADCs without sacrificing accuracy. This is achieved by leveraging sparsity in the weights and input activations and dynamically adjusting the number of input activations and distributing MAC operations across multiple cycles during runtime. We perform our evaluations using a commercial 7nm FinFET PDK and show that AR-PIM offers an appealing trade-off, delivering 1.7× higher energy efficiency and 4.3× better area benefits without losing accuracy. The latency overhead is modest, only 10% over a baseline PIM architecture.**

Index Terms—**Processing in memory, deep neural network.**

I. Introduction

Deep neural network (DNN) has enjoyed exceptional successes in many applications and it has become one primary workload for modern computing hardware. Many works have demonstrated substantial acceleration for DNN workloads using a large amount of compute resources and power by GPUs, CPUs, FPGAs, and digital ASICs [1]–[4]. In mobile and IoT devices, the energy budget is severely limited, which requires hardware solutions of higher energy efficiency [5]–[7] to enable DNN processing on these devices.

Crossbar-based processing in memory (PIM) architecture, also known as compute in memory, is a promising candidate for DNN inference computation to improve performance and energy efficiency. In essence, PIM removes the memory wall by eliminating data movement between memory units and processing units and by performing multiply-accumulate (MAC) operations at the cross-point locations. However, PIM architecture requires complex analog circuits that must be carefully designed or else they can dominate the area and energy of the entire design [8]. These include digital-to-analog converters (DACs) for the digital inputs and analog-to-digital converters (ADCs) for digitizing the outputs between layers. The complexity of ADC and DAC grow exponentially with their resolution. Lowering the resolution of ADC and DAC is desirable for area- and energy-efficient solutions. However, an insufficient resolution generally results in accuracy degradation. Without further network engineering, the accuracy can drop dramatically for sub-8b ADC resolution.

SRAM and NVM have been used as PIM memory devices. Nonvolatile memory (NVM) devices such as resistive RAM (RRAM), magnetoresistive RAM (MRAM), phase-change RAM (PCRAM), ferroelectric RAM (FeRAM), are suitable for PIM thanks to their nonvolatility, low standby power, and high density. However, the commercially available NVM devices are still at 22nm [9]–[12], lagging the scaling of logic devices. The process, voltage, temperature (PVT) variations of the NVM devices also require a diligent control. Although both SRAM and NVM suffer from variability, SRAM has no drift issues, infinite endurance, and always leads technology scaling. Comparing a 7nm SRAM to a 22nm RRAM or MRAM, a 7nm SRAM is a better candidate for PIM because of its energy efficiency if nonvolatility is not of concern.

In this work, we investigate PIM based on a 7nm SRAM and explore practical analog PIM design choices. The investigation points to a promising direction of utilizing data sparsity in an adaptive design to relax the high-resolution ADC requirements without accuracy loss. The contributions of this work are summarized below:

1) We provide an analysis of the practical design choices for 7nm SRAM PIM, accounting for the noise and non-idealities derived from the intrinsic nature of the SRAM cell and analog accumulation in the bitline.
2) We present runtime range detection and adaptive-range PIM (AR-PIM) to achieve high accuracy with minimal latency overhead even using a low-resolution ADC.
3) We benchmark AR-PIM against a baseline on multiple DNN workloads using MNIST dataset and ImageNet dataset, showing the energy efficiency and area improvement at minimal latency increment.

II. Preliminary and Related Work

PIM architectures can reduce data movement by adopting a weight-stationary approach. The weights are stored in a memory array and the input activations are passed to the array to perform computation. The weights are stored in a bit-parallel way across columns as in [13]–[15]. In computation, a vector of inputs is driven, one per wordline (WL). The cells along a column are turned on, and the currents are summed on the bitline (BL), accomplishing the dot product between the input vector and the weight vector stored on the column of the

979-8-3503-1176-1/23 $31.00 © 2023 IEEE

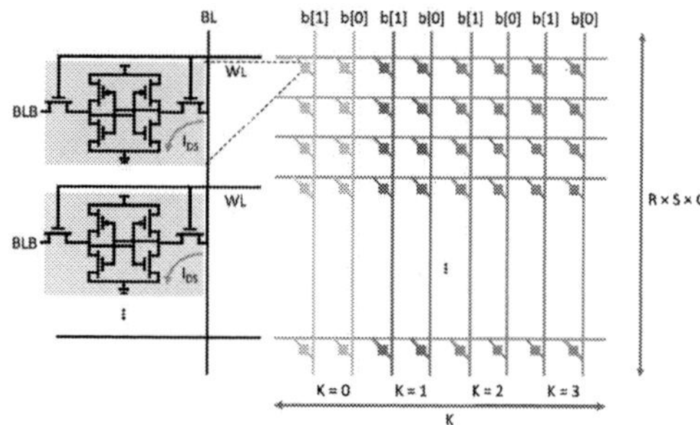

Fig. 1: The PIM mapping of the convolution operation where $R \times S$ is the kernel size, C is the input channel and K is the output channel. The 2b weight in the example is stored in two columns. The zoom-in view shows the 6T SRAM and current I_{DS} discharged by each bitcell. Both baseline PIM and AR-PIM adopt this mapping for convolution operation.

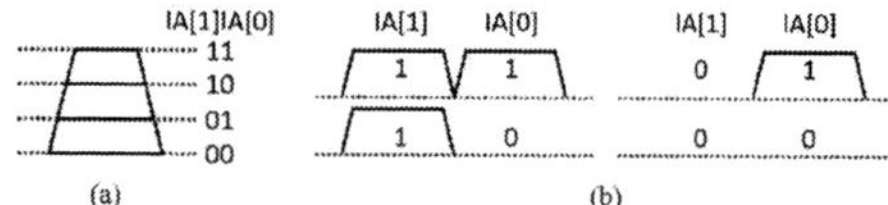

Fig. 2: The multi-bit (2b in the example) input representation of (a) WL pulse amplitude and (b) WL pulse train.

memory array. Across the columns, dot products are conducted in parallel, realizing vector-matrix multiplication (VMM).

An SRAM bitcell stores a value and its complement. When its WL turns on, the stored value drives BL and the complement drives BLB. In SRAM-based PIM, either BL or BLB can be taken as the output (Fig. 1). In this work, BL is taken as the output. The key design parameters are considered below.

Array Size. Using a larger array, more cells are activated in parallel, achieving higher performance; and the row and column peripheral circuits are amortized more effectively, leading to a higher compute density. A drawback of a larger array is the lower utilization in mapping smaller VMMs, leaving unused cells. A larger array also presents higher capacitive loading on WL and BL, resulting in a longer delay. Finally, a larger array implies the potential activation of more cells contributing current to the same BL, and thus the accumulation of more noise that may degrade the signal-to-noise ratio (SNR).

Input Encoding. The input activations can be encoded in two forms as shown in Fig. 2: (a) pulse amplitude or (b) pulse train, i.e., each bit of the multi-bit input is represented by a 1b pulse. The pulse train is more linear compared to pulse amplitude or width, and it can be better controlled [16], but the latency increases with the bitwidth. Past designs have combined pulse amplitude and pulse train [8], [17], [18].

BL Resolution. The BL resolution depends on the WL resolution (b_{WL}), the memory cell resolution, and the maximum allowable activated memory cells (N_{cells}) in a column. Since SRAM is a digital (1b) memory, the BL resolution is $b_{BL} = b_{WL} + \log_2 N_{cells}$. A higher BL resolution requires a higher-resolution ADC, which in turn significantly impacts power and area [8], [19]. A higher BL resolution also exacerbates PIM's variation and reduces the capability of error tolerance

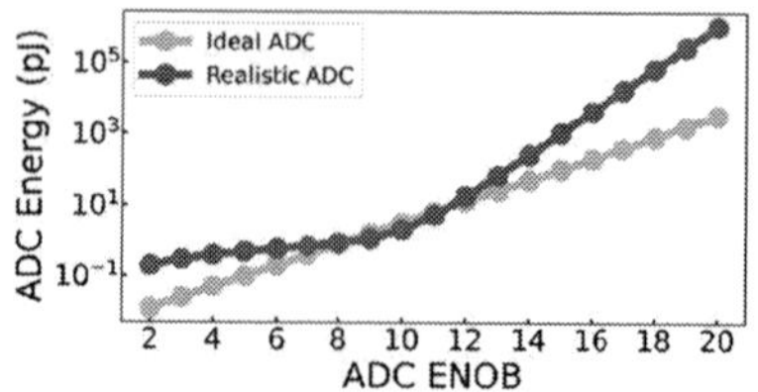

Fig. 3: ADC energy consumption with effective number of bits (ENOB) from [19].

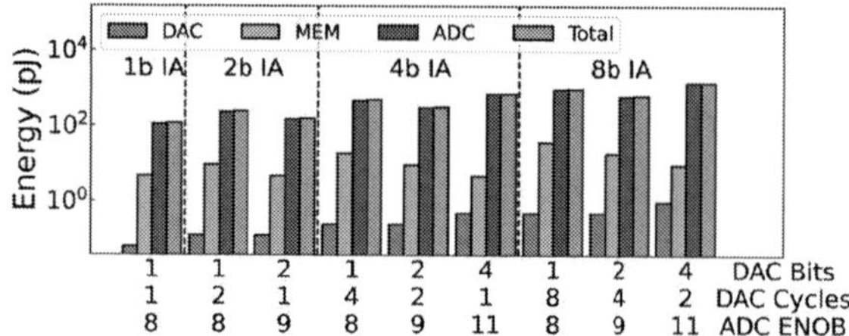

Fig. 4: Energy efficiency of 1b/2b/4b/8b input activations. The input encoding for each configuration is implemented by DAC bits and DAC cycles with the corresponding ADC resolution.

[20]. Recently, an all-digital PIM design [21] is proposed, the power and area of peripheral circuits increase with the BL resolution.

ADC Sharing. The ADC area can be significantly larger compared to the SRAM bitcell pitch. Placing one ADC per BL is difficult due to the physical layout constraint. Since the conversion time of ADC can be much shorter than the time it takes for the SRAM BL current to develop, sharing an ADC between BLs becomes a necessity, e.g., 1 ADC is shared by 4 BLs in [16]. ADC sharing requires extra circuits such as sample-and-hold circuit [8] or weighted capacitors [16] to store the BL value before the conversion starts.

III. PIM Design Considerations and Challenges

The following evaluations are based on the SPICE simulation of a 128×128 SRAM array in 7nm FinFET technology. The ADC energy is extracted from [19] and shown in Fig. 3. DACs are adopted from [22] and most circuit component models are adopted from [8]. The array size of 128 is chosen to obtain high utilization for DNN workloads as in [23].

A. Input Encoding

The energy with various input encoding choices is investigated. If the input activations are 1b, they can be encoded in 1b WL pulses. Since a BL is connected to 128 bitcells, the BL resolution is $b_{BL} = 8$. An 8b ADC consumes 96% of the total energy (Fig. 4), significantly higher than the energy of the memory access or the 1b DAC. For a 2b input activation, two input encoding options are available: 1b pulses over 2 cycles (with partial sums scaled and added digitally post-ADC) or a single-cycle 2b pulse, resulting in 8b and 9b BL resolution, respectively. A 9b ADC consumes 35% more energy than an 8b ADC (Fig. 3). Hence, two 8b analog-to-digital (A/D) conversions with DAC bits = 1 and DAC cycles = 2 result in 49% higher energy than one 9b A/D conversion with DAC

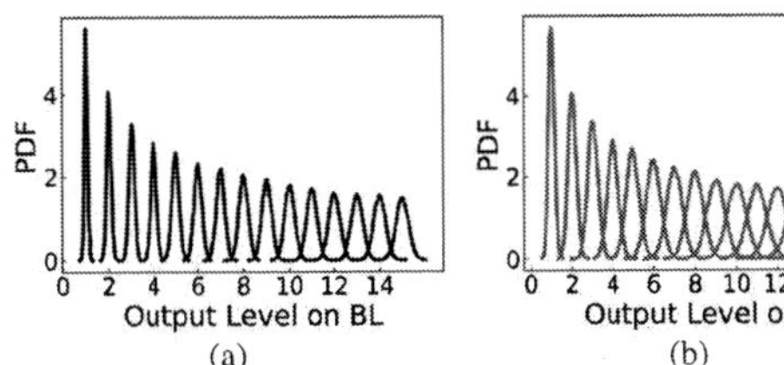

Fig. 5: SRAM BL current levels from the SPICE simulation for WL voltage of (a) 0.8V and (b) 0.6V. Each Gaussian distribution represents one output level. The less overlap between two distributions means the better sensing margin for distinguishing two output levels.

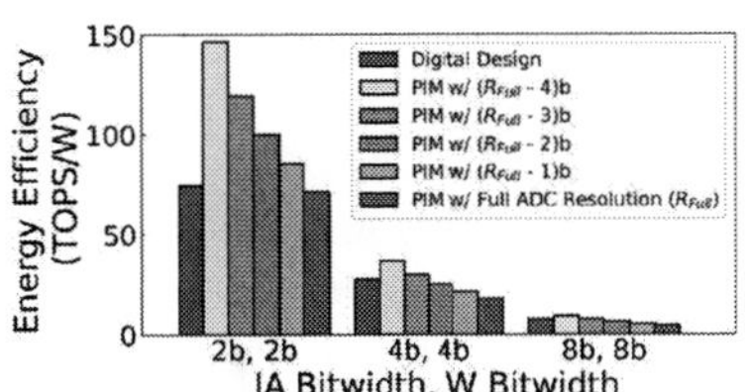

Fig. 6: The energy efficiency comparison of synthesized digital design and PIM designs with various ADC resolution settings and 3 Input Activation (IA) and Weight (W) bitwidth combinations.

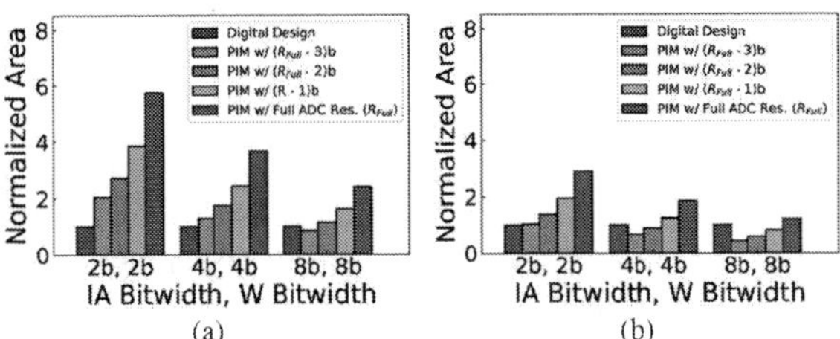

Fig. 7: The area comparison for synthesized digital design with the SIMD architecture and PIM designs with various ADC resolution settings for (a) 1 ADC per BL and (b) 1 ADC shared by 2 BLs.

bits = 2 and DAC cycles = 1, so the single-cycle 2b pulse encoding is preferable in terms of energy (Fig. 4).

Moving to 4b and 8b input activations, more input encoding options are available. The 2b-pulse encoding was found to be the sweet spot in energy consumption. However, regardless of the input encoding choice, the ADC dominates the energy consumption. A WL resolution of higher than 2b is not practical due to the significant escalation of ADC energy.

B. BL Current Levels under PVT Variations

If a bitcell storing a 1 is activated, the bitcell discharges one unit of current from BL. However, process variations complicate the picture. As more discharging bitcells on the same BL are activated, the distribution of current gets wider as in Fig. 5(a). The wide distribution makes it challenging to decode as few as 16 current levels. This insight suggests that the degradation of SNR due to process variations may make using a very high-resolution ADC inconsequential.

As the WL voltage level is reduced, e.g., in supporting WL pulse-amplitude input encoding, the current level boundaries are further obscured as seen in Fig. 5(b).

C. ADC Resolution and ADC Sharing

For this investigation, a reference SRAM-based PIM design is adopted: a 128×128 SRAM array; the input is provided bit serially; WL is encoded in 1b pulses. For example, an 8b input activation is passed to WL by a 1b DAC in 8 cycles. Therefore, the full resolution required at each BL is 8b. An 8b weight is stored in 8 SRAM bitcells in a row. A weight-stationary digital design was synthesized using an array of multiply-accumulates (MACs) with weight storage to mimic PIM. The partial sums are accumulated along a column of MACs. The digital design also follows the same bitwidth as the PIM design for comparison at the MAC level.

The energy efficiency of the PIM design is compared to the digital SIMD architecture based on [24] (with matched BL bitwidth to the PIM design for a fair comparison) in Fig. 6, assuming one ADC per BL. PIM achieves the best energy when the input activation and weight bitwidth are low. Also, note that PIM with an ADC that supports the full BL resolution ($R_{Full} = b_{BL}$) fares worse than the digital design. For the energy of PIM to be competitive, the ADC resolution needs to be reduced to 3b or 4b below the full BL resolution.

The area of PIM is compared to the synthesized digital design in Fig. 7 for one ADC per BL. The ADC area is based on existing IPs and extrapolations. Due to the relatively large area of ADC, especially a high-resolution ADC, the area of PIM easily exceeds the digital design by up to 4.3×. Even with the short bitwidth of 2b input activation and 2b weight, the full BL resolution still requires the use of relatively high-resolution ADCs that consume a large area. When the ADC resolution is reduced to 3b below the full BL resolution ($R_{Full} - 3$), the area becomes comparable. Fig. 7(b) shows the configurations of one ADC shared by 2 BLs to reduce the PIM area. The results highlight the importance of reducing ADC resolution and increasing ADC sharing to keep the PIM area competitive.

D. Necessity to Control Resolution

The above sections highlight the challenges behind a high BL resolution and its feasibility due to the sensing margin. Reducing the ADC resolution is a must to make PIM more competitive in energy efficiency and area.

To control the BL resolution, WL resolution can be reduced by employing 1b-pulse encoding over multiple cycles and activating only a subset of rows at a time such as in [25]. However, these approaches requires more cycles to complete the computation. To address this operation limitation we proposed AR-PIM, described in Section IV.

IV. Adaptive Range-PIM (AR-PIM) Architecture

AR-PIM leverages data sparsity by controlling the BL range at runtime. This approach can prevent sacrificing the inference accuracy incurred by direct truncation on BL values with the reduced ADC resolution. For a bitcell to contribute to the BL current and increase the BL range, the bitcell needs to store a 1 and it needs to be activated. This implies that both the weight value and the input activation value are 1 (at the bit

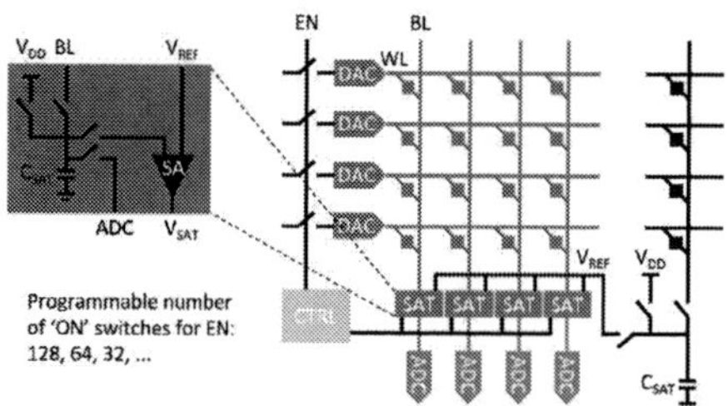

Fig. 8: AR-PIM architecture with BL saturation detection circuitry (SAT). The light-weight SAs and control with programmable activated number of WLs enable runtime range detection.

level). If either the weight value or the input activation value is 0, the bitcell does not contribute to the BL current or the BL range. Therefore, the ADC resolution quoted in the previous sections is the maximum resolution, while the effective BL range can be lower with the bit-level sparsity.

The presence of zeros in weights and input activations is referred to as sparsity. Sparsity exists even in unpruned models, especially with the rectified linear unit (ReLU) activation function that generates zero activations. Sparsity can be further increased using pruning algorithms to remove weights. In addition to word-level sparsity, plenty of bit-level sparsity exists in weights and input activations as identified by [26].

In DNN inference, a model is given and the weight sparsity is static. However, the activation sparsity is dynamic, namely input-dependent, and determined in runtime. As a result, when the computation of DNN inference is mapped to PIM, the BL range can vary due to the dynamic activation sparsity. We propose a technique to detect the runtime sparsity (or density) for the computation of DNN inference. If the density is low, an energy-inexpensive low-resolution ADC can be used; and if the density is high, the BL range can be adjusted by activating only a portion of the bitcells.

A. Runtime Range Detection

The runtime range detection can be implemented by reusing the SRAM sense amplifier (SA) in the readout circuitry and a reference column as shown in Fig. 8. The reference column stores a preset number of 1s to correspond to a given density level, e.g., 25% of the reference column storing 1 to represent a density of 25%. Prior to the detection, a readout from the reference column is performed by applying unit pulses on all WLs. The reference column's BL current is integrated on a sampling capacitor as the threshold voltage.

The range detection is done by an SRAM readout. The BL current is integrated on a sampling capacitor to be the BL voltage. The SA compares the BL voltage to the threshold voltage generated by the reference column. If the BL voltage is below the threshold voltage, the value of the column is higher than the reference value, and the SA sets $EN = 1$ to the controller.

The controller checks all BLs' SA outputs. If an SA signals $EN = 1$, the controller activates only 50% of the WLs and another round of SRAM readout follows for only a small number of saturated columns. In the next round, if one

TABLE I: Array utilization, mean of BL value, and standard deviation of BL value. Note that the maximum of BL values is 128 for a 128×128 array.

Model	LeNet		AlexNet				
Dataset	MNIST		ImageNet				
Layer	CONV1	CONV2	CONV1	CONV2	CONV3	CONV4	CONV5
Array Utilization	2.64%	42.2%	94.5%	96.2%	96.4%	100%	100%
BL Value Mean	0.383	4.630	27.957	14.478	6.932	2.945	2.764
BL Value Std	1.038	4.277	17.429	12.804	6.123	2.977	2.999

Model	VGG11							
Dataset	ImageNet							
Layer	CONV1	CONV2	CONV3	CONV4	CONV5	CONV6	CONV7	CONV8
Array Utilization	21.1%	90%	100%	100%	100%	100%	100%	100%
BL Value Mean	6.420	9.030	9.906	5.907	6.798	3.905	3.463	2.925
BL Value Std	4.250	9.604	8.650	5.468	5.710	3.750	3.420	3.110

$EN = 1$, the controller activates just 25% of the WLs with the columns at which the BL value is above the reference value in subsequent SRAM readouts. The process continues until the BL value is reduced to the reference level or below, thereby controlling the BL range.

Fig. 9 illustrates BL range control. Assume that prior to the detection, the threshold voltage is set to represent a 25% density. In the example, in cycle 0, the first and the second BL density exceed the 25% threshold, and the controller only activates 50% of the rows in the subsequent cycle 1 and cycle 2. In cycle 1, the first BL density still exceeds the threshold, and the controller activates only 25% of the rows in the subsequent cycle 3 and cycle 4. By actively limiting the density below 25%, the effective BL resolution is reduced by 2b. The proposed range control adapts to the effective BL resolution by activating more or fewer bitcells, thus we call it adaptive-range PIM or AR-PIM.

B. Energy Minimization

To reduce the ADC resolution and improve the sensing margin, low-resolution ADCs are necessary. When adopting low-resolution ADCs, only a portion of all the rows in the array at a time can be activated and the BL values are read out sequentially. Therefore, it may result in more processing cycles, which in turn costs more energy and a longer latency. To amortize this overhead, AR-PIM exploits the lower effectual BL range in runtime originating from data density levels of input activations and weights.

The lowest energy is investigated by sweeping the input activation and weight density each from 5% to 75%. If the ADC resolution is set based on the effective BL range (using the IA/W density levels as the proxy indicator), the number of processing cycles and energy consumption can be minimized. Fig. 10 shows the result of choosing the appropriate ADC resolution represented as ENOB for optimal energy consumption.

V. Evaluation of Energy and Performance

The energy consumption of the AR-PIM architecture is evaluated using DNN workloads based on the bit-level sparsity of activations and weights. The DNN workloads include LeNet with MNIST dataset as well as AlexNet and VGG11 with ImageNet dataset. The energy of AR-PIM is highly dependent on two factors, the runtime data sparsity and the ADC resolution.

For LeNet with MNIST dataset, both activations and weights are quantized to 8b for evaluations. Table I shows

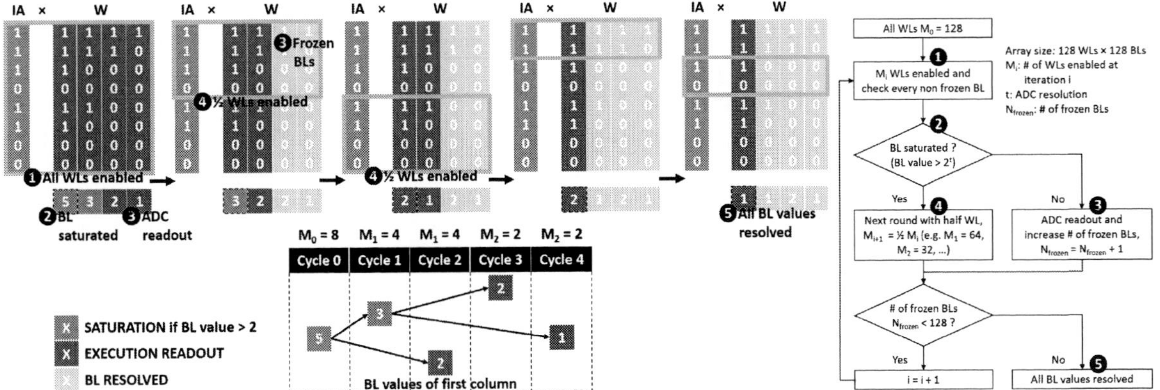
Fig. 9: Runtime range detection and adaptation of a simple 8×4 array example and the general flowchart for a 128×128 array.

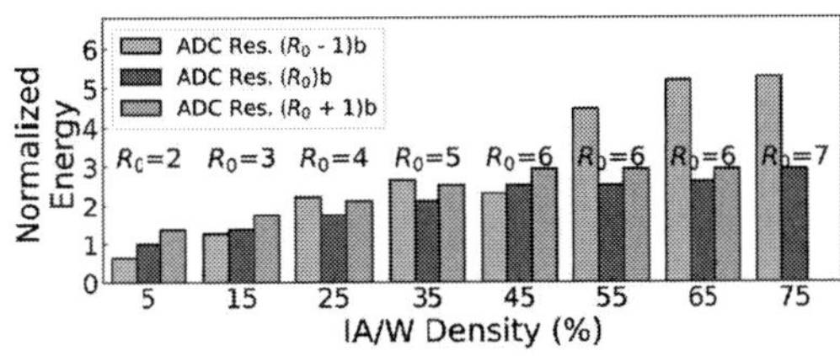
Fig. 10: Energy of AR-PIM for IA/W density each from 5% to 75% with conditions of ADC resolution R_0 - 1, R_0, R_0 + 1, where R_0 is the nominal value of ADC resolution settings depending on the product of array size, IA density, and W density.

the average BL values when running inference in the first and the second convolutional layers. Each layer is mapped to a 128×128 PIM module [8] as a baseline. In both layers, the average BL densities stay below 10%, making AR-PIM suitable as the effective BL range is low and consistent between columns.

Fig. 11(a) shows the normalized energy with different ADC resolution settings for the first and the second convolutional layers of LeNet. In the first convolutional layer, a lower ADC resolution reduces the energy consumption. The low utilization of cells along each column leads to a consistently low effective BL range. The low average BL value and the narrow distribution (Table I) allow the setting of a low ADC resolution to aggressively reduce the BL resolution to save the most energy.

In the second convolutional layer, a different trend is observed: when the ADC resolution is too low, the energy consumption increases. Different from the first layer, the utilization in the second layer of mapping is higher. The higher utilization results in a broader BL value distribution (Table I). The lowest ADC resolution could result in a large number of extra cycles and more energy. The energy-optimal ADC resolution can be set to capture most of the BLs, leaving only a small number of extra cycles to capture the remaining BLs.

Fig. 11(b) shows the latency implications of different ADC resolution settings. Generally speaking, the lower the ADC resolution, the higher the latency. The energy-optimal points tend to be low-resolution points where the latency does not increase excessively.

Fig. 11(c) and Fig. 11(d) show the normalized energy and latency of AR-PIM with different ADC resolution settings for the first, middle, and last convolutional layers in AlexNet with ImageNet dataset. The activations and weights are quantized to 16b and 12b in evaluations while maintaining the inference accuracy. The mean of BL values decreases and the distribution gets narrower in deep layers as in Table I.

Similar behavior can be observed in VGG11 as in Fig. 11(e) and Fig. 11(f). Fig. 12 shows the accuracy and energy tradeoff between with and without AR-PIM. The accuracy can be recovered in lower-resolution ADCs with minimal energy increase. Across different DNN workloads, AR-PIM can minimize the energy consumption while maintaining the inference accuracy over the baseline PIM with 7b ADC resolution. As a result, AR-PIM improves the energy efficiency over the baseline PIM. By using AR-PIM, latency-constrained applications can achieve up to 1.7× higher energy efficiency.

VI. Conclusion

This work explores the design boundary for analog PIM using SRAM in a 7nm process. The input encoding, BL range, ADC resolution, and ADC sharing are studied to analyze their impacts on the energy efficiency and the area cost of analog PIM. From the analyses, we conclude that low-bitwidth quantized NNs are more suitable to be deployed on analog PIM to save energy on power-constrained mobile devices. Addressing the challenges behind the deployment of multi-bit matrices in analog accelerators, AR-PIM is presented with a runtime BL range detection mechanism to adapt to a lower effective BL range.

AR-PIM eliminates the need for high-resolution ADCs and reduces the energy consumption of the ADCs. By adapting to the lower effectual BL range, AR-PIM also enhances the variation tolerance and the sensing margin. Considering the energy gain and latency overhead together, our evaluations show that AR-PIM provides 1.7× higher energy efficiency over the baseline PIM with 4.3× area reduction while maintaining the inference accuracy.

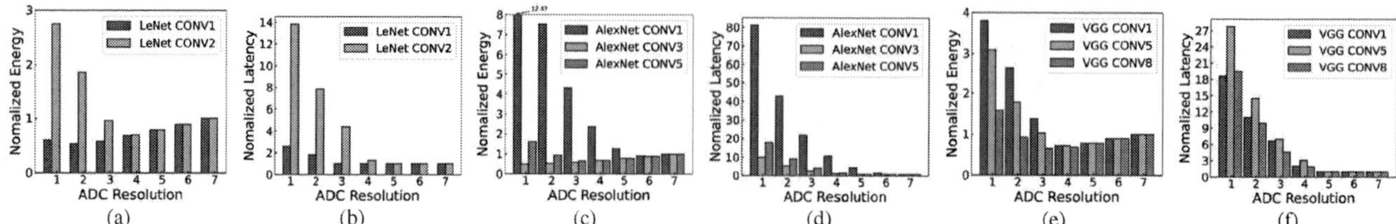

Fig. 11: Energy consumption and latency of AR-PIM compared and normalized to the baseline PIM with 7b ADC resolution (the rightmost bar in each figure) for each layer running (a)(b) LeNet using MNIST dataset, (c)(d) AlexNet using ImageNet dataset, and (e)(f) VGG11 using ImageNet dataset.

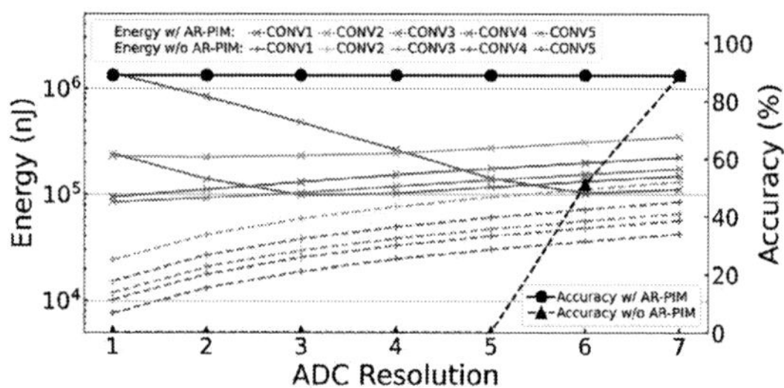

Fig. 12: Accuracy and energy trade-off with ADC resolution settings.

ACKNOWLEDGMENT

We thank Mudit Bhargava for advice and discussion. The work at the University of Michigan was supported in part by NSF CCF-1900675.

REFERENCES

[1] D. Ciregan, U. Meier *et al.*, "Multi-column deep neural networks for image classification," in *IEEE Conference on Computer Vision and Pattern Recognition (CVPR)*, 2012, pp. 3642–3649.

[2] "NVIDIA Grace CPU." [Online]. Available: https://www.nvidia.com/en-us/data-center/grace-cpu/, 2021.

[3] J. Fowers, K. Ovtcharov *et al.*, "A configurable cloud-scale DNN processor for real-time AI," in *Proceedings of the International Symposium on Computer Architecture*, 2018, pp. 1–14.

[4] N. P. Jouppi, C. Young *et al.*, "In-datacenter performance analysis of a tensor processing unit," in *Proceedings of the International Symposium on Computer Architecture*, 2017, pp. 1–12.

[5] T. Chen, Z. Du *et al.*, "DianNao: A small-footprint high-throughput accelerator for ubiquitous machine-learning," in *Proceedings of the International Conference on Architectural Support for Programming Languages and Operating Systems*, 2014, pp. 269–284.

[6] Y.-H. Chen, J. Emer *et al.*, "Eyeriss: A spatial architecture for energy-efficient dataflow for convolutional neural networks," in *Proceedings of the International Symposium on Computer Architecture*, 2016, pp. 367–379.

[7] Y. Chen, T. Luo *et al.*, "DaDianNao: A Machine-Learning Supercomputer," in *Proceedings of the IEEE/ACM International Symposium on Microarchitecture*, 2014, pp. 609–622.

[8] A. Shafiee, A. Nag *et al.*, "ISAAC: A convolutional neural network accelerator with in-situ analog arithmetic in crossbars," in *Proceedings of the International Symposium on Computer Architecture*, 2016, pp. 14–26.

[9] P. Jain, U. Arslan *et al.*, "A 3.6 Mb 10.1 Mb/mm^2 embedded non-volatile ReRAM macro in 22nm FinFET technology with adaptive forming/set/reset schemes yielding down to 0.5 V with sensing time of 5ns at 0.7 V," in *IEEE International Solid-State Circuits Conference (ISSCC)*, 2019, pp. 212–214.

[10] L. Wei, J. G. Alzate *et al.*, "A 7Mb STT-MRAM in 22FFL FinFET technology with 4ns read sensing time at 0.9 V using write-verify-write scheme and offset-cancellation sensing technique," in *IEEE International Solid-State Circuits Conference (ISSCC)*, 2019, pp. 214–216.

[11] Y.-D. Chih, Y.-C. Shih *et al.*, "A 22nm 32Mb embedded STT-MRAM with 10ns read speed, 1M cycle write endurance, 10 years retention at 150°C and high immunity to magnetic field interference," in *IEEE International Solid-State Circuits Conference (ISSCC)*, 2020, pp. 222–224.

[12] C.-C. Chou, Z.-J. Lin *et al.*, "A 22nm 96KX144 RRAM macro with a self-tracking reference and a low ripple charge pump to achieve a configurable read window and a wide operating voltage range," in *IEEE Symposium on VLSI Circuits*, 2020, pp. 1–2.

[13] J. Zhang, Z. Wang *et al.*, "A machine-learning classifier implemented in a standard 6T SRAM array," in *IEEE Symposium on VLSI Circuits*, 2016, pp. 1–2.

[14] S. K. Gonugondla, M. Kang *et al.*, "A 42pJ/decision 3.12TOPS/W robust in-memory machine learning classifier with on-chip training," in *IEEE International Solid-State Circuits Conference (ISSCC)*, Feb 2018, pp. 490–492.

[15] A. Biswas and A. P. Chandrakasan, "Conv-RAM: An energy-efficient SRAM with embedded convolution computation for low-power CNN-based machine learning applications," in *IEEE International Solid-State Circuits Conference (ISSCC)*, Feb 2018, pp. 488–490.

[16] Q. Dong, M. E. Sinangil *et al.*, "A 351TOPS/W and 372.4 GOPS compute-in-memory SRAM macro in 7nm FinFET CMOS for machine-learning applications," in *IEEE International Solid-State Circuits Conference (ISSCC)*, 2020, pp. 242–244.

[17] P. Chi, S. Li *et al.*, "PRIME: A novel processing-in-memory architecture for neural network computation in ReRAM-based main memory," in *Proceedings of the International Symposium on Computer Architecture*, 2016, pp. 27–39.

[18] L. Song, X. Qian *et al.*, "PipeLayer: A pipelined ReRAM-based accelerator for deep learning," in *IEEE International Symposium on High Performance Computer Architecture*, 2017, pp. 541–552.

[19] B. Murmann, "ADC Performance Survey 1997-2021 (ISSCC & VLSI Symposium)." [Online]. Available: http://web.stanford.edu/ murmann/adcsurvey.html, 2021.

[20] S. K. Gonugondla, M. Kang *et al.*, "A variation-tolerant in-memory machine learning classifier via on-chip training," *IEEE Journal of Solid-State Circuits*, vol. 53, no. 11, pp. 3163–3173, Nov 2018.

[21] Y.-D. Chih, P.-H. Lee *et al.*, "An 89TOPS/W and 16.3 TOPS/mm^2 all-digital sram-based full-precision compute-in memory macro in 22nm for machine-learning edge applications," in *IEEE International Solid-State Circuits Conference (ISSCC)*, 2021, pp. 252–254.

[22] S. Cosemans, B. Verhoef *et al.*, "Towards 10000TOPS/W DNN inference with analog in-memory computing–A circuit blueprint, device options and requirements," in *IEEE International Electron Devices Meeting (IEDM)*, 2019, pp. 22–2.

[23] T.-J. Yang and V. Sze, "Design considerations for efficient deep neural networks on processing-in-memory accelerators," in *IEEE International Electron Devices Meeting (IEDM)*, 2019, pp. 22–1.

[24] B. Moons, R. Uytterhoeven *et al.*, "Envision: A 0.26-to-10TOPS/W subword-parallel dynamic-voltage-accuracy-frequency-scalable Convolutional Neural Network processor in 28nm FDSOI," in *IEEE International Solid-State Circuits Conference (ISSCC)*, 2017, pp. 246–247.

[25] W.-H. Chen, K.-X. Li *et al.*, "A 65nm 1Mb nonvolatile computing-in-memory ReRAM macro with sub-16ns multiply-and-accumulate for binary DNN AI edge processors," in *IEEE International Solid-State Circuits Conference (ISSCC)*, 2018, pp. 494–496.

[26] J. Albericio, A. Delmás *et al.*, "Bit-pragmatic deep neural network computing," in *Proceedings of the IEEE/ACM International Symposium on Microarchitecture*, 2017, pp. 382–394.

Low Power Logic Obfuscation Through System Level Clock Gating

Daniel Xing, Yuntao Liu, Ankur Srivastava
University of Maryland, College Park, MD, USA
{dxing97, ytliu, ankurs}@umd.edu

***Abstract*—Logic locking methods such as Stripped Functionality Logic Locking (SFLL) tend to yield high overheads. SFLL only corrupts a small part of the input space by design in order to maintain good SAT resilience and in doing so selects high frequency inputs to corrupt (protect) and therefore increases locking's impact on system level error. This implies that much of the time stripped modules are doing unnecessary work while the restore units are correcting the computations. We propose taking advantage of this fact to selectively clock gate the modules when protected inputs are being processed. Under the highest possible level of attack resilience, this alone can yield up to 24.5% dynamic power savings when protected inputs are applied to synthesized MediaBench benchmarks. We also propose a system-level design approach that utilizes the data-flow graph to also gate operations that fully depend on other gated operations. In conjunction with modifying operation binding, this increases power savings to 32.9% under the same strict security constraints.**

***Index Terms*—clock gating, logic locking**

I. Introduction

As the cost of maintaining advanced technology IC foundries continues to rise, many chip designers have chosen to become fabless and rely on offshore foundries for fabrication. However, this outsourcing can jeopardize the security of the IC supply chain since the foundries are not controlled by the designer.

Logic locking has emerged as a protection of the intellectual properties in chip designs against untrusted fabs [1]. Logic locking involves a secret key input, known only to the designer, that must be correctly applied to the circuit for it to function properly. Various types of logic locking mechanisms have been proposed, starting with inserting XOR/XNOR gates in the design netlist [2] and progressing to more advanced techniques based on VLSI testing principles that produce high corruption at the output bits when an incorrect key is applied [3], [4].

The logic locking field was transformed by the Boolean satisfiability-based attack, also known as the SAT attack [5]. SAT offers a robust mathematical approach for identifying the correct locking key of a logic locked IC by progressively eliminating incorrect keys. In response, point function (PF)-based logic locking, such as SARLock [6] and Anti-SAT [7], limits the number of wrong keys pruned out in each iteration, resulting in an exponential increase in the number of SAT iterations required relative to the key size. However, such logic locking techniques were proven vulnerable to approximate SAT attacks [8], [9] and removal attacks [10].

In a more recent development, stripped functionality logic locking (SFLL) was proposed which empowers designers to choose a group of protected input patterns (PIPs) that are impacted by a significant proportion of incorrect keys, while other input patterns are affected by only a small percentage of incorrect keys [11]. Details of SFLL is introduced in Sec. II-A. Robust Strong Anti-SAT (RSAS) achieves the same level of security by also stripping the PIPs' functionality from the original circuit and using improved Anti-SAT infrastructure to restore the functionality when the correct key is applied [12].

Our proposed method gives SFLL the ability to lower power overhead. While this method can be applied to modules in isolation, we show that a system-level approach can enable even greater power savings for the same level of SAT attack resilience. Other system-level logic locking methods have been proposed in past works [13]–[17], including one work that proposes a system-level sharing method for reducing locking-related overhead [18]. However, our work is the first to propose clock gating entire functional modules locked with SFLL while incorporating a high level design perspective. Treating the look-up tables (LUTs) in SFLL as a high-level power-saving resource grants system designers greater design flexibility when considering power overheads while still maintaining granular control over SAT attack resilience.

A. Contributions

In this work, we propose utilizing SFLL's LUTs in a new way: enabling clock gating of entire modules by performing the lookup in an earlier clock cycle. While this technique can be readily applied at the module level, we demonstrate that a system-level design view enabled by high-level synthesis (HLS) can enable even greater power savings, all while keeping SAT attack resilience and LUT sizes in check. Our contributions can be summarized as follows:

1) We formally define finding the power-optimal operation binding and locking configuration as a 0-1 integer linear program (ILP). This formulation jointly determines which operations to protect (and therefore clock gate) with SFLL and operation-to-hardware binding so that power savings are maximized without compromising SAT resilience.
2) We also present a post-binding greedy heuristic for selecting protected operations and PIPs using system-level information, but implemented at a module level. This simplifies the control circuitry needed to implement clock gating at the expense of sub-optimal power reduction.
3) We evaluate our clock gating techniques on MediaBench benchmarks. Our heuristic and optimal methods on average reduce dynamic power by 24.5% and 32.9%,

This work was supported by the NSF under Grant 1953285.

979-8-3503-1176-1/23 $31.00 © 2023 IEEE

respectively, while maintaining maximal SAT attack resilience. Further power reductions are possible under relaxed attack resilience requirements.

II. Preliminaries

A. SAT Attack and SFLL

The SAT attack provides a strong mathematical formulation and was able to defeat all the logic locking techniques that predated it. The potential adversary could be either an untrusted foundry or user with the capability to reverse engineer the produced chip and acquire the locked gate-level netlist. The SAT attack requires two resources: (1) the locked netlist and (2) an activated chip (one that has the correct key loaded) from which the adversary can query the correct output of selected input vectors. Details of the mathematical formulation of the SAT attack can be found in [5]. Simply put, in each iteration of the SAT attack, a Boolean satisfiability problem is solved, and a *distinguishing input (DI)* is found. The DI is capable of eliminating the set of wrong keys that produce incorrect output for the DI. When all the wrong keys are eliminated by DIs, a correct key will be found.

The advanced technique of Stripped Functionality Logic Locking (SFLL) comprises of two main components: a Functionality Stripped Circuit (FSC) and a Restore Unit (RU). The FSC is the original circuit, but with its functionality altered for a selected set of PIPs, rendering SFLL resistant to removal attack. Removing the RU renders the PIPs' functionality different from that of the original circuit, making the attack futile. The RU is often implemented with an LUT that stores the key, verifies the input of the circuit against the key, and produces a restore vector that is XOR-ed with the FSC output. If the key is correct, the restore vector will correct the FSC's output, resulting in a correct output for the circuit. SFLL comes in two types: SFLL-HD and SFLL-flex. SFLL-HD produces specific structural traces in the FSC, which can be captured through functional analysis-based attacks [19]. On the other hand, SFLL-flex leaves minimal structural traces in the FSC, thanks to a fault-injection-based approach used for stripping the functionality [20]. Of note is SFLL's inherent direct inverse relationship between the number of PIPs and the expected time required by SAT attack [21], [22]. More more PIPs lead to higher impact that SFLL has on the overall functionality of a locked circuit, doing so comes at the cost of decreased SAT attack resilience.

B. High-Level Synthesis

HLS, or high-level synthesis, is a process that involves transforming a high-level description of functionality, such as a behavioral description in a high-level language, e.g. C or SystemC, into a register-transfer level (RTL) design. During HLS, there are generally three main design optimizations: resource allocation, scheduling, and resource binding. Resource allocation involves determining the quantity and type of hardware resources, such as functional units (FUs), that are necessary for the design. Scheduling imposes clock-cycle boundaries on the target behavioral code to resolve data dependencies. This produces a scheduled data flow graph (DFG), a directed acyclic graph whose nodes and edges represent operations and dependencies between them, respectively. Resource binding maps operations in the scheduled DFG to the allocated FUs from resource allocation. The binding must meet the minimal timing and performance requirements of the design while trying to optimize for area or power. Common binding schemes aim to minimize area [23] or switching power [24], [25]. During binding, the expected input space for a circuit is generally known. This enables switching power estimation to inform power-aware binding decisions. Power-aware binding techniques aim to minimize the switching activity of FUs by selecting those that have a lower expected switching activity, thus reducing the overall power consumption of the design.

HLS-based technques have been exploited to strengthen SFLL. For example, the intermediate representation during the compilation of the high-level design code can be analyzed to identify suitable combinational logic cones to insert SFLL [26]. Furthermore, security-aware binding was proposed in [13] where the operations are selected to be bound to locked FUs in order to maximize the occurrence of PIPs in the locked FUs. In our work, we show that the binding step, i.e. assigning operations to SFLL-locked FUs, provides an opportunity to reduce the switching power for the entire design, and the security of logic locking and power savings can be achieved simultaneously without compromise.

III. Gating, Graphs, and You: Saving Dynamic Power, Securely

Conventional logic locking methods always require some amount of additional circuitry to implement. SFLL-based methods in particular require a RU (typically implemented as a LUT) to correct stripped functionality when the correct key is applied. However, only a small number of protected inputs are chosen in order to strengthen SFLL against SAT attacks and to keep LUT sizes small. By moving LUT lookup into an earlier clock cycle (i.e. after module inputs are known but before inputs are applied), we can store precomputed output values and save them in the LUT, allowing the system to opportunistically clock gate entire locked modules.

A. Module-Constrained Gating

To illustrate this idea, consider the locked single-module datapath shown in Fig. 1a. A subset of the locked module's input is compared against a stored key value that characterizes the PIP. If they match, then the output of the FSC will be inverted according to the stored flip vector. Since the functionality of SAT-secure stripped modules only differs from the original design for a few inputs, it is feasible to implement correction functionality using a small LUT, and indeed this practice is adopted by PIP-based locking schemes such as SFLL and RSAS. Note that during this operation, when the PIP are being processed by the LUT, the module itself is doing useless work and can therefore be gated.

Our proposed modification to the RU is illustrated in Fig. 1b. Instead of storing a flip vector, we store the module's output value corresponding to the protected input. Note that the RU is moved one clock step earlier and connected to the output of the module feeding the input FFs of the stripped module. Whenever the input to the RU matches the key, instead of correcting the corrupted output of the FSC, we clock gate the entire FSC and supply a precomputed output value to any modules or registers dependent on the FSC's output. Since the FSC's correct output value is completely known at lookup

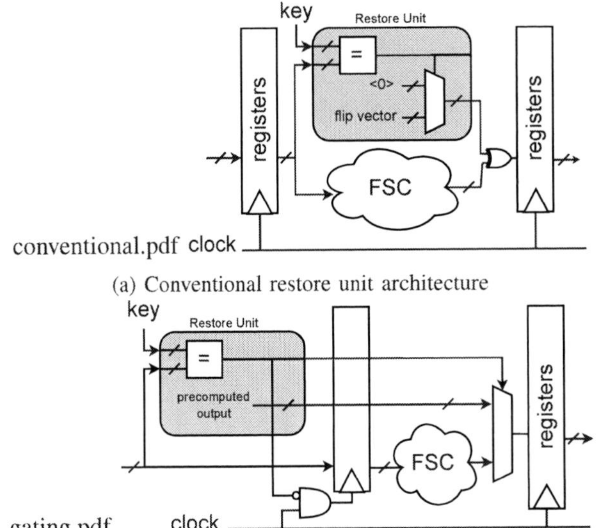

Fig. 1: Two single-module datapaths locked with SFLL. 1a places the RU in the conventional way, while 1b configures the RU to clock gate the FSC when protected inputs are applied.

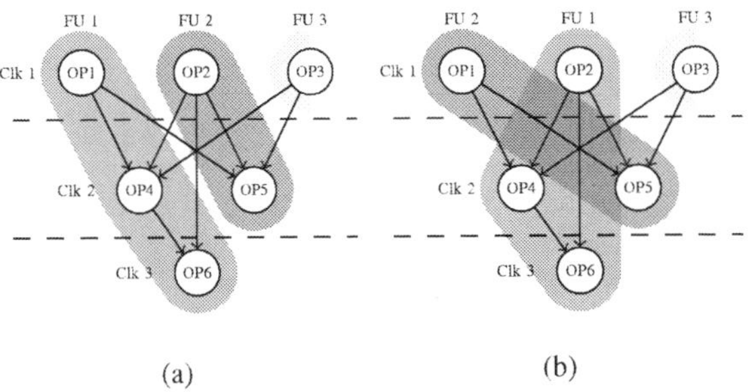

Fig. 2: Two different resource bindings for the same scheduled DFG.

time, the overall functionality of the system is not affected. We therefore perform the LUT lookup operation before the locked circuit is scheduled to operate, but after its input value is available. Since a SAT-secure locked circuit will protect only a few inputs, the LUT only needs to contain a few entries, minimally impacting timing. Provided the correct key values are loaded into the LUT, the locked module will operate correctly. Also, if the PIP are chosen such that they have high frequency of occurrence, the clock gating will be active often leading to large power savings.

B. System-Level Gating with DFGs

While this LUT look-ahead method can be readily applied within the confines of a single module on any synthesized design, we show that system-level modifications can enable more gating opportunities and provide the same guarantees against the SAT attack by considering data dependencies in the DFG and resource binding respectively.

Consider the scheduled DFG in Fig. 2a. We assume that a particular DFG input has been chosen for protection, so each operation in the DFG will already have a candidate protected input. Suppose the designer is limited to placing just two LUTs (e.g. due to area limitations). If PIPs for operations 2 and 4 are chosen to be placed in LUTs, then the system can clock gate up to *three* operations: operations 2, 4, and 6. The protected inputs of operations 2 and 4 are both stored by LUTs, and can be directly gated. Operation 6 can be gated if both operations 2 and 4 are also gated because the input of operation 6 is fully determined by the outputs of operations 2 and 4, which are also stored in our modified LUTs. Therefore, the output value for operation 6 when both operation 2 and 4 receive protected inputs is fully known at design time, and can be saved on-chip. On the other hand, if inputs of operations 1 and 2 are chosen for protection, then the number of gated operations drops from three to two when the protected DFG input is applied, since no operations in the DFG depend on only operations 1 and 2. Hence judicious allocation of limited restore LUT resources to operations impacts how many operations can be gated.

Binding of operations to hardware modules also affects the SAT-secureness of each locked module because each locked module's PIPs are chosen based on which operations are bound to each module. For example, consider the binding shown in Fig. 2a. Again assuming operations 2 and 4 are protected, modules FU1 and FU2 will only need to strip one input each, and therefore are maximally secure against SAT attack. If the binding is instead what's shown in Fig. 2b, then FU1 will need to strip the protected inputs of both operation 2 and 4, worsening its resilience against SAT attack (since we will need to protect two PIPs instead of one), while FU2 will not be locked at all. Therefore, binding should be carefully done to ensure that each module does not have too many protected inputs and remains sufficiently secure against SAT attacks.

This HLS-driven system-level approach is the one we take when clock gating modules. We select operations to protect and map protected operations to locked functional modules such we maximize the number of gated operations under protected inputs and ensure locked modules remain sufficiently secure against SAT attack.

IV. Finding Power-Optimal LUT and Binding Configurations with ILP

Finding locking configurations that maximize power saved in this way can be expressed as a 0-1 integer linear program (ILP). To help make our formulation easier to follow, Table I lists constants, variables, and some notation used in our formulation.

First, the constraints. Besides operations with LUT-protected inputs, we also want to gate operations that depend only on other gated operations. To start, we first describe operations that only depend on other operations and not DFG inputs ($NI(V)$). The two constraints calculates whether a node's predecessors are all gated and are not all gated respectively:

$$\prod_{i \in pre(k)} g_i \leq p_v \quad \forall v \in NI(V) \tag{1}$$

$$1 - p_v \leq n_v \quad \forall v \in NI(V) \tag{2}$$

Note that while constraint 1 is not linear, since all variables are constrainted to be binary, it can be rewritten as a set of linear constraints.

If every operation that v depends on is gated, then the inputs of v can be fully determined from stored LUT entries, and

TABLE I: ILP variables, constants, and definitions

Notation	Definition
$G(V, E)$	a DAG representing the scheduled DFG, where nodes represent operations and edges represent data dependencies
$pre(v)$	predecessors/data dependencies of a node/operation $v \in V$
$NI(V)$	Operations in G that depend only on other operations (i.e. don't depend on DFG inputs)
$start(v)$	The first clock cycle that operation v is active for
$end(v)$	The last clock cycle that operation v is active for
$w(v)$	Expected dynamic power used by operation v
M	set of functional modules
K	set of clock cycles in the schedule
l	User-defined limit on the total number of instantiated LUTs
C_m	User-defined limit on the number of LUTs allowed for each module m, this represents the number of PIPs stripped from m which is decided by the level of SAT attack resilience desired
g_v	Variable that encodes if an operation v is gated. g_v is 1 if operation v is gated, 0 otherwise.
p_v	Variable that encodes whether or not predecessors of v are gated. p_v is 1 if all predecessors are gated, 0 otherwise.
n_v	Logical inverse of p_v
L_v	Variable for keeping track of what operation a LUT is protecting. L_v is 1 if operation v's input is protected by a LUT, 0 otherwise.
$b_{v,m}$	Variable associated operation binding. $b_{v,m}$ is 1 if operation v is bound to functional module m, 0 otherwise.
$busy_{v,m,k}$	Variable for keeping track of the schedule during binding. $busy_{v,m,k}$ is 1 if operation v is scheduled during clock k and is bound to module m.

therefore v itself can also be gated. This can be expressed as the following constraint:

$$p_v \leq g_v \quad \forall v \in NI(V) \tag{3}$$

If not every parent operation of v is gated, then the output of v cannot be determined from LUTs assigned to parent operations alone. However, it can still be gated if a LUT is assigned to protect it. We can express this as the following constraint, again noting that it can be rewritten as a set of linear constraints:

$$g_v n_v \leq L_v \quad \forall v \in NI(V) \tag{4}$$

For operations v that do depend on DFG inputs (i.e. $V \setminus NI(V)$) then the inputs of v cannot be fully known by examining inputs and outputs of other operations alone. Therefore, v can only be gated if a LUT is assigned to it:

$$g_v \leq L_v \quad \forall v \in V \setminus NI(V) \tag{5}$$

For area-limited designs, we can constrain the total number of LUTs to some user-defined limit l with the following inequality:

$$\textstyle\sum_{v \in V} L_v \leq l \tag{6}$$

We use the following constraints to ensure that operations are properly bound to functional modules. Constraint 7 determines which which operations at what clock cycles each module is bound to:

$$b_{v,m} \leq busy_{v,m,k} \quad \forall v \in V, m \in M, k \in [start(v), end(v)] \tag{7}$$

Constraint 8 ensures that there can only be at most one operation bound to a module in any clock cycle:

$$\textstyle\sum_{v \in V} busy_{v,m,k} \leq 1 \quad \forall m \in M, k \in K \tag{8}$$

Constraint 9 ensures that every operation is bound to exactly one module:

$$\textstyle\sum_{m \in M} b_{v,m} = 1 \quad \forall v \in V \tag{9}$$

To ensure each locked functional module is still secure against SAT attack, we limit the number of LUTs (and therefore the number of PIPs per module) in each locked module. Since each LUT resource only protects a single input value, limiting the number of LUTs per module will also limit the number of PIPs, and therefore ensures locked modules will be sufficiently resilient against expected SAT attacks:

$$sum_{v \in V} b_{v,m} L_v \leq C_m \quad \forall m \in M \tag{10}$$

The overall optimization objective is to maximize expected dynamic power savings due to gated operations

$$\max_{g,p,n,L,b,busy} \sum_{i \in V} w_i g_i \tag{11}$$

subject to the constraints given.

A. ILP Usage

A system designer seeking to secure a design using lookahead LUT-based RUs will first need to perform the scheduling and resource allocation steps of HLS so that the scheduled DFG $G(V, E)$ and available functional modules M are known. One or more DFG inputs will need to be selected for protection, and should be propagated through the DFG so that each operation has one or more candidate protected inputs. The designer will also need to determine the maximum number of protected inputs that each module can protect (C_m) before SAT attack resilience degrades excessively. If the design is overhead constrained, then the maximum number of LUTs that can be instantiated l without exceeding area or static power limits will need to be determined.

Once these system parameters have been found, they can be encoded into the ILP described previously, and solved using one of the many available academic or commercial ILP solvers. Optimal solutions of g_v, L_v, and $b_{v,m}$ indicates which operations should be gated, which operations should have its input protected by an LUT, and which operations should be bound to which modules respectively.

V. Post-Binding Module-level Heuristic

While in practice ILP solvers can efficiently solve some kinds of large ILP instances, the worst-case runtime is still NP-hard. To avoid this worst-case time complexity and to simplify the control circuitry needed to implement optimal gating of downstream operations, we present a greedy PIP selection heuristic that can be implemented after operation binding has occurred.

While gating downstream operations increases the number of clock gating events without adding additional LUTs, proper implementation requires additional control circuitry to coordinate LUTs across different modules. As an example, again consider the scheduled and bound DFG shown in Fig. 2a. Suppose a particular DFG input is selected for obfuscation. While it may be the designer's intention to protect particular inputs of operation 2 and 4 (denoted as PIP_2 and PIP_4 respectively) calculated from the protected DFG input, in practice other DFG inputs may result in PIPs occuring at other clock cycles besides the ones operations 2 and 4 are scheduled for. For example, there may be a DFG input where PIP_2 occurs at FU2 in clock 1 but PIP_4 does *not* occur at FU1 in clock 2, or PIP_4 occurs in clock 1 of FU1 instead of clock 2. We cannot assume PIPs will occur at only in their intended scheduled operation, so some additional control logic is needed to make sure downstream operations with unprotected inputs

Algorithm 1 Greedy heuristic algorithm

Input: $G(V, E), w(\cdot), M, l, b, C$
Output: $L(v, m)$
Initialization:
1: $candidates = V$
2: $L(v, m)$ is initialized to 0 for all $v \in V$ and $m \in M$
3: $lcount = 0$
Loop:
4: **while** $lcount < l$ **do**
5: v_{sel} = highest weighted op in $candidates$
6: **if** $\sum_{w \in V} L(w, b(v_{sel})) < C_{b(v_{sel})}$ **then**
7: $L(v_{sel}, b(v_{sel})) = 1$
8: $lcount = lcount + 1$
9: **end if**
10: **end while**
11: **return** L

are only gated when all upstream operations are gated. Not gating downstream operations removes the need for control circuitry beyond what is already required by a single module at the expense of decreased power savings.

If operation binding is fixed and clock gating is confined to just the protected operation, then only protected inputs for each module need to be selected. This can be done efficiently with the greedy algorithm shown in Alg. 1. The algorithm first sets the pool of candidate gatable operations to V, (line 1) initializes the binding matrix $L(v, m)$ to 0 (line 2), and sets the total number of allocated LUTs to 0 (line 3). In each loop iteration, the operation v_{sel} with the highest dynamic power consumption is chosen (line 5). If the module that v_{sel} is bound to $b(v_{sel}) \in M$ has not exceeded its SAT-attack resiliency requirement $C_{b(v_{sel})}$ (line 6), then v_{sel} can be added to the set of operations protected by $b(v_{sel})$ (line 7). The loop terminates when all LUTs have been assigned (line 4).

The initialization requires sorting all operations in V by their dynamic power consumption, which takes $O(|V| \log |V|)$ time. The algorithm loop runs at most $|V|$ times (the highest value l can be set to) and the summation in line 6 can be implemented in $O(1)$ time if the length of each row of L is stored and updated separately from L. Therefore the heuristic's runtime is $O(|V| \log |V|)$.

VI. Results

To evaluate our optimal and heuristic clock gating approaches, we locked designs synthesized from C functions used in the MediaBench suite [27]. Each function's DFG was extracted using SUIF and scheduled using a path-based scheduler [28]. Up to three adders and three multipliers were allocated for each benchmark.

Locking was performed with SFLL-flex, although our method can be applied to any locking method that uses restore units. While SFLL PIPs can be of any bit width, we set the width to be the same as the input vector size of each functional module to maximize SAT attack resilience and simplify LUT design. For similar security reasons, we limited each locked module to strip at most one input (i.e. $C_m = 1 \forall m$). To evaluate the heuristic method, we used a security-aware binding method [13] to bind each operation to functional modules.

We implemented the ILP using Gurobi [29]. All experiments were performed on a desktop machine equipped with a 2.5 GHz Intel processor and 16 GB of system memory. Dynamic power was calculated for designs built using FreePDK45 [30] and synthesized with Cadence Genus tools.

To compare our techniques with the conventional no-gating approach, Fig. 4 shows dynamic power consumed by a locked datapath when a protected DFG input is applied to the system, normalized to the conventional no-gating method. Our optimal system-level approach reduces dynamic power consumed by the locked datapath under protected DFG inputs on average by 32.9% while ensuring that each locked module only strips one input, maximizing SAT attack resilience. Performing clock gating using our module-constrained heuristic at the same security level yields 24.5% dynamic power savings, although the advantage our heuristic holds over the ILP binding approach strongly depends on the structure of the DFG. For example, all operations in `jctrans2`'s DFG depend directly on a DFG input, so operations can only be gated via a LUT lookup, and cannot depend exclusively on preceding operations alone. However, both `fir` and `dct` have tree-like DFGs with multiple operations that do not directly depend on DFG inputs, so a system-level DFG-aware approach is able to save more power than a module-level one.

We can also explore the security-power tradeoff by varying C_m of each allocated module, shown in Fig. 3. Here, we assume that the system design allows enough area overhead to add any additional LUTs needed to implement restore units. When a small number of PIPs protects each module, the opportunities for clock gating decrease and so the dynamic power savings are less. Conversely, increasing the number of protected inputs per locked module increases the number of protected operations, and therefore the number of gated operations. At very high LUT allocation limits, every operation that depends on DFG inputs can be protected, and therefore gated, which in conjunction with gating downstream operations, brings dynamic power down to near-zero for protected inputs. Naturally, this is an extreme case since for such a scenario, much of the power will be dissipated in the LUTs themselves. For the heuristic module-limited gating method, this will happen when every operation in the DFG has a LUT protecting it. However, since there is a direct inverse relationship between the number of locked inputs and expected SAT iterations required, these additional power savings comes at the cost of decreased SAT attack resilience and increased LUT power dissipation.

VII. Conclusion

In this work we propose using SFLL restore unit lookup tables to store locked module outputs and therefore clock gate entire locked modules to reduce dynamic power consumption. We show that a system-level design approach guided by operation data dependencies can yield power savings beyond what a module-level approach can save. We demonstrate both a module-constrained and a system-level clock gating approach on synthesized designs from the MediaBench [27] suite. Our proposed module-constrained and system-level approach reduces dynamic power for protected inputs by 24.5% and 32.9% respectively compared to the non-gated locking method.

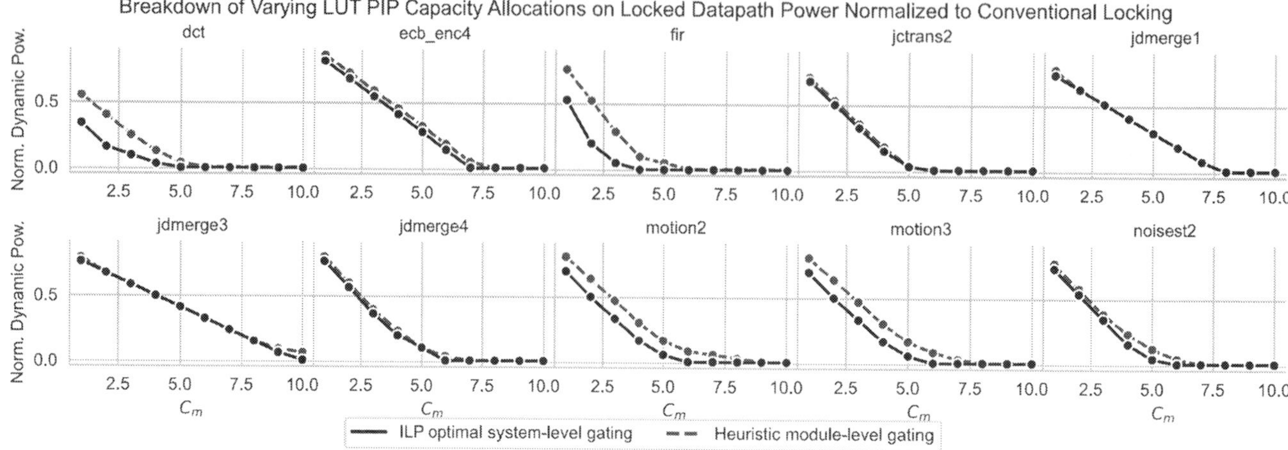

Fig. 3: Breakdown of how system dynamic power decreases as limits on the number of PIPs per locked module (C_m) are raised for both module-level heuristic and system-level ILP methods. The same C_m value is set for all allocated functional modules. Power is normalized to a system locked with a conventional no-gating appraoch.

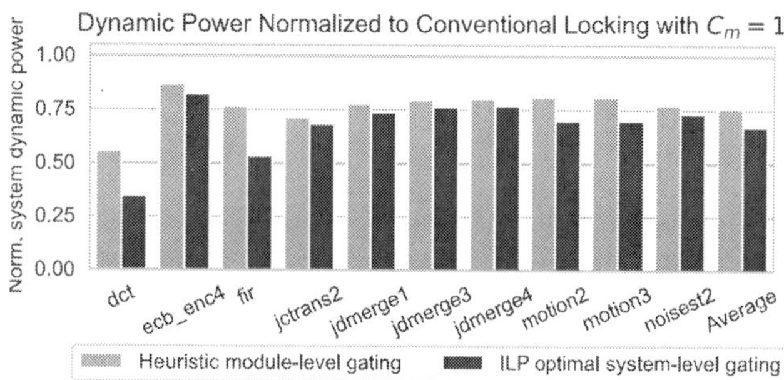

Fig. 4: Locked datapath dynamic power for benchmarks tested, normalized to a conventionally locked datapath. Data is shown for locking solutions found with $C_m = 1$ for all modules.

REFERENCES

[1] A. Chakraborty *et al.*, "Keynote: A disquisition on logic locking," *IEEE Transactions on Computer-Aided Design of Integrated Circuits and Systems*, 2019.

[2] J. A. Roy *et al.*, "Epic: Ending piracy of integrated circuits," in *Conference on Design, automation and test in Europe*, 2008.

[3] J. Rajendran *et al.*, "Security analysis of logic obfuscation," in *Proceedings of Design Automation Conference*, 2012.

[4] ——, "Fault analysis-based logic encryption," *IEEE Transactions on computers*, vol. 64, no. 2, pp. 410–424, 2015.

[5] P. Subramanyan *et al.*, "Evaluating the security of logic encryption algorithms," in *2015 IEEE International Symposium on Hardware Oriented Security and Trust (HOST)*. IEEE, 2015, pp. 137–143.

[6] M. Yasin *et al.*, "Sarlock: Sat attack resistant logic locking," in *Intl. Symposium on Hardware Oriented Security and Trust*, 2016.

[7] Y. Xie and A. Srivastava, "Anti-sat: Mitigating sat attack on logic locking," *IEEE Transactions on Computer-Aided Design of Integrated Circuits and Systems*, vol. 38, no. 2, pp. 199–207, 2018.

[8] K. Shamsi *et al.*, "Appsat: Approximately deobfuscating integrated circuits," in *2017 IEEE International Symposium on Hardware Oriented Security and Trust (HOST)*. IEEE, 2017, pp. 95–100.

[9] Y. Shen and H. Zhou, "Double dip: Re-evaluating security of logic encryption algorithms," in *Great Lakes Symposium on VLSI 2017*, 2017.

[10] M. Yasin *et al.*, "Removal attacks on logic locking and camouflaging techniques," *Transactions on Emerging Topics in Computing*, 2017.

[11] ——, "Provably-secure logic locking: From theory to practice," in *Conference on Computer and Communications Security*, 2017.

[12] Y. Liu *et al.*, "Robust and attack resilient logic locking with a high application-level impact," *ACM Journal on Emerging Technologies in Computing Systems*, 2021.

[13] M. Zuzak *et al.*, "A resource binding approach to logic obfuscation," in *2021 58th ACM/IEEE Design Automation Conference (DAC)*. IEEE, 2021, pp. 235–240.

[14] C. Pilato *et al.*, "On the optimization of behavioral logic locking for high-level synthesis," *arXiv preprint arXiv:2105.09666*, 2021.

[15] M. R. Muttaki *et al.*, "Hlock: Locking ips at the high-level language," in *2021 58th ACM/IEEE Design Automation Conference (DAC)*. IEEE, 2021, pp. 79–84.

[16] C. Pilato *et al.*, "Assure: Rtl locking against an untrusted foundry," *IEEE Transactions on Very Large Scale Integration (VLSI) Systems*, vol. 29, no. 7, pp. 1306–1318, 2021.

[17] N. Limaye *et al.*, "Fortifying rtl locking against oracle-less (untrusted foundry) and oracle-guided attacks," in *2021 58th ACM/IEEE Design Automation Conference (DAC)*. IEEE, 2021, pp. 91–96.

[18] D. Xing *et al.*, "Low overhead system-level obfuscation through hardware resource sharing," in *2023 24th International Symposium on Quality Electronic Design (ISQED)*, 2023, pp. 1–8, ISSN: 1948-3295.

[19] D. Sirone and P. Subramanyan, "Functional analysis attacks on logic locking," *IEEE Transactions on Information Forensics and Security*, vol. 15, pp. 2514–2527, 2020.

[20] A. Sengupta *et al.*, "Truly stripping functionality for logic locking: A fault-based perspective," *IEEE Transactions on Computer-Aided Design of Integrated Circuits and Systems*, 2020.

[21] M. Zuzak *et al.*, "Trace logic locking: Improving the parametric space of logic locking," *IEEE Transactions on Computer-Aided Design of Integrated Circuits and Systems*, 2020.

[22] H. Zhou *et al.*, "Resolving the trilemma in logic encryption," in *International Conference on Computer-Aided Design (ICCAD)*, 2019.

[23] C.-Y. Huang *et al.*, "Data path allocation based on bipartite weighted matching," in *Design Automation Conference*, 1991.

[24] J.-M. Chang and M. Pedram, "Register allocation and binding for low power," in *ACM/IEEE Design Automation Conference (DAC)*, 1995.

[25] A. Stammermann *et al.*, "Binding allocation and floorplanning in low power high-level synthesis," in *International Conference on Computer Aided Design*. IEEE, 2003.

[26] M. Yasin *et al.*, "Sfll-hls: Stripped-functionality logic locking meets high-level synthesis," in *Intl. Conf. on Computer-Aided Design*, 2019.

[27] C. Lee *et al.*, "Mediabench: A tool for evaluating and synthesizing multimedia and communications systems," in *International Symposium on Microarchitecture*. IEEE, 1997.

[28] S. Ogrenci Memik *et al.*, "A super-scheduler for embedded reconfigurable systems," in *IEEE/ACM International Conference on Computer Aided Design. ICCAD 2001. IEEE/ACM Digest of Technical Papers (Cat. No.01CH37281)*, Nov. 2001, pp. 391–394, iSSN: 1092-3152.

[29] Gurobi Optimization, LLC, "Gurobi Optimizer Reference Manual," 2023. [Online]. Available: https://www.gurobi.com

[30] J. E. Stine *et al.*, "FreePDK: An Open-Source Variation-Aware Design Kit," in *2007 IEEE International Conference on Microelectronic Systems Education (MSE'07)*, Jun. 2007, pp. 173–174.

FPGA-Patch: Mitigating Remote Side-Channel Attacks on FPGAs using Dynamic Patch Generation

Mahya Morid Ahmadi
Faculty of Informatics
Vienna University of Technology (TU Wien)
Vienna, Austria
mahya.ahmadi@tuwien.ac.at

Lilas Alrahis
Division of Engineering
New York University Abu Dhabi (NYUAD)
Abu Dhabi, United Arab Emirates
lma387@nyu.edu

Ozgur Sinanoglu
Division of Engineering
New York University Abu Dhabi (NYUAD)
Abu Dhabi, United Arab Emirates
ozgursin@nyu.edu

Muhammad Shafique
Division of Engineering
New York University Abu Dhabi (NYUAD)
Abu Dhabi, United Arab Emirates
muhammad.shafique@nyu.edu

***Abstract*—We propose *FPGA-Patch*, the first-of-its-kind defense that leverages automated program repair concepts to thwart power side-channel attacks on cloud FPGAs. FPGA-Patch generates isofunctional variants of the target hardware by injecting faults and finding transformations that eliminate failure. The obtained variants display different hardware characteristics, ensuring a maximal diversity in power traces once dynamically swapped at run-time. Yet, FPGA-Patch forces the variants to have enough similarity, enabling bitstream compression and minimizing dynamic exchange costs. Considering AES running on AMD/Xilinx FPGA, FPGA-Patch increases the attacker's effort by three orders of magnitude, while preserving the performance of AES and a minimal area overhead of 14.2%.**

I. Introduction

Field programmable gate arrays (FPGAs) are gaining increasing attention for their accelerated and low-power computations, leading cloud service providers (CSPs) to deploy FPGA platform-as-a-service [1], [2]. However, CSPs consider FPGA virtualization for multi-tenancy, which poses security risks in cloud FPGAs [3], [4].

New security vulnerabilities and attack surfaces arise due to the shared resources among the tenants. Various attacks are reported in the literature on remote FPGAs, such as fault injection, covert channel, denial-of-service, and side-channel attacks (SCAs) [5], [6]. For example, an attacker can leak critical information of co-located victim tenants via the shared power distribution network (PDN) [7], i.e., a power SCA (see Fig. 1). In such an attack, an adversary (i.e., malicious tenant) inserts on-chip power sensors, such as ring oscillators (RO) [8] or time to digital converters (TDC) [4] to record the power trace of a victim tenant and extract secret assets.[1]

Besides the conventional mitigation techniques against power SCAs, it is imperative to investigate FPGA-assisted

[1]We study a remote side-channel attack scenario where the attacker extracts information from the cloud without physical access to the platform [8].

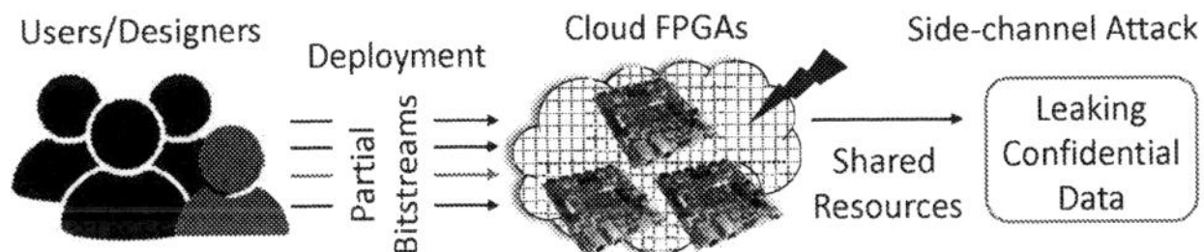

Fig. 1. Shared resources in cloud FPGAs leak secret data to malicious users.

defense techniques in this new remote attack surface [9]–[11]. The state-of-the-art (SOTA) defenses and their limitations are discussed next and summarized in Table I.

A. State-of-the-art and Their Limitations

Offline Bitstream Scanning: Pioneer CSPs, e.g., Amazon® [12], check the FPGA bitstream and prevent the deployment of combinational loops on the cloud as power measurement sensors [13]–[15]. While these techniques detect some attack circuits, e.g., combinational loops in ROs [14], [16], they cannot identify stealthy attack sensors, e.g., arithmetic-logic units [17]. Also, restricting the hardware design limits the implementation of essential security primitives, e.g., true random number generators.

Run-time Monitoring detects power fluctuations in shared FPGA platforms (indicating a co-located malicious tenant) [18]. Such techniques mitigate active attacks (e.g., fault-injecting) and cannot detect passive remote SCAs [19].

Noise Addition: Defense mechanisms against SCAs decrease the signal-to-noise ratio to hide the process of the critical assets. By introducing noise to the run-time power trace, attackers need more power traces to extract the secret asset (e.g., crypto key). However, these techniques suffer from extremely high area/power overhead (e.g., 100% area overhead in [20]), making them impractical in real-world applications.

Implementation Diversity: Recently, researchers have proposed diversity-based techniques to obfuscate the power trace by exploiting modern FPGA's dynamic partial reconfiguration (DPR) feature [21], [22]. These methods use isofunctional variant swapping to hide the correlation

979-8-3503-1176-1/23 $31.00

TABLE I
COMPARISON OF THE DEFENSE METHODS AGAINST REMOTE SCAS

Defense	Unrestrained Design	SCAS Prevention	Low Overhead (Area)
Offline Bitstream Scanning [13]	No	Yes	–
Run-time Monitoring [18]	Yes	No	No (>30%)
Noise Addition [20]	Yes	No	No (100%)
Implementation Diversity [21]	No	Yes	No (>2x)
Proposed FPGA-Patch	**Yes**	**Yes**	**Yes** (14%)

between device power consumption and the target core's intermediate values. However, while diversity-based solutions offer a strong defense, they currently have the following limitations that need to be addressed.

- SOTA offer minimal SCA security due to the limited number of deployed variants. Asghar *et al.* [22] proposed netlist randomization, generating 3 classes of 9 netlist variants, with increasing the attacker's effort only by $\sim$3.5×.
- The SOTA techniques modify the entire design randomly. Thus, each variant is as large as an entire design bitstream, e.g., in [21], 790 MB is required to store 128 variants. In addition, large bitstream sizes impose large reconfiguration latency and resource utilization overheads. For example, in [21], increasing the resistance against power SCAs by two factors caused resource utilization to double.
- SOTA solutions have been demonstrated via simulations only. The hardware implementation and physical properties of the target platform add technical challenges, such as noise due to process variation in the chip, which cannot be accurately considered in only-simulation results [23].

There is a gap in designing lightweight and effective diversity-based obfuscation defenses. Toward that end, we propose *FPGA-Patch*, the first-of-its-kind concept that employs automated program repair methods to generate design variants and thwart power SCA in shared FPGAs. FPGA-Patch is a proactive lightweight defense that cloud-FPGA users can employ at the design time prior to bitstream generation. Moreover, FPGA-Patch is generic and can protect any given hardware design. Our novel contributions are as follows.

B. Our Novel Contributions

1) **Hardware patching for variants generation (Sec. III-A).** FPGA-Patch injects errors into the design to generate faulty netlists, which are then recovered using equivalent checker methods to create diverse power traces at run time.
2) **Eradicating large storage overhead (Sec. III-C).** FPGA-Patch adopts a difference-based bitstream generation technique in FPGA design tools to decrease the storage overhead and reconfiguration latency.
   By limiting the location of modified fault points, FPGA-Patch enables compressed bitstream generation.
3) **Dynamic FPGA hardware implementation (Sec. III-D).** FPGA-Patch is implemented on an AMD/Xilinx ZYNQ FPGA using dynamic function exchange (DFX) technology. We evaluate the security of an advanced encryption standard (AES) application core protected by FPGA-Patch against a correlation power analysis (CPA) attack using remote power measurement techniques.

Key Results. FPGA-Patch employs 128 variants to thwart the CPA attack, resulting in an increase in the minimum traces to disclosure (MTD) by over 1,000× and a 2.25× decrease in the CPA value. These improvements significantly reduce the attacker's confidence in key detection. Our experiments show that this enhanced security comes at a minimal area overhead of 14.2% while preserving the performance of AES.

II. THREAT MODEL OF PROPOSED FPGA-PATCH

In power SCAs on multi-tenant FPGAs, the attacker is a malicious tenant that exploits the shared PDN [8]. Consistent with SOTA work, we assume that the attacker knows the target application running on the victim's partial region. This assumption enables a security assessment of FPGA-Patch in the absence of obscurity. Further, we assume the worst-case scenario for the defender, i.e., when there is no noise added by other tenants' computation in the attacker's observed traces.

Considering crypto cores, the attacker requests the victim to encrypt plain text and observes the produced power traces.

III. FPGA-PATCH METHODOLOGY

Power SCAs collect power trace samples during multiple executions using power measurement circuitry. These traces are then aligned offline and statistically analyzed to isolate key-correlated parts of each trace. Adding uncertainty to the power profile of the circuit makes trace alignment challenging, reducing the correlation between consumed power and the secret key, thus hampering the SCA. FPGA-Patch tackles this issue by frequently switching between isofunctional variants with diverse power profiles, which enforces heterogeneity in power traces. Fig. 2 illustrates the integration of FPGA-Patch into the standard FPGA design flow.[2]

A. Design of Equivalent Function Variants

FPGA-Patch uses LEC[3] for automatic program repair to generate variants. Fig. 2 **A** shows the standard FPGA design flow, where the hardware's register transfer level (RTL) description is simulated for functional testing, synthesized, and mapped to look up tables (LUT)s. To create design-equivalent heterogeneous variants, FPGA-Patch selects candidate nets in the target design netlist for fault injection (See Fig. 2 **B**).

1) Net selection: We consider two net selection methods, discussed below. The number of selected nets is a parameter to evaluate the diversity of variants.

Random Selection. For our initial exploration, we consider random net selection. However, some selected nets could not change the run-time power characteristics effectively.

Critical Net Selection. We modify FPGA-Patch to select "effective" candidate nets, increasing the diversity of variants.

[2]FPGA-Patch is an effective and versatile defense against SCAs targeting the key-dependent execution part of cryptography cores. However, it has limitations, as it only applies to designs where correlation power analysis is valid and does not defend against attacks on machine learning systems, such as side-channel-based model stealing attacks on deep neural networks.

[3]The LEC tool compares the netlists of the two designs to identify differences and generates a report of faults in the revised logic [24].

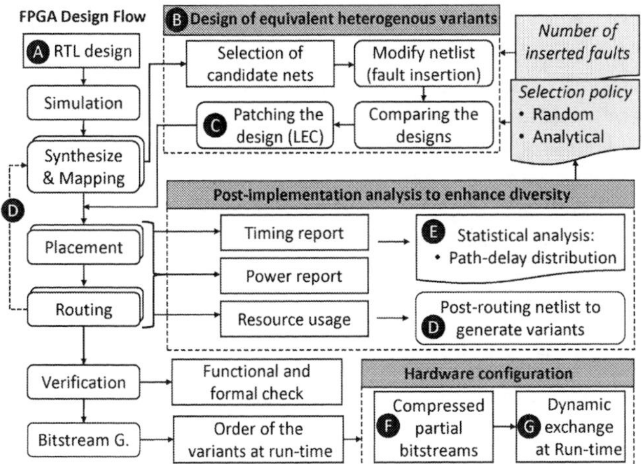

Fig. 2. Proposed FPGA-Patch methodology for mitigating power SCAs.

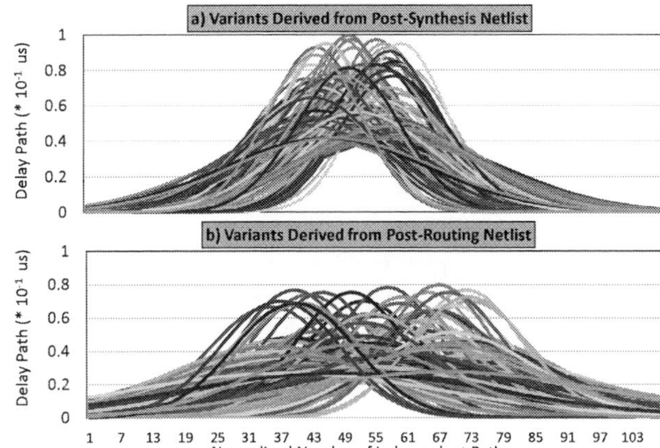

Fig. 3. Timing comparison of 128 variants generated by FPGA-Patch regarding the delay of independent paths in one clock cycle. The lines represent the best Gaussian fit of the estimated path delays of routed design.

Selecting nets on the critical timing path, maximizes the probability of run-time power obfuscation, see Fig. 2 E.

2) Fault Insertion: Selected nets are connected to the "0" and "1" logical ports, resembling stuck-at-0 (SA0) and stuck-at-1 (SA1) faults, respectively. After fault injection, the design tool optimizes the netlist and removes unnecessary logic gates, unused ports, and unconnected wires. These manipulations are intended to alter design functionality.

3) Design-Level for Fault Insertion: We explore the difference between injecting the faults at the post-synthesis netlist versus the post-routing netlist. The proposed flow is the same in both cases. More details are provided in Sec. III-B.

4) Patching the Faulty Netlists: Each faulty netlist should be patched for missing logic using the LEC methods, as shown in Fig. 2 C. To recover the target design's functionality, each faulty netlist is compared with the netlist of the original design. Formal equivalence checking can catch any formal errors in combinational logic, known as Nonequivalents (NEQs). To repair the faulty netlist, the LEC traces back the NEQs to find the unmapped or incorrectly mapped nets. After detecting all NEQs, the LEC adds the logic required to recover each missing state to the design steadily until all NEQs are resolved.

B. More on Fault Injection

1) Post-Synthesis: For the initial exploration, we inject the faults into the post-synthesis netlist and study the resource utilization of the different variants, as shown in Fig. 2 D. Fig. 3(a) indicates the delay distribution between 128 generated variants, considering AES as a target application. As can be observed, the diversity between the variants is limited. Our investigation indicates that the altered nets were different bits of the same data bus. Therefore, they all had a similar effect on the hardware implementation. Further, passing the recovered netlists through the rest of the FPGA design flow steps, i.e., routing, etc., applies logic optimization methods, eliminating redundant paths and leading to lower diversity.

2) Post-Routing: Next, we investigate the effect of injecting the faults into the post-routing netlist (Fig. 2 D) and present the path delay distribution for the 128 variants in Fig. 3 (b).

We can infer that the distribution of maximum delay of independent paths for post-routing is more diverse compared to post-synthesis. This diversity leads to variable dynamic power when switching between the variants at run time and hiding the secret key's power trace. *Therefore, we recommend that FPGA-Patch gets adopted post-routing.*

C. Hardware Configuration File

When the difference between two hardware versions is minimal and localized, we can use difference-based bitstream generation to generate configuration files; see Fig. 2 F.

This compressed partial bitstreams program only the difference between two given variants. Switching the configuration of a module from one implementation to another is fast because the bitstream differences are smaller than the entire partial reconfiguration region (PRR) bitstream.

Initially, on device power-up, a complete bitstream must be loaded into the device prior to any partial bitstreams. Therefore, a full bitstream configuring all PRRs to initial configurations needs to be loaded. After the first reconfiguration, a partial bitstream will be loaded to each PRR. During the reconfiguration process, the states of the flip-flops are preserved, and the fixed parts of the design remain fully operational. This ensures that reconfiguration does not affect the performance of the target application and FPGA-patch does not impose performance overhead, in terms of the throughput of the encryption core.

D. Dynamic Exchange at run-time

FPGA-Patch obfuscates confidential applications' dynamic power trace by changing the underlying hardware dynamically. A difference-based method is deployed to generate the configuration file of the variants, where the distance vector of two consecutive designs is stored in memory. Therefore, the order of deployed bitstreams in a specific PRR is determined at design time. We divide the 128 diverse designs into 8 categories, and each category is assigned to one PRR. Hence, we can avoid stalls in the execution of the encryption algorithm during the reconfiguration via cycling around the

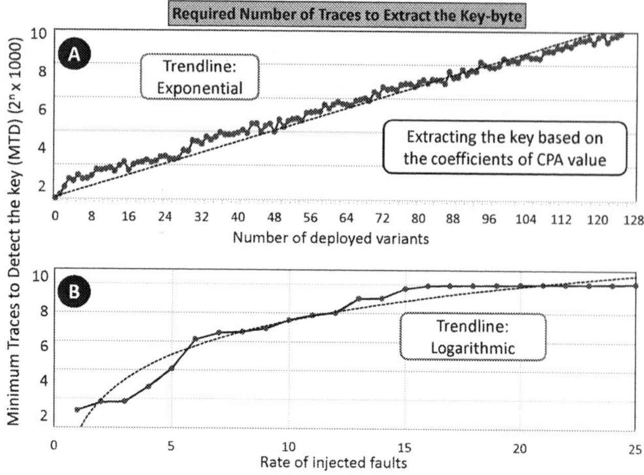

Fig. 4. Analysis of the number of deployed variants on attacker's effort in terms of MTD, when the injected fault rate is 10%.

PRRs and inside the categories per each encryption. The partial bitstreams in PRR use a fixed order of configurations independently (C1 to C2 to C4,..., to C8).

The target module activates one PRR at a time while the other PRRs undergo reconfiguration with the next variant. To ensure the required time to finish the next module reconfiguration and preserve the throughput, the active time of each variant should be less than MTD for unprotected design. The DFX module in Xilinx/AMD FPGAs automatically programs the next PRR to be used in the reconfiguration process, as illustrated in Fig. 2 Ⓖ.

IV. Experimental Investigations

A. Experimental Setup and Tool Flow

Target Hardware. Following SOTA works [21], [25], we use a serial AES-128 encryption core, employed by a security application encrypting a test dataset. The S-box operation which is the focus of design diversity is implemented using Gallois field implementation and taken from opencores platform [26].

After the first round (SubBytes), the state is a function of input data and the encryption key. Hence, the SubByte is considered as the operation-to-be-protected by FPGA-Patch.

Implementation. The AES is implemented on the ZedBoard Zynq-7000 ARM/FPGA SoC development board, using Vivado Design Suite'21, requiring 256 LUTs and 260 registers. Enabling the I/O interfaces makes the static region require 422 CLBs in Zedboard (1277 LUTs, 2024 registers).

For **Fault Injection**, we adopt the engineering change orders (ECOs) flow. ECOs are modifications to the design netlist to implement the changes with minimal impact on the original design. Modern FPGA design methods, i.e., Vivado, provide an ECO flow, which allows modification of a design checkpoint, implements the changes, runs reports on the changed netlist, and generates the final programming files. FPGA-Patch leverages ECO to select and modify the netlist to ensure the semantics of the design netlist after fault injection.

Variants. To study the security of FPGA-Patch, 128 variants are generated for each experiment, which are divided into 8 categories for exchanging at run time. Following the design constraints, the operating frequency of AES is set to 10 MHz.

Dynamic Exchange. The FPGA processor hosts a Linux OS to control the I/O interface, reconfiguration application, and the AES IP core data transfer. To enable high-speed reconfiguration of PRR, we adopt the internal configuration access port implemented for DFX procedure in Vivado.

The **Order of Variants** generated by FPGA-Patch does not impact its security performance. The correlation equation considers all collected traces, ensuring that the distribution of variant usage is fixed. Even if an attacker obtains and re-organizes all the traces, the security remains intact as long as this distribution is maintained.

Attack Measurement. We collect the power traces of FPGA-Patch using a TDC [27] with 2 channels, running in an independent PRR. A single encryption run corresponds to one measured trace. The output is periodically sent to a host workstation application for further analysis via the CPA technique. As an evaluation metric, we perform a CPA [28], with the aforementioned Hamming Distance power model targeting two consecutive S-box outputs during the first round. Then, we compute the CPA value for each key-byte guess and for each power trace based on the number of traces.

B. Security Analysis

We analyze the security of FPGA-Patch against CPA in terms of (i) MTD (see Fig. 4) and (ii) the correlation coefficient of all key-bytes compared to the correct key-byte (see Fig. 5).

Attacker's Effort. The attacker attempts to find the correct key-byte by correlating power samples with the hypothetical power model of each possible byte value for the first byte of the key. The values that are $1.5\times$ higher than the rest are considered the correct key. The MTD of AES with respect to the #variants deployed at runtime is shown in Fig. 5 Ⓐ. Assuming that the correct key-byte in a CPA against AES can be detected in $\approx$ 1000 traces, deploying 1 to 128 variants in FPGA-Patch results in an exponential growth of the MTD for various numbers of traces. By applying 128 variants generated with 10% injected faults to the post-routing netlist while selecting nets analytically (as explained in Sec. IV-C), the MTD has increased by at least three orders of magnitude.

Variant diversity depends on the number of injected faults. Our analysis reported in Fig. 5 Ⓑ, shows that injecting more than 10% of faults increases the MTD only by 4x, where the threshold of circuit recovery by LEC, has been observed to be $\sim$25%. Therefore, the effective injected fault rate for the random net selection policy is identified to be 10%.

Attacker's Confidence. The maximum correlation coefficient of each key-byte guess and assumed power model is shown in Fig. 5. Details of the results are explained below.

Effect of the Injected Fault Rate. The rate of injected faults is defined as the number of candidate nets to assign the fault divided by the total number of nets in the DPR module (SubByte). The CPA value for 128 variants deployed at run time is compared for different fault injection rates. The

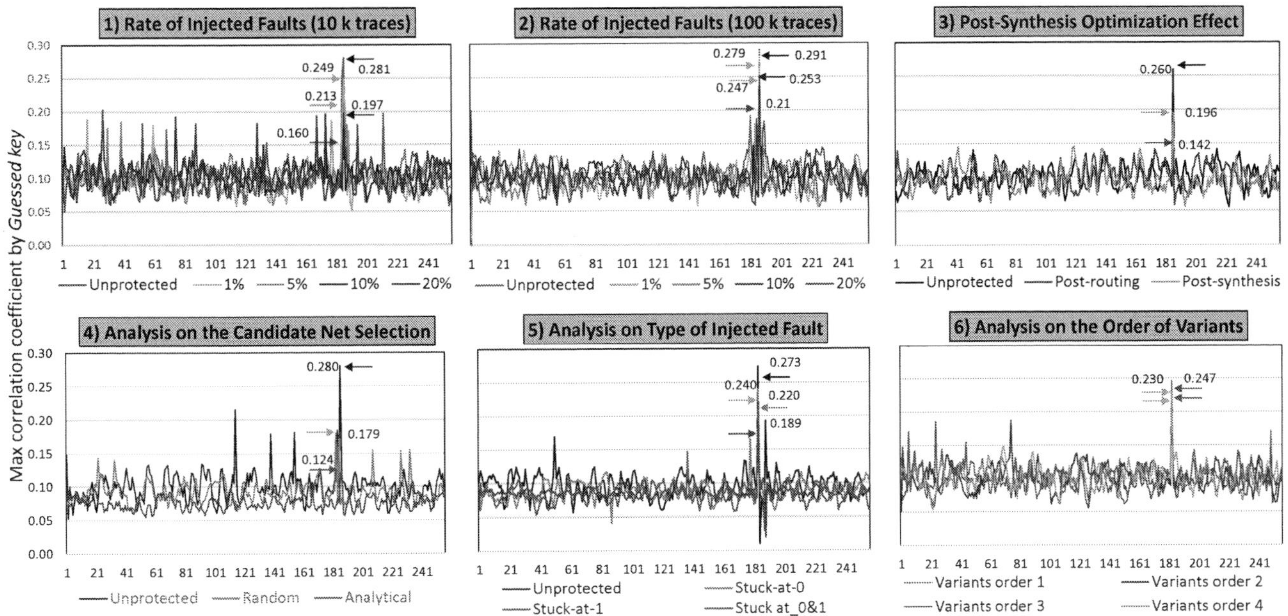

Fig. 5. Max correlation coefficient analysis by guessed key, with correct key-byte of 185 (0xB9). The red line highlights the best-obfuscated results.

TABLE II
OVERHEAD COMPARISON OF THE INJECTED FAULT RATE IN FPGA-PATCH METHODOLOGY

Fault Rate (%)	1	2	3	4	5	6	7	8	9	10	11	12	13	14	15	16	17	18	19	20	21	22	23	24	25
Area (LUT) (%)	3.3	4.3	6.6	7.3	8.1	12.1	12.5	13	13.2	14.2	14.7	15.2	18.5	22.4	23.2	25.9	31.5	32	35.7	37.5	47.4	50.9	51.5	52.3	54
Power (%)	0.6	0.8	1.2	1.3	2.1	5.2	6.7	6.9	7	7.3	7.5	7.6	8.4	9.1	9.4	10	10.2	10.3	10.5	12.2	15.1	17.4	17.7	17.8	18
Path Delay (%)	0.18	0.19	0.2	0.2	0.2	0.21	0.22	0.22	0.22	0.24	0.24	0.26	0.27	0.27	0.29	0.29	0.3	0.32	0.33	0.33	0.34	0.38	0.39	0.39	0.39

comparison indicates that increasing the number of randomly injected faults decreases the CPA value by 1.75× when the most repairable nodes are faulty.

Increasing #Traces. Attacker's confidence in extracting the secret key-byte, is defined by CPA value for a specific number of traces to analyze. As shown in Fig. 5 1 and 2, collecting and analyzing 10x more power traces, eliminates the false positive in results and increases CPA value by 0.27x.

Post-Synthesis Fault Injection vs. Post-Routing. The experiments on the adopted netlist from post-synthesis and post-routing, show that applying faults to the routed design is 1.83× more effective than only synthesized netlist, in terms of decreasing the CPA. The optimization in re-synthesis stage eliminated the effect of patch logic on power obfuscation.

Effect of Node Selection. The candidate nets to inject the faults are selected randomly and analytically (based on the critical path delay order). Selection of the nets randomly decreases the CPA value for 1.56×, while selecting nodes based on the timing reports decreases the CPA peak for 2.25×.

Effect of Fault Type. To study the impact of the type of injected faults, the CPA value for SA0 and SA1 and both, are shown in Fig. 5.5. It shows the value of CPA for both SA0 and SA1 are in similar trends, while when the both are deployed, the peak of CPA value decreases by 1.44×.

C. Performance Overhead

We have explored a fault rate in the range from 1% to 25%, which is the maximum repairable threshold for LEC. The trade-off between security and overhead shows that the area overhead, from 10% to 25% increases by 2.8×, while increasing MTD by only 4×. Therefore 10% is selected as an effective fault rate in our experiments.

Area Overhead. Exploiting the optimized netlist and adding logic to repair the faults increases the number of logic gates and the used area. As shown in Table II, the area overhead increases with the number of faults injected into the design. In our security analysis, considering the effective fault rate as 10%, the area overhead is 14.2% in FPGA-Patch.

Delay Overhead. FPGA-Patch affects delay only if the candidate nets affect the critical paths. The effect of FPGA-Patch on the average path delay is minimal. Considering 10% fault rate, the path delay overhead is 0.24%.

Power Overhead. The overhead of dynamic power in FPGA-Patch includes the reconfiguration controller and switching activity of the hardware. Moreover, our analysis indicates that when candidate nets are selected based on the timing report, the increase in dynamic power is 12% more than when selected randomly.

Memory Storage Size. A full bitstream of AES is ∼460 kB, and each partial bitstream depending on the injected fault rate is ∼ 10 to 50 kB. The total storage size required for full bitstream and 128 variants is 1.8 to 7 MB, whereas, in module-based bitstream generation, the required storage is ∼60 MB.

Performance overhead. FPGA-Patch's small variant bitstream size enables fast reconfiguration. Further, the cyclic

TABLE III
COMPARISON WITH RELATED WORKS

Comparison Category	[21]	[22]	[29]	FPGA-Patch
Increased MTD	2.2×	3.54×	95×	1000×
Number of variants	128	9	16	128
Throughput overhead	NA	∼5×	0	0

selection of PRRs guarantees an active configuration for encryption, preserving AES's throughput and performance.

D. Comparison with State-of-the-Art

Recently, design diversity enabled by DPR, has been studied as a defense method against SCA. In this section, we compare our findings with three approaches [21], [22], [29] that advocate for the dynamic exchange of hardware variants, summarized in Table III.

In [21], the variants are diversified in placement and routing strategies. This work deploys 128 AES variants to exchange at runtime, and the gained security in terms of confidence of attacker, in detecting the key is limited to ∼2.2×. Similarly, Asghar *et al.* [22] have proposed netlist randomization techniques that randomly enable delay elements in the design. By developing 9 netlist variants from three classes of diversity, they show the peak of CPA value has decreased by ∼3.54×. Moreover, [22] employs long chains of delay elements and dummy modules, which increases the hardware resource overhead by ∼5×. Furthermore, in [29], several reconfiguration regions are reserved in the FPGA, and the design is moved among these regions in each reconfiguration, in addition to noise generators to further obfuscate the power traces. While in [29] the authors show MTD is increased by ∼95×, the security of this technique is bounded by the process variation on the FPGA SoC.

In contrast, the design size in FPGA-Patch exhibits negotiable changes (i.e., $<14.2\%$) for resource overhead (due to adding patch logic) to AES, increasing the attacker's effort to extract the correct key-byte by $> 3\times$.

V. CONCLUSION

CSPs are highly concerned about securing shared platform applications against remote SCA. To address this concern, we propose *FPGA-Patch*, a novel defense that utilizes program repair algorithms to prevent SCAs on cloud FPGAs. By generating isofunctional variants of the target hardware and dynamically exchanging them at runtime, we can obfuscate power traces and hinder SCA. The variant generation policy of FPGA-Patch allows for bitstream compression and minimizes dynamic exchange costs, making it a lightweight and effective diversity-based obfuscation defense against remote SCAs. Overall, FPGA-Patch provides a promising solution for CSPs seeking to enhance the security of their shared platforms.

VI. ACKNOWLEDGEMENT

This work is supported by the Doctoral College Resilient Embedded Systems, which is run jointly by the TU Wien's Faculty of Informatics and the UAS Technikum Wien and is part of the Moore4Medical project funded by the ECSEL Joint Undertaking under grant number H2020-ECSEL-2019-IA-876190. It is also jointly supported by the Center for Cyber Security (CCS) at New York University Abu Dhabi.

REFERENCES

[1] C. Jin et al., "Security of Cloud FPGAs: A Survey," *arXiv:2005.04867 [cs]*, 2020. [Online]. Available: http://arxiv.org/abs/2005.04867
[2] C. Ramesh et al., "FPGA Side Channel Attacks without Physical Access," in *FCCM*, Apr. 2018, pp. 45–52, iSSN: 2576-2621.
[3] A. Vaishnav et al., "A survey on FPGA virtualization," in *FPL*, 2018, pp. 131–1317.
[4] J. Krautter et al., "CPAmap: On the Complexity of Secure FPGA Virtualization, Multi-Tenancy, and Physical Design," *TCHES*, vol. 2020, no. 3, pp. 121–146, 2020.
[5] G. Provelengios et al., "Power distribution attacks in multitenant FPGAs," *TVLSI*, vol. 28, pp. 2685–2698, 2020.
[6] R. Elnaggar et al., "Multi-Tenant FPGA-based Reconfigurable Systems: Attacks and Defenses," in *DATE*, 2019, pp. 7–12.
[7] S. Moini et al., "Voltage sensor implementations for remote power attacks on FPGAs," *TRETS*, 2022.
[8] M. Zhao et al., "FPGA-Based Remote Power Side-Channel Attacks," in *IEEE S&P*, 2018, pp. 229–244.
[9] S. S. Mirzargar et al., "Physical side-channel attacks and covert communication on FPGAs: A survey," in *FPL*, 2019, pp. 202–210.
[10] S. Duan et al., "A survey of recent attacks and mitigation on FPGA systems," in *ISVLSI*. IEEE, 2021, pp. 284–289.
[11] M. Morid Ahmadi et al., "ShapeShifter: Protecting FPGAs from side-channel attacks with isofunctional heterogeneous modules," in *IOLTS*. IEEE, 2023, pp. 1–6.
[12] Amazon. (2021) Aws ec2 FPGA hdk+sdk errata. [Online]. Available: https://github.com/aws/aws-fpga/blob/master/ERRATA.md
[13] D. R. E. Gnad et al., "Checking for Electrical Level Security Threats in Bitstreams for Multi-tenant FPGAs," in *FPT*, 2018, pp. 286–289.
[14] J. Chaudhuri et al., "Detection of malicious FPGA bitstreams using cnn-based learning," in *ETS*. IEEE, 2022, pp. 1–2.
[15] S. Tian et al., "Fingerprinting Cloud FPGA Infrastructures," in *FPGA*, 2020, pp. 58–64.
[16] R. Elnaggar et al., "Learning malicious circuits in FPGA bitstreams," *TCAD*, pp. 1–1, 2022.
[17] D. R. Gnad et al., "Stealthy logic misuse for power analysis attacks in multi-tenant FPGAs," in *DATE*. IEEE, 2021, pp. 1012–1015.
[18] D. Jayasinghe et al., "RFTC: Runtime frequency tuning countermeasure using FPGA dynamic reconfiguration to mitigate power analysis attacks," in *DAC*, 2019, pp. 1–6.
[19] J. Gravellier et al., "High-speed ring oscillator based sensors for remote side-channel attacks on FPGAs," in *ReConFig*, 2019, pp. 1–8.
[20] J. Krautter et al., "Active Fences against Voltage-based Side Channels in Multi-Tenant FPGAs," in *ICCAD*, 2019, pp. 1–8.
[21] B. Hettwer et al., "Securing cryptographic circuits by exploiting implementation diversity and partial reconfiguration on FPGAs," in *DATE*, 2019, pp. 260–263.
[22] A. Asghar et al., "Increasing Side-Channel Resistance by Netlist Randomization and FPGA-Based Reconfiguration," in *Applied Reconfigurable Computing. Architectures, Tools, and Applications*, Cham, 2021, pp. 173–187.
[23] N. Khan et al., "Moving target and implementation diversity based countermeasures against side-channel attacks," in *ARC*. Springer, 2021, pp. 188–202.
[24] D. Krishnegowda, "A primer on logical equivalence checking (lec) using conformal," 10 2021.
[25] I. Bow et al., "Side-channel power resistance for encryption algorithms using implementation diversity," vol. 4, no. 2, 2020, p. 13.
[26] J. Villar. (2019) Opencores 128/192 aes. [Online]. Available: https://opencores.org/projects/systemcaes
[27] M. Adamič and A. Trost, "A fast high-resolution time-to-digital converter implemented in a zynq 7010 soc," in *Austrochip*, 2019, pp. 29–34.
[28] E. Brier et al., "Correlation power analysis with a leakage model," in *CHES*. Springer, 2004, pp. 16–29.
[29] N. Khan et al., "Moving target and implementation diversity based countermeasures against side-channel attacks," in *Applied Reconfigurable Computing. Architectures, Tools, and Applications*. Cham: Springer International Publishing, 2021, pp. 188–202.

Enabling DVFS Side-Channel Attacks for Neural Network Fingerprinting in Edge Inference Services

Erich Malan, Valentino Peluso, Andrea Calimera, Enrico Macii*
Department of Control and Computer Engineering, Politecnico di Torino, Turin, Italy
*Interuniversity Department of Regional and Urban Studies and Planning, Politecnico di Torino, Turin, Italy
{erich.malan, valentino.peluso, andrea.calimera, enrico.macii}@polito.it

***Abstract*—The Inference-as-a-Service (IaaS) delivery model provides users access to pre-trained deep neural networks while safeguarding network code and weights. However, IaaS is not immune to security threats, like side-channel attacks (SCAs), that exploit unintended information leakage from the physical characteristics of the target device. Exposure to such threats grows when IaaS is deployed on distributed computing nodes at the edge. This work identifies a p otential v ulnerability of low-power CPUs that facilitates stealing the deep neural network architecture without physical access to the hardware or interference with the execution flow. O ur a pproach r elies o n a Dynamic Voltage and Frequency Scaling (DVFS) side-channel attack, which monitors the CPU frequency state during the inference stages. Specifically, w e i ntroduce a d edicated load-testing methodology that imprints distinguishable signatures of the network on the frequency traces. A machine learning classifier is then used to infer the victim architecture. Experimental results on two commercial ARM Cortex-A CPUs, the A72 and A57, demonstrate the attack can identify the target architecture from a pool of 12 convolutional neural networks with an average accuracy of 98.7% and 92.4%**

***Index Terms*—Inference-as-a-Service, Convolutional Neural Networks, Side-Channel Attacks, DVFS.**

I. Introduction

Inference-as-a-Service (IaaS) is a licensing and delivery model that lets connected clients get access to pre-trained deep neural networks while preserving the intellectual property rights of the service provider. Originally conceived for the cloud, the IaaS paradigm can be implemented on edge nodes, like IoT gateways, to improve response times, scalability, and geographical coverage [1]. The edge-IaaS infrastructure relies on the microservice architectural approach. Microservices leverage virtualization technologies to build hardware-agnostic inference engines deployable on a multitude of computing platforms, including embedded systems commonly equipped with low-power CPUs, such as the ARM Cortex-A cores. The clients query the inference service through an Application Programming Interface (API) that prevents access to the neural network, protecting code and weights. Unfortunately, microservices deployed on distributed edge nodes are more vulnerable to malicious acts [2], like Side-Channel Attacks (SCAs), which can extract sensitive information about the neural network from the physical characteristics of the hosting device.

TABLE I
Overview of black-box SCAs against IaaS.

Reference	Access	Target Device	Side-Channel Signal	Information
[3]	Physical	ARM Cortex-A	Power consumption	Architecture
[4]	Physical	NVIDIA GPU	Memory and PCIe bus	Architecture
[5]	Physical	NVIDIA GPU	PCIe bus	Weights
[6]	Physical	ARM Cortex-M	EM emission	Weights
[7]*	Remote	Intel x86	Cache timing	Hyperparams
[8]	Remote	Intel x86	Latency	# of Layers
[9]	Remote	Intel x86	Performance counters	Layers Type
[10]‡	Remote	NVIDIA Tegra	Resource utilization	Family
[11]*	Remote	Intel x86	Cache timing	Architecture
[12]†	Remote	NVIDIA GPU	Power consumption	Architecture
This work	Remote	ARM Cortex-A	DVFS state	Architecture

* Manipulate the execution flow. †Require integrated power sensors.
‡ Neural networks belonging to the same family show the same topology and core organization but differ in the number of layers and/or filters.

Previous studies demonstrated various types of SCAs that target different hardware platforms and can extract different details. Table I summarizes state-of-the-art SCAs, and in particular, black-box SCAs, namely, methods that assume no prior knowledge of the deep neural network. The table offers a taxonomy based on four main features: the required access, the target device, the source of side-channel leakage, and the recovered information. Physical attacks require direct access to the target device for measuring physical signals, like power consumption [3], memory and PCIe bus traces [4], [5], or electromagnetic emissions (EM) [6]. Those signals contain fine-grained information from which the attacker can retrieve the neural network architecture and even reconstruct the network weights, enabling the replica of the neural network. However, physical probing and costly laboratory equipment limit applicability in real-life scenarios. On the other hand, remote attacks are software methods based on monitoring system metrics and performance counters accessible from operating system logs without special privileges. However, they can simply retrieve partial information, like a subset of architectural hyperparameters (e.g., stride, pooling, padding in case of convolutional neural networks) [7], the number of layers [8], the type of layers (without details about the number of filters) [9], the family to which the neural architecture belongs [10]. Very few remote attacks allow for the identification of the neural architecture [11], [12]. Moreover, remote attacks show several weaknesses that limit their detection capability (64.17% Top-1 accuracy, as reported in [12]). For example, cache attacks [11] require the ability to modify the execution flow through additional cache instructions, which

979-8-3503-1176-1/23 $31.00 © 2023 IEEE

may interfere with the network execution. Alternatively, attacks that probe power consumption using software-readable on-chip current sensors are affected by low sampling rates and low resolution of the sensors [13]. To notice that those current sensors are mounted on development kits but are often unavailable in production-grade boards.

This work uncovers a vulnerability in ARM Cortex-A CPUs that can be exploited to bypass IaaS protection via remote SCA. Specifically, our study investigates the use of Dynamic Voltage and Frequency Scaling (DVFS) state as a signal for neural network fingerprinting. We demonstrate that existing DVFS SCAs are ineffective at extracting signatures that could lead to the reconstruction of the neural architecture. Hence, we introduce a novel method based on a load-testing procedure that sends a sequence of queries to the prediction API according to pre-defined schemes. The procedure regulates the service load inducing variations in the voltage-frequency state of the device under attack. The generated frequency traces are processed through a machine learning (ML) classifier trained offline to infer the neural architecture. The proposed methodology is validated on two embedded CPUs from the ARM Cortex-A family: the A72 and A57. The collected results show that our approach achieves an average accuracy of 98.7% (A72) and 92.4% (A57) in detecting the neural architecture from a pool of 12 state-of-the-art convolutional neural networks (CNNs) suited for embedded applications. These findings suggest that DVFS can be exploited as a source of leakage for neural architecture theft, highlighting the need for security measures against these types of attacks.

II. Background & Related Works

A. DVFS & Governors

DVFS is a runtime power and thermal management technique for digital cores. Low-power ARM CPUs based on the Cortex-A architecture have a predefined set of voltage-frequency pairs available, and each voltage level is associated with one specific operating frequency. The *CPUFreq* is the software interface integrated into the Linux kernel in charge of controlling DVFS. It consists of two main components: the scaling governors and the scaling driver. Scaling governors implement algorithms to set up the right CPU frequency based on the system's workload. The scaling driver is responsible for accessing the hardware interfaces to change the CPU frequency as requested by the scaling governors. The standard workflow of a governor involves continuous monitoring of the CPU utilization, computed as the ratio between the number of busy (non-idle) CPU time and the total time elapsed since the last evaluation. The time interleaving two consecutive workload estimations defines the sampling period, which can range from microseconds to milliseconds.

The Linux kernel embeds four standard scaling governors, i.e., *conservative, ondemand, schedutil, interactive* [14], differently available depending on the hosting platform. The *conservative* policy gradually adjusts the frequency level following a hysteresis scheme. When the CPU utilization exceeds a predefined upper threshold (e.g., 80%), the operating frequency

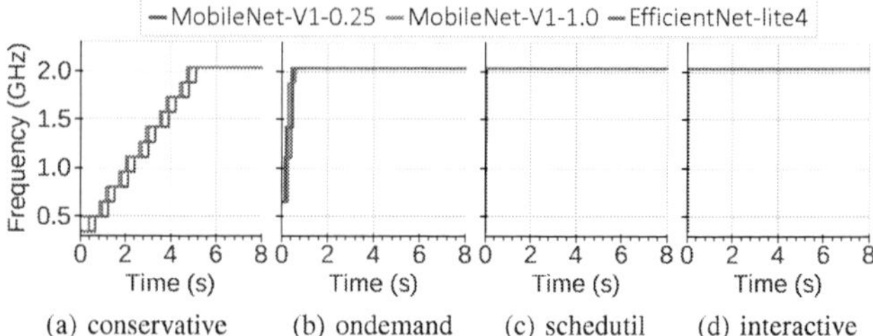

(a) conservative (b) ondemand (c) schedutil (d) interactive

Fig. 1. Examples of frequency traces collected on A57 with the standard procedure for DVFS SCAs.

is increased to the next available level; when it falls below the lower threshold (e.g., 20%), the frequency is decreased to the previous level. The *ondemand, schedutil,* and *interactive* governors implement a workload-proportional scaling approach, where the frequency is set to values linearly proportional to the CPU utilization. The *ondemand* policy maps 0%/100% CPU utilization to the lowest/highest frequency available, whereas the *schedutil* policy applies an over-scaling factor of 1.25× to prevent the frequency from being too low, which could lead to poor performance. Furthermore, *schedutil* boosts the frequency to the maximum level in case of a deadline event. The *interactive* governor favours frequency up-scaling and penalizes down-scaling. It reacts to sudden increases in CPU utilization by instantly forcing the maximum frequency level, and it waits a short period before decreasing the frequency to prevent costly performance fluctuations. That makes the *interactive* governor more responsive to intensive workloads but also more conservative in saving power.

Users with root privileges can alter the governor settings and parameters, e.g. the thresholds. Generic users without special permissions are limited to reading the governor policy and monitoring the operating frequency.

B. DVFS Side-Channel Attacks

DVFS SCAs collect frequency traces from the victim device and use ML models trained offline to steal sensitive data or to detect the applications in use. Extensive investigations on the efficacy of DVFS SCAs for data theft have been conducted in [15], showing that it is possible to identify the websites visited by users and even retrieve passwords entered on their smartphones. In [16] and [17], the authors showed the use of DVFS SCAs in the context of application detection. The general idea relies on profiling DVFS states during continuous loops of the same application. The approach was validated in [16] for a collection of 22 Android benchmarks running on an ARM Cortex-A CPU. A more exhaustive assessment conducted in [17] presented a benchmarking over two representative application suites: *PARSEC 3.0* and *SGX-bench.* The report proved that DVFS SCAs are effective for applications with long execution times and varying CPU usage (PARSEC 3.0) but are prone to fail for short and intensive workloads (SGX-bench), where the F1 score reaches a mere 10% over ten applications. As discussed in the next section, neural networks fall into the latter category, which motivates this work.

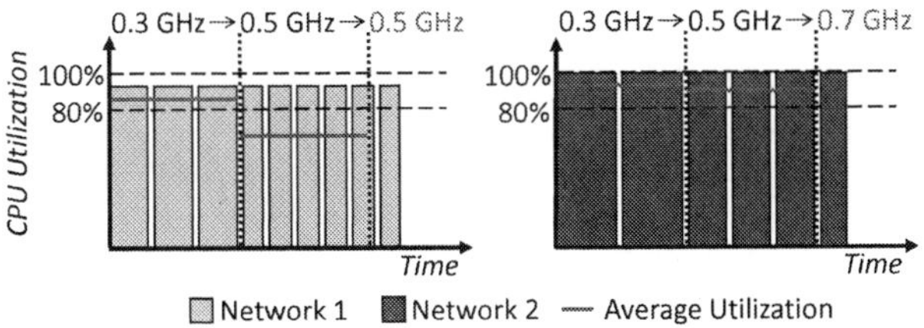

Fig. 2. CPU Frequency evolution during the execution of two different networks with the proposed approach.

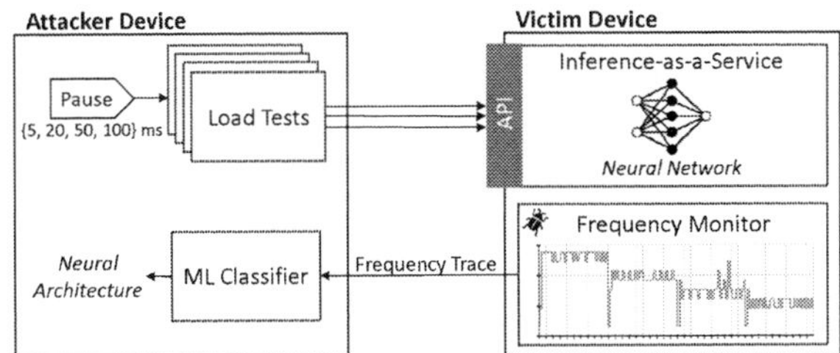

Fig. 3. Extraction phase overview.

III. Methodology

A. Understanding the Limitations of DVFS SCAs

In standard DVFS SCAs, the target application is run repeatedly while monitoring how the CPU frequency evolves. The same approach can be applied to IaaS by making continuous requests to the prediction API. Unfortunately, the deep neural networks backed by embedded applications are optimized through extensive algorithmic and code optimizations [18], resulting in dense workloads that are short in time and highly resource-demanding. This forces DVFS governors to set the highest frequency available and keep it constant until the end of the inference, no matter the network architecture. We deployed three networks of different sizes and algorithmic complexity (i.e., MobileNet-V1-0.25, MobileNet-V1-1.0, EfficientNet-lite4) on an A57 CPU (more technical details in Sec. IV-A) and we profiled the frequency traces during 8 seconds of continuous queries using the four governors introduced in Sec. II-A. The results reported in the line plots of Fig. 1 show overlaps that prevent discriminating the neural architecture under attack. The high resource demand quickly drives and keeps the frequency to the maximum level (2.0 GHz) even for the *conservative* governor.

B. Altering the Frequency Profiles

We propose a simple yet effective strategy to alter the CPU frequency updates driven by DVFS governors. The idea is to introduce pause intervals between consecutive API requests. Longer (shorter) idle periods result in lower (higher) CPU utilization and eventually lower (higher) frequency levels. The number of such pauses will be inversely proportional to the response time, a metric tightly coupled with (i) the neural architecture deployed for serving the request and (ii) the CPU frequency. Different frequency profiles over long observation periods serve as a signature to discriminate the neural architecture.

Fig. 2 graphically depicts how different frequency states could originate from two different neural architectures (Network 1 and Network 2, with Network 1 less resource-demanding than Network 2). For the sake of simplicity, the example assumes a *conservative* governor: the frequency is increased when the average CPU utilization during the observation period exceeds 80%. The coloured bars (light grey for Network 1, dark grey for Network 2) refer to the processing stage of an inference request. The bars' width reflects the CPU time to process the request, which is a function of the neural architecture and the CPU frequency. The bars' height reflects the CPU utilization during the request processing, which is another metric dependent on the neural architecture. Both response time and CPU utilization of each inference request affect the workload estimation during an observation window. At the end of each window, indicated by the vertical dotted lines, the governor evaluates the average CPU utilization (red lines in the plots) and decides on the frequency update. In the first window, the CPU operates at the lowest frequency of 0.3 GHz for both networks. Since the average CPU utilization exceeds the 80% threshold, the governor increases the frequency to 0.5 GHz for both networks. In the second window, the pause intervals (white spaces in the plots) inserted between the requests alter the workload estimation. With the increased frequency, the requests are processed faster (thinner bars for both networks). The number of pauses also increases (more white spaces), leading to a decrease in average CPU utilization, potentially falling below the threshold. This is the case for the lighter network (Network 1), for which the CPU utilization falls below the threshold (<80%) and the governor does not change the frequency. The same is not true for the more complex network (Network 2), where the average utilization is still over-threshold (>80%) and the governor drives a frequency increase to 0.7 GHz. Therefore, the frequency traces for the two networks exhibit distinct patterns. Without the pause intervals, the governor would have made the same frequency update for both networks in the third window, resulting in no discernible difference between their frequency traces. It is worth noticing that the length of the pause intervals is a hyper-parameter that can be leveraged to enhance the differentiation capability among different neural networks and CPUs. Further implementation details are provided in the following subsections.

C. Load testing

Load testing is a type of performance testing that aims at assessing how a system behaves under (i) specific workloads or (ii) requests volume. A standard approach is to generate a large amount of traffic by creating a pool of concurrent virtual users that query the service according to a predefined request rate. The number of virtual users and the request rate are parameters that can be fine-tuned to emulate different scenarios and test the stability and scalability of the service.

We borrowed a typical load testing infrastructure to implement the pausing mechanism introduced in the previous

TABLE II
HARDWARE PLATFORMS, OPERATING SYSTEM (OS), MAXIMUM (MAX. FREQ.) AND MINIMUM (MIN. FREQ.) FREQUENCY, AND NUMBER OF FREQUENCY LEVELS OF THE TARGET CPUS.

CPU	Platform	OS	Min. Freq.	Max. Freq.	Levels
A72	Raspberry Pi 4	Ubuntu 22.04	600 MHz	1.500 GHz	10
A57	NVIDIA TX2	Ubuntu 18.04	345 MHz	2.035 GHz	12

TABLE III
GOVERNORS AND THEIR SAMPLING PERIOD (ms) ON THE TARGET CPUS.

CPU	conservative	ondemand	schedutil	interactive
A72	10.0	10.0	10.0	N/A
A57	300.0	300.0	0.5	20.0

TABLE IV
POOL OF NEURAL NETWORK ARCHITECTURES.

Architecture	Layers	Size (MB)	Latency (ms) A72	Latency (ms) A57	Util. (%) A72	Util. (%) A57
MobileNet-V1-0.25	28	0.49	4.71	3.57	85.6	92.0
MobileNet-V1-0.50	28	1.31	9.68	7.44	91.9	95.1
MobileNet-V1-0.75	28	2.51	17.80	13.53	95.0	97.0
MobileNet-V1-1.0	28	4.09	27.86	21.03	96.8	98.0
MobileNet-V2	53	3.42	23.39	17.85	96.2	97.7
MobileNet-V3-small	54	2.52	10.20	8.06	92.5	95.6
MobileNet-V3-large	64	5.35	25.80	20.26	96.4	97.9
EfficientNet-lite0	50	5.18	21.62	16.48	90.7	90.3
EfficientNet-lite1	65	6.12	32.32	23.87	92.4	92.6
EfficientNet-lite2	65	6.83	44.96	33.37	93.6	93.6
EfficientNet-lite3	74	9.15	68.97	51.46	95.2	95.2
EfficientNet-lite4	92	14.34	119.60	89.04	96.8	96.9

subsection, specifically a single virtual-user setup with the pause interval as the only testing parameter. The pause interval is defined as the time distance between the service response and the next client request and can be used as a knob to improve the SCA efficiency. More in detail, it is a knob to increase and reduce CPU utilization. The optimal values, i.e., those which produce enough frequency variations, may depend on non-controllable factors, such as the governor in use at the endpoint hosting the inference service and the neural network characteristics (latency and utilization). We thereby resorted to multiple load tests where the pause length is swept over a predefined set of values, as detailed in the next subsection.

D. Attack Overview

The SCA plan involves performing four load tests in sequence, each with a duration of 10 s and a growing pause length of 5, 20, 50, and 100 ms.

We consider a threat model based on the following assumptions. First, the attacker owns a device, referred to as the template device, with characteristics and specifications identical to those of the victim device; this falls into the broad category of *profiled attacks* [19]. Second, the attacker has access to the victim's device to install malware for monitoring and transmitting the CPU frequency traces. This kind of malware can gather raw data without requiring root permissions. Third, the target neural network belongs to a pool of well-known neural network architectures (although network weights may differ). This assumption is quite realistic in most use cases. It is standard practice today to leverage transfer learning [20], a deep learning strategy where a neural architecture is taken from a collection of publicly available networks and fine-tuned on a proprietary dataset.

Following the standard organization of a profiled attack, our method consists of a *profiling* phase and an *extraction* phase.

In the profiling phase, we collect frequency traces of the target neural architectures to train an ML classifier capable of inferring the target architecture from traces. To collect the dataset, we execute the load tests described in Sec. III-C on each target neural architecture deployed on the template device. During the execution of the load tests, a background process records the frequency profiles by reading the frequency level every 10 ms using the *CPUFreq* interface. The same load test is processed 50 times to account for small fluctuations in the frequency traces caused by operating system routines running in the background. This procedure is repeated four times, one for each pause length. The resulting dataset thus includes multiple frequency traces annotated with a label corresponding to one neural architecture. The classifier consists of a MiniRocket [21] feature extractor, commonly used for time-series data, followed by a multinomial logistic regressor; it is trained in a supervised manner using the *lbfgs* solver and 5-fold stratified cross-validation for hyper-parameters optimization. The training loop consists of 200 iterations. It is worth emphasizing that the data collection campaign and the training pipeline are repeated for all governors and template devices considered in our experiments, resulting in a dedicated ML classifier for each governor and hardware configuration.

In the extraction phase, the attack is deployed to steal the target neural architecture, as illustrated in Fig. 3. The victim device is one of the template devices analyzed in the profiling phase, but the target neural architecture is unknown. The attacker runs the four load tests on the prediction endpoint, recording the victim device's CPU frequency through the pre-installed malware. After the load tests, the collected frequency trace and the active governor are sent to the attacker and fed into the ML classifier trained for that specific governor. The ML classifier returns the neural architecture used by the victim inference service.

IV. EXPERIMENTAL RESULTS

A. Experimental Setup

We operated two commercial platforms as test benches: the Raspberry Pi 4 and the NVIDIA Jetson TX2. These platforms are powered by different versions of the ARM Cortex-A CPU, i.e., A72 and A57 respectively, whose technical specifications are reported in Table II. Both CPUs are quad-core but differ in terms of frequency range, frequency levels, and governors' setups. Table III shows the available governors and the sampling period (defined by the CPU vendor). Notice that the sampling period of a governor may vary by one order of magnitude depending on the CPU architecture, affecting the speed of the DVFS state update.

TABLE V
RESULTS ON A72.

	Architecture	conservative	ondemand	schedutil
1	MobileNet-V1-0.25	100.0%	100.0%	99.0%
2	MobileNet-V1-0.50	99.0%	93.8%	84.5%
3	MobileNet-V1-0.75	99.0%	100.0%	100.0%
4	MobileNet-V1-1.0	99.0%	100.0%	100.0%
5	MobileNet-V2	100.0%	100.0%	100.0%
6	MobileNet-V3-small	100.0%	94.2%	86.5%
7	MobileNet-V3-large	100.0%	100.0%	100.0%
8	EfficientNet-lite0	100.0%	100.0%	100.0%
9	EfficientNet-lite1	99.0%	100.0%	100.0%
10	EfficientNet-lite2	100.0%	100.0%	100.0%
11	EfficientNet-lite3	100.0%	100.0%	100.0%
12	EfficientNet-lite4	100.0%	100.0%	100.0%
	Macro-F1	99.7%	99.0%	97.5%
	Top-1 Accuracy	99.7%	99.0%	97.5%
	Top-2 Accuracy	100.0%	100.0%	99.8%
	Misclassifications	-	2-6	2-6

TABLE VI
RESULTS ON A57.

	Architecture	conservative	ondemand	schedutil	interactive
1	MobileNet-V1-0.25	99.0%	100.0%	100.0%	100.0%
2	MobileNet-V1-0.50	62.6%	100.0%	100.0%	100.0%
3	MobileNet-V1-0.75	93.9%	100.0%	100.0%	99.0%
4	MobileNet-V1-1.0	80.8%	59.3%	95.9%	90.7%
5	MobileNet-V2	68.0%	99.0%	99.0%	99.0%
6	MobileNet-V3-small	63.4%	100.0%	100.0%	100.0%
7	MobileNet-V3-large	77.9%	72.9%	95.1%	94.0%
8	EfficientNet-lite0	67.9%	100.0%	100.0%	99.0%
9	EfficientNet-lite1	93.2%	100.0%	100.0%	98.0%
10	EfficientNet-lite2	95.9%	100.0%	100.0%	98.0%
11	EfficientNet-lite3	100.0%	100.0%	100.0%	65.4%
12	EfficientNet-lite4	100.0%	100.0%	100.0%	65.3%
	Macro-F1	83.6%	94.3%	99.2%	92.4%
	Top-1 Accuracy	83.5%	94.5%	99.2%	92.3%
	Top-2 Accuracy	97.3%	99.7%	99.8%	99.8%
	Misclassifications	2-6, 5-8, 4-7	4-7	4-7	11-12

Table IV collects the CNNs used as benchmarks, all taken from TensorFlow HuB [22] in TFLite format; it shows the average values of latency and CPU utilization over 100 consecutive inference runs processed at the maximum frequency with 4-thread execution. The picked networks, pre-trained on the ImageNet dataset and quantized to 8-bit, belong to four distinct families: MobileNet-V1 [23], MobileNet-V2 [24], MobileNet-V3 [25], and EfficientNet [26]. The pool includes architectures with very similar characteristics. For example, networks of the MobileNet-V1 family have the same number of layers but differ in the number of filters. Furthermore, some architectures exhibit similar latency and utilization, such as Mobilenet-V1-0.50 and MobileNet-V3-small. These similarities can affect the detection accuracy of the attack. Our experiments aim to evaluate the capability of our methodology to identify the different networks, even under these challenging conditions.

B. Results

We evaluate the performance of the attacker classifier using standard metrics for multi-class classification. First, we report the F1 score, which is defined as the harmonic mean of precision and recall. Precision measures the fraction of correctly predicted positive instances (true positives) over the total number of instances predicted as positive (true positives plus false positives). Recall measures the fraction of correctly predicted positive instances over the total number of positive instances (true positives plus false negatives). Higher F1 scores indicate more accurate classifications. We also compute per-class F1 scores, which measure the performance of the model for each class individually. The per-class F1 score is computed by considering each class as positive and the remaining classes as negative, resulting in a score for each class. The average F1 score (Macro-F1) is computed by taking the arithmetic mean of the per-class F1 scores, providing an overall measure of the model's performance across all classes. Additionally, we report the Top-1 Accuracy, which is the ratio between the number of correct predictions and the total number of instances in the dataset. With the Top-2 Accuracy, we consider a prediction correct if any of the two classes with the highest prediction probabilities matches the expected label. All these metrics were evaluated on a test set having the same size as the training set, i.e. 600 traces (50 traces per network) for each CPU/governor configuration.

The experimental results presented in Tables V and VI prove the efficacy of the proposed SCA method. The reported Macro-F1 scores range from 97.5% to 99.7% for the A72 and from 83.6% to 99.2% for the A57, depending on the selected governor. These results indicate that the classifier deployed on the attacker side can accurately identify the target network, despite the different CPU architectures and the governor settings.

The target network and the governor sampling period primarily affect the prediction quality. Concerning the target network, we observed that most misclassifications occur between network pairs with similar characteristics, as highlighted in the last row of the results tables. This observation is consistent with the Top-2 accuracy, which reveals that even in cases where the Top-1 accuracy is low, the Top-2 accuracy is significantly higher. For instance, the A57 with a *conservative* governor has the lowest Top-1 accuracy at 83.5%, yet its Top-2 accuracy is much higher at 97.3%. Overall, the Top-2 accuracy is close to 100% in all cases, and misclassifications are limited to only a few governors and network pairs.

Regarding the sampling period, higher values lead to fewer CPU frequency switches, resulting in less information leakage in the traces. This effect is evident for the *ondemand* and the *conservative* governors, which show prediction scores on the A57 (sampling period 300 ms) lower than that on the A72 (sampling period 10 ms). In *conservative* mode, the Macro-F1 score is 99.7% for the A72, compared to 83.6% for the A57. For the *schedutil* governor instead, the scores on the A57 (sampling period 0.5 ms) get higher than that recorded on the A72 (sampling period 10 ms). Intuitively, under larger sampling periods, the probability of frequency switches reduces, and variations in CPU utilization might get missed, making it more challenging to profile the neural architecture.

Fig. 4 provides visual evidence of the effectiveness of the proposed methodology. The figure shows frequency traces

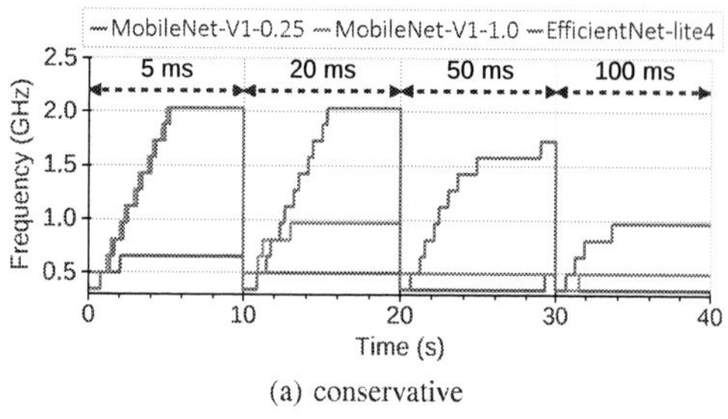

(a) conservative

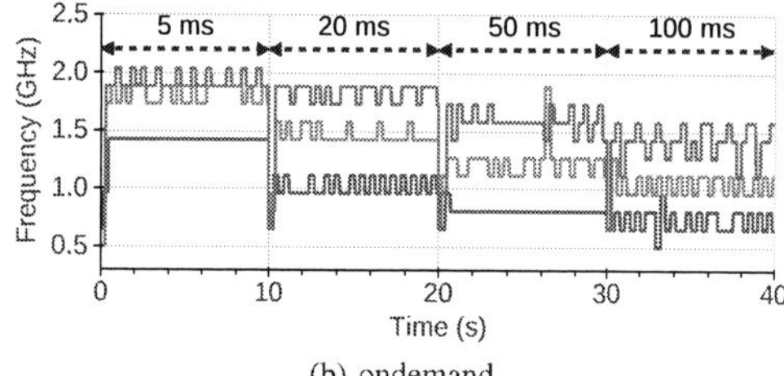

(b) ondemand

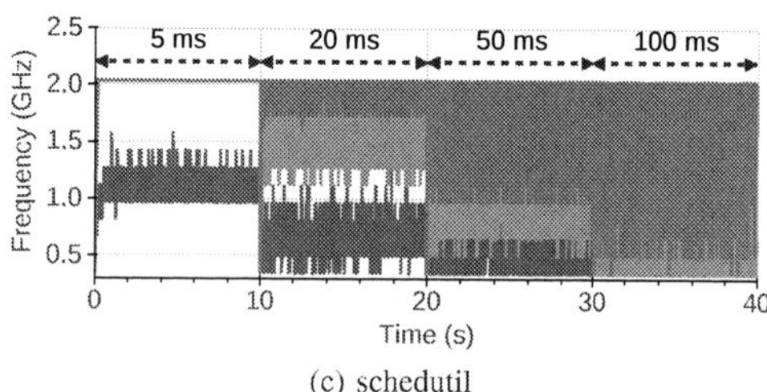

(c) schedutil

Fig. 4. Examples of frequency traces collected on the A57 with the proposed load testing methodology. The four load tests are delimited by dashed arrows, with pause lengths indicated above the arrows.

collected for three networks deployed on the A57 CPU, i.e., MobileNet-V1-0.25, MobileNet-V1-1.0, and EfficientNet-lite4, with the *conservative*, *ondemand*, and *schedutil* governors (*interactive* omitted for the sake of space). These are the same networks of Fig. 1. Unlike standard DVFS SCAs, the frequency traces are distinguishable, enabling a successful attack. The improvement is due to the testing procedure, which emphasizes the characteristics of the governors. With *conservative*, the frequency gradually increases and stops at different levels. With *ondemand*, it switches to a value proportional to the workload with small fluctuations due to system noise. With *schedutil*, it oscillates with sudden surges due to the low governor sampling period (0.5 ms).

Moreover, the plots demonstrate the need for multiple load tests with different pause intervals. This can be seen in the traces collected during the load test with a pause of 5 ms for the *conservative* and *schedutil* governors (Figs. 4a and 4c). In these cases, the traces of MobileNet-V1-1.0 and EfficientNet-lite4 do overlap, while they get distinguishable with longer pause intervals (20 ms, 50 ms, and 100 ms). We observed similar cases for other instances not reported in the figure. In summary, only a combination of multiple load tests can create distinguishable traces for all networks and governors.

V. Conclusion

This work has shown that DVFS SCAs can perform network fingerprinting in edge inference services, bypassing IaaS protection via software-based procedures. Our methodology based on load testing has proven capable of generating frequency traces that serve as signatures, enabling the detection of neural architectures with high accuracy. Our findings highlight the need for security measures to counteract such attacks, which can expose a concrete source of leakage for neural architecture stealing with severe implications for security and privacy in the context of IaaS. In particular, malicious actors with knowledge of the neural architecture can launch more efficient adversarial attacks to mislead the output prediction or white-box attacks aiming to replicate the network functionality.

References

[1] V. Peluso *et al.*, "Inference on the edge: Performance analysis of an image classification task using off-the-shelf cpus and open-source convnets," in *SNAMS*, 2019.

[2] J. Hou *et al.*, "Model protection: Real-time privacy-preserving inference service for model privacy at the edge," *IEEE Trans. Dependable Secur. Comput.*, vol. 19, no. 6, pp. 4270–4284, 2022.

[3] Y. Xiang *et al.*, "Open DNN box by power side-channel attack," *IEEE Trans. Circuits Syst.*, vol. 67-II, no. 11, pp. 2717–2721, 2020.

[4] X. Hu *et al.*, "Deepsniffer: A DNN model extraction framework based on learning architectural hints," in *ASPLOS*, 2020.

[5] Y. Zhu *et al.*, "Hermes attack: Steal DNN models with lossless inference accuracy," in *USENIX Security Symposium*, 2021.

[6] L. Batina *et al.*, "SCA strikes back: Reverse-engineering neural network architectures using side channels," *IEEE Des. Test*, vol. 39, no. 4, pp. 7–14, 2022.

[7] M. Yan, C. W. Fletcher, and J. Torrellas, "Cache telepathy: Leveraging shared resource attacks to learn DNN architectures," in *USENIX Security Symposium*, 2020.

[8] V. Duddu *et al.*, "Stealing neural networks via timing side channels," *CoRR*, vol. abs/1812.11720, 2018.

[9] B. A. D. Kumar *et al.*, "Inferring DNN layer-types through a hardware performance counters based side channel attack," in *AIMLSystems*, 2021.

[10] K. Patwari *et al.*, "DNN model architecture fingerprinting attack on CPU-GPU edge devices," in *EuroS&P*, 2022.

[11] Y. Liu and A. Srivastava, "GANRED: gan-based reverse engineering of dnns via cache side-channel," in *CCSW*, 2020.

[12] S. E. Arefin and A. Serwadda, "Deep neural exposure: You can run, but not hide your neural network architecture!" in *IH&MMSec*, 2021.

[13] A. Büşra and Y.-M. Ayse, "A study on power and energy measurement of nvidia jetson embedded gpus using built-in sensor," in *UBMK*, 2022.

[14] Linux Governors. Accessed on 2023/03/20. [Online]. Available: https://docs.kernel.org/admin-guide/pm/cpufreq.html

[15] D. R. Dipta and B. Gülmezoglu, "DF-SCA: dynamic frequency side channel attacks are practical," in *ACSAC*, 2022.

[16] N. Chawla *et al.*, "Application inference using machine learning based side channel analysis," in *IJCNN*, 2019.

[17] C. Liu *et al.*, "Methodology of assessing information leakage through software-accessible telemetries," in *HOST*, 2021.

[18] C. Wu *et al.*, "Machine learning at facebook: Understanding inference at the edge," in *HPCA*, 2019.

[19] M. Taouil, A. Aljuffri, and S. Hamdioui, "Power side channel attacks: Where are we standing?" in *DTIS*, 2021.

[20] A. Kolesnikov *et al.*, "Big transfer (bit): General visual representation learning," in *ECCV*, 2020.

[21] A. Dempster, D. F. Schmidt, and G. I. Webb, "Minirocket: A very fast (almost) deterministic transform for time series classification," in *KDD*, 2021.

[22] TensorFlow Hub. Accessed on 2023/03/20. [Online]. Available: https://tfhub.dev

[23] A. G. Howard *et al.*, "Mobilenets: Efficient convolutional neural networks for mobile vision applications," *CoRR*, vol. abs/1704.04861, 2017.

[24] M. Sandler *et al.*, "Mobilenetv2: Inverted residuals and linear bottlenecks," in *CVPR*, 2018.

[25] A. Howard *et al.*, "Searching for mobilenetv3," in *ICCV*, 2019.

[26] M. Tan and Q. V. Le, "Efficientnet: Rethinking model scaling for convolutional neural networks," in *ICML*, 2019.

Hardware Trojans in fdSOI

Christian Lanius, Florian Freye, Shutao Zhang, Tobias Gemmeke
RWTH Aachen University
Aachen, Germany
{lanius,freye,zhang,gemmeke}@ids.rwth-aachen.de

***Abstract*—With shortening turn-around times and increasing complexity for digital circuits, design reuse, third party IP and today even physical chiplets has increased. Malicious actors have more options to introduce hardware backdoors to packaged systems, which will leak data if triggered. In this work, we show two novel approaches to introduce such backdoors, that are possible due to the specifics of fully depleted silicon on insulator (fdSOI) technology. The first method relies on modifying the doping profile of an antenna cell to introduce a covert short between the back gate and logic signals. The second method constructs specific illegal states which are latched when the clock is running with the trigger frequency. Basic test structures have been designed such that they are DRC and STA clean. LVS does not reveal the hidden structure, while measurements in silicon confirm their operation.**

***Index Terms*—hardware trojans, fully depleted silicon on insulator, static timing analysis, antenna effect**

I. Introduction

Hardware trojans have entered the discourse in the last 15 years, as a danger due to more federated chip developments: Buying IP from vendors and fabricating ICs at independent fabs has broken the chain of trust. In the most extreme case, the specification, post package testing and in-field-monitoring are the only trusted parts of the lifetime cycle [1].

With this loss of control over the circuit, the danger of hardware trojans has increased: Unwanted covert functionality which can lead to a variety of undesired effects, such as leakage of private data or denial of service. Hardware trojans can be inserted by various actors, from IP houses, to implementation services and foundries.

Because of the reliance of the modern world on security sensitive applications, such as medical or military, on modern ICs and processors, a lot of work has been done to investigate ways malicious actors might introduce hardware trojans and how they can be detected. In this paper, we present two novel approaches to implement hardware trojans which are specific to fdSOI devices, due to their superior tuning ability and the introduction of an isolator underneath the channel.

In summary, our contributions in this paper are as follows:

- Design of a modified antenna cell shorting signals to the well by only changing doping
- Design of a construction approach to make intermediate states in digital logic predictable
- Assessment of their use as hardware trojans
- Simulative and measurement results of the proposed methods in a 22nm fdSOI technology

This work was partially funded by the Federal Ministry of Education and Research (BMBF, Germany) in the project NEUROTEC II (project number 16ME0399).

II. Background and Previous Work

A. Hardware Trojans and their Detection

A challenge in assessing hardware trojans is the wide range in terms of implementation (how is it implemented), trigger (when does it act) and payload (what does it do). In general terms, a hardware trojan is an unwanted part of circuitry, which will change the behavior of the circuit when the trigger is being executed. A lot of work has gone into analyzing hardware trojans and dividing them into meaningful taxonomies [1], [2].

Trojan analysis and detection is a field as large as the development of the trojans in order to provide sufficient defense against them. They can be roughly split into two types: Detecting the trojan during validation and detecting them in-field. As trojans should only be triggered in-field, integrated trojan detection circuitry [3], [4] can only detect them during real operation.

Traditional verification methods can be used to evaluate RTL code for hardware trojans [5]: Coverage indicates possible lines of code which are not executed or FSM states not reached [6] and property definition and their assertion show deviations from the specification. RTL code not executed during test points towards possible trojans and deviations from the specification indicate hardware trojans or bugs. For gate-level netlists, toggle coverage can be used to highlight untested parts of the circuit.

One downside is the "high level" modeling that is used: Even if a netlist is investigated, it has to be assumed that the design is behaving synchronously w.r.t. the clocked registers. Any intermediate state on the inputs of the registers can be safely discarded and undetermined states might appear, due to the logic implemented, if the clock is not enabled. As they do not get latched by the register though, their value does not matter. Similarly, it is obvious that any analysis on gate level netlist is dependent on the model of the gates to be correct.

The authors in [7] propose to replace the p-doping of the pMOS source and drain region with n-doping, thus creating a short to the n-well, which is tied to V_{dd} in bulk technologies. This way they generate a "stuck-at" fault. While the work was done using the FreePDK45 [8] PDK and is not directly applicable to modern technologies, it shows that doping based trojans are especially hard to detect and are a realistic scenario for foundry-side attacks.

979-8-3503-1176-1/23 $31.00 © 2023 IEEE

Foundry side attacks could also use modified dummy contacts, as seen in [9], to alter gate functionality. This method does however increase manufacturing complexity.

B. fdSOI Technology

Fully depleted silicon on insulator (fdSOI) devices provide better power-performance trade-off than classic bulk technologies. They reduce leakage, provide larger tunability of the threshold voltage via the backgate and remove parasitic bipolar junction transistors, preventing latch-up [10]. The cross section of a typical fdSOI transistor is shown in Fig. 1: By fabricating the devices on a thin buried-oxide, the channel can be undoped (i.e. fully depleted) and it is isolated from the well.

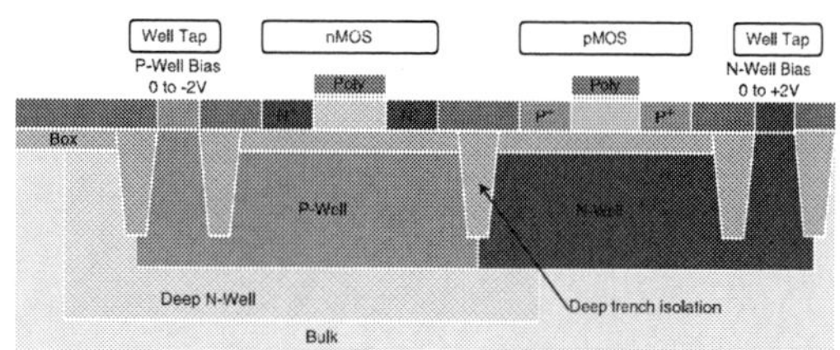

Fig. 1: Flipped well fdSOI transistors. Adapted from [11].

If an nMOS transistor is placed in an n-well, it can be forward body biased by applying a positive back gate voltage, decreasing the threshold voltage and increasing speed at the cost of higher leakage. If placed in a p-well, negative body bias can be applied, decreasing leakage and speed. In traditional bulk technologies the body bias potential is limited, as a too high forward voltage between the bulk and source contact $|V_{BS}|$, will lead to unacceptable power loss over the corresponding pn-diode. This diode is not present in fdSOI technologies, as the buried oxide isolates the diffusion areas from the wells. Thus, the body bias voltage can be selected over a larger range, even beyond the supply voltage.

During fabrication, before the entire metal stack is connected, large continuous pieces of metal act as "antennas" and accumulate charge [12], [13]. This can lead to the destruction of the devices, as the electrical field over the gate oxide can become too large. To prevent device breakdown during manufacturing, antenna cells are inserted during the digital flow, which contain two diodes in opposite directions, thus clamping the voltage to the breakdown voltage (plus the forwards bias voltage), as shown in Fig. 2.a.

III. Proposed Trojan Techniques

A. Fake Antenna Cell

Antenna cells are commonly used in digital designs, resulting from the physical implementation, but with supposedly no impact on the logic function implemented, (cf. cross section of the antenna cell in Fig. 2.a). The principle of this first trojan is to realize a hidden connection between two signals, where one of the two can selectively override the other. This connection is realized as well connections to look like an antenna diode, as shown in Fig. 2.b.

By doing such a modification of the doping profile only, the logic signal carried by the metal can be controlled by the body bias voltage. A visual analysis, even with a scanning electron microscope, does not discover the change in doping profile. Only a time consuming and expensive scanning capacitance microscopy analysis would discover the modified doping profile for an antenna cell [14]. However, such an approach is destructive and not scalable to larger circuits [2], [7].

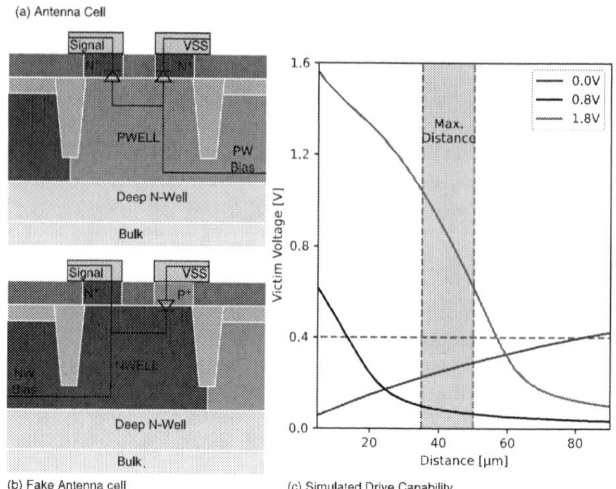

Fig. 2: (a) Real antenna cell. (b) Proposed fake antenna cell. (c) Simulated voltage as a function of the distance between welltap and fake antenna cell.

The well resistance and $R_{DS,on}$ of the overpowered driver cell form a voltage divider between 0 V and V_{bias} or between V_{dd} and 0 V, as the well can not be negatively biased in the proposed scheme without forwards biasing diodes and increasing leakage. A positive V_{bias} can be as high as 1.8 V. Fig. 2 shows the resulting voltage as a function of the distance to the welltap, if the victim cell is a tie cell (tie-high for the 0 V case, tie-low otherwise). The victims can be easily overpowered up to a distance of 15 µm. If a higher bias voltage is used, even longer distances are possible, though high voltages for V_{GS} can become a concern. An uppper bound on distance to the welltaps is given by the design rules anyways.

B. Suitable signals for the fake antenna cell

One downside to consider in the fake antenna cell is that it is an electrical short: Normal signals which have to toggle for the system to work correctly in the normal state are not suitable to be overwritten by the antenna cell. Only constant signals are suitable to be overwritten.

Clock enable cells are a prime candidate: They provide a secondary clock enable pin, which is tied to ground if the registers are not part of the scan chain. At the same time, they provide a mechanism to impact the state of registers. The signal on the D pin of the register is a "don't care" if the enable of the clock gating cell is not set. The fake antenna cell now forces the cell to latch this state, with a setup as shown in Fig. 3. Oversights in the RTL description can make this value predictable, either by being explicitly coded or by being synthesized to a specific value, based on the logic function.

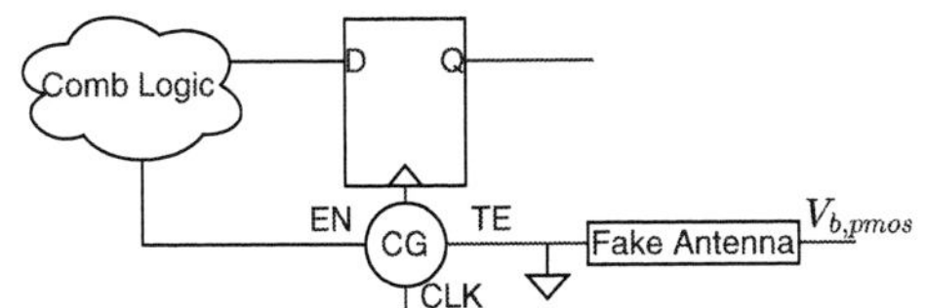

Fig. 3: The fake antenna cell can overpower the tie low of the TE pin to make the register latch invalid states.

Formally, all values on pin D are equivalent if enable is not set (as they will never be latched during normal operation), thus logic analysis of a malicious circuit will never raise any issues. According to formal equivalence checks, the code snippet below is equivalent to the same circuit with the mux statement removed (replaced with a direct assignment of a to d).

```
wire [3:0] a,d;
assign d = en ? a : 4'b1010;
always @(posedge clk) if (en) q <= d;
```

By setting a DONT_TOUCH constraint on d, the synthesis tool may not optimize away the statement in line 2. As the muxed value is constant (here 4b1010), it will be implemented as a simple OR/AND operation with the EN signal.

While the invalid state transition can be forced by this construction, it also shows up naturally in synthesized code, which is not specifically made to exhibit a weakness to this kind of trojan.

A scan-chain could undermine security of sensitive parts of a circuit (such as RNG or encryption/decryption cores) [15] and recommendations from industry indicate the non-inclusion of security critical components in the scan chain [16]. Hence, a tied-off TE pin can be considered a common practice in security cores, making the proposed method less suspicious.

C. Timing Based Trojan

Digital circuits are traditionally characterized by the static register transfer function: Based on the state of the registers, combinational logic computes the next state, which will be latched by the registers at the next clock cycle. Most importantly, the functionality is independent of the clock frequency.

This means that analysis can be reduced down to the transfer function T, which evolves the current state S_k at step k into $S_{k+1} = T(S_k)$. Static timing analysis will guarantee that this "static" assumption is valid, i.e. that the specified clock frequency f_{SO} is slow enough for the digital logic. Setup checks guarantee that the longest path is shorter than the clock period and hold checks guarantee that no old data is latched by the registers.

If those checks fail, the logic function of the circuit is not defined. Anyway, the timing critical path is in typical designs only rarely triggered, such that the silicon still operates properly most of the time. In this work, we propose to modify an implemented design to purposely change the functionality as function of the time window a signal is observed in, i.e. latched in a register. Formally, we modify the digital circuit such that

$$S_{k+1} = \begin{cases} T^*(S_k) \text{ if } f \approx f_t \\ T(S_k) \text{ if } f < f_{\text{SO}} \\ \text{undefined else} \end{cases} \tag{1}$$

In the context of hardware trojans, the modified transfer function $T^*(S_k)$ leaks secrets or weakens encryption, when the operating frequency is near the trigger frequency f_t.

The required modifications can be constructed and validated with special timing constraints. As an input to the construction, a one-to-one mapping between target nets and victim registers is required. The victim register will latch the value of the target net instead of the "correct" data.

The design without any modifications to its functionality is implemented with a traditional digital flow. At this stage, post layout extraction can be done and parasitics and propagation delay is accurately known. A mux is inserted in front of the victim register, with its second input connected to the target. If the mux control signal is set, the target value will propagate to the victim register, otherwise no change in functionality is observed.

The mux control signal must, under static analysis, be always 0. However it must also switch to 1. The most simple circuit would be a toggling signal like the LSB of a counter, which is XOR'd with a delayed version of itself. This will generate a 1 pulse with its width determined by the inserted delay, but will statically always be 0. A minimal functional example of such circuit is shown in Fig. 4.

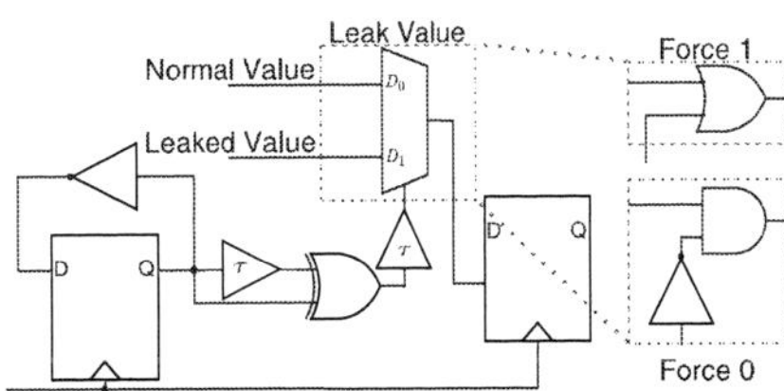

Fig. 4: A pulsed mux can replace the correct value. Logic gates can be used to force the register to latch a specific value.

By appropriately delaying this pulse, so that its high phase is aligned to the trigger frequency's clock edge, the hardware trojan can be implemented. In normal operation, the clock edge will lie outside of the pulse and the "correct" signal will propagate through the mux and get latched by the victim, as shown as "normal operation" in Fig. 5.

Static timing analysis can be used to ensure correct timing of the trojan by validating these constraints (from the toggling register's clock pin $\text{REG}_{\text{tog,CLK}}$ through the pulse generating XOR gate's Z pin XOR_Z to the victim register's D pin $\text{REG}_{\text{vic,D}}$:

$$\begin{aligned}
\text{REG}_{\text{tog,CLK}} \uparrow\rightarrow \text{XOR}_Z \uparrow\rightarrow \text{REG}_{\text{vic,D}} &< T_t - \tau_{\text{setup}} \\
\text{REG}_{\text{tog,CLK}} \uparrow\rightarrow \text{XOR}_Z \uparrow\rightarrow \text{REG}_{\text{vic,D}} &> \tau_{\text{hold}} \\
\text{REG}_{\text{tog,CLK}} \uparrow\rightarrow \text{XOR}_Z \downarrow\rightarrow \text{REG}_{\text{vic,D}} &> T_t + \tau_{\text{hold}} \\
\text{REG}_{\text{tog,CLK}} \uparrow\rightarrow \text{XOR}_Z \downarrow\rightarrow \text{REG}_{\text{vic,D}} &< 2T_t - \tau_{\text{setup}}
\end{aligned}$$

While static timing analysis is well suited to analyze these constraints and make sure they are fulfilled, setup and hold fixing of standard implementation tools does not generate appropriate structures to fulfill them. We instead can construct the appropriate delays by increasing the delay before the XOR and MUX.

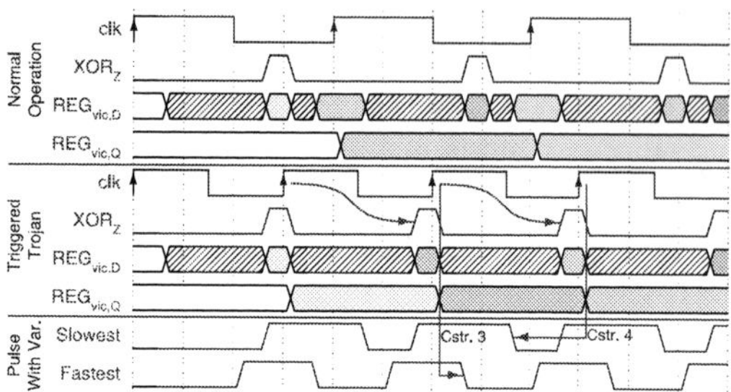

Fig. 5: The pulse changes the data on the D pin of the victim register.

The insertion is done on a fully placed and routed database, with all other parts being kept fixed, enabling the implementation tool to only reroute the minimum number of signals for the inserted/removed delays and perform incremental extraction. It is visualized in Fig. 6.a.

After inserting the structure with minimal delays (left most delays in Fig. 6.a), we iteratively increase the number of delays in front of the mux until the first constraint is only barely fulfilled. In Fig. 6.a, this corresponds to the point where the difference in delays for all corners starts to increase. Now, the rising edge of the pulse is at the closest possible position (time wise) left of the next clock edge (in the figure aligned to the dotted line). This also fixes the trivial second constraint.

In the slowest corner, the rising edge is barely in front of the rising edge of the clock. In the fast corner, the pulse arrives earlier and is therefore already gone before the clock edge appears. This violates constraint 3, i.e. hold slack for the path is negative (the slower signal in the fast corner is still inside the gray box in Fig. 6.a). By inserting delays in front of the XOR gate, the pulse is widened until the hold constraint is fulfilled. For illustrative purposes, we continued the delay insertion in Fig. 6.a until constraint 4 is no longer fulfilled. The smallest window is observed as soon as constraint 3 is fulfilled and would be used in practice.

All delays inserted are necessary to fulfill constraints 2 to 3. Constraint 4 can not be explicitly fulfilled, as the delay is determined by the other constraints: If the delay variations between SS and FF is too large, constraint 4 is violated before constraint 3 is fulfilled, as can be seen in Fig. 6.a, in the upper most row, where constraint 4 is violated before constrained 3 is fulfilled.

fdSOI technology can be trimmed far more via back body biasing than traditional bulk technology. Adaptive body biasing is a technique unique to fdSOI processes, where the body bias is controlled to counteract PVT variations, as well as other time dependent effects like aging, by adaptively setting the body bias [17]. Reference ring oscillators are distributed on the chip, whose nominal frequency is known. Based on their measured frequency, the body bias can be adapted until the ring oscillators, and thus the digital logic, exhibits delays at worst similar to TT. The cost for this improved performance in SS is the increase in leakage power: ABB increases leakage in SS corner and high temperature by 10x.

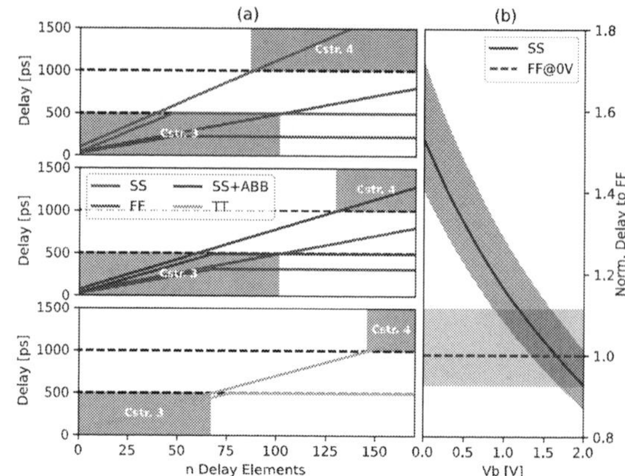

Fig. 6: (a) Visualization of proposed construction of pulse with traditional corners, SS with ABB and with a single corner. (b) Delay in SS (+OCV), as a function of bias voltage.

Because of process variation, traditional corners (upper graph) exhibit too much variations: Constraint 3 and 4 can not be fulfilled at the same time. If ABB is activated in the SS corner (center graph), a window opens up with a well defined trigger frequency range. If ABB is used to only counteract process variation, it reduces down to a single corner (bottom graph), where the trigger frequency window can be selected in a wide range. Fig. 6.b shows that the body biasing's impact is large enough to trim SS devices to FF performance, thus such an assumption of a single corner is valid.

D. Detectability

The fake antenna cell is invisible for methods based on netlist analysis: Any method will assume the "antenna cell" to be just that. As it has no effect on the operation, no approach can flag it as a suspicious gate.

The presented modification of the circuit to force the correct idle state for the register is logically equivalent to the non-modified netlist. The same is true for the logic inserted for the timing based trojan. Thus, any method based on logic equivalence checking will not flag the inserted logic.

IV. Applications

A. Privilege Escalation in RISC–V Processor

Modern processors with multiple privilege levels and support for operating systems are complex and are thus one of the common elements to be bought off-the-shelf. Using open source implementations, like the RISC–V processor CVA6 [18], is becoming more popular as well.

The CVA6 core is vulnerable to the fake antenna cell trojan without any modification: The register holding the privilege level is clock gated and is set in its default state to machine mode[1]. This means that the test enable pin of the clock gating cell can be attached to the fake antenna cell and tied to ground. If the bias voltage is driven to high, the test enable pin will allow the clock to propagate to the privilege register, always latching machine mode. The mode is going to change for a single cycle if an mret/sret instruction is issued and the MPP register is set appropriately but will instantly change back to machine mode the next cycle.

To validate the approach, we crafted a minimal example program, which will boot the processor, switch into user mode and will then attempt to access a CSR which is only accessible in machine mode. In normal operation mode, this access will fail as an illegal instruction [19]. If the fake antenna cell is triggered, the processor will consider itself to be running in machine mode and will thus willingly execute the supposed invalid instruction.

This shows that open source code is not a defense against this attack: A malicious actor can provide an implementation, which is logically equivalent to the freely available source code. Even if the authors of the original core did not include a malicious trojan, a purely "physical" cell will alter the functionality to completely break the execution security. Even having the malicious IP house only perform the implementation for a given netlist, such HW trojans might be implemented.

Similarily, the timing based attack can replace the privilege level to be latched with machine mode: In front of the D pin of the register, an OR gate can be inserted, which OR's together the pulse and the true next state. Inside the pulse window, the privilege level latched is always going to be machine mode, but in normal operation the correct privilege level will be used.

B. AES128 Encryption

Looking at AES128, the encryption is based on multiple iterations of the same function blocks, including: Substitution of bytes, transposition and mixing [20]. Area efficient implementations reuse the same hardware blocks and sequence the operations with a simple state machine. Each iteration through the operations increases the possible key space.

Fig. 7 shows an example AES implementation. By modifying the state-machine, such that the D pin of the register holding n idles at the penultimate value (which has no impact on the normal operation, as the enable for it is not set), the fake antenna cell trojan can be employed. When triggered, n will be loaded with the wrong value while the circuit is idling. When the start signal comes, instead of performing 10 rounds, it will instead go from the COMP state back to the idle state after a single cycle, only applying the key once. By providing the plain text and obtaining the weakly ciphered data, the attacker can trivially compute the private key.

An alternative approach is to fully bypass the encryption step and directly output the secret. The extended key can be muxed into the output register (requiring the red highlighted mux to be inserted).

[1]Tested version was commit 6deffb2

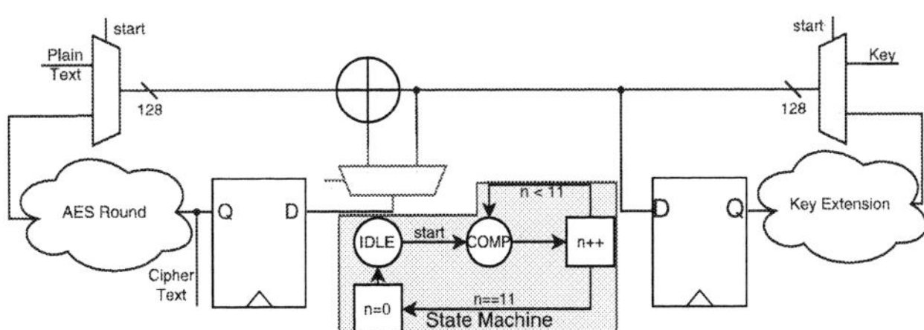

Fig. 7: Example AES implementation and its state machine. The red mux is inserted and driven by the timing based attack.

C. RNG Generation

Another application is reducing the entropy of randomly generated numbers, which is necessary e.g. generating secure private keys. As both presented trojans can act as a tie-cell, the same application as presented in [7] can be used: Their doping modified standard cells act as tie-high/tie-low cells. The registers holding the random key K are overwritten by a constant value, introduced by the proposed hardware trojans. The random input c has most of its values (128 - n) fixed to a constant value by the hardware trojans. This reduces the entropy of the numbers to n bits. However, for an unsuspecting observer, the numbers look random, due to the good properties of AES. Outside of the realization of the tie implementation, all the same points as the original work apply.

D. Tapeout Results

We have validated both trojans in a 22nm fdSOI technology. The micrograph and the actual circuits are shown in Fig. 8.

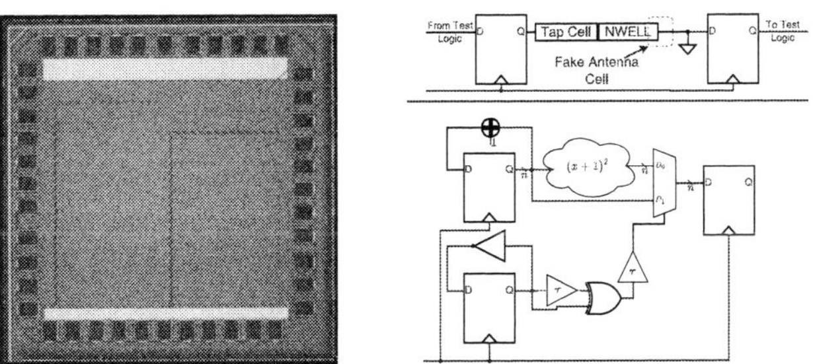

Fig. 8: Micrograph and implemented circuits

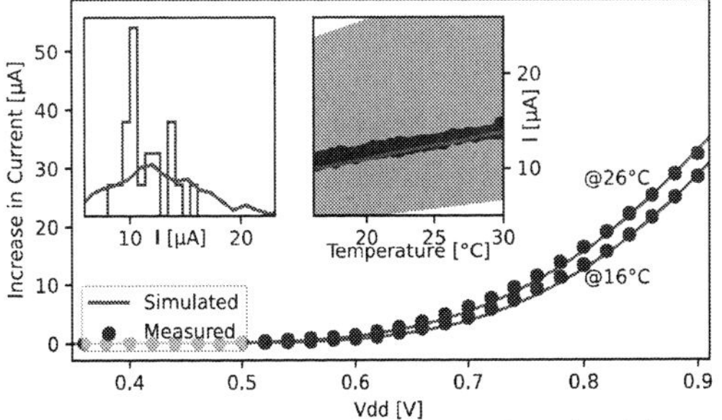

Fig. 9: Short circuit current introduced by the fake antenna cell overpowering a tie low cell

Fig. 9 shows the increase in current as a function of the supply voltage. The increase in current consumption because

979-8-3503-1176-1/23 $31.00 © 2023 IEEE

of the well driving against the tie cell is less than 35 µA, which is a lot less than typical leakage currents at room temperature. Top left shows the chip-to-chip variation of the short circuit current. Top center compares the simulated and measured temperature dependency. The overlap of these two effects leads to high variance in the short circuit current, making the detection by analyzing the power consumption difficult. Due to the low number of samples (only 24 chips are available) and the better than normal process control for the engineering samples, the measured C2C variation spreads only across a smaller range of the simulated results.

A normal Shmoo plot shows the functionality of a circuit as a binary decision (or as a probability) as a function of supply voltage and frequency. For the proposed timing based trojan, three results can occur: The true result of the computation can be obtained, in that case the circuit is performing normally, the secret that is muxed in by the timing based trojan is leaked, or the circuit outputs a wrong result.

In Fig. 10 we show a Shmoo plot with color coded performance of the chip. There is a clear region in green where the secret is leaked. If the period is outside that region, normal operation resumes. At the boundary between these domains, wrong results are computed, as some of the registers already latch the secret, but some still latch the computation result.

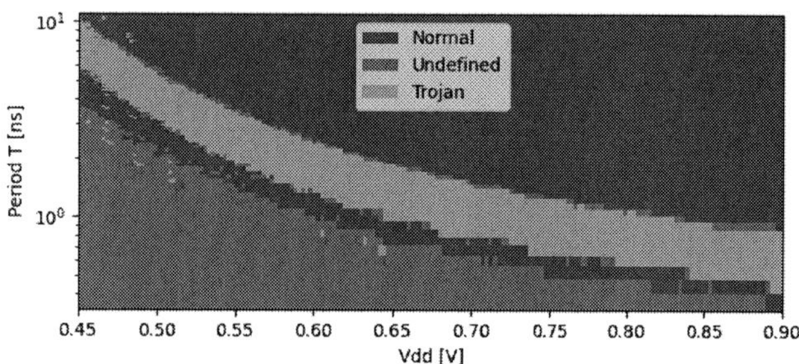

Fig. 10: Extended Shmoo plot for timing based trojan

We have opted to insert the trojan considering both the ABB and FF corner, making the frequency window larger than what would be possible if a single corner setup, as discussed in Sec. III-C, was used. At 0.8 V, for a clock with a period between 564 ps and 1012 ps, the hardware trojan triggers.

V. Summary and Outlook

In this paper, we have presented two fdSOI enabled hardware trojans, which are triggered either by the body bias voltage or by a specific frequency respectively. Both approaches have been validated in silicon. Multiple applications for the trojans have been presented.

Covert functionality can be used to implement hardware trojans, but also for less nefarious uses: IP protection is a possible other field to apply these approaches. The presented approaches allow obfuscation of functionality which can be used to prevent reverse engineering. It also allows limitation of functionality (such as frequency range), depending on agreed upon performance specification.

A possible extension of this work is to investigate trigger frequencies for the timing based trojan, which are dependent on PVT variations: This would reduce the need for the timing constraints to be fulfilled over all corners. Different trigger frequencies could be choosen in different corners, making the pulse widths only dependent on the on-chip and cycle-to-cycle variation.

References

[1] R. Chakraborty, S. Narasimhan, and S. Bhunia, "Hardware Trojan: Threats and emerging solutions," in *IEEE Int. High Level Design Validation and Test Workshop*, 2009, pp. 166–171.

[2] M. Rijoy, R. Sree Ranjani, and C. Rajat Subhra, "A comprehensive survey of physical and logic testing techniques for Hardware Trojan detection and prevention," in *Journal of Cryptographic Engineering*, vol. 12, no. 4, 2022, pp. 495–522.

[3] R. Fani and M. Saheb Zamani, "Runtime hardware Trojan detection by reconfigurable monitoring circuits," in *The Journal of Supercomputing*, vol. 78, no. 10, 2022.

[4] G. Bloom, B. Narahari, and R. Simha, "OS support for detecting Trojan circuit attacks," in *IEEE Int. Workshop on Hardware-Oriented Security and Trust*, 2009, pp. 100–103.

[5] X. Zhang and M. Tehranipoor, "Case study: Detecting hardware Trojans in third-party digital IP cores," in *IEEE Int. Symp. on Hardware-Oriented Security and Trust*, 2011, pp. 67–70.

[6] M. Hicks, M. Finnicum *et al.*, "Overcoming an Untrusted Computing Base: Detecting and Removing Malicious Hardware Automatically," in *IEEE Symp. on Security and Privacy*, 2010, pp. 159–172.

[7] G. Becker, F. Regazzoni *et al.*, "Stealthy Dopant-Level Hardware Trojans," in *Cryptographic Hardware and Embedded Systems - CHES*. Berlin, Heidelberg: Springer, 2013, pp. 197–214.

[8] J. Stine, I. Castellanos *et al.*, "FreePDK: An Open-Source Variation-Aware Design Kit," in *IEEE Int. Conf. on Microelectronic Systems Education (MSE'07)*, 2007, pp. 173–174.

[9] B. Shakya, H. Shen *et al.*, "Covert Gates: Protecting Integrated Circuits with Undetectable Camouflaging," *IACR Transactions on Cryptographic Hardware and Embedded Systems*, 2019.

[10] F. Assaderaghi, D. Sinitsky *et al.*, "A dynamic threshold voltage MOSFET (DTMOS) for ultra-low voltage operation," in *IEEE Int. Electron Devices Meeting*, 1994, pp. 809–812.

[11] GlobalFoundries, "22FDX for Cost-Effective Low Energy Designs," *Tensilica Days*, 2018.

[12] T. Watanabe and Y. Yoshida, "Dielectric breakdown of gate insulator due to reactive ion etching," in *Solid State Technology*, vol. 26, no. 4, 1984, p. 263.

[13] H. Shin, C. King, and C. Hu, in *30th Annual Proceedings Reliability Physics*.

[14] R. Torrance and D. James, "The State-of-the-Art in IC Reverse Engineering," in *Cryptographic Hardware and Embedded Systems - CHES 2009*, ser. Lecture Notes in Computer Science. Berlin, Heidelberg: Springer, 2009, vol. 5747, pp. 363–381.

[15] Intel, "Intel® Digital Random Number Digital Random Number Generator Generator(DRNG), Software Implementation Guide], pages=16,," 2012.

[16] J. Walker, "Conceptual Foundations of the Ivy Bridge Random Number Generator], journal = Presentation at ISTS Computer Science Department Colloquium at Dartmouth College," 11 2012.

[17] S. Höppner, H. Eisenreich *et al.*, "How to Achieve World-Leading Energy Efficiency using 22FDX with Adaptive Body Biasing on an Arm Cortex-M4 IoT SoC," in *49th European Solid-State Device Research Conf. (ESSDERC)*, 2019, pp. 66–69.

[18] F. Zaruba and L. Benini, "The Cost of Application-Class Processing: Energy and Performance Analysis of a Linux-Ready 1.7-GHz 64-Bit RISC-V Core in 22-nm FDSOI Technology," *IEEE Trans on VLSI Systems*, vol. 27, no. 11, pp. 2629–2640, Nov 2019.

[19] A. Waterman, K. Asanovic, and J. Hauser, "The RISC-V Instruction Set Manual, Volume II: Privileged Architecture, Document Version 20211203," *RISC-V Int.*, 2021.

[20] N. I. of Standards and Technology, "Advanced Encryption Standard (AES)," *NIST FIPS PUB 197*, 2001.

Bridging the Gap between Spiking Neural Networks & LSTMs for Latency & Energy Efficiency

Gourav Datta, Haoqin Deng, Robert Aviles, Zeyu Liu, Peter A. Beerel
Ming Hsieh Department of Electrical and Computer Engineering, University of Southern California, USA
{gdatta, haoqinde, rsaviles, liuzeyu, pabeerel}@usc.edu

***Abstract*—Spiking Neural Networks (SNNs) have emerged as an attractive spatio-temporal computing paradigm for complex vision tasks. However, most existing works yield models that require many time steps and do not leverage the inherent temporal dynamics of spiking neural networks, even for sequential tasks. Motivated by this observation, we propose an optimized spiking long short-term memory networks (LSTM) training framework that involves a novel ANN-to-SNN conversion framework, followed by SNN fine-tuning v ia backpropagation through time (BPTT). In particular, we propose novel activation functions in the source LSTM architecture and convert a judiciously selected subset of them to leaky-integrate-and-fire (LIF) activations with optimal bias shifts. Moreover, we propose a pipelined parallel processing scheme that hides the SNN time steps, significantly i mproving s ystem l atency, e specially f or long sequences. The resulting SNNs have high activation sparsity and require only accumulate operations (AC), in contrast to expensive multiply-and-accumulates (MAC) needed for ANNs, except for the input layer when using direct encoding, yielding significant improvements in energy efficiency. W e e valuate o ur framework on sequential learning tasks including temporal MNIST, Google Speech Commands (GSC), and UCI Smartphone datasets on different LSTM architectures. We obtain test accuracy of 94.75% with only 2 time steps on the GSC dataset with $\sim 4.1\times$ lower energy than an iso-architecture standard LSTM.**

I. Introduction & Related Work

In contrast to the neurons in ANNs, the neurons in Spiking Neural Networks (SNNs) are biologically inspired, receiving and transmitting information via spikes. SNNs promise higher energy-efficiency t han A NNs d ue t o t heir h igh activation sparsity and event-driven spike-based computation [1] which helps avoid the costly multiplication operations that dominate ANNs. To handle multi-bit inputs, such as typical in traditional datasets and real-life sensor-based applications, the inputs are often spike encoded in the temporal domain using rate coding [1], temporal coding [2], or rank-order coding [3]. Alternatively, instead of spike encoding the inputs, some researchers explored directly feeding the analog pixel values in the first c onvolutional l ayer, a nd t hereby, e mitting spikes only in the subsequent layers [4]. This can dramatically reduce the number of time steps needed to achieve the state-of-the-art accuracy, but comes at the cost that the first l ayer now requires MACs [4]. However, all these encoding techniques increase the end-to-end latency (proportional to the number of time steps) compared to their non-spiking counterparts.

In addition to accommodating various forms of spike encoding, supervised learning algorithms for SNNs, such as surrogate gradient learning (SGL) have overcome various roadblocks associated with the discontinuous derivative of the spike activation function [5]–[7]. SNNs with integrate-and-fire (IF) neurons can be converted from ANNs with low error by approximating the activation value of ReLU neurons with the firing rate of spiking neurons [1], [8]–[10]. Previous works proposed shifting the ReLU activation functions in CNNs for both identical and independent distributions (IID) [11], [12] as well as skewed distributions [10], the latter proposing to learn the optimal shift via SGL. The resulting SNNs have been able to perform similar to SOTA CNNs in traditional image recognition tasks [4], [9] and dynamic vision tasks [13], [14] with significant advantages in compute efficiency.

Interestingly, there has been relatively little research that target SNNs for sequence learning tasks. Some proposed spiking LSTM works are limited to binary inputs [15], [16], which might not represent several real-world use cases, and others [1], [11], [17] are obtaining from vanilla RNNs, which can lead to large accuracy drops. Other works [18] explored converting SNNs from LSTMs but incurred large drops in accuracy and lacked architectural parallelism that led to long latencies. A more recent work [16] uses a more complex neuron model than the leaky-integrate-and-fire (LIF) model to improve the recurrence dynamics and multi-bit activation maps to improve training at the cost of requiring multiplication and the associated higher energy consumption. Finally, we note that although transformers have recently attracted significant attention for temporal tasks [19], existing works on spiking transformers only target static vision tasks [20] and come at the cost of a very high compute and memory footprint. Moreover, LSTMs rather can ingest naturally temporal inputs, unlike transformers that can start processing only after the entire input sequence arrives.

Our work leverages both the efficient temporal and sparse dynamics of SNNs to reduce the inference latency and energy of large-scale streaming workloads while achieving close to SOTA accuracy. In particular, we adopt an LSTM which provides memorization using an unrolled architecture over a sequence length and is independent of the number of SNN time steps. We aim to reduce the number of time steps to reduce latency and energy consumption and improve the resulting inference accuracy through novel training techniques. The key contributions of our work are summarized below.

- We propose a training framework that starts with a spiking LSTM that is converted from a pre-trained non-spiking LSTM. Our framework has four steps. i) Convert the traditional sigmoid and tanh activation functions in the

This work was supported in part by the NSF CCF-1763747 award, the DARPA HR00112190120 award, and a gift from Intel.

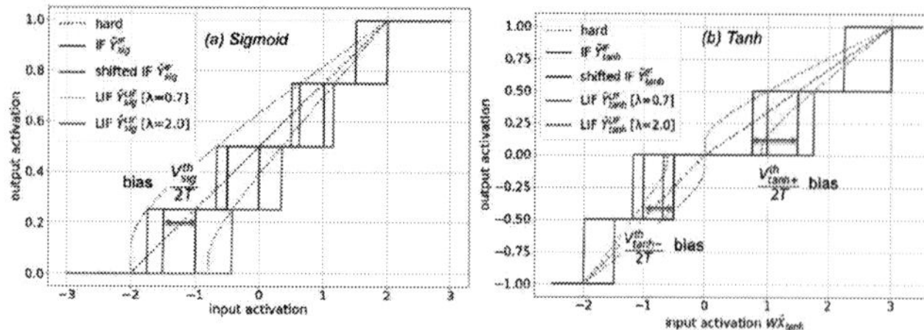

Fig. 1: Spiking (hard) and non-spiking activation (IF and LIF) functions corresponding to (a) sigmoid and (b) tanh activation for $T = 4$, and $V^{th}_{sig} = 4$, $V^{th}_{tanh+} = 3$, $V^{th}_{tanh-} = -2$. We show the proposed bias shifts for IF activations. The green and red dotted lines show the continuous versions of the discrete LIF activation functions.

source LSTM to clipped versions, ii) judiciously convert a subset of these functions to IF activation functions such that the SNN does not require the expensive MAC operations, iii) find the optimal shifts of the IF activation functions for IID inputs, and iv) fine-tune the SNN with LIF activation functions, updating the shifts for non-IID training inputs.

- We propose a high-level parallel and pipelined implementation of the resulting SNN-based models that incurs negligible latency overhead compared to the baseline LSTM and improves hardware utilization over existing spiking LSTMs.
- We demonstrate the energy-latency-accuracy trade-off benefits of our framework through FPGA synthesis and place-and-route and algorithmic experiments with different LSTM architectures on sequential tasks from computer vision (temporal MNIST), spoken term classification (Google Speech Commands), and human activity recogniton (UCI Smartphone) applications, and comparisons with existing spiking and non-spiking LSTMs.

II. Preliminaries

In this work, we adopt the popular IF [8] and LIF [21] models to capture the computation dynamics of an SNN, where each neuron has an internal state called its membrane potential $U(t)$ which captures the bias B and sum of the weight W modulated incoming spikes $S(t)$ from each neuron in the previous layer. In the LIF model, $U(t)$ leaks with a fixed time constant, denoted as λ. With the spiking threshold denoted as V^{th}, the LIF neuron dynamics are expressed as

$$U^{temp}(t) = \lambda U(t-1) + WS(t) + B \tag{1}$$

$$S(t) = \begin{cases} V^{th}, & \text{if } U^{temp}(t) > V^{th} \\ 0, & \text{otherwise} \end{cases} \tag{2}$$

$$U(t) = U^{temp}(t) - S(t) \tag{3}$$

The IF model is simply the LIF model with $\lambda = 1$.

III. Proposed Training Framework

A. Non-spiking LSTM

For traditional LSTM-based architectures, the non-linear activation functions are tanh and sigmoid. In order to yield accurate LSTM-based SNN models, we first replace them with *hard* (clipped) versions, as illustrated in the blue dotted lines in Fig. 1(a-b). In particular, we clip the hard sigmoid function to 1 at $x = \frac{V^{th}_{sig}}{2}$ and to 0 at $x = -\frac{V^{th}_{sig}}{2}$, where V^{th}_{sig} is the intended threshold value of the converted SNNs. To increase the representative power, we model the hard tanh function with two hard sigmoid functions, one for when its input $x \geq 0$ clipped at $x = V^{th}_{tanh+}$ to 1 and the other for when $x < 0$ clipped at $x = V^{th}_{tanh-}$ to -1. Note that in our training framework the SNN threshold values are initialized with their values in trained non-spiking LSTM models.

B. Conversion from non-spiking to spiking LSTMs

Let $U^{temp}_{sig}(t)$ and $U^{temp}_{tanh}(t)$ denote the accumulated membrane potentials at time step t associated with our novel sigmoid and tanh SNN modules. As per our proposed clipping, the associated spike outputs $S_{sig}(t)$ and $S_{tanh}(t)$ are

$$S_{sig}(t) = \begin{cases} 1, & \text{if } U^{temp}_{sig}(t) > \frac{V^{th}_{sig}}{2} \\ 0, & \text{otherwise,} \end{cases} \tag{4}$$

$$S_{tanh}(t) = \begin{cases} 1, & \text{if } U^{temp}_{tanh}(t) > V^{th}_{tanh+} \\ -1, & \text{if } U^{temp}_{tanh}(t) < V^{th}_{tanh-} \\ 0, & \text{otherwise.} \end{cases} \tag{5}$$

Note that the two types of spikes (positive and negative) we propose for $S_{tanh}(t)$ may be less bio-plausibile but can be easily implemented in hardware as illustrated in Fig. 2(a)-(b) and does not significantly degrade the energy-efficiency.

Our SNN training framework starts by training a baseline non-spiking LSTM model that we convert into a spiking LSTM. To minimize the error during this conversion, inspired [11], [12], [22] that focuses on CNNs, we first define a notion of IF activation functions $\bar{Y}^{IF}_{sig}$ and $\bar{Y}^{IF}_{tanh}$ that represent the average of IF activation outputs over all time steps.

$$\bar{Y}^{IF}_{sig} = \frac{1}{T}\text{clip}\left(\left\lfloor \frac{T}{V^{th}_{sig}}\left(W\bar{X}_{sig} + B + \frac{V^{th}_{sig}}{2}\right)\right\rfloor, 0, T\right)$$

$$\bar{Y}^{IF}_{tanh} = \begin{cases} \frac{1}{T}\text{clip}\left(\left\lfloor \frac{T}{V^{th}_{tanh+}}(W\bar{X}_{tanh} + B)\right\rfloor, 0, T\right), \text{if } A \\ \frac{1}{T}\text{clip}\left(\left\lfloor \frac{T}{V^{th}_{tanh-}}(W\bar{X}_{tanh} + B)\right\rfloor, -T, 0\right), \text{otherwise} \end{cases}$$

Here, T denotes the total number of SNN time steps, $\bar{X}_{sig}$ and $\bar{X}_{tanh}$ denote the time-averaged inputs, and A corresponds to $W\bar{X}_{tanh} > 0$. $\text{clip}(x, y, z) = y$ when $x \leq y$, $\text{clip}(x, y, z) = x$ when $y < x < z$, and $\text{clip}(x, y, z) = z$ when $x \geq z$.

We then observe, as illustrated in Fig. 1(a), the IF activation $\bar{Y}^{IF}_{sig}$ when $B = 0$ is always less than its sigmoid counterpart. Hence the error accumulates over the multiple time steps and input elements in the sequence. To mitigate this error, we propose to shift the IF activation curve to the left by appropriately shifting the bias term B as shown in Fig. 1(a). Under the assumption that $\bar{X}_{sig}$ is IID and the converted spiking bias value is 0 (i.e., we do not include the bias term in the pre-trained non-spiking LSTM), the optimal value of B is $V^{th}_{sig}/2T$ as shown in Fig 1(a). Similarly, we can show that, under the assumption that $\bar{X}_{tanh}$ is IID for $\bar{X}_{tanh} \geq 0$ and $\bar{X}_{tanh} < 0$ separately, the optimal value of $B = V^{th}_{tanh+}/2T$ for $\bar{X}_{tanh} \geq 0$ and $B = V^{th}_{tanh-}/2T$ for $\bar{X}_{tanh} < 0$.

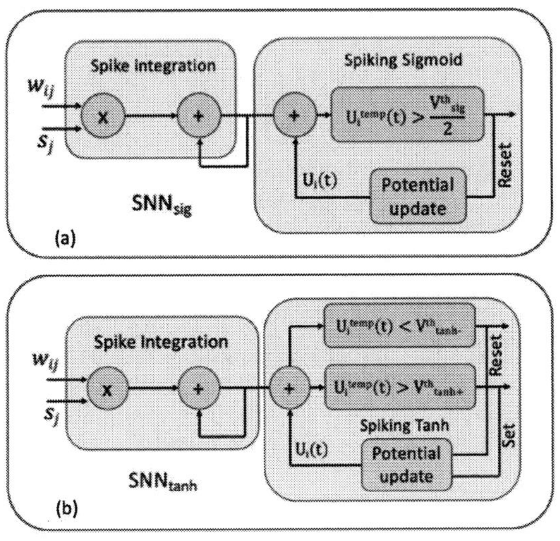

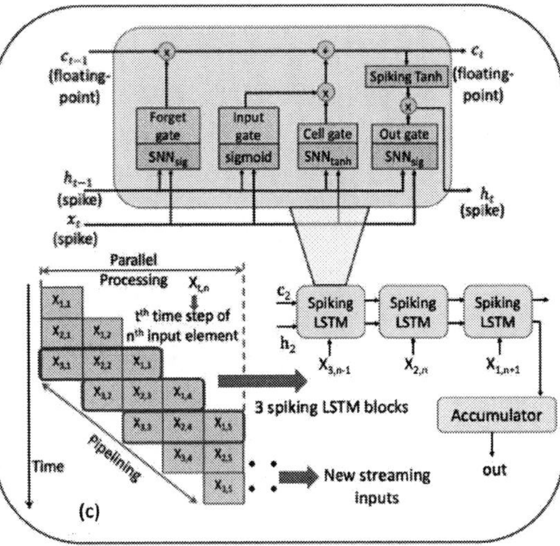

Fig. 2: LIF activation function corresponding to the (a) sigmoid and (b) tanh activation function used in the spiking LSTM architecture, and (c) Proposed spiking LSTM architecture and dataflow with the parallel pipelined execution for the example of 5 time steps and 3 input elements in the sequence.

C. SNN Training with LIF Activation Shifts

After initial conversion to a spiking LSTM, we aim to further optimize the error between the outputs of the IF and non-spiking LSTM activations in order to reduce the number of time steps and resulting energy consumption. Towards this goal, we convert the IF model to its LIF counterpart by incorporating the leak term that provides a tunable control knob to minimize this error. The LIF activations converted from the hard sigmoid and hard tanh functions can be computed (derivation not shown due to space constraints) as

$$\bar{Y}^{LIF}_{sig} = \begin{cases} \frac{1}{T}\left\lfloor\frac{T}{t_{sig}}\right\rfloor, \text{if } W\bar{X}_{sig} > V^{th}_{sig}(1-\lambda) \\ 0, \text{otherwise}, \end{cases} \tag{6}$$

$$\bar{Y}^{LIF}_{tanh} = \begin{cases} \frac{1}{T}\left\lfloor\frac{T}{t_{tanh+}}\right\rfloor, \text{if } W\bar{X}_{tanh} > V^{th}_{tanh+}(1-\lambda) \\ -\frac{1}{T}\left\lfloor\frac{T}{t_{tanh-}}\right\rfloor, \text{else if } W\bar{X}_{tanh} < V^{th}_{tanh-}(1-\lambda) \\ 0, \text{otherwise} \end{cases}$$

The functions have been illustrated for two different settings of λ in Fig. 1(a), one for which $\lambda > 1$ and one for which $\lambda < 1$. Assuming $K = \frac{(1-\lambda)}{W}$,

$$t_{sig} = \left\lceil \frac{log\left(1 - \frac{KV^{th}_{sig}}{X_{sig}}\right)}{log(\lambda)} \right\rceil \tag{7}$$

$$t_{tanh+} = \left\lceil \frac{log\left(1 - \frac{2KV^{th}_{tanh+}}{X_{tanh}}\right)}{log(\lambda)} \right\rceil \quad t_{tanh-} = \left\lceil \frac{log\left(1 + \frac{2KV^{th}_{tanh-}}{X_{tanh}}\right)}{log(\lambda)} \right\rceil \tag{8}$$

In both cases, the optimal value of λ can minimize the difference between the LIF and non-spiking LSTM output activations, but depends on the input distributions which are difficult to model. We thus propose to optimize both leak term along with the threshold terms and weights via SGL after conversion during the proposed SNN fine-tuning phase.

D. Selective conversion of LSTM activation functions

Instead of converting all the sigmoid and tanh activation functions in the non-spiking LSTM architecture to spiking counterparts, we judiciously select a subset of them such that only one of the two inputs of each LSTM multiplication operation is spiking. This ensures that all multiplications are replaced with the conditional addition or subtraction of the weights. This also avoids the unnecessary accumulated error due to the spiking gradients and improves the inference accuracy at low time steps.

To motivate this conversion, let us review the equations governing the LSTM architecture [23] in Eqs. 9 10 11 where h_t and c_t denote the hidden and cell state tensors respectively. We denote f_t, i_t, g_t, and o_t as outputs of the forget, input, cell, and output gates respectively. We assume the weight tensors $W_{a,h}$, $W_{a,x}$, $W_{g,h}$, $W_{g,x}$ to be multi-bit as per typical SNN setups [9].

$$\mathbf{a}_t = \text{sig}_{\mathbf{a}}(W_{\mathbf{a},h}h_{t-1} + W_{\mathbf{a},x}x_t) \ \forall \mathbf{a} \in \{f, i, o\} \tag{9}$$

$$g_t = \tanh_g(W_{g,h}h_{t-1} + W_{g,x}x_t) \tag{10}$$

$$c_t = f_t \odot c_{t-1} + i_t \odot g_t, \ h_t = o_t \odot \tanh_c(c_t) \tag{11}$$

We propose to encode x_t using a spike tensor (of length equal to the number of hidden neurons in a layer), as otherwise the MAC operation with the weight tensors would require costly multiplications. Similarly, h_t also needs to be a spike tensor, which implies that o_t should be a spike tensor and $\tanh_c$ should be converted to a LIF activation. A spiking o_t necessitates conversion of the sig_o to LIF (see Eq. (9)). On the other hand, Eq. (10) implies that either f_t or c_{t-1} and i_t or g_t need to be a spike tensor for multiplier-less operations. Between f_t and c_{t-1}, we have to choose f_t as the spike tensor because c_{t-1} is the sum of two tensors which is not naturally a spike tensor. Moreover, sig_f can be easily converted to LIF activation using our framework. Between i_t and g_t, we can arbitrarily choose one to spike.

IV. Pipelined Parallel SNN Processing

The operation of each LSTM unit (both spiking and non-spiking) is repeated, i.e., unrolled, as many times as the sequence length of each input, which we denote as N. In each rolling, the LSTM operation (both spiking and non-spiking) works with a different element of the sequence. Each such element is direct encoded and repeatedly inputted to the spiking LSTM unit T times. To hide the latency of the LIF dynamics, we propose a pipelined and parallel processing scheme exemplified in Fig. 2(c) for N=5 and T=3.

The LSTM states h_t and c_t for element n are updated every time step t, modulated by the weights, and processed by the LIF activation function. This state update allows us to update states for element n for time $t+1$ as well as start processing the t time step of the $n+1$ input tensor in the sequence, provided we have sufficient LSTM hardware to pipeline/parallelize these operations.

Because, the LSTM algorithm is limited to processing a maximum of T input elements at the same time and because we achieve relatively small T, this level of hardware pipelining is quite manageable. By doing so, in each time step, a new input element in the sequence will begin to be processed, the first spike input of the Nth input element will be processed at the Nth time step. To process the remaining $T-1$ spike inputs of its encoding, we need an additional $T-1$ time steps. Hence, the total number of time steps required to process the whole input sequence with our spiking LSTM is $(T+N-1)$.

For hardware with built-in parallel processing capability such as GPUs, our approach improves the hardware utilization compared to non-spiking LSTMs that are sequential in nature. Note that previous research on LSTM-based SNNs [18] accumulates the spike outputs of the different gates over all the time steps for processing a single input element. As a result, it uses $T \times N$ time steps to process the entire input sequence. Moreover, the hidden state input to the next unrolled LSTM block becomes multi-bit which necessitates using energy-expensive multiplications.

V. Experimental Results

We validate our proposed techniques on temporal MNIST [24], Google speech commands (GSC) with 11 classes [25], and UCI smartphone datasets [26]. For temporal MNIST (T-MNIST), we use row-wise sequential inputs, resulting in 32 image pixels each over a sequence of 32 frames [18], [27]. For GSC, we pre-process the raw audio inputs using log-mel spectrograms resulting in 20 frequency features over a sequence of 81 frames [28]. For UCI smartphone, we pre-process the sensor signals obtained from the smartphone worn on the waist of the participating humans by applying butterworth low-pass filters within a fixed-width sliding windows of 2.56 seconds and 50% overlap (128 readings per window). For all the three datasets, we use both one and two-layer LSTMs with 128 hidden neurons in each layer. While we use a single fully-connected (FC) classifier layer for the T-MNIST and UCI datasets, we use two FC layers of 32 and 11 neurons each, with softmax output for the GSC dataset, following [28]. We do not convert the FC layers to spiking counterparts as they consume $<0.03\%$ of total energy.

A. Inference Accuracy

Our results for single layer LSTMs are illustrated in Table II for 2 time steps, both with and without a pre-trained non-spiking LSTM model. Each of our proposed techniques improve the test accuracy for large-scale tasks such as GSC, with an overall improvement of 4.1% (90.59% to 94.75). For T-MNIST, which is an easier task with less room for improvement, our techniques lead to a 0.31% improvement in accuracy, while the threshold and leak optimizations yielding hardly any benefit. For the relatively more challenging UCI dataset, the leak optimization leads to a maximum accuracy increase of +0.49% with a pre-trained model, while the total increase due to all our techniques is 1.23%. While the UCI accuracy can be further increased with the use of bi-directional and stacked LSTMs [29] (+1.5%), it increases energy consumption. We also evaluate the scalability of our approach on the bi-directional and two-layer LSTM architectures for the large-scale tasks, as shown in Table III. The results indicate that our approach can yield stacked and bi-directional spiking LSTMs with a $0.4-0.5\%$ drop in accuracy compared to non-spiking counterparts. Note that the accuracies obtained without pre-trained models are lower, particularly for the more complex applications.

We also compare the impact of direct and Poisson encoding on the SNN accuracy in Fig. 3(a-b). Note that the test accuracies obtained after only ANN-to-SNN conversion shown in Fig. 3(b) are significantly lower than those obtained after SNN training, especially for more complex tasks. In particular, our approach, with SNN training yields close to state-of-the-art (SOTA) accuracy with only 2 time steps, providing more than $7\times$ reduction in the latency compared to our conversion-only approach for the GSC dataset.

Our conversion framework provides the optimal value of the SNN threshold (before SNN fine-tuning) backed by our theoretical insights. Thus, our conversion framework acts as a good initializer for the weights and the membrane potential, without which the test accuracy of more complex tasks, such as GSC, drops by $>3.4\%$. Note that our conversion framework alone also outperforms the existing ANN-to-SNN conversion frameworks for LSTMs. For example, our conversion-only spiking LSTM models yield 94.3% test accuracy with 15 time steps; the conversion-only baseline spiking models (without our proposed threshold shifts) require 32 time steps to obtain similar accuracy. On the other hand, the baseline models can attain only 86% accuracy with 15 time steps.

We compare the test accuracies obtained by our training framework with that of existing works in Table III. Our accuracies are close to SOTA (within 0.6%) obtained by the non-spiking models, while yielding significant energy and latency savings as shown in Fig. 4. We surpass the SOTA spiking models in terms of accuracy for both the T-MNIST and GSC datasets.[1]

[1]Note that we were unable to find any deep SNN architectures, classifying the UCI dataset for comparison.

TABLE I: Test accuracy on temporal MNIST, GSC, and UCI datasets obtained by proposed approaches with direct encoding for 2 time steps. S and NS denote the spiking and non-spiking LSTM variants respectively. On the other hand, P and NP denotes the accuracies with and without a pre-trained non-spiking LSTM model respectively.

LSTM	V^{th}	V^{th}	λ	NS	NS	T-MNIST Acc. (%)		GSC Acc. (%)		UCI Acc. (%)	
Model	Shift	Train	Train	sig$_i$	tanh$_g$	P	NP	P	NP	P	NP
NS	×	×	–	–	–	–	98.6±0.2	–	95.42±0.1	–	90.37±0.2
S	×	×	×	×	×	97.84±0.2	97.74±0.3	90.59±0.2	63.45±0.2	88.17±0.2	87.63±0.1
	✓	×	×	×	×	97.98±0.2	97.87±0.1	92.05±0.1	91.45±0.2	88.60±0.2	88.13±0.3
	✓	✓	×	×	×	97.92±0.1	97.84±0.2	92.87±0.2	91.33±0.3	88.64±0.3	86.87±0.2
	✓	✓	✓	×	×	98.0±0.2	97.95±0.2	93.57±0.1	92.14±0.1	89.13±0.4	87.50±0.3
	✓	✓	✓	✓	×	98.1±0.3	97.98±0.1	94.75±0.1	92.63±0.2	89.23±0.2	89.20±0.2
	✓	✓	✓	×	✓	98.15±0.1	98.12±0.2	94.53±0.2	92.61±0.3	89.40±0.3	89.12±0.1

TABLE II: Test accuracy on GSC and UCI datasets obtained by proposed approaches with direct encoding for 4 time steps on bi-directional and stacked LSTMs. 'St.' denotes a two-layer stacked LSTM, with both layers having 128 nodes each. 'Bi-St.' denotes a two-layer LSTM, with the first layer being bi-directional having 128 nodes.

LSTM	V^{th}	V^{th}	λ	NS	NS	GSC Acc. (%)		UCI Acc. (%)	
Model	Shift	Train	Train	sig$_i$	tanh$_g$	St.	Bi-St.	St.	Bi-St.
NS	×	×	–	–	–	95.03±0.3	94.72±0.3	91.42±0.1	91.90±0.1
S	×	×	×	×	×	63.24±0.2	62.87±0.3	86.24±0.2	87.51±0.1
	✓	×	×	×	×	92.77±0.2	89.75±0.2	87.80±0.2	88.46±0.2
	✓	✓	×	×	×	93.12±0.2	91.51±0.1	87.72±0.3	88.80±0.2
	✓	✓	✓	×	×	93.23±0.3	91.88±0.2	89.26±0.1	89.65±0.3
	✓	✓	✓	✓	×	94.61±0.1	94.22±0.3	91.00±0.2	91.68±0.2
	✓	✓	✓	×	✓	94.39±0.2	94.16±0.1	89.53±0.2	91.07±0.3

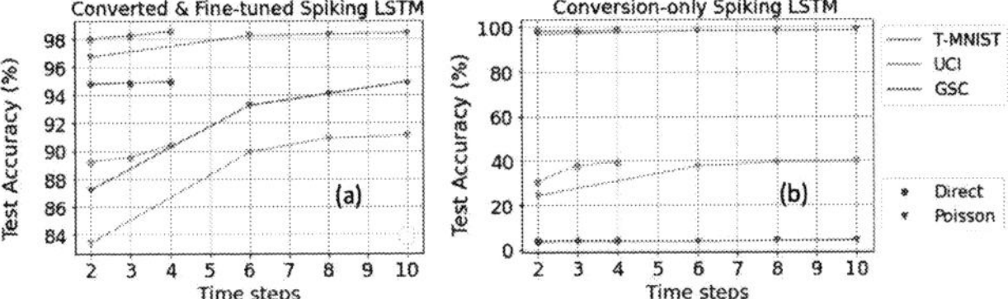

Fig. 3: Comparison between the accuracies obtained by our direct and Poisson encoded spiking LSTMs (a) with both conversion and SNN fine-tuning and (b) with only conversion.

B. Inference Energy Efficiency

The inference energy is dominated by the total number of floating point operations (FLOPs) and memory accesses. For non-spiking LSTMs, the FLOPs consists of the MAC, AC, hard sigmoid and hard tanh operations required in the four gates, and the memory cost includes the weight accesses for each unrolled LSTM operation. On the contrary, for spiking LSTMs, each emitted spike indicates which weights need to be accumulated at the post-synaptic neurons and results in a fixed number of AC operations. This, coupled with the comparison operations for the membrane potential in each time step dominates the spiking compute energy. However, spiking LSTMs incur higher memory energy compared to non-spiking counterparts, due to the repetitive membrane potential and weight accesses at each time step. Our framework can reduce this memory cost compared to existing solutions, thanks to the use of only a few number of time steps.

We use custom RTL specifications and 28 nm Kintex-7 FPGA platform to estimate the post place-and-route energy

TABLE III: Accuracy comparison of the best performing models obtained by our training framework with SOTA spiking and non-spiking LSTM models on different datasets.

Ref.	Model	Training technique	Architecture	Accuracy (%)
		Dataset : temporal MNIST		
[33]	Spiking	SGD	LSTM(220)	96.4
[15]	Spiking	SGD	LSTM(1000)	98.23
[18]	Spiking	ANN-SNN conv.	LSTM(128)	98.72 (T=64)
[27]	Spiking	SGD	RNN(64-256-256)	98.7
[28]	Non-spiking	RC-BPTT	LSTM (320)	98.14
This work	Spiking	Hybrid training	LSTM(128)	**98.93 (T=8)**
		Dataset : GSC		
[34]	Spiking	BPTT	CNN(64-64-64)	94.5
[28]	Partly spiking	RC-BPTT	LSTM(128)	**95.6**
[35]	Spiking	BPTT	LSTM(128)	91.2
This work	Spiking	Hybrid training	LSTM(128)	**95.02 (T=4)**
		Dataset : UCI		
[34]	Non-Spiking	SGD	Bi-dirLSTM(-)	**91.1**
This work	Spiking	Hybrid training	LSTM(128)	**90.78 (T=4)**

consumption of the hardware implementations of the spiking and non-spiking networks. In particular, we develop Verilog RTL block-level models to design, simulate, and synthesize an inference pipeline that captures the LSTM processing including the writing and reading of the weights (and membrane potentials for the spiking) for the LSTMs on our target FPGA device. In addition, for comparison purposes, we develop a similar synthesizable RTL design for the non-spiking LSTMs.

Fig. 4(a) and (b) illustrate the energy consumption for our spiking and non-spiking LSTM architectures used for classifying the three datasets, along with the SOTA spiking LSTM implementation [18]. As we can see, we obtain 2.8-5.1× and 10.1-13.2× lower energy than the non-spiking and SOTA spiking implementations respectively for direct coding. The reductions obtained by Poisson encoding are a little lower (1.8-3.5× compared to non-spiking and 6.6-9.0× compared to SOTA spiking) due to the degraded trade-off between more time steps and less energy due to ACs. On custom neuromorphic architectures, such as TrueNorth [30], and SpiNNaker [31], the total energy is estimated as $FLOPs{*}E_{compute}{+}T{*}E_{static}$ [32], where $(E_{compute}, E_{static})$ can be normalized to $(0.4, 0.6)$ and $(0.64, 0.36)$ for TrueNorth and SpiNNaker, respectively [32]. Since the FLOPs for the LSTM architectures are generally several orders of magnitude higher than T [18], [28], we expect to see similar compute energy improvements on them.

C. Inference Latency

The latency incurred by the non-spiking LSTM architecture depends on the latency of our RTL-implemented MAC,

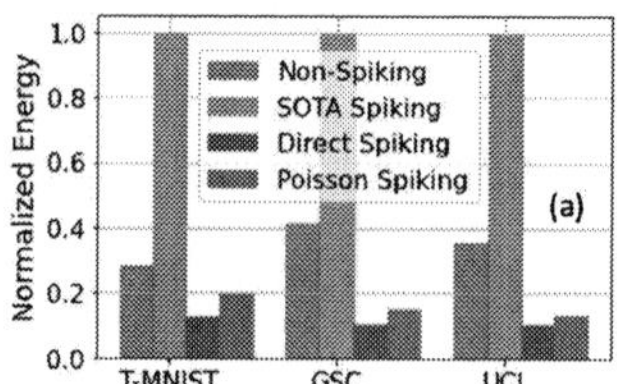

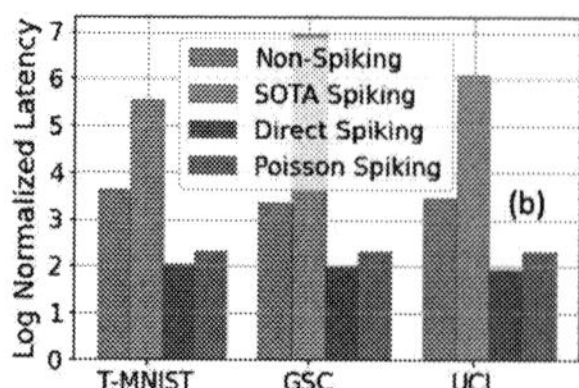

Fig. 4: (a) Energy and (b) Delay comparisons between the non-spiking LSTM, proposed direct and Poisson encoded spiking LSTM, and the SOTA spiking LSTM model [18], that does not include any of our proposed approaches.

AC, multiplication, hard sigmoid, hard tanh modules and the memory accesses. Depending on the number of the compute units available in each LSTM block, we parallelize the MAC operations, followed by the AC and activation functions in the four LSTM gates. We incur additional AC and hard tanh delay to produce c_t and h_t respectively. In contrast, with our proposed implementation, both the direct and Poisson encoded spiking LSTM architectures can process T different input elements simultaneously by internally pipelining the RTL models. Note that we use the similar RTL models and FPGA evaluation setup illustrated above to evaluate the latency of the LSTM implementations.

As shown in Fig. 4(b), our processing scheme, coupled with the low time steps and accumulate-only operations in SNN, results in $\sim 4\times$ and 25.9-105.7$\times$ reduction in latency compared to the non-spiking LSTM and SOTA spiking LSTM implementation respectively. This significant improvement over the SOTA spiking implementation can be attributed to three factors. Firstly, the SOTA spiking LSTM architecture require more time steps (3-8$\times$) to encode the original multi-bit input tensor than ours to achieve similar test accuracy. Secondly, while our proposed spiking architecture requires a total of $(T+N-1)$ time steps to process the whole input sequence, the existing spiking counterpart requires $T' \times N$ time steps where T and T' denote the total number of SNN time steps for the proposed and existing networks respectively. Lastly, since the hidden and cell state tensors are multi-bit tensors, the LSTM block requires MACs for certain computations, which also increases the latency by 5.1$\times$ obtained from our FPGA simulations compared to our AC-only approach.

VI. Conclusions & Broader Impact

ML models for large-scale streaming tasks are typically compute-intensive and often demand significant processing power that have large carbon footprints. This work proposes a spiking LSTM training framework which reduces the inference latency by up to 4$\times$ and energy efficiency by up to 3.2$\times$ when implemented on FPGA hardware compared to existing works with minimal ($< 0.6\%$) accuracy drop for diverse large-scale streaming use cases. Our models can thus reduce the carbon footprint, helping enable smart home assistants to efficiently perform on-device action and audio recognition. Finally, our LIF derivations approximating sigmoid and tanh activation functions may bridge the gap between traditional deep learning and SNNs for other use cases of these forms of non-linearity.

References

[1] P. U. Diehl *et al.*, "Conversion of artificial recurrent neural networks to spiking neural networks for low-power neuromorphic hardware," in *ICRC*, 2016, pp. 1–8.
[2] I. M. Comsa *et al.*, "Temporal coding in spiking neural networks with alpha synaptic function," in *ICASSP*, 2020, pp. 8529–8533.
[3] S. R. Kheradpisheh *et al.*, "Temporal backpropagation for spiking neural networks with one spike per neuron," *IJCNN*, 2020.
[4] N. Rathi *et al.*, "DIET-SNN: Direct input encoding with leakage and threshold optimization in deep spiking neural networks," *TNNLS*, 2021.
[5] Y. Kim and P. Panda, "Revisiting batch normalization for training low-latency deep spiking neural networks from scratch," *Frontiers in Neuroscience*, 2021.
[6] E. O. Neftci *et al.*, "Surrogate gradient learning in spiking neural networks: Bringing the power of gradient-based optimization to spiking neural networks," *IEEE SPM*, 2019.
[7] P. Panda *et al.*, "Toward scalable, efficient, and accurate deep spiking neural networks with backward residual connections, stochastic softmax, and hybridization," *Frontiers in Neuroscience*, 2020.
[8] A. Sengupta *et al.*, "Going deeper in spiking neural networks: VGG and residual architectures," *Frontiers in Neuroscience*, 2019.
[9] N. Rathi *et al.*, "Enabling deep spiking neural networks with hybrid conversion and spike timing dependent backpropagation," in *ICLR*, 2020.
[10] G. Datta and P. A. Beerel, "Can deep neural networks be converted to ultra low-latency spiking neural networks?" *DATE*, 2022.
[11] S. Deng and S. Gu, "Optimal conversion of conventional artificial neural networks to spiking neural networks," in *ICLR*, 2021.
[12] T. Bu *et al.*, "Optimized potential initialization for low-latency spiking neural networks," *AAAI*, 2022.
[13] Y. Kim and P. Panda, "Optimizing deeper spiking neural networks for dynamic vision sensing," *Neural Networks*, 2021.
[14] Y. Li *et al.*, "Neuromorphic data augmentation for training spiking neural networks," *ECCV*, 2022.
[15] A. L. Rezaabad and S. Vishwanath, "Long short-term memory spiking networks and their applications," in *ICONS*, 2020.
[16] W. Ponghiran *et al.*, "Spiking neural networks with improved inherent recurrence dynamics for sequential learning," *arXiv:2109.01905*, 2021.
[17] N. Moritz *et al.*, "Unidirectional neural network architectures for end-to-end automatic speech recognition," in *INTERSPEECH*, 2019.
[18] W. Ponghiran *et al.*, "Hybrid analog-spiking long short-term memory for energy efficient computing on edge devices," in *DATE*, 2021.
[19] A. Vaswani *et al.*, "Attention is all you need," in *NeurIPS*, vol. 30, 2017.
[20] Z. Zhou *et al.*, "Spikformer: When spiking neural network meets transformer," in *ICLR*, 2023.
[21] C. Lee *et al.*, "Enabling spike-based backpropagation for training deep neural network architectures," *Frontiers in Neuroscience*, vol. 14, 2020.
[22] Y. Li *et al.*, "A free lunch from ANN: Towards efficient, accurate spiking neural networks calibration," *arXiv preprint arXiv:2106.06984*, 2021.
[23] A. L. Rezaabad and S. Vishwanath, "Long short-term memory spiking networks and their applications," *arXiv:2007.04779*, 2020.
[24] Y. Lecun *et al.*, "Gradient-based learning applied to document recognition," *Proceedings of the IEEE*, 1998.
[25] P. Warden, "Speech commands: A dataset for limited-vocabulary speech recognition," *arXiv preprint arXiv:1804.03209*, 2018.
[26] D. Anguita *et al.*, "A public domain dataset for human activity recognition using smartphones," in *European Symposium on Artificial Neural Networks, Computational Intelligence and Machine Learning*, 2013.
[27] B. Yin *et al.*, "Accurate and efficient time-domain classification with adaptive spiking recurrent neural networks," *arXiv:2103.12593*, 2021.
[28] A. Jeffares *et al.*, "Spike-inspired rank coding for fast and accurate recurrent neural networks," in *ICLR*, 2022.
[29] Y. Zhao *et al.*, "Deep residual bidir-lstm for human activity recognition using wearable sensors," *arXiv:1708.08989*, 2017.
[30] P. Merolla *et al.*, "A million spiking-neuron integrated circuit with a scalable communication network and interface," *Science*, 2014.
[31] S. B. Furber *et al.*, "The SpiNNaker project," *Proceedings of the IEEE*, 2014.
[32] S. Park *et al.*, "T2FSNN: Deep spiking neural networks with time-to-first-spike coding," *DAC*, 2020.
[33] G. Bellec *et al.*, "Long short-term memory and learning-to-learn in networks of spiking neurons," in *NeurIPS*, 2018.
[34] T. Pellegrini *et al.*, "Low-activity supervised convolutional spiking neural networks applied to speech commands recognition," *arXiv:2011.06846*, 2020.
[35] D. Salaj *et al.*, "Spike frequency adaptation supports network computations on temporally dispersed information," *eLife*, 2021.

Partial-sum Quantization for near ADC-Less Compute-In-Memory Accelerators

Utkarsh Saxena, Kaushik Roy
Elmore School of Electrical and Computer Engineering
Purdue University
West Lafayette, USA
{saxenau, kaushik}@purdue.edu

***Abstract*—Resistive Crossbar (Xbar) Array based Compute-in-Memory (CiM) accelerators form an attractive hardware substrate for acceleration of Deep Neural Networks (DNNs) on edge devices. They perform highly efficient M atrix Vector Multiplication (MVM) operation, employing the power of analog compute. However, efficiency g ains w ith C iM a ccelerators are limited due to the overhead posed by peripheral circuits, primarily the Analog-to-Digital Converters (ADCs). In this work, we improve efficiency o f C iM a ccelerators b y d eveloping ADC-Less and near ADC-Less CiM accelerators which either eliminate or minimize the ADC overhead. More specifically, w e leverage partial-sum quantization to reduce ADC precision to binary (1-bit) or ternary (1.5-bit) values. Xbars with binary partial sums require a sense amplifier f or a nalog-to-digital conversion leading to ADC-Less design. Xbars with ternary partial-sums require two comparators for the conversion process leading to a near ADC-Less design. We develop a CiM hardware aware DNN quantization methodology to mitigate accuracy degradation with partial-sum quantization. We show the effectiveness of our training methodology by achieving high accuracies and minimal accuracy degradation on CIFAR-10 and Imagenet datasets. Consequently, we achieve 14x, 178x and 131x improvements over baseline (8-bit ADC) in Energy, Latency and Compute Efficiency (TOPS/mm^2), respectively on Resnet-20 (CIFAR-10) with ADC-Less design and 11x, 55x and 36x improvements over baseline (8-bit ADC) in Energy, Latency and TOPS/mm^2, respectively on Resnet-18 (ImageNet) with Near ADC-Less design.**

***Index Terms*—Compute-In-Memory, Quantization, Hardware-Software Co-design, Non-Volatile Memories, Analog Compute**

I. Introduction

Compute in Memory (CiM) accelerators enabled by Resistive Crossbar (Xbar) Arrays have shown tremendous potential for efficient a cceleration o f D eep N eural N etworks (DNNs) [1]–[3]. Xbars fabricated with various non-volatile memories (NVM) provide high density and low power storage and have been shown to efficiently execute Matrix-Vector Multiplication (MVM) operation [4], a key kernel in DNNs. Xbar based CiM accelerators can achieve high energy efficiency by performing computations in the analog domain within the memory array in addition to significantly reducing data movement, a bottleneck in DNN execution [5].

Analyses of the energy and area breakdown of Xbar based CiM accelerators reveal the peripheral circuits to be the primary bottleneck in limiting performance. Digital to Analog Converters (DAC) and Analog to Digital Converters (ADC) consume as much as 75% energy within an Xbar macro [6]. Note, ADCs alone consume as much as 60% energy and occupy as much as 80% area in Xbar macros [6]. Additionally, high ADC area is amortized by sharing an ADC across multiple Xbar columns, thereby, limiting throughput of the hardware. Further, each ADC conversion takes multiple clock cycles deteriorating the latency of Xbar based compute [2].

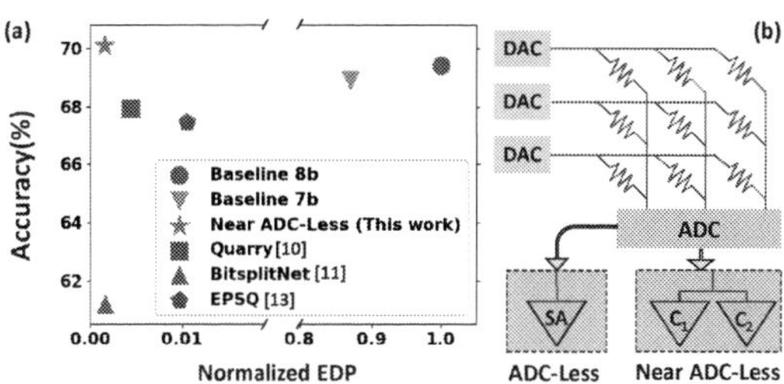

Fig. 1. (a) Energy-Delay-Product (EDP) vs Accuracy comparison of our work with baselines on (a) Imagenet, (b) Schematic of Xbar with ADC-Less and Near ADC-Less design.

A straightforward way to reduce ADC overhead is to reduce its precision, correspondingly improving its area, power and latency. However, reducing ADC precision quantizes the partial-sums. In this work we developed a partial-sum quantization aware training to achieve high accuracy at extremely low ADC precision (1-bit or 1.5-bit). Results on CIFAR-10 dataset show minimal accuracy degradation with 1-bit ADC precision. To scale to larger datasets like Imagenet, we leverage ternary partial-sum quantization which reduces ADC precision to 1.5-bits. 1-bit ADCs can be implemented using just a sense amplifier (Fig. 1(b)), while ADCs with 1.5-bit precision require two comparators for analog-to-digital conversion leading to Near ADC-Less Xbar design (Fig. 1(b)). While there have been several works using partial-sum quantization to reduce ADC precision [7]–[12], our work achieves the highest accuracy on Imagenet with lowest Energy-Delay-Product (EDP) (Fig.1 (a)).

We make the following contributions,

- We develop a CiM hardware aware DNN weight, activation and partial-sum quantization framework incorporating key architectural aspects of CiM hardware like tiling, bit-slicing and layer mapping.
- We develop ADC-Less (1-bit partial-sum) and near ADC-Less (1.5-bit partial-sum) Xbar based CiM accelerators which eliminate or minimize the ADC overhead in con-

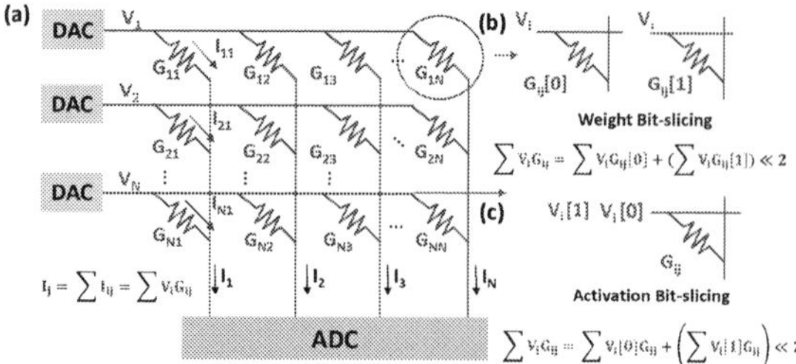

Fig. 2. (a) In-memory MVM operation in Xbars, (b) Weight bit-slicing, (c) Activation bit-slicing.

TABLE I
RELATED WORKS ON PARTIAL-SUM QUANTIZATION.

	Multibit W/A	**Bit-Slicing**	**Large Datasets**	**High Accuracy with low ADC precision**
BNN RRAM [12]	✗	-	✗	✓
Saxena et al. [8]	✓	✓	✗	✓
Kim et al. [9]	✗	-	✗	✗
Bit-SplitNet [11]	✓	✓	✓	✗
Quarry [10]	✓	✗	✓	✗
EPSQ [13]	✓	✓	✓	✗
This work	✓	✓	✓	✓

ventional Xbar based CiM accelerators.

- We implement ADC-Less (1-bit partial-sum) and Near ADC-Less (1.5-bit partial-sum) Xbar based CiM accelerator design on a system level DNN performance estimator and evaluate the improvements in performance metrics.
- We provide an evaluation of the impact of Xbar based non-idealities within ADC-Less and Near ADC-Less designs.

II. BACKGROUND

A. Matrix Vector Multiplication in Xbars

Xbars are augmented with peripheral circuits (DAC, ADC) to perform matrix vector multiplication operation within the memory array in the analog domain. The weight matrix is stored within the memory array as conductance of the memory device element. While, activation vectors are applied as voltages on the wordlines using DAC. Consequently, the signal developed on the bitline is read using ADCs and it corresponds to the matrix vector multiplication output (Fig.2(a)).

Often the precision of memory devices and DAC is lower than the precision demanded by the weight matrix and activation vectors. In such a case, bit-slicing is utilized to scale the hardware to higher precision of weights and activations [2], [6]. Bit-slicing involves decomposing the MVM computation into multiple bit-wise MVM operations which can be supported by the Xbar hardware. Activations are bit-sliced and serially applied on the Word Lines (WLs) across multiple compute cycles (Fig2.(c)) while weights are bit-sliced and stored in multiple Xbars (Fig.2(b)). For a given bit-slice precision of weights and activations, s_w and s_a, respectively, and an Xbar size of N, the ADC precision to capture entire dynamic range of analog signals is given as,

$$B_{ADC} = log(N) + s_w + s_a - 1, \tag{1}$$

Bit-slicing weights and activations to lower precision bit-slices reduces the required precision for ADCs. Reducing ADC precision below B_{ADC} (eq. 1) leads to inaccuracies in capturing entire dynamic range of analog signals. Consequently, it leads to quantization of read partial-sums.

III. RELATED WORKS

Comparison of our approach with related works is listed in Table 1. Authors in [7] develop a reinforcement learning based search framework to find weight, activation and partial-sum precision with minimal accuracy degradation. However, they do not train the workload with quantization and hence require high partial-sum precision. BNN-RRAM [12] trains a binary neural network with 1-bit partial-sum quantization to achieve extreme ADC precision reduction. Their approach requires a batch normalization operation to be implemented in analog domain which exacerbates the impact of non-idealities and is not considered in their analysis. Authors in [9] show high accuracy results with 1-bit partial-sum without analog batch normalization but are only limited to binary neural networks. Authors in [8] improve this approach to multi-bit weight, activation precision while still keeping 1-bit partial-sum quantization. However, the results in [8], [9], [12] are only limited to small scale datasets (CIFAR-10). Bit-Split-Net [11] shows results on large scale datasets like Imagenet with multiple bits of weights and activations but suffers from high accuracy degradation (6%) with 1-bit partial-sum quantization. Additionally, Bit-Split-Net modifies the bit-wise execution of MVM operation in CiM hardware demanding specific hardware for deploying their trained models [11]. Quarry [10] shows impressive results on ImageNet with low ADC precision but bit-slicing is not incorporated in their approach. Note, considering bit-slicing in the methodology is essential to scale the algorithm for different NVM devices which offer different precision of storage and different DAC precision. While EPSQ [13] incorporates bit-slicing, both EPSQ and Quarry suffer from accuracy degradation with very heavy partial-sum quantization. To the best of our knowledge, our work is the first work to achieve highest accuracy on Imagenet dataset with the least ADC precision (1.5-bits). Improved results in our work compared to related works can be attributed to three aspects *1) Superior training methodology employing state of the art weight and activation quantization, 2) Bit-slicing aware DNN training, 3) Partial-sum quantization with trainable scaling factor.*

IV. QUANTIZATION METHODOLOGY

A. Weight and Activation Quantization

1) Integer Quantization: We follow the methodology described in LSQ [14] for integer quantization of weight and activation. Given a full precision value v, scaling factor s and positive and negative quantization levels Q_P, Q_N, respectively, the scaled integer representation v_{int} and the quantized representation v_q are given by,

$$v_{int} = \lfloor clip(\frac{v}{s}, -Q_N, Q_P) \rceil, \quad v_q = v_{int} \times s \tag{2}$$

For signed weights : $Q_N : 2^{B_{w-1}}$, $Q_P : 2^{B_{w-1}} - 1$ while for unsigned activations : $Q_N : 0$, $Q_P : 2^{B_a} - 1$ where B_w and B_a are the weight and the activation precision, respectively. For weight and activation quantization, we use a per layer scaling factor to minimize parameter overhead.

Algorithm 1 Bit-Slicing Algorithm

Input: Integer matrix $\mathbf{v_{int}}$, precision $\mathbf{B_v}$, bit-slice precision $\mathbf{s_v}$

Output: Bit-sliced matrix $\mathbf{v_b}$

$n_b \leftarrow \lceil \mathbf{B_v}/\mathbf{s_v} \rceil$

for $i \in [0, n_b)$ **do**

$\mathbf{v_{b_i}} \leftarrow floor(\mathrm{v}_{int}/(2^{\mathbf{s_v}})^i)$

end for

$\mathbf{v_b} \leftarrow remainder(\mathrm{v_b}, 2^{\mathbf{s_v}})$

2) Bit-slicing: The bit-slicing function divides the integer matrices to multiple low precision bit slices according to Algorithm 1. The bit-slicing algorithm involves floor function which has zero gradients and needs to be approximated during the backward pass. We approximate gradients through the floor and remainder function using straight through estimator. Consequently, the gradient output through the bit-slicing function ($\nabla_{v_{int}}$) is given by,

$$v_b = BitSlice(v_{int}) \quad \nabla_{v_{int}} = \frac{1}{n_b} \times \sum_{i=0}^{n_b - 1} \frac{\nabla_{v_{bi}}}{(2^{s_v})^i}, \tag{3}$$

where n_b is the number of bit-slices. For signed matrices, we apply bit-slicing function on the magnitude of matrix values and then apply the sign.

B. Partial sum quantization

1) Quantization Function: We consider two variants of partial-sum quantization: ADC-Less quantization which quantizes partial-sums to a binary value (1-bit) and Near ADC-Less which quantizes partial-sums to ternary values (1.5-bit). ps being the high precision partial-sum and α being the scaling factor, the binary partial-sum ps_b is given by,

$$ps_b = \alpha * \begin{cases} 1 & \text{if } ps \geq 0 \\ -1 & \text{if } ps < 0 \end{cases} \tag{4}$$

Training with the quantization function involves a backward pass through the quantization function. This involves computing gradients for the full precision partial-sum ps and the scale parameter α. The gradient for the full precision partial-sum (∇_{ps}) and the scale parameter (∇_α) is derived from gradients for binary partial sums (∇_{ps_b}) as follows,

$$\nabla_{ps} = \nabla_{ps_b} \times \begin{cases} 1 & \text{if } -1 \leq \frac{ps}{\alpha} \leq 1 \\ 0 & \text{otherwise,} \end{cases} \tag{5}$$

$$\nabla_\alpha = \nabla_{ps_b} \times sign(\frac{ps}{\alpha}) \tag{6}$$

The ternary partial-sum ps_t with Near ADC-Less design is given by,

$$ps_t = \alpha \times \begin{cases} 1 & \text{if } ps \geq \frac{\alpha}{2} \\ 0 & \frac{-\alpha}{2} < ps < \frac{\alpha}{2} \\ -1 & \text{if } ps \leq \frac{-\alpha}{2} \end{cases} \tag{7}$$

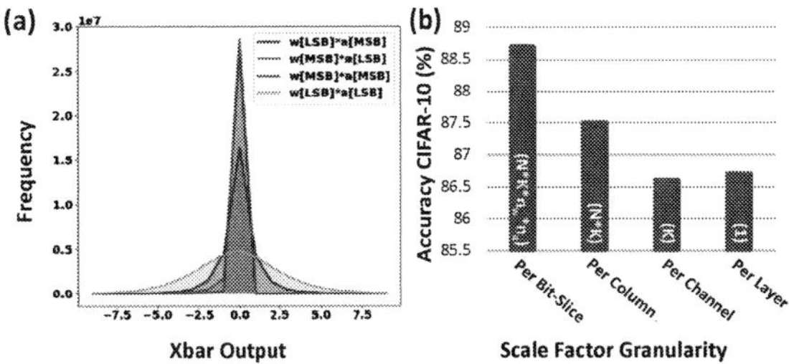

Fig. 3. (a)Distribution of partial-sums for MVM between different bit-positions of weights and activations. (b) Accuracy on CIFAR-10 with different granularity of partial-sum quantization scaling factor (text in bar graphs correspond to number of scaling factors in a layer) .

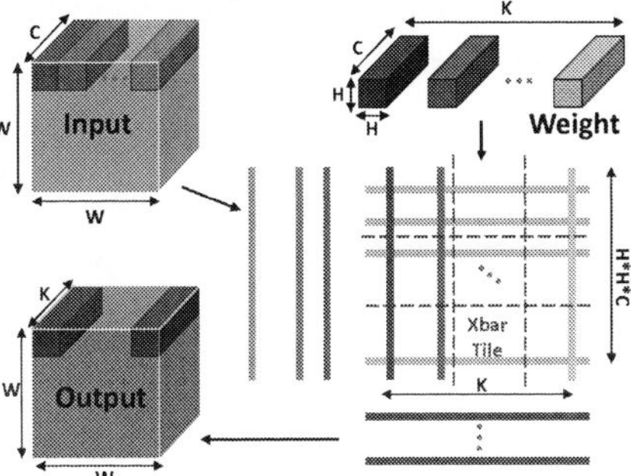

Fig. 4. Convolution layer mapping for Xbar based CiM accelerators.

The gradients of full-precision partial sums (∇_{ps}) are used to obtain gradients for ternary partial-sums (∇_{ps_t}) and the scale parameter (∇_α). They are given by,

$$\nabla_{ps_t} = \nabla_{ps} \times \begin{cases} 1 & \text{if } -1 \leq \frac{ps}{\alpha} \leq 1 \\ 0 & \text{otherwise,} \end{cases} \tag{8}$$

$$\nabla_\alpha = \nabla_{ps} \times \begin{cases} 1 & \text{if } \frac{ps}{\alpha} \geq 1 \\ round(\frac{ps}{\alpha}) - \frac{ps}{\alpha} & -1 < \frac{ps}{\alpha} < 1 \\ -1 & \text{if } \frac{ps}{\alpha} \leq 1 \end{cases} \tag{9}$$

2) Partial-sum quantization Scaling factor (α): Finer granularity of scaling factor (per channel, per layer) [14] helps in capturing the distribution of quantized values with a much higher fidelity. For partial-sum quantization, we evaluated several choices of granularity of quantization as shown in Fig.3(b). We use per-bit slice scaling factor where each Xbar column has a different scale factor for each bit-slice. This corresponds to $N * n_a$ scaling factors for Xbar of size N and n_a number of activation bit-slices. As shown in Fig3(a), partial-sums corresponding to MVM between LSBs of weights and activations have a very different distribution than corresponding operation between MSBs. Per bit-slice scaling factor helps in capturing the variation in partial-sum distribution across different bit-positions. There is an additional 25% parameter overhead for Xbar size 128 when using per bit-slice full precision (32-bit) scaling factors. Scaling factors can further be quantized to reduce the parameter overhead.

C. Convolutional Layer Mapping on Xbar based CiM accelerators

Convolution (Conv) operation between layer inputs and weights is converted to MVM operation for execution on

Xbar based CiM accelerators as shown in Fig. 4. It involves flattening of the 4-dimensional weight kernel along height, width and channel dimension to generate the weight matrix. Similarly, each activation window is flattened to form the activation vector. Since flattened tensors are much larger than a single Xbar, the activation vectors and weight matrix are tiled according to the Xbar size. Further, each tiled tensor is bit-sliced according to the precision supported by the hardware. Each Xbar generates a partial-sum for bitwise MVM operation which is scaled and accumulated spatially and temporally to generate the layer output. While we show results for the layer mapping described here, our methodology is able to support any different kind of layer mapping.

V. Accuracy Evaluation

In this section we present details on the DNN training process with our quantization methodology. We show accuracy results obtained with ADC-Less and Near ADC-Less on CIFAR-10 and Imagenet dataset. Table 2,3 show the compiled accuracy results and comparison with related works. W/A represent weight and activation precision, while s_w and s_a represent bit-slice precision of weights and activations, respectively. Baseline accuracy involves accuracy obtained without any partial-sum quantization but with weights and activation quantized.

A. Two Stage Training

We follow a two stage training process to obtain the final DNN model with weight, activation and partial-sums quantized. In the first step, we train the DNN model with only weights and activation quantization while partial-sums are not quantized. Then, in the second step, the partial-sum quantization along with bit-slicing is considered during training and the model is trained again. We adopt this two step quantization aware training process to incrementally introduce quantization of data-structures during training.

B. CIFAR-10 Accuracy Results

We show results on Binary Resnet20, Binary Wide Resnet20 and Quantized Resnet20. The compiled results on CIFAR-10 are shown in Table 2. We train the model for 400 epochs using SGD Optimizer during each training stage for the two stage training using CosineAnnealing learning rate scheduler with initial learning rate of 0.01. We use The first and last layers are kept to be full precision following [8], [9].

1) Results on Binary Wide Resnet20 and Resnet20: : Comparing with Kim *et al.* [9], our methodology with ADC-Less quantization outperforms their 1-bit partial-sum quantization. With respect to 2-bit partial-sum quantization in [9], we achieve higher absolute accuracy and comparable gap to baseline with Near ADC-Less design (1.5-bit). Comparing with Saxena *et al.* [8], we achieve much higher accuracy and much lower gap to baseline with our methodology. Additionally, we are able to close the accuracy gap with baseline to be within 1% with near ADC-Less.

TABLE II
Accuracy Results on CIFAR-10. (Xbar size 128)

CIFAR-10						
	W/A	s_w/s_a	ADC precision	baseline	Accuracy	Gap
Binary Wide-Resnet-20 (4x-wide)						
Kim *et al.* [9]	1/1	1/1	1	90.96	86.3	4.66
Kim *et al.* [9]	1/1	1/1	2	90.96	90.5	0.46
ADC-Less	**1/1**	**1/1**	**1**	**92.48**	**91.1**	**1.38**
Near ADC-Less	**1/1**	**1/1**	**1.5**	**92.48**	**91.7**	**0.78**
Binary Resnet-20						
Saxena *et al.* [8]	1/1	1/1	1	85.5	82.7	3.3
ADC-Less	**1/1**	**1/1**	**1**	**85.5**	**83.7**	**1.8**
Near ADC-Less	**1/1**	**1/1**	**1.5**	**85.5**	**85.0**	**0.5**
Quantized Resnet-20						
Saxena *et al.* [8]	3/3	1/1	1	89.3	87.5	1.8
ADC-Less	**3/3**	**1/1**	**1**	**89.6**	**88.2**	**1.4**
Near ADC-Less	**3/3**	**1/1**	**1.5**	**89.6**	**88.7**	**0.9**

2) Results on Quantized Resnet-20: : Comparing with Saxena *et al.* [8], we achieve higher accuracy with lower gap to baseline with both ADC-Less and Near ADC-Less design. These results highlight the effectiveness of our training methodology when bit-slicing is applied to multi-bit weight and activation.

C. ImageNet Accuracy Results

Table III contains the results obtained on Imagenet and comparison with related works. The model is trained for 90 epochs using SGD optimizer during the first stage and for 30 epochs during the second stage. During both stages, we use CosineAnnealing learning rate scheduler with initial learning rate of 0.01. Following Quarry [10], the partial-sums of first and last layer are not quantized. With Near ADC-Less design, we are able to get no accuracy loss when compared with the baseline. Getting higher accuracy than baseline can be attributed to more training effort in the case of partial-sum quantization as a consequence of two-stage training.

Quarry [10] does not consider bit-slicing in their training methodology and suffer from high accuracy degradation with heavy partial-sum quantization (1-bit). By incorporating bit-slicing in our training, we are able to achieve no accuracy loss with ternary partial-sums. Bit-split-Net [11] does consider bit-slicing of activations but still suffers from high accuracy loss. Bit-Split net modifies the bit-wise execution within CiM hardware by considering each bit-path separately for parallel execution. Since accumulation across bit-slices do not occur until the final layer of DNN model, there is a considerable information loss which manifests as a large accuracy degradation. Our approach is based on conventional CiM hardware which performs accumulation across bit-slices after every MVM operation [2], [3]. EPSQ [13] does consider bit-slicing of both weights and activations yet achieves substantial accuracy degradation with 3-bit ADCs (2%). This can be attributed to the short-comings of their training methodology, particularly, the partial-sum quantization function. EPSQ uses heuristic based optimization for evaluating scaling factors for their partial-sum quantization, while in our methodology, the scaling factors are optimized using gradient descent during training.

D. Impact of Xbar size and Bit-slicing

Larger Xbar size (N) requires higher precision ADC to capture full dynamic range of partial-sums according to eq

TABLE III
ACCURACY RESULTS ON IMAGENET.(XBAR SIZE 128)

	ImageNet					
	W/A	s_w/s_a	ADC precision	Baseline	Accuracy	Gap
	Quantized Resnet-18					
Bit-Split-Net [11]	3/3	3/1	1	67.6	61.2	6.4
Quarry [10]	3/3	3/3	1	67.73	62.93	4.8
Quarry [10]	3/3	3/3	4	67.73	67.93	-0.2
EPSQ [13]	4/4	1/1	3	69.71	67.45	2.26
EPSQ [13]	4/4	1/1	4	69.71	69.06	0.65
Near ADC-Less	**3/3**	**1/1**	**1.5**	**69.4**	**70.1**	**-0.7**

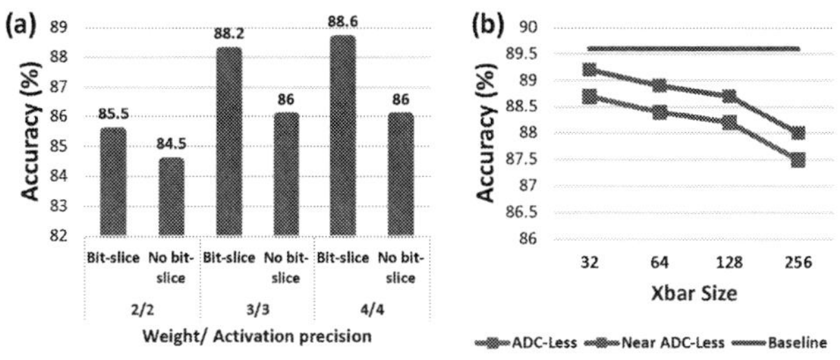

Fig. 5. (a) Impact of bit-slicing on Quantized Resnet-20 accuracy on CIFAR-10 for different weight and activation precision. (b) Impact of Xbar size on Quantized Resnet-20 (3-bit weight, 3-bit activation) CIFAR-10 accuracy.

(1). Consequently, larger xbar sizes cause much severe quantization error with ADC-Less and near ADC-Less design. It is observed that larger xbar sizes achieve lower accuracy with both ADC-Less and near ADC-Less design as shown in Fig. 5(b). Similarly, bit-slicing lowers required ADC precision (eq. 1) causing lesser quantization error with partial-sum quantization. Therefore, higher accuracy is achieved when bit-slicing is incorporated in the training methodology (Fig. 5(a)).

E. Impact of Variations

Device variations impacting workload accuracy is a common issue in Xbar based CiM accelerators [16]. Device variations influence the value of weight bit-slice affecting workload accuracy. Device variations can be spatial and temporal impacting resistance of NVM device. Variations in temperature and operating voltages induce cycle to cycle variations in NVM device resistance. Variation and imperfections in process technology cause spatial variation in NVM devices. Authors in [16] show that NVM device resistance variation can be modelled as a log-normal distribution with zero mean. We add variations to weight bit-slices according to the equation 10. Let w be the weight bit-slice. The weight bit-slice after variation w_{var} is given as,

$$w_{var} \leftarrow e^{\theta} \times w, \tag{10}$$

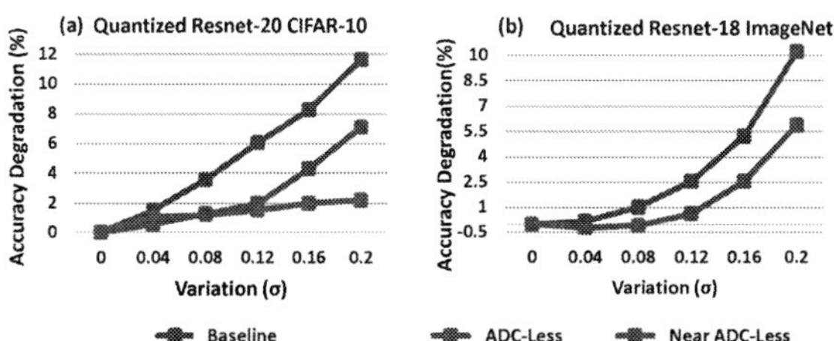

Fig. 6. Impact of NVM device variation on accuracy obtained with ADC-Less and Near ADC-Less design compared with baseline with no partial-sum quantization.

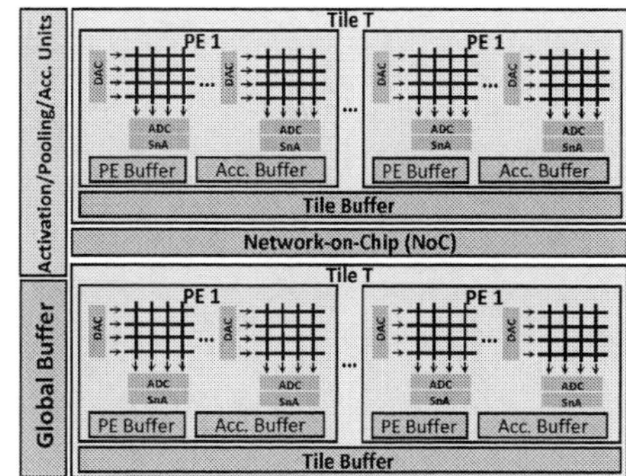

Fig. 7. Spatial architecture for system level evaluation of DNN workload [15]

TABLE IV
ADC PRECISION AND SHARING FOR BASELINES.

	ADC precision	ADC type	#columns sharing an ADC
Baseline 8-bit ADC	8b	SAR	128
Baseline 7-bit ADC	7b	SAR	128
ADC-Less	1b	Flash	1
Near ADC-Less	1.5b	Flash	1
EPSQ	3b	Flash	8
Quarry	4b	Flash	16
Bitsplitnet	1b	Flash	1

where $\theta \sim \mathcal{N}(0, \sigma^2)$, $\mathcal{N}$ is normal distribution with zero mean and standard deviation σ. We vary σ from 0 (no variation) to 0.2 (high variation) and observe the impact on accuracy degradation as shown in Fig. 6. We observe that quantizing partial-sums to very low precision values makes the DNN model resilient to small variations in weight bit-slice values and reduces the impact of high variations.

VI. HARDWARE PERFORMANCE

In this section we evaluate the hardware performance improvement obtained with ADC-Less and Near ADC-Less CiM accelerators. We implement the designs in DNN+Neurosim V1.3 Simulator [15] and report the performance metrics.

A. Architecture Overview

The spatial architecture (Fig.7) consists of mutiple tiles connected via network-on-chip. The top level of the spatial architecture has an array of tiles, global buffer for storing activations, accumulation units, activation units and pooling units. A tile consists of multiple processing elements (PEs), tile buffer for storing input and output activations. Each PE has multiple Xbar Units with buffers for storing inputs and partial-sums. Each Xbar Units consits of a DAC, ADC, Xbar and Shift and Add (SnA) modules. Depending on relative area of ADCs and Xbars, an ADC might be shared across multiple xbar columns requiring a multiplexer for routing between Xbar and ADC. We consider that the on-chip memory is sufficient to store entire DNN weights, thus only off-chip access is to fetch the input activation.

B. Performance Metrics

We analyze three performance metrics: Energy, Latency and Compute Efficiency (TOPS/mm^2) for Quantized Resnet-18 (3-bit weight and activation) on Imagenet and Quantized Resnet-20 (3-bit weight and activation) on CIFAR-10 (Fig.8).The baselines involve higher ADC precision which requires sharing

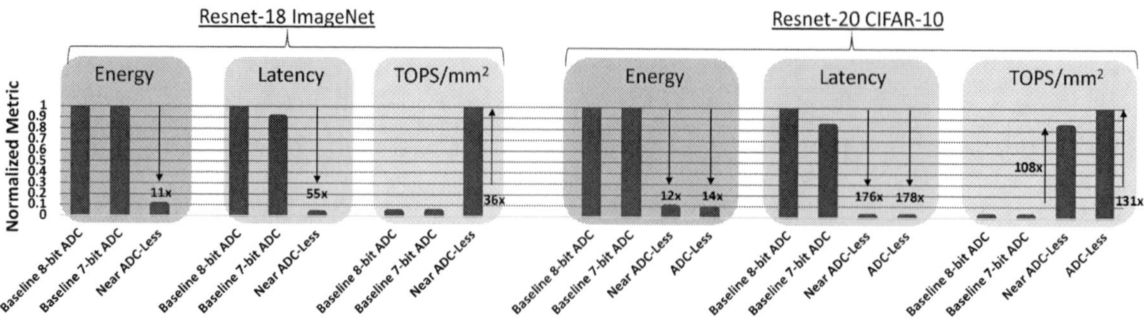

Fig. 8. Normalized Performance metrics for Resnet-20 on CIFAR-10 and Resnet-18 on Imagenet.

an ADC across multiple Xbar columns. The ADC precision, ADC type and ADC sharing for our design and the baselines is listed in Table IV. Baseline 8-bit ADC correspond to the design without partial-sum quantization and uses a single Successive-approximation-register (SAR) ADC for all Xbar columns following [2]. We observe that reducing ADC precision by 1-bit causes accuracy degradation of less than 1%, hence, we compare with another baseline with 7-bit SAR ADC.

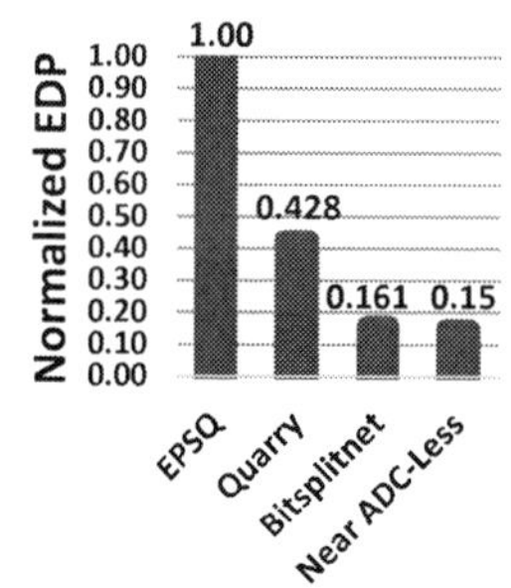

Fig. 9. Normalized EDP for Resnet-18 on Imagement.

Compared to 8-bit/7-bit baseline, we get 12x/12x, 176x/146x, 108x/90x improvements in Energy, Latency and TOPS/mm^2 with Near ADC-Less design, and, 14x/14x, 178x/148x, 131x/109x improvements in Energy, Latency and TOPS/mm^2 with ADC-Less design on CIFAR-10. On ImageNet, we get 11x/11x, 55x/49x, 36x/32x improvements in Energy, Latency and TOPS/mm^2 with Near ADC-Less design. Additionally, we compared the EDP improvements with Near ADC-Less design with related partial-sum quantization works (Fig. 9). The ADC sharing in related works is determined by the relative area of Flash ADC compared to the Xbar area [15]. We get 7% improvement in Energy-Delay-Product (EDP) over the most competitive partial-sum quantization baseline, Bitsplitnet, while getting 9% improvement in accuracy on Imagenet with Near ADC-Less design. Bit-Split net modifies the bit-wise execution within CiM hardware by considering each activation bit-path independently. Since, each activation bit in BitSplitNet is independent, energy and latency for 3-bit activations is obtained by scaling the energy and latency obtained with 1-bit activation by 3.

VII. Conclusion

In this work, we present a partial-sum quantization approach to ADC-Less (1-bit ADC) and Near ADC-Less (1.5-bit ADC) Xbar based CiM accelerators. We achieve accuracies close to the baseline with quantized partial-sums on a variety of workloads and datasets. We are able to achieve 14x, 178x and 131x improvements over 8-bit ADC baseline in Energy, Latency and Compute Efficiency (TOPS/mm2) with ADC-Less CiM accelerators on Resnet20 CIFAR-10 and 11x, 55x and 36x improvements, respectively, over baseline in Energy, Latency and TOPS/mm2 with Near ADC-Less CiM accelerators on Resnet18 Imagenet.

Acknowledgement

This work was supported in part by the National Science Foundation, Intel Corporation, Semiconductor Research Corporation (SRC), IARPA MicroE4AI program, and by the Center for Brain-Inspired Computing (C-BRIC), one of six centers in JUMP, funded by Semiconductor Research Corporation (SRC) and DARPA.

References

[1] A. Biswas *et al.*, "Conv-sram: An energy-efficient sram with in-memory dot-product computation for low-power convolutional neural networks," *IEEE JSSC*, 2019.

[2] A. Ankit *et al.*, "Puma: A programmable ultra-efficient memristor-based accelerator for machine learning inference," in *ASPLOS*, 2019.

[3] A. Shafiee *et al.*, "Isaac: A convolutional neural network accelerator with in-situ analog arithmetic in crossbars," in *ISCA*, 2016.

[4] X. Sun *et al.*, "Xnor-rram: A scalable and parallel resistive synaptic architecture for binary neural networks," in *DATE*, 2018.

[5] N. Verma *et al.*, "In-memory computing: Advances and prospects," *IEEE Solid-State Circuits Magazine*, 2019.

[6] K. Roy *et al.*, "In-memory computing in emerging memory technologies for machine learning: An overview," in *DAC*, 2020.

[7] S. Huang *et al.*, "Mixed precision quantization for reram-based dnn inference accelerators," in *ASP-DAC*, 2021.

[8] U. Saxena *et al.*, "Towards adc-less compute-in-memory accelerators for energy efficient deep learning," in *DATE*, 2022.

[9] Y. Kim *et al.*, "Mapping binary resnets on computing-in-memory hardware with low-bit adcs," in *DATE)*, 2021.

[10] A. Azamat *et al.*, "Quarry: Quantization-based adc reduction for reram-based deep neural network accelerators," in *ICCAD*, 2021.

[11] H. Kim *et al.*, "Algorithm/hardware co-design for in-memory neural network computing with minimal peripheral circuit overhead," in *DAC*, 2020.

[12] Y. Kim *et al.*, "Neural network-hardware co-design for scalable rram-based BNN accelerators."

[13] ——, "Extreme partial-sum quantization for analog computing-in-memory neural network accelerators," *J. Emerg. Technol. Comput. Syst.*, 2022.

[14] S. K. Esser *et al.*, "Learned step size quantization," 2019.

[15] X. Peng *et al.*, "Dnn+neurosim: An end-to-end benchmarking framework for compute-in-memory accelerators with versatile device technologies," in *IEDM*, 2019.

[16] G. Charan *et al.*, "Accurate inference with inaccurate rram devices: Statistical data, model transfer, and on-line adaptation," in *DAC*, 2020.

Efficient Multi-Objective Optimization for PVT Variation-Aware Circuit Sizing using Surrogate Models and Smart Corner Sampling

Octavian Pascu*, Catalin Visan*, Georgian Nicolae*, Mihai Boldeanu*, Horia Cucu*
Cristian Diaconu†, Andi Buzo†, Georg Pelz†
*University "Politehnica" of Bucharest, †Infineon Technologies

Abstract—**Circuit sizing for designs with many design variables and responses is a complex task that requires highly experienced and creative designers to invest precious time in trial and error, routine work. In addition, sizing the circuit while also taking into account PVT (process, voltage, temperature) variation corners increases the complexity further. To simplify such tasks, designers select the most unfavorable PVT corner in advance (leveraging their expertise), perform circuit sizing for this condition, and finally verify the resulting design in all PVT corners. This procedure might generate designs that fail the specifications in other PVT corners leading to more design-verification iterative loops. Recent years brought machine learning (ML) and optimization techniques to the field of circuit design, with evolutionary algorithms and Bayesian models showing good results for automated circuit sizing. However, these methods can still require an unfeasibly large number of simulations, especially if taking into account several PVT corners. In this context, we introduce a methodology that uses surrogate ML models to perform PVT variation-aware circuit sizing. We propose to dynamically select the worst PVT corners and take them into account when sizing the circuit. In addition, we explore the best ways to model process corners with Gaussian Processes, leading to more than 10x improvements for such surrogate models. We evaluate the proposed corner management method on two voltage regulators showing different levels of complexity and highlight that it enables finding feasible solutions 2x faster when compared to baseline algorithms which optimize in all PVT corners. In addition, the quality and diversity of the proposed solutions are significantly higher by one to three orders of magnitude in terms of population hypervolume.**

Index Terms—**Evolutionary Algorithms; Gaussian Processes; Circuit Design; Multi-objective Optimization; PVT corners.**

I. Introduction

In order to design circuits, experienced and skilled engineers use complex CAD tools to find suitable configurations that respect a certain specification. An important step in circuit design that relies on designer experience is circuit sizing, which can be very time-consuming. In recent times, Artificial Intelligence techniques have started to be used for circuit sizing which enhances this experience-based tuning. This approach does not only reduce significantly the necessary time for circuit sizing, but also reduces R&D costs.

Historically two approaches were used to automatize circuit sizing. The quantitative approach uses only data from circuit simulations and is usually based on algorithms such as Particle Swarm Optimization (PSO) [1], Bayesian Optimization (BO) [2] or Evolutionary Algorithms (EAs). The drawback of this approach is that it relies only on circuit simulations which results in a high computational cost. The qualitative approach implies creating a model of the circuit using existing circuit data and finding the optimal configuration by using that model. Common techniques for circuit modeling are based on geometric programming [3]. While the main advantage of this approach is the low computational cost, the model can deviate from the real circuit, leading to sub-optimal solutions.

To get the advantages of both approaches and mitigate the drawbacks, hybrid circuit sizing methods have been proposed. Hybrid methods commonly use algorithms mentioned in the quantitative approach but are enhanced using a surrogate model to reduce the number of necessary circuit simulations. The surrogate model is trained using previous simulations, and it is used to evaluate circuit configurations proposed by the optimization algorithm [2] [4]. This way only the most promising circuit configurations are evaluated with the circuit simulator.

Reducing the number of design-verification loops is crucial to assuring a good time to market for new circuits. One way to reduce the number of design-verification loops, involves taking into account process-voltage-temperature (PVT) variation corners during circuit sizing. Typically, designers select the most unfavorable PVT corner in advance, leveraging their expertise. Going further, they perform circuit sizing for this condition and, in the end, the design is verified in all PVT corners to demonstrate its robustness. This minimizes the number of circuit simulations required for circuit sizing, but often leads to multiple design-verification loops because the solution which was feasible for the PVT corner chosen up-front turns out not to feasible in all other corners. Alternatively, the sizing can be done while verifying all PVT corners for each sizing configuration proposed by the algorithm. However, this is computationally expensive (ex. requires 27x more simulations for a usual case of 3 operating conditions with 3 levels each) and leads to exhausting the simulation budget without being able to find feasible solutions. In this context, we propose integrating PVT verification in the circuit sizing phase in a more intelligent manner by periodically selecting the worst PVT corners and performing the sizing only in those. This allows obtaining similar results as if optimizing in all corners

979-8-3503-1176-1/23 $31.00 © 2023 IEEE

while respecting the simulation budget. Moreover, we pay close attention to the way process corners can be modeled and we adapt the surrogate model to take them into account in the most convenient way. We compare the performance of multiple approaches: single surrogate model with categorical axes; single surrogate model with multiple axes; multiple surrogate models.

The main contributions of this paper are the following: (i) we propose an algorithm that takes into account PVT variation during circuit sizing, (ii) we integrate this algorithm in a state of the art solution for circuit sizing, based on multi-objective optimization, (iii) we demonstrate that the proposed algorithm finds feasible solutions using a reduced simulation budget compared to existing methods.

II. Background

In this section, we explain the problem, metrics, and state-of-the-art algorithm that we use.

We treat the circuit sizing task as a Multi-Objective Optimization Problem (MOOP) where the design parameters (DPs) represent the inputs and the circuit responses represent the outputs which need to satisfy some constraints. For solutions that already satisfy the constraints, some responses were chosen as optimization objectives for further improvement.

In the particular case of circuit sizing, one must take into account not only the DPs that can be modified by an optimization algorithm in order to find the optimal configuration, but also the operating conditions (OCs) of the circuit. They are externally imposed by the designer and reflect the PVT corners that have to be taken into consideration while simulating the circuit.

The focus of any global optimizer is to find the input configuration that gives the best value for the objective function. In single-objective optimization there is always a global optimum, represented by a certain point on the output function. So, the goal of the optimizer is to find that point by avoiding getting stuck in local optima. However, in MOOPs the objectives are usually conflicting with each other. In consequence, there is no single point globally optimal, but a front in the objective's hyperspace called the Pareto front.

a) Metrics: There are multiple ways of handling constraints [5]. The priority of this work is to first find feasible solutions (which satisfy all constraints). The importance of the constraints is considered equal. The normalized violation of the specification for each circuit response is calculated resulting in a number of violation components equal to the number of constraints. For a circuit response that meets the specification the violation component is 0, for the other cases the component has a value in the interval $(0, 1]$. The reference worst-case value for each response is taken from the initial population. The violation components are averaged [6] resulting in a metric called Constraint Violation (CV).

$$CV(x) = \frac{1}{N_c} \sum_{i=1}^{N_c} \frac{cv_i(x)}{\max\limits_{x \in P0} \{cv_i(x)\}}, \quad (1)$$

where N_c is the number of constraints, $cv_i(x)$ is the violation component of the response i of solution x and $P0$ is the initial population. CV is used to compare unfeasible solutions in terms of performance.

To compare feasible solutions (for which the CV is 0) we employ a different metric. The Hypervolume (HV) is bounded by a virtual worst-case point and represents the space occupied by one or more solutions. A solution dominates a certain space if all possible solutions inside the space have worse objective values across all objectives' axis. A high HV value corresponds to a high performance solution. The HV can be also used to compare groups of solutions or entire populations by considering the space dominated by at least one of the solutions in the group. However, HV has an important drawback: it can be computationally expensive. The time needed to compute HV grows rapidly with the number of objectives of the optimization problem and with the number of considered solutions.

b) Algorithm: Evolutionary computation is often used for optimization purposes due to its characteristics that make it suitable for global search. In circuit sizing applications one commonly used algorithm is NSGAII [7]. It is successfully used for analog IC design in [8] or with additional machine learning (ML) techniques in [9]. Also, Differential Studies in the literature comparing the most promising five evolutionary algorithms for circuit sizing show that GDE3 is the most promising one [10] [11]. It has a slower but steady convergence rate which is desirable in complex problems with high numbers of objectives. It maintains a high degree of diversity in the population, and it is less impacted by randomness. These traits give it a higher chance to find the global optimum in a large variety of problems. EAs require a high number of simulations in order to find the best circuit configuration. Unfortunately, for complex circuits the circuit simulations are quite computationally expensive, therefore time-consuming. In order to reduce the number of simulations, a surrogate model can be embedded into the algorithm. Probably the most common surrogate model in the context of circuit design is represented by Gaussian Processes (GPs) since this type of model works quite well with a low number of training data points. One solution is to train the model once and then use it for the rest of the optimization with very few updates. This "offline training" method is used for GPs in [12]. Recent methods incline to use "online training" of the surrogate model making the algorithm more computationally expensive but also more efficient in terms of convergence. This approach is used on top of EAs in [13].

MODEBI is an optimization algorithm inspired by GDE3 that uses GPs as surrogate models. At each iteration, the differential evolution engine proposes ten times more offspring than the population size. The surrogate model is retrained each time on the circuit simulator data before selecting from the proposed offspring only the most promising candidates, equal to 25% of the population. Those candidates are evaluated on the real simulator. The final population is chosen based on a survival mechanism. In [14] there are two selection mechanisms and two survival mechanisms detailed, but for the scope of this work we consider the combination which shows the best performance, using Pool Selection (PSel) and Improved Survival (ISv). PSel does not take into account the information about the parents of the offspring; it just selects the best ones in terms of performance. In case of feasible solutions, it uses the HV metric, while for unfeasible ones it uses a combination of CV and the Distribution Metric (DM) [15]. Similarly, ISv does not take into account the information about the parents either. It starts from an extended population including the current population and the offspring evaluated on the real simulator. Then it builds the next population step by step always adding the best performing solution that was not selected yet. It uses CV and DM to compare unfeasible solutions, and HV or CD to compare feasible ones. The major drawback of MODEBI, which we address in this work, is that it evaluates all solutions in all PVT corners. This leads to a relatively long optimization time, which makes MODEBI unusable in some cases.

III. Method

In this section, we propose and present two methods that aim to improve the performance and reduce the number of simulations needed in a circuit sizing task. These advancements can generally be applied to any such algorithm that uses surrogate models. We chose to add them to the *MODEBI* algorithm, which was briefly described in section II. In subsection III-A we describe different training strategies for fitting GPs in order to better model the PVT corners and in subsection III-B we propose an algorithm that evaluates solutions in fewer PVT corners while maintaining its performance.

A. Process corners representation in Gaussian Process Training

In *MODEBI* the authors represented process corners as discrete values on a new dimension for the purpose of GP training. This method provides an easy way of including process corners in the surrogate model training process, but it is not representative of the real-world circuit behavior. In order to improve process corner modeling and the accuracy of the GPs, we propose two different training methods.

a) Representing process corners on multiple dimensions: In circuit design, process corners are commonly represented with two-letter designators (ex. *fs*). The first letter refers to the N-channel MOSFET ("fast") and the second one refers to the P channel ("slow").

We propose a method that improves GP fitting performance by representing the designators on multiple dimensions, with values between 0 and 1. The *slow* corner corresponds to 0, the *fast* corner corresponds to 1 and *typical*, considered to be in the middle corresponds to 0.5.

As an example, the *fs* designator is short for *fast slow* which would be represented as [1,0] for GP training in 2 dimensions. This technique takes advantage of the *slow-fast* ordering behavior and improves the surrogate model performance.

b) Training one GP for each process corner: Training one GP with all available data becomes costly over the optimization process. In order to reduce the training time and since solutions evaluated in different process corners can return very different responses, the second method we propose is training a separate GP for each process corner.

As an example, if in the course of the optimization, the solutions are evaluated in 4 process corners (such as *ss*, *sf*, *fs* and *ff*), 4 GPs would be employed, each trained using 25% of the data.

Using this method each process corner is modeled by a GP, which in turn guarantees a fast and accurate prediction given sufficient data. One significant drawback is that data from other corners are not used in the learning and prediction process. In theory, this data should help the GP better model the circuit because the circuit responses from different corners of the same solution are correlated.

B. Corner Management

In MODEBI the authors proposed evaluating all the corners of all solutions at each iteration in an effort to minimize the number of design-verification loops that take place before an adequate solution is found. Usually in circuit sizing tasks, finding feasible solutions in some corners can take significantly more simulations than in other corners. In essence, the optimization process comes to a point where most solutions in the population are already feasible in one or more PVT corners, but it is struggling to find feasible solutions in other, more difficult ones.

We propose the MODEBI-CM algorithm, an improvement over the MODEBI algorithm which uses a *worst corner selection* mechanism so that simulations are not wasted on easy corners which are likely within specifications already if the *worst corners* meet them. The MODEBI-CM algorithm is described in Algorithm 1. The P worst corners are selected in two ways:

- If there are less than 10 feasible solutions in the population, the P worst corners are chosen based on CV. The mean CV of each corner is computed, and the highest P are selected.
- If there are more than 10 feasible solutions, the P worst corners are chosen based on HV. The mean HV of each corner is computed, and the lowest P are selected.

In our algorithm, the *P worst corners* are updated after an arbitrary number of iterations. Once this number is met, a *worst corners update* simulation budget is assigned, and while the budget is available, the most promising generated solutions (with the lowest CV/highest HV on evaluated corners) are evaluated in all missing corners until the budget is depleted. After the budget is depleted, all solutions that have been evaluated in all corners are put together, and the P worst corners are chosen again. This process is repeated until the optimization stops.

By using the previously mentioned technique, the MODEBI-CM algorithm reduces the simulations needed to reach feasible solutions in the first optimization part. In the second part, where feasible solutions have been reached, by selecting the worst corners based on HV we speed up the optimization process even further. This way we find better solutions (with higher HV) faster.

Algorithm 1: Proposed optimization procedure that performs the evaluations only in the P-worst corners

Input: A random population **POP** with **N** individual DP configurations
Input: A problem with O Operating Conditions (OC) and M PVT corners

Evaluate the starting population in all possible corners
Select the worst P corners
while *Optimization simulation budget available* **do**
  (re)Train the metamodel on all simulation results available so far
  Generate **N*10** new candidate solutions
  Predict circuit responses using the metamodel on the **N*10** new candidate solutions
  Select the best **N/4** solutions out of the **N*10** predicted ones
  Evaluate the **N/4** candidates in the P **corners** using the real simulator
  Select the best **N** individuals out of the $\{\mathbf{N} + \mathbf{N/4}\}$ individuals as the new population **POP**
  if *OC corners should be updated* **then**
    Sort the population based on performance
    while *Worst corners update simulation budget available* **do**
      Evaluate the best candidates in all missing corners using the real circuit simulator
    end while
    Use the above simulation results to select the worst P OC corners
  end if
end while

IV. Experimental Results

In this section, we present the circuits used for the experiments (subsection IV-A) as well as the results obtained with various GP training methods (subsection IV-B) and the proposed CM algorithm (subsection IV-C).

A. Circuits description

We used two circuits to run the experiments proposed in this paper. *Circuit one* is a voltage regulator with 27 design parameters and 11 responses, while *circuit two* is a voltage regulator having eight design parameters and six circuit responses. Both circuits are described in detail in [14]. For the proposed process corner GP training methodologies we have experimented only on *circuit two* (on *circuit one* the designers didn't include process corners) and for Corner Management (CM) we have experimented on both circuits.

B. Evaluation of different GP training methodologies for process corners

The proposed methodologies (subsection III-A) were evaluated on *circuit two* in terms of error, timing and experimental results.

In order to evaluate the accuracy of the different GP training methodologies we used data obtained from a previous optimization with a budget of 9000 simulations. For each iteration the solution responses evaluated by the simulator were compared to the responses predicted by the proposed GP training methods. Table I depicts the Symmetric mean absolute percentage error (SMAPE) of the different GP training methodologies at different points in the optimization process. The *MODEBI GP* training methodology quickly reaches 1.5 and does not achieve a SMAPE lower than 1.43. The proposed methods reach an error of 0.02 after 9000 simulations, which is more than 50 times lower than the *MODEBI* method. Between the 2 proposed methods, representing the process corners on 2 dimensions has the lowest SMAPE, followed by training a GP for each process corner. Overall both proposed GP training methodologies show much better performance than the *MODEBI* method.

TABLE I
GP Symmetric mean absolute percentage error comparison between the *MODEBI* technique and our proposed methods: one GP per process corner (GP per PC) and 2 dimensional process corners (2 dim PC)

GP type	Simulations			
	1000	*2000*	*4000*	*9000*
MODEBI GP	2.43	1.85	1.58	1.43
GP per PC	0.97	0.19	0.047	0.023
2 dim PC	0.93	0.18	0.044	0.020

In terms of timing, the proposed methods are divided in 2 categories. Since the *MODEBI* technique and 1 of our proposed methods (2 dimensional PC) use all data for GP training, they have no significant difference in training and prediction timing. One category represents training a GP with all available data while the other represents training multiple GPs with a fraction of the total data. Since *circuit two* contains 5 process corners, it means that when training with one GP per process corner, each GP will be trained with 20% of the total data. Table II presents the total (training plus prediction) time needed for both categories of GP training methodologies. The timing results for the second method are obtained by training the GPs sequentially. All the GPs can also be trained in parallel and the total timing would be reduced by a factor of 5 in our case. When training with low amounts of data, training *One GP* on all data is faster. As more simulations are gathered, the total time needed for one iteration increases much faster for the *One GP* version.

Taking both SMAPE and timing into consideration, representing the process corners on 2 dimensions and training one GP for each process corners are both promising methods.

If the high GP training time is only a low portion of the simulation time, then training one GP with all data would

TABLE II
GP TIMING (TRAINING+PREDICTION) COMPARISON BETWEEN TRAINING ALL DATA WITH ONE GP (*One GP*) AND TRAINING ONE GP FOR EACH PROCESS CORNER (*Five GPs* SINCE *circuit two* CONTAINS 5 PROCESS CORNERS)

Number Of GPs	Simulations			
	1000	*5000*	*10000*	*15000*
One GP	1.5s	71.3s	202.8s	1092.5s
Five GPs	3.3s	6.1s	28.3s	75.1 s

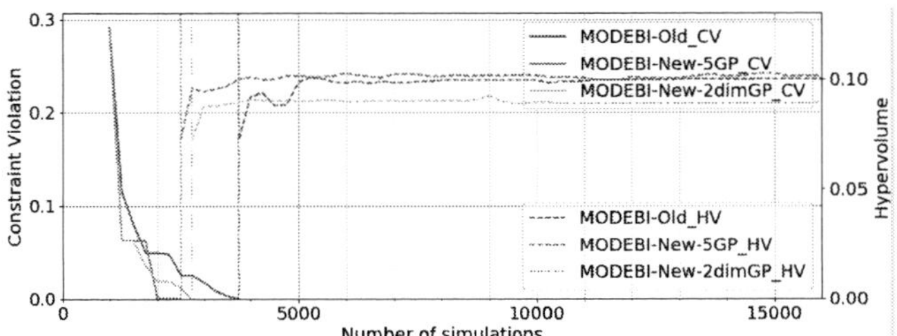

Fig. 1. Circuit two / Contraint Violation (CV) of the best solution represented with continuous line / Hypervolume (HV) of the best 10 feasible solutions represented with horizontal dotted line. Please note that the absence of process corners in Circuit one resulted in the omission of this experiment for that particular circuit.

theoretically be better, as it has a lower SMAPE. If timing is important and simulations are not very costly then training one GP per process corner would make the optimization process significantly faster. Because of the previously mentioned conclusions, we chose to perform design optimization experiments with both proposed methods.

Figure 1 depicts with a continuous line the Constraint Violation (CV) of the best solution in each population obtained when applying the algorithms on *circuit two*. The vertical dotted line represents finding the first feasible solution while the horizontal dotted line represents the Hypervolume (HV) evolution during the optimization process. With blue line the original *MODEBI* algorithm is represented while with red and yellow lines the *MODEBI* algorithm with the new proposed methods are represented. The new methods find feasible solutions faster (2500 for one GP per process corner, 2750 for 2 dimensional process corners) than *MODEBI* (3750). The *One GP per process corner* method finds feasible solutions faster than *2 dimensional process corners* and reaches a higher HV over the optimization process.

In conclusion the proposed GP training methodologies are a significant improvement over the original MODEBI algorithm [14].

C. Evaluation of the proposed CM algorithm

MODEBI algorithm [14] evaluates all solutions in all PVT corners. In section III-A we proposed an improvement of this algorithm with a different GP training methodology for process corners. In section III-B we proposed a Corner Management (CM) algorithm which aims to further improve the performance of MODEBI algorithm by reducing the number of corners in which each solution is evaluated. In order to better present the advantages of the CM algorithm we have increased the number of corners for both circuits by adding extra verification points for both circuits operating conditions. The testing is done on circuit one having 27 corners and circuit two having 20 corners.

Figure 2 depict the results of the proposed *MODEBI-CM* algorithm on both circuits. Two variations of this algorithm are presented, with 2 and 4 worst corners, and are represented in the figure as MODEBI-CM2 and MODEBI-CM4. With blue line, the MODEBI results with *2 dimensional process corners* and no CM are presented.

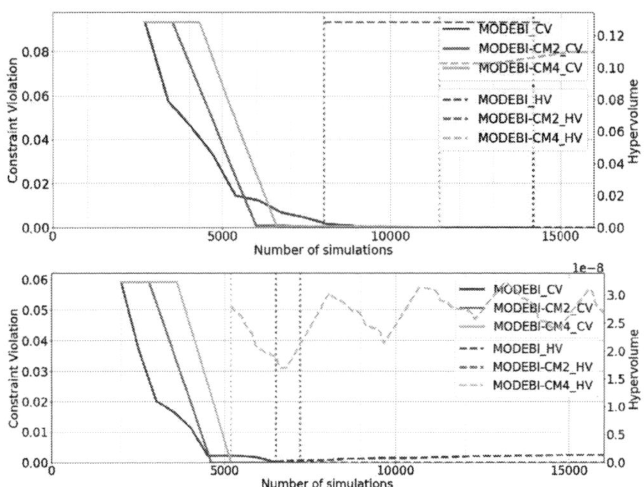

Fig. 2. *Circuit one*(top) and *Circuit two* (bottom)/ Constraint Violation (CV) of the best solution represented with continuous line / Hypervolume (HV) of the best 10 feasible solutions represented with horizontal dotted line/ Only solutions evaluated in all corners

On the first circuit, which has 27 PVT corners, the MODEBI algorithm finds feasible solutions after 14200 simulations, while the fastest MODEBI-CM variant (MODEBI-CM2) finds feasible solutions after 8000 simulations. On the second circuit, which has 20 PVT corners, similar results can be seen. The MODEBI algorithm finds feasible solutions after 6500 simulations, while MODEBI-CM4 finds feasible solutions after 4500 simulations, faster than the MODEBI-CM2 variant.

This happens is because for the second circuit, on the MODEBI-CM2 algorithm, the two worst PVT corners keep switching frequently. Focusing only on 2 PVT corners reduces the performance on other corners, which results in the worst corners changing at the next iteration. When using the CM4 algorithm, the four worst PVT corners are more stable. On the other hand, when applying MODEBI-CM2 on the first circuit, the two worst corners do not change as much and that results in a significant improvement over the MODEBI-CM4 algorithm.

Besides searching for feasible solutions, the MODEBI algorithm tries to maintain a high diversity throughout the optimization process. After finding feasible solutions, the performance (and diversity) of the solutions are represented with the HV dotted line. The CM algorithm takes advantage of this, and because of the high number of evolutions throughout the optimization, the MODEBI-CM variants manage to maintain a much higher diversity of the solutions. On circuit one, MODEBI reaches a HV of $1.4*10^{-4}$ while MODEBI-CM2 reaches a HV of $1.3*10^{-1}$. On circuit two, MODEBI reaches a HV of $1.3*10^{-9}$ while MODEBI-CM4 reaches a HV of $3.3*10^{-8}$.

In conclusion, the proposed MODEBI-CM algorithm advancement significantly improved upon the MODEBI algorithm, offering a smaller amount of necessary corner evaluations at each iteration. On average, an optimized MODEBI-CM finds feasible solutions with 37% less simulations than MODEBI and with a much better HV.

For the best possible results, the number of worst PVT corners must be chosen on a circuit by circuit basis, as it depends on the worst corners consistency.

V. Conclusions and Future Work

In this paper, we proposed two methods for approaching circuit sizing problems of higher complexity using state-of-the-art evolutionary algorithms. These advancements can generally be applied to any such algorithm that uses surrogate models, and aim to improve the performance and reduce the number of evaluations needed in the circuit sizing process.

Within the first method, we have shown two different GP training techniques that take advantage of the *slow-fast* process corners ordering correlation for the surrogate model. Both techniques resulted in significantly lower number of simulations needed for the algorithm to find feasible solutions (33% for the first technique, 27% for the second technique). For our second method, we proposed a Corner Management extension of the algorithm, which reduces the number of necessary simulations at each iteration by choosing the *worst PVT corners* for evaluation. This lowers the number of required simulations even further, resulting in an overall 54% reduction in the number of simulations necessary to reach feasible solutions, when compared to the original MODEBI algorithm. This method also finds better and more diverse solutions, reaching a higher HV compared to the original MODEBI algorithm.

We have observed that the performance of Corner Management algorithm is sensitive to the number of worst corners chosen and how frequently those worst corners are updated, as outlined in subsection IV-C. To ensure the highest performance, adequate hyperparameters must be chosen for each circuit. Another possibility is to assign a small part of the iteration budget to evaluate the missing corners of the top solutions in the population. This would ensure that a minimum of fully evaluated solutions are present at each iteration in the population, and that the *worst corners* are updated at each step. As an advantage of this possible future work, this method provides easier hyperparameter tuning and an overall more robust optimization process.

In subsection III-A we have described the process of GP modeling when process corners are involved, by representing them in multiple dimensions depending on specific *slow-fast* ordering. Another interesting avenue to mixed variable optimization problems in the literature involves integrating a novel latent-variable approach for mixed-variable GP modeling (LVGP) [16]. How the LVGP procedure performs in a circuit sizing task will be investigated in future work.

Acknowledgment

This work was partly supported by a grant of the Ministry of Research, Innovation and Digitization, CCCDI - UEFISCDI, project no. PN-III-P2-2.1-PTE-2021-0460, within PNCDI III.

References

[1] R. Phelps, M. Krasnicki, R. Rutenbar, L. Carley, and J. Hellums, "Anaconda: simulation-based synthesis of analog circuits via stochastic pattern search," *IEEE Transactions on Computer-Aided Design of Integrated Circuits and Systems*, vol. 19, no. 6, pp. 703–717, 2000.

[2] W. Lyu, F. Yang, C. Yan, D. Zhou, and X. Zeng, "Multi-objective bayesian optimization for analog/rf circuit synthesis," in *2018 55th ACM/ESDA/IEEE Design Automation Conference (DAC)*, 2018, pp. 1–6.

[3] S. P. Boyd and S. J. Kim, "Geometric programming for circuit optimization," in *Proceedings of the 2005 International Symposium on Physical Design*, ser. ISPD '05. New York, NY, USA: Association for Computing Machinery, 2005, p. 44–46.

[4] W. Lyu, P. Xue, F. Yang, C. Yan, Z. Hong, X. Zeng, and D. Zhou, "An efficient bayesian optimization approach for automated optimization of analog circuits," *IEEE Transactions on Circuits and Systems I: Regular Papers*, vol. 65, no. 6, pp. 1954–1967, 2018.

[5] K. Li, R. Chen, G. Fu, and X. Yao, "Two-archive evolutionary algorithm for constrained multiobjective optimization," *IEEE Transactions on Evolutionary Computation*, vol. 23, no. 2, pp. 303–315, 2018.

[6] S. Zeng, R. Jiao, C. Li, X. Li, and J. S. Alkasassbeh, "A general framework of dynamic constrained multiobjective evolutionary algorithms for constrained optimization," *IEEE transactions on Cybernetics*, vol. 47, no. 9, pp. 2678–2688, 2017.

[7] K. Deb, A. Pratap, S. Agarwal, and T. Meyarivan, "A fast and elitist multiobjective genetic algorithm: NSGA-II," *IEEE Trans. Evol. Comput.*, vol. 6, no. 2, pp. 182–197, Apr. 2002.

[8] N. Lourenço, R. Martins, A. Canelas, R. Póvoa, and N. Horta, "AIDA: Layout-aware analog circuit-level sizing with in-loop layout generation," *Integration*, vol. 55, pp. 316–329, Sep. 2016.

[9] A. Canelas, R. Martins, R. Povoa, N. Lourenco, and N. Horta, "Efficient yield optimization method using a variable K-Means algorithm for analog IC sizing," in *Design, Automation Test in Europe Conference Exhibition (DATE), 2017*, Mar. 2017, pp. 1201–1206.

[10] M. Stanescu, C. Visan, G. Sandu, H. Cucu, C. Diaconu, A. Buzo, and G. Pelz, "Multi-objective optimization algorithms for automated circuit sizing of analog/ mixed-signal circuits," in *2021 International Semiconductor Conference (CAS)*. IEEE, Oct. 2021, pp. 1–4.

[11] C. Visan, O. Pascu, M. Stanescu, H. Cucu, C. Diaconu, A. Buzo, and G. Pelz, "Versatility and population diversity of evolutionary algorithms in automated circuit sizing applications," in *2021 International Conference on Speech Technology and Human-Computer Dialogue (SpeD)*. IEEE, Oct. 2021, pp. 1–6.

[12] O. Okobiah, S. Mohanty, and E. Kougianos, "Fast design optimization through simple kriging metamodeling: A sense amplifier case study," *IEEE Transactions on Very Large Scale Integration (VLSI) Systems*, vol. 22, no. 4, pp. 932–937, 2014.

[13] B. Liu and A. Nikolaeva, "Efficient global optimization of mems based on surrogate model assisted evolutionary algorithm," in *2016 Design, Automation Test in Europe Conference Exhibition (DATE)*, 2016, pp. 555–558.

[14] C. Vişan, O. Pascu, M. Stănescu, E.-D. Şandru, C. Diaconu, A. Buzo, G. Pelz, and H. Cucu, "Automated circuit sizing with multi-objective optimization based on differential evolution and bayesian inference," *Knowledge-Based Systems*, vol. 258, p. 109987, 2022.

[15] K. Zheng, R.-J. Yang, H. Xu, and J. Hu, "A new distribution metric for comparing pareto optimal solutions," *Struct. Multidiscip. Optim.*, vol. 55, no. 1, pp. 53–62, Jan. 2017.

[16] Y. Zhang, D. W. Apley, and W. Chen, "Bayesian optimization for materials design with mixed quantitative and qualitative variables," *Scientific reports*, vol. 10, no. 1, pp. 1–13, 2020.

Model-Driven Dataset Generation for Data-Driven Battery SOH Models

Khaled Sidahmed Sidahmed Alamin*, Francesco Daghero*, Giovanni Pollo*,
Daniele Jahier Pagliari*, Yukai Chen†, Enrico Macii*, Massimo Poncino*, Sara Vinco*
*Politecnico Di Torino, Turin, Italy {name.surname@polito.it} †IMEC, Leuven, Belgium {yukai.chen@imec.be}

***Abstract*—Estimating the State of Health (SOH) of batteries is crucial for ensuring the reliable operation of battery systems. Since there is no practical way to instantaneously measure it at run time, a model is required for its estimation. Recently, several data-driven SOH models have been proposed, whose accuracy heavily relies on the quality of the datasets used for their training. Since these datasets are obtained from measurements, they are limited in the variety of the charge/discharge profiles.**

To address this scarcity issue, we propose generating datasets by simulating a traditional battery model (e.g., a circuit-equivalent one). The primary advantage of this approach is the ability to use a simulatable battery model to evaluate a potentially infinite number of workload profiles for training the data-driven model. Furthermore, this general concept can be applied using any simulatable battery model, providing a fine spectrum of accuracy/complexity tradeoffs. Our results indicate that using simulated data achieves reasonable accuracy in SOH estimation, with a 7.2% error relative to the simulated model, in exchange for a 27X memory reduction and a ≈2000X speedup.

***Index Terms*—Battery modeling, digital twin, automotive**

I. Introduction

The accuracy of onboard State of Health (SOH) estimation in Battery Management Systems (BMS) is essential for ensuring the safety and reliability of battery systems of a battery-powered device, and in particular for Electric Vehicles (EVs). As there is no practical physical way to instantaneously measure the SOH, such tracking inevitably requires a model.

The literature about SOH models is extremely vast, including electrochemical, equivalent circuits, semi-empirical, analytical, and statistical models [1]. More recently, on the wave of the Machine Learning (ML) hype, a new category of *data-driven models* has emerged, in which a set of instantaneously measurable battery parameters (typically, voltage, current, and temperature) relative to a charge or discharge session of a battery is labeled with corresponding SOH values calculated at session's end [2]. These labeled measures are then used as a dataset to train appropriate ML models [3], [4].

Data-driven models essentially solve the two main drawbacks of traditional models: (i) they are more general, as models for different battery types can be naturally obtained by training them with measurements on different devices, thus also covering variability aspects; and (ii) they do not require any kind of simulation, thus resulting in significant reductions in time and space complexity when deploying the model on resource-constrained devices (i.e., no need for a large amount of simulation operations to estimate battery dynamics and/or for a simulation engine).

On the other hand, the quality of data-driven models is strongly dependent on the size and the variety of the dataset. As these datasets are obtained by experimental measurements, it is *materially unfeasible to provide an acceptable coverage of the design space*: datasets are generated at specific working conditions, determined by the application domain, and in limited time, thus restricting the variety of explored charge/discharge/rest patterns, discharge current profiles, and load currents. Last but not least, such a large exploration space should possibly be repeated on multiple battery instances in order to account for the intrinsic variability of the devices. The consequence is that datasets obtained by measurements are by definition very accurate, but accuracy is guaranteed only in the few points of the experiment space that have been measured.

The key motivation of this paper is to fundamentally swap this asymmetry, i.e., to *sacrifice some accuracy while extending the coverage of the design space*. We propose *to use measurements* (possibly much fewer than those required to generate the whole dataset) *to build a simulatable battery model* that incorporates the desired effects. This model can then be used to generate arbitrarily large datasets, which will serve as the training set for constructing more lightweight and flexible data-driven models.

Several are the advantages of this approach:

- **Exploration of virtually unlimited data points:** once the battery model is built, any kind of current and/or temperature workload can be simulated to generate as many data points as needed, at a lower cost and in less time than would be required for experimental measurement setups;
- **Tunable accuracy/complexity tradeoff:** depending on the quality of the available measurement data, more or less accurate battery models can be built, providing a flexible range of datasets for the data-driven model;
- **Possibility of model combination:** multiple types of battery models can be built and integrated to cover different aspects of battery dynamics, increasing the comprehensiveness of the final battery model;
- **The final SOH model is still a data-driven SOH model:** its execution does not require simulation and it is essentially a callable function of "live" parameters that can be deployed on a target resource-constrained device.

Our results show that using data obtained from the chosen simulation models [5], we achieved an error of about 7% with respect to the SOH data while reducing memory usage by 27X and speeding up calculations by ≈2000X.

II. Background and Related Work

A. Battery Aging

Battery aging is the effect of (i) *calendar aging* (L_{cal}), reflecting battery intrinsic degradation when in rest conditions as an effect of temperature, State of Charge (SOC), and elapsed time; and (ii) *cycle aging*, representing capacity loss during each charge/discharge cycle (L_{cyc}), depending on average values of current I, SOC, cell temperature T and Depth of Discharge (DoD, i.e., difference between final and initial SOC) [6]–[8]. Overall capacity loss (L_C) is thus the sum of a global term for calendar aging plus the sum of the degradation in each cycle [9]:

$$L_C(t, SOC, DoD, I, T) = L_{cal}(t, SOC, T) + \sum_{i=1}^{N} L_{cyc}(I_i, SOC_i, DoD_i, T_i) \quad (1)$$

where N represents the number of charge/discharge cycles, SOC and T are average over an interval of length t in L_{cal}, and refer to each individual cycle i in L_{cyc}.

State-of-the-art models for L_{cal} and L_{cyc} leverage either the similarities of fatigue process of materials subjected to cyclic loading [6] or incorporate electro-chemical properties of the charge/discharge process [9]. It is outside the objective of this work to provide further details about the models themselves; an exhaustive overview of these models is available in [10].

B. Data-Driven SOH Estimation

The relative simplicity of casting the estimation of the SOH as the problem of building a predictive model has spurred a number of datasets available online [11] and a vast literature about models [3]. The approaches used to estimate SOH differ in two key aspects: how SOH is measured and the ML model used for estimation.

Concerning the former aspect, SOH is measured either in terms of the loss of capacity (which is more typical) or in terms of the increase of the internal resistance. For the latter aspect, conversely, the spectrum of options is definitely much wider: models range from various types of Neural Networks (NNs, feed-forward or recurrent ones) to simpler models like random forests, Support Vector Machines (SVMs), or Bayesian networks. Many of these approaches claim to estimate capacity or resistance with high accuracy, making them promising candidates for SOH estimation.

However, as emphasized by the authors of [3], comparing different approaches and establishing reference models is challenging. One reason is the *quality* of the datasets. As a matter of fact, most of these datasets are too limited in size. Besides the obvious impact that small datasets have on the accuracy of data-driven models (in particular NNs), there is also the problem of the *variety* of the dataset points. As they are obtained from lab measurements, there are some intrinsic limitations in generating some specific data points (e.g., very low load currents, which will require prohibitive runtimes) and are susceptible to measurement errors and noise.

III. Proposed Methodology

A. Workflow of the Proposed Methodology

Our idea is to use the datasets mentioned in the former section (or possibly a small portion thereof) to build a full-fledged, simulatable battery model including the SOH *together with the entire battery dynamics*, and then use this simulatable model to generate additional data points, that become the training set for a higher quality SOH data-driven model. Figure 1 sketches the envisioned flow to implement this approach.

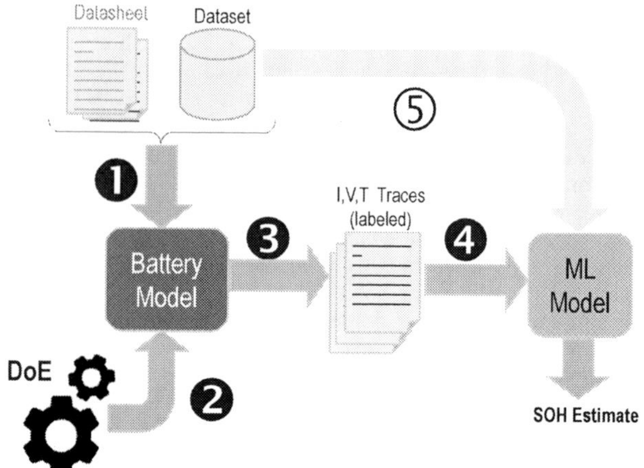

Fig. 1: Conceptual flow of the proposed methodology

The flow starts (❶) with an *available set of battery data*, which typically includes measurements collected at various operating conditions and can be generated afresh or obtained from public datasets. While datasheets can be used to obtain battery information, they generally result in poorer accuracy in the construction of the model [12].

Battery datasets are then used to *identify the parameters of a battery model* (hereafter the *simulation model*), which tracks (at least) the desired target quantity (in our case, SOH). Depending on the type of model, an appropriate procedure is followed for the identification of the model parameters [12]–[14]. Section III-B elaborates on the requirements for the simulation model, and Section III-C surveys the main ones available in the literature that comply with these requirements.

Once the simulation model is built, *an exhaustive set of synthetic traces is generated* in a Design-of-Experiment (DoE) step (❷), exercising as many points as possible in the space of the model inputs. For each of these design points, a simulation of the battery model is run to yield one output trace (❸).

Finally each trace generated by the simulation model is used for training the SOH ML model (hereafter the *data-driven model*, ❹). Section III-D will describe the various options for the data-driven models and their impact on a BMS. If the format of the traces ❶ and ❸ are the same, we can use a mix of real (measured) and simulated (model-driven) traces to train or test the model (⑤). This step is not explicitly depicted in the flow, but it showcases the flexibility of the approach.

B. General Requirements for Battery Simulation Models

The need for a simulatable model raises the issue of identifying a common *interface of the model*, defining the requirements in terms of modeled quantities. The general interface used in our framework is shown in Figure 2.

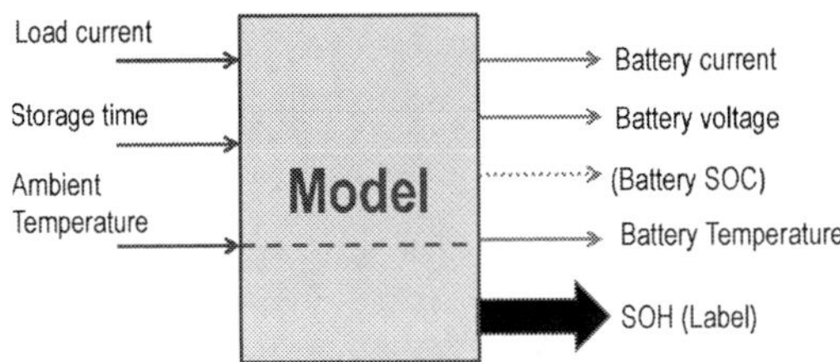

Fig. 2: General interface of the battery simulation model.

The inputs of the models are the independent variables that represent the external conditions under which the battery is used, and must be measurable at runtime. Inputs are load (charge/discharge) current, ambient temperature, and total storage time (needed for calendar aging). Notice that although the general aging model (Equation 1) contains the number of cycles N, this is not strictly required as an input of the battery model; N can in fact be simply obtained from the dataset by using the timestamp of each data point: a new cycle is counted anytime we complete a charge/discharge sequence.

The outputs of the model are typically arbitrary and can include a range of quantities typical of most public datasets [11]. However, two constraints must be considered. Firstly, traces generated by the battery model in ❸ **must be labeled**, meaning that they must include a label that represents SOH, so that the traces can be used to train the ML battery model effectively. Secondly, if possible, the simulation model traces should *follow the same format of the datasets used in* ❶. This will allow the synthetic dataset to be used as an extension of the original dataset for training and/or testing the resulting data-driven battery model, thereby enabling a wider scenario.

Finally, it is worth emphasizing the importance of temperature in modeling SOH and highlighting the distinction between *ambient* and *battery* temperature. If the model is chosen carefully, it will include a thermal model that, based on the electrical quantities **and** the external temperature, can accurately predict the battery's internal temperature. However, if the thermal model is not available, a rough approximation is to use the ambient temperature as a proxy for the internal temperature (the dashed line). This is essential because SOH is highly sensitive to temperature.

C. Choice of the Battery Simulation Model

The requirements implied by the model interface defined in the previous section might result in a possible difficulty in our approach. Fortunately, models with these characteristics are available in the literature [15], with different ranges of accuracy and complexity. Electrochemical models represent the chemical reactions as differential equations [16], but they are too complex to be executed on board in real-time. Circuit-equivalent models like [17], [18] require an RC network solver or a state-space computation environment and may result in being very heavy; improving their performance on the other hand implies a reduction in accuracy. Kalman filters are a good compromise: they allow to specify an error bound and, being an iterative process, they tend to reduce the estimation error to zero [19].

For the aforementioned reasons, we chose the Kalman filter-based simulation model of [5] as a reference for this work (Figure 3). This pre-defined model is designed to describe a 27Ah battery and uses Simscape to represent both the thermal and electrical dynamics of the battery. The model reproduces capacity fading due to thermal cycling and uses an unscented Kalman filter to estimate battery SOC and internal resistance based on values of voltage, current and internal temperature at runtime. These estimates are then used as inputs of a SOH estimation block built with measurement-based lookup tables.

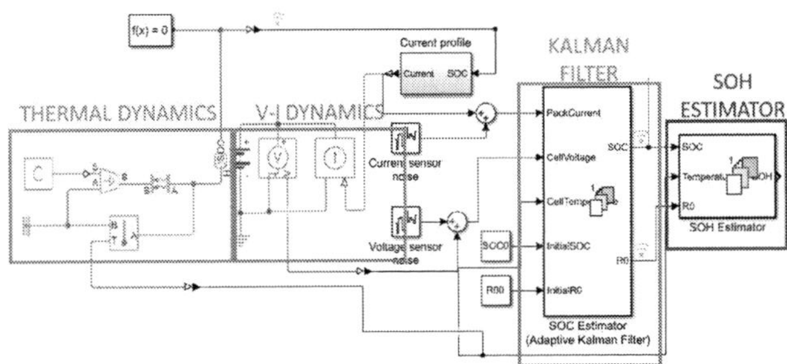

Fig. 3: Adopted battery model based on Kalman filters.

D. Choice of the Battery Data-Driven Model

The last step of the flow is the training of a data-driven model on the dataset generated by the simulatable model. The proposed flow is *independent of the target data-driven model*: in principle, any type of regressor can be embedded in the "ML Model" block of Figure 1, ranging from a simple linear model to a complex deep NNs: this offers a much finer-grain tradeoff between accuracy and computational complexity than traditional physics-based models.

We focus on a scenario in which SOH estimation must be implemented on a low-power Microcontroller (MCU), possibly hosted in the BMS, to enable onboard SOH estimation. To achieve this, we selected two lightweight data-driven models: (i) a Gradient Boosted Tree (GBT) regressor, i.e., an ensemble of decision trees selected for its very fast prediction consisting of a small number of branching operations [20], and (ii) a Multi-Layer Perceptron (MLP), i.e., a simple fully-connected neural network appropriate for dealing with pre-aggregated features. These models are just examples to prove the flexibility of our proposed approach, and the most suitable data-driven model should be selected based on the desired SOH estimation accuracy and the constraints of the target platform.

1) Model Training: The training dataset is populated from the raw samples of battery current (I), voltage (V), and temperature (T) generated by the simulation model at the frequency f_s that we expect onboard measurements to occur ($f_s = 1$ Hz in our experiments, but this highly depends on the time constants of the considered system). The ML regressors are then trained on 12 statistical features of I, V and T, aggregated over a *time-window* of configurable duration W (2 hours in our experiments): *mean*, *variance*, *min*, and *max* of each quantity. Given that we target deployment on a highly-constrained, low-performance MCU, we avoid more complex features (e.g., Skewness or Kurtosis) since the computational cost for their calculation could outweigh the resulting accuracy improvement. Given that windows have a constant duration, it is not necessary to include also the elapsed time.

ML models are trained to predict $\Delta SOH = SOH_{initial} - SOH_{final}$ in each window, normalized to [0:1] for numerical stability. The training loss function for both models is the Mean Squared Error (MSE) between their prediction and the ΔSOH estimated by the simulation model. Note that the raw SOH prediction can be recovered by de-normalizing and accumulating predictions over consecutive windows (1 multiplication and 1 sum per window).

2) Model Space Exploration: A large number of design points can be obtained by (1) exploring model hyperparameters, and (2) selecting model features. Concerning hyperparameters exploration, we vary the number of estimators (decision trees) of the GBT and their maximum depths in the sets [5, 10, 20, 50, 100, 200] and [1, 2, 3, 4, 5, 10, 30, 50] respectively, for a total of 48 configurations. Similarly, we consider 42 MLP variants, with 1 to 2 hidden layers of sizes in [4, 8, 16, 32, 64, 128]. We explore these options with grid search and 50/20/30% train/validation/test split to obtain a Pareto-frontier in the MSE vs. execution time and MSE vs. memory occupation spaces. For the MLP, we use the Adam optimizer with a batch size of 64 and a learning rate of 0.001, training for 50 epochs. The remaining hyper-parameters are kept at the default values of the respective training libraries (see Section IV-A).

A selection of the best feature set, down to a minimum of 3 features, is applied by using a Recursive Feature Elimination (RFE) algorithm. Thus, the grid search described above is repeated 10 times, once for each feature sub-set, resulting in a total of 480 GBT and 420 MLP models being evaluated.

Since the deployment of such a large number of model variants would be impractical, we use simple mathematical models in the exploration phase to estimate the time/energy and memory complexity of each point.

For time and energy estimation, we count the operations required for the two steps involved when using the model:

- **Feature Extraction**: Since all considered features can be extracted with $O(N_{samples})$ operations, where $N_{samples} = \frac{W}{f_s}$ is the number of samples in a window, we estimate feature extraction time as $\frac{W}{f_s} \cdot N_f$, where N_f is the number of features.
- **Model Evaluation**: MLP evaluation cost is obtained by counting the total number of multiply-and-accumulate (MAC) operations, while GBT cost is obtained by counting the number of branch operations.

Since both phases consist mainly of arithmetic operations (with no division), we use the above two time quantities also as proxies of energy consumption: this is reasonable as the exploration relies on relative comparisons and it is architecture-independent, so we can reasonably assume that the energy cost is roughly equivalent to the execution time times the average power cost of an arithmetic operation. With these models, we note that feature extraction time is one or two orders of magnitudes higher than model evaluation time, especially for GBTs.

Concerning memory, we estimate the cost of an MLP configuration as the number of bytes required to store the network weights and the two largest activation buffers [21]. For the GBT, we count the bytes required to store the ensemble data structure and the input feature buffer [20].

While simplified, these models are effective in preserving complexity rankings among different configurations, especially for simple hardware like an MCU. On the other hand, using them allows us to perform the entire training and hyper-parameters search process in less than 4 hours on a laptop.

IV. Experiment Results

A. Experiment Setup

We train data-driven models using the Scikit-Learn and Keras Python libraries for GBT and MLP respectively. As the target embedded device, we consider the ultra-low-power RISC-V MCU PULPissimo [22], onto which we deploy both GBTs and MLPs using optimized libraries written in C. For GBTs, we leverage an in-house implementation similar to the one described in [20] for random forests, whereas for the MLPs we use a single-core version of the PULP-NN library [21]. Time and energy results refer to a 22nm realization of PULPissimo working at 205.1 MHz [22].

B. Dataset Generation

Simulation data is generated by configuring the model of [5] with the default parameters. We run a set of simulations, each lasting a maximum of 1,000 hours, or until battery SOH reaches 0. Each simulation uses either a different temperature or a different current profile. Specifically, we consider ambient temperature values in the [10°C : 40°C] range with a step of 5°C. We stimulate the battery model with constant, square-wave, and "random walk" load current profiles for both charge and discharge cycles, with values ranging from ± 0.25A to ± 2A. Each current pulse in a random walk has a duration of ≈ 1 minute, whereas square waves have a period of ≈ 30 minutes. Combining these conditions, we obtained 110 simulations that, once aggregated in non-overlapping windows of length $W = 2$ hours, gave us 17,842 samples.

We split those data into training, validation, and test sets with 50/20/30% proportions for all our experiments. Importantly, the split is performed *at the simulation level* (i.e., not at the window level), meaning that windows belonging to the same simulation *cannot* be simultaneously present, for instance, in the training and test sets. This is the most realistic

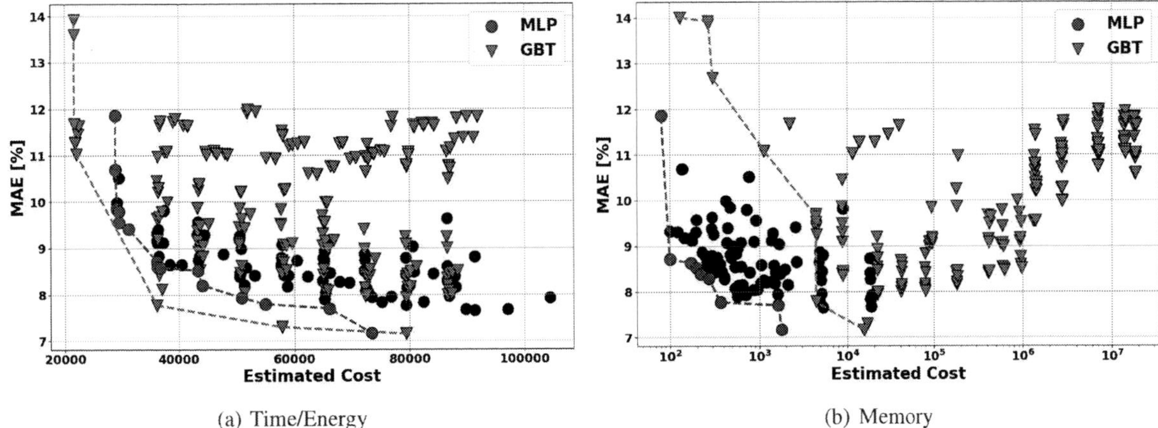

(a) Time/Energy

(b) Memory

Fig. 4: Pareto-fronts obtained from the hyper-parameters exploration of data-driven models

scenario, since in order to perform well, the data-driven models must learn to extrapolate the ΔSOH for different simulation conditions w.r.t. those seen during training.

Notably, this setup would allow us to easily generate more data, for example conducting an error analysis to identify the T and I conditions in which our GBT/MLP models perform worse and enhancing the dataset accordingly.

C. Pareto Analysis

Figure 4 shows the results of model space exploration for data-driven models. The x-axes of the two plots report the time/energy and memory estimated cost respectively, according to the models of Sec. III-D, whereas the y-axis reports the ΔSOH MAE with respect to the simulation model, in percentage. In both charts, each dot/triangle refers to one MLP/GBT hyper-parameter and input features configuration, and the blue/red points highlight the respective Pareto fronts.

Tuning models configurations, we obtain Pareto-optimal solutions spanning a 4x range in estimated time/energy and more than 2 order of magnitudes in memory occupation, with MAEs ranging between 7% and 14%. GBT models achieve superior results in terms of error vs. time/energy trade-off, but have higher estimated memory than MLPs. This demonstrates the flexibility achievable by selecting from the rich spectrum of data-driven model architectures.

Table I reports the detailed results of the *extremes* of the two Pareto curves. Namely, for each model type, we report the configuration achieving the lowest error (-E suffix), the lowest estimated time (-T), and the lowest estimated memory (-M). Note that the latter two coincide with the MLP. Besides the precise number of features, the hyper-parameter setting, and the MAE, we also report two additional error metrics, i.e., the Mean Squared Error (MSE) and the R^2 score. These results show that both hyper-parameter tuning and feature selection are important to find optimal data-driven model configurations.

D. Deployment Results

Figure 5 reports the results of deploying all Pareto-optimal models from Figure 4 on PULPissimo, thus replacing the cost

TABLE I: Extremes of the Pareto-curve.

GBT						
Model	# Feat.	# Trees	Max Depth	MAE [%]	MSE [%]	R^2
GBT-E	11	50	5	7.15	0.96	0.729
GBT-T	3	5	2	13.92	2.87	0.182
GBT-M	4	5	1	14.00	3.14	0.107
MLP						
Model	# Feat.	# Layers	Hidden Size	MAE [%]	MSE [%]	R^2
MLP-E	10	3	128	7.16	0.93	0.736
MLP-T	4	3	8	11.86	2.04	0.420
MLP-M	4	3	8	11.86	2.04	0.420

estimates with actual latency and memory (data plus code) measures on the target. The detailed deployment results for the extremes of the MLP and GBT Pareto fronts (same models of Table I) are also reported in the first six rows of Table II, in terms of the number of clock cycles, latency, total memory occupation, and energy consumption per regression.

In order to compare the time and memory costs of our data-driven models against those incurred by deploying a simulation-based model on-device, we compiled our ground truth reference from [5] to a binary executable through the Simulink coder toolbox. We targeted a laptop-class CPU (Apple M1 Pro), since [5] turned out to be impossible to compile for our RISC-V embedded target, as the Simulink coder relies on pre-compiled, x86-only support libraries, and the memory required by this model exceeds the 512kB L2 available on the target.

To perform a fair comparison, we also compiled our lowest-MAE data-driven model (GBT-E) for the M1 Pro, using the C-based library of [20]. The results are presented in the two rows marked with † in Table II. The last row estimates the figures of the simulation model on PULPissimo by re-scaling the simulation model results on M1 Pro to the RISC-V, using the ratio of the results obtained by GBT-E on the two platforms as proportionality factors. The corresponding latency and memory values are also represented as light-blue dots in Figure 5.

The results demonstrate the flexibility of a data-driven approach: we obtain configurations with latency, energy, and memory values that vary approximately by a factor of 3 (e.g., from 25ms/0.94μJ to 68ms/2.6μJ per regression and from 13.1

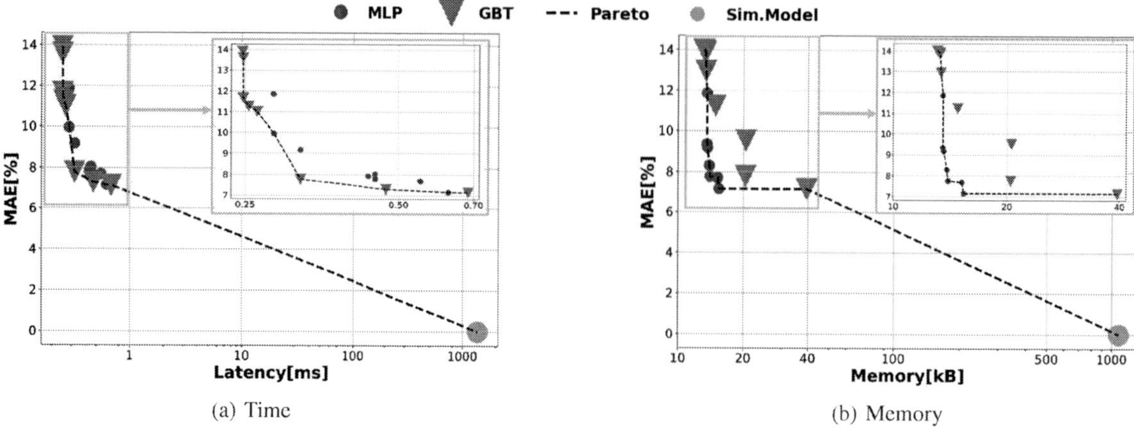

(a) Time

(b) Memory

Fig. 5: Deployed data-driven models and comparison with the simulation-based model.

TABLE II: Deployment Results.

Model	MAE [%]	Cycles	Latency [ms]	Memory [kB]	Energy [μJ]
GBT-E	7.15	$140 \cdot 10^3$	0.68	39.46	2.6
GBT-T	13.92	$51 \cdot 10^3$	0.25	13.26	0.94
GBT-M	14.00	$58 \cdot 10^3$	0.28	13.06	1.04
MLP-E	7.16	$122 \cdot 10^3$	0.6	15.38	2.36
MLP-T	11.86	$58 \cdot 10^3$	0.28	13.54	1.01
MLP-M	11.86	$58 \cdot 10^3$	0.28	13.54	1.01
Simulation-model Comparison					
GBT-E†	7.2	$560 \cdot 10^3$	0.20	49.15	n.a.
Simul.†	0	$127 \cdot 10^7$	395	1343.49	n.a.
Simul.*	0	$31 \cdot 10^6$	1377	1078.57	n.a.

† Results collected on Apple M1 Pro
* Results scaled from those on Apple M1 Pro

to 39.5 kB), with corresponding MAE values ranging from 7.2 to 14%. Considering that we use a window of 2 hours as input to the models, the energy consumption values obtained by the data-driven solutions can be considered completely negligible.

Furthermore, on the M1 Pro CPU, our lowest error model pays a 7.2% MAE in exchange for a striking 2,000X reduction in latency and 27X reduction in memory, with respect to directly executing a simulation-based SOH model. The latter, when scaled to the RISC-V platform, would require more than 1s to execute, implying a much higher energy overhead and more than 1MB of total memory, which would exceed the available space on most embedded microcontrollers.

V. Conclusions

Data-driven models are the most suitable option for a digital twin of a battery to be hosted on-board a BMS, but their fidelity strongly depends on the quality of the training dataset. We have shown that it is possible to use a simulation model to generate an arbitrarily large dataset in a much smaller time than that required by datasets obtained through measurements. As results showed, this option also allows a tradeoff between model accuracy and model execution time or memory.

References

[1] S. Tamilselvi *et al.*, "A review on battery modelling techniques," *Sustainability*, vol. 13, no. 18, 2021.

[2] S. A. Hasib *et al.*, "A comprehensive review of available battery datasets, rul prediction approaches, and advanced battery management," *IEEE Access*, vol. 9, pp. 86 166–86 193, 2021.

[3] C. Vidal *et al.*, "Machine learning applied to electrified vehicle battery state of charge and state of health estimation: State-of-the-art," *IEEE Access*, vol. 8, pp. 52 796–52 814, 2020.

[4] F. Heinrich *et al.*, "A comprehensive study on battery electric modeling approaches based on machine learning," *Energy Inform*, vol. 4, no. 17, 2021.

[5] MathWorks, "Battery state-of-health estimation," https://it.mathworks.com/help/simscape-battery/ug/battery-state-of-health-estimation.html, 2023.

[6] A. Millner, "Modeling lithium ion battery degradation in electric vehicles," in *IEEE CITRES*, 2010, pp. 349–356.

[7] A. Bocca *et al.*, "An aging-aware battery charge scheme for mobile devices exploiting plug-in time patterns," in *33rd IEEE ICCD*, 2015, pp. 407–410.

[8] J. Vetter *et al.*, "Ageing mechanisms in lithium-ion batteries," *Journal of Power Sources*, vol. 147, no. 1, pp. 269–281, 2005.

[9] B. Xu *et al.*, "Modeling of lithium-ion battery degradation for cell life assessment," *IEEE Trans Smart Grid*, vol. 9, no. 2, pp. 1131–1140, 2018.

[10] G. Vennam *et al.*, "A survey on lithium-ion battery internal and external degradation modeling and state of health estimation," *Journal of Energy Storage*, vol. 52, p. 104720, 2022.

[11] G. dos Reis *et al.*, "Lithium-ion battery data and where to find it," *Energy and AI*, vol. 5, no. 1, 2021.

[12] M. Petricca *et al.*, "An automated framework for generating variable-accuracy battery models from datasheet information," in *IEEE/ACM ISLPED*, 2013, pp. 365–370.

[13] V. Barreras *et al.*, "Datasheet-based modeling of li-ion batteries," in *2012 IEEE Vehicle Power and Propulsion Conference*, 2012.

[14] A. Bocca *et al.*, "Composable battery model templates based on manufacturers' data," *IEEE Design & Test*, vol. 35, no. 3, 2017.

[15] B. Balagopal and M.-Y. Chow, "The state of the art approaches to estimate the state of health (soh) and state of function (sof) of lithium ion batteries," in *13th IEEE INDIN*, 2015, pp. 1302–1307.

[16] K. A. Smith *et al.*, "Model-based electrochemical estimation and constraint management for pulse operation of lithium ion batteries," *IEEE Trans. on Control Systems Technology*, vol. 18, no. 3, pp. 654–663, 2010.

[17] O. Erdinc *et al.*, "A dynamic lithium-ion battery model considering the effects of temperature and capacity fading," in *2009 ICCEP*, 2009, pp. 383–386.

[18] M. Cacciato *et al.*, "Real-time model-based estimation of soc and soh for energy storage systems," *IEEE Trans. Power Electron.*, vol. 32, no. 1, pp. 794–803, 2017.

[19] D. Haifeng *et al.*, "A new soh prediction concept for the power lithium-ion battery used on hevs," in *2009 IEEE VPPC*, 2009, pp. 1649–1653.

[20] F. Daghero *et al.*, "Adaptive random forests for energy-efficient inference on microcontrollers," in *IFIP/IEEE 29th VLSI-SoC*, 2021, pp. 1–6.

[21] A. Garofalo *et al.*, "PULP-NN: accelerating quantized neural networks on parallel ultra-low-power RISC-V processors," *Philos. Trans. Royal Soc. A*, vol. 378, no. 2164, p. 20190155, 2 2020.

[22] P. D. Schiavone *et al.*, "Quentin: an ultra-low-power pulpissimo soc in 22nm fdx," in *IEEE S3S*, 2018, pp. 1–3.

Ocellus: Highly Parallel Convolution-in-Pixel Scheme Realizing Power-Delay-Efficient Edge Intelligence

Sepehr Tabrizchi*, Shaahin Angizi†, Arman Roohi*
*School of Computing, University of Nebraska–Lincoln, Lincoln NE, USA
†Department of Electrical and Computer Engineering, New Jersey Institute of Technology, Newark, NJ, USA
aroohi@unl.edu, shaahin.angizi@njit.edu

***Abstract*—With the advent of Edge Intelligence (EI) devices, always-on intelligent and self-powered visual perception systems are receiving considerable attention. These emerging systems require continuous sensing and instant processing; however, the high energy data conversion/transmission of raw data and the limited available energy and computation resources make designing energy-efficient and low bandwidth CMOS vision sensors vital but challenging. This paper proposes a low-power integrated sensing and computing engine, namely `Ocellus`, which considerably decreases power costs of data movement/conversion and enables data/compute -intensive neural network tasks. Ocellus offers several unique features, including a highly parallel analog convolution-in-pixel scheme and reconfigurable filtering modes with filter pruning capability. These features realize low-precision ternary weight neural networks to mitigate the overhead of analog-to-digital converters and analog buffers. Moreover, the proposed structure supports a zero-skipping scheme to further reduce power consumption. Our circuit-to-application co-simulation results demonstrate comparable, even better, accuracy to the full-precision baseline on object classification tasks, while it achieves a frame rate of 1000 and efficiency of ~1.45 TOp/s/W.**

I. Introduction

Nowadays, several serious challenges are associated with cloud-based communication and computation, including high latency, questionable scalability, quality of service (QoS), privacy, and security. As the Internet of Things (IoT) or the Internet of Everything (IoE) advances, these issues might be addressed by shifting computing architecture from a cloud-centric to a thing-centric perspective. It is projected that the use of IoTs will reach $1100 billion by 2025, with an interconnected web of 75+ billion IoTs, including smart homes, smart cities, smart industries, vehicle-to-everything technologies (V2X), wearables, and implantables systems for healthcare. Although IoT devices have become commonplace and academia and industry have worked to make them smaller, more powerful, and more energy efficient, their intelligence and decision-making still rely on the cloud. Dumb IoTs with sensory systems capable of collecting massive amounts of data from the environment are classified as the Sense-Communicate-Decide-Action paradigm, where communication is still a performance bottleneck. It is worth noting that one bit's communication energy (e.g., Bluetooth low energy (BLE) needs 1 nJ/bit [1]) is usually orders of magnitude higher than its computation energy (e.g., Multiply-Accumulate (MAC) < 2 pJ/6-bit) [2]–[4]. Due to the insufficient computing ability and constraints in energy resources and area, nearly 90% of the data generated by the IoT is not analyzed or processed [5]. Therefore, the conventional paradigm is revised to Sense-Decide-Action, proposing Edge Intelligence (EI) devices, which can analyze data locally rather than sending it to the cloud to reduce the amount of communication. These emerging systems require both continuous sensing and instant processing. Nonetheless, the high energy data conversion/transmission of raw data and the limited available energy from ambient energy sources make designing energy-efficient and low bandwidth CMOS vision sensors challenging. Moreover, even using low-power sensors to realize artificial intelligence tasks such as object classification faces serious challenges for their tractability in computational and storage resources. Effective techniques in both the software and the hardware domains have been developed to improve the performance and computing efficiency of EIs by alleviating the power, memory, instruction-level parallelism, and, recently, network walls bottleneck. Nevertheless, leveraging current von Neumann computing architectures with separate computing and memory, which imposes serious problems, including long memory access latency, limited memory bandwidth, and energy-hungry data transfer, the mentioned challenges are hardly addressable [6]. The concept of instant image pre-processing with smart image sensors has therefore been extensively investigated [6]–[9] as a potential remedy. Using an on-chip processor, pixels' digital outputs can be accelerated where the sensor is located, paving the way for enhanced sensor paradigms such as Processing-Near-Sensor (PNS). Other promising alternatives are a Process-in-Sensor (PIS) platform [10] that processes pre-Analog-to-Digital Converter data and a hybrid PIS-PNS [11] platform to incorporate vision sensors and eliminate redundant data output. Conventional designs rely on data transfer from CMOS image sensors to memory, which reduces the speed of feature extraction.

In this paper, advancements from both algorithm and hardware architecture perspectives are deployed to accelerate compute-intensive applications making real-time and decision-making EI devices a reality. In summary, our major contributions to this paper can be listed as follows: **(1)** We introduce a reconfigurable channel technique in which an RGB input image is converted to grayscale, considered as pseudo-channel pruning, to alleviate the power costs of data conversion and transmission. **(2)** Further improvement in the performance efficiency of systems, such as power and delay reduction, is achieved by redesigning conventional pixels to enable low-precision ternary weight neural networks (TWNN), mitigating the overhead of analog buffers and analog-to-digital converters. **(3)** We propose a novel highly efficient and parallel analog convolution-in-pixel scheme, namely, **`Ocellus`**. Compared to the full-precision model, it can massively reduce the required storage and computational resources in the inference paths.

979-8-3503-1176-1/23 $31.00 © 2023 IEEE

(4) Finally, the evaluation of system accuracy in classification tasks and the system performance in speed and energy are carried out. Applying 1 to 3 leads to a significant reduction in power and delay with acceptable accuracy, making Ocellus a promising solution for emerging agriculture technologies.

II. Background

Systematic integration of computing and sensor arrays has been widely studied to eliminate off-chip data transmission and reduce ADC bandwidth, known as a processing-near-sensor (PNS) [9], [12], combining sensor and processing element so-called processing-in-sensor (PIS) [8], and integrating pixels and computation unit, known as a processing-in-pixel (PIP) [13]. In [9], photocurrents are converted into pulse-width modulation signals, and a dedicated analog processor is used to perform feature extraction. To run spatiotemporal image processing, 3D-stacked column-parallel ADCs, and processing elements are implemented and utilized in [6]. The CMOS image sensor with dual-mode delta-sigma ADCs described in [14] is designed to process 1st-conv. layer of binarized-weight neural networks (BWNN). Charge-sharing tunable capacitors are used by RedEye [15] to implement the convolution operation. By sacrificing accuracy in favor of energy savings, this design reduces energy consumption compared to a CPU/GPU. However, for high-accuracy computation, the required energy per frame increases dramatically by 100×. As a PIS platform, MACSen [8] processes the 1st-conv. layer of BWNNs with the correlated double sampling procedure and achieves speeds of 1000fps in computation mode. This method, due primarily to the SRAM-based PIS, however, suffers from a humongous area overhead and high power consumption. An example of a pulse-domain algorithm is [16], which optimizes near-sensor image processing by using photodiode arrays and an ADC to minimize design complexity and increase cost and speed. Similar to the previous state-of-the-art works, we mainly focus on the first layer for the following reasons and observations. Regarding accuracy, in the most quantized neural network accelerators, the first and the last layers of the networks remain in the full-precision, floating-point domain. Although the authors in [17], showed that in vision-based applications, the input feature map (ifmap) generally includes fewer channels (e.g., red, green, and blue) compared to the internal layers (e.g., 512), communications are relatively high. Moreover, for efficiency, because raw image data is full precision, the first layer's convolution operations are the performance bottleneck in different hardware/software co-design accelerators and require a lot of memory and processing resources [18]. On the other hand, in conventional image sensors, most of the power (>96% [19]) is consumed by processing and converting pixel values. That means pixel circuits consume only four percent of power to perform photovoltaic conversions, whereas signal amplification, digital-to-analog conversion, and data transmission consume most of the power.

III. Proposed Ocellus Architecture

We propose Ocellus as an efficient and reconfigurable sensory architecture to address the aforementioned challenges and limitations. Ocellus consists of a 32×32 Compute Focal Plane (CFP) array, row and column controllers (Ctrl), command decoder, sensor timing Ctrl, global and weight buffers, and switch/ADC circuitry, as shown in Fig. 1. The CFP is designed to co-integrate sensing and processing for low-power but high classification accuracy applications. The output, i.e., 1st-layer of DNN, is transmitted to an on-chip deep learning accelerator to accelerate further.

A. Pixel Structure

The proposed reconfigured pixel, shown in Fig. 2(b), contains four sensors for Red, Green (×2), and Blue colors, as depicted in Fig. 2(c). Sensors are equipped with two inputs, Rst and Discharge signals, to produce positive (PC) and negative (NC) currents. All the PC (NC) outputs within a pixel are integrated and connected to multiplexer (MUX) inputs. Then, based on the value of weight stored in the weight buffers, one of them is connected to the bit line (BL). This pixel structure offers two beneficial features: (1) supports ternary values (i.e., -1, 0, +1), which enables TWNNs. (2) All identical sensors, e.g., Red (R), have a common voltage source, leading to four separate sources for three colors that facilitate the pseudo-channel pruning scheme. It is worth mentioning that in our design, two V_{DD} sources are assigned to each green sensor (G). Ocellus can dynamically turn ON/OFF multiple sensors (colors) within a pixel, which reduces the total power consumption.

1) Ternary Values Enabling TWNN: Figure 2(b) shows the modified version of a conventional 4-T (Transistor) sensor to support negative current as well. In this case, T_3 and T_4 are responsible for creating positive current (PC) and negative current (NC), respectively, based on the voltage of the photodiode capacitor (V_{CPD}). In every clock, the V_{CPD} is charged to V_{DD} using T_1. Then in the evaluation phase and based on the light intensity, it starts to discharge through T_2 and the photodiode (PD). After evaluation, the value of the V_{CPD} remains unchanged for following convolution operations. In

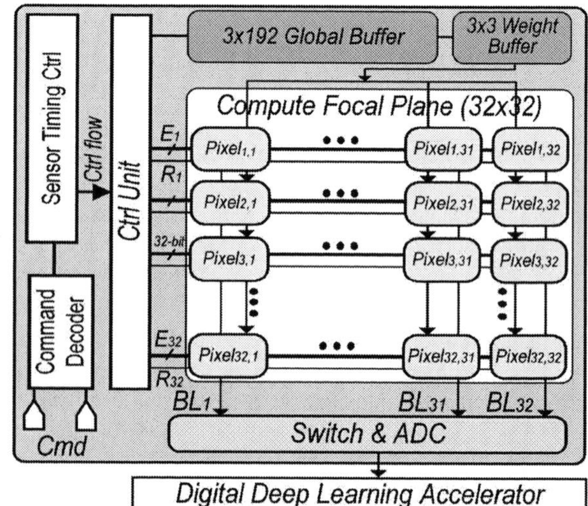

Fig. 1: Proposed Ocellus architecture.

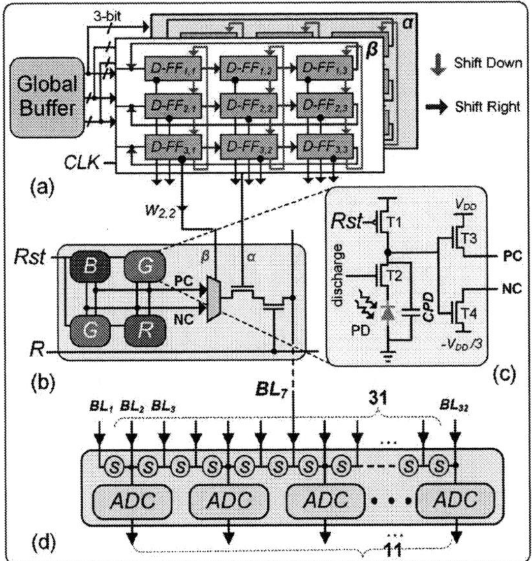

Fig. 2: The main components of Ocellus includes (a) the global buffer and weight buffer with down and right shift capabilities, (b) the proposed pixel design includes four sensors, depicted in (c), and switch & ADC peripheral circuits.

the pixel structure, to choose which PC or NC should be connected to the BL, a multiplexer is leveraged with weights as its selector. Disabling the α path disconnects the current path, preventing the pixel from producing positive or negative current. Otherwise, the current of sensor output directly relates to the V_{CPD}. Combinations of the (α) signal and stored weights (β) provide us with ternary values, summarized in Table I. The functionality of one proposed pixel is conducted using the HSpice simulator for a 45nm CMOS technology, and the power consumption results are reported in Table II. As expected, activating more sensors in a pixel increases the power consumption of the pixel, whereas the accuracy does not necessarily increase, as elaborated in the accuracy section. The simulation results for one pixel are shown in Fig. 3, where all the inputs and outputs are demonstrated with red and green colors, respectively. In ❶, the pixel is initialized/enabled. Then α is set to V_{DD} to disable the sensor. So V_{CPD} cannot charge, and the produced current on BL is approximately zero ❷. On the other hand, by changing the EN value to zero, with the first Rst clock ('0') in ❸, V_{CPD} is charged to V_{DD}. Now by setting Rst to '1' again and enabling $discharge$, V_{CPD} starts discharging. ❹ follows a similar behavior but for a different weight value, i.e., '1.' The value of the V_{CPD} has remained the same as steps ❸ and ❹, shown in ❺, where the pixel generates a positive current on BL.

2) Pseudo-Channel Pruning Scheme: Ocellus is capable of channel pruning to reduce data movement overhead considerably. Each of the sensors within a pixel can be justified

TABLE I: Ternary Values regarding the stored weights.

Weight Buffer 0 (α)	Weight Buffer 1 (β)	Represented Weight	Output Current
0	×	0	0
1	0	-1	CPD
1	1	1	$-CPD$

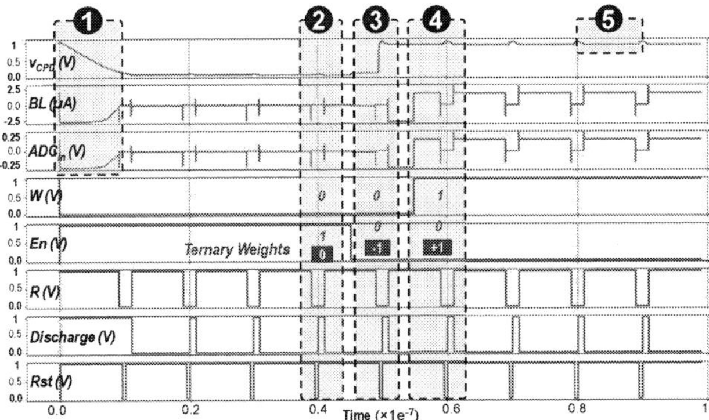

Fig. 3: Transient waveforms of a pixel regarding ternary weights.

individually to meet the desired performance, e.g., power consumption or/and accuracy. Due to its structure, Ocellus instantly reduces the number of input channels from three (RGB) to one (grayscale). Therefore, Ocellus can support ten legal combinations of RGB sensors. In the evaluation section, an extensive study is performed to examine the effectiveness of RGB sensors on various neural network architectures.

B. Peripheral Circuits

Additionally, the Ocellus architecture incorporates two Global/Weight Buffers, as well as Switch and ADC circuitry. In accordance with Table I, two bits are needed to generate ternary weights, which are stored in separate weight buffers. The global buffer has less than 1KB of storage to hold all the pre-trained quantized weights. This capacity is efficient for all neural network models with 1 up to 64, 3×3 filters in their 1st-layer. To avoid logic failure, based on our Monte Carlo simulation, 64 different filters can be applied to the captured inputs; however, we should mention that the capacitors (CPDs) can keep the charges at least 64ms, similar to a conventional DRAM [20], which is sufficient for our analog convolution-in-pixel. The global buffer passes a desired 3×3 filter for convolution to the weight buffer, shown in Fig. 2(a). The weight buffer is capable of shifting in two directions, left-to-right (blue arrow) and top-to-bottom (red arrow). Because the Ocellus specifically targets filters with a 3×3 spatial dimension, the connections between the weight buffer and pixels are hardwired, which provides an efficient convolution-in-pixel scheme. Our proposed accelerator reuses both input feature map (ifmap) and weights data types to form input stationery and weight stationary dataflows, which reduces the overall data transfer significantly. Furthermore, herein, the number of ADCs is much smaller than in the previous PIS/PNS designs, where they require a separate ADC for each column; thus, for instance, for a 32×32 CFP, they

TABLE II: Power (W) of a pixel w.r.t number of active sensors and different weights.

#Active Sensor	Ternary Weight -1	0	1
1	1.086e-07	6.061e-09	1.196e-07
2	1.849e-07	1.075e-08	1.919e-07
3	2.594e-07	1.526e-08	2.684e-07
4	3.348e-07	1.917e-08	3.428e-07

need 32 ADCs. While in our design, due to the 3×3 pixel boxes, $\lceil 32/3 \rceil = 11$ ADCs, and 31 switches are required, depicted in Fig.2 (d). The switches are readily implemented by transmission gates with a near-optimal full-swing switching behavior. Two switches accumulate three sets of integrated currents generated by three pixels in the same column. Finally, the summation is passed to an ADC to produce a 1-bit output feature map (ofmap).

C. Convolution-in-Pixel

The proposed Ocellus performs 1st-layer's convolution operations in the analog domain instantly after capturing an image that (1) increases MACs throughput[1], and (2) decreases the ADC overheads[2]. The operation principle of the Convolution-in-Pixel is explained in Algorithm 1.

Ocellus implements the input stationary (IS) dataflow to minimize the reuse distance of ifmaps, maximizing the convolutional and ifmap data reuse. All capacitors within the 32×32 pixel array are written regarding the light intensity of a target image. The stored values remain almost constant for a specific time interval (τ), which is determined by the capacitance of the capacitors. The τ value is long enough and set to a greater value than delays of ADCs and MUXs, to implement an efficient IS. The weight buffer (WB) in the Ocellus architecture has the capability of shifting values to the right and to down to implement stride behavior. The stride window is considered 1. Moreover, the WB supports the weight stationary (WS) dataflow for specific timeframes. The loop with variable K, which indexes filter weights, is placed as the outermost loop, shown in 1 (line 6). It activates three rows (line 10) and performs convolutions for all selected columns. Then shifts weights in the same filter to the right and continues the previous process. After considering all possible movements based on the stride window, weights are shifted down (line 13), and so on. Since the connections between WB's blocks and 1024 pixels are hardwired, different weights of a $R \times S$ filter are unicast to a group of 9 pixels, while the same filter is broadcast to other groups of pixels in different columns. The spatial dimensions of a filter are represented by R and S, height and width, respectively. Thus, Ocellus can compute $R \times S \times n$ MAC operations (i.e. 27) in only one clock cycle, as shown in Fig 4 Clk **1**, where $n = \lfloor \frac{W}{3} \rfloor$ and W is input's width, i.e. 32. Regardless of size S, all the columns are traversed after a maximum of three cycles. However, applying the same filter to other R input rows needs an extra cycle, Clk **2** in Fig 4. Therefore, in the worst-case scenario and to maximize weight data reuse, $\lfloor \frac{H}{3} \rfloor$ cycles are required before shifting the filter's values, where H is the input's height, i.e., 32. Figure 4 depicts several steps of convolutions using stored weights $\in \{1, 2, 3, \ldots, 9\}$ in WB. As it can be seen, each block of WB, e.g., 11, is hardwired to multiple pixels, which offers highly parallel convolution-in-pixel. It means three cycles with no shift operation are required to produce 9 ofmaps, red boxes.

[1]The main portions (> 99%) of DNNs, are occupied by MACs [21].

[2]In some cases, its hardware cost can go beyond 90% [22].

Algorithm 1 Convolution-in-Pixel (`CiP`) Algorithm

1: **Input:** Captured image via 32×32 pixel array
2: **K:** Number of filters ▷ Filters' 3D-dimension: $K \times 3 \times 3$
3: **WB:** A 3×3 filter
4: **Output:** 1st-layer convolution ▷ Produces the complete ofmap
5: **procedure** `CiP`
6:   **for** $(k = 1; k \leq K; k++)$ **do** ▷ WS dataflow
7:     **for** $(i = 1; i \leq 3; i++)$ **do**
8:       **for** $(r = i + 1; r \leq 32; r+= 3)$ **do**
9:         **for** $(j = 1; k \leq 3; j++)$ **do**
10:           `Active_Row` (r-1, r, r+1)
11:           `Calculate_CONV` ()
12:         `Shift_Right` (WB$\rightarrow$)
13:       `Shift_Down` (WB$\downarrow$)
14:     `Load_New_Weight` (GB $\Rightarrow$ WB)

To implement stride, all the weights are shifted one bit to the right, shown as a blue arrow in Fig. 4. Then Ocellus performs the same process as before, traversing all the columns in a cycle and visiting all the rows in three cycles (Clk **4**, **5**, and **6**) using the same filters. The right shift operations continue till all the columns are visited, Clk **9**. In cycle **10**, the WB shifts the weights to the right without any computation, which is considered as an idle cycle. We must do that to reset all the shifts and reproduce the original filter. Then, the WB shifts down the weights once and passes them through a different group of pixels, cycles 11 and 12. From this time forth, three right shifts are required before another shift down operation. Therefore, the total number of right shifts and down shifts are three and $\lceil \frac{H}{3} \rceil$, respectively.

Figure 5 exhibits the circuit-level schematic of Fig. 4 in Clk **11**, where the original filter is shifted down to perform convolutions properly. The second row of WB is connected to pixels $a, b, c, j, \ldots$ located in the second row of the pixel array. Due to the hardwired scheme, each WB's block is connected to $S + x$ inputs, where S is the width of the filter, i.e., 3, and x is the column index of the selected element. Moreover, the same WB's row is connected to all $R + y$, where R is the height of the filter, i.e., 3, and y is the number of the selected filter's row. For example, the $W_{2,1}$ element of WB is shared and connected to inputs with indices, $\{\{2,1\},\{2,4\},\{2,7\},\ldots\}$ (in a column-wise way), and $\{\{2,1\},\{5,1\},\{8,1\},\ldots\}$ (in a row-wise manner). Mathematically, let $G_{j,i}$ be the conductance of the synapse connecting i^{th} to the jth node, the current through that synapse is $G_{j,i}V_i$ and the collection of the current through each BL represents the MAC result ($I_{sum,j}=\sum_i G_{j,i}V_i$), according to Kirchhoff's law. This is readily calculated by measuring the voltage across a sensing resistor and proper signaling of the switches. This mechanism converts every input pixel value to a weighted current according to the stored weight that is interpreted as the multiplication in DNNs. For instance, in cycle **11**, for a shown portion in Fig. 5, by enabling the first and second switches, nine MAC operations are performed, and the result is passed to the ADC_l for further calculations.

IV. Simulation Results

A. Performance Evaluation

Table III compares the structural and performance characteristics of selective PIS and PIP designs. For an impartial

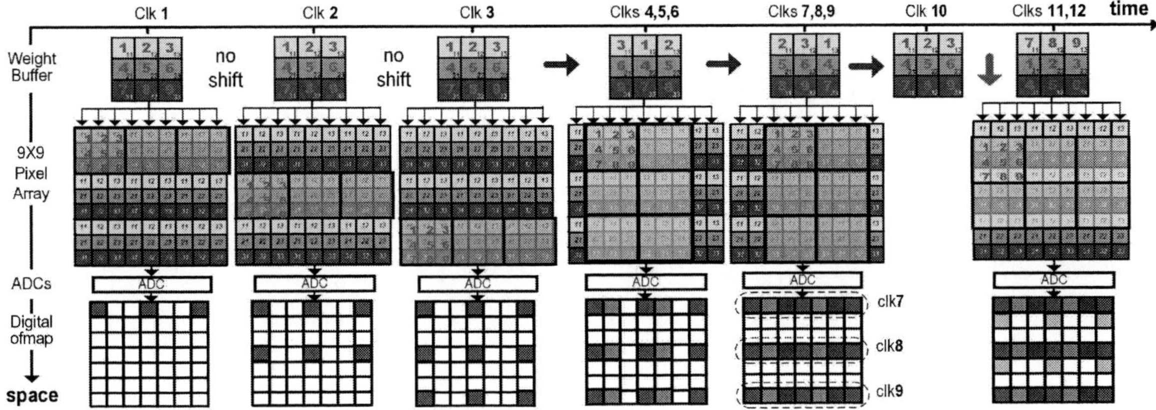

Fig. 4: An example of Convolution-in-Pixel approach with 3×3 filter size and 9×9 pixel array as inputs for 12 clk cycles.

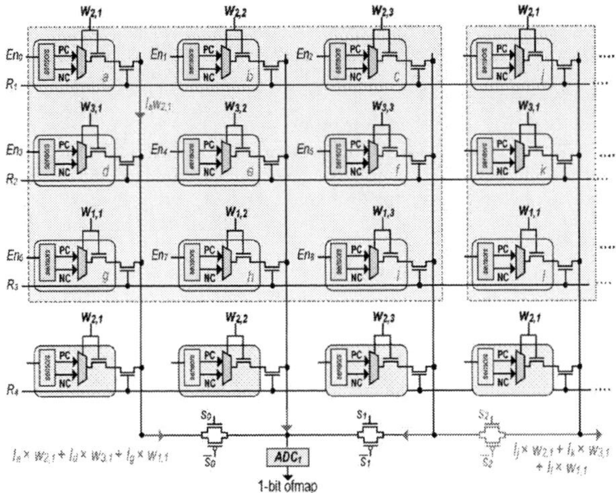

Fig. 5: Convolution steps in the cycle 11 of Fig. 4.

comparison, as different designs are developed for specific domains, the power consumption of PIS units executing the similar task of processing the 1st-layer of DNN is estimated. With its frame rate of 1000 and efficiency of $\sim$1.45 TOp/s/W, the Ocellus has been assessed as the most efficient design. However, it is the structure in [23] that achieves the highest frame rate. Due to the lack of access to other layouts' configurations, comparing area overheads is difficult. However, we believe a ballpark assessment can be made by comparing the number of transistors in previous SRAM-based designs and Ocellus's lower-overhead compute add-on.

B. Accuracy

To evaluate Ocellus's accuracy, we implemented several prominent models in the PyTorch framework including: VGG11, ResNet11, DenseNet121, MobileNetV1, MobileNetV2, DPN92, and EfficientNetB0. Each model has different specifications, which improve our assessment process. We conduct experiments of Ocellus on Corn Leaf Infection (CLI) dataset [25]. The CLI dataset is an agricultural dataset gathered from corn farms. It consists of RGB images, which are distributed in two classes to distinguish healthy corn leaves from infected ones. The number of training and test images are 3378 and 845, respectively. We selected this dataset to show an example application for Ocellus and how it can be utilized inside edge sensors for agricultural manners. To satisfy our implementation purposes, we applied two preprocessing steps to the dataset. First, we mapped RGB channels to a single grayscale channel by multiplying each channel by a constant factor, shown in Table IV, **Channel Scaling Factor**. These constant values are selected based on legal mapping and Ocellus hardware design and constraints. Then, we reshaped the size of input images to 32×32. In Table IV, **FP** columns summarize the accuracy of each corresponding model concerning constant scaling factors of RGB channels. In the channel scaling factor column, the order of the sensor is [R, G, B]. For example, [0.0, 0.67, 0.33] means the Red sensor is OFF, and two Green sensors and one Blue sensor are ON. Herein, the first scaling factor, [0.29, 0.59, 0.12], relates to the conventional RGB to grayscale factor, which is not hardware-friendly. As shown in the table, using all sensors is not necessarily cause better results, e.g., accuracy results for [0.0, 1.0, 0.0] *vs.* [0.29, 0.59, 0.12]).

TABLE III: Performance comparison of various sensor units.

Designs	**Technology** (nm)	**Purpose**	**Frame Rate** ($frame/s$)	**Power** (mW)	**Efficiency** ($TOp/s/W$)
[24]	180	2D optic flow est.	30	0.029	0.0041
[9]	180	edge*/blur/sharpen/ 1st layer DNN	480	sensing: 0.077 processing: 0.091	0.777
[6]	60/90	STP†	1000	sensing: 230 processing:363	0.386
[8]	180	1st layer BNN	1000	0.0121	1.32
[23]	180	edge*/TMF‡	100,000	1230	0.535
Ocellus	45	edge 1st layer DNN	1000	sensing: 0.017 processing: 0.075	1.45

* Edge extraction. † Spatial Temporal Processing. ‡ Thresholding Median Filter.

Next, to perform convolution inside Ocellus architecture and to further increase the power efficiency, we quantized the weights of the 1st-conv. layer in our ternary fashion by sorting and partitioning them into three regions. Then we replaced the weights of each region with a specific constant value. We always set the weights of the middle region to zero. Still, to specify the length of each region and replacement values of upper and lower regions, we partitioned the weights using their percentiles and selected the replacement values from these percentiles. Table IV, **Q1**, and **Q2** show the corresponding accuracy for two different percentile approaches that we utilized to partition weights and quantize them. For

TABLE IV: Accuracy of different models on CLI dataset, w.r.t the constant channel scaling factor for converting RGB inputs to grayscale.

Channel Scaling Factor	VGG			ResNet18			DenseNet121			MobileNetV1			MobileNetV2			DPN92			EfficientNetB0		
	FP	Q1	Q2	FP	Q1	Q2	FP	Q1	Q2	FP	Q1	Q2	FP	Q1	Q2	FP	Q1	Q2	FP	Q1	Q2
[0.29, 0.59, 0.12]*	90.53	75.27	49.23	94.56	85.80	83.31	94.91	85.68	90.77	90.30	77.28	68.64	92.54	85.09	81.66	92.31	73.25	56.45	92.90	69.94	67.10
[1.0, 0.0, 0.0]	92.19	67.57	61.89	94.32	83.31	89.23	94.08	77.63	85.68	91.36	62.01	83.20	91.36	69.47	87.34	91.12	57.63	75.74	92.78	77.75	84.97
[0.5, 0.5, 0.0]	91.01	60.47	55.98	94.67	81.07	86.86	95.15	82.37	91.24	88.40	49.70	47.81	91.95	80.95	82.72	92.66	53.73	62.13	93.37	80.36	74.79
[0.5, 0.0, 0.5]	92.78	50.65	57.63	92.19	68.17	83.55	94.79	55.03	81.18	90.06	69.94	81.54	93.14	58.22	75.50	92.07	55.86	69.70	91.60	54.08	82.25
[0.33, 0.67, 0.0]	91.01	71.36	49.11	94.44	71.36	70.77	94.79	85.92	86.27	91.12	48.76	47.34	93.25	73.37	55.98	91.95	75.15	69.94	92.78	77.40	73.61
[0.33, 0.33, 0.33]	91.01	64.62	76.33	94.32	82.72	84.62	94.44	86.04	84.26	88.28	50.30	55.74	92.07	77.75	82.72	89.70	71.60	66.27	91.95	78.22	83.79
[0.25, 0.5, 0.25]	91.36	77.75	82.84	94.44	87.69	91.48	94.56	85.92	90.53	88.40	76.92	55.86	92.07	81.30	80.83	91.24	81.89	86.04	92.43	75.50	52.90
[0.0, 1.0, 0.0]	91.83	64.38	47.34	94.79	79.64	88.28	95.27	81.54	90.06	90.30	55.74	52.31	92.31	69.47	84.73	92.07	48.76	64.97	91.83	54.44	41.78
[0.0, 0.67, 0.33]	89.70	76.33	60.95	91.60	79.64	81.18	93.96	83.20	89.47	91.72	63.79	48.40	93.02	83.91	83.20	92.31	81.54	88.17	92.19	79.17	82.37
[0.0, 0.5, 0.5]	92.19	66.15	68.88	93.25	75.50	78.11	95.03	86.15	86.63	86.75	62.49	74.08	91.36	78.34	81.54	90.30	76.09	68.40	91.72	68.52	65.21
[0.0, 0.0, 1.0]	91.48	57.04	65.09	93.14	53.14	83.67	94.44	72.31	74.79	88.76	59.17	56.80	91.24	50.06	58.58	91.48	49.47	69.47	90.06	50.77	63.43

*Although it is impossible to represent this scaling factor within sensors, herein, it is considered as a baseline.
The accuracy is listed for using floating-point operation (FP) and using two different quantization parameters, Q1(10, 20, 80, 90), and Q2(10, 30, 70, 90).

instance, in Table IV, **Q2** "(10, 30, 70, 90)" indicates that we replaced the values of weights less than the 30th percentile with the value of the 10th percentile; and from the 30th percentile to the 70th percentile, we replaced the values of weights with zero; and finally, for the values larger than the 70th percentile, we replaced them with the value of 90th percentile. Herein, we only showed two sets of suitable percentiles for our architecture (Q1 and Q2). While quantizing the weights of the 1st-conv. layer and reducing the computation cost, based in Table II, we further improve power consumption by eliminating multiplied-by-zero operations inside Ocellus architecture. For example, we replaced 40% and 60% of the weights in the 1st-conv. layer with zero after adopting our "(10, 30, 70, 90)" and "(10, 20, 80, 90)" percentile approaches, respectively. As shown in Table IV, except for a few particular models and Channel Scaling Factors, accuracy degradation is near 5%, on average.

V. Conclusion

This paper proposed Ocellus, as an intelligent visual perception architecture that realizes a low-power convolution-in-pixel scheme with filter pruning capability. Ocellus supports analog convolutions enabling low-precision TWNN to mitigate the overhead of analog buffer and analog-to-digital converters. The proposed structure supports a zero-skipping scheme to further reduce power consumption. Our circuit-to-application co-simulation results demonstrate comparable accuracy to the full-precision baseline on the classification tasks. In some configurations, Ocellus shows better accuracy, while it achieves a frame rate of 1000 and efficiency of ~1.45 TOp/s/W.

Acknowledgments

This work is supported in part by the National Science Foundation under Grant No. 2216772 and 2216773.

References

[1] S. Sen, "Context-aware energy-efficient communication for iot sensor nodes," in *2016 53nd DAC*. IEEE, 2016, pp. 1–6.

[2] T. Bouguera *et al.*, "Energy consumption model for sensor nodes based on lora and lorawan," *Sensors*, vol. 18, no. 7, p. 2104, 2018.

[3] I. F. Akyildiz *et al.*, "Wireless sensor networks: a survey," *Computer networks*, vol. 38, no. 4, pp. 393–422, 2002.

[4] K. C. Barr and K. Asanović, "Energy-aware lossless data compression," *ACM Transactions on Computer Systems (TOCS)*, vol. 24, no. 3, pp. 250–291, 2006.

[5] "Five ways cognitive computing will power the internet of things, url = https://www.forbes.com/sites/ibm/2015/12/15/five-ways-cognitive-computing-will-power-the-internet-of-things/?sh=4b0282b6724a."

[6] T. Yamazaki *et al.*, "4.9 a 1ms high-speed vision chip with 3d-stacked 140gops column-parallel pes for spatio-temporal image processing," in *IEEE ISSCC*. IEEE, 2017, pp. 82–83.

[7] S. Tabrizchi *et al.*, "Tizbin: A low-power image sensor with event and object detection using efficient processing-in-pixel schemes," in *IEEE ICCD*. IEEE, 2022, pp. 770–777.

[8] H. Xu *et al.*, "Macsen: A processing-in-sensor architecture integrating mac operations into image sensor for ultra-low-power bnn-based intelligent visual perception," *IEEE TCAS II: Express Briefs*, vol. 68, no. 2, pp. 627–631, 2020.

[9] T.-H. Hsu *et al.*, "A 0.5-v real-time computational cmos image sensor with programmable kernel for feature extraction," *IEEE Journal of Solid-State Circuits*, vol. 56, no. 5, pp. 1588–1596, 2020.

[10] H. Xu *et al.*, "Senputing: An ultra-low-power always-on vision perception chip featuring the deep fusion of sensing and computing," *IEEE TCAS I: Regular Papers*, 2021.

[11] M. Abedin *et al.*, "Mr-pipa: An integrated multilevel rram (hfo x)-based processing-in-pixel accelerator," *IEEE JxCDC*, vol. 8, no. 2, pp. 59–67, 2022.

[12] S. Tabrizchi *et al.*, "Ocelli: Efficient processing-in-pixel array enabling edge inference of ternary neural networks," *Journal of Low Power Electronics and Applications*, vol. 12, no. 4, p. 57, 2022.

[13] ——, "Appcip: Energy-efficient approximate convolution-in-pixel scheme for neural network acceleration," *IEEE JETCAS*, vol. 13, no. 1, pp. 225–236, 2023.

[14] W.-T. Kim *et al.*, "An on-chip binary-weight convolution cmos image sensor for neural networks," *IEEE Transactions on Industrial Electronics*, vol. 68, no. 8, pp. 7567–7576, 2020.

[15] R. LiKamWa *et al.*, "Redeye: analog convnet image sensor architecture for continuous mobile vision," *ACM SIGARCH Computer Architecture News*, vol. 44, no. 3, pp. 255–266, 2016.

[16] F. Taherian and D. Asemani, "Design and implementation of digital image processing techniques in pulse-domain," in *IEEE Asia Pacific Conference on Circuits and Systems*. IEEE, 2010, pp. 895–898.

[17] I. Hubara *et al.*, "Binarized neural networks," *Advances in neural information processing systems*, vol. 29, 2016.

[18] F. Muñoz-Martínez *et al.*, "Stonne: Enabling cycle-level microarchitectural simulation for dnn inference accelerators," in *IEEE IISWC*, 2021, pp. 201–213.

[19] J. Choi *et al.*, "An energy/illumination-adaptive cmos image sensor with reconfigurable modes of operations," *IEEE Journal of Solid-State Circuits*, vol. 50, no. 6, pp. 1438–1450, 2015.

[20] H. Kwon *et al.*, "Reducing refresh overhead with in-dram error correction codes," in *18th ISOCC*. IEEE, 2021, pp. 211–214.

[21] H. Sharma *et al.*, "isc: Bit-level dynamically composable architecture for accelerating deep neural network," in *ACM/IEEE 45th ISCA*. IEEE, 2018, pp. 764–775.

[22] M. Imani *et al.*, "Floatpim: In-memory acceleration of deep neural network training with high precision," in *2019 ACM/IEEE ISCA*. IEEE, 2019, pp. 802–815.

[23] S. J. Carey *et al.*, "A 100,000 fps vision sensor with embedded 535gops/w 256× 256 simd processor array," in *2013 Symposium on VLSI Circuits*. IEEE, 2013, pp. C182–C183.

[24] S. Park *et al.*, "7.2 243.3 pj/pixel bio-inspired time-stamp-based 2d optic flow sensor for artificial compound eyes," in *2014 IEEE ISSCC*. IEEE, 2014, pp. 126–127.

[25] R. Acharya, "Corn leaf infection dataset," 2020. [Online]. Available: https://www.kaggle.com/datasets/qramkrishna/corn-leaf-infection-dataset

Sky-NN: Enabling Efficient Neural Network Data Processing with Skyrmion Racetrack Memory

Yong-Cheng Liaw
National Yang Ming Chiao Tung University
Taipei, Taiwan
tomhot246@gmail.com

Shuo-Han Chen
National Yang Ming Chiao Tung University
Hsinchu, Taiwan
shchen.nycu@gmail.com

Yuan-Hao Chang
Academia Sinica
Taipei, Taiwan
johnson@iis.sinica.edu.tw

Yu-Pei Liang
National Chung Cheng University
Chiayi, Taiwan
ypliang@cs.ccu.edu.tw

***Abstract*—The thriving of artificial i ntelligence h as brought numerous efforts to build strengthened and sophisticated neural network models to resolve almost all kinds of problems in different academic fields. O wing t o t he g rowing c omplexity a nd s ize of neural networks, nonvolatile random access memory (NVRAM) has been utilized to avoid excessive data movements between volatile memory and persistent storage. Among various NVRAM alternatives, skyrmion racetrack memory (SK-RM) is regarded as a promising candidate owing to its high memory density and efficient r eads a nd w rites. N evertheless, d ue t o t he distinct shift operation of SK-RM, directly applying existing data process methods of neural networks on SK-RM hinders the benefits and performance of both SK-RM and neural networks. To resolve this issue, this paper proposes *Sky-NN* to enable efficient N N data processing methods on SK-RM by utilizing the distinct shift and re-assemblability capability of skyrmions. A series of experiments were conducted to demonstrate the capability of Sky-NN.**

***Index Terms*—skyrmion racetrack memory, SK-RM, efficient, data processing, neural networks**

I. Introduction

Neural network (NN) has been widely utilized in the area of artificial i ntelligence, a nd v arious N N m odels h ave been proposed to advance the techniques of object detection [2], natural language processing (NLP) [6], and recommendation systems [4]. Nevertheless, with the huge datasets and complex problems to solve, the size and computation requirements of NN models have grown rapidly. For instance, within Google's Bidirectional Encoder Representations from Transformers (BERT) [6], which is a transformer-based NN model designed to perform NLP pre-training, the number and the size of *weights* in the large BERT model are roughly 3.4E08 and 1 GB, respectively. This size will become much larger during the runtime since NN processing engines (*i.e.*, Tensorflow [1]) require extra metadata (*i.e.*, tensor) and memory to facilitate the forward propagations and deal with neural activations.

In the conventional computer architecture, the growing size of modern NN models has led to prolonged processing latency and excessive energy consumption. This is because the conventional architecture relies on fast but small memory and slow but large storage to strike a balance between cost and performance. Owing to the limited size of memory in computer systems, swap operations between memory and storage are required, thus aggravating the processing performance of already-computation-intensive training and inference of NN models. Meanwhile, dynamic random access memory (DRAM), which is typically used as the volatile memory, faces the issue of high leakage power [8] and causes energy efficiency concerns. To resolve the above issues, the non-volatile random access memory (NVRAM), such as spin-transfer torque RAM (STT-RAM) [3] and skyrmion racetrack memory (SK-RM) [7, 12, 14], has become a great alternative to resolve the latency and energy efficiency concerns of large NN models.

Among various NVRAMs, racetrack memory (RM) has received growing attention owing to its near hard disk drive data density (*i.e.*, 1-$4F^2$ per cell), close performance to DRAM, and no endurance concerns when compared with other NVRAM alternatives. The first generation of RM is demonstrated by IBM [13] and stores data bits on a tape-like nanowire through the presence or absence of magnetic domains. Since each magnetic domain is separated by physical domain walls, this type of RM is known as domain-wall racetrack memory (DW-RM). Later, skyrmion racetrack memory (SK-RM) [7, 12, 14] is proposed to further enhance memory density by eliminating domain walls and achieves better energy efficiency with its ultra-low shift current [11]. Notably, unlike DRAM or other NVRAMs, the bit addressability of SK-RM is achieved through shifting skyrmions to access ports to detect/insert/remove skyrmions for data reads and writes. Owing to this distinct shift requirement during reading/writing data bits on SK-RM, *directly applying NN models on SK-RM could hinder the overall performance if excessive shifts are induced by the data processing methods of NN models.*

In this paper, the TensorFlow engine is utilized to demonstrate

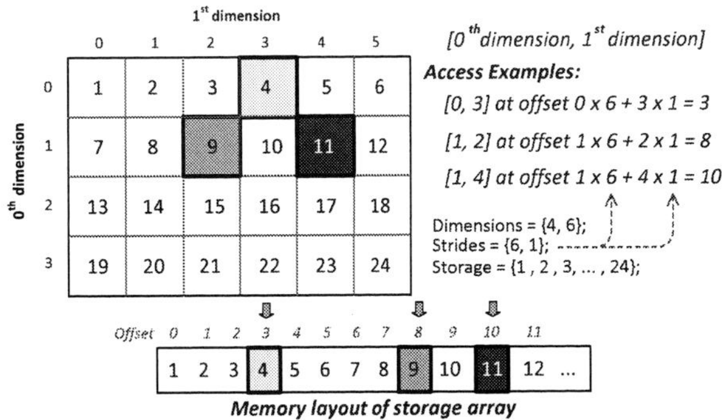

Fig. 1. Data memory layout of matrix with TensorFlow engine, which relies on the strides array and the access indexs to compute the offset of the to-be-access elements in the *storage* array.

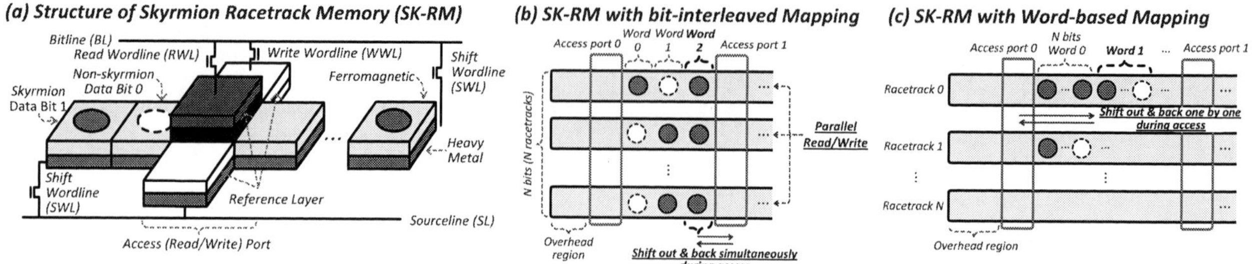

Fig. 2. Structure and data mapping methods of SK-RM, which shows that the data stream is divided into fix-sized words (*i.e.*, 32 or 64 bytes) and stored across multiple racetracks or on the same racetrack according to the bit-interleaved or word-based mapping methods. While bit-interleaved mapping allows parallel operations after aligning words to access ports, the word-based mapping requires shifting the whole word accross the access port for skyrmion operations.

the data memory layout and data processing methods of NN models. Although NN models are usually visualized as matrices with arbitrary dimensions, matrices are stored on memory in three different arrays, which are the *dimensions*, *strides*, and *storage* arrays, during the runtime of the TensorFlow engine. An example is shown in Figure 1 for a better explanation. As shown in the figure, to access the *weight* at *[0, 3]*, the row and column indexes are multiplied by the corresponding values in *strides* array to derive the offset of the to-be-accessed *weights* in *storage* array. Notably, the values within the *strides* array can be interpreted as the number of elements it requires to skip in each dimension on the *storage* array for accessing the values at certain indexes. Even though the above data accessing works well on DRAM, accessing those *weights* in the *storage* array on SK-RM becomes complicated due to the needs of shift operations. Although, based on SK-RM, numerous studies have been proposed to lower the number of shifts or injections by innovative racetrack designs [18], permutating existing skyrmions to compose new data pattern [16], and buffering existing skyrmions on buffer tracks for future reuse [17], *the repetitiveness and randomness of the NN data processing methods have received much less attention.*

To resolve the above concerns, this paper proposes, Sky-NN, a new set of SK-RM optimized NN data processing methods, including (1) the circular block multiplication, (2) the fixed-skyrmions approximation, and (3) the zero-shift matrix transpose. The main contribution of this study lies in properly utilizing the distinct in- and inter-track shift operations and the re-assemblability of skyrmions for efficient NN data processing. Notably, re-assemblability refers to the ability to compose new data patterns through previously-injected skyrmions. First, since block matrix multiplication induces multiple and random reads within the to-be-multiplied metrics, the circular block multiplication is proposed to lower the number of shifts during block matrix multiplication via utilizing the inter-track shift capabilities of SK-RM. Second, based on NN quantization, the fixed-skyrmions approximation eliminates the need for skyrmion inserts and deletes via the re-assemblability of skyrmions during the updates of *weights* and activations in NN models. Finally, instead of rewriting the whole metrics for matrix transpose, the zero-shift matrix transpose achieves the effect of transpose via a few pieces of additional metadata and the checkerboard-pattern layout of SK-RM. With the aforementioned three components, Sky-NN can achieve up to 41.06% and 43.39% latency and energy reduction, respectively, when using the long short-term memory (LSTM) and BERT as case studies.

II. Background and Motivation

A. *Skyrmion Racetrack Memory*

Skyrmion racetrack memory (SK-RM) utilizes the skyrmion, uniformly-distributed and topologically-protected magnetization, on racetracks to represent data bits 1 and 0 via the presence and absence of skyrmions. As shown in Figure 2(a), a skyrmion racetrack comprises a ferromagnetic layer and a heavy metal layer. Then, on the racetrack, access ports are attached with a fixed interval to generate skyrmions and non-skyrmions with different applied voltages on the sourceline (SL) and bitline (BL). Those generated skyrmions and non-skyrmions can then be shifted along the racetrack to specific offsets for representing different data patterns. Notably, since access ports induce additional area overhead, the number of access ports per racetrack is limited. On the other hand, to support detecting, injecting, removing, and shifting skyrmions, wordline (WL) is split into read wordline (RWL), write wordline (WWL), and shift wordline (SWL). For each connected line, different voltage combinations will be generated to read and write skyrmions at access ports or to shift skyrmions to access ports for multiplication.

Based on the structure of a racetrack, SK-RM can be formed by a group of racetracks with shared SL, BL, and WL. Meanwhile, access ports are connected between each racetrack to allow inter-track shifts [17, 18] for moving skyrmions between racetracks. Owing to the racetrack layout and the limited number of access ports, data mapping methods are proposed to store data bits on SK-RM and enable random data reads and writes. Figures 2(b) and (c) summarize the bit-interleaved and the word-based mapping methods. In both mapping methods, the to-be-stored data stream is divided into fix-sized words. The difference is that while the bit-interleaved mapping stores data bits of a word vertically across multiple tracks, the word-based mapping stores one or multiple words horizontally between two access ports on the same racetrack. During data reads and writes, the bit-interleaved mapping allows skyrmions of a single word to be shifted to the access port simultaneously and allows parallel detection/deletion/injection on skyrmions. For instance, as shown in Figure 2(b), if *Word 2* is going to be accessed, the whole word will be shifted left to *Access Port 2*. Then, all skyrmions and non-skyrmions within *Word 2* can be accessed in parallel. On the other hand, the word-based mapping requires shifting skyrmions of a word across the access port one at a time and induces a higher number of shift operations. As shown in Figure 2(c), if *Word 1* is going to be accessed, skyrmions of *Word 0* need to be first shifted to the overhead region through *Access Port 0*. Then, the skyrmions of *Word 1* can be shifted through *Access Port 0* one by one for accesses.

Even though the bit-interleaved mapping only requires a certain number of shift operations to align the to-be-access word to an access port and allows parallel skyrmion operations, the bit-interleaved mapping method is less preferable in terms of energy efficiency. This is because, while skyrmion injection induces the highest energy consumption among skyrmion operations, bit-interleaved mapping removes all skyrmions of previous data patterns and injects new skyrmions for composing future data patterns without exploiting the re-assemblability of skyrmions. On the other hand, the word-based mapping has better energy efficiency since the permutation write (PW) technique [16] can be enforced to reuse existing skyrmions for composing new data patterns via shifting skyrmions to the overhead region, thus lowering the number of skyrmion injections. However, the word-based mapping induces higher shift operations due to the requirement of shifting a whole word across access ports during reads and writes. In summary, *both the data memory layout and processing methods have limited aspects, and should be considered carefully to avoid excessive shift or inject operations.*

B. Data Processing Methods of NN

During the training and inference of NN models, one of the most widely-used operations is matrix multiplication. Within the Tensorflow engine, multiplying matrices is carried out by following the concept of block matrix multiplication. As shown in Figure 3(a), block matrix multiplication divides large matrices into smaller blocks, multiplies each block, and combines the block products to compose the product of the original large matrices. In this example, $BlockA_1$ is multiplied by $BlockB_1$ and $BlockB_2$, then followed by $BlockA_2$ to be multiplied by $BlockB_3$ and $BlockB_4$, and so on. Block matrix multiplication is utilized because it allows optimization through different block partition methods. However, due to the repetitiveness of matrix multiplication, blocks of the original larger matrices are read multiple times during the multiplication. Meanwhile, as the memory layout of matrices is composed of *dimensions*, *strides*, and *storage* arrays (See Figure 1), accessing the elements of blocks ends up as random and repetitive accesses to the *storage* array, as illustrated in Figure 3(b). While accessing an element of matrices on SK-RM involves shifting skyrmions out and back, random accesses further enlarge the number of shift operations due to the non-sequential access manner.

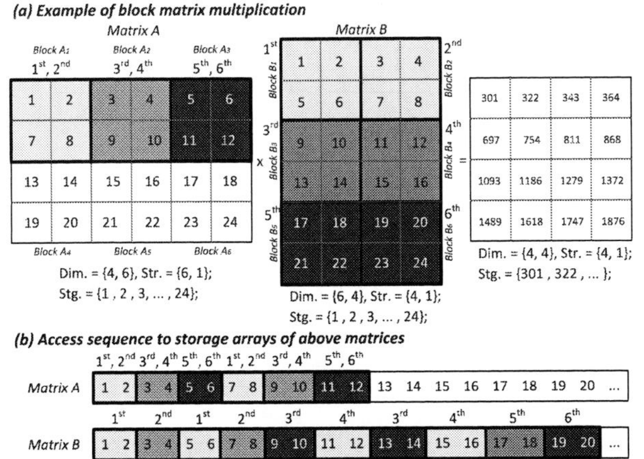

Fig. 3. Example of block matrix multiplication, which induces random access to the storage arrays of to-be-multiplied matrices. The access of blocks and elements in the *storage* array are denoted with sequence number to illustrate their accessing sequence.

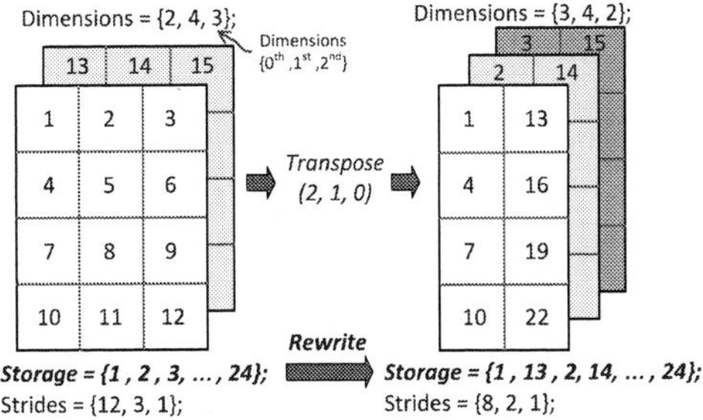

Fig. 4. Transposing from dimension [2, 4, 3] to dimension [3, 4, 2], which leads to completely rewrite the storage array for reordering matrix elements.

In addition to matrix multiplication, the matrix transpose is widely used to switch the indices of specified dimensions. Transpose is widely used with NN computations to make *weights* and *inputs* comply with matrix multiplication requirements. For instance, within the multi-head attention phase of Transformer models, transpose is utilized after matrix reshape to prepare matrices as the multi-head form and before output attention results. Figure 4 shows an example of matrix transpose, in which the specified $transpose(2, 1, 0)$ rotates indices of the 0th and the 2nd dimensions of the input matrix. The existing implementation of matrix transpose simply rewrites all indices within the *storage* array. Although rewriting is simple and effective, rewriting on SK-RM induces multiple insert/delete/shift/detect operations. In addition, while skyrmion injections induce the highest latency and energy consumption, frequent transpose degrades the benefits of SK-RM and worsens the data processing performance. In summary, *the existing data processing methods of NN are unfriendly to SK-RM due to the random accesses of block multiplication, the rewrites of matrix transpose, and SK-RM characteristics.*

C. Motivation

To investigate the excessive shifts issue of directly hosting NN models on SK-RM, experiments are conducted to investigate the number of shift operations during block matrix multiplication for NN models of different sizes. In this experiment, the word-based mapping method with the permutation write capability is chosen for its energy efficiency, and two words (*i.e.*, *weights*) are placed between access ports. Accordingly, as shown in Figure 5(a), during block matrix multiplication, each *weight* access induces 64 shift operations if 32 bits are used to represent each *weight*. In the example, shift current can be applied from *Access ports 1* to *0* to shift *Word 0* across the *Access port 0*. Meanwhile, even though parallel access at different access ports are feasible via shifting all skyrmions on the racetrack from the rightmost end to the leftmost end, parallel accesses require to-be-accessed words located at offset with an equal interval, as shown in Figure 5(b), and the randomness of accesses in the *storage* array, as suggested in Figure 3(b), prevents parallel accesses from being utilized.

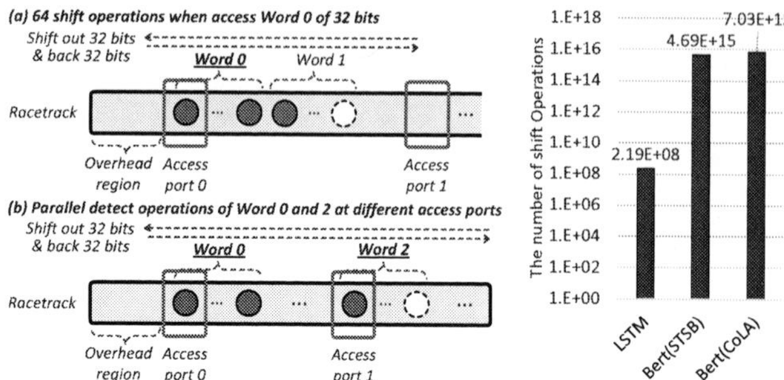

Fig. 5. Accessing words (*i.e., weights*) on SK-RM, which shows that accessing a word requires shift all data bits out and back.

Fig. 6. The number of shift operations.

The experimental results are summarized as Figure 6, and experiments are conducted by extending the Tensorflow engine to simulate the behavior of SK-RM, while the long short-term memory (LSTM) and the small-sized BERT are selected as the examples of large and small size NN models. The total numbers of *weights* in the LSTM and the small-sized BERT models are 419 and 3.63E07, respectively. The CoLA and STSB datasets are used for BERT, while the CoLA dataset is 1.5 times larger than the STSB dataset, and the 365 days stock market dataset is used for LSTM. Figure 6 shows the number of shift operations of a training epoch after convergence. Accordingly, it can be observed that the number of shift operations becomes larger as the size of the datasets or the NN models grows, thus leading to degraded efficiency when utilizing SK-RM as the main memory for NN models.

III. Sky-NN: SK-RM Optimized NN Data Processing

A. Overview

To enable efficient NN data processing on SK-RM, this study presents Sky-NN to reconsider NN computations with the awareness of SK-RM characteristics. This design goal of Sky-NN is achieved through redesigning the data processing methods of NN models based on the shift capability and re-assemblability of skyrmions. Meanwhile, to facilitate those redesigned methods, Sky-NN extends the existing data memory layout of matrices based on the checkerboard-pattern layout of SK-RM. The main rationale behind the checkerboard-pattern layout is to utilize the multi-racetracks and multi-access ports structure of SK-RM for efficient NN data processing. Figure 7 demonstrates the two-dimension checkerboard-pattern layout and indexes-to-offset conversion of Sky-NN.

As shown in Figure 7, a three-dimension matrix is transferred into a two-dimension layout by placing the matrices of 0th dimension side by side onto the SK-RM. Accordingly, owing to the two-dimension layout, the *strides* array is decomposed into vertical and horizontal ones, which indicates how many racetracks and how many elements need to be skipped during index accesses. Horizontally, since the 1st of the original three-dimension matrix is retained vertically on SK-RM, *Strides_H* array only comprises the number of elements that need to be skipped in the original 0th and 2nd dimensions after fitting onto SK-RM. Vertically, in the *strides_V* array, only the 1st

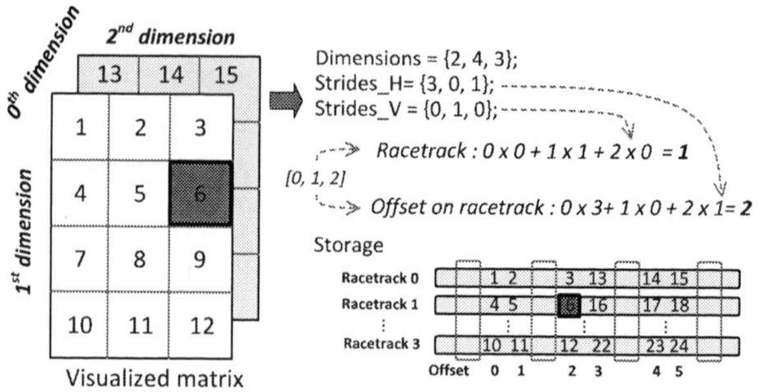

Fig. 7. Two-dimension data memory layout, which demonstrates the necessary metadata (*i.e., Strides_H* and *Stride_V* arrays) to support two-dimensions checkerboard-pattern layout on SK-RM. A conversion example of an access with index [0, 1, 2] to the racetrack and offset is given.

dimension need to be skipped since the original 0th and 2nd dimensions are already considered in the *Strides_H* array. Then, based on the above metadata, indexes can be converted into the number of racetracks and offset on the racetrack. Based on the two-dimension checkerboard-pattern layout of matrices on SK-RM, the proposed Sky-NN introduces (1) the circular block multiplication to lower shifts through vertical shifts and an additional racetrack, (2) the fixed-skyrmions approximation to recompose data pattern of different floating-point numbers via the same group of skyrmions, and (3) the zero-shift matrix transpose to eliminate the matrices rewrites.

B. Circular Block Multiplication

To lower the number of shift operations when hosting NN models on SK-RM, Sky-NN introduces a circular block multiplication for the goal of applying one while accessing stored *weights* for block matrix multiplication. To achieve this goal, an additional racetrack is included on the top of matrices, and the vertical shift is utilized to move skyrmions between racetracks. Accordingly, during a block matrix multiplication (See Figure 3), those stored *weights* in one block are shifted across access ports for detection and onto another racetrack for temporally buffering. Then, followed by the next block matrix multiplication, those *weights* are moved onto different racetracks and across the access ports again for detection. In other words, as those *weights* are shifted circularly on racetracks, each *weight* only needs to be shifted once during each block access, rather than the original two shifts (See Figure 5). The process can be summarized as Figure 8.

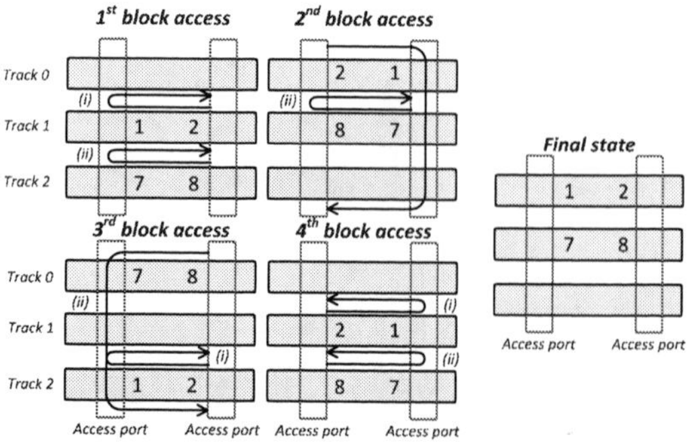

Fig. 8. Example of circular-shift matrix multiplication, which follows up the block matrix multiplication example in Figure 3. (i) and (ii) denotes the shift ordering during each block access and the final state after 4 block accesses is moving the whole block upward by a track.

C. Fixed-Skyrmions Approximation

To enhance the efficiency of SK-RM-based NN data processing, Sky-NN proposes the concept of fixed-skyrmions approximation to completely remove the need for skyrmion injections and deletions after the first write of NN models on SK-RM. The core concept is to inject a fixed number of skyrmions only once when matrices are first written on SK-RM, and those skyrmions are reused repetitively to compose the data bits pattern of updated *weights*. This concept is made possible by the re-assemblability of skyrmions, the quantization capability of NN models, and the IEEE-754 floating-point format. Notably, re-assemblability refers to the fact that injected skyrmions are stable enough to reuse for composing different data patterns through shift operations. Meanwhile, unlike the previous permutation write mechanism that needs to inject or remove skyrmions when the number of skyrmions is inconsistent between two contiguous updates, the proposed fixed-skyrmions approximation induces no skyrmions injections and deletions.

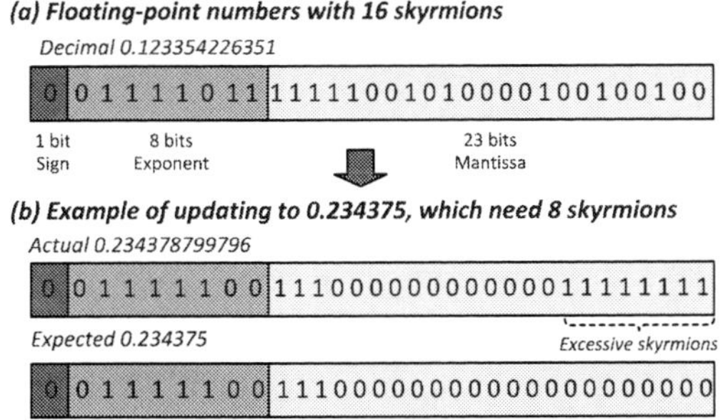

Fig. 9. The fixed-skyrmions approximation, which relies on a fixed number of skyrmions to represent different floating-point numbers without the need of injections or removals.

Meanwhile, the quantization capability allows NN models to lower the precision of *weights* by using fewer data bits. Differently, in Sky-NN, the number of data bits for each *weight* update remains identical. In other words, there could be insufficient or excessive skyrmions after each *weight* update. While insufficient skyrmions only leads to a smaller value, those excessive skyrmions are buffered as mantissa bits in IEEE-754 floating-point format. Above approach is made possibly by the IEEE-754 floating-point format without incurring negative impacts on the precision of NN computation. This is because, as shown in Figure 9(a), a floating-point number is composed of a *sign* bit, 8 *exponent* bits, and 23 *mantissa* bits. Based on this format, the value of storage floating-point number can be calculated as $sign \cdot 2^{exponent} \cdot (1.mantissa)$. In this format, the *sign* bit, *exponent* bits, and the former portion of *mantissa* bits are more important in terms of representing *weights* precisely. Therefore, as shown in Figure 9(b), even if there are excessive skyrmions left after composing the whole floating-point number, those excessive skyrmions have limited precision impacts. In this study, we assume 16 skyrmions are injected to represent different floating-point numbers in the 32-bits word space.

D. Zero-Shift Matrix Transpose

Another source of skyrmion injections and removals during NN data processing is the matrix transpose. Matrix transpose is widely utilized in NN computation to transform inputs of different sizes to meet the dimension requirement of matrix multiplication. In Sky-NN, instead of rewriting the whole *storage* array to comply with the indexing mechanism of the TensorFlow engine, the zero-shift matrix transpose includes an additional *inverted* array to derive the correct offset when accessing indexes in the transposed matrices. Notably, this approach can be applied to transposed matrices that results in switching the horizontal and vertical directions of the checkerboard-pattern layout of SK-RM. In other words, the transpose can be simply achieved by rotating the indexing calculation of rows and columns with the labeled indexes in the *inverted* array to denote which index of the *stride* array should be used after matrices transpose.

IV. Performance Evaluation

A. Experiment Setup

To evaluate the effectiveness of Sky-NN, a series of experiments are conducted based on widely-used and well-known LSTM and BERT models. While the 365 days stock market dataset [9] is chosen for the LSTM model, the CoLA and STSB datasets of Glue Benchmark [15] are chosen for the BERT model. These three datasets are chosen based on their size differences, and the sizes of CoLA, STSB, and stock market datasets are 8551, 5750, and 365 rows, respectively. Mechanisms of the proposed Sky-NN are integrated within the TensorFlow engine on top of simulated SK-RM. Meanwhile, as the main goal of Sky-NN is to enhance the efficiency of NN data processing on SK-RM, the number of shift and inject operation, the accumulated latency, the energy consumption, and the NN model accuracy are reported. The latency and energy consumption parameters of each skyrmion operation are summarized in Table I. In this paper, since the word-based mapping method is chosen as the foundation of Sky-NN for its energy efficiency, the latency and energy consumption of Sky-NN are compared with the word-based mapping methods with permutation write techniques.

TABLE I
Latency and Energy Consumption of SK-RM Operations [17].

Operations	Detect	Shift	Remove	Inject
Energy	2 fJ	20 fJ	20 fJ	200 fJ
Latency	0.1 ns	0.5 ns	0.8 ns	1 ns

Notably, based on racetrack memory, most of the existing works accelerate NN computations through special processing-in-memory (PIM) computing circuits. For instance, Liu et al. [10] and Choong et al. [5] propose efficient in-memory arithmetic circuits based on racetrack memory. However, as this study mainly relies on the distinct in- and inter-track shift operations of SK-RM to enable efficient NN data processing without special PIM circuits, prior works relying on PIM circuits have different design goals and implementations. Therefore, experimental comparisons are not conducted.

B. Experimental Results

As the design goal of the proposed Sky-NN is to enable efficient NN data processing on SK-RM, the number of shift

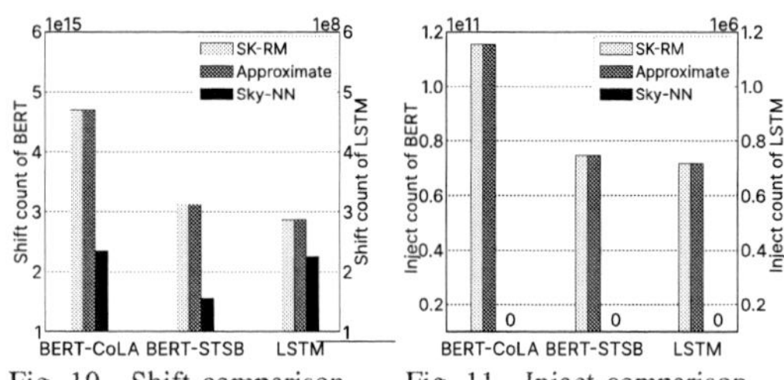

Fig. 10. Shift comparison.

Fig. 11. Inject comparison.

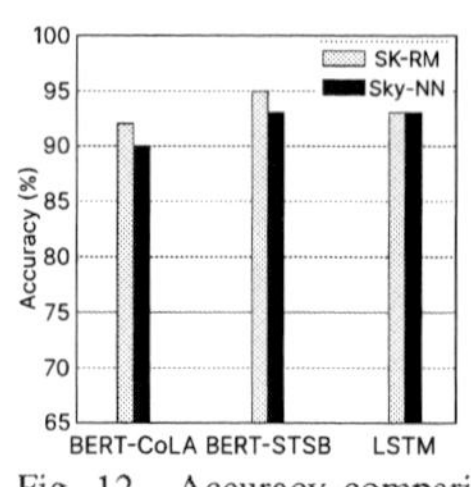

Fig. 12. Accuracy comparison.

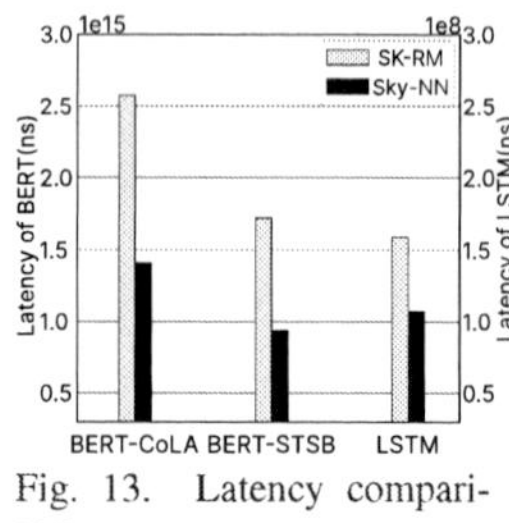

Fig. 13. Latency comparison.

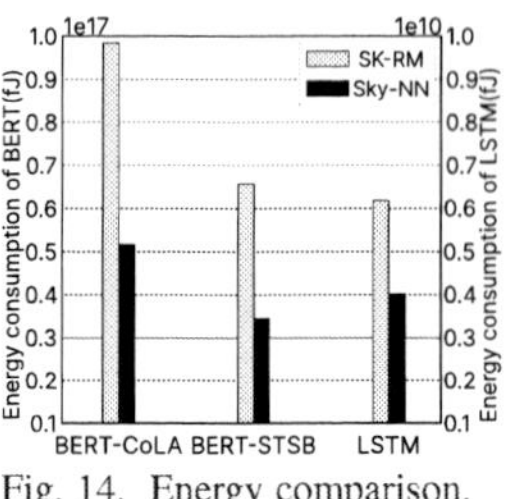

Fig. 14. Energy comparison.

and inject operations are compared in Figures 10 and 11. In the following figures, the legend, SK-RM, refers to the word-based mapping SK-RM with permutation write capability, while *Circular* and *Approximate* corresponds to enabling a single component in the proposed Sky-NN. The shift and inject operations of singly enabling the fixed-skyrmions approximation and the circular block multiplication are included to investigate whether additional shift or inject operations are induced while enforcing these mechanisms. Accordingly, when compared with SK-RM, Sky-NN with the circular block multiplication can effectively reduce the shift operations by an average of 44.95%, while inducing no additional inject operations. Meanwhile, Sky-NN with the fixed-skyrmions approximation can effectively eliminate the need for injecting skyrmions by allowing approximation on the stored floating-point numbers without extra shifts.

According to the model precision investigation shown in Figure 12, the overall precision difference is only 0.66% on average. In other words, the fixed-skyrmions approximation does not incur significant negative impacts on NN precision while greatly lowering the data processing overhead. To further understand the latency and energy consumption of Sky-NN, Figures 13 to 14 summarizes and compares the results. Accordingly, the overall latency and energy consumption can be reduced by up to 41.06% and 43.39%, respectively. This is because Sky-NN can effectively lower the shifts by the circular block multiplication, eliminate the injections by the concept of approximation with a fixed number of skyrmions, and avoid completely rewriting the whole matrix during matrix transposes.

V. Conclusion

To resolve the energy efficiency and performance issue of utilizing SK-RM as the memory for NN models, this study presents a pioneering design, Sky-NN, as a set of efficient and SK-RM-optimized NN data processing methods. The foundation of Sky-NN is to renovate the data memory layout of matrices from a single-dimension array to a two-dimensions checkerboard-pattern layout on SK-RM. Then, based on the distinct in- and inter-track shift capability, the circular block multiplication reduces the number of shifts during accessing matrices. Meanwhile, the fixed-skyrmions approximation eliminates the need for skyrmions injections and removals by repetitively using the same group of skyrmions to represent different floating-point numbers. Furthermore, the inefficient matrix rewrite during the matrix transpose is replaced by a set of new index methods. With these components, the Sky-NN can reduce the processing latency and the energy consumption by up to 41.06% and 43.39%, respectively, when compared with the word-based mapping SK-RM with permutation write. Future work of this study lies in extending the proposed Sky-NN to other NN models and tuning the fixed-skyrmions approximation for a balance between skyrmion injections and NN accuracy.

Acknowledgements

This work was supported in part by National Science and Technology Council under grant no. 109-2222-E-027-007-MY3, and 111-2221-E-027-076-MY3, 111-2223-E-001-001, 111-2923-E-002-014-MY3, 111-2221-E-001-013-MY3, and 112-2927-I-001-508 and Academia Sinica under grant nos. AS-IA-111-M01 and AS-GCS-110-08.

References

[1] M. Abadi, P. Barham, J. Chen, Z. Chen, A. Davis, J. Dean, M. Devin, S. Ghemawat, G. Irving, M. Isard, M. Kudlur, J. Levenberg, R. Monga, S. Moore, D. G. Murray, B. Steiner, P. Tucker, V. Vasudevan, P. Warden, M. Wicke, Y. Yu, and X. Zheng, "TensorFlow: A system for large-scale machine learning," in *12th USENIX Symposium on Operating Systems Design and Implementation (OSDI 16)*, 2016.

[2] A. Bochkovskiy, C.-Y. Wang, and H.-Y. M. Liao, "YOLOv4: Optimal speed and accuracy of object detection," 2020-04-23.

[3] E. Chen, D. Lottis, A. Driskill-Smith, D. Druist, V. Nikitin, S. Watts, X. Tang, and D. Apalkov, "Non-volatile spin-transfer torque RAM (STT-RAM)," in *68th Device Research Conference*, 2010, pp. 249–252.

[4] H.-T. Cheng, L. Koc, J. Harmsen, T. Shaked, T. Chandra, H. Aradhye, G. Anderson, G. Corrado, W. Chai, M. Ispir, R. Anil, Z. Haque, L. Hong, V. Jain, X. Liu, and H. Shah, "Wide & deep learning for recommender systems," in *Proceedings of the 1st Workshop on Deep Learning for Recommender Systems*, ser. DLRS 2016. Association for Computing Machinery, 2016, pp. 7–10.

[5] B. C. M. Choong, T. Luo, C. Liu, B. He, W. Zhang, and J. T. Zhou, "Hardware-software co-exploration with racetrack memory based in-memory computing for cnn inference in embedded systems," *Journal of Systems Architecture*, vol. 128, 2022.

[6] J. Devlin, M.-W. Chang, K. Lee, and K. Toutanova, "BERT: Pre-training of deep bidirectional transformers for language understanding," in *Proceedings of the 2019 Conference of the North American Chapter of the Association for Computational Linguistics: Human Language Technologies, Volume 1 (Long and Short Papers)*. Association for Computational Linguistics, 2019, pp. 4171–4186.

[7] W. Kang, Y. Huang, X. Zhang, Y. Zhou, and W. Zhao, "Skyrmion-electronics: An overview and outlook," *Proceedings of the IEEE*, pp. 2040–2061, 2016.

[8] Z. Lan, M. Chen, S. Goodman, K. Gimpel, P. Sharma, and R. Soricut, "ALBERT: A lite BERT for self-supervised learning of language representations," 2020.

[9] Liao, "365days stock klines dataset." [Online]. Available: https://github.com/DandinPower/SKY-NN-Dataset/blob/main/lstm/stocks.csv

[10] B. Liu, S. Gu, M. Chen, W. Kang, J. Hu, Q. Zhuge, and E. H.-M. Sha, "An efficient racetrack memory-based processing-in-memory architecture for convolutional neural networks," in *2017 IEEE International Symposium on Parallel and Distributed Processing with Applications and 2017 IEEE International Conference on Ubiquitous Computing and Communications (ISPA/IUCC)*, 2017, pp. 383–390.

[11] Y. Luo, S.-Z. Lin, M. Leroux, N. Wakeham, D. M. Fobes, E. D. Bauer, J. B. Betts, J. D. Thompson, A. Migliori, M. Janoschek, and B. Maiorov, "Skyrmion lattice creep at ultra-low current densities," vol. 1, no. 1, pp. 1–7, 2020.

[12] S. Mühlbauer, B. Binz, F. Jonietz, C. Pfleiderer, A. Rosch, A. Neubauer, R. Georgii, and P. Böni, "Skyrmion lattice in a chiral magnet," no. 5916, pp. 915–919, 2009.

[13] S. S. P. Parkin, M. Hayashi, and L. Thomas, "Magnetic domain-wall racetrack memory," vol. 320, no. 5873, pp. 190–194, 2008.

[14] R. Tomasello, E. Martinez, R. Zivieri, L. Torres, M. Carpentieri, and G. Finocchio, "A strategy for the design of skyrmion racetrack memories," vol. 4, no. 1, p. 6784, 2014.

[15] A. Wang, A. Singh, J. Michael, F. Hill, O. Levy, and S. Bowman, "GLUE: A multi-task benchmark and analysis platform for natural language understanding," in *Proceedings of the 2018 EMNLP Workshop BlackboxNLP: Analyzing and Interpreting Neural Networks for NLP*. Association for Computational Linguistics, 2018.

[16] T.-Y. Yang, M.-C. Yang, J. Li, and W. Kang, "Permutation-write: Optimizing write performance and energy for skyrmion racetrack memory," in *2020 57th ACM/IEEE Design Automation Conference (DAC)*, 2020, pp. 1–6.

[17] Y.-H. Yang, S.-H. Chen, and Y.-H. Chang, "Evolving skyrmion racetrack memory as energy-efficient last-level cache devices," in *Proceedings of the ACM/IEEE International Symposium on Low Power Electronics and Design*, 2022, pp. 1–6.

[18] Y.-H. Yang, Y.-P. Liang, C.-H. Tseng, and S.-H. Chen, "Exploring skyrmion racetrack memory for high performance full-nonvolatile ftl," in *2021 IEEE 10th Non-Volatile Memory Systems and Applications Symposium (NVMSA)*, 2021.

RF2P: A Lightweight RISC Processor Optimized for Rapid Migration from IEEE-754 to Posit

Hyun Woo Oh, Seongmo An, Won Sik Jeong, and Seung Eun Lee*
Department of Electronic Engineering
Seoul National University of Science and Technology, 01811, Seoul, Republic of Korea
Email: {ohhyunwoo, ahnseongmo, jeongwonsik, seung.lee}@seoultech.ac.kr

***Abstract*—This paper presents a lightweight processor and evaluation platform for migrating from IEEE-754 to posit arithmetic, with an optimized posit arithmetic unit (PAU) supporting existing floating-point instructions. The PAU features a reconfigurable divider architecture for diverse operating conditions and lightweight square root logic. The platform includes a posit-optimized compiler, divider generator, JTAG environment builder, and programmable logic controller. The experimental results demonstrate the successful execution of legacy IEEE-754 code with a small additional workload and up to 60.09 times the performance improvement through hardware acceleration. Additionally, the PAU and divider consume 11.00% and 57.87% fewer LUTs, respectively, compared to the best prior works.**

***Index Terms*—real number arithmetic, processor core, hardware acceleration, reduced instruction set computer, lightweight embedded systems**

I. Introduction

Nowadays, computation for real number arithmetic is a widespread technique in digital applications by reason of ongoing advances in semiconductor technology. As numerous applications are based on real number arithmetic, choosing the appropriate number system for applications is one of the major points in digital systems. In general, the IEEE-754 floating-point (FP) standard has been widely adopted for real number operations due to the practicality and flexibility originating from a wider dynamic range than fixed-point arithmetic since first established in 1985. However, emerging applications that require a higher precision and wider dynamic range prompted the development of novel number systems [1], [2].

Posit is a number format that has been proposed as a potential replacement for the FP [3]. Fig. 1 illustrates the main differences between the FP and posit. The key distinction of posit from IEEE-754 is the dynamic bit-width of the exponent and mantissa parts, which is facilitated by the use of a separate encoding format for the exponent field. This architecture encodes data through the run-length method for the regime field and raw least-significant bits for the exponent field, enabling posit numbers to achieve a wider dynamic range and higher precision in certain regions that are frequently used in computation [3]. Indeed, many studies adopting posit were held and reported the performance enhancements of adopting posit arithmetic in a variety of applications [4]–[6].

Hardware acceleration of posit arithmetic is one of the main topics because of the low throughput of software-only (SW-only) implementation [7]. In lightweight embedded systems where area usage and energy consumption are major concerns, optimizing the microarchitecture to minimize hardware resource usage is a major challenge in the design process [8], [9]. In the case of systems that adopted the FP acceleration, the floating-point unit (FPU) takes a major share of resource usage [10]. As a consequence, many researchers regard posit as an appropriate replacement for the FP because posit generally requires fewer resources, derived by such factors including the abandonment of the extraordinary representations such as subnormal numbers and adoption of the geometric rounding [3], [10], [11]. However, optimizing the resources of the posit arithmetic hardware still has complications originating from the decoding and encoding algorithms and general arithmetic algorithms for posit that are more complex than those for IEEE-754 due to differences in the representation of the exponent part and higher maximum bit-width of the mantissa part. This attribute enforces the entire system to have a lower operating frequency due to the increment of the critical path, slowing down the overall performance. On the other hand, the additional workload for porting legacy codes to the posit hardware is one of the impediments to the prevalent diffusion of posit arithmetic. To address this issue, preserving the former methodology for SW development is essential. For this reason, fostering a legacy-friendly environment by optimizing both hardware design and the compiler should be pursued.

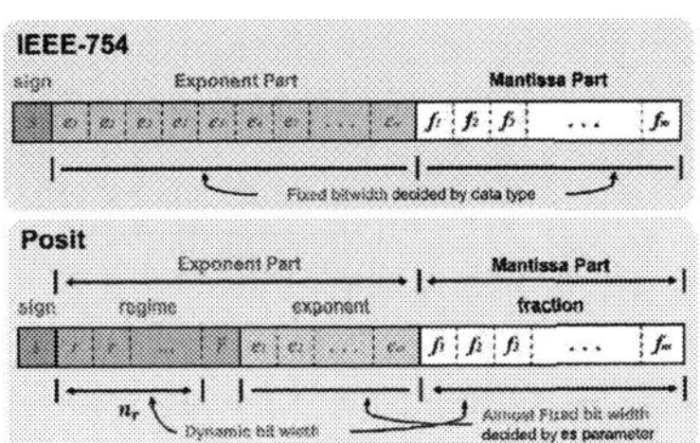

Fig. 1. Representation architecture of both IEEE-754 and posit.

In this paper, we propose a lightweight processor with an adequate platform optimized for migrating from IEEE-754 to posit. The processor architecture is built to support the posit arithmetic operation by utilizing the legacy FP extensions. To provide a wide range of resource optimization for numerous operating conditions, we designed the reconfigurable posit arithmetic unit (PAU). Next, we constructed the integrated platform for the development and evaluation based on the modern system-on-chip field-programmable gate array (SoC FPGA) environment composed of a posit-optimized GNU C compiler (GCC), programmable logic (PL) controller, and JTAG environment builder. We verified and evaluated the PAU as well as the processor through the PL implementation on a number of operating conditions and execution of the functions on the PL utilizing the evaluation platform.

II. Related Works and Motivations

The works on [12] and [13] focus on adding the posit arithmetic support running with custom instruction set extensions (ISE), providing faster execution than previous accelerators connected to the system bus. However, these works do not provide compiler support for a high-level language, which leads to a significant amount of additional workload required for utilizing posit hardware due to the writing of the assembly language. The work on [14] adds the 32-bit and 64-bit posit support on RISC-V utilizing the FP extensions. Through this, a major share of the additional workload reduces because the compiler generates the FP instruction codes from the FP variables and operators written in previous source codes. However, as this work excludes compiler modification, the variable initialization does not create the appropriate binary representation for posit. Therefore, the additional workload to fix the representation still exists. One of the key research that shares a similar aspect in fostering the legacy-friendly environment is explained in [15], which handles this issue through ISA-compatible core integration and compiler optimization. This work presents the reconfigurable PAU configured by the *es* parameter and integrates the PAU to the RISC-V core while providing compatibility for the FP extensions in RISC-V specification. Despite these solutions, applying posit to lightweight systems is still distant as the PAU design in [15] is optimized for a complex processor running on limited operating conditions. Meanwhile, research in [16] suggests a method for reducing the area usage and energy consumption by fixing the configurable parameters of the posit format specifications, i.e., *es*, and unit bit-width N. The key aspects of our work differentiate from previous works are as follows:

- Designing a reconfigurable PAU that parameterizes the internal architecture of the arithmetic unit, supporting optimization for a variety of lightweight operating conditions while reducing resource usage by fixing the configurable parameters on the posit specifications.
- Providing the rapid evaluation platform for the proposed processor utilizing the SoC FPGA environment, including compiler optimization and subsidiary SW running on the pre-fabricated processor (PS) block in the SoC FPGA.

III. Processor Architecture

Fig. 2 presents the proposed processor architecture. To design a processor providing posit arithmetic with swift evaluation and versatility while using minimal resources, the processor architecture consists of only the essential components: central processing unit (CPU), system bus with nested JTAG interconnect, boot ROM with boot mode configuration register (BMCR), on-chip memory, controller for external non-volatile memory (NVM), communication interfaces. Extendability support for additional peripheral devices, e.g., domain-specific accelerators, is provided through the bus topology.

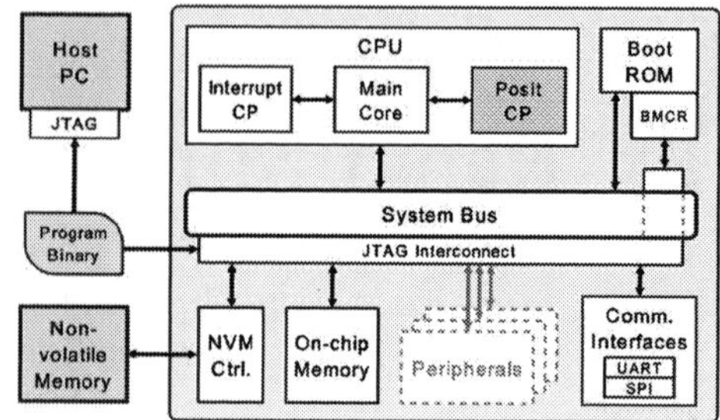

Fig. 2. The architecture of the proposed lightweight processor

The CPU comprises the main core, the posit coprocessor, and the interrupt coprocessor (CP) to handle the basic ISA, FP extensions, and trap conditions. The JTAG interconnect provides direct and precise access to the peripherals organizing bus topology from the host device by passing the signals for manipulating the system bus. This characteristic simplifies and improves the accuracy of the evaluation process, as the peripherals are operated and monitored cycle-accurate.

One crucial consideration for applying to the embedded application is the independent operability without any host device. For this reason, the processor supports two separate boot modes called JTAG boot and NVM boot. While the processor is in the booting state, the program in the boot ROM firstly checks the BMCR to select the mode. As the BMCR, similar to the other peripherals, is configured through JTAG interconnect, changing the boot mode can be done from the external host device. Through this, the processor can be used in both the evaluation process and distribution-level applications.

IV. Posit Arithmetic Unit Architecture

Fig. 3 shows the architecture of the posit coprocessor. The coprocessor is composed of the six pipeline stages similar to the main core: instruction fetch, decode, execute, memory 1, memory 2, and write back. The coprocessor maintains the pipeline logic separate from the main core except for the instruction fetch stage, which is shared with the main core and other coprocessors due to the in-order instruction queue. The PAU is embedded in the execute stage and operates through three stages: decoding, calculation, and encoding. The temporary store registers are inserted between each stage to reduce critical path delay. The decoding stage, which is processed by the posit decoder, extracts the two sets of sign, exponent, and mantissa parts from the two operands. These extracted sets are used as input for the calculation stage. The calculation stage contains the arithmetic logic responsible for every operation that is present in MIPS FP extensions: add/subtract (ADD), multiply (MUL), divide (DIV), integer to posit conversion (I2P), square root (SQRT), absolute/negate (ABS/NEG), posit to integer conversion (P2I), and compare

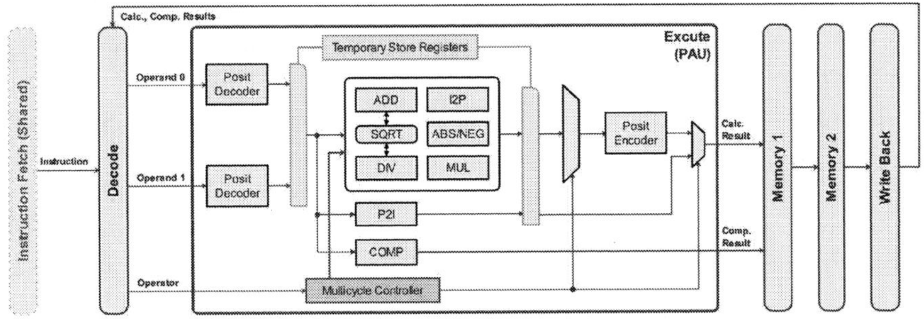

Fig. 3. The architecture of the posit coprocessor

(COMP). These logics begin their calculations using the extracted sets simultaneously and save the output results to temporary store registers, except for the I2P, which also operates simultaneously but uses the raw operand inputs. The encoding stage selects the appropriate result from the temporary store registers and compresses the selected result to posit representation by using the multiplexer (MUX) and the posit encoder. As the results of the operations are not limited to the data encoded to posit, the final calculation result is selected by another MUX, which selects the output data from the result of the P2I or the output from the encoder. Considering that the required cycles for each operation differ, the multicycle controller, not part of the stages, exists to handle operations that are processed over multiple cycles by sending the control signals for each logic and triggering the stall signals for the current stage while waiting for the operation completion.

Typically, the division and the square root are regarded as one of the most intensive operations in real number arithmetic. Reducing the critical path delay of these operations to less than that of the main core preserves the operating frequency. Therefore, we focused on two major concerns to enable flexible adoption in a wide range of systems: designing an SW-generated scalable divider architecture reflecting the operating frequency and developing a minimal square root architecture.

A. Decoder and Encoder

Algorithm 1 represent the decoding process of the regime field. The posit decoder uses concatenated 2-input MUXs to extract the exponent and mantissa from any regime and exponent field cases in a single clock cycle. The number of required MUXs in this architecture is derived by equations (1), (2), and (3). The w refers to the bit-width of any data.

$$w_{exp\_val} = \lceil \log_2 max(v_{exp}) \rceil + 1 \quad (1)$$

$$w_{m\_val} = w_{raw\_data} - 3 - es \quad (2)$$

$$N_{MUX} = (w_{raw\_data} - 1) \times (w_{exp\_val} + w_{m\_val}) \quad (3)$$

The encoding process is performed by the reversed execution of the decoding process.

As our design targets lightweight systems operating at low frequency, the maximum path delay of the concatenated MUXs does not violate the timing constraints, which means that the throughput performance is maximized without degrading energy efficiency by avoiding extra stall conditions.

Algorithm 1 Decoding the regime field

Input:
A_r : Bit array storing the regime field data
N : Bit-width of the data type
es : Configured parameter for exponent calculation
Output:
$regime$: Decoded exponent value

$r \leftarrow A_r[0]$
$n_r \leftarrow 1$
for $(i = 1; A_r[i] == r \ \&\& \ i < N; i = i + 1)$ **do**
  $n_r \leftarrow n_r + 1$
end for
if $r == 0$ **then**
  $regime \leftarrow -2^{es} \times n_r$
else
  $regime \leftarrow 2^{es} \times (n_r - 1)$
end if

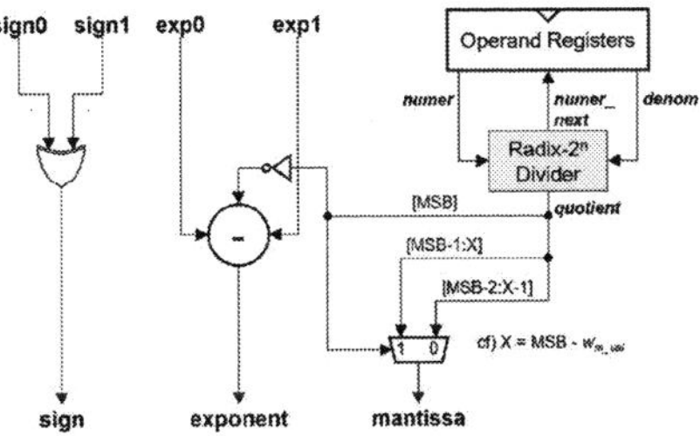

Fig. 4. The architecture of the divider logic

B. Division

Fig. 4 shows the divider logic architecture. The divider logic consists of the operand registers, scalable radix-2^n divider, subtractor, and ancillary components. The calculation flow of the division is as follows:

1) Store the mantissa values extracted by the decoder to the operand registers.
2) Calculate the division for certain clock cycles through the radix-2^n divider and the operand registers.
3) Derive the output exponent value and mantissa value using the subtractor and the MUX.

The radix-2^n division is the algorithm that parameterizes the number of subtractions performed in one clock period. Fig.

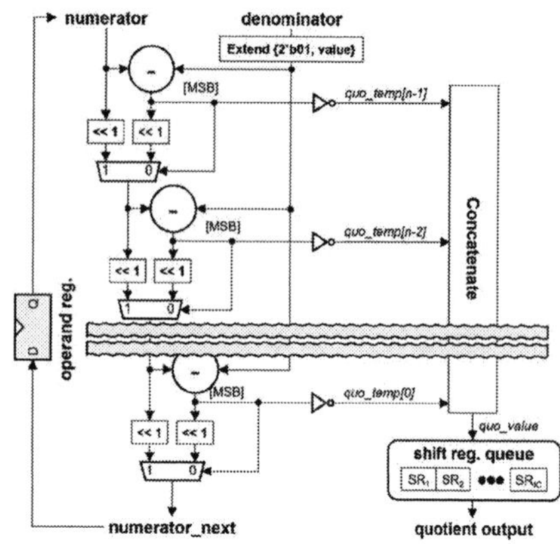

Fig. 5. The architecture of the proposed radix-2^n divider

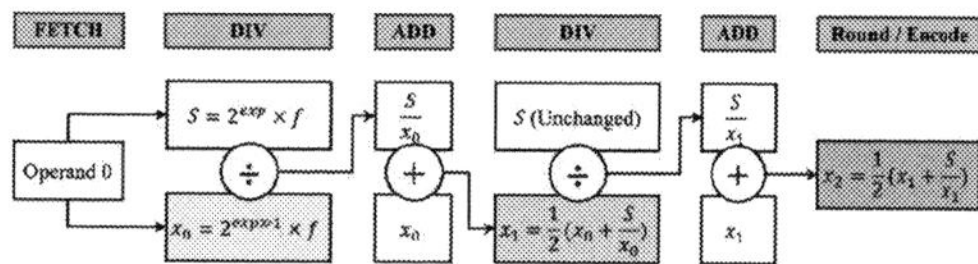

Fig. 6. The sequence to calculate square root

5 illustrates the architecture of the radix-2^n divider. In this architecture, the division is performed by iterating the divider sequence for certain times decided by the w_{m_val} constant and the n parameter. The iteration count is calculated through the following equation.

$$i = \lceil \frac{w_{m\_val} + 2}{n} \rceil \tag{4}$$

We composed the memory to temporarily store the output as the shift register (SR) queue to eliminate the path delay for searching the block entry to be stored.

According to the architecture, the maximum path delay is proportionate to the number of subtractions specified by the n parameter. Thus, the scalable radix-2^n divider provides versatility for numerous operating conditions. Furthermore, we developed the radix-2^n divider generator to provide the divider reflecting the n parameter swiftly.

C. Square Root

The square root operation is performed by repeating the Babylonian method twice times. The Babylonian method is organized by basic arithmetic operations: multiply, add, and divide, as shown in equation (5). The S refers to the input value to acquire the square root.

$$x_{n+1} = \frac{1}{2}\left(x_n + \frac{S}{x_n}\right) \tag{5}$$

This attribute eliminates the necessity of extra arithmetic logic because the algorithm is executed by utilizing the maintained logic. Thus, we designed only the routing logic steered by the multicycle controller (see Fig. 6) to compose the equation (5).

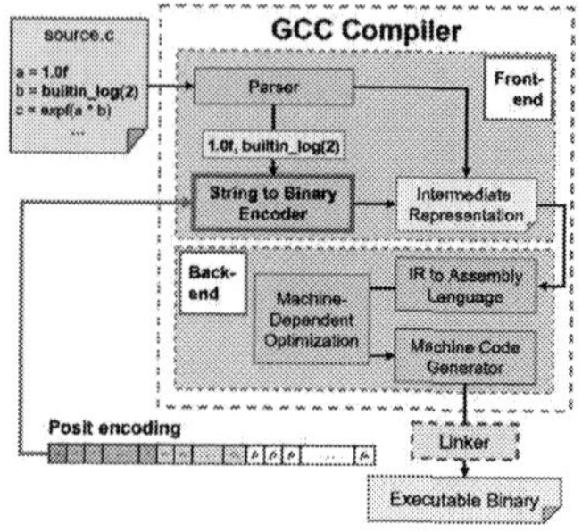

Fig. 7. Compiler operation optimized for posit arithmetic

V. Evaluation Platform

We construct the evaluation platform utilizing the SoC FPGA environment with three additional SW components: posit-optimized compiler, JTAG environment builder, and PL controller. The posit-optimized compiler, which is developed by modifying the string to the binary encoder in the GCC front-end (see Fig. 7), enables swift conversion from IEEE-754-based operation to posit-based operation. Through this compiler, the binary representation converted from the variable initialization codes replaced from the FP to the posit. The JTAG environment builder generates the driver that directly controls the peripherals linked to the system bus with nested JTAG interconnect (see Fig. 2), supporting accessibility from the PS block. The PL controller, which is executed on the Linux operating system, provides user-friendly interfaces to manipulate the designed processor.

VI. Implementation & Evaluation

A. Radix-2^n Divider

We verified and evaluated the scalable radix-2^n divider architecture through register-transfer level (RTL) simulation running on the Vivado simulator. Additionally, we conducted FPGA synthesis through Vivado 2022.2 targeting the Xilinx xc7z020 SoC FPGA for the entire divider logic with several different configurations. Each divider logic was designed for posit(N=32, es=3). Fig. 8 and 9 present the performance of each configuration. Except for the $n = 1$ configuration, the maximum frequency of the divider is approximately inversely proportional to the n parameter, and the throughput performance tends to maintain similar values as expected. The abnormality of the $n = 1$ configuration originates from the other circuits because the path delay of the radix-2^1 divider is already reduced enough. In the case of a relative analysis between area usage, i.e., look-up tables (LUTs) required for constructing the logic, and throughput performance (see Fig. 9), reducing the n parameter makes the area efficiency higher as the required LUTs per throughput decrease except for radix-2^1. These results are in line with the previous abnormality: the radix-2^n divider is relatively small to the other circuits in the entire divider logic. The overall result shows that the designed radix-2^n divider with generation SW provides flexible division logic for a variety of operating conditions.

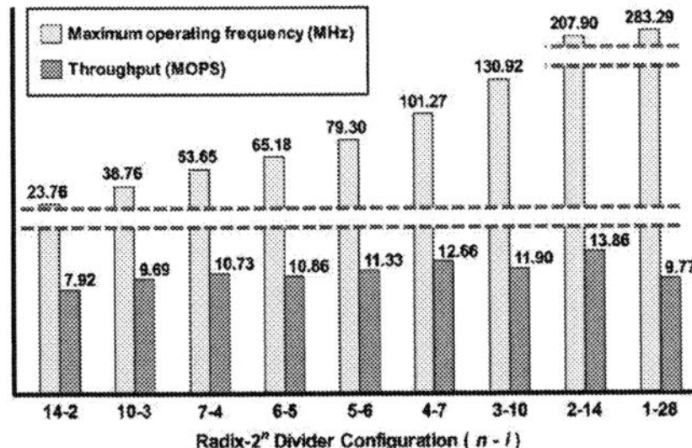

Fig. 8. Maximum operating frequency and throughput on FPGA

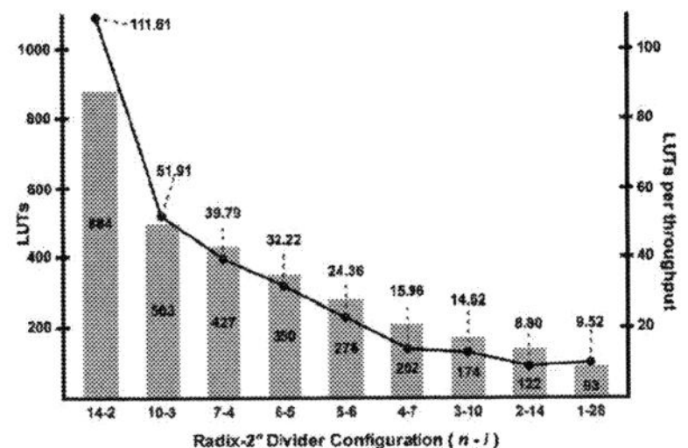

Fig. 9. Throughput performance by area usage on FPGA

To analyze the divider logic more precisely, we synthesized the logic with separated area-first (AF) and throughput-first (TF) constraints using Synopsys Design Compiler and TSMC 180nm technology. Fig. 10 presents the performance analysis for each configuration. As shown in Fig. 10 (a), the area usage varies from 4322.35 μm^2 to 69561.50 μm^2 across the different configurations. This represents a dynamic range of 16.09 and demonstrates versatility. In the case of power and energy (see Fig. 10 (b), and (c)), the result shows that the divider consumes the power of less than 15.53 mW and the energy of less than 273.94 pJ for one division operation, ensuring the aptness for lightweight systems. The maximum throughput of the divider is 56.02 mega operations per second (MOPS) (see Fig. 10 (e). The overall result indicates that the divider architecture is suitable for both lightweight systems and high-performance systems due to the variation in area usage and throughput.

Table I shows the power per throughput (PPT) and energy per throughput (EPT) of each configuration, regarded as the breakdown for each power and energy efficiency. One concern is that the most common configurations, $n = 2$ and $n = 4$, are not always the most efficient in terms of power and energy. The most energy-efficient configuration for the AF is $n = 7$, while $n = 2$ shows the second-worst performance. Though $n = 2$ or $n = 4$ configuration shows the best power efficiency with the TF, embedding those circuits in lightweight systems is not appropriate in terms of power and area. Therefore, utilizing the scalable divider architecture to adopt the adequate configuration with respect to the specifications, e.g., power, energy, and throughput, enhances the overall system efficiency.

B. Processor Architecture & Evaluation Platform

We implemented the proposed processor architecture with our evaluation platform by designing the processor written in

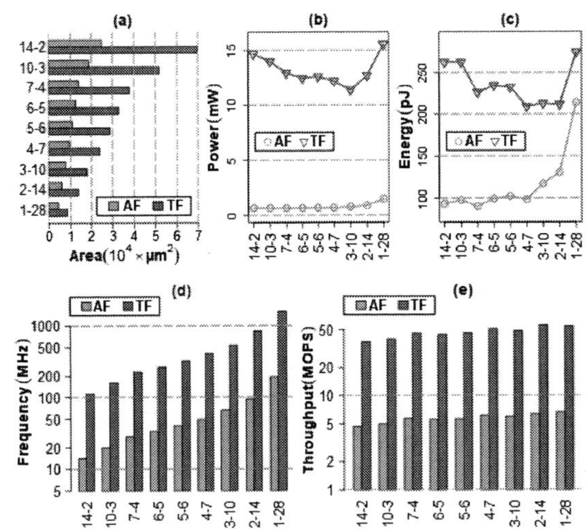

Fig. 10. ASIC synthesis results. **(a)** Area usage. **(b)** Maximum power consumption. **(c)** Energy dissipation for one operation. **(d)** Maximum operating frequency. **(e)** Throughput running on maximum frequency.

TABLE I
EFFICIENCY BREAKDOWNS ON POWER AND ENERGY

	Area-first synthesis results		Throughput-first synthesis results	
	PPT ($\mu W/MOPS$)	EPT ($pJ/MOPS$)	PPT ($\mu W/MOPS$)	EPT ($pJ/MOPS$)
14-2	13.99	19.94	39.34	7.03
10-3	13.02	19.55	34.95	6.55
7-4	11.29	**15.81**	28.21	4.92
6-5	11.91	17.93	28.09	5.28
5-6	11.95	18.12	27.10	4.99
4-7	**11.24**	16.05	23.89	4.10
3-10	12.91	19.61	23.39	4.37
2-14	14.01	20.64	**22.66**	**3.77**
1-28	22.15	32.07	28.37	5.00

Verilog HDL and testing the processor utilizing the platform. Operating at 100 MHz, we selected the $n = 3$ configuration for the divider to maximize performance. The DIV instruction requires three more cycles for division as the temporary registers exist in between the three stages (see Fig. 3).

We verified the proposed PAU and the processor with our evaluation platform through the following processes:

1) Generate the JTAG driver through the builder.
2) Develop the PL controller utilizing the JTAG driver.
3) Build the testing SW with the posit-optimized GCC.
4) Upload the testing SW to the main memory of the designed processor through the PL controller.
5) Check the results through the communication SW.

The testing SW executes the exponential function included in the standard math library. According to the result, the hardware acceleration with the posit coprocessor achieved up to approximately 60.09 times the performance improvement compared to the SW-only implementation.

We evaluated the throughput and area performance through comparative analysis of latencies and FPGA synthesis results of state-of-the-art works and our work. Every PAU presented in the tables is for handling 32-bit data type. The work on [13], in which DIV and SQRT operations are based on approximate

TABLE II
COMPARISONS OF THROUGHPUT PERFORMANCE OF THE POSIT INSTRUCTIONS BETWEEN THE PREVIOUS WORKS

	This work	[13]	[14]	[15]
Required cycles for each instruction				
ADD/SUB	5	3	1	6
MUL	5	2	1	6
DIV	13	2[a]	32	3-31[b]
SQRT	27	2[a]	32	3-30[b]
I2P	4	1	1	3
P2I	2	1	1	3
Others	1	1	1	1
FPGA implementation environment				
Target FPGA	xc7z020-1	xc7k325t-2	xc7s75t-1	xc7a100t-1
Target clock	100 MHz	50 MHz	24.5 MHz[c]	100 MHz

[a]Based on the logarithm-approximate multiplier.
[b]Proportionate to the bit-width of the mantissa part of numerators.
[c]Based on the critical path indicated in the timing report.

TABLE III
COMPARISONS OF AREA USAGE BETWEEN THE PREVIOUS WORKS

	This work		[13]		[14]		[15]	
	LUTs	FFs	LUTs	FFs	LUTs	FFs	LUTs	FFs
PAU*	2856	289	4753	255	4046	145	3209	184
DIV	174	66	413	43	1033	145	794	184

*Only for the arithmetic logics.

logic with a maximum relative error of 11.11%, suggests the best throughput despite the target clock being relatively lower than our work, except for ADD and SUB (see Table II). Concerning only the exact arithmetic, our processor has advantages in DIV instruction as the number of required cycles is lower when the bit-width of the mantissa part is higher, compared to the best prior work, [15] (see Table II). In terms of area usage, our PAU and divider consume 11.00% and 57.87% fewer LUTs compared to the best prior works, the PAU of [15] and the divider logic of [13], respectively, demonstrating the compactness of our design (see Table III). Though evaluating the performance of different works with varying specifications, such as operating clock and ISA, is subjective, we consider our work as a promising approach to adopting posit arithmetic in lightweight systems due to the reduced area usage.

VII. CONCLUSION

This paper presented a lightweight processor and evaluation platform for rapid migration from IEEE-754 to posit. The key aspects we considered, versatility and usability, were provided by the following studies: designing the reconfigurable PAU focusing on the most intensive operations such as division and square root, designing the processor architecture supporting posit responsible for existing FP extensions, and constructing the platform to supply the swift conversion of the previous applications and fast evaluation. The experimental results showed that our work is suitable for lightweight systems due to the variable operating frequency and feasibility and, moreover, the reduced area usage compared to the previous studies. In future work, we plan to port a variety of applications to evaluate our designs in detail. Ultimately, we aim to fabricate the processor into a chip to confirm feasibility and compactness.

ACKNOWLEDGMENT

This research was supported by the MSIT (Ministry of Science and ICT), Korea, under the ITRC(Information Technology Research Center) support program (IITP-2023-RS-2022-00156295) supervised by the IITP(Institute for Information & Communications Technology Planning & Evaluation). This work was partly supported by Institute of Information & Communications Technology Planning & Evaluation (IITP) grant funded by the Korea government(MSIT) (2022-0-01013, Development of DRAM PIM semiconductor technology for enhanced computing function for edge, 50%).

REFERENCES

[1] J. Choquette, W. Gandhi, O. Giroux, N. Stam, and R. Krashinsky, "NVIDIA A100 tensor core GPU: Performance and innovation," *IEEE Micro*, vol. 41, no. 2, pp. 29–35, 2021.

[2] V. Popescu, M. Nassar, X. Wang, E. Tumer, and T. Webb, "Flexpoint: Predictive numerics for deep learning," in *2018 IEEE 25th Symposium on Computer Arithmetic (ARITH)*, 2018, pp. 1–4.

[3] Gustafson and Yonemoto, "Beating floating point at its own game: Posit arithmetic," *Supercomput. Front. Innov.: Int. J.*, vol. 4, no. 2, p. 71–86, jun 2017. [Online]. Available: https://doi.org/10.14529/jsfi170206

[4] Z. Carmichael, S. H. Fatemi Langroudi, C. Khazanov, J. Lillie, J. Gustafson, and D. Kudithipudi, "Deep positron: A deep neural network using the posit number system," in *2019 Design, Automation Test in Europe Conference Exhibition (DATE)*, 03 2019, pp. 1421–1426.

[5] J. Lu, C. Fang, M. Xu, J. Lin, and Z. Wang, "Evaluations on deep neural networks training using posit number system," *IEEE Transactions on Computers*, vol. 70, no. 2, pp. 174–187, 2021.

[6] N. Buoncristiani, S. Shah, D. Donofrio, and J. Shalf, "Evaluating the numerical stability of posit arithmetic," in *2020 IEEE International Parallel and Distributed Processing Symposium (IPDPS)*, 2020, pp. 612–621.

[7] H. W. Oh, W. S. Jeong, and S. E. Lee, "Evaluation of posit arithmetic on machine learning based on approximate exponential functions," in *2022 19th International SoC Design Conference (ISOCC)*, 2022, pp. 358–359.

[8] S. Moini, B. Alizadeh, M. Emad, and R. Ebrahimpour, "A resource-limited hardware accelerator for convolutional neural networks in embedded vision applications," *IEEE Transactions on Circuits and Systems II: Express Briefs*, vol. 64, no. 10, pp. 1217–1221, 2017.

[9] I. Pérez and M. Figueroa, "A heterogeneous hardware accelerator for image classification in embedded systems," *Sensors*, vol. 21, no. 8, 2021. [Online]. Available: https://www.mdpi.com/1424-8220/21/8/2637

[10] R. Chaurasiya, J. Gustafson, R. Shrestha, J. Neudorfer, S. Nambiar, K. Niyogi, F. Merchant, and R. Leupers, "Parameterized posit arithmetic hardware generator," in *2018 IEEE 36th International Conference on Computer Design (ICCD)*, 2018, pp. 334–341.

[11] A. A. Esmaeel, S. Abed, B. J. Mohd, and A. A. Fairouz, "Posit vs. floating point in implementing IIR notch filter by enhancing radix-4 modified booth multiplier," *Electronics*, vol. 11, no. 1, 2022. [Online]. Available: https://www.mdpi.com/2079-9292/11/1/163

[12] M. Cococcioni, F. Rossi, E. Ruffaldi, and S. Saponara, "A lightweight posit processing unit for RISC-V processors in deep neural network applications," *IEEE Transactions on Emerging Topics in Computing*, vol. 10, no. 4, pp. 1898–1908, 2022.

[13] D. Mallasén, R. Murillo, A. A. D. Barrio, G. Botella, L. Piñuel, and M. Prieto-Matias, "PERCIVAL: Open-source posit RISC-V core with quire capability," *IEEE Transactions on Emerging Topics in Computing*, vol. 10, no. 3, pp. 1241–1252, 2022.

[14] A. M Vaidyanathan, G. Bhairathi, and H. Hayatnagarkar, "PERC: Posit enhanced rocket chip," in *4th Workshop on RISC-V for Computer Architecture Research (CARRV)*, no. 8, May 2020.

[15] S. Tiwari, N. Gala, C. Rebeiro, and V. Kamakoti, "PERI: A configurable posit enabled RISC-V core," *ACM Trans. Archit. Code Optim.*, vol. 18, no. 3, Apr 2021. [Online]. Available: https://doi.org/10.1145/3446210

[16] V. Gohil, S. Walia, J. Mekie, and M. Awasthi, "Fixed-posit: A floating-point representation for error-resilient applications," *IEEE Transactions on Circuits and Systems II: Express Briefs*, vol. 68, no. 10, pp. 3341–3345, 2021.

Scaled Population Division for Approximate Computing

Kunal Bharathi*, Sunil P. Khatri† and Jiang Hu‡
Department of ECE, Texas A&M University
College Station, TX, USA
Email: *kunal-bharathi@tamu.edu, †sunilkhatri@tamu.edu, ‡jianghu@tamu.edu

***Abstract*—In this paper we present an approximate division scheme for Scaled Population (SP) arithmetic, a technique that improves on the limitations of stochastic computing (SC). SP arithmetic circuits are designed (a) to perform all operations with a constant delay, and (b) they use scaling operations to help reduce errors compared to SC circuits. As part of this work, we also present a method to correlate two SP numbers with a constant delay. We compare our SP divider with SC dividers, as well as fixed-point dividers (in terms of area, power and delay). Our 512-bit SP divider has a delay (power) that is $0.08\times$ ($0.06\times$) that of the equivalent fixed-point binary divider. Compared to a equivalent SC divider, our power-delay-product is $13\times$ better.**

***Index Terms*—Approximate Arithmetic, Stochastic Computing, Computer Arithmetic, Approximate Division, Fast Division**

I. Introduction

Applications in signal processing [1], machine learning [2] and real-time systems [3] require fast arithmetic circuits and can tolerate small errors. Approximate computing techniques take advantage of this tolerance to errors, trading computational accuracy for more power-efficient operations. A popular approximate arithmetic scheme is Stochastic Computing (SC) [4] [5]. SC has seen uses in high-throughput decoding of LDPC codes [6], low area/power Deep Neural Networks [7].

Stochastic computing (SC) uses simple logic circuits for arithmetic operations. However, SC has several limitations. SC has a runtime complexity of O(Π) (Π is the length of the SC *bit stream*). SC accuracy also depends on the number and distribution of '1's and '0's in the input bit streams. The range of values that the bit streams are able to represent is limited to [0, 1]. The authors of [8] address these limitations of SC in a scheme called Scaled Population (SP) arithmetic. SP is a low area/power overhead scheme that has a constant delay (in terms of gates), wider range of values and smaller errors than SC. SP addition and multiplication schemes were presented in [8], and an SP subtraction scheme was presented in [9]. However, no technique for SP division has been reported to date. The key contributions of our work are:

- We exploit the correlation of the bits in the SP-based input operands to perform division with logarithmic delay. As part of this work, we also present a novel approach to correlate two SP bit vectors with a constant delay (using the notion of "strong correlation" which means maximum overlap of the '1' bits of the two SP numbers).
- We also compare our SP division method to fixed-point dividers and SC dividers with respect to the 4 metrics. Our 512-bit SP divider has a delay (power) that is $0.08\times$ ($0.06\times$) times that of the equivalent (same numerical precision) fixed-point binary divider, and compared to a equivalent SC divider, our power-delay-product is $13\times$ better.
- We evaluate the error performance of approximate dividers using a novel metric called ENOB (Estimated Number of Bits)

This work is partially supported by the RTML program of the National Science Foundation, project CCF-1937396.

II. Background and Previous Work

SP supports a wider range of numbers and mathematical operations are designed to have a constant delay. A scaling term (exponent) is used by SP to represent a wider ranger of numbers. SP numbers are expressed as a 2-term tuple: $\{\sigma, \pi\}$, where σ is a Σ-bit term (σ is a binary number) and π is a Π-bit vector [8]. Next, we present an overview of existing division techniques in SC.

A. Division in Stochastic Computing

The authors of [10] proposed the first scheme for dividing two SC numbers. Their method exploited the logic of a JK Flip-Flop to perform division. The accuracy of the method is dependant on the operands (better results are obtained when the dividend is significantly smaller than the divisor). The design was improved by introducing a feedback element, resulting in more accurate results. However, they still require significantly long bit streams, resulting in a large delay. The authors of [11] present CORDIV, a scheme that relies on the correlation of bits in the inputs to perform division. Compared to previous approaches, CORDIV achieves better accuracy but the delay is still proportional to the length of the operand bit streams. DFSM-DIV [12] and CBDIV [13] are improvements to the CORDIV technique. However, in addition to having a delay that is proportional to the length of the input bit streams, another drawback of these schemes is that CORDIV, DFSM-DIV and CBDIV require binary inputs, and the result for DFSM-DIV and CBDIV are also in binary. These dividers, therefore, cannot be directly cascaded with other SC operators. Next, we will look at CORDIV in more detail, and subsequently present our idea.

979-8-3503-1176-1/23 $31.00 © 2023 IEEE

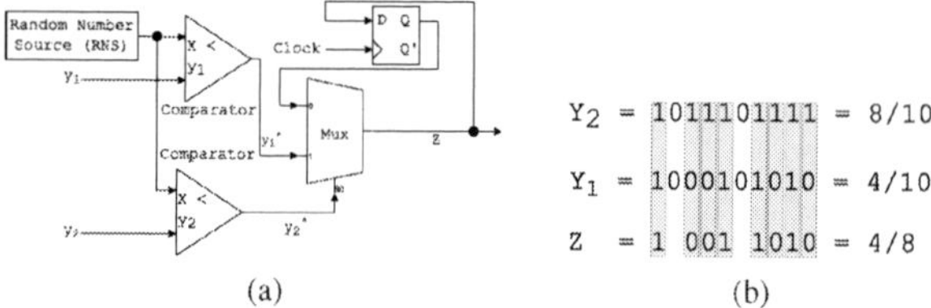

Fig. 1: CORDIV [11]

1) CORDIV: CORDIV [11] relies on *strong correlation* between two SC bit streams (the divisor (Y_2) and dividend (Y_1)) to perform division. Since we are dealing with SC numbers, a number is simply a bit stream. As shown in Figure 1a, the CORDIV unit reads one bit every clock cycle. Let $P(y_i)$ be the probability of observing a '1' at a given clock cycle in the SC bit stream Y_i. $P(y_1, y_2)$ is the joint probability of observing a '1' in both bit streams in the same clock cycle. From probability theory [14] we know that: $P(y_1|y_2) = \frac{P(y_1,y_2)}{P(y_2)}$. When $Y_1 < Y_2$ and both SC numbers are strongly correlated, the probability relation simplifies to $P(y_1|y_2) = \frac{P(y_1)}{P(y_2)}$. CORDIV achieves division by constructing a new SC bit stream Z, whose value is equal to $P(y_1|y_2)$, and therefore also equal to $\frac{P(y_1)}{P(y_2)}$. In Figure 1b illustrates the construction of output Z. From Y_1, we select only those bits whose corresponding bit in Y_2 is a '1' (the shaded columns represent Z, whose value is $\frac{y_1}{y_2}$). There are three major drawbacks that prevent CORDIV from being used in SP arithmetic:

1) The construction of Z violates the constant delay requirement of SP arithmetic.
2) Strong correlation negatively affects the error of addition and multiplication for SC or SP.
3) The output is of variable size (as shown in Figure 1a). The size of the output Z is determined by the number of '1's in Y_2.

In the next section, we present our SP design that builds on CORDIV, and overcomes these drawbacks.

III. An Overview of SP Division

In this section we will describe our SP divider. Figure 2 gives an overview of the SP division technique. Our method is divided into 4 *stages*. Each stage is designed to eliminate the drawbacks of CORDIV:

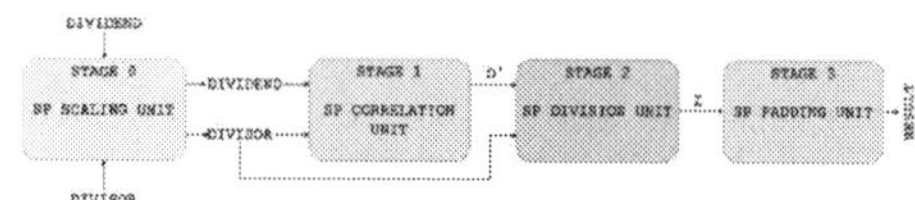

Fig. 2: SP Division

1) **Stage 0: The SP Scaling Unit** is used to modify the exponent of the input SP operands to ensure that the mantissa of the dividend $\pi_{dividend} \leq \pi_{divisor}$, the mantissa of the divisor.
2) **Stage 1: The SP Correlation Unit** is used to take the dividend and divisor SP bit streams as input, and output D', a SP bit vector which has the same value as the original dividend, but is now strongly correlated with the divisor.
3) **Stage 2: The SP Division Unit** takes as input D' and the divisor, and outputs Z, the variable length result.
4) **Stage 3: SP Padding Unit**. SP arithmetic requires the result to be of a fixed length. The SP Padding Unit takes the variable length Z and outputs a division result that has the same number of bits as the divisor and dividend.

In the next sections, we will study the design of each of the stages in more detail.

A. Stage 0: SP Scaling Unit

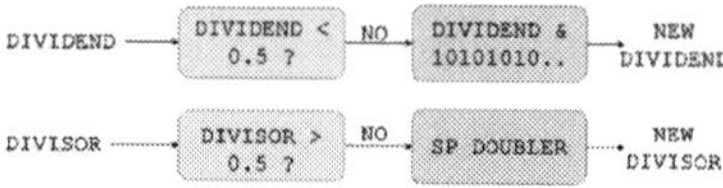

Fig. 3: Stage 0: SP Scaling Unit

CORDIV requires $\pi_{dividend} \leq \pi_{divisor}$. However, because SP has an exponent term, it is possible to have $dividend \leq divisor$, but $\pi_{dividend} \geq \pi_{divisor}$. We use stage 0 to adjust the π and σ of the 2 operands to ensure $\pi_{dividend} \leq \pi_{divisor}$. The logic of Stage 0 is described in Figure 3. SP arithmetic [8] has units that can compare numbers to fixed constants. We use these to implement Stage 0. If $\pi_{dividend} \geq 0.5$, we halve the value of the mantissa. This is easily achieved in constant time by using bit-wise logical AND as shown in Figure 3. To ensure that the value of the dividend is unchanged, the exponent is incremented by one if $dividend \geq 0.5$. Next, we want to modify the $\pi_{divisor}$ so that $\pi_{divisor} \geq 0.5$. To achieve this, we use SP population vector doublers from [9]. This unit doubles the numbers of ones in the π of its input. Therefore, we subtract one from the exponent, to keep the value of the divisor unchanged. The divisor needs to be doubled until $\pi_{divisor} \geq 0.5$. In the worst case, this process can take $(\log_2(\Pi) - 1)$ cycles. While all other stages are constant delay, this scaling of the divisor results in a logarithmic delay for our SP divider.

B. Stage 1: SP Correlation Unit

Given two SP numbers Y_1 and Y_2, where $Y_1 < Y_2$, we need to transform Y_1 such that its '1's are maximally overlapped (correlated) with the '1's in Y_2. The SP Correlation Unit is described in Algorithm 1. We next describe its steps.

1) Step 1 ($r = Y_1 \wedge Y_2$): If we perform a bit-wise logical AND operation between the SP numbers Y_1 and Y_2, then the result r will be strongly correlated to Y_2. However, numerically, $Y_1 \geq r$. The further r is from Y_1, the greater will be the error in our computation. For example, if $Y_1 = 11000$ and $Y_2 = 10011$, then $r = 10000$.

2) Step 2 (Fixing the Error): In the above example, the reason $r \leq Y_1$ is because Y_1 can have '1's in positions that Y_2 does not. We need to "collect" these ones and "insert" them back into to r. We achieve this with the help of the bit-wise difference operator ('$\setminus$') or bit-wise AND-NOT operation ($a \setminus b \Rightarrow a \wedge (\neg b)$). Let $t_1 = Y_1 \setminus r$. Continuing our example,

$t_1 = 01000$. t_1 represents the '1's in Y_1 that were "left behind" in step 1. Next, we shuffle (using hardwired connections to randomly transpose the bits) the bits of t_1 to obtain t'_1. We want to shuffle the bits in order to try and align the "left behind" '1's with '1's in Y_2. We obtain t_2 using: $t_2 = Y_2 \wedge t'_1$. The '1's in t_2 will always overlap with the '1's in Y_2. We now insert the '1's of t_2 back to r using the operation: $r = r \vee t_2$. We repeat Step 2 a certain number of times ($Rounds$) until $r \approx Y_1$. When $Rounds = 1$, we perform only Step 1. For $Rounds = 2$, we perform Step 1 once, and Step 2 once. The loop represents Step 2. In the circuit implementation, we unroll this loop to improve the circuit delay.

Algorithm 1 SP Correlation Unit

Require: $Y_2 \geq Y_1$, $\{Y_1, Y_2\} \in SP\ Numbers$
  $r \leftarrow bit\ vector\ of\ all\ 0s$ ▷ Output, Initially 0
  $Rounds \leftarrow From\ Circuit\ Designer$
  $r \leftarrow Y_1 \wedge Y_2$ ▷ Step 1
  for $i \in [2....Rounds]$ **do** ▷ Step 2
    $t_1 = Y_1 \setminus r$
    $t'_1 = shuffle(t_1)$
    $t_2 = Y_2 \wedge t'_1$
    $r = r \vee t_2$
  end for

C. Stage 2: SP Division Unit

Stage 1 provides us two operands that are now strongly correlated. In Stage 2 we perform our CORDIV-based division but with a key improvement over [11]: we will construct the same output, but with a constant delay. The logic for Stage 2 is depicted in Figure 4. In Figure 4 the SP bit vectors are of length 10 ($\Pi = 10$). The divisor (Y_2) has a numerical value of $\frac{8}{10}$, while the dividend (Y_1) has a numerical value of $\frac{4}{10}$. Since the numbers are strongly correlated, we can find the result of division by selecting the bits Y_1^i, whose corresponding i^{th} bit in Y_2, Y_2^i is a '1'. We accomplish this using a set of Π MUXes. As shown in the figure, the select line for the ith MUX is the Y_2^i bit (indicted by the blue connections), and the input selected when the select line is high is the Y_1^i (indicated by the red connections). The connection of the other input of the MUX will be discussed in the next section. The output of each MUX forms the Z^i bits. In Figure 4, the 8 bits of Z, corresponding to the 8 '1' bits in Y_2, are assigned the bit value from Y_1 via the MUXes. If we consider just these 8 bits, then the result of division is as expected ($\frac{4/10}{8/10} = \frac{4}{8} = \frac{1}{2}$). However, in SP, we cannot have variable length outputs, and the remaining bits need to be "padded" or filled in, with minimal affect to the number represented by Z. We present our scheme of doing this in the next section.

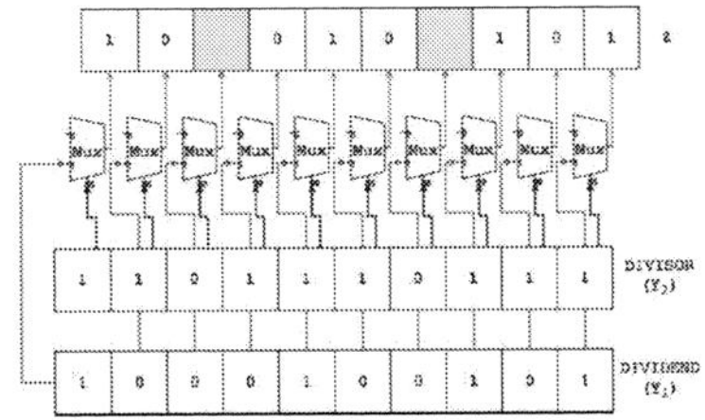

Fig. 4: Stage 2 - Unrolled CORDIV

D. Stage 3: SP Padding Unit

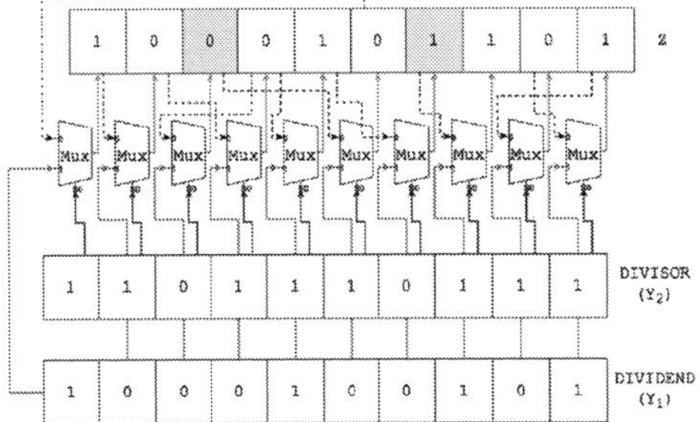

Fig. 5: Stage 2 + Stage 3 (Padding Circuit)

The easiest way to pad Z, is to fill the yellow entries of Figure 4 randomly with '1' bits or '0' bits. But, doing so will alter the value represented by the SP number Z. We want to fill the yellow entries of Figure 4 with a bit that has a probability $\frac{Y_1}{Y_2}$ of being a 1. If we sample a bit from an SP bit vector at random, the probability of sampling a '1' bit is equal to the value represented by the SP number. We use this property to pad the missing (yellow) bits of Z in Figure 4. In Figure 5, the previously unused MUX connections are randomly hard-wired using the dashed black lines. Now, if a bit is missing (yellow) from Z (the corresponding bit in Y_2 is a '0'), then this circuit *randomly* (with a probability $\approx \frac{Y_1}{Y_2}$ of being 1) selects a bit from a different part of Z. This way, we pad the result of stage 2, without affecting the value of the final padded result appreciably. In Figure 5, the "padded" bits are indicated using a yellow color. Therefore, the padded result Z (final result) now has a value $\frac{5}{10} = \frac{1}{2}$. We construct the hardwired feedback connections in a manner that guarantees the absence of cyclic dependencies whenever at most $\frac{\Pi}{2}$ divisor bits are '0'. Since the logic of Stage 0 ensures that $\pi_{divisor} \geq 0.5$, we avoid any cyclical dependencies in Stage 3. In the next sections we will study the performance of our divider using different metrics obtained via experimental simulations.

IV. Simulation Experiments and Results

A. Experimental Methodology

In Section IV-A1 we discuss our method of computing error, while Section IV-A2 describes our circuit implementation approach.

1) ENOB: The Effective-Number-of-Bits (ENOB) is used in analog circuits like ADCs or DACs [15], and represents the precision of any analog circuit scheme in terms of binary bits. We adopt this metric in the context of SP. Every Π-bit SP number has a precision that is at most $\log_2 \Pi$ binary bits. Therefore, the maximum ENOB for a Π bit SP number is $\log_2 \Pi$. The approximate nature of SP computations introduce errors in the output. If the RMSE (Root-Mean-Square Error) of the SP arithmetic operation is ϵ, then the number of SP bits affected by the error is $\epsilon * \Pi$. Therefore, the effective number of accurate SP bits available after the SP arithmetic operation is $\Gamma = \Pi - \epsilon * \Pi$. The new ENOB is therefore: $\log_2 \Gamma = \log_2(\Pi - \epsilon * \Pi)$. In the discussion that follows, we

979-8-3503-1176-1/23 $31.00 © 2023 IEEE

use the ENOB as one of the metrics to quantify the errors due to the approximate nature of our divider. The total number of *whole* bits of accuracy after accounting for approximation errors is $ENOB' = \lfloor (ENOB) \rfloor$.

2) Hardware Implementation: In order to find the circuit area, power, delay and energy of our SP divider design, we expressed the logic of our divider in Verilog [16]. The logic for all 4 stages consists of only combinational circuits (no flip-flops/latches). The design was then synthesized using Synopsys DC using the ASAP7 [17] technology library. Prior to synthesis, we allow Synopsys DC to optimize the logic of the top-module if it can. We used "compile_ultra" since we have tight timing constraints (we want fast circuits, since speed is a key driving philosophy of SP arithmetic). The area reported by DC is the cell area. The ASAP7 technology library has a well known 4x scale factor [18] and we apply this correction for all designs before reporting all our results. In order to compute the power numbers, we first create a switching activity file using Synopsys VCS for every design using 10000 test vectors. The activity file is then used by Synopsys DC to report the final power numbers. The delay is obtained from the timing report generated by Synopsys DC. The power report provides us with the total power consumption data.

B. Experimental Results

1) SP Divider Error and Hardware Implementation: In this section we will study the error performance of the entire SP divider and the hardware implementation of the SP divider logic.

Error Simulations: In order to study the error of the entire SP divider, we implemented the logic in python, and performed 100,000 simulations. For each simulation we pick a *dividend* randomly from a uniform distribution with range (0, 1) and a *divisor* randomly from a uniform distribution with range (dividend, 1).

ENOB		ROUNDS					Ideal ENOB
		1	2	3	4	5	
Π	8	2.52	2.62	2.68	2.71	2.71	3
	16	3.59	3.69	3.75	3.79	3.80	4
	32	4.63	4.72	4.79	4.83	4.84	5
	64	5.65	5.74	5.81	5.85	5.86	6
	128	6.66	6.75	6.81	6.86	6.88	7
	256	7.66	7.75	7.82	7.86	7.88	8
	512	8.66	8.75	8.82	8.87	8.89	9

TABLE I: ENOB for SP Divider

We perform these simulations for $\Pi = \{8, 16, 32, 64, 128, 256, 512\}$. For each Π, we simulated division for $Rounds \in \{1, 2, 3, 4, 5\}$. We omit the plots for $Rounds = 2$ for brevity.

Figure 6 plots the RMSE v/s Exact result for different Πs and *Rounds* values. The Exact Result here refers to the result assuming 0 errors, in other words, it is the division result calculated using an exact (not approximate) method. We use IEEE-754 floating point arithmetic for our exact results. In each of the figures, for a given population size, we plot 9 data points. Consider the last black diamond ($\Pi = 512$) in Figure 6a. This data point indicates that in our experiments, for the exact values in the range (0.9, 1], we obtained an average RMSE of 0.38. The behavior of these figures is as expected. The error performance improves with more *Rounds* (due to more accurate correlation) and also improves with Π (due to more mantissa bits). The only outliers are the plots of $\Pi = 8$, which we attribute to noise due to a very short mantissa. Observe that for $\Pi = 8$ the error is 0 when the exact result is close to 0 (which is much better than the error of larger Π). This is because the smallest value that can be represented by an SP number with $\Pi = 8$ is 0.125. Therefore, any division that is expected to result in a value less than 0.125 will always result in 0. For the same reason, any error in the range [0, 0.125), results in 0 error. An important point to note is that we are reporting average RMSE values, and approximate computations will always have certain inputs for which the errors will be significantly greater. However, in applications like ML, with millions of computations, the average error is more important than a few outliers.

For each Π and *Rounds* value, we use the average RMSE across the entire simulation to compute the ENOB values, using the formulas discussed in Section IV-A1. Table I summarizes the ENOB for all our simulations. We note that a reasonable value of *Rounds* beyond which the ENOB doesn't appreciably increase is 2 or 3.

Circuit Implementation: In the previous section we described the error performance for different configurations of the SP divider. Now we will study the area, power and delay of our SP divider's circuit implementation. Figure 7a plots the area performance of our SP divider. The graph behaves as expected, where the designs with larger Π and/or larger *Rounds* have a larger area footprint. The user can use the data of Figures 6 and 7 to choose a design based on the area-error trade-off. The circuits with larger area have a smaller error, as indicated by the corresponding ENOB value (↑ ENOB ⇒ ↓ Error) in Table I. The delay performance of the circuit is shown in Figure 7b. Our SP divider has a logarithmic delay due to Stage 0. In Figure 7b, we report the delay assuming that the scaling of the divisor requires a single iteration. With this assumption, the delay numbers should be independent of Π, and only depend on the *Rounds*. In Figure 7b we see that this holds true for $Rounds = \{1, 2, 3\}$. For $Rounds = \{4, 5\}$ this property is no longer precisely valid. This can be explained based on our implementation methodology. Since we allow Synopsys DC to optimize the overall Verilog of our design, it finds optimizations and chooses slightly different cells for the designs in question, leading to slightly different delays. Therefore, we observe some variations. Figure 7c reports the delay assuming the worst case scenario of requiring $(\log_2(\Pi) - 1)$ cycles to scale the divisor (we call this *Delay Max*). Observe that the plots for different Π no longer overlap due to the logarithmic dependency of the delay on Π. Again, based on these plots and the ENOB values (Table I), a user can choose the best design based on the delay-error (or Delay Max-error) trade-offs.

Figure 7d plots the power performance of our SP divider. The total power consumption of the circuit is shown. This graph behaves as expected, where the designs with larger

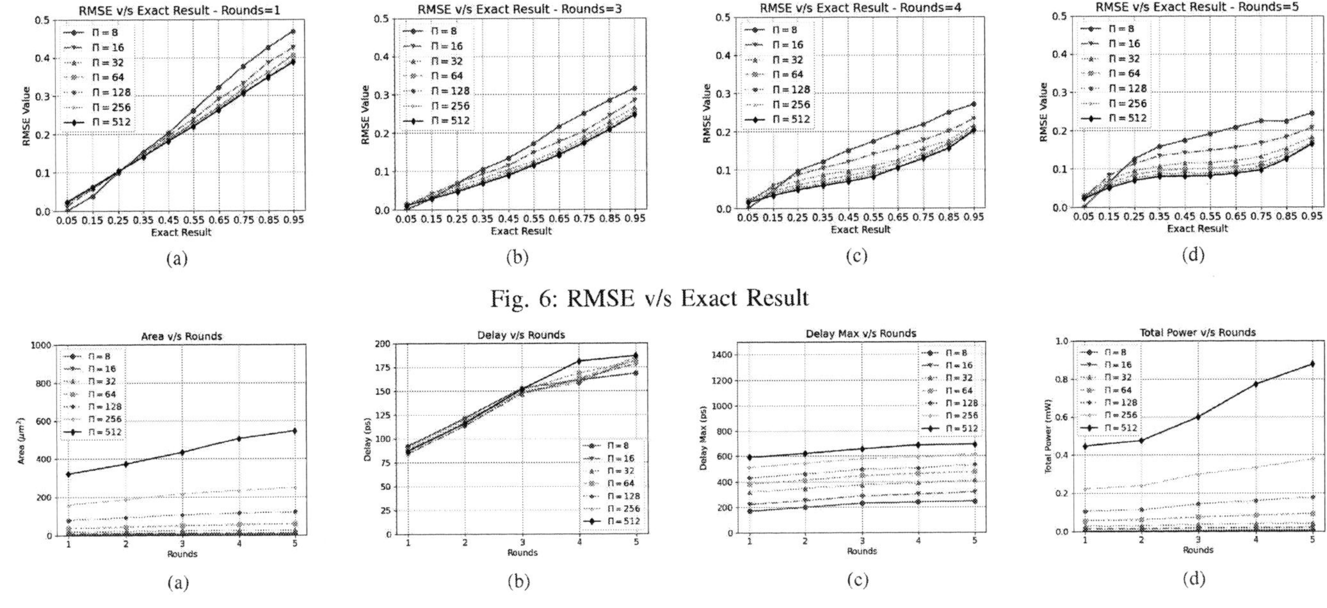

(a) (b) (c) (d)

Fig. 6: RMSE v/s Exact Result

(a) (b) (c) (d)

Fig. 7: Circuit Implementation of SP Divider

Π and/or larger *Rounds* have a larger power consumption. The user can use this data along with the data in Figure 6, to choose a design based on the power-error trade-off. The circuits with larger power have a smaller error, as indicated by the corresponding ENOB value (Table I).

Note that the incremental ENOB gains beyond $Rounds =$ 2 or 3 is small (Table I), and the area and power increase linearly with *Rounds*. This suggests that a practical value of *Rounds* is 2 or 3.

In the next section, we will see how our SP divider compares with other approximate and non-approximate dividers, reported in the literature.

2) Comparisons with Alternate Methods: In this section we will study how our SP divider performs relative to other approximate and non-approximate (precise) dividers. First, we will benchmark the area, power and delay of the SP divider against a fixed-point precise divider. Next, we compare the SP divider against dividers in the SC domain.

CFPD (Combinational-Fixed-Point-Divider)			
Bit Width	Delay(ps)	Area(μm^2)	Total Power(mW)
2	76.43	0.73	0.42
3	201.73	2.08	1.06
4	505.10	3.66	2.11
5	749.43	5.92	3.53
6	1012.58	9.20	5.46
7	1348.88	12.16	7.56
8	1785.82	17.50	10.81
9	2099.32	21.51	13.17

TABLE II: CFPD Circuit Implementation

Fixed-Point Divider: In order to compare our divider against a "precise" divider, we chose a fixed point divider design from the Synopsys DesignWare Library. The division logic used in these precise dividers relies on the Newton-Raphson [19] approximation technique. As a result, a purely combinational divider design will require cascaded CPAs (Carry-Propagation-Adders) and the area, power and delay numbers increase super-linearly with operand bit-width. Synopsys therefore provides both combinational and sequential dividers, and recommends the use of sequential dividers for operand bit-width ≥ 16.

For a fair comparison, we compare dividers of *equal precision*. The precision of an SP divider in terms of binary bits is determined by ENOB', and the precision of a precise divider is determined by its bit-width. The highest ENOB value across our designs is 8.89 (referring to the values in Table I). Therefore, we synthesize fixed-point dividers with operand bit-widths of ≤ 9, and use the Combinational-Fixed-Point-Divider (*CFPD*) component in the Synopsys DesignWare library, to compare our work with.

Table II reports the delay, area and total power for the CFPD for input operand bit-widths in the range [2, 9]. Table III presents the ratio the area, power and delay for SP dividers with an ENOB' of 8, with the corresponding CFPD design (with a bit-width 8). Each row in the table corresponds to a SP divider design (with a different *Round* value). The values reported are ratios of the quantity for the SP divider to the corresponding quantity for the CFPD. The SP divider is clearly much faster than the CFPD, on average having a delay (delay max) that is just $0.08\times$ ($0.36\times$) that of the CFPD delay. The CFPD has a large delay because of a long critical path through the CPAs in the design. This speed-up is at the cost of higher area than the CFPD, with the SP divider being on average $25\times$ larger. A key reason for this increase is that the SP design is required to have an $O(1)$ delay (in terms of gates) per round. The power for the SP divider is again much better ($0.06\times$ on average) than the power for the CFPD. The main reason for such a large power difference is the large dynamic power component required in the CFPD.

SP Divider / CFPD Performance Ratios (Equivalent ENOB)						
Π	ENOB'	Rounds	Delay Ratio	Delay Max Ratio	Area Ratio	Power Ratio
512	8	1	0.05	0.33	18.37	0.04
512	8	2	0.07	0.35	21.36	0.04
512	8	3	0.09	0.37	24.79	0.06
512	8	4	0.10	0.38	29.00	0.07
512	8	5	0.10	0.39	31.30	0.08
Average:			**0.08**	**0.36**	**24.97**	**0.06**

TABLE III: SP Divider v/s CFPD (for $ENOB' = 8$)

CBDIV				
Binary Input Size	Area(um^2)	Delay(ps)	Latency(ps)	Total Power (mW)
4	4.58	74.83	1.20E+03	0.15
8	11.39	87.38	2.24E+04	0.25
16	20.76	104.83	6.87E+06	0.31
32	44.89	127.60	5.48E+11	0.49
64	92.22	154.69	2.85E+21	0.79

TABLE IV: CBDIV Circuit Implementation

SC Dividers: Next, we compare the performance of our SP divider against an approximate divider from the SC domain. We chose CBDIV [13], a very recent work that builds on CORDIV in the SC domain, as our benchmark circuit. Using the same methodology as before, we implemented the circuits for CBDIV, in order to get the area, delay and power numbers. The result of our synthesis is shown in Table IV. The inputs to CBDIV are binary numbers (they convert from binary to SC as part of the design). Therefore, the first column in the table refers to the size of input operands in the binary number system. The corresponding SP divider is one with an ENOB' that matches the binary input size. CBDIV is a sequential circuit that uses an FSM to perform the correlation that is required for the underlying CORDIV logic. *Latency* here refers to the total amount of time required to generate the result of division. Latency is calculated using the formula: $Delay * Cycles$. The number of cycles required by the sequential circuit of CBDIV is 2^N, where N is the number of binary bits. Table V compares our SP divider with CBDIV (for $ENOB' = 8$). Comparing equivalent CBDIV and SP division circuits we see that SP division has much better delay max and power-delay-max-product numbers, and CBDIV has better area and power numbers. This is expected, as CBDIV is designed to be a serial circuit that achieves small area (and power) at the cost of long computation time. Note that the CBDIV design includes the circuitry to convert binary numbers to SC numbers and back. Chaining together successive division operations in CBDIV incurs expensive conversions to-and-from the binary number system.

V. Conclusion

In this paper we presented an approximate division technique called SP division. Prior to this work, there was no division scheme in SP arithmetic. We provide the user/designer different SP division design variants, allowing the designer to chose based on the specific area-power-delay-error trade-offs of the target application. We have shown that on average, and SP divider with ENOB' of 8, is 12× faster, and consumes 16× less power, than the corresponding Fixed-Point-Combinational-Divider. We also developed a constant delay method to correlate two SP numbers.

SP Divider / CBDIV Performance Ratios (Equivalent ENOB)							
Π	ENOB'	Rounds	Delay Ratio	Delay Max Ratio	Area Ratio	Power Ratio	Power Delay Max Product (PDP) Ratio
512	8	1	<0.01	0.03	28.22	1.79	0.053
512	8	2	0.01	0.03	32.81	1.91	0.057
512	8	3	0.01	0.03	38.09	2.40	0.072
512	8	4	0.01	0.03	44.55	3.09	0.092
512	8	5	0.01	0.03	48.08	3.51	0.105
Average:			**0.01**	**0.03**	**38.35**	**2.54**	**0.075**

TABLE V: SP Divider v/s CBDIV (for $ENOB' = 8$)

References

[1] R. St Amant, A. Yazdanbakhsh, J. Park, B. Thwaites, H. Esmaeilzadeh, A. Hassibi, L. Ceze, and D. Burger, "General-purpose code acceleration with limited-precision analog computation," *ACM SIGARCH Computer Architecture News*, vol. 42, no. 3, pp. 505–516, 2014.

[2] D. S. Khudia, B. Zamirai, M. Samadi, and S. Mahlke, "Rumba: An online quality management system for approximate computing," in *ISCA*. IEEE, 2015, pp. 554–566.

[3] Z. Wang, S. Mohajer, and K. Bazargan, "Low latency parallel implementation of traditionally-called stochastic circuits using deterministic shuffling networks," in *ASPDAC*, 2018, pp. 337–342.

[4] B. R. Gaines, "Stochastic Computing," in *Proceedings of the Joint Computer Conference*, 1967, pp. 149–156.

[5] A. Alaghi and J. P. Hayes, "Survey of Stochastic Computing," *ACM Trans. Embed. Comput. Syst.*, vol. 12, no. 2s, may 2013. [Online]. Available: https://doi.org/10.1145/2465787.2465794

[6] S. Sharifi Tehrani, W. Gross, and S. Mannor, "Stochastic decoding of LDPC codes," *IEEE Communications Letters*, vol. 10, no. 10, pp. 716–718, 2006.

[7] A. Ardakani, F. Leduc-Primeau, N. Onizawa, T. Hanyu, and W. J. Gross, "VLSI Implementation of Deep Neural Network Using Integral Stochastic Computing," *IEEE Transactions on Very Large Scale Integration (VLSI) Systems*, vol. 25, no. 10, pp. 2688–2699, 2017.

[8] H. Zhou, S. P. Khatri, J. Hu, and F. Liu, "Scaled Population Arithmetic for Efficient Stochastic Computing," in *2020 25th Asia and South Pacific Design Automation Conference (ASP-DAC)*, 2020, pp. 611–616.

[9] K. Bharathi, J. Hu, and S. P. Khatri, "Scaled Population Subtraction for Approximate Computing," in *2020 IEEE 38th International Conference on Computer Design (ICCD)*, 2020, pp. 348–355.

[10] B. R. Gaines, "Stochastic computing systems," in *Advances in Information Systems Science*. Springer, 1969, pp. 37–172.

[11] S.-I. Chu, "New Divider Design for Stochastic Computing," *IEEE Transactions on Circuits and Systems II: Express Briefs*, vol. 67, no. 1, pp. 147–151, 2020.

[12] N. Temenos and P. P. Sotiriadis, "Deterministic Finite State Machines for Stochastic Division in Unipolar Format," in *2020 IEEE International Symposium on Circuits and Systems (ISCAS)*, 2020, pp. 1–5.

[13] S. Yu, Y. Liu, and S. X.-D. Tan, "Approximate Divider Design Based on Counting-Based Stochastic Computing Division," in *2021 ACM/IEEE 3rd Workshop on Machine Learning for CAD (MLCAD)*, 2021, pp. 1–6.

[14] F. Dekking, C. Kraaikamp, H. Lopuhaä, and L. Meester, *A Modern Introduction to Probability and Statistics: Understanding Why and How*, ser. Springer Texts in Statistics. Springer, 2005. [Online]. Available: https://books.google.com/books?id=XLUMIIombgQC

[15] K. Scott and S. P. Khatri, "Flash-based Digital to Analog Conversion," in *2022 IEEE 40th International Conference on Computer Design (ICCD)*, 2022.

[16] "IEEE Standard Verilog Hardware Description Language," *IEEE Std 1364-2001*, pp. 1–792, 2001.

[17] L. Clark, V. Vashishtha, L. Shifren, A. Gujja, S. Sinha, B. Cline, C. Ramamurthy, and G. Yeric, "ASAP7: A 7-nm FinFET predictive process design kit," *Microelectronics*, vol. 53, pp. 105–115, Jul. 2016.

[18] L. T. Clark, V. Vashishtha, D. M. Harris, S. Dietrich, and Z. Wang, "Design flows and collateral for the ASAP7 7nm FinFET predictive process design kit," in *2017 IEEE International Conference on Microelectronic Systems Education (MSE)*, 2017, pp. 1–4.

[19] Wikipedia contributors, "Newton's method — Wikipedia, the free encyclopedia," 2022, [Online; accessed 19-May-2022]. [Online]. Available: https://en.wikipedia.org/w/index.php?title=Newton%27s_method&oldid=1087753056

Cryogenic CMOS as an Enabler for Low Power Dynamic Logic

Rakshith Saligram
School of Electrical & Computer Engineering
Georgia Institute of Technology
Atlanta, USA
rakshith.saligram@gatech.edu

Suman Datta
School of Electrical and Computer Engineering
Georgia Institute of Technology
Atlanta, USA
sdatta68@gatech.edu

Arijit Raychowdhury
School of Electrical and Computer Engineering
Georgia Institute of Technology
Atlanta, USA
arijit.raychowdhury@ece.gatech.edu

***Abstract*—Cryogenic High-Performance Computing (HPC) has gained traction for server and cloud systems which demand large scale, energy efficient and fast computing systems. Dynamic logic satisfies these goals and at cryogenic temperature, its inherent problems of charge leakage are readily addressed thanks to the exponential reduction in subthreshold leakage currents. Fully Depleted Silicon on Insulator (FDSOI) devices present an additional "dial" of back gate biasing which opens multiple design options and solutions to further enhance the circuit power-performance metrics. In this paper we present a solution – selective back gate biasing – applied to dynamic and domino logic circuits, to increase their energy efficiency and/or performance. With the proposed method, we show up to 48% decrease in delay at constant energy and 41% decrease in energy at constant delay at 77K compared to 300K. We further scale up the circuit to a radix-4 sparse-2 64 bit adder where the proposed technique increases energy efficiency by 53% and/or performance by 56% going from 300K to 77K.**

***Keywords*— Adder, Back Gate Biasing, Cryogenic CMOS, Dynamic Logic, High Performance, Low Power, Threshold Voltage.**

I. Introduction

Low Temperature CMOS or Cryogenic CMOS is a promising technology for high performance server and cloud computing applications [1-2] due to performance boosters such as enhanced carrier mobility, steeper subthreshold slope [3], improved reliability, reduced interconnect resistance [4] and lower self-heating. Cryogenic CMOS also plays a key role in several other applications including but not limited to space electronics, astronomical detectors, metrology and interface circuits to quantum computers [5-9]. High frequency circuits can utilize the dynamic CMOS logic for better performance. Compared to static CMOS circuits, dynamic CMOS circuits consume lesser area, have lower delays and are more power efficient. Despite these advantages, room temperature dynamic circuits have high charge leakage which along with lower noise margins makes their usage difficult for reliable operation. Also, since they need to be periodically precharged at a minimum rate (which is again dictated by leakage currents), they are set to operate at a lower cut-off frequency which makes their testing significantly difficult. However, dynamic circuits at cryogenic temperature can address these issues thanks to the ultra-low leakage.

Many technologies have been deployed at cryogenic temperatures including bulk CMOS, FinFETS, SiGe HBT etc., In these, the increase in the drive current arising from enhanced carrier mobility is partially deteriorated due to increase in threshold voltage. One particular technology where this increase in threshold voltage can be compensated by applying body/back gate/substrate bias is the Fully Depleted Silicon on Insulator (FDSOI). In this paper, we characterize production grade foundry 22nm FDSOI devices from Global Foundries at room and cryogenic temperatures and study the effect of back gate bias at 77K. BSIM4 models are calibrated to this device data to accurately simulate the behavior at low temperature.

We next propose a technique of applying selective back gate bias to the device in the pull down network of the dynamic CMOS logic in order to tune their threshold voltage and enhance the performance. We extend the proposed technique to a radix-4 sparse-2 64 bit carry chain and demonstrate energy and performance benefits at 77K. The rest of the paper is organized as follows: in section II FDSOI device behavior and effect of back gate biasing is studied. Section III introduces the proposed technique of selective back gate biasing and analyzes the energy/performance trade-offs. Section IV shown the implementation of the 64 bit adder with domino logic under the proposed technique followed by conclusions.

II. Cryogenic CMOS

A. 22nm FDSOI Device I-V Characteristics and Modelling

The transistors are fabricated in 22nm-FDX [10] Fully Depleted Silicon on Insulator (FDSOI) technology from Global Foundries using standard foundry process. The test chips are measured using Lakeshore CPX-VF Cryogenic Probe Station from 300K to 4K. The dc characterization was performed using Keithley 4200 SCS parameter analyzer. The *measured* device I_{DS}-V_{GS} characteristics for 300K and 77K for both NMOS and PMOS type devices in the linear and saturation regions is shown in Fig. 1 (Symbols). The I_{DS}-V_{GS} curves clearly indicate several orders of magnitude reduction ($> 10^4\times$) in the sub-threshold leakage current (I_{OFF}) as well as improvement in the device ON current (I_{ON}) for both NMOS and PMOS (~30%) due to enhanced carrier mobilities resulting from reduced phonon scattering.

The subthreshold slope (*SS*) factor reduces (from ~80mV/dec (76mV/dec) at 300K to ~20mV/dec (20 mV/dec) at 77K for PMOS (NMOS). Going lower in temperature does not yield better *SS* due to higher interface trap charges and band tail effects [3]. Thus, 77K promises to be the ideal sweet spot temperature for reaping low temperature benefits and apply them to high performance computing. However, the threshold

This work is sponsored by Defense Advanced Research Project Agency (DARPA) Low Temperature Logic Technology (LTLT) project. Any opinions, findings, conclusions expressed in this material are those of the authors and do not necessarily reflect the views of DARPA.

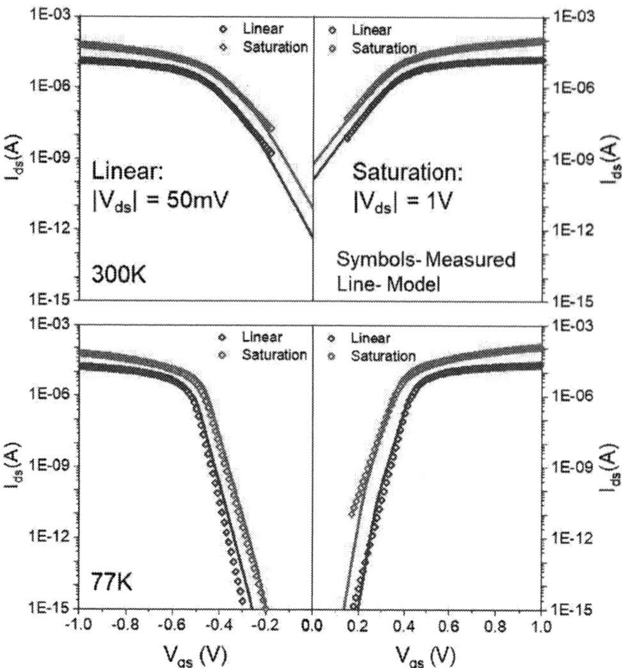

Fig. 1. Calibrated BSIM models for 22nm FDSOI measured device I_{DS}-V_{GS} for linear and saturation regions at 300K and 77K for NMOS and PMOS. Symbols indicate measurement, line indicates models. The voltage levels for V_{DS} are 50mV for linear region and 1V for saturation region.

voltage (V_{th}) of the transistors show an upward shift of approximately 120mV and 160mV for NMOS and PMOS going from 300K to 77K due to increase in the Fermi potential at lower temperature. This increase in V_{th} leads to reduced voltage headroom for operating the circuits at low temperature diminishing the benefits of increased I_{ON}. In the next sub-section, we shall see how to mitigate this effect for FDSOI devices in particular.

First, in order to be able to simulate the circuits at low temperature, we calibrate BSIM models to the measured device characteristics using BSIM ProPlus and flow explained in [11]. The parameters are individually fit to the temperature points so as to reduce the error of fit. We tune a minimal number of crucial parameters like threshold voltage (*VTH0*), saturation velocity (*VSAT*), low field carrier mobility (*U0*), first order body bias coefficient (*K1*), Drain Induced Barrier Lowering (DIBL) coefficient (*ETA0*), DIBL coefficient in subthreshold region (*DSUB*), interface trap capacitance (*CIT*), and drain source resistance (*RDSW*) that control the device behavior. The percentage of fitting error is < 5.8%. The calibrated BSIM models are also shown in Fig. 1 (Lines).

B. Effect of Substrate Biasing

FDSOI devices present with an excellent substrate/body control which helps to steer the threshold voltage of the transistors adding more flexibility to circuit design allowing to switch from between high performance (forward body bias – FBB) and ultra-low leakage (reverse body bias – RBB) modes on the fly. For an NMOS device, applying a positive body bias ($V_{BS} > 0$) will reduce the depletion charge and hence the threshold voltage as given by a common expression in (1). Here, γ is called the body bias coefficient/back gate coefficient, ϕ_F is

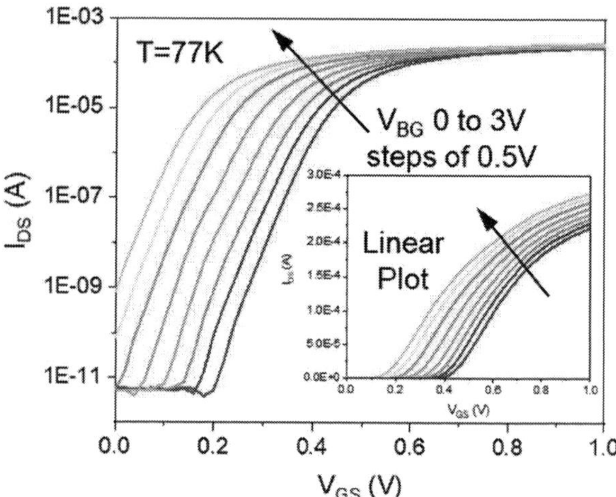

Fig. 2. Measured effect of back gate bias V_{BG} on current at cryogenic temperature of 77K- the saturation current increases while the subthreshold leakage current I_{OFF} increases exponentially with increasing V_{BG}

the Fermi potential and V_{th0} is the threshold voltage at zero body bias.

$$V_{th} = V_{tho} + \gamma.\left(\sqrt{|-2\phi_F + V_{SB}|} - \sqrt{|-2\phi_F|}\right) \qquad (1)$$

The resultant reduction in the V_{th} with applied body bias (denoted by V_{BG} henceforth, which is same as V_{BS}), will lead to higher gate overdrive for the same value of supply voltage V_{DD}. This causes the device I_{ON} to increase. Shown in Fig. 2 is the *measured* transistor I_{DS}-V_{GS} for an NMOS in the saturation regime at 77K for different values of back gate voltage V_{BG}. As seen from the linear plot inset, the current increases. However, the decrease of V_{th} with increasing V_{BG} has another effect – the increase in the subthreshold leakage current. This can be seen from the increase in the Y-intercept (I_{OFF}) on the logarithmic scale. The I_{OFF} for lower value of V_{BG} do not show variation which is mainly because of the measurement setup sensitivity. Further, the BSIM models are again calibrated to incorporate the effects of back gate biasing. One needs to strike a balance between leakage current and device ON current in order to achieve best performance.

III. Dynamic Logic at Cryogenic Temperatures

Dynamic logic has been known for its high performance capabilities albeit higher power consumption due to pre-charge event at every clock cycle. It also suffers from charge leakage problems which have been mitigated by use of weak keeper transistors and other methods. Before diving into the details of cryogenic behavior of dynamic logic, we need to understand the relationship between back gate bias, threshold voltage and leakage current.

A. V_{th} tuning at Cryogenic temperature with back gate bias

As seen from section II, by applying back gate bias V_{BG} to the transistor, the threshold voltage can be varied. We *experimentally* applied different V_{BG} to the NMOS transistor in saturation and extracted the V_{th} at multiple temperature points from the I_{DS}-V_{GS} curves. This is plotted in Fig. 3. A line parallel to X-axis indicates a constant Y (V_{th}) value and is denoted as *Iso-V_{th}*. This line's intersection with the different temperature curves

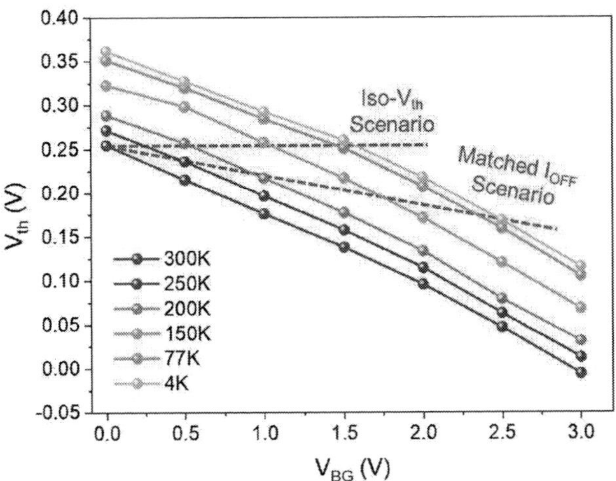

Fig. 3. Measured variation of threshold voltage V_{th} with applied back gate bias V_{BG} across temperature. Intercepts of curves with Iso-V_{th} and matched I_{OFF} lines at a given temperature indicate value of V_{BG} that needs to be applied to achieve the corresponding scenario.

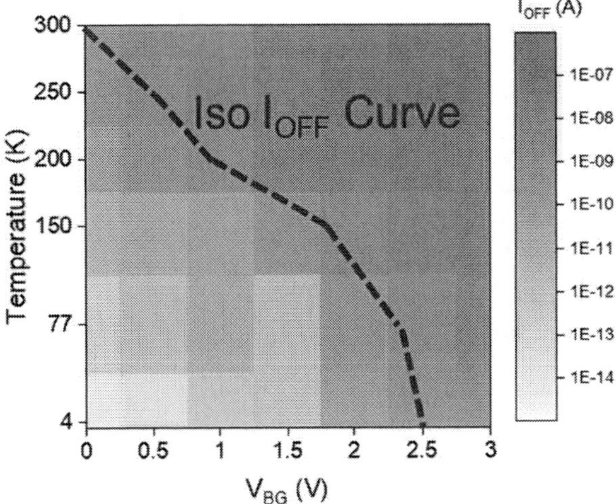

Fig. 4. Thermal plot of subthreshold leakage current as a function of back gate bias V_{BG} across temperature and intercept curve drawn to indicate *Iso-I_{OFF}* scenario.

indicate the amount of back gate bias required to achieve the threshold voltage same as that of room temperature (300K). For instance, applying V_{BG} of 1.5V at 77K will yield same V_{th} at 77K as that of a 300K device.

We can also envision another scenario where instead of matching the V_{th} of the devices across temperature, we can match their subthreshold leakage currents. To do this, we use the calibrated models to evaluate the leakage current (since the measurement setup has lower sensitivity to I_{OFF} at lower V_{BG} at cryogenic temperatures) as a function of temperature and V_{BG} to obtain the thermal plot shown in Fig. 4. If we trace the points with same I_{OFF} values across temperature at various V_{BG} values, we get the curve indicated. To better understand, these V_{BG} values are connected on plot in Fig. 3 to get the matched I_{OFF} scenario curve. We can notice that the corresponding threshold voltages decrease with decrease in temperature compared to 300K. Thus, the V_{th} can be decreased at 77K (say for example) compared to 300K, without increase in subthreshold leakage yielding higher voltage headroom and hence higher device performance for iso gate overdrive voltage.

B. Achieving balance in Dynamic Logic

Dynamic CMOS logic offers many advantages including lower area (arising from lower transistor count), higher speed of operation, glitch free behavior etc., However, the charge leakage is one of the major problems which makes circuit realization harder. Another drawback is that dynamic circuits cannot operate at lower frequencies making testability harder. However, addition of keeper/bleeder transistors to restore the leaked charge has proven to be an effective technique but requires additional area. These problems can be easily addressed at cryogenic temperatures due to the near absence of leakage as seen from the previous section.

Consider a simple 2 input NAND gate realized in dynamic logic as shown in Fig. 5(a). It consists of two precharge transistors (1 PMOS and bottom most NMOS) connected to the precharge clock signal φ. During the precharge phase (PCH), φ is low, turning ON the PMOS and charging the output node to V_{DD}. During the evaluate phase (EVAL), φ is high, turning the pull down NMOS ON and allowing the inputs to set the output. Since the temperature change affects all transistors, the leakage of NMOS transistors which have signal inputs also decreases with increasing V_{th} degrading the evaluation time (at a given V_{DD} – due to lower gate overdrive) which is undesirable. We propose a method of selectively applying back gate (V_{BG}) bias to these transistors as shown in Fig. 5(b) (highlighted) to increase their performance. In this way, the header and footer transistors are still having considerably low leakage at 77K alleviating the problem of charge leakage and hence removing the requirement of bleeder/keeper transistor, while the active devices which are driven by signal inputs can still help achieve better performance. A generic Boolean logic implementation with the proposed method is shown in Fig. 5(c).

C. Energy – Performance Analysis of Proposed Circuits

The proposed technique of applying back gate bias voltage creates an effect similar to using Dual-Threshold or Multi-Threshold CMOS (DTCMOS/MTCMOS), and it can be argued that one can use different V_{th} flavor cells during the design. However, the proposed technique is advantageously distinguishable due to the following reason – (i) the solution is temperature scalable in the sense we can tune the threshold voltage to the value that is required now making V_{th} a control knob instead of a design parameter; (ii) we can tweak additional performance thanks to the higher gate overdrive voltage available at disposal; (iii) We can make the circuit more energy efficient by operating at lower V_{DD} while meeting the speed requirement; (iv) leakage control can be possible at room temperature with reverse body biasing.

Consider a simple 2 input NAND gate driving a load of 1fF output capacitance. The transistors in the pull up and pull down network have been sized to achieve equal strengths. However, the PMOS sizing can be optimized depending on the time available for precharging (subject to pulse width and cycle time of φ). The two transistors with signal inputs have back gate voltage V_{BG} applied to tune their V_{th}. Next, we parametrically

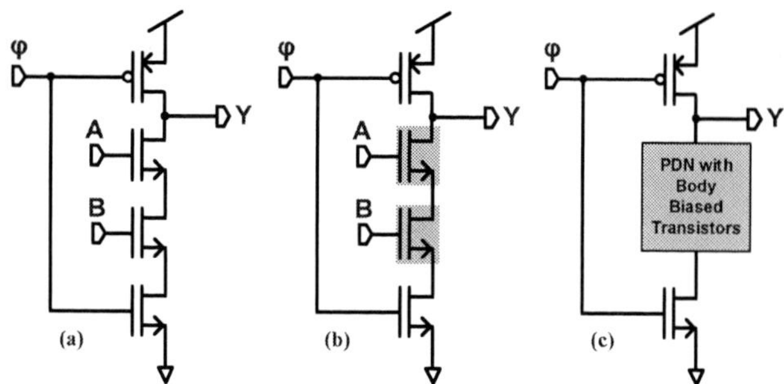

Fig. 5. (a) Conventional (b) Proposed – Transistor implementation of dynamic 2 input NAND gate with footer device. The highlighted devices are V_{th} tuned by applying V_{BG} to have *Iso-*I_{OFF} as 300K devices. (c) Generic dynamic logic realization of boolean function using proposed scheme.

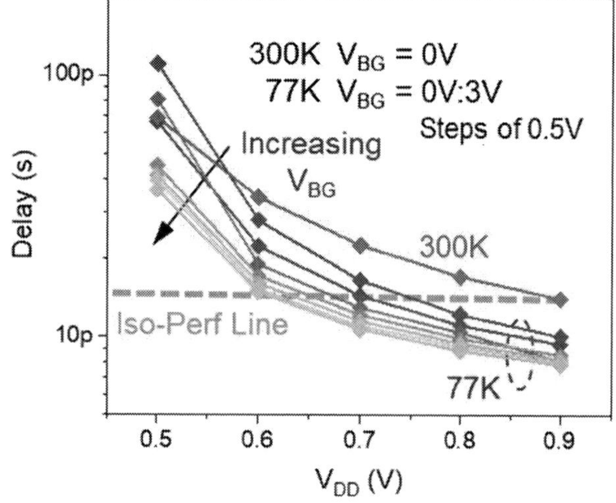

Fig. 6. Simulated variation of delay with supply voltage V_{DD} for different back gate bias voltages V_{BG} at 77K compared with 300K at zero V_{BG}. The Iso-Performance line indicates that for similar delay, we can reduce V_{DD} by ~33%.

sweep the supply voltage V_{DD} and measure the rise/fall delay which is the delay between the rise/fall of input at its 50% value to rise/fall of output at its 50% value. The simulated blue curves at 77K in Fig. 6 show an increased delay with decreasing V_{DD} due to lower ON current for discharging the output capacitance. With increasing V_{BG} this delay value reduces due to reduction in V_{th} thereby creating higher voltage headroom, while maintaining the trajectory with V_{DD}. We also plot the delay vs supply voltage for 300K with zero V_{BG} for reference. An Iso-performance line drawn parallel to the X-axis indicates that we can reduce the supply voltage at 77K by more than 30% compared to 300K while still achieving same delay metrics. At 77K, under iso-V_{DD}, we can obtain 26% improvement in delay with zero V_{BG}, 53 % with V_{BG} of 3.0V and 47% with V_{BG} tuned for iso-I_{OFF} compared to 300K.

We next calculate the energy expended in the switching activity for each of these V_{DD}/V_{BG} points and obtain the energy vs delay plot as shown in Fig. 7. For a given V_{BG}, as we decrease the supply voltage V_{DD}, the delay increases exponentially because the discharge current decreases. Also, since the switching energy is proportional to the square of supply voltage ($E_{sw} \propto C.V_{DD}^2$), the energy decreases. With increasing V_{BG}, the V_{th} decreases, so for the same V_{DD}, we create higher overdrive

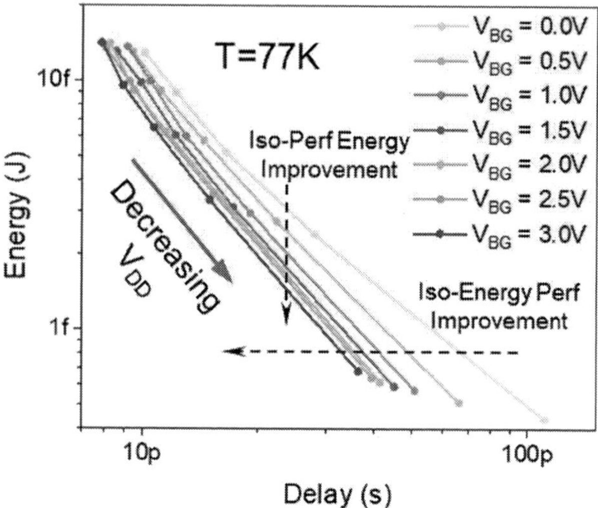

Fig. 7. Simulated energy delay plot of dynamic 2 input NAND gate at 77K at different back gate bias voltages V_{BG} showing performance improvement at iso energy and energy reduction at iso performance with increasing V_{BG}.

voltage which increases the current, decreases the delay. However, there is a small increase in the capacitance that is switched due to additional drain-body, source-body capacitances (C_{BS}, C_{DB}) that are developed when the back gate bias is applied which slightly increases the switching energy. This can be noticed as the slight upward drift of the data points with increasing V_{BG}. However, this is very small compared to the energy benefits that are obtained by operating at lower V_{DD} seen as iso-performance energy improvement line in Fig. 7. Similarly, we can also obtain performance benefits by operating at iso-energy. We see 48% improvement in delay under iso energy and 41% improvement in energy at iso delay. We need to note that the cooling cost of the system has not been considered in calculating these improvements.

IV. 64 Bit Domino Logic Adder

In this last section, we will examine the domino logic implementation of a 64 bit adder. Domino logic is a variation of dynamic logic which is used to overcome the cascading problem wherein the concurrent evaluation of subsequent stages might lead to erroneous output at the last stage due to propagation delays from the previous stage. In domino logic, this problem is averted by two techniques – (i) addition of inverter after every dynamic stage (ii) driving cascaded stages with delayed precharge signals. Using skewed precharge signals also minimizes the cross-over current between the two dynamic stages. One disadvantage of domino logic is that it can implement only true functions but not complements.

Carry-look-ahead techniques involve parallel calculation of groups of carries (in this case, 4 at a time) in a modular fashion that reduces the carry calculation time to $\log_r[n]$ +2 gate delays where r is the group size and n is the width of the adder. The potential advantage of using Ling's equations comes after unrolling the recursions. Ling's pseudo-carry equations [12,13] reduce transistor stack height in the first stage of the carry tree however increasing the complexity of sum pre-computation. The radix/valency of a carry tree is the number of carries merged

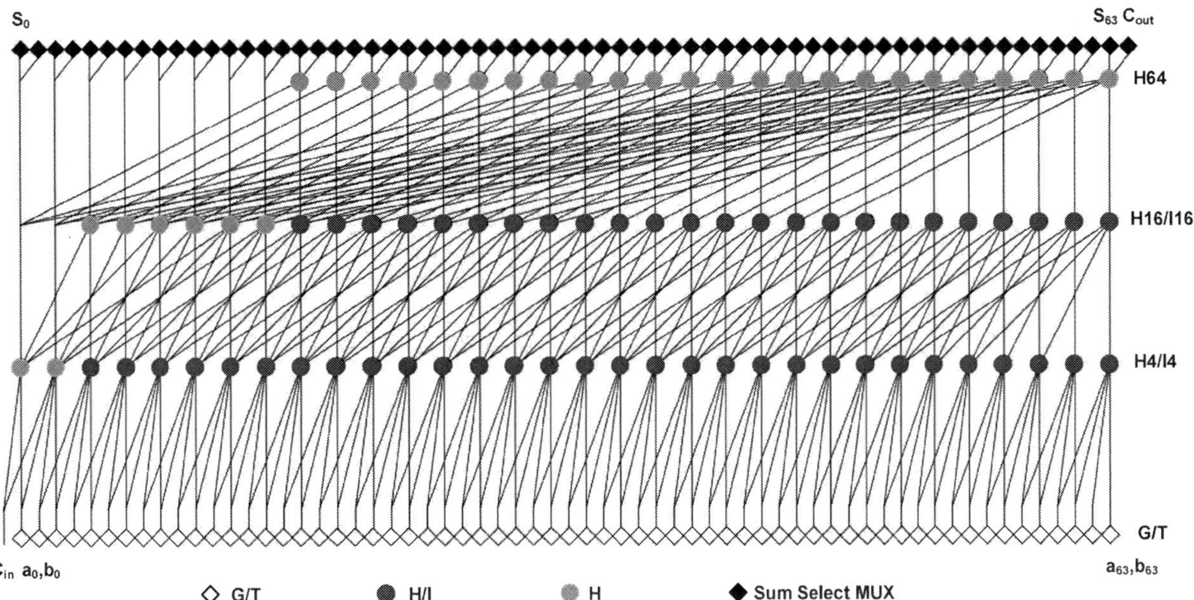

Fig. 8. Radix-4 Sparse-2 64 bit Carry Tree

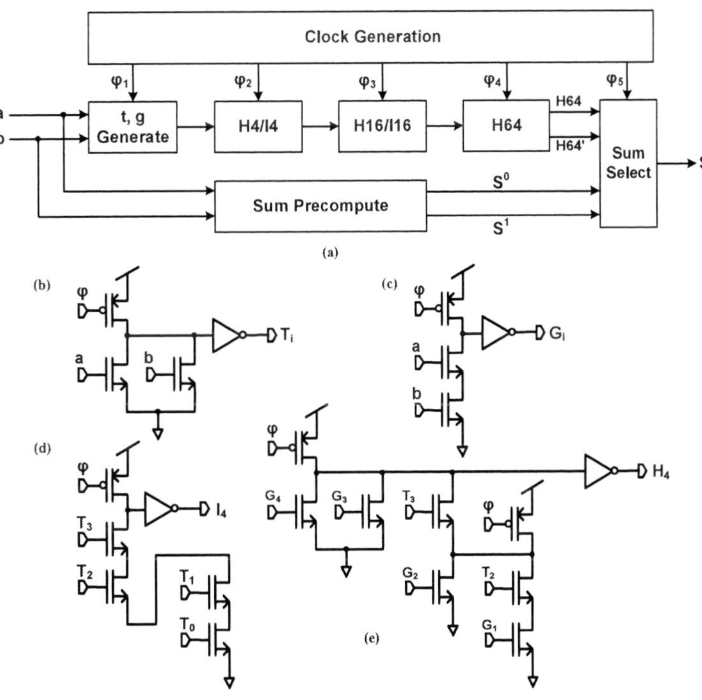

Fig. 9. (a) Block diagram of the radix-4 sparse-2 64bit adder showing various blocks and clocking, (b) Domino logic implememtation of T gate, (c) Domino logic implementation of G gate, Domino logic implementation of Group 4 (d) I and (e) H gates.

in each step. A 64 bit adder requires 3 radix 4 stages or 6 radix 2 stages. Radix-4 adders thus have fewer stages in the critical path composed of more complex gates with more branching than radix-2. Further, while full trees compute the final carry at every bit, it is possible to compute only some of the carries and select the sum based only on the available carry bits. In a sparse-2 tree, for instance, we can compute only the even carries in the CLA block and use them to select the multiplexers in the sum-select stage. The gates and wires corresponding to the eliminated carries are pruned, dramatically reducing the complexity of the tree. However, the sum-precomputation block complexity will increase even more.

We implement a radix-4 sparse-2 64 bit carry tree (Fig. 8) domino logic adder based on Ling's equation similar to one proposed in [14] as shown in Fig. 9. The well-known equations for the propagate, generate, and sum are given below:

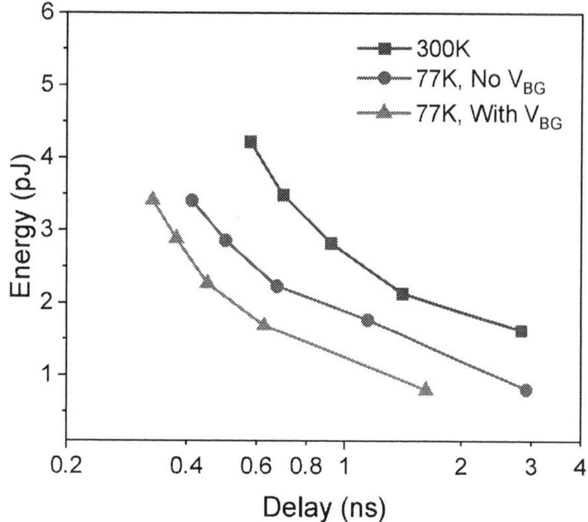

Fig. 10. Simulated energy vs delay plot using calibrated BSIM models for Radix-4 Sparse-2 64 bit Domino adder at 300K and 77K with & without selective back gate biasing showing energy efficiency and performance improvements.

$$g_i = a_i . b_i \tag{2}$$

$$t_i = a_i + b_i \tag{3}$$

$$H_i = g_i + t_{i-1} . H_{i-1} \tag{4}$$

Even numbered sum bits:

$$S_i^0 = a_i \oplus b_i \tag{5}$$

$$S_i^1 = a_i \oplus b_i \oplus (a_{i-1} + b_{i-1}) \tag{6}$$

Odd numbered sum bits:

$$S_i^0 = a_i \oplus b_i \oplus (a_{i-1} . b_{i-1}) \tag{7}$$

$$S_i^1 = a_i \oplus b_i \oplus \ [a_{i-1} . b_{i-1} + (a_{i-1} + b_{i-1}) . (a_{i-2} + b_{i-2})] \tag{8}$$

The different blocks and their domino implementation are shown in Fig. 8. Since the delayed clocks are used for precharging subsequent phases of the domino logic, and we assume the inputs to be monotonically available for the evaluation, the stages can be implemented without the footer transistors. The transistors in the pull down network with signal inputs are V_{th} tuned by applying V_{BG} to have iso-I_{OFF} as that of 300K devices. The clock generator will generate the required skewed precharge signals to the 5 stages of the carry propagation chain and the output of the MUX will provide the correct sum bits.

This design is simulated by varying the supply voltage V_{DD} and calculating the delay and energy of operation for 300K with no back gate bias, 77K with no back gate bias and 77K with back gate bias for matched I_{OFF}. The resultant energy delay plots are shown in Fig. 9. We can clearly see that we can achieve energy improvement at iso performance or have higher performance at same energy expended. We see up to 56% improvement in performance at constant energy and 53% improvement in energy at iso-performance (however, without cooling cost included) under application of V_{BG}. This slightly greater value of improvement in metrics as compared to section III can be accredited to the absence of footer device in the latter implementation. All these make the proposed dynamic logic with selective back gate bias a viable option for high performance and/or energy efficient cryogenic computing.

V. Conclusion

In this paper we examine the cryogenic behavior of FDSOI devices by examining their transfer characteristics and their behavior under the effect of back gate bias. The device behavior is accurately modelled by the calibrated BSIM4 models. The unique capability of V_{th} tunability in FDSOI through back gate bias allows us to improve performance and/or energy per operation. We employ selective back gate biasing of the pull down devices in the dynamic/domino logic gates to increase their performance while the precharge transistors act as high V_{th} devices at cryogenic temperature limiting the leakage problems. We apply this design technique to a radix-4 sparse-2 64 bit adder to show 56% improvement in performance and/or 53% improvement in energy. Thus, the proposed selective back gate bias cryogenic dynamic logic can address the issues of charge leakage while achieving energy efficiency making it a feasible and competitive candidate for cryogenic high performance computing.

References

[1] H. L. Chiang et al., "Cold CMOS as a Power-Performance-Reliability Booster for Advanced FinFETs," 2020 IEEE Symposium on VLSI Technology, Honolulu, HI, USA, 2020, pp. 1-2.

[2] R. Saligram et al, "A 64-Bit Arm CPU at Cryogenic temperatures: Design Technology Co-Optimization for Power and Performance," IEEE Custom Integrated Circuits Conference (CICC), Austin, TX, USA, 2021, pp. 1-2.

[3] A. Beckers, F. Jazaeri and C. Enz, "Theoretical Limit of Low Temperature Subthreshold Swing in Field-Effect Transistors," in IEEE Electron Device Letters, vol. 41, no. 2, pp. 276-279, Feb. 2020.

[4] R. Saligram, S. Datta and A. Raychowdhury, "Scaled Back End of Line Interconnects at Cryogenic Temperatures," in IEEE Electron Device Letters, vol. 42, no. 11, pp. 1674-1677, Nov. 2021.

[5] X. Xue et al., ''CMOS-based cryogenic control of silicon quantum circuits,'' Nature, vol. 593, no. 7858, pp. 205–210, May 2021.

[6] F. Zocca et al, ''Setup of cryogenic front-end electronic systems for germanium detectors readout,'' in Proc. IEEE Nucl. Sci. Symp. Conf. Rec. (NSS/MIC), Oct. 2009, pp. 368–372

[7] C. L. Degen, F. Reinhard, and P. Cappellaro, ''Quantum sensing,'' Rev. Mod. Phys., vol. 89, no. 3, pp. 1–39, Jul. 2017,

[8] T. Chen et al., ''CMOS reliability issues for emerging cryogenic Lunar electronics applications,'' Solid-State Electron., vol. 50, no. 6, pp. 959–963, Jun. 2006

[9] W. Kuhnet al., "A microtransceiver for UHF proximity links including Mars surface-to-orbit applications," Proc. IEEE, vol. 95, no. 10,pp. 2019–2044, Oct. 2007.

[10] R. Carteret al., "22nm FDSOI technology for emerging mobile, Internet-of-Things, and RF applications," inIEDM Tech. Dig., Dec. 2016, p. 2.

[11] R. Saligram et al., "Power Performance Analysis of Digital Standard Cells for 28 nm Bulk CMOS at Cryogenic Temperature Using BSIM Models," in IEEE Journal on Exploratory Solid-State Computational Devices and Circuits, vol. 7, no. 2, pp. 193-200, Dec. 2021.

[12] H. Ling, "High Speed Binary Adder," IBM J. R&D, vol. 25, no. 3, pp. 156-166, May, 1981.

[13] S. Naffziger, "A Sub-Nanosecond 0.5μm 64b Adder Design," ISSCC Dig. Tech. Papers, pp. 362-363, Feb., 1996.

[14] S. Kao, R. Zlatanovici and B. Nikolic, "A 240ps 64b carry-lookahead adder in 90nm CMOS," 2006 IEEE ISSCC pp. 1735-1744.

Quantifying the Overheads of Modular Multiplication

Deepraj Soni*, Mohammed Nabeel†, Negar Neda*, Ramesh Karri*, Michail Maniatakos†, Brandon Reagen*
*New York University, New York, USA
†Center for Cyber Security, New York University Abu Dhabi, Abu Dhabi, UAE

***Abstract*— As security and privacy continue to grow in importance, new techniques, including fully homomorphic encryption (FHE) and post-quantum cryptography (PQC), have emerged to provide new capabilities. Many of these techniques are based on the ring learning with errors problem and operate over rings. Elements of a ring are computed using modular arithmetic, with modular multiplication being a primary component. These components are far more complex than standard integer computing, especially when working with large bit widths. As FHE and PQC become increasingly popular, the need for well-designed and optimized modular multipliers also grows in importance. In this paper, we analyze the power, area, performance, energy, and thermal characteristics of two commonly used modular multipliers: Barrett (bit parallel) and Interleaved (bit parallel). To understand these multipliers' characteristics, this study provides necessary insights into the sources of area, power, frequency, and energy overhead, considering a range of different bit widths (16-256). This paper rigorously analyzes the sub-blocks of modular multipliers and their contributions to overall power, performance, and area (PPA).**

I. Introduction

Modular arithmetic is used in many fields of computer science, mainly in the fields of cryptography (e.g., RSA and Diffie-Hellman key exchange), information theory (e.g., Reed-Solomon codes), digital signal processing (e.g., digital filters), etc. Among all modular arithmetic operations, multiplication is one of the most complex. The naive way of implementing modular multiplication includes costly division operations for the remainder calculation. Fortunately, optimizations have been proposed to eliminate the division. Despite this, software emulation is still too slow. Most main-stream architectures do not natively support modular arithmetic, resulting is significant slowdown. Today, many are looking at hardware implementations.

Hardware efforts have been fueled by the recent attention modular multiplication has recieved due to its use in Post-Quantum Cryptography (PQC) and Fully Homomorphic encryption (FHE), which are based on the Ring Learning with Errors (RLWE) problem. Popular FHE schemes including CKKS [1] and BGV [2], and PQC schemes, such as CRYSTALS-Dilithium [3] and CRYSTALS-Kyber [4], are based on RLWE. Here, the underlying operation is modular arithmetic. Recent work on hardware accelerators for FHE and PQC schemes [5]–[10] shows that the modular multiplier is a key component for high-performance and efficient solutions.

Most prior work focuses on overall modular multiplier area and performance. Only a few studies have published numbers on modular multiplier power estimation [5], [11], [12]. As analyzing multiplier building blocks can help identify opportunities for optimization and improvement, which can lead to better PPA, there is a need for a holistic, detailed study on the impact of sub-blocks employed in these multipliers to overall power, performance, and area (PPA). Another important aspect is understanding the energy requirement of modular multipliers. Such a study helps select the right modular multiplier for the task, e.g., in a battery-operated devices verses the cloud. In this paper, we perform an analysis of power, energy, performance, and the area of two popular modular multiplier implementations: bit-serial and bit-parallel. Specifically, we use Barrett Modular Multiplier (BM) and Interleaved Modular Multiplier (ILM) as representatives. Montgomery multipliers have also been proposed in the literature, but the requirement of domain transformations between the original domain and Montgomery domain must be considered. These domain transformations can lead to high overhead [13] and need to be studied in the context of a full system. Therefore, we focus on Barrett and Interleaved as representatives of bit-parallel and bit-serial designs, respectively.

This paper provides a thorough analysis of modular multipliers by quantifying the effects of modulus size on area, frequency, power, energy, and thermal dissipation. For this purpose, we examined all the sub-blocks of the selected modular multipliers, which is achieved via structural RTL implementation where each sub-module is a separate component. We observe that energy requirements are higher for smaller ILM compared to the BM. We also observe that area and power numbers increase differently in BM and ILM when the modulus size increases. We also examine the contribution of combinational and sequential logic to the area and power consumption and the impact of clock network power. Finally, we also perform thermal analysis.

This paper makes the following contributions:

- We analyze and quantify the PPA results of bit-parallel and bit-serial modular multipliers.
- We perform energy and thermal analysis of ASIC modular multiplication.
- We dissect the modular multipliers and analyze power and area results for their sub-modules.

979-8-3503-1176-1/23 $31.00 © 2023 IEEE

Algorithm 1 Barrett Reduction

Input: q, A, B, R, n
Conditions: $q \geq 3,\ q \neq 2^i$; A, B $\in \mathbb{Z}_q$; $n = \lceil \log_2 q \rceil$; $R = \lfloor 4^n/q \rfloor$
Output: $t = (A*B) \bmod q \in \mathbb{Z}_q$

```
s ← (A*B)                          ▷ MULT1
s_h ← s ≫ n − 1
c ← (s_h*R) ≫ n+1                  ▷ MULT2
t' ← s − c*q                       ▷ MULT3 & SUB1
if t' > 2*q then t ← t' − 2*q      ▷ SUB2
else if t' > q then t ← t' − q     ▷ SUB3
else t ← t'
end if
return t
```

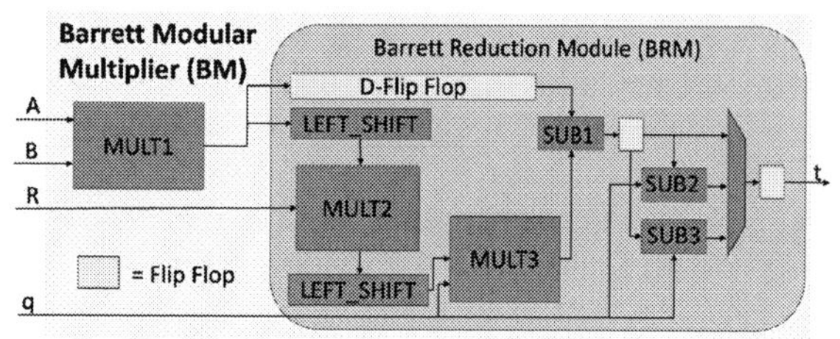

Fig. 1. Barrett Modular Multiplier (BM).

II. Modular Multiplication

Modular multiplication multiplies two numbers and performs modulo reduction on the multiplication result. Modular reduction requires costly division operations. Different algorithms have been proposed to elide the division operation. Two popular ways to do modular multiplication include: (1) bit-serial multiplication combined with reduction (i.e., Interleaved modular multiplication (ILM)), and (2) bit-parallel full multiplication followed by a reduction (i.e., Barrett Modular Multiplication (BM)).

A. Barrett Modular Multiplier (BM)

The Barrett modular reduction avoids division operation by approximating the modulus q to R = 2^n/q; this replaces the division with bit-shifts and multiplications.

1) s = A * B (result can be 2n bits)
2) p = s * R (result can be 3n+1 bits)
3) c = $(p >> 2^{2n})$*q (result can be max 2n+1 bits)

Note only the upper n+1 bits of the second multiplier output are used. The operation $(p >> 2^{2n})$ can be rewritten as: $(s_h*2^{n-1} + s_l)*R/2^{2n}$ where s_h is the upper n+1 bits of s and s_l is the lower n-1 bits. Final equation will be: $(s_h*R/2^{n+1} + s_l*R/2^{2n})$. Here, the second part of the equation is 0, as s_l*R can be a maximum of 2n bits. So it is enough to perform only s_h*R multiplication, which is (n+1) bits * (n+1) bits instead of s*R multiplication which is (2n) bits * (n+1) bits. Based on this observation, Figure 1 and Algorithm 1 show the optimized BM. We implement each multiplier using the Wallace tree algorithm [14] and we use four-stage pipeline multiplier.

B. Interleaved Modular Multiplication (ILM)

The Interleaved modular multiplication has three inputs: Two operands and a modulus. Based on a few bits of the first

Algorithm 2 Interleaved radix-2 multiplication (IL_R2)

Input: A, B, q
Conditions: $A < q$, $B < q$, $q = [0, 2^n)$, $n = \lceil \log_2 q \rceil$;
Output: $AB \bmod q$

```
1: procedure IL_R2 (A,B,q)
2:     B_lut[j = 0 : 1] ← (jB mod q)
3:     q_lut[j = 0 : 2] ← jq
4:     P ← 0
5:     for i = n − 1 : 0 do
6:         P ← P ≪ 1
7:         P ← P + B_lut[A[i]]                ▷ ADD & MUX
8:         for parallel j = 0 : 2 do
9:             P_j ← P − q_lut[j]             ▷ SUB1 & SUB2
10:        end for
11:        P ← select(0 ≤ P_j < q)            ▷ MUX
12:    end for
13:    return P
14: end procedure
```

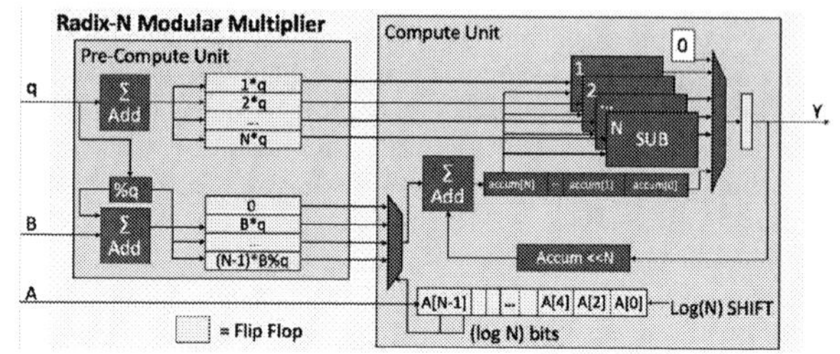

Fig. 2. Interleaved Modular Multiplier (ILM).

operand, interleaved modular multiplication adds the second operand to the partial product (initialized to zero). This partial product is reduced to the range [0, q) where q is modulus. The reduced partial product is shifted right. Thus, ILM performs multiplication and reduction simultaneously for a few bits of the first operand. These steps are repeated with different bits of the first operand until all the bits of the first operand are consumed. Radix decides how many bits of first operand are operated on in one iteration. For radix-N, $\log_2$(N) bits are selected for one iteration. A higher radix requires more hardware but less latency for modular multiplication. Radix-2 implementation of interleaved modular multiplication is shown in Algorithm 2. Lines 3 and 2 forms the pre-computation unit, where lookup tables (LUT) are created iteratively for the multiples of the operand B, which is used for iterative addition in line 7 and for the multiples of the modulus which is used to reduce the partial product in line 9. As the size of the radix increases, the pre-compute unit needs to compute more modulus multiples and this increases the latency.

III. Methodology

Here we provide details on the synthesis, verification, and power estimation process used for the analyzed modular multipliers. We use Synopsys Design Compiler (DC) with the TSMC 28nm technology library to synthesize all designs, with worst-case corner characterization for parasitic resistance, capacitance, 0.81V voltage, and 125°C temperature. We synthesized the multipliers with a 0.5ns clock period, to achieve highest frequency for all the designs. We use the DesignWare library for the multipliers, adders, and subtractors.

For each modulus size, BM and ILM are verified for one million random test cases using Synopsys VCS for functional simulation. We selected random numbers for modulus with 75% of numbers in the range $[2^{n-1} - 1,\ 2^n - 1)$ where n is modulus size. We estimate the power using Synopsys DC and Prime Time Average power analysis mode, which derives power from user-defined switching activity and switching derived from Verilog simulation. For accurate power measurements, we use Synopsys Prime Time tool time-based power analysis, taking the Value Change Dump (VCD) file from gate-level simulation (GLS) of 1000 random test cases on the modular multiplier netlist.

To compare with previous work on 14/12nm technology nodes, we use scaling and report power & area estimates using foundry-reported scaling factors. Specifically, we scale 0.4× for power and 0.5× for area from 28nm to 14/12nm [15].

IV. Results and Analysis

Here we evaluate and characterize the modular multipliers by rigorously analyzing their area, latency, power, energy, and thermal requirements. We explore the performance of different components of the modular multiplier, such as internal multipliers, adders, subtractors, multiplexers, and flip-flops used to store intermediate results. Furthermore, we delve into the area, and power breakdown of these building blocks to gain a detailed understanding of how they contribute to the overall performance of the modular multiplier.

Table I reports the area, frequency, power, energy, and power density for both the Barrett modular multiplier (BM) and the Interleaved modular multiplier (ILM) Radix-2. There are several key observations to be made from the results. First, BM achieves higher frequency than ILM. This is because BM's three multipliers are implemented using the Wallace tree algorithm [14] and are pipelined with four stages to reduce the critical path. However, the compute unit of the interleaved multiplier, which limits its frequency, cannot be broken down efficiently into multiple clock cycles using pipelining. By adding just one pipeline stage to the ILM, the frequency can be increased but this also nearly doubles the initiation interval (II). As a result, the overall throughput is reduced. Second, the relative area of the designs is highly modulus dependent. BM has a 3× larger area than ILM for a modulus size of 16. Barrett multiplier has 46× large area than ILM for a modulus size of 256. As the bitwidth doubles, the BM area almost triples, since the BM has three multiplication operations which increase their area by 3× with doubling the modulus size. As the bitwidth doubles, the ILM area increases by 1.5×, as its internal component sizes increase sublinearly with modulus size. For 256-bits modulus size, BM and ILM require 1292 KGates and 28 KGates, respectively. Third, the BM design requires 7× more power than the interleaved multiplier for a modulus size of 16. PTPX time-based analysis estimates power as reported in Table I. BM design needs 124× more power than the ILM for a modulus size of 256. This is due to the

TABLE I
Performance, Power, and Area (PPA) of Modular Multipliers. $|q|$: Modulus Size, PD: Power Density.

Multiplier	$\|q\|$	Freq. (GHz)	Latency (cc)	II (cc)	Area (μm^2)	Power (mW)	Energy (pJ)	PD (W/mm^2)
BM	16	2.17	11	1	3,330	7.63	3.59	2.29
	32	2.12	11	1	10,182	22.6	10.9	2.22
	64	2.17	11	1	39,983	91.2	42.9	2.28
	128	2.17	11	1	156,412	358	168	2.28
	256	2.08	11	1	488,505	1,004	493	2.05
ILM	16	2.08	18	16	925	1.14	9.8	1.23
	32	1.78	34	32	1,580	1.71	32.6	1.08
	64	1.56	66	64	2,977	2.93	124	0.984
	128	1.38	130	128	5,341	4.37	437	0.874
	256	1.23	258	256	10,647	8.27	1,730	0.777

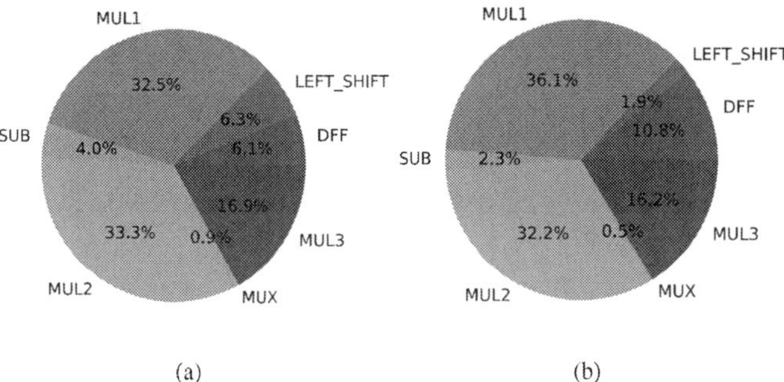

(a) (b)

Fig. 3. **32-bit Barrett Multiplier (a) area and (b) power breakdown.**

fact that the BM has a high area and has all pipeline stages being active simultaneously. As the bitwidth doubles, the BM has over 3× power increase, and the ILM has an average power increase of 1.5×. However, the energy required by the BM to calculate one multiplication is *less* compared to the interleaved multiplier, despite having less area and power. This is because energy is directly proportional to execution time; the interleaved multiplier takes more time due to more clock cycles and a longer clock period. Finally, BM and ILM have a maximum power density of 2.29 W/mm^2 and 1.7 W/mm^2, respectively. The BM has a higher power density compared to the ILM. Power density in modern chips with larger die sizes can achieve an average chip power density of 4.25W/mm^2 or lower with adequate cooling [16]. If a modular multiplier is causing excessive heat, a designer can switch to a larger modulus size because a larger modulus size has lower heat generation due to lower power density. Smaller designs have high power density as they run on high frequency.

A. Barrett Multiplier

Figure 3 provides a breakdown of the area and power consumption of a 32-bit Barrett multiplier and several key observations can be made from this data. First, the majority of the area and power numbers are attributed to the three multipliers. Among the multipliers, MULT2 has the highest area as it is a (N+1)-bit multiplier. MULT3 has less area because Barrett uses only the lower half of its output. Hence, the synthesis tool optimizes MULT3 heavily. MULT1 has the highest power despite having less area compared to MULT2. This is because both operands of MULT1 change with each input, resulting in more switching activity compared to MULT2, where only one operand changes, causing less switching activity. Although LEFT_SHIFT takes up 6.3% of the area, it makes up only 1.9% of the power. This is because the shift value rarely changes, resulting in minimal switching activity. Flip-flops, on the other hand, take 10.8% of the power despite taking up only 6.1% of the area as sequential logic is power hungry.

The area breakdown of the Barrett multiplier for various modulus sizes ($|q|$) is illustrated in Figure 4. As the modulus size increases, the three multipliers require a larger proportion of the design area. Specifically, for 16-bit and 256-bit modulus sizes, the three multipliers need 76% and 94% of the total area, respectively. As $|q|$ doubles, the multiplier area triples and other modules' area double.

Similarly, Figure 5 displays the power breakdown of the Barrett multiplier for various modulus sizes. As the modulus size increases, the three multipliers require a greater proportion of the design power. For instance, for 16-bit and 256-bit modulus sizes, the three multipliers require 77% and 96% of the total power, respectively. Because multiplier power triples with each doubling of the modulus size, unlike other modules that double the area. Additionally, Figure 4(a) and Figure 5(a) show that the area and power estimates of N-bit modular multiplier is similar to 2*N-bit integer multiplication.

Figure 6 shows how the percentage of area and power of combinational logic, sequential logic, and clock network are distributed. This analysis is based on the post-synthesis (pre-routed) design, where the clock tree is considered as an ideal network, and no optimization is performed on it. Therefore, the clock network is not taken into account in the area estimation. As the modulus size increases, the area percentage and power percentage of the combinational logic increase. This is mainly because of the full adders used in the three Wallace tree multipliers, whose count is more than the number of pipeline flops used within the multiplier. The number of full adders used in the Wallace tree can be calculated from the equation $3log_2N+2N-1$, and the maximum number of pipeline stages possible is by adding flip flops at the output of each full adder;

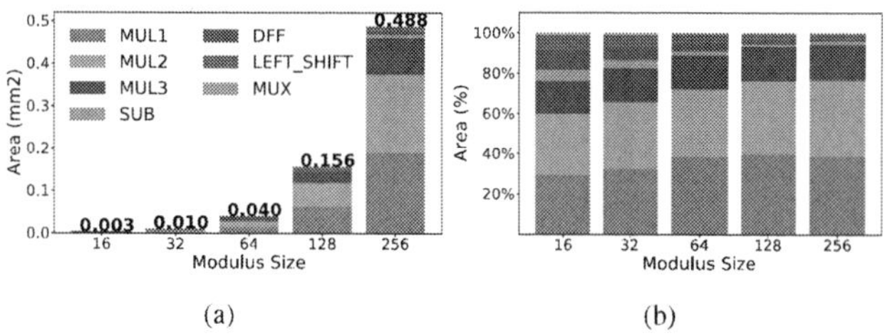

(a) (b)

Fig. 4. **Barrett Multiplier area breakdown.**

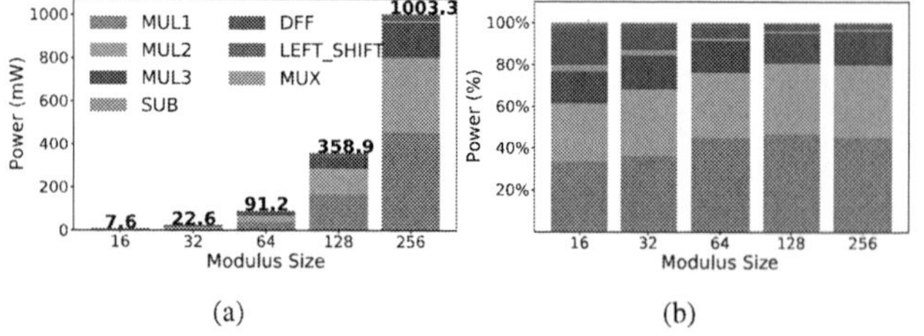

(a) (b)

Fig. 5. **Barrett Multiplier power breakdown.**

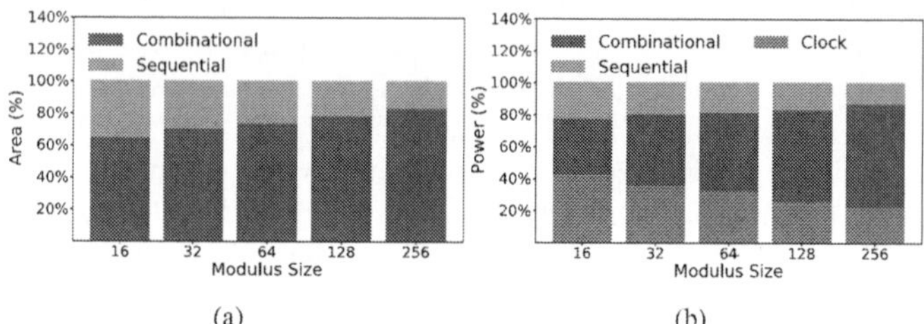

(a) (b)

Fig. 6. **Barrett area and power breakdown for combinational and sequential logic.**

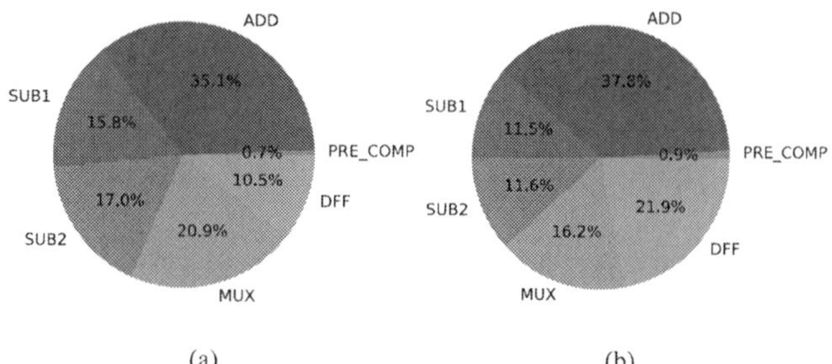

(a) (b)

Fig. 7. **32-bit Interleaved Multiplier (a) area and (b) power breakdown.**

hence, the area of combinational logic will always be greater than the sequential logic as the full adder standard cell has more area than the D flipflop ($3.654um^2$ versus $2.268um^2$ in TSMC $28nm$ library). The sequential logic affects the power consumption of the clock network, which decreases with the increase in modulus size. As $|q|$ increases the percentage of sequential logic area decreases and frequency decreases.

Internal power, which is any power dissipated within the boundary of a standard cell, takes most of the power in the Barrett multiplier. This includes power dissipated charging and discharging internal load capacitance in a standard cell and short-circuit power caused by the momentary short circuit between the P and N transistors of a gate during the signal transition. The major contributor to the internal power is the internal power of registers due to toggling clock pins. Switching power, which depends on the charging of the output loads of each cell, increases with the modulus size as the overall number of logic cells increases with the modulus size. The leakage power also increases because of the increase in the number of logic cells.

Finally, we include thermal analysis from Hotspot [16] for BM. We did a backend implementation of a 32-bit Barrett multiplier and did power analysis on the post-routed netlist to get the power traces of each sub-blocks. We give following input parameters: sub-block power traces, along with their location coordinates, the size of the design ($.15mm \times 1.5mm$), heat sink dimension (minimum - $2mmX2mm$), heat sink material (copper), and thermal conductivity (400 W/mK). With these input parameters, HotSpot reported that the temperature rises over the ambient temperature of $25°C$ is $10°C$. BM can tolerate till $125°C$ as per the techonology library. Thus, heat dissipation from BM designs should not damage the chip.

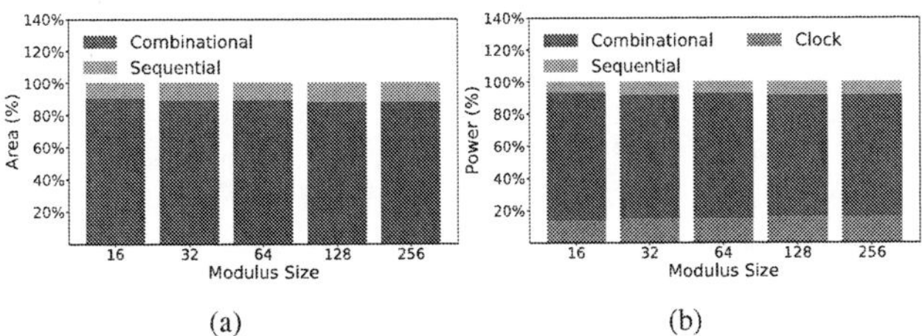

Fig. 8. **Interleaved Multiplier area and power breakdown for combinational and sequential logic.**

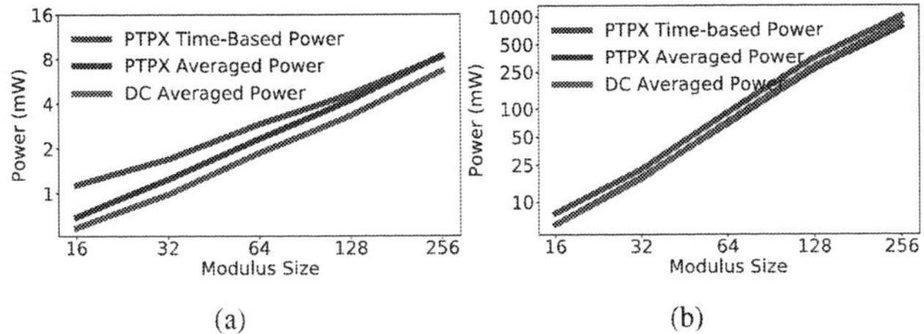

Fig. 9. **PTPX and DC Power analysis for (a) Interleaved Multiplier and (b) Barrett Multiplier.**

B. Interleaved Multiplier

Figure 7 shows the area and power breakdown of 32bit Interleaved multiplier. There are two key observations. First, adders and two subtractors require 35.1%, 15.8%, and 17.0% of ILM area, respectively. As the power is correlated with area, adders and two subtractors require 37.8%, 11.5%, and 11.6% of ILM area, respectively. Power estimates of the adder is higher compared to other modules because both the input operands are changing every clock cycle for the adder, while only one operand changes each clock cycle for subtractors. Though adder and subtractors have the same size, the adder requires higher power and area because Synopsys uses a faster implementation of the adder to achieve greater frequency. Second, DFFs need 21.9% power, even though it takes only 10.5% of the total area. Each modulus size has a similar area and power breakdown as each of the module area and power increases linearly with modulus size.

The distribution of area and power consumption for the combinational logic, sequential logic, and clock network is presented in Figure 8. We observe that the combinational logic accounts for 88% of the total area, while the sequential logic occupies the remaining area. For power consumption, the combinational logic, sequential logic, and clock network consume 76%, 8%, and 16% of the total power, respectively. The percentage of power and area requirements for each component remains similar across different modulus sizes. This is due to the linear increase in area and power consumption of each component with the modulus size, which leads to a constant percentage of area and power for the combinational logic, sequential logic, and clock network.

For thermal analysis, the same flow previously mentioned for the Barrett design is followed. Design dimension after backend implementation is $.05mm \times .05$. We observe that the temperature rise above ambient is $16.5°C$, which is $6.5°C$ more than the Barrett design. This can be attributed to the higher energy requirement of ILM designs compared to BM.

C. Power Estimates

To estimate the power of BM and ILM design, we utilized conventional DC averaged power analysis, and PrimeTime PX (PTPX) averaged power analysis. As the power requirement depends upon many parameters, such as input test vectors, the switching activity, and frequency, averaged power analysis, which supports propagation of switching activity based on user-defined toggle rate and switching activity, may not give the most accurate power estimate. To estimate the power more accurately, we used PTPX time-based analysis. PTPX time-based analysis evaluates power based on RTL/gate-level simulation activity over time, using an event-driven algorithm to generate detailed time-based power waveform for more accurate power estimates.

Figure 9 depicts that conventional power estimate methods, such as configuring the Synopsys DC tool with the typically used setting of toggle rate with 0.1 and switching activity with 0.5, are less than perfect for these complex designs. To obtain accurate power estimates, we ran the PTPX feeding in the switching activity file acquired from the Gate level simulation (GLS) and clock period defined based on the critical path of the design. Hence, PTPX emerges as a reliable tool for accurately estimating the power consumption of complex designs. For Interleaved multipliers, PTPX time-based analysis reports 50% and 30% more power compared to DC power analysis for small modulus size (16) and large modulus size (256), respectively. For Barrett multiplier, PTPX time-based analysis reports 25% and 15% more power compared to DC power analysis for small modulus size (16) and large modulus size (256), respectively. This result shows that, for complex designs, for accurate power estimates, one should rely on flows like PTPX time-based analysis instead of the commonly used flow of deriving power based on defining static probability for the switching activity.

D. Comparison

Table II compares power, area, frequency, and power density of the BM design reported in this study and the design presented in F1 [5]. The proposed Barrett design requires $0.5\times$ power and operates on $3\times$ frequency compared to F1, while occupying the similar area. Table III compares the area and latency of the ILM design with the state-of-the-art FPGA design [13], which implements an iterative Barrett design. For a fair comparison, we ported the ILM design to a FPGA by replacing optimized IPs from Designware library with RTL modules. Table III reports the ILM design has $2.5\times$ less area and $3.5\times$ high clock period compared to the state-of-the-art, which uses carry-save compression and a modified booth algorithm. Comparison in Tables II and III demonstrate that the designs developed and analyzed in this study has comparable performance with the current state-of-the-art designs.

V. Related Work

Recently, many researchers have proposed solutions for modular multipliers on alternative platforms [17]–[20] with a focus on NTT and polynomial multiplication. Ozturk et

TABLE II
COMPARISON WITH STATE-OF-THE-ART (32-BITS) ASIC DESIGN.

Design	$\|q\|$	Frequency (GHz)	Area (μm^2)	Power (mW)	Energy (pJ)	PD (W/mm^2)
F1 [5]	32	0.76	10,542	46	-	3.49
Proposed	32	2.12	10,182	22.6	10.9	2.22

TABLE III
COMPARISON WITH STATE-OF-THE-ART (1024 BITS) FPGA DESIGN.

Design	Period (ns)	Area			Latency (cc)
		LUT	FF	DSP	
[13]	4.3	12,872	3,078	0	1029
Proposed*	14.35	5,445	2,067	0	1026

*: FPGA design where ASIC DesignWare IPs are replaced with RTL.

al. [21] proposed a co-processor to offload computationally expensive operations, using a hardware polynomial multiplier. The design implemented on a Virtex-7 FPGA is about 100× faster than in software. FPGA implementations on cutting-edge technological platform promise large performance gain compared to GPU implementations [12], [22], [23].

Axel et al. [5] implemented ASIC 32-bit Barrett Modular Multiplier and Montgomery Montgomery Multiplier for 14/12 nm technology node. Moreover, Nabeel et al. [11] fabricated a co-processor on 55 nm technology for FHE using modular multiplier, modular adder, and modular subtractors. It utilizes 128-bit Barrett Modular Multiplier for efficient implementation of FHE applications.

VI. CONCLUSION

This paper thoroughly examines the area, power, energy, performance, and thermal characteristics of optimized Barrett Modular Multiplier (BM) and Interleaved Modular Multiplier (ILM). We have observed that the BM offers faster implementation with higher frequency compared to ILM at the expense of high area and power. We breakdown the design in sub-modules to understand the power and area complexity of modular multipliers. Though ILM requires comparatively low area and power, ILM draws high energy to perform single multiplication compared to BM.

VII. ACKNOWLEDGEMENT

This work is supported in part by NSF award 2016650. Additionally, this research was developed with funding from the Defense Advanced Research Projects Agency (DARPA), under the Data Protection in Virtual Environments (DPRIVE) program, contract HR0011-21-9-0003. Finally, we are grateful to the reviewers for their valuable input, which has significantly enhanced the quality of the paper.

REFERENCES

[1] J. H. Cheon, A. Kim, M. Kim, and Y. Song, "Homomorphic encryption for arithmetic of approximate numbers," in *Advances in Cryptology – ASIACRYPT*, 2017.

[2] Z. Brakerski, C. Gentry, and V. Vaikuntanathan, "Fully homomorphic encryption without bootstrapping," *Cryptology ePrint Archive*, 2011.

[3] L. Ducas, E. Kiltz, T. Lepoint, V. Lyubashevsky, P. Schwabe, G. Seiler, and D. Stehlé, "CRYSTALS-Dilithium: A lattice-based digital signature scheme," *IACR Transactions on Cryptographic Hardware and Embedded Systems*, 2018.

[4] J. Bos, L. Ducas, E. Kiltz, V. Lyubashevsky, J. M. Schanck, P. Schwabe, G. Seiler, and D. Stehlé, "CRYSTALS-Kyber: A CCA-secure module-lattice-based kem," in *2018 IEEE European Symposium on Security and Privacy (EuroS&P)*, IEEE, 2018.

[5] N. Samardzic, A. Feldmann, A. Krastev, S. Devadas, R. Dreslinski, C. Peikert, and D. Sanchez, "F1: A fast and programmable accelerator for fully homomorphic encryption," in *MICRO-54: 54th Annual IEEE/ACM International Symposium on Microarchitecture*, 2021.

[6] N. Samardzic, A. Feldmann, A. Krastev, N. Manohar, N. Genise, S. Devadas, K. Eldefrawy, C. Peikert, and D. Sanchez, "Craterlake: a hardware accelerator for efficient unbounded computation on encrypted data," in *Proceedings of the 49th Annual International Symposium on Computer Architecture (ISCA)*, 2022.

[7] R. Geelen, M. Van Beirendonck, H. V. Pereira, B. Huffman, T. McAuley, B. Selfridge, D. Wagner, G. Dimou, I. Verbauwhede, F. Vercauteren, *et al.*, "Basalisc: Flexible asynchronous hardware accelerator for fully homomorphic encryption," *arXiv preprint arXiv:2205.14017*, 2022.

[8] L. Beckwith, D. T. Nguyen, and K. Gaj, "High-performance hardware implementation of crystals-dilithium," in *2021 International Conference on Field-Programmable Technology (ICFPT)*, 2021.

[9] D. Soni, K. Basu, M. Nabeel, N. Aaraj, M. Manzano, and R. Karri, *Hardware Architectures for Post-Quantum Digital Signature Schemes*. Springer, 2020.

[10] D. Soni, N. Neda, N. Zhang, B. Reynwar, B. Heyman, M. Nabeel, A. A. Badawi, Y. Polyakov, K. Canida, *et al.*, "RPU: The ring processing unit," *2023 IEEE International Symposium on Performance Analysis of Systems and Software (ISPASS)*, 2023.

[11] M. Nabeel, D. Soni, M. Ashraf, M. A. Gebremichael, H. Gamil, E. Chielle, R. Karri, M. Sanduleanu, and M. Maniatakos, "CoFHEE: A co-processor for fully homomorphic encryption execution," *Design, Automation and Test in Europe (DATE) Conference*, 2023.

[12] X. Cao, C. Moore, M. O'Neill, E. O'Sullivan, and N. Hanley, "Accelerating fully homomorphic encryption over the integers with super-size hardware multiplier and modular reduction," *Cryptology ePrint Archive*, 2013.

[13] B. Zhang, Z. Cheng, and M. Pedram, "A high-performance low-power Barrett modular multiplier for cryptosystems," in *2021 IEEE/ACM International Symposium on Low Power Electronics and Design (ISLPED)*, IEEE, 2021.

[14] K. Abbas, *Handbook of Digital CMOS Technology, Circuits, and Systems*. Springer Nature, 2020.

[15] T. S. M. C. Limited, "1612nm technology." https://www.tsmc.com/english/dedicatedFoundry/technology/logic/l_16_12nm Accessed on 2023-Jun-11.

[16] W. Huang, M. R. Stan, S. Gurumurthi, R. J. Ribando, and K. Skadron, "Interaction of scaling trends in processor architecture and cooling," in *2010 26th Annual IEEE Semiconductor Thermal Measurement and Management Symposium (SEMI-THERM)*, 2010.

[17] B. Zhang, Z. Cheng, and M. Pedram, "High-radix design of a scalable montgomery modular multiplier with low latency," *IEEE Transactions on Computers*, 2021.

[18] B. Reagen, W.-S. Choi, Y. Ko, V. T. Lee, H.-H. S. Lee, G.-Y. Wei, and D. Brooks, "Cheetah: Optimizing and accelerating homomorphic encryption for private inference," in *2021 IEEE International Symposium on High-Performance Computer Architecture (HPCA)*, 2021.

[19] K. Javeed, D. Irwin, and X. Wang, "Design and performance comparison of modular multipliers implemented on FPGA platform," in *Cloud Computing and Security: Second International Conference, ICCCS 2016*, Springer, 2016.

[20] R. Liu and S. Li, "A design and implementation of montgomery modular multiplier," in *2019 IEEE International Symposium on Circuits and Systems (ISCAS)*, IEEE, 2019.

[21] E. Öztürk, Y. Doröz, E. Savas, and B. Sunar, "A custom accelerator for homomorphic encryption applications," *IEEE Transactions on Computers*, 2016.

[22] M. S. Riazi, K. Laine, B. Pelton, and W. Dai, "HEAX: An architecture for computing on encrypted data," in *Proceedings of the Twenty-Fifth International Conference on Architectural Support for Programming Languages and Operating Systems (ASPLOS)*, 2020.

[23] W. Wang, Y. Hu, L. Chen, X. Huang, and B. Sunar, "Accelerating fully homomorphic encryption using GPU," in *2012 IEEE Conference on High Performance Extreme Computing*, 2012.

Multi-Source Transfer Learning for Design Technology Co-Optimization

Jakang Lee, Jaeseung Lee, Seonghyeon Park and Seokhyeong Kang
Department of Electrical Engineering,
Pohang University of Science and Technology, Pohang, South Korea
{wkrkd95, jae2seung, seonghyeon98, shkang}@postech.ac.kr

***Abstract*—In advanced technology nodes, pitch scaling have not kept up with the Moore's Law. To continue progression, the design technology co-optimization (DTCO) has been proposed. However, implementing DTCO requires significant time cost and resources due to iterative trials. In addition, optimal design and technology option depend on each design, thus it should start from scratch whenever the target design changes. We present a DTCO framework based on Bayesian optimization that efficiently explores design feedback for optimization. In addition, our framework incorporates a multi-source transfer Gaussian process (MTGP) that ensures robust optimization even for unseen designs. MTGP significantly improves prediction and generalization performance by integrating multiple single source transfer Gaussian processes. Our framework, on average, reduced the mean absolute error of power and area by 47.3% and 24.1%, respectively, and power and area by 37.3% and 19.9%, respectively, compared to the reference, in 7nm technology nodes.**

***Index Terms*—Design Technology Co-Optimization, Bayesian Optimization, Gaussian Process, Transfer Learning**

I. Introduction

In recent decades, improvements in CMOS technology have largely been achieved through device scaling. However, as technology nodes shrink to sub-10nm, achieving the same level of power, performance, and area (PPA) improvements through scaling has become increasingly challenging. To overcome this issue, a new approach called Design Technology Co-Optimization (DTCO) has been introduced.

DTCO is an optimization methodology that encompasses both design and technology. The general flow of DTCO is depicted in Fig. 1, which highlights how foundries can customize technology to meet specific design requirements. This approach provides chip designers with an assortment of process design kits (PDK) that are optimized for PPA [1], [2]. As technology continues to develop, it becomes increasingly difficult to improve PPA, highlighting the crucial role that DTCO plays in advanced nodes [2].

Continuing with this trend, several studies have recently introduced DTCO technologies to enhance the design quality in advanced nodes [3]–[5]. All of these studies achieved significant design quality improvement by modifying the standard cell layout and various design and technology options. Nevertheless, these studies did not address the issue of design feedback that explores the optimal technology option again based on the PPA of the design. Hence, to achieve optimization using these approaches, the designer should evaluate the PPA of the design and select a suitable solution based on their expertise, making the optimization performance heavily reliant on the designer's knowledge. Therefore, there is an urgent need for a method that can effectively integrate design feedback into DTCO to promote efficient optimization.

The problem of reflecting design feedback effectively is one of the design space exploration problems that selects the design and technology option that is predicted to be optimal. In the electronic design automation (EDA), several recent works have proposed techniques for efficiently exploring large design spaces [6]–[9]. These works demonstrate the effectiveness of Bayesian optimization for exploring vast design spaces. However, a major limitation of these methods is their inability to reuse data since PPA to be predicted are

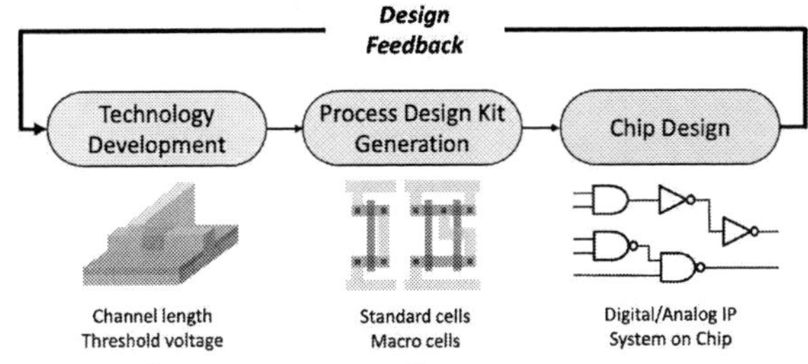

Fig. 1. Common DTCO flow

highly dependent on the target design. Consequently, it is challenging to apply these methods when sufficient data is not available due to design time constraints. Therefore, there is a growing need for techniques that enable the reuse of existing design data and facilitate the training of predictive models for new designs.

Transfer learning emerged as an optimal solution for reusing existing data [10], [11]. This approach uses source design data to construct a predictive model for target design data. Source design data means the PPA data of the previous design, and target design data means the PPA data of the design to be optimized. In general, transfer learning proceeds using a large amount of source design data compared to target design data. The aforementioned studies [10], [11] have demonstrated that transfer learning can enhance the accuracy of predictions and enable more efficient optimization search, even when there is a scarcity of target design data. However, one of the limitations of this method is that it can only utilize a single design as the source design data, thus it cannot incorporate multiple source design. As the performance of the target model is highly dependent by the specific source design used for transfer, this can lead to considerable variability in performance, which, in turn, results in a significant reduction in the generalization performance of the model.

In this study, we introduce a multi-source transfer Gaussian process (MTGP) based DTCO framework. Our framework can effectively explore the near-optimal design and technology options within a large design space. We incorporated Bayesian optimization to efficiently reflect design feedback, which is widely used due to its strong optimization performance, even with noise like the results of EDA tools [6], [7]. Furthermore, we utilized the GP as a surrogate model for Bayesian optimization and applied the MTGP method to overcome the drawbacks of the single-source transfer Gaussian process (STGP) approach [10]. MTGP is a method that utilizes data from multiple source designs for transfer to improve prediction performance and reduce generalization errors in target design. The major contributions of this study is summarized as follows.

- We propose a framework exploiting MTGP based Bayesian optimization to efficiently reflect design feedback in DTCO flow, thereby reducing the number of optimization iterations.
- We demonstrate that small designs can be utilized to improve the PPA predictive model performance of large designs, indicating the possibility of scalability for larger designs.

- By transferring data from several designs, the predictive performance dependency on source design is reduced, resulting in robust optimization performance.

II. Preliminaries

A. DTCO knobs

At the system level, achieving a balance between performance and power consumption is a challenge, as it is difficult to achieve both high performance and low power simultaneously. Therefore, designers need to prioritize and optimize designs based on their intended purpose. Our approach focuses on reducing power and area, while ensuring that the design meets the required timing constraints. To achieve this, we adjust the three key elements of DTCO knobs, which are described in detail in [1].

1) Supply Voltage: The power consumption is primarily dependent on the supply voltage. Specifically, the switching power is proportional to the square of the supply voltage, which implies that scaling the supply voltage can significantly reduce dynamic power consumption. To achieve energy efficiency, modern designs often employ voltage scaling and multiple supply voltages [12]. However, reducing the supply voltage negatively impact the timing performance of the device. Hence, setting an appropriate supply voltage that meets the target specifications is crucial to maintaining the performance.

2) Threshold Voltage: This is related to the sub-threshold leakage power. As the threshold voltage increases, the leakage power decreases exponentially. However, as with the supply voltage, increasing the threshold voltage degrades the timing performance of the device. Therefore, achieving an optimal balance between power consumption and performance is crucial. To balance leakage power and timing performance, modern chip designs use Multi-V_{th} technique to selectively place the low-V_{th} in the timing-critical path and selectively place the high-V_{th} cell in the non-critical path.

3) Standard Cell Height: The height of a standard cell is often expressed as a horizontal routing track. Figure 2 shows cells with 7.5 and 6 routing tracks, respectively. Here, the two cells differ in the number of fins and their height. Increasing the number of tracks improves the device's timing performance, but also increases power consumption and area. Therefore, high-track cells are used where high performance is required, and low-track cells are used where low power and high density are required.

The three factors discussed earlier have a significant impact on power consumption, timing performance, and chip area. Moreover, the complex interdependency among these factors requires designers to search for an optimal balance between them by considering various design and technology options carefully.

B. Threshold Voltage Engineering

The threshold voltage formula is given as:

$$V_{th0} = V_{fb} + \phi_S + \left.\frac{\sqrt{qN_a 2\epsilon_s \phi_S}}{C_{ox}}\right|_{\phi_S = 2\phi_B}, \tag{1}$$

$$V_{fb} = \Phi_G - \Phi_S - \frac{Q_{ox}}{C_{ox}}, \qquad C_{ox} = \frac{\epsilon_{ox}}{T_{ox}}, \tag{2}$$

where V_{th0} is a threshold voltage, ϕ_S is the surface potential of the channel, q is the charge of the electron, N_a is acceptor concentration of the channel, ϵ_s is the dielectric constant of semiconductor, C_{ox} is the gate capacitance, V_{fb} is a work function difference between gate electrode and the silicon.

According to the equations (1)-(2), there are four methods for modifying the threshold voltage of a transistor. Traditional methods that control the doping concentration of the channel (N_a) are not suitable for advanced nodes due to the random fluctuations of dopants. Similarly, modifying the gate dielectric constant (ϵ_{ox}) is difficult to implement at advanced nodes that use metal gates, as only metal and silicon-compatible materials can be used as dielectrics.

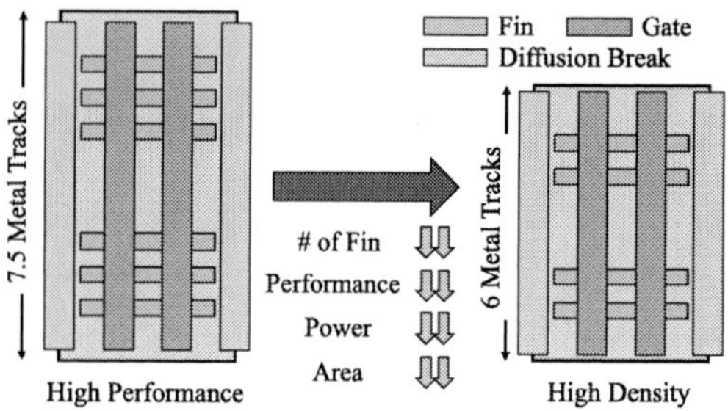

Fig. 2. Standard cell height

Therefore, dielectric thickness tuning (T_{ox}) and gate work function (Φ_G) engineering are commonly used techniques for threshold voltage control.

Increasing the thickness of the dielectric raises the threshold voltage and reduces the gate oxide tunneling current, which greatly reduces leakage power. However, if the thickness is reduced too much, the leakage power increases excessively. Therefore, this approach is mainly used for manufacturing high V_{th} cells. Gate work function engineering is considered the most effective approach for providing multiple threshold voltages.

In light of the aforementioned information, we adopt gate work function engineering to fine-tune the threshold voltage. By directly manipulating the technology, this method also enables us to consider potential side effects that may arise from changes to the process parameters, such as the gate fringing capacitance. Consequently, the reliability of our study can be significantly enhanced.

C. Gaussian Process

The field of EDA often employs heuristic algorithms to generate results, which can be challenging to define and may contain noise. To address this issue, the EDA industry has undertaken numerous studies utilizing GP to model noisy black box problems. GP is a powerful technique estimates both predicted values and their corresponding uncertainties using the covariance matrix. The posterior mean and variance of the GP are calculated using conditional probabilities in the following manner:

$$\mu_* = k(X_*, X)^\top (k(X,X) + \sigma^2 I)^{-1} Y,$$

$$\sigma_*^2 = k(X_*, X_*) - k(X_*, X)^\top (k(X,X) + \sigma^2 I)^{-1} k(X_*, X), \tag{3}$$

where the new input values for a prediction are denoted by X_*, while the previous inputs are represented by X. Additionally, the predicted value of the objective function for the new input is denoted by μ_*, while σ_*^2 denotes the uncertainty of the prediction. The previous output values are denoted by Y, while $k(X, X)$ represents the covariance matrix. In most cases, the covariance matrix is replaced with a kernel function.

A widely employed approach for training GP models is to identify a suitable kernel function and associated parameters that accurately represent the given data. To accomplish this, maximum likelihood estimation is commonly utilized. The objective of maximum likelihood estimation is to find a kernel function that maximizes the likelihood of observed data given a GP model. Based on this, kernel parameters and data noise are adjusted to achieve the best fit for the model. The kernel function is of paramount importance in GP learning as it determines the shape and continuity of the GP. The radial basis function (RBF) is frequently used as a kernel function in GP learning because of its smoothness and infinite differentiability, making it a popular choice.

D. Transfer Kernel

GP is a powerful modeling technique, but efficiency is reduced because a new model must be created every time the target design changes. To overcome this limitation, transfer learning has emerged as a promising approach. In this section, we introduce the transfer kernel method, which is applicable to GP. This method learns the similarity between different designs as a kernel parameter and utilizes data from the different design to learn the GP of the target design. The modified kernel function for transfer learning can be expressed as:

$$K_{nm}(x_n, x_m) = \begin{cases} \lambda k(x_n, x_m), x_n \in S \ \& \ x_m \in T, \\ k(x_n, x_m), otherwise, \end{cases} \tag{4}$$

where S is one source design, λ is the coefficient for the relatedness of the source and target designs, and $k(x, x')$ is a kernel function.

Geng *et al.* [10] constrained λ within the bounds of -1 to 1 to facilitate acquisition of both positive and negative correlations. Specifically, when $|\lambda| = 1$, $\lambda k(x, x')$ indicates that the source and target designs are closely related, and all source design data is transferred to the target design. Conversely, when $\lambda = 0$, $\lambda k(x, x')$ implies that only the target design data is used for training, as the source and target designs are entirely dissimilar. According to Equation (4), the covariance matrix of the target design's GP changes as:

$$\tilde{\mathbf{K}} = \begin{pmatrix} K_{SS} & \lambda K_{ST} \\ \lambda K_{TS} & K_{TT} \end{pmatrix}, \tag{5}$$

where K_{SS} refers to the covariance matrix generated exclusively from the source design data, while K_{TT} denotes the covariance matrix produced solely from the target design data. Furthermore, λK_{ST}(=$\lambda K_{TS}^\top$) is the covariance matrix constructed to capture the resemblance between the source and target designs. Utilizing the covariance matrix at hand, Equation (3) may be adapted as follows.

$$\mu_* = \mathbf{k}_*(X_*, X)^\top (\tilde{\mathbf{K}}(X, X) + \mathbf{\Sigma})^{-1} Y,$$

$$\sigma_*^2 = k(X_*, X_*) - \mathbf{k}_*(X_*, X)^\top (\tilde{\mathbf{K}}(X, X) + \mathbf{\Sigma})^{-1} \mathbf{k}_*(X_*, X), \tag{6}$$

where the matrix $\mathbf{\Sigma}$ incorporates the noise present in the source and target design data, and the matrix $\mathbf{k}_*$ is used to estimate the existing data by dividing it into source and target components.

$$\mathbf{\Sigma} = \begin{pmatrix} \sigma_s^2 \mathrm{I_s} & 0 \\ 0 & \sigma_t^2 \mathrm{I_t}, \end{pmatrix} \quad \mathbf{k}_*(X_*, X) = \begin{pmatrix} \lambda k(X_*, X), X \in S \\ k(X_*, X), X \in T \end{pmatrix} \tag{7}$$

Through Equations (6)-(7), the posterior distribution of the target design's GP given the source design data can be obtained. It could enhance the performance of the model by leveraging correlations (λ) between the source and target design datasets. However, this method may increase the likelihood of overfitting the target GP with respect to specific source design data because of limited capacity.

III. Proposed Method

A. Overall Framework

The objective of this research is to explore the design and technology options that can achieve power and area optimization while meeting the timing constraints. The iterative process from RTL to GDS incurs significant costs and time consumption. To mitigate this challenge, we concentrate on cost-effective optimization following logic synthesis. Our proposed framework encompasses five major steps, including the selection of the design and technology options, generation of PDK, logic synthesis, establishment of multi-source transfer GP (MTGP), measurement of PPA and utilization of Bayesian optimization for selecting the optimal design and technology options. Fig. 3 shows the overall framework.

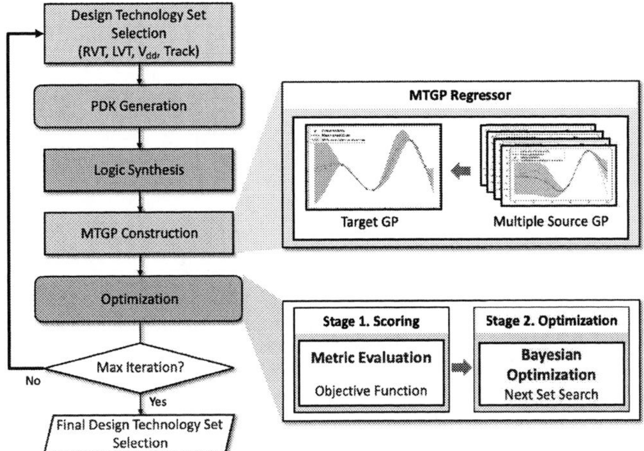

Fig. 3. Overall framework

B. Multi-source Transfer Gaussian Process

We introduce MTGP to overcome the overfitting problem due to the capacity limitations of STGP mentioned in Section II-D. MTGP has been proposed to reduce generalization errors in the target design by leveraging knowledge from multiple source designs [15]. The formula for MTGP is:

$$k_\lambda(x, x') = \begin{cases} \lambda_{D_i, D_j} k(x, x'), x \in D_i \ \& \ x' \in D_j, i \neq j \\ k(x, x'), x \ \& \ x' \ in \ the \ same \ domain \end{cases}, \tag{8}$$

where D is defined as a set containing both source and target designs, while kernel function is represented by $k(x, x')$. As explained in II-D, each domain is assigned a similarity metric λ_{D_i, D_j}, which is re-parameterized as $\lambda_{D_i, D_j} = \alpha_{D_i} \alpha_{D_j}$. The parameter α_{D_i} represents the similarity between the i-th domain and the target domain. As MTGP is trained, $\lambda_{T,S}$ and $\lambda_{S,S}$ could gradually acquire information about the relationship between the target and source designs and the source and source designs, respectively. The covariance matrix employed in this method is expressed as follows.

$$\tilde{\mathbf{K}} = \begin{pmatrix} K_{S_1,S_1} & \lambda_{S_1,S_2} K_{S_1,S_2} & \ldots & \lambda_{S_1,T} K_{S_1,T} \\ \lambda_{S_2,S_1} K_{S_2,S_1} & K_{S_2,S_2} & \ldots & \lambda_{S_2,T} K_{S_2,T} \\ \ldots & \ldots & \ldots & \ldots \\ \lambda_{T,S_1} K_{T,S_1} & \lambda_{T,S_2} K_{T,S_2} & \ldots & K_{T,T} \end{pmatrix} \tag{9}$$

MTGP can effectively consider similarity between all source and target designs by Equation (9) and has improved generalization error compared to STGP. The process of learning MTGP is accomplished by utilizing maximum likelihood estimation, a method similar to that of conventional GP. However, since the source design data induces modifications in the kernel, adjustments are necessary. The adapted log-likelihood function for MTGP is presented below:

$$L_\Theta = \log p(Y_T | X_T, X_S, Y_S; \Theta)$$
$$= -\frac{1}{2}(\log |\sigma_*{}^2| + (Y_T - \mu_*)^\top \sigma_*{}^{-2} (Y_T - \mu_*) + N \log(2\pi)), \tag{10}$$

where the values of μ_* and σ_*^2 can be obtained by extending the source part in Equation (6), where N represents the number of data used for training. The set of parameters that need to be learned is represented by Θ, which includes kernel parameters, data noise (σ_{s1}^2, ..., σ_{s2}^2, σ_t^2), and the similarity of the domain (α_D).

The benefits of MTGP can be summarized as follows.

- The utilization of multiple sources for training in MTGP results in an enhancement of its generalization performance.

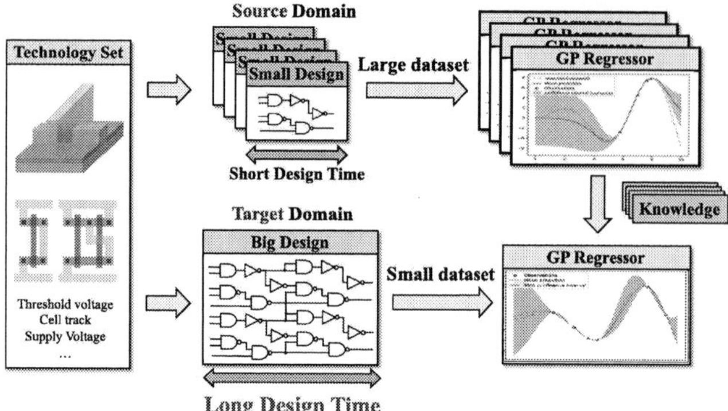

Fig. 4. The necessity of MTGP for efficient DTCO: multiple small design designs can be reused for larger design prediction models.

- MTGP can enhance the accuracy of the model in the target design even in situations where there is a limited amount of target design data available.
- MTGP achieve greater expressiveness by taking into account not only the similarity between the source and target designs but also the similarity between different source designs.

MTGP can leverage these capabilities to enhance the efficiency of DTCO by improving its optimization performance.

C. Cost Function

Our objective is to minimize both power and area while meeting the timing constraints. Due to the inherent trade-offs among PPA, a flexible black-box optimization approach is necessary. To address this, we referred to [9] to construct a weighted sum cost function that accounts for multiple metrics. Given that each metric has distinct units, we normalized each metric to a value determined using the reference foundry-set provided.

$$P_{nor} = \frac{P_{cur} - P_{ref}}{P_{ref}}, \quad A_{nor} = \frac{A_{cur} - A_{ref}}{A_{ref}}, \quad S_{nor} = \frac{S_{cur}}{P_{clock}}, \tag{11}$$

where the normalized power, area, and worst negative slack (WNS) are denoted as P_{nor}, A_{nor}, and S_{nor}, respectively. The power, area, and WNS of the current design using the modified design and technology options are denoted as P_{cur}, A_{cur}, and S_{cur}, while the power and area obtained from the reference foundry-set are represented by P_{ref} and A_{ref}, respectively. The clock period used for that design is denoted as P_{clock}.

We then utilized a single-objective Bayesian optimization framework to optimize the weighted-sum cost function. The specific form of the cost function we employed is as follows:

$$f = \begin{cases} \alpha \cdot P_{nor} + \beta \cdot A_{nor}, & S_{cur} > 0 \\ \alpha \cdot P_{nor} + \beta \cdot A_{nor} - \gamma \cdot S_{nor}, & S_{cur} < 0 \end{cases}, \tag{12}$$

where the weights for power and area are denoted by α and β, respectively. Furthermore, the penalty weight for WNS is represented by γ, since power and area improvements are performed under the assumption that timing violations are avoided. Consequently, a penalty is imposed when the timing constraints are not met. The designer can adjust these values flexibly based on the optimization goals of the framework.

D. Bayesian Optimization

Bayesian optimization is a promising approach for optimizing black-box problems, which consists of two main components: a surrogate model and an acquisition function. The surrogate model is utilized to estimate the cost function of a black box, while the acquisition function is responsible for searching for a point where the cost function is expected to be minimized, based on the surrogate model. At this stage, it is crucial to balance the trade-off between exploration and exploitation. Exploration involves searching for points with high variance (σ^2), which corresponds to identifying regions of uncertainty. In contrast, exploitation involves selecting a value with a smaller predicted average (μ) from previously searched points, with the aim of identifying a lower value around a known point. To achieve a good balance between exploration and exploitation, we introduce a lower confidence bound as the acquisition function. The lower confidence bound is calculated as the sum of the mean of the values and the confidence interval, and recommends the best place for further exploration. The formula for the lower confidence bound is as follows.

$$\begin{aligned} a(x; \lambda) &= \mu(x) - \kappa\sigma(x), \\ x_* &= \arg\min(a(x; \kappa)), \end{aligned} \tag{13}$$

where the κ parameter denotes the sampling propensity of the model, whereby higher values of κ correspond to an increase in the number of searches performed by the model. κ is set to 1 in normal. x_* denotes the subsequent point of sampling that is anticipated to yield a lower cost.

TABLE I
TESTCASES SPECIFICATIONS

Bench	# of cells	Design time	Domain
aes	12k	4 min	Source
keccak	28k	12 min	Source
ldpc	42k	35 min	Source
des3	68k	20 min	Source
nova	144k	40 min	Target
rocket_chip	691k	180 min	Target

IV. Experimental Setup & Result

A. Experimental Setup

The experiments were conducted on a computing system consisting of an AMD EPYC-Rome CPU with 15 cores operating at 2.8 GHz and 200 GB RAM, and an NVIDIA Geforce 1080 Ti graphics processing unit. The operating system used was CentOS Linux 7.9. The ASAP7 PDK [16] with BSIM-CMG model [17] was employed, with *OpenCores* circuit designs [18] and *RocketChip* [19] serving as test cases. The characteristics of the test cases are presented in TABLE I.

TABLE II
DESIGN & TECHNOLOGY PARAMETERS

Parameter	min	max	Step size	# of combinations
LVT	4.250	4.345	0.005	20
RVT	4.350	4.450	0.005	21
VDD	0.50	0.90	0.05	9
TRACK	6T	7.5T	1.5T	2
# of all combinations: 7,560				

* LVT and RVT refer to the corresponding gate work function, respectively.
* Units for LVT and RVT are in eV, while VDD is in V.

B. Process Design kits Generation

To generate process design kits (PDK), we modified the design and technology options described in Section II-A. To adjust the threshold voltage, we employed the BSIM-CMG model, which is capable of simulating device characteristics based on process changes. Specifically, we adjusted the gate work function parameter *phig* to modify

TABLE III
PREDICTION: COMPARISON RESULTS OF PREDICTING POWER, AREA, AND PERFORMANCE (WNS)

Metric		Target 1						Target 2					
		GP	STGP [10]				MTGP	GP	STGP [10]				MTGP
			ldpc	keccak	aes	des3			ldpc	keccak	aes	des3	
MAE	power	655	532	775	532	491	**386**	29,513	16,108	21,535	26,232	16,046	**13,691**
	area	172	126	160	183	185	**117**	801	812	880	862	793	**673**
	wns	185.60	209.97	171.62	173.49	168.64	**120.56**	**11.42**	11.75	12.21	12.06	11.99	11.59
R2	power	0.8673	0.9149	0.8693	0.9189	0.9279	**0.9570**	0.5789	0.9040	0.7839	0.6774	0.9015	**0.9196**
	area	0.9863	0.9922	0.9880	0.9843	0.9830	**0.9935**	0.9891	0.9841	0.9903	0.9899	0.9909	**0.9934**
	wns	0.3408	0.1821	0.4473	0.4309	0.4550	**0.7334**	0.5175	0.5431	0.4698	0.4842	0.4592	**0.5749**
95% CI ACC	power	78.84%	85.06%	80.50%	80.08%	83.40%	**100.00%**	65.98%	80.91%	70.54%	64.73%	78.84%	**96.68%**
	area	**100.00%**	99.17%	98.76%	**100.00%**	**100.00%**	**100.00%**	99.17%	98.76%	**100.00%**	**100.00%**	99.17%	**100.00%**
	wns	36.51%	58.09%	9.96%	11.62%	15.35%	**93.36%**	70.12%	16.18%	52.28%	55.60%	60.17%	**83.40%**

* The Target 1 is nova and the Target 2 is rocket_chip.
* The best results are shown in bold.
* 95% CI ACC measures the percentage of test data that fell within the 95% confidence interval.

the threshold voltage. Since modern designs employ Multi-V_{th} technique to minimize leakage power, we utilized two V_{th} cell libraries, LVT and RVT. In this context, LVT and RVT refer to low and regular threshold voltage cells respectively. Therefore, we specified each set of data points in the range of [4.250, 4.345] and [4.350, 4.450] in units of eV as input parameters for a pair of NMOS gate work functions. The endpoints of this range corresponded to the values of the foundry-set SLVT and HVT, respectively, and the step length was set to 0.005 eV. In this context, the foundry-set SLVT and HVT refer to the super-low and high threshold voltage cell libraries provided by the foundry as standard, respectively. To ensure symmetric operation of CMOS, we first modified the NMOS gate work function, found the NMOS drain-source current i_{ds} through SPICE simulation, and then adjusted the PMOS gate work function to match the PMOS i_{ds} to 90% of the NMOS i_{ds}. We also considered the effect of supply voltage scaling using library characterization tools. We specified each data point in the range of [0.50, 0.90] with a step length of 0.05 V. Finally, we used 7.5T and 6T standard cells to represent high-performance and high-density cells, respectively, resulting in a total design space of 7,560. A summary of the input parameters is presented in TABLE II. Subsequently, logic synthesis was performed to assess the PPA of the modified design and technology options. The clock period for synthesis was determined as a value between the maximum clock periods that could be achieved using each foundry-set LVT and foundry-set RVT. For our experiments, *HSPICE v21.9* was used for SPICE simulation, *Silicon Smart v21.6* for library characterization, and *Design Compiler v18.6* for logic synthesis.

C. Prediction Model Evaluation

In our study, we selected three evaluation metrics to assess the performance of our predictive model: mean absolute error (MAE), R2 score, and 95% confidence interval accuracy. Since GP predicts both the expected value (μ) and the confidence interval (σ^2), we needed to include an evaluation indicator for the predicted confidence interval. To do this, we measured the accuracy of the test data within the 95% confidence interval predicted by the model. To evaluate the predictive model's performance, we utilized 250 grid data points that were evenly spaced in the design space. We used the ADAM [20] optimizer with 1,000 iterations to train each predictive model. To reduce learning time, multi-source transfer GP (MTGP) set $\alpha_{S,T}$ learned through single-source trasfer GP (STGP) as the initial value of MTGP, and then only learned 100 times. All of our models were implemented in *Python* using *PyTorch*.

We conducted an experiment to compare transfer performance using 200 of source design data and 10 of target design data. The experimental results are presented in the TABLE III. In our study, we compared the performance of MTGP and reference GP in predicting target design data. We found that MTGP outperformed reference GP by reducing both MAE of power and MAE of area by 47.33% and 24.07%, respectively, as well as reducing MAE of WNS by 16.75%. MTGP also showed improved R2 scores for power, area, and WNS, with the R2 score of WNS showing the most significant improvement at 63.14%. To illustrate the performance of our predictive model, we present the power prediction results for Target 2 as a representative example in Fig. 5. We further demonstrated that using multiple source design data for learning can increase the accuracy of the 95% confidence interval and prevent overfitting to a specific source design. These results suggest that building a predictive model using multiple sources is more effective than relying on data from a single source design.

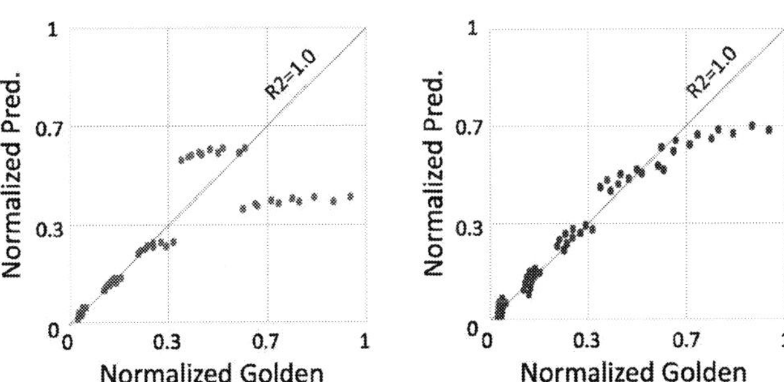

Fig. 5. Performance comparison of GP and MTGP for power prediction of Target 2: The power prediction plots in the left and right panels show the results obtained using the reference GP and MTGP models, respectively.

D. Optimization Result

We utilized MTGP for DTCO optimization, building on the model obtained in Section IV-C. We performed optimization 30 times using Bayesian optimization, with α, β, and γ set to 1.0, 1.0, and 3.0, respectively. The reference design used the results using foundry-set LVT, foundry-set RVT, 0.70V and 7.5T. We conducted a Bayesian optimization on multiple GP models and analyzed their convergence performance. Fig. 6 shows the results, where the vertical axis represents cost, as per Equation (12), and the horizontal axis represents the number of optimization steps. To better capture trends, we plotted cost values as 11-step-sized moving averages. The plot clearly shows that MTGP outperforms other models in terms of optimization performance such as the convergence and stability for both targets.

The results presented in TABLE IV illustrate the minimum cost points that each model explored while meeting the timing constraints. These results demonstrate that Bayesian optimization using MTGP

TABLE IV
OPTIMIZATION: COMPARISON RESULTS OF OPTIMIZING POWER AND AREA WHILE MEETING TIMING CONSTRAINTS

Metric	Target 1						Target 2					
	GP	STGP [10]				MTGP	GP	STGP [10]				MTGP
		ldpc	keccak	aes	des3			ldpc	keccak	aes	des3	
power	-28.10%	30.36%	2.68%	28.91%	3.15%	**34.06%**	14.52%	14.52%	32.03%	3.65%	39.10%	**40.51%**
area	0.83%	-0.86%	0.45%	-0.72%	0.08%	**17.80%**	-0.18%	-0.18%	**22.09%**	22.07%	-0.27%	21.93%
wns	0.00	0.00	0.00	0.00	0.00	0.00	0.00	0.00	0.00	0.00	0.00	0.00

* The Target 1 is nova and the Target 2 is rocket_chip.
* The best results are shown in bold.

outperforms conventional Bayesian optimization in terms of both optimization performance. Additionally, while the optimization results achieved through STGP can vary depending on the source and target design, MTGP consistently exhibits excellent performance across various conditions. For example, as shown in Fig. 6, STGP transferred *ldpc* outperforms STGP transferred *keccak* for Target 1, while the opposite is true for Target 2. As shown in Table III, the WNS's MAE of STGP transferred *ldpc* at Target 1 was observed to be inferior to that of STGP transferred keccak, whereas the 95% confidence interval accuracy is significantly higher for STGP transferred *ldpc*. These observations can be explained by the fact that in Bayesian optimization, the selection of the next sampling point takes into account not only the predicted value but also the confidence interval, as per Equation (13). Hence, as MTGP outperforms STGP in terms of both MAE and accuracy of 95% confidence interval, it can be concluded that the optimization performance is robust and effective, without any overfitting issues. Our experimental results demonstrate that the proposed methodology improves power and area by an average of 37.29% and 19.87%, respectively, for both target designs compared to the reference design, while meeting the imposed timing constraints.

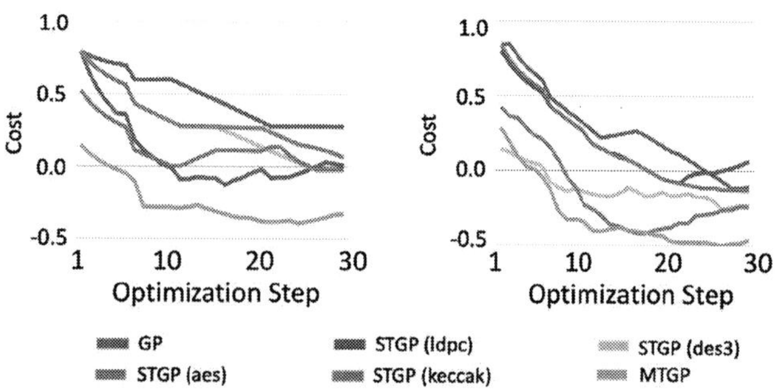

Fig. 6. Convergence comparison of GP, STGP, and MTGP. The left plot shows the convergence tendency for Target 1, and the right plot shows the convergence tendency for Target 2.

V. CONCLUSION

In this study, we propose a DTCO framework based on the Multi-source Transfer Gaussian Process (MTGP) that effectively explores a vast design and technology space. Our approach transfers small design data to a larger design, improving the performance of the predictive model and demonstrating its scalability. Moreover, we introduce MTGP to integrate and transfer multiple source designs into the framework, achieving a lower generalization error than only transferring single source design. In our experiments, we achieved an average reduction of 47.33% and 24.07% in the average mean absolute error for power and area, respectively, compared to the reference GP. Furthermore, in the 7nm technology node, we achieved a average reduction of 37.29% and 19.87% in power and area, respectively, while meeting imposed timing constraints, compared to the reference design. Our experimental results demonstrate that our method reduces the optimization iterations required for DTCO and enables efficient design feedback.

VI. ACKNOWLEDGMENTS

This work was partly supported by Institute for Information & communications Technology Promotion (IITP) grant funded by the Korea government (MSIT) (No.RS-2023-00222085, Development of memory module and memory compiler for non-volatile PIM optimized for data characteristics and data access characteristics of AI processor / No.2021-0-00754, Software Systems for AI Semiconductor Design / No.2022-0-01172, DRAM PIM Design Base Technology Development). Furthermore, this research was supported in part by Samsung Electronics Co., Ltd.

REFERENCES

[1] T. Song, *et al.*, "3nm Gate-All-Around (GAA) Design-Technology Co-Optimization (DTCO) for succeeding PPA by Technology", *Proc. CICC*, 2022.
[2] V. Moroz, *et al.*, "DTCO Launches Moore's Law Over the Feature Scaling Wall", *Proc. IEDM*, 2020.
[3] CK, Cheng, *et al.*, "Complementary-FET (CFET) standard cell synthesis framework for design and system technology co-optimization using SMT", *IEEE Transactions on VLSI*, 2021, pp. 1178-1191.
[4] CK, Cheng, *et al.*, "PROBE2.0: A systematic framework for routability assessment from technology to design in advanced nodes", *IEEE Transactions on CAD*, 2021, pp. 1495-1508.
[5] J. Jeong, *et al.*, "A Study on Optimizing Pin Accessibility of Standard Cells in the Post-3 nm Node", *Proc. ISLPED*, 2022.
[6] B. Reagen, *et al.* "A case for efficient accelerator design space exploration via bayesian optimization", *Proc. ISLPED*, 2017.
[7] Y. Ma, *et al.*, "CAD tool design space exploration via Bayesian optimization", *Proc. MLCAD*, 2019.
[8] H. Geng, *et al.*, "PTPT: physical design tool parameter tuning via multi-objective Bayesian optimization", *IEEE TCAD*, 2022, pp. 178-189.
[9] J. Jung, et al, "METRICS2.1 and Flow Tuning in the IEEE CEDA Robust Design Flow and OpenROAD", *Proc. ICCAD*, 2021, pp. 1-9.
[10] H. Geng, *et al.*, "PPATuner: pareto-driven tool parameter auto-tuning in physical design via gaussian process transfer learning", *Proc. DAC*, 2022.
[11] Z. Zhang, *et al.*, "A fast parameter tuning framework via transfer learning and multi-objective bayesian optimization", *Proc. DAC*, 2022.
[12] V. De, *et al.*, "Near threshold voltage (NTV) computing: Computing in the dark silicon era", *IEEE Design & Test*, 2017, pp. 24-30.
[13] M. Mustafa, T. et al, "Threshold Voltage Sensitivity to Metal Gate Work-Function Based Performance Evaluation of Double-Gate n-FinFET Structures for LSTP Technology", *World Journal of Nano Science and Engineering*, 2013, pp. 17-22.
[14] P. Wei, *et al.* "Adaptive Transfer Kernel Learning for Transfer Gaussian Process Regression", *IEEE TPAMI*, 2022.
[15] P. Wei, *et al.*, "Transfer Kernel Learning for Multi-Source Transfer Gaussian Process Regression", *IEEE TPAMI*, 2022, pp. 10.1109. 2016.
[16] L. T. Clark, *et al.*, "ASAP7: A 7-nm finFET predictive process design kit", *Microelectronics Journal*, 2016, pp. 105-115.
[17] S. Khandelwal, *et al.*, "BSIM-CMG 110.0.0: Multi-gate MOSFET compact model: technical manual", 2014
[18] OpenCores: "Open Source IP-Cores", http://www.opencores.org.
[19] Asanovic, Krste, *et al.*, "The rocket chip generator", EECS Department, University of California, Berkeley, Tech., 2016.
[20] Kingma, D. P, *et al.*, "Adam: A method for stochastic optimization", arXiv preprint arXiv:1412.6980, 2014.

Enabling Highly-Efficient DNA Sequence Mapping via ReRAM-based TCAM

Yu-Shao Lai
National Taipei Univ. of Tech.
Taipei, Taiwan
cxi197@gmail.com

Shuo-Han Chen
National Yang Ming Chiao Tung Univ.
Hsinchu, Taiwan
shchen.nycu@gmail.com

Yuan-Hao Chang
Academia Sinica
Taipei, Taiwan
johnson@iis.sinica.edu.tw

***Abstract*—In the post-pandemic era, third-generation DNA sequencing (TGS) has received increasing attention from both academics and industries. As TGS technologies have become a requisite for extracting DNA sequences, the DNA sequence mapping, which is the most basic bioinformatics application and the core of polymerase chain reaction (PCR) tests, receives great challenges, due to the large size and noisy nature of TGS technologies. In addition, the ever-increasing data volume of DNA sequences also induces the issue of memory wall while large datasets are moved between the memory and the computing units. However, much less effort has been devoted to DNA sequence mapping acceleration while considering both the memory wall issue and the challenges of TGS technologies. To enable highly-efficient DNA sequence mapping, this study proposes a novel resistive random-access memory (ReRAM)-based ternary content-addressable memory (TCAM) and exploits the intrinsic parallelity of ReRAM crossbar for efficient mapping acceleration. Promising results have been demonstrated through a series of experiments with different scales of datasets.**

***Index Terms*—DNA, sequence mapping, ReRAM, TCAM**

I. Introduction

The booming of biotechnology applications and studies have facilitated the evolving of DNA sequencing. The goal of DNA sequencing is to determine the nucleic acid sequence, which is the order of nucleotides (*i.e.*, Adenine, Thymine, Cytosine and Guanine) in DNA, for measuring gene expression. The latest third-generation DNA sequencing (TGS) [8] have proven to be cost-effective and capable of producing accurate genome assemblies. For instance, the Oxford Nanopore Technologies (ONT) [13] has proposed a nanopore-based TGS platform for producing high-quality genome assemblies at an affordable cost. Practically, instead of producing a complete genome assembly at once, DNA sequencing produces multiple fixed-length genes, which are known as *DNA reads* or *reads* and are used as the basic unit for bioinformatics applications. The strength of TGS is the capability of producing substantially long reads at the size over 10 kb in length. Unlike those small reads produced by the first a nd s econd g eneration of DNA sequencing technologies, large-sized reads lower the duplications between each read and enhance the efficiency of DNA sequencing. However, *the error rate also increases and complicates the process of DNA sequence mapping.*

Among various bioinformatics applications, the DNA sequence mapping is the first s tep f or v arious a nalysis a nd is utilized to align DNA reads to a reference genome. Figure 1

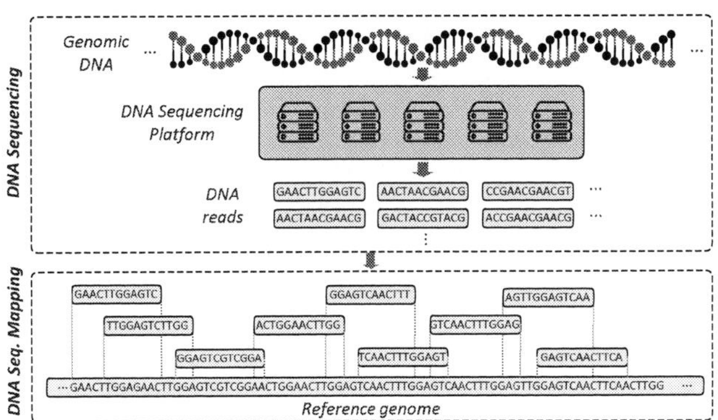

Fig. 1. Workflow of DNA sequencing and DNA sequence mapping, *in which genomic DNA are sampled via sequencing platforms and digitized as DNA reads. DNA reads are then mapped to a reference genome.*

illustrates the process of DNA sequencing and DNA sequence mapping. As shown in the figure, *reads* produced by DNA sequencing platforms are aligned to the reference genome based on the nucleotides order; thus, extensive read and compare operations are involved. In other words, DNA sequence mapping is usually the most computationally intensive part of bioinformatics applications [12]. To facilitate the process of DNA sequence mapping, various read-mapping tools [4, 9, 16], software optimizations [3, 14], and hardware accelerations [1, 10] are proposed. Nevertheless, due to the large-sized and noise-prone-nature of TGS technologies, previous deigns have become less effective and could lead to inaccurate genome assemblies [9]. In addition, while large-sized reads of TGS technologies are moved between memory and computing units for DNA sequence mapping, the energy overhead and inefficiency increases as the issue of memory wall emerges.

To alleviate the issue of memory wall, in-memory computing (IMC) accelerations have been proposed to perform computation directly with memory cells. For instance, based on short reads, Huangfu et al. [10] propose facilitating the existing BLASTN mapping tool through content-addressable memory (CAM) cells. The circuitry of CAM allows parallel match/mismatch operations to be conducted within one memory cycle and is regarded as a good IMC candidate for DNA sequence mapping. Nevertheless, instead of exploiting IMC circuitry characteristics for pattern matching, the previous CAM-based design includes additional transistors and resistors in each memory cell for pattern comparison, thus increasing areal overhead. In addition, since the CAM-based

979-8-3503-1176-1/23 $31.00 © 2023 IEEE

design demands for complete match between DNA reads and reference genome, *CAM-based accelerations could lead to lowered accuracy when the length and error rate of DNA sequence increase.*

To resolve aforementioned challenges of TGS technologies, this paper propose a pioneering resistive random-access memory (ReRAM)-based ternary content-addressable memory (ReTCAM) to produce the mapping results of DNA reads and reference genome. Unlike the previous CAM that requires complete match between DNA reads and reference genome, TCAM allows *don't care* bits in the to-be-compared sequence and skips the comparison while performing match/mismatch operations. To exploit the benefit of *don't care* feature, the proposed ReTCAM marks nucleotide as *don't care* based on quality scores provided by TGS technologies. The quality score is an integer value that indicates the estimated probability of an error when extracting a nucleotide from DNA molecules. In other words, the proposed ReTCAM is able to skip the comparison of those possible-inaccurate nucleotides within each DNA read. Notably, the proposed ReTCAM implements the functionality of TCAM within ReRAM crossbar circuitry without including new transistors and resistors, thus achieving high energy efficiency and low area overhead. Meanwhile, to support TCAM within ReRAM crossbar, the encoding method of ReTCAM is carefully designed to support the *don't care* feature of TCAM. Lastly, the accumulated current of each ReTCAM comparison operations are utilized to aid the subsequent DNA sequence mapping steps. The evaluation results show that the proposed ReTCAM can outperform conventional CPU-based DNA sequence mapping by 99.72% and 99.76% in terms of energy and latency.

The rest of this paper is organized as follows. The background and research motivation are summarized in Section II. Then, Section III describes the proposed ReTCAM. Then, a series of experiments are conducted in Section IV. Finally, Section V concludes this paper with research remarks.

II. Background and Motivation

A. DNA Sequence Mapping

While DNA sequencing refers to the laboratory technique for extracting the nucleotide sequences (*i.e.*, reads) from DNA

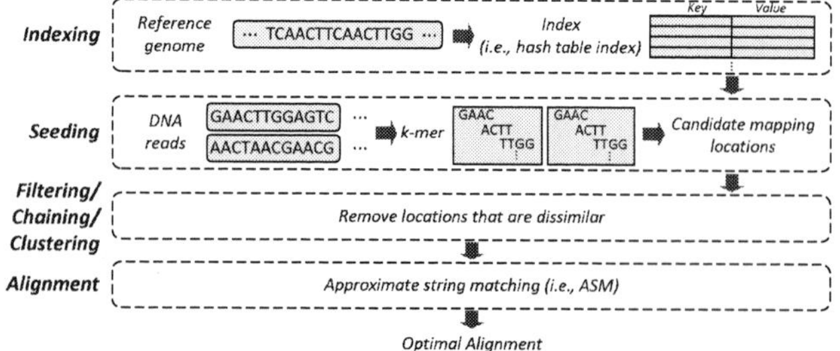

Fig. 2. Steps of typical seed-and-extend mapping tools, *in which the reference genome is preprocessed as index and the DNA reads are sub-sequenced as k-mer. Then, alignment results are derived through approximate string matching (i.e., approximate string matching, ASM).*

Position	1	2	3	4	5	6
Sequence	G	A	A	C	T	T
k-mer	G	A	A			
		A	A	C		
			A	C	T	
				C	T	T

Fig. 3. Generating k-mers by sub-sequencing and pushing position back by one at a time. *Notably, k-mers are utilized as the unit of seeding and refers to all the sub-sequences with the length of k in DNA reads. k-mers with length of 3 can be noted as 3-mers.*

molecules, the DNA sequence mapping specifies the process of aligning the reads produced by DNA sequencing techniques to a reference genome for further bioinformatics analysis. To perform DNA sequence mapping, most modern mapping tools follow the *seed-and-extend* approach, which is composed of the seeding and the extension stages. Seeding refers to the process of identifying identical or highly-matching patterns between the sub-sequences of DNA reads and the reference genome. Then, during extension stage, the reads are aligned to reference genome based on comparison results of seeding stage. Typically, as summarized in Figure 2, mapping tools that adopt the seed-and-extend approach can be further divided into four steps, including indexing, seeding, optional filtering/chaining/clustering, and alignment.

Indexing pre-processes the reference genome and generates an index of the reference. Then, seeding generates a set of *k-mers* (See Figure 3), which are k-length sub-sequences of DNA reads, and finds the exact matching locations of these k-mers in the reference genome. Notably, these mapped locations are referred to as *seeds*. Next, filtering/chaining/clustering is an optional step to lower the number of candidate mapping regions (*i.e.,* seeds) by removing dissimilar locations. The goal of this step is to enhancing the alignment performance by decreasing the number of alignment operations. Finally, alignment finds the similarities and differences between the read and the reference regions from those filtered candidate mapping locations from the previous step. Following above steps, Minimap [14] has been proposed as a TGS-suitable mapping tool with excellent performance. Minimap outperforms other mapping tools by discovering that long reads can be processed without the post-assembled error correction and genome assemblies can still be derived with similar accuracy, when compared with other long-read mapping tools.

Apart from software-based tools, various studies also attempt to accelerate read-mapping through hardware-based computation. For instance, GateKeeper [1], BWA-MEM [9], CloudBurst [16] and MapReduce [4] aim to enhance read-mapping algorithms by exploiting the parallelism through FPGA, GPU, cloud and distributed computing. Even though these hardware accelerations are effective, the issue of memory wall has emerged due to the sheer amount of DNA nucleotides and the increased read length of TGC technologies. Among the steps of DNA mapping tools, the seeding step is the most computational-intensive part and is responsible for most of the memory space requirement and the memory accesses. Such circumstances has made in-memory computing (IMC) become an ideal acceleration method for the seeding step to lower the amount of data movement between memory and computing units. For instance, GRIM-Filter [11] focuses on improving hash-table based filters using the shifted hamming distance algorithm, while Huangfu et al. [10] directly apply the existing BLASTN mapping toolto content-addressable memory (CAM) cells for in-memory computation. Nevertheless, as most previous accelerators can only support short reads, *previous designs could have lower mapping accuracy or inefficiency due to the large-sized reads of TGC methods.*

B. CAM and ReRAM Corssbar

To resolve the performance issue and bandwidth limitation of data-intensive workloads, a promising and effect solution is to access memory via content directly, rather than memory address. This type of memory are often referred to as the associative memory or the content addressable memory (CAM) [15] and is optimized for performing parallel searches through data. CAM are a collection of basic CAM cells laying out in a two dimensional array. Basic CAM cells are composed of SRAM and additional associated comparison circuitry, which are utilized to conduct data search in the whole CAM in a single clock cycle. Conventionally, the term, CAM, usually refers to the binary CAM (BiCAM), which can only store '0' and '1', while ternary CAM (TCAM) can store '0', '1' and 'x' (*i.e.,* don't care). Figure 4 illustrates the working flow of two-dimensional CAM cells. During each comparison cycle, the match line is first precharged and the input data arrives at each cell through the search lines. Meanwhile, those to-be-compared data are stored in the SRAM of each CAM cell. If the pattern matches, the current on the match line flows through the cell and arrive at the sense amplifier. Otherwise, the current on the match line is redirected to ground. In summary, with the ability to perform searches in one memory cycle, both CAM and TCAM are highly suitable for fast searching applications and are widely utilized for network packet forwarding, packet classification, and lookup tables.

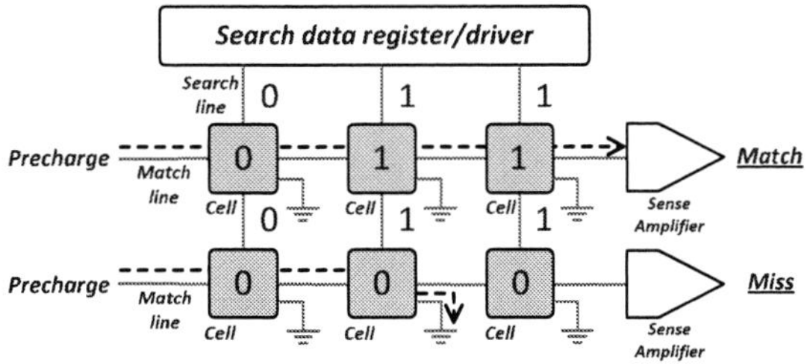

Fig. 4. Example of CAM-based comparison with 3 data bits, *which shows that the precharge current flows till the sense amplifier if pattern matches. Otherwise, the precharge current is grounded by CAM cells if mismatch.*

Although CAM and TCAM can perform fast searching through its specialized hardware, conventional CAM cells still rely on volatile SRAM for storing to-be-compared data patterns. In other words, even though SRAM-based CAM functions well in its current application fields, such as network packet forwarding, with relative smaller data volume, the efficiency of SRAM-based CAM decreases as the data volume grows, owing to the issue of memory wall. To resolve above concerns, CAM designs that rely on nonvolatile memory, such resistive memory [6] and STT-RAM [7], are proposed. In addition, owing to the high structural similarity between CAM and ReRAM crossbar, ReRAM crossbar has also be exploited to construct CAM with additional transistors and resistors for data comparison [6]. The structure and internal details of ReRAM corssbar are illustrated in Figure 5.

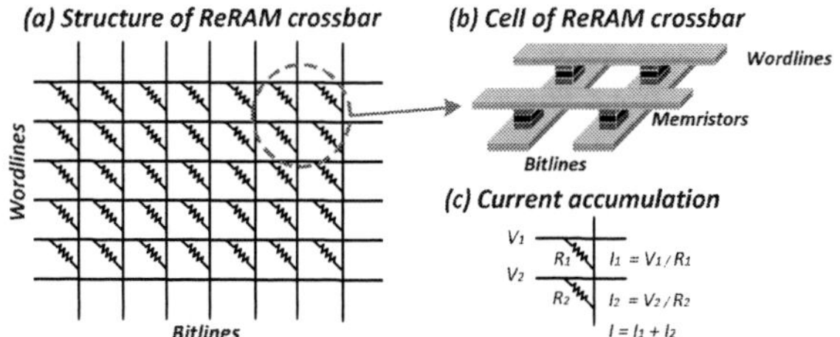

Fig. 5. ReRAM crossbar, *which shows the 2-dimension ReRAM structure, the cell structure, and the accumulated current if two cells are activated.*

As shown in Figure 5(a), ReRAM crossbar interconnects multiples memristors at the intersection of horizontal wordlines and vertical bitlines. To store information within ReRAM crossbar, the resistance of memristors is switched between the low resistance state (LRS) and high resistance state (HRS) through *SET* and *RESET* current. Figure 5(b) further shows that each memristor is fabricated by sandwiching a metal oxide layer between the top and bottom metal layers, and the cell size could be as small as $4F^2$, which is basically the size of overlapped area between the wordline and bitline. Meanwhile, by applying proper voltages to both the worldline and bitline, cells can be accessed independently or at the same time and those accessed cells are denoted as the *full-selected cells.* Notably, when multiple cells on the same bitline are accessed a the same time, the generated accumulated current, as shown in Figure 5(c) on the same bitline can exploited for further applications, such as matrix multiplication [17] or CAM [9] with additional transistors and resistors. However, *the possibility of utilizing ReRAM crossbar for constructing TCAM without additional transistors and resistors has received much less attention.*

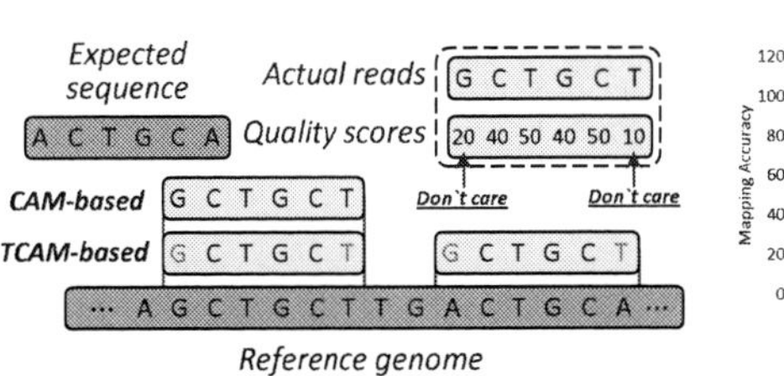

Fig. 6. Motivational example, *in which the CAM-based mapping produces an inaccurate mapping location due to incorrect actual read and TCAM-based derives two possible locations with a correct one.*

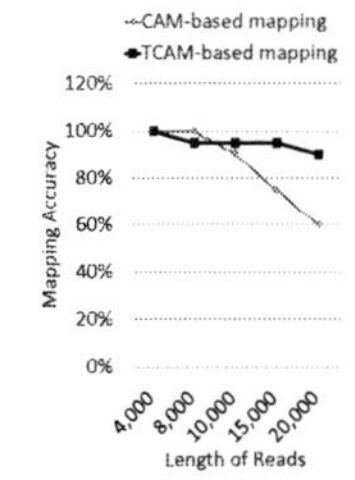

Fig. 7. Accuracy comparison, *which shows the accuracy of CAM-based mapping drops.*

C. Motivation

In this section, both a motivational example and the experiment results are included to explain how the long-sized reads of TSG technologies affect the accuracy of DNA sequence mapping. As shown in Figure 6, the DNA reads produced by TGS methods are labeled with quality scores for each nucleotide. If the quality score is smaller than certain value (*i.e.,* 30), the sampled nucleotide is highly likely to be an incorrect one. These incorrect nucleotides could lead to inaccurate candidate mapping locations in the reference genome and ultimately lower the mapping accuracy. Accordingly, the issue of lowered accuracy aggravates as the size of DNA reads grows. This is because the growing size of DNA reads also increases the number of nucleotides with low quality value in each DNA read. To better understand the accuracy difference, experiments are conducted based on the TGS-suitable Minimap design and Figure 7 summarizes the results.

The CAM-based mapping represents that the DNA reads is mapped to the reference genome only if the patterns are exactly identical. On the other hand, the TCAM-based mapping marks those nucleotides with quality score lower than 30 as *don't care* and skips the comparison for those nucleotides (See Figure 6). The investigated length of DNA reads is set between 4 and 20 thousand nucleotides and DNA reads are typically regarded as long reads if the length is longer than 10 thousand nucleotides. Accordingly, the accuracy drop varies from 5% to 33% when the length is between 10 and 20 thousand. In other words, according to our investigation, the growing read length does lower the mapping accuracy of the TGS-suitable Minimap design and exploiting TCAM with *don't care* feature for DNA sequence mapping seems maintaining good accuracy, when compared with CAM-based mapping.

III. ReRAM-based TCAM

A. Overview

Motivated by the worsening accuracy of previous TGS-suitable DNA mapping tools, this study proposes ReTCAM to utilize the *don't care* feature of TCAM for efficient DNA sequence mapping while maintaining good accuracy, regardless of the DNA read sizes. First, the proposed ReTCAM relies on the ReRAM crossbar for its nonvolatility, thus eliminating the need to move data between the computation units and the main memory. Second, the accumulated current of activated memristors on bitlines is utilized as an indication for match, mismatch, and *don't care* condition directly. This approach is made possible by the encoding method of the proposed ReTCAM, and the accumulated current intensity can also be utilized for adding other stages of DNA sequence mapping algorithms. Accordingly, the system architecture of the proposed ReTCAM is summarized in Figure 8.

As shown in the figure, the proposed ReTACM is a basic component in larger hardware for enforcing in-memory computation. The overall hardware comprises multiple process engine (PE), and, in a single PE, there are peripheral circuits

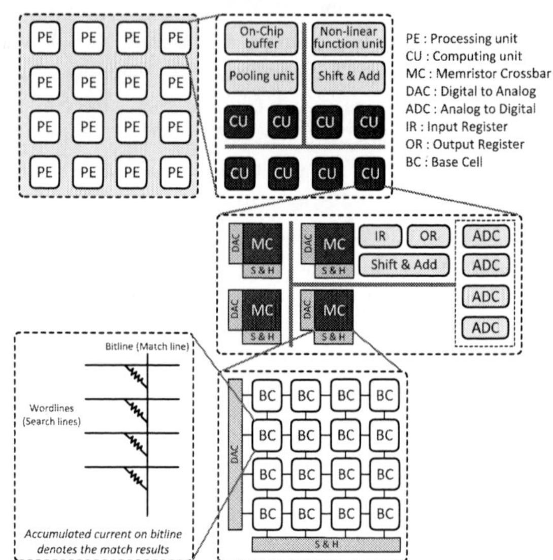

Fig. 8. System architecture of ReTCAM, *which shows that the proposed ReTCAM is part of larger in-memory computing hardware.*

and many computing units (CU), which is composed of several memristor crossbars (*i.e.,* ReRAM) (MC), where the proposed ReTCAM is implemented. In the proposed ReTCAM, the reference genome is stored within MC, and each base unit (BC) represents a nucleotide. For comparison, the input DNA read is divided into k-length subsequence, k-mer, for comparison. Within each MC, the reference genome is also divided into k-mer, each of which is stored in multiple BC vertically. In other words, the number of BC in the vertical direction is based on the length of k-mer. Then, the horizontal direction in each MC is the number of k-mers of the reference genome.

B. Nucleotide Encoding

Conventionally, four nucleotides can be represented by 2 bits after the nucleotide is digitized. However, to utilize memristors for comparing the input bit pattern and the stored bit pattern, the generated current should be consistent and should be able to indicate the match or mismatch result without additional comparison. If 2-bit encoding is utilized, the number of 1 bit in each nucleotide would be either 1 or 2, as shown in Figure 9, and the generated current varies according to different input and stored nucleotides. Although it seems to be manageable from the software perspective, it

Base	2 bits	3 bits	4 bits
A	00	000	0001
C	01	001	0010
G	10	010	0100
T	11	100	1000
don't care	-	111	1111

Fig. 9. Encoding methods with different number of bits, *which shows that 4-bits pattern with one-hot encoding allows consistent current intensity for match and mismatch results.*

induces additional overheads if it is implemented as hardware since additional current intensity comparisons are required. A similar condition can be found for 3-bit encoding. To enable consistent current intensity for match and mismatch results, ReTCAM proposes utilizing the 4-bit pattern with one-hot encoding method and denotes the *don't care* with four 1 data bits. In other words, for match and mismatch conditions, the generated current of each base cell would be identical, while *don't care* generates the highest current on the bitline (See Section III-C).

C. Circuitry

With the goal of deriving match and mismatch results directly from memristors without additional transistors and resistors, ReTCAM utilizes the 4-bit pattern with the one-hot encoding method for each nucleotide. Then, within each based cell of ReTCAM, different current intensities would be generated according to three conditions, including match, mismatch, and *don't care*, regardless of the input and stored nucleotide. Examples are given in Figure 10 for a better explanation. For example (a), the input nucleotide is A with the bit pattern 1000, and the stored nucleotide is C with the bit pattern 0100. It can be inferred that there are bit differences and therefore is a mismatch condition. The final generated current intensity can be calculated by dividing input voltage by high resistance, denoted as $1/H$, assuming the voltage is 1 and resistance is H. Same current intensity can be derived for different nucleotide mismatches with the help of the one-hot encoding method.

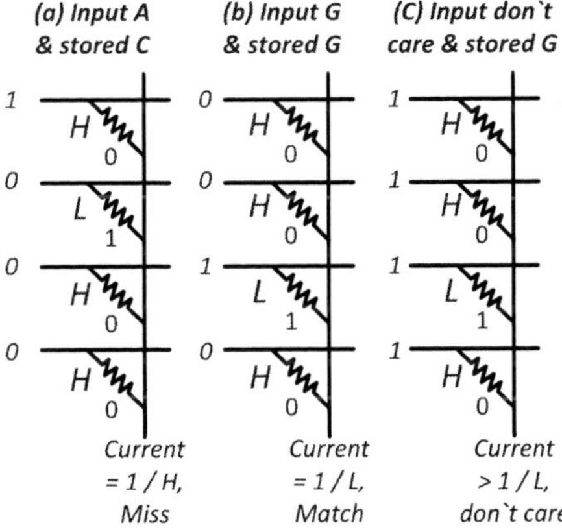

Fig. 10. Examples of generated different current intensities, *in which the high and low resistance states of memristors represents of the stored 0 and 1 data bits.*

On the other hand, in example (b), the current of the match condition can be calculated as dividing input voltage by low resistance, denoted as $1/L$. Similarly, the same current intensity applies to all the other match nucleotide pairs. Lastly, as the *don't care* is represented by 1111 bit pattern, all four wordlines are activated, and the current would be at least larger than $1/L$ after calculation. In summary, the proposed ReTCAM can properly utilize the memristors to derive comparison results with the 4-bit pattern with the one-hot encoding method.

D. Accumulated Current on Bitlines

As shown in Figure 8, each base cell of ReTCAM produces the matching result between an input nucleotide and the stored nucleotide. The comparison results are indicated through current, as the memristor is an analog circuitry. If we combine all the generated currents when comparing a DNA read with the reference genome, the current intensity would be different according to the number of match, mismatch, and *don't care* results. As mentioned above, the bit patterns of *don't care* are full of 1 and generates the highest current among all nucleotide comparisons. In other words, if all the currents are combined to represent the comparison result of a DNA read, more *don't care* ends up with higher current intensity. Therefore, the current intensity can be used as an indication of both the quality of the DNA reads and the credibility of a DNA read comparison.

As those comparison results are used as seeds in DNA sequence mapping algorithm, ReTCAM proposes utilizing the current intensity of a DNA read comparison as a credibility indication of the following extend stage to first align seeds with lower current intensity. When compared with the original randomly aligning approach, utilizing the current intensity indication can better arrange the order of seeds for alignment and enhance the efficiency of the extend stage. In other words, ReTCAM not only enhances the seeding stage while performing mapping within the ReRAM corssbar but also utilizes the current intensity to aid other stages within DNA sequence mapping algorithms.

IV. Performance Evaluation

A. Experiment Setup

To evaluate the effectiveness of the proposed ReTCAM, a series of experiments are conducted and compared to the TGS-suitable Minimap tool. While the proposed ReTCAM is simulated by following the ReRAM parameters of NVSim [5], the minimap tool is executed through gem5 [2] simulator for counting the required CPU cycles. Accordingly, datasets of 3 different sizes are chosen, and the information is summarized in Table I. Then, the SimLord [18] tool is utilized to generate long reads by mimicking the actual TGS platform, and the total number of generated DNA reads is 500. The length of each read is between 9000 and 20000 nucleotides. This is because the SimLord tool can simulate the actual TGS platform and produce nonuniform reads. The k-mer size of 15 (See Figure 3). In other words, reads are divided into subsequences with a length of 15 nucleotides. While ReTCAM performs parallel comparisons between the input DNA reads and the stored reference genome, the sizes of required ReRAM space are 0.39, 0.92, and 28.23 MB. Evaluation matrices of this experiment include the energy and latency comparison for a better understanding of the efficiency of the proposed ReTCAM.

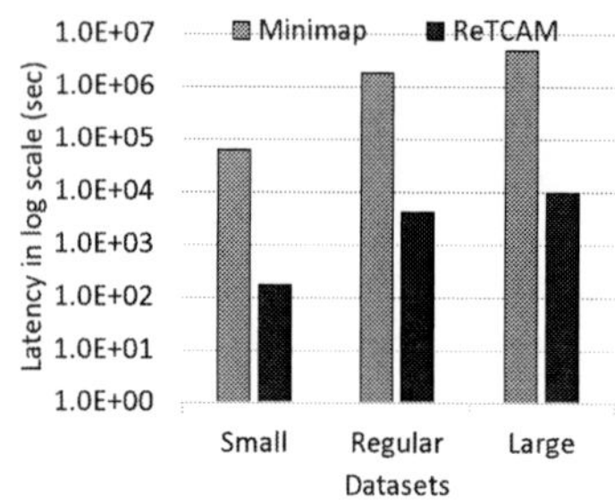

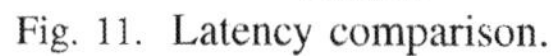

Fig. 11. Latency comparison.

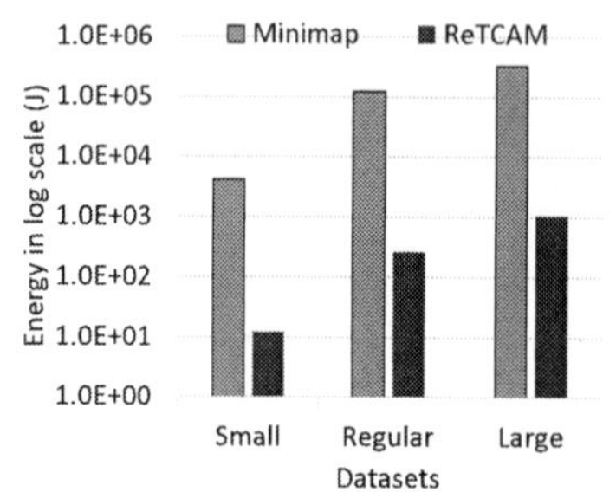

Fig. 12. Energy consumption comparison.

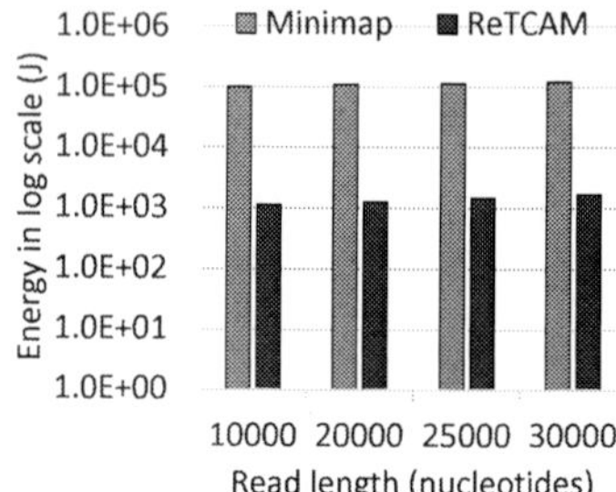

Fig. 13. Energy comparison with different read length.

TABLE I
DATASETS OF DIFFERENT SCALES.

Scales	Name	Species	Size (nucleotides)
Small	ERS646601	Haemophilus influenzae	1,882,235
Medium	ERS544009	Yersinia pseudotuberculosis	5,110,789
Large	PE-ce-10X	Caenorhabditis elegans	10,195,8257

B. Experimental Results

As the goal of the proposed ReTCAM is to resolve the inefficiency of current DNA sequence mapping, the latency and energy comparisons are first reported in Figures 11 and 12. For latency, the proposed ReTCAM can effectively achieve 99.76% reduction on average. This is because the original time-consuming seeding step is replaced with the proposed ReTCAM to perform the in-memory comparison. Meanwhile, a similar reduction can be found in the consumed energy during performing mapping through the CPU-based Minimap tool and the proposed ReTCAM, and the difference is 96.72% on average. Notably, according to the accuracy comparison as shown in Figure 7, the proposed ReTCAM can also enhance the accuracy by marking low-quality DNA as *don't care* during sequence comparison. Lastly, to investigate the energy consumption of different DNA read lengths, Figure 13 is included. Accordingly, the proposed ReTCAM can achieve lower energy consumption regardless of the DNA read length and the energy reduction is 98.77% on average.

V. CONCLUSION

To resolve the inefficiency and degraded accuracy of DNA sequence mapping, this study proposes the ReTCAM to construct an efficient in-memory sequence DNA mapping without including additional transistors and resistors when compared with the conventional ReRAM. In addition, the accuracy of the TGS-suitable mapping tool is also enhanced with the *don't care* feature of TCAM and eliminates the comparison for low-quality nucleotides. A new encoding method is then included for nucleotides, and the accumulated current of each comparison within the proposed ReTCAM is utilized to aid the subsequent mapping steps (*i.e.*, extension). The evaluation results show that, when compared with the conventional CPU-based Minimap tool, ReTCAM can effectively reduce the energy consumption and latency of long-read DNA sequence mapping by an average of 99.72% and 99.76%, respectively.

ACKNOWLEDGEMENTS

This work was supported in part by National Science and Technology Council under grant no. 109-2222-E-027-007-MY3, and 111-2221-E-027-076-MY3, 111-2223-E-001-001, 111-2923-E-002-014-MY3, 111-2221-E-001-013-MY3, and 112-2927-I-001-508 and Academia Sinica under grant nos. AS-IA-111-M01 and AS-GCS-110-08.

REFERENCES

[1] M. Alser, H. Hassan, H. Xin, O. Ergin, O. Mutlu, and C. Alkan, "GateKeeper: a new hardware architecture for accelerating pre-alignment in DNA short read mapping," *Bioinformatics*, vol. 33, no. 21, pp. 3355–3363, may 2017. [Online]. Available: https://doi.org/10.1093%2Fbioinformatics%2Fbtx342

[2] N. Binkert, B. Beckmann, G. Black, S. K. Reinhardt, A. Saidi, A. Basu, J. Hestness, D. R. Hower, T. Krishna, S. Sardashti, R. Sen, K. Sewell, M. Shoaib, N. Vaish, M. D. Hill, and D. A. Wood, "The gem5 simulator," *SIGARCH Comput. Archit. News*, vol. 39, no. 2, p. 1–7, aug 2011.

[3] M. J. Chaisson and G. Tesler, "Mapping single molecule sequencing reads using basic local alignment with successive refinement (blasr): application and theory," *BMC Bioinformatics*, vol. 13, 2012.

[4] J. Dean and S. Ghemawat, "Mapreduce: Simplified data processing on large clusters," *Commun. ACM*, vol. 51, no. 1, p. 107–113, jan 2008.

[5] X. Dong, C. Xu, Y. Xie, and N. P. Jouppi, "Nvsim: A circuit-level performance, energy, and area model for emerging nonvolatile memory," *IEEE Transactions on Computer-Aided Design of Integrated Circuits and Systems*, vol. 31, no. 7, pp. 994–1007, 2012.

[6] Q. Guo, X. Guo, Y. Bai, and E. İpek, "A resistive tcam accelerator for data-intensive computing," in *2011 44th Annual IEEE/ACM International Symposium on Microarchitecture (MICRO)*, 2011, pp. 339–350.

[7] Q. Guo, X. Guo, R. Patel, E. Ipek, and E. G. Friedman, "Ac-dimm: Associative computing with stt-mram," vol. 41, no. 3, p. 189–200, jun 2013.

[8] J. M. Heather and B. Chain, "The sequence of sequencers: The history of sequencing dna," *Genomics*, vol. 107, no. 1, pp. 1–8, 2016.

[9] E. J. Houtgast, V.-M. Sima, K. Bertels, and Z. Al-Ars, "Hardware acceleration of bwa-mem genomic short read mapping for longer read lengths," *Computational Biology and Chemistry*, vol. 75, pp. 54–64, 2018.

[10] W. Huangfu, S. Li, X. Hu, and Y. Xie, "Radar: A 3d-reram based dna alignment accelerator architecture," in *2018 55th ACM/ESDA/IEEE Design Automation Conference (DAC)*, 2018, pp. 1–6.

[11] J. S. Kim, D. S. Cali, H. Xin, D. Lee, S. Ghose, M. Alser, H. Hassan, O. Ergin, C. Alkan, and O. Mutlu, "Grim-filter: Fast seed location filtering in dna read mapping using processing-in-memory technologies," *BMC Genomics*, vol. 19, no. S2, may 2018.

[12] D. Lavenier, J.-F. Roy, and D. Furodet, "Dna mapping using processor-in-memory architecture," in *2016 IEEE International Conference on Bioinformatics and Biomedicine (BIBM)*, 2016, pp. 1429–1435.

[13] T. Laver, J. Harrison, P. O'Neill, K. Moore, A. Farbos, K. Paszkiewicz, and D. Studholme, "Assessing the performance of the oxford nanopore technologies minion," *Biomolecular Detection and Quantification*, vol. 3, pp. 1–8, 2015.

[14] H. Li, "Minimap and miniasm: fast mapping and de novo assembly for noisy long sequences," *Bioinformatics*, vol. 32, no. 14, pp. 2103–2110, 2016.

[15] K. Pagiamtzis and A. Sheikholeslami, "Content-addressable memory (cam) circuits and architectures: a tutorial and survey," *IEEE Journal of Solid-State Circuits*, vol. 41, no. 3, pp. 712–727, 2006.

[16] M. C. Schatz, "CloudBurst: highly sensitive read mapping with MapReduce," *Bioinformatics*, vol. 25, no. 11, pp. 1363–1369, 04 2009.

[17] A. Shafiee, A. Nag, N. Muralimanohar, R. Balasubramonian, J. P. Strachan, M. Hu, R. S. Williams, and V. Srikumar, "Isaac: A convolutional neural network accelerator with in-situ analog arithmetic in crossbars," in *2016 ACM/IEEE 43rd Annual International Symposium on Computer Architecture (ISCA)*, 2016, pp. 14–26.

[18] B. K. Stöcker, J. Köster, and S. Rahmann, "SimLoRD: Simulation of Long Read Data," *Bioinformatics*, vol. 32, no. 17, pp. 2704–2706, 05 2016.

A Self-powered Predictive Maintenance System Based on Piezoelectric Energy Harvesting and TinyML

Zijie Chen, Yiming Gao and Junrui Liang
School of Information Science and Technology, ShanghaiTech University, Shanghai 201210, China
Email: {chenzj1, gaoym, liangjr}@shanghaitech.edu.cn

Abstract*—Nowadays, the Industrial Internet of Things (IIoT) plays a more and more significant role in smart manufacturing. Predictive Maintenance (PdM) is one of the essential applications, recognizing the current status of the machine and preventing disastrous breakdowns. End-point sensors for such monitoring systems are powered mainly by batteries. As IIoT grows, constantly replacing batteries across thousands of devices is cost-prohibitive. In addition, tremendous original sensing data are wirelessly transmitted to the server for data analysis, causing colossal energy consumption. In this paper, we propose the first self-powered on-device PdM system based on piezoelectric energy harvesting and tiny machine learning (TinyML). A trained TinyML model is deployed on the low-cost microcontroller (MCU) for on-device inferring; only the diagnosis result is transmitted. With an emphasis on ultra-low-power demands, a piezoelectric energy harvester is utilized as an energy source and self-powered sensor (SPS) simultaneously. The energy-aware circuit provides reconfigurable on/off threshold voltages for efficient and robust intermittent operation. The balance between energy supply and demand in the battery-free system has been achieved by a handy design. A rich SPS dataset has been collected in a simulated vibration environment and analyzed by five well-known machine-learning models. Random forest stands out given ultra-small data length and sampling rate with accuracy up to 99% for four-class similar vibration diagnosis. Lab test validates the feasibility and performance. As a cyber-electro-mechanical co-design, the system provides a promising solution to the ubiquitous artificial intelligence of things (AIoT).

I. Introduction

With the development of the Industrial Internet of Things (IIOT), factories are placing higher demands on pervasive machine monitoring, and maintenance [1]. Since the impact of maintenance represents a total of 15 to 60% of all operational costs in the manufacturing industry [2], an efficient maintenance strategy proves significant in solving machine anomalies. Predictive maintenance (PdM) is a superior technique that triggers the corresponding necessary maintenance measures only when needed or just before, avoiding shutdown in the production processes and minimizing maintenance costs [3]. The most fundamental part of a PdM system is to identify the current state of the equipment in real time.

Machine learning (ML) models have proved the outstanding performance in PdM [4]. However, most end-point sensors are regarded as data tubes transmitting all the original data to the remote server, which relies on the resource-rich and power-hungry device for artificial intelligence (AI) model training and inferring. Such systems may lead to undesired data privacy issues and large energy consumption. Researchers have focused on deploying ML models at the edge in recent years [5]. Tiny machine learning (TinyML), a booming branch of state-of-the-art ML techniques, enables the low-cost MCU to run on-device ML models at a milliwatt-level power consumption without real-time support of large servers [6]. Yet, TinyML is still at a very early age. More application-specific studies based on TinyML are needed.

Replacing and repairing batteries for the exponentially increasing number of IIoT end-point devices is labor-intensive and environmental-unfriendly [7]. Energy Harvesting (EH), trying to transform the ambient energy from wasted to useful, proves a promising technology for battery-free IoT (Internet of Things) [8]. As the kinetic energy harvesting (KEH) based solution booms, some cyber-electro-mechanical co-design have been proposed for ubiquitous IoT, ignoring the effects of the volatile light radiation and radio frequency signals [9]. However, systematic KEH-IoT designs are few in academia and industry due to the interdisciplinarity of mechanical, electrical, and computer knowledge [10]. In addition, many KEH-IoT approaches are proposed for easy application, such as beacon counting and temperature sensing. To further increase the efficiency of the self-sustained systems, some studies paid attention to simultaneously energy harvesting and sensing (SEHS) [11], which means the harvester serves a self-powered sensor (SPS) to substitute the commercial sensor, such as the inertial measurement unit (IMU).

To the best of our knowledge, an interdisciplinary study for PdM with TinyML, energy harvesting, and SEHS remains unexplored. In this paper, we propose a self-powered TinyML-based PdM system where the harvester works as both an energy source and an SPS, emphasizing ultra-low-power and low-cost end-point intelligence. It brings a promising solution for pervasive sensing and ubiquitous AI.

II. Related Work

PdM is a cutting-edge strategy to predict severe malfunctions in factory machines, ensuring the smooth operation of production line [3], thus decreasing the high costs resulting from unpredictable machine downtime and defective products [2]. It is based on the fact that many devices have a detectable

979-8-3503-1176-1/23 $31.00 © 2023 IEEE

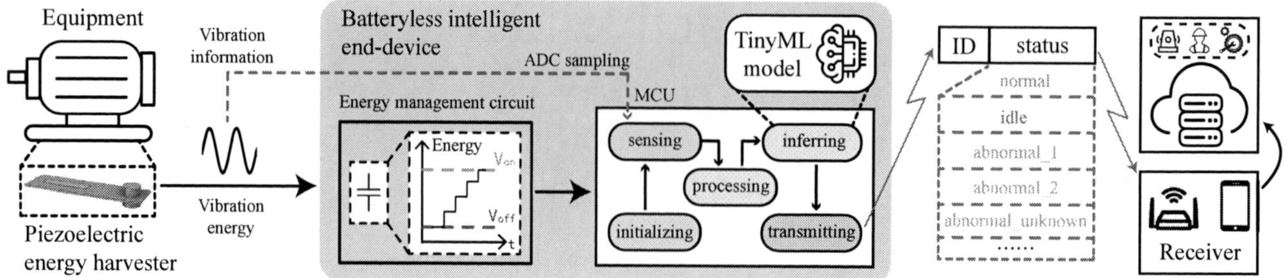

Fig. 1. Architecture of the proposed self-powered predictive maintenance system embedded with TinyML.

latency period from potential failure to performance failure [12]. ML has attracted increasing academic and industry interest for its robust performance. The authors in [13] trained a deep neural network with IMU data for tool wear monitoring and predictive maintenance. For further low-power demand, some studies utilized SPS to replace IMU. In [14], a piezoelectric MEMS-based vibration sensor was designed for surface roughness prediction using an ML algorithm.

TinyML is a burgeoning field that empowers ultra-low power devices like MCUs to infer original data at the endpoint with an ML algorithm [6]. The efficiency of TinyML enables a plethora of battery-powered, always-on applications that can revolutionize the real-time collection and processing of data [15]. In [16], an intelligent rail vehicle running states monitoring system based on TinyML and IMU is proposed.

Energy harvesting is an emerging technology enabling self-maintained IoT devices. As a cyber-electro-mechanical co-design, leveraging the energy harvester as a power source and a sensor has recently attracted increasing interest. ES-ETHG [17], a system based on triboelectric nanogenerators and electromagnetic generators, is proposed to realize self-powered IoT nodes for remote collection of wind speed and direction. In [18], a KEH-based pavement roughness estimation system is presented with SEHS ability. Yet, many KEH-based SEHS systems use different harvesters as energy and information sources, respectively, with increased cost or use the relationship between energy and information to achieve perception with low sensitivity.

Given the challenges above, the proposed self-sustained PdM system utilized only one energy harvester for powering and sensing with an embedded TinyML technique for on-device inferring and similar vibration diagnosis.

III. System Overview

Fig. 1 shows the architecture of the proposed self-powered TinyML-based PdM system. A piezoelectric cantilever as an energy source, as well as SPS, is utilized to capture vibration energy, substitute commercial IMU, and form an intelligent end device with a well-designed energy management circuit and a low-cost MCU. Since the energy generated from the tiny vibration is still insufficient to support the MCU's stable operation, the capacitor inside the circuit first stores the converted energy from the harvester. While embedded with the UVLO (under-voltage lockout) function, the energy management circuit can sense the voltage of the capacitor and turn on the MCU when the stored energy is sufficient, ensuring all operations are completed.

A voltage signal is generated from the based-excited SPS, reflecting the current vibration information. First, The real-world data for different machine statuses are collected with a low sampling rate for saving energy and used for pre-training a TinyML model. When the specific abnormal information can be obtained handily in advance, the PdM system can not only identify both the occurrence of the anomaly and the specific abnormal situation. After deploying the compressed model on an MCU, the end device can carry out the intelligent evaluation. A low-power BLE SoC-based minimum system with a transceiver and CPU is chosen as the MCU. It starts running when the energy is sufficient. Given the awareness of intermittent operation, the software program is set to allow the MCU only to run the necessary parts, including initializing, sensing, processing, inferring, and transmitting. After on-device inference, the current machine state is identified and reported to the near receiver within 1 second with low power consumption. Subsequently, when the circuit senses that the energy consumption has reached a certain level, the MCU would be cut off until the next running cycle. The machine information can be sent to the cloud server for remote monitoring according to the application requirements. The vibration energy and information are fully utilized in this system with EH and TinyML techniques for self-contained end-point intelligence.

IV. Working Principle

To realize a cyber-electro-mechanical co-design, a lost-cost, well-rounded energy management circuit is designed for stable intermittent operation and SEHS. Utilizing the piezoelectric energy harvester as a data source and TinyML as a data inferring method, the system achieves end-point intelligence efficiently.

A. Simultaneously energy harvesting and sensing

Fig. 2(a) shows a piezoelectric cantilever that generates a voltage according to the vibration. One end of the cantilevered beam is fixed on the vibrating structure, usually called the base. The other end is free and mounted with a tip mass.

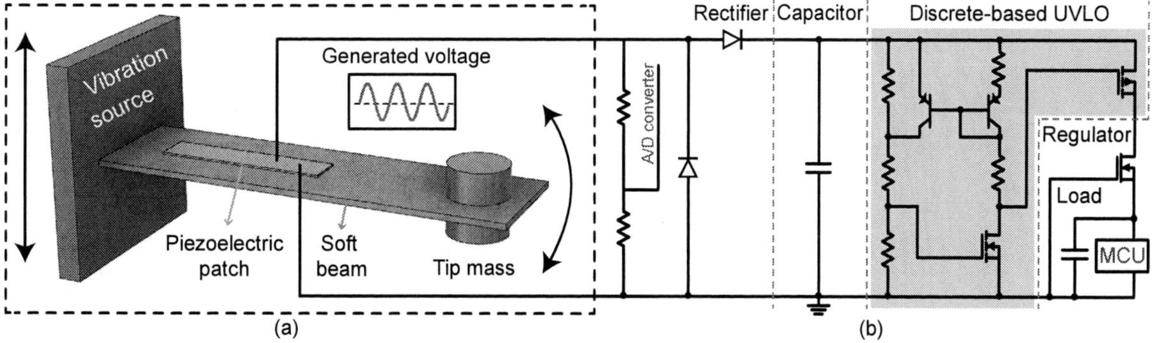

Fig. 2. The proposed circuit for simultaneously energy harvesting and sensing. (a) Base-excited piezoelectric cantilever. (b) Well-rounded circuit for energy management and SEHS.

As the machine vibrates, the beam is simultaneously excited by the vibrating base. At the open-circuit condition, the piezoelectric patch subsequently generates a voltage $v_p(t)$, which is proportional to the beam deflection $x(t)$ with a ratio of α, i.e.,

$$v_p(t) = \alpha x(t). \tag{1}$$

Intermittent operation is proven to be the better executing mode for KEH-based IoT systems [10]. A comprehensive board-level energy-aware circuit, incorporating rectification, energy storage, and voltage regulation functions, is designed and utilized to manage the energy and realize the intermittency in computing regarding [19], as shown in Fig. 2(b). Two threshold voltages, V_{start} and V_{close}, can be conveniently changed by adjusting the resistor network, realizing low-power analog UVLO. Moreover, this circuit offers a comparably stable regulated output voltage to power the IoT devices while requiring no additional quiescent current skillfully using a depletion-mode MOSFET [19].

A piezoelectric transducer can be considered an SPS. For a common IMU-based sensing method, the generated three-axis data is transmitted to an MCU through an I^2C bus. For a self-powered sensing approach, one-axial vibration data is generated without any extra power supply, thus decreasing the power consumption for sensing. For realizing simultaneously energy harvesting and sensing, there exist three challenges:

1) Common ground problem during sensing: We use the analog-to-digital converter (ADC) inside the MCU for voltage sensing. Since the piezoelectric sheet generates alternating current, a common choice for rectification is the full bridge rectifier circuit. However, the harvester and MCU are not co-grounded, which leads to huge signal distortion. Thus in this design, we use a diode for half-wave rectifying and a diode for the continuity of the equivalent capacitance current of the piezoelectric sheet, as shown in Fig. 2(b). We intend to sacrifice some energy harvesting efficiency for more significant sensing accuracy.

2) Connection between SPS and ADC: For SEHS, the ADC is connected to the SPS for original voltage sensing. Due to the characteristic of ADC, half-wave voltage is cut in this way. Since we have used a half-wave rectifier circuit, this problem is acceptable. However, if the ADC is connected directly to the harvester, it will take away part of the generated energy, decreasing energy harvesting efficiency. We connected two same large megohm resistors in parallel, as shown in Fig. 2(b), allowing only small signals to flow into the ADC.

3) Signal distortion: Considering the extra cost and power consumption of adding an auxiliary circuit for sensing both positive and negative half-cycle signals, in this study, we simply drop the negative half-cycle signals. Furthermore, we intend to sense the signal at a low sampling rate with small data for energy saving, which leads to considerable data distortion. With advanced ML models, the distortion can be well compensated.

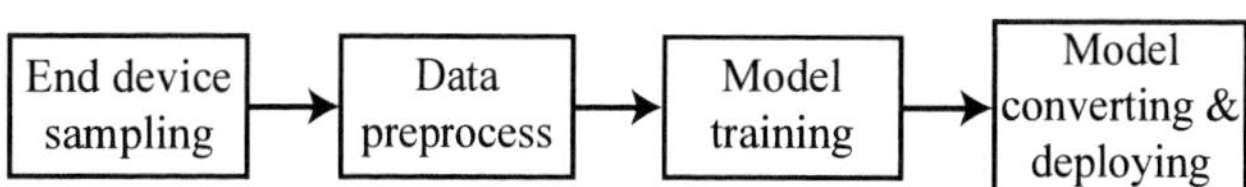

Fig. 3. The TinyML pipeline of the proposed system.

B. TinyML Development and Deployment

As mentioned above, TinyML equips the IIoT end devices with basic intelligence. Fig. 3 illustrates the TinyML pipeline of the proposed system, including end device sampling, data preprocessing, model training, and model deploying. The SPS is used to capture the vibration information from the equipment. To further reduce the power consumption, the lower the sampling rate of MCU for data acquisition, the better. Subsequently, the real-world data goes through some preprocessing methods, such as data normalization and missing value imputation. For traditional ML, the model is built with the open-source scikit-learn library and trained with the processed data. Then the model is converted by advanced inference frameworks, such as micromlgen. After being converted into a document suitable for embedded systems, the lightweight models are easily deployed to the TinyML-supported MCUs for on-device monitoring from the real-time SPS signals.

C. On-device Intermittent Operation

EH-based IoT devices primarily work intermittently. Fig. 4 demonstrates the energy picture for the proposed battery-

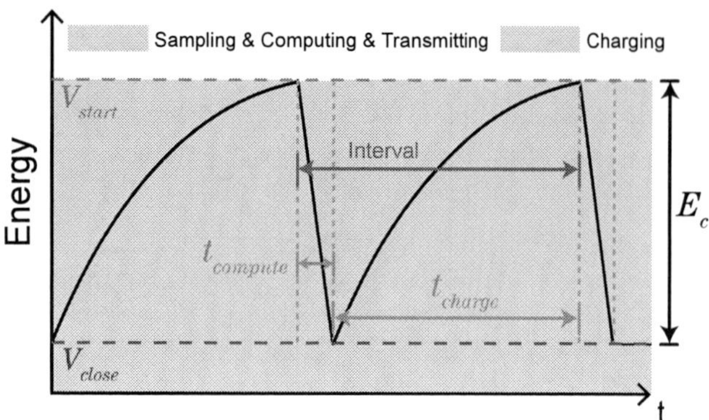

Fig. 4. Energy picture of the system during intermittent operation mode.

free IoT system during the normal intermittent working cycle. There are two basic phases: 1) charging; and 2) sensing, computing, and transmitting. In the beginning, the stored energy is close to the off threshold. As the energy keeps flowing into the storage capacitor, the stored energy accumulates. It is the charging period when the IoT device is turned off. When the energy reaches the turn-on threshold, the IoT device is turned on to execute necessary tasks. The energy consumption of the IoT device can be measured, denoted as E_c. As the MCU finishes the scheduled tasks, the stored energy drops quickly and reaches the off threshold. Thus t_{compute} is generally much smaller than t_{charge}. If the machine stays in a stable frequency, the interval of one round operation is a rough constant. Repeatedly, the IoT device would be turned off and wait until the energy is sufficient.

After measuring the total energy consumed by the MCU for one round of necessary operation, we can derive the minimal energy required for each interval, i.e., E_c. According to the energy formula

$$E_c = \frac{1}{2}C\left(V_{\text{start}}^2 - V_{\text{close}}^2\right), \tag{2}$$

the harvested energy in each round can be set as small as possible by taking E_c as a reference and handily adjusting the on/off threshold voltages of the energy management circuit. This balance of supply and demand is critical for highly-efficient battery-free IoT systems.

V. Experiment

To comprehensively evaluate the performance of the proposed system, we have conducted several experiments. A rich SPS data set is built and analyzed by five well-known supervised ML models, including Decision Tree (DT), Random Forest (RF), Support Vector Machine (SVM), Logistic Regression (LR), and K-Nearest Neighbor (KNN). The proposed system is prototyped, and the performance is validated. The specifications of the system are listed in Table I.

A. Data acquisition

The anomaly information is particular in each maintenance problem [20]. In addition, there is no publicly available SPS-based dataset for simulation experiments. Therefore, in this study, we use a vibration platform to simulate a working machine and collect multiple data sets. The experimental setup for data acquisition and performance analysis is shown in Fig. 5. A cantilevered piezoelectric energy harvester is installed on the vibration source, controlled by a computer-based vibration controller (computer #1). The test PCB integrates the circuit shown in Fig. 2(b) and an off-the-shelf nRF52832 minimal system board. As the platform oscillates, the generated energy flows into the energy management circuit. For data acquisition, the MCU is separated and battery-powered, transmitting real-time digital values to another computer (computer #2) at 40 Hz, 60 Hz, and 80 Hz with 12-bit precision through a CH340 module. Since the voltage of the capacitor is set to the threshold voltage of 3.3 V every time the MCU is powered on, we control the amplitude of the piezoelectric energy harvester at 3.3 V and start sampling data after the capacitor voltage is stabilized. A piezoelectric sheet collects the most energy when the vibration reaches its resonant frequency. However, as mentioned, every machine may have different vibration status, so it is not fair and general to simulate a vibration in resonant frequency. The prototyped harvester's resonant frequency is 24.8 Hz. So we choose 20 Hz vibration as a normal state, while a 0 Hz one is regarded as in the idle state. Vibration signals at 19 and 21 Hz are regarded as two specific abnormal conditions. Note that We strictly control the same amplitude of the voltage generated by the piezoelectric sheet for the three vibration cases, i.e., 4.2 V. Thus, we can only identify these data according to the frequency information.

TABLE I
Specifications of the proposed system.

Unit	Component	Specification	
Energy harvester	Soft beam	Size	10×80×0.3 mm^3
		Material	Copper
	Proof mass	Weight	13g
Circuit	Energy-aware circuit	Resistance	1, 18, 31, 20, 0.1 MΩ
		Capacitance	470, 1 μF
	Sampling circuit	Resistance	3, 3 MΩ
MCU	SoC	nRF52832	
	Software	nRF5 SDK V17.1	

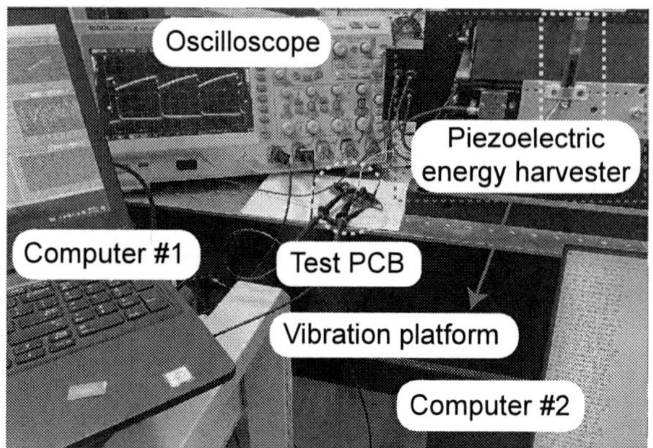

Fig. 5. Experimental setup for data acquisition and performance analysis.

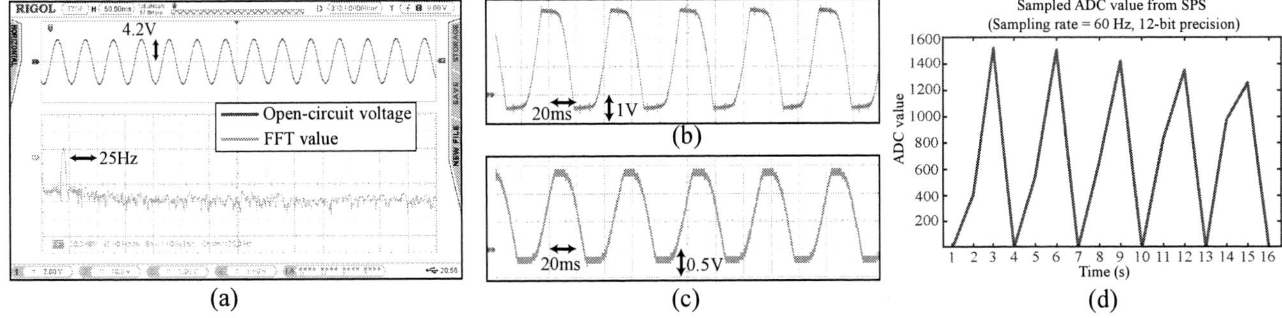

Fig. 6. Data illustration. (a) Open-circuit voltage signal generated by the SPS under a 20 Hz vibration environment with some inevitable noise. (b) A voltage signal of the harvester when it is connected to the circuit for energy charging under the 20 Hz vibration environment. (c) Corresponding sensing voltage on the resistance. (d) 16-point digitized values after ADC conversion at a 60 Hz sampling rate.

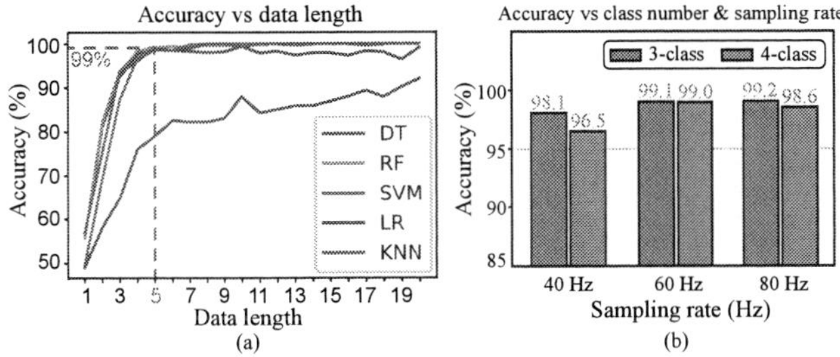

Fig. 7. Machine learning analysis. (a) Model accuracy of different models for data length ranging from 1 to 20 on a quadruple classification case. (b) Performance of RF model at different sampling rates in triple and quadruple classifications

The whole test PCB is batteryless and embedded with the MCU for performance analysis. An oscilloscope is used to measure the voltage of different units. A dataset of more than 2000 seconds of vibration data against different frequencies and sampling rates has been collected and used for data analysis.

Fig. 6(a) shows the open-circuit voltage signal generated by the SPS under a 20 Hz vibration environment with some inevitable noise, such as electromagnetic interference. Fig. 6(b) shows the voltage signal of the harvester when connected to the circuit for energy charging under the same vibration environment. And Fig. 6(c) shows the corresponding sensing voltage on the resistance, whose voltage is half of the SPS voltage due to two same resistances. As we can see, the negative part of the SPS voltage is nearly dropped due to the half-wave rectifier and the direct connection of ADC. The less required sensing data, the better for energy saving. The digital 16-point values are shown in Fig. 6(d), where the voltage is sampled at 60 Hz with the MCU. Large distortions are caused by discarding the negative values and sampling at such a low sampling rate.

Due to the space limit, the fast Fourier transform (FFT) figures of 16-point ADC values in 19, 20, and 21 Hz are not shown. The result shows that they all look nearly the same, which means an expert system building for the small SPS data size will cost a certain amount of effort. Although this distorted signal is hard to analyze using a simple algorithm, we can still use the efficient TinyML technology for realizing the on-device inferring.

B. Machine Learning Analysis

Given the challenges of the distorted SPS signal, we intend to use five ML algorithms for analysis, including DT, RF, SVM, LR, and KNN. ML models are based on the default settings of the sklearn library, and the results are based on 5-fold cross validation.

Fig. 7(a) shows the accuracy of different models for data length ranging from 1 to 20 on a quadruple classification problem with 0, 19, 20, and 21 Hz SPS vibration signals in a 60 Hz sampling rate. Except for the LR algorithm, all models reach a high accuracy when the data length is only 5, which means the window size is less than 0.1 seconds. RF achieves accuracy up to 99%, whose model size is under 100 KB. Therefore, AI methods can satisfy high performance in identifying SPS signals with ultra-small window sizes.

Fig. 7(b) illustrates the performance of the RF model at different sampling rates in triple and quadruple classifications. 0, 19, and 20 Hz SPS signals are used for analyzing the triple classification case. RF model is set to be ten decision trees with gini criterion. Considering the tradeoff between accuracy and power consumption, the 60 Hz sampling rate is preferred. Since the model has an ultra-high recognition ability for different machine states, when unknown anomalies occur, the MCU will have a relatively low confidence rate, which means the MCU can also accurately discern the occurrence of unknown anomalies with a simple threshold setting.

C. Performance validation

Fig. 8(a), (b), and (c) show the waveform under the frequencies of 19 Hz, 20 Hz, and 21 Hz, respectively. The on and off threshold voltages of the system are set to 3.1 V and 2 V, respectively, while the regulated output voltage is set to 2.8 V. From the waveform, we can see the system is working intermittently as originally designed. The software code of the EUU is optimized to only broadcast two BLE beacon packets in one round after completing initializing, sensing, preprocessing, and inferring. As we can see, the difference

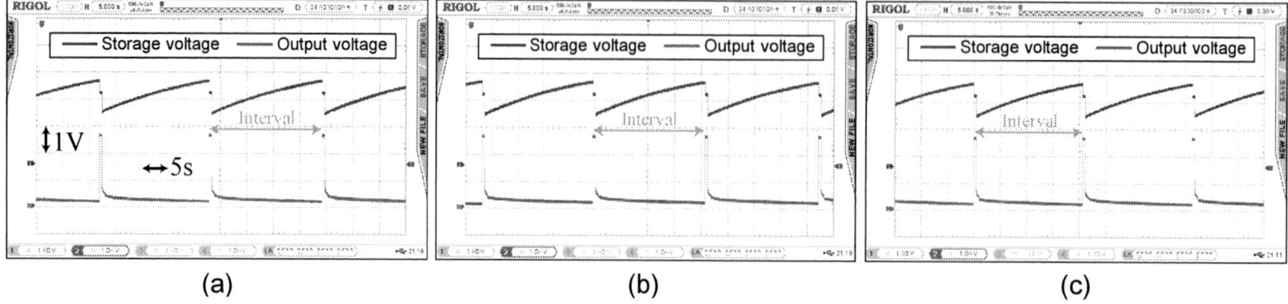

Fig. 8. Results of the experimental test. (a)-(c) The waveform showing the voltage of the storage capacitor and MCU when the environmental frequency is set to 19 Hz, 20 Hz, and 21 Hz, respectively.

between the beacon intervals in three similar vibration statuses is ultra-small. By simply counting the interval between received wireless packets, the current machine status is hard to deduce. With TinyML, we successfully receive the correct beacon in different vibration statuses within 20 meters.

VI. Conclusion

This paper proposed a self-powered TinyML-based PdM system with an emphasis on ultra-low power and low-cost end-point intelligence. TinyML equipped the end machine with intelligence for perceiving its situation in real time, thus relieving the tight dependency on high-end servers. By knowing the minimal power consumption of one round of computational tasks, we have set the on/off thresholds of the energy-aware circuit to realize a balance between supply and demand. To further reduce power consumption, the piezoelectric energy harvester was utilized as both an energy and information source, realizing SEHS. A rich SPS dataset was collected with a vibration platform and analyzed by five well-known ML models. Random forest algorithm identified raw SPS signal with 99% accuracy when the required data length and sampling rate were 5 and 60 Hz, respectively. Experiments were carried out to validate the availability and performance. After a thorough design from EH, sensing, analyzing, intermittent operation mode, etc., the system can reliably fulfill condition-based PdM. This cyber-electro-mechanical co-design is hoped to provide valuable inspiration for future ubiquitous AIoT studies.

References

[1] A. G. Frank, L. S. Dalenogare, and N. F. Ayala, "Industry 4.0 technologies: Implementation patterns in manufacturing companies," *International Journal of Production Economics*, vol. 210, pp. 15–26, 2019.

[2] R. K. Mobley, *An Introduction to Predictive Maintenance*. Elsevier, 2002.

[3] G. A. Susto, A. Schirru, S. Pampuri, S. McLoone, and A. Beghi, "Machine learning for predictive maintenance: A multiple classifier approach," *IEEE transactions on industrial informatics*, vol. 11, no. 3, pp. 812–820, 2014.

[4] Q. Cao, C. Zanni-Merk, A. Samet, C. Reich, F. d. B. de Beuvron, A. Beckmann, and C. Giannetti, "Kspmi: a knowledge-based system for predictive maintenance in industry 4.0," *Robotics and Computer-Integrated Manufacturing*, vol. 74, p. 102281, 2022.

[5] S. Duan, D. Wang, J. Ren, F. Lyu, Y. Zhang, H. Wu, and X. Shen, "Distributed artificial intelligence empowered by end-edge-cloud computing: A survey," *IEEE Communications Surveys & Tutorials*, 2022.

[6] V. J. Reddi, B. Plancher, S. Kennedy, L. Moroney, P. Warden, A. Agarwal, C. Banbury, M. Banzi, M. Bennett, B. Brown *et al.*, "Widening access to applied machine learning with tinyml," *arXiv preprint arXiv:2106.04008*, 2021.

[7] Z. Chen, F. Gao, and J. Liang, "Kinetic energy harvesting based sensing and iot systems: a review," *Frontiers in Electronics*, vol. 3, p. 1017511, 2022.

[8] M. M. Sandhu, S. Khalifa, K. Geissdoerfer, R. Jurdak, and M. Portmann, "SolAR: Energy positive human activity recognition using solar cells," in *2021 IEEE International Conference on Pervasive Computing and Communications (PerCom)*. ieeexplore.ieee.org, Mar. 2021, pp. 1–10.

[9] X. Li, H. Tang, G. Hu, B. Zhao, and J. Liang, "ViPSN-pluck: A transient-motion-powered motion detector," *IEEE Internet of Things Journal*, vol. 9, no. 5, pp. 3372–3382, 2021.

[10] Liang, Li, and Yang, "Kinetic energy harvesting toward Battery-Free IoT: Fundamentals, Co-Design necessity and prospects," *ZTE Communications*, 2021.

[11] D. Ma, G. Lan, W. Xu, M. Hassan, and W. Hu, "Simultaneous energy harvesting and gait recognition using piezoelectric energy harvester," *IEEE Trans. Mob. Comput.*, pp. 1–1, 2020.

[12] R. Ahmad and S. Kamaruddin, "An overview of time-based and condition-based maintenance in industrial application," *Computers & industrial engineering*, vol. 63, no. 1, pp. 135–149, 2012.

[13] D. F. Hesser and B. Markert, "Tool wear monitoring of a retrofitted cnc milling machine using artificial neural networks," *Manufacturing letters*, vol. 19, pp. 1–4, 2019.

[14] S. Trivedi, R. H. Ganesh, T. Shen, P.-W. Huang, and S.-S. Li, "Piezoelectric mems vibration sensor module for machining quality prediction," in *2020 IEEE SENSORS*. IEEE, 2020, pp. 1–4.

[15] C. Banbury, C. Zhou, I. Fedorov, R. Matas, U. Thakker, D. Gope, V. Janapa Reddi, M. Mattina, and P. Whatmough, "Micronets: Neural network architectures for deploying tinyml applications on commodity microcontrollers," *Proceedings of Machine Learning and Systems*, vol. 3, pp. 517–532, 2021.

[16] S. Zhou, Y. Du, B. Chen, Y. Li, and X. Luan, "An intelligent iot sensing system for rail vehicle running states based on tinyml," *IEEE Access*, vol. 10, pp. 98 860–98 871, 2022.

[17] B. Zhang, S. Zhang, W. Li, Q. Gao, D. Zhao, Z. L. Wang, and T. Cheng, "Self-Powered sensing for smart agriculture by Electromagnetic–Triboelectric hybrid generator," *ACS Nano*, vol. 15, no. 12, pp. 20 278–20 286, Dec. 2021.

[18] H. Yang, L. Teng, and J. Liang, "A battery-free pavement roughness estimation system based on kinetic energy harvesting," in *2022 IEEE International Symposium on Circuits and Systems (ISCAS)*. IEEE, 2022, pp. 2763–2767.

[19] L. Teng, J. Liang, and S. Du, "A nano-power wake-up circuit for energy-driven iot applications," in *2022 IEEE International Symposium on Circuits and Systems (ISCAS)*. IEEE, 2022, pp. 2383–2387.

[20] M. Luo, Z. Xu, H. L. Chan, and M. Alavi, "Online predictive maintenance approach for semiconductor equipment," in *IECON 2013-39th Annual Conference of the IEEE Industrial Electronics Society*, 2013, pp. 3662–3667.

Temperature-Aware Memory Mapping and Active Cooling of Neural Processing Units

Vahidreza Moghaddas[*], Hammam Kattan[†], Tim Bücher[†], Mikail Yayla[*], Jian-Jia Chen[*§], and Hussam Amrouch[†‡¶]
TU Dortmund University[*], *University of Stuttgart*[†], *Lamarr Institute for Machine Learning and Artificial Intelligence*[§]
AI Processor Design, Technical University of Munich[‡], *Munich Institute of Robotics and Machine Intelligence*[¶]
Corresponding author: vahidreza.moghaddas@tu-dortmund.de

***Abstract*—Neural processing units (NPUs) have become indispensable for meeting the high computational demands of deep neural networks (DNNs). They provide a very efficient solution, thanks to having a huge MAC array that enables massive parallelism. Nevertheless, such an architecture exhibits excessive on-chip power densities leading to a localized hot-spot that seriously heats its surroundings. This work demonstrates how the on-chip temperatures induced by the MAC array create a spatial thermal gradient through the on-chip SRAM memory. This makes the memory regions sensitive to different error probabilities (Perror), leading to significant accuracy drops when DNNs are being executed. To surmount this challenge, we employ on-chip superlattice thermoelectric (TEC) cooling devices that effectively reduce the memory temperature. Although scaling the memory voltage makes SRAM cells more sensitive to errors, it significantly decreases the leakage power, which compensates for the power consumed by the incorporated TEC devices. Furthermore, operating the SRAM at a lower voltage and temperature substantially increases its lifetime because voltage and temperature are key stimuli of transistor aging. By running multi-physics simulations using commercial finite-element tools and SPICE simulations for the 14nm FinFET technology, we accurately derive the relation between the Perror in different memory regions and the corresponding cooling cost. We then propose a three-stage temperature-aware layer-wise memory mapping that exploits different degrees of the sensitivity of NN layers to errors towards maximizing the DNN accuracy while minimizing the cooling cost. Experimental results reveal that our method notably improves the DNN accuracy compared to existing temperature-oblivious memory mapping.**

***Index Terms*—Thermal management, Neural processing unit (NPU), Thermoelectric cooling (TEC), On-chip memory**

I. Introduction

In the past decade, deep neural networks (DNNs) have significantly improved the accuracy of neural networks (NNs). However, running NN applications on general-purpose CPUs is inefficient (in terms of energy and performance) due to the large number of multiplication-accumulation (MAC) operations required to calculate the sum of products of the neurons' inputs. Custom neural network accelerators are therefore necessary to meet the increasing computational demands. This prompted Google to design and fabricate its first generation of NPUs known as tensor processing unit (TPU), which were deployed in its data centers in 2015, revolutionizing DNN accelerators. TPUv1 can speed up DNN inference by 15-30X with 30-80X less energy consumption than conventional CPUs and GPUs [1].

This breakthrough has motivated other companies to develop their own NPUs, such as Tesla's full self-driving (FSD) computer, which comprises two NPU chips in dual configuration, each with 32MB of on-chip SRAM and 96×96 MAC units [2]. TPUv1 features 256×256 MAC units with 24MB SRAM, while subsequent generations of TPUs have varying MAC units and/or different on-chip SRAM sizes [3]. Nevertheless, having a large number of MAC units in a limited and confined area of the die has reliability consequences. Amrouch et al. in [4] showed that even with maximum air cooling, the power density of a 128×128 MAC array can skyrocket to even above 300W/cm^2, violating the critical temperature of the chip. Furthermore, in 2018, Google reported that they switched to liquid cooling for their TPUv3 due to its extreme on-chip power density [3]. However, liquid cooling necessitates its own maintenance, cooling infrastructure, and consumes a significant amount of power [3].

With the advent of thin-film superlattice thermoelectric cooling (TEC) devices that can be integrated on top of the silicon die, fine-grained *active cooling* is now possible [5]. Controlled active cooling allows cooling specific region(s) of the die only when needed. These devices can pump heat out of the die with significant heat dissipation of up to 1300W/cm^2, as experimentally demonstrated [6]. In [4], a hybrid method is employed for thermal management in NPUs using active on-chip TEC, frequency scaling, and precision scaling. However, the method is only used for the inference of DNNs. Considering NPU for both inference and training, [7] explores the design space of Google TPUv3 using Negative Capacitance FET (NCFET) for the MAC array to minimize the TEC cost. Nevertheless, in both papers TEC is considered for the entire die, and the effect of the high power density of the MAC array on the on-chip SRAM memory is not considered.

Reagen et al. in [8] developed a fault injection framework and explored the resilience of DNNs to errors across different models. Another study proposed software-level error-aware training to tolerate voltage overscaling SRAM errors [9]. The robustness of different quantization schemes and error training against low-voltage induced random bit errors has been studied in [10] for DNN accelerators with quantized weights stored in SRAM. However, this work assumes the same probability of bit error for the entire memory, while in another work the voltage of different regions of DRAM is scaled down non-uniformly

979-8-3503-1176-1/23 $31.00 © 2023 IEEE

to save energy [11]. Notably, neither study considered the impact of high power density of the MAC array on the on-chip memory, nor did they explore the use of active cooling.

Transistor aging due to Bias Temperature Instability (BTI) and Hot-carrier Injection (HCI) phenomena is the major reliability concern for circuit designers. This is due to the profound impact on the circuit performance degradation over time, which may unpredictably, lead to catastrophic timing errors. Both BTI and HCI have a very strong dependency on voltage and temperature. On-chip SRAM memories are one of the prime sources of failures in SoCs because they are powered on continuously throughout the entire lifetime and hence they considerably suffer from aging. Transistor aging manifests itself as drift in the electrical parameters such as Threshold Voltage (V_T) [12]. This reduces the resiliency of SRAMs against noise and hence errors during read operations start to appear.

Problem and key goals: In this work, we address the problem of efficiently mapping the NN parameters in on-chip memory for NPUs. We aim to investigate the impact of the high power density of the NPU's MAC array on the nearby SRAM memory, which results in different probabilities of errors (P_{error}) across the memory regions. This exposure to varying error probabilities poses a significant challenge. To tackle this, we employ localized on-chip TEC specifically targeting the SRAM memory regions. However, employing such active cooling comes with additional power costs. Nevertheless, operating the SRAM memory at a scaled voltage of 0.3V instead of 0.7V provides substantial power savings, which can compensate for, if not entirely offset, the extra cooling costs. Note that: (1) Apart from power savings, there is also a significant mitigation of transistor aging, leading to a significant improvement in SRAM reliability. (2) Operating the SRAM at a lower voltage, however, results in higher P_{error}. Therefore, our research explores the effective utilization of TECs to compensate for accuracy loss caused by temperature-induced memory errors while minimizing the associated cooling costs.

In short: We propose a three-stage layer-wise on-chip memory mapping approach for NPUs that maximizes accuracy and minimizes cooling cost by taking into account the varying degrees of robustness of different NN layers to errors. Our approach is orthogonal to any quantization schemes and/or NPUs, and similar results are anticipated.

Our novel contributions within this paper are as follows:
(1) Investigating the impact of MAC array of NPU on the nearby on-chip SRAM memory that creates a spatial thermal gradient through the memory.
(2) Integrating active on-chip cooling and memory mapping to maximize the accuracy and minimize the cooling cost.

II. System Model and Assumptions

A. NPU Thermal Model

In this work, for the NPU model, we adopt the Google TPUv1 chip. The floorplan is obtained from [1] and consists of a systolic array that occupies around 24% of the total chip, on-chip SRAM memory occupies around 29% and the remaining part is used for auxiliary management area (I/O, PCIe, peripheral, control, etc.) which occupies around 47%. The thermal

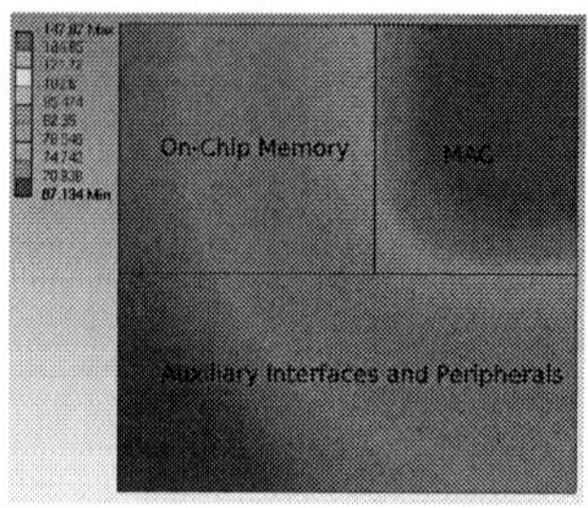

Fig. 1: Thermal gradient of TPUv1 when MAC array is hot.

model is performed using commercial ANSYS tool flows that offer accurate multi-physics simulations using finite element methods. In the assumed thermal model, the TPU chip is mounted on top of a printed circuit board (PCB) and is located underneath a cooling unit that consists of a heatspreader, a heatsink, and a fan. A thin layer of thermal interface materials is sandwiched between the TPU silicon die and the heatspreader. Fig. 1, shows how the high temperature of the MAC array creates spatial thermal gradient and affects the on-chip memory (see Section IV-A for simulation setup). Since different memory regions operate under different temperatures, this makes the memory regions susceptible to different probability of errors.

To model the superlattice thin-film TEC, we adopt the proposed structure in [5], [6]. Each TEC module consists of 3×3 thermoelectric couples and has a total surface area of 1.75mm $\times$ 1.75mm. Several TEC modules are implemented to fully cover the entire on-chip SRAM area. The TEC modules have been integrated inside the thermal interface layer to provide effective cooling to the silicon die while avoiding an increase in thermal resistance. When the TEC device is fed with current, a temperature difference between the upper and lower side is formed. The complex interaction between the full system (i.e., PCB, silicon die, heat spreader, heat sink, etc.) and the cooling effects induced by the TEC are accurately captured by the finite element simulations performed by ANSYS. Furthermore, any side effects caused by Joule heat induced inside the TEC will also be captured by the finite element simulations. In order to model the relation between the cooling cost and the corresponding temperature reduction, we sweep the current from 0A all the way up to 6A with a step of 0.5A. For every current step, we obtain the entire thermal map of the die from the thermal simulation. Then, we calculate the temperature reduction and the corresponding cooling power.

B. Modeling Error Probability in SRAMs under Voltage and Temperature Effects

To accurately model the probability of errors in SRAM cells, we employed SPICE simulations. First, we calibrated the industry-compact model for the FinFET technology (BSIM-CMG) to reproduce measurement data from Intel 14nm FinFET. Then, we built a 6-T SRAM cell consisting of two coupled-inverters and two access transistors. To capture the impact of process variation of n-FinFET and p-FinFET transistors, we also calibrated the variation against variability data from the same technology node (i.e., Intel 14nm FinFET). Then, we performed Monte-Carlo simulations for the SRAM circuit, in which we measured the transfer characteristics to extract the so-called *butterfly* curve. This enabled us to compute the static noise margins for both hold and read operations. To obtain sufficient statistical information, we performed $10,000$ SPICE simulations, providing us with the Hold Noise Margin (HNM) and Read Noise Margin (RNM) distributions. When the distribution crosses the thermal noise which is defined as 26mV, a bit-flip in the SRAM cell occurs. In practice, using the probability density function of the obtained HNM and RNM distributions along with the threshold at which failures occur, the P_{error} of SRAM is then calculated.

The impact of temperature on the underlying transistors and hence on the SRAM circuits is accurately captured by the physics-based models within the industry-standard compact model (BSIM-CMG) that we have calibrated against measurements. We repeated the performed Monte-Carlo simulations for different temperature steps, starting from room temperature (25°C) all the way up to the maximum temperature of 125°C. Afterwards, we did the aforementioned Monte-Carlo analysis for the reduced/scaled voltage (0.3V) instead of the nominal voltage (0.7V). In practice, decreasing the voltage reduces the ON current of pFinFET and nFinFET transistors, affecting the HNM and RNM distributions by shifting them considerably towards the thermal noise limit where errors occur. This, in turn, manifests itself as a significant increase in the probability of errors that the SRAM cells will exhibit.

C. Neural Network Model

The proposed method in this paper is not limited to any specific NN, and we provide a general approach to deal with high temperatures that affect memory and degrade the performance of NNs. However, since in most NPUs, quantization is normally applied, in our work, we also consider quantization. Notably, we do not intend to propose a novel quantization and/or exercise state-of-the-art quantization schemes; rather, we use a straightforward quantization to present our method.

Quantization: In this work, we use a uniform-asymmetric quantization scheme. Assume a quantization scheme $\mathbb{R} \rightarrow Q$, where Q is a set with a finite number of levels, which we denote here as ordered $Q = \{q_1, \ldots, q_L\}$, where $q_v < q_{v+1}$ for $1 \leq v \leq L$. We define the number of bits for quantization as n_q, which allows $L = 2^{n_q}$ quantization levels. Let the value to be quantized be x_f, then the quantized output value x_q is obtained as:

$$x_q = round\left((x_f - min_{x_f})\frac{2^{n_q} - 1}{max_{x_f} - min_{x_f}}\right) \quad (1)$$

The values min_{x_f} and max_{x_f} denote the minimum and maximum values that require quantization. In evaluations, we quantize all parameters of the NN to 4 bits, similar to previous studies that also included low-bit weight quantization [10], [11].

Training: For regular NNs, stochastic gradient descent (SGD) is employed with mini-batches. Assume that the training data is described as $\mathcal{D} = \{(x_1, y_1), \ldots, (x_I, y_I)\}$ with I samples, $x_i \in \mathcal{X}$ as the inputs, $y_i \in \mathcal{Y}$ as the labels, and $\mathcal{L}$ as the loss function. The objective of training is to find a solution for the optimization problem: $\arg\min_W \frac{1}{I}\sum_{(x,y)\in\mathcal{D}} \mathcal{L}(\mathcal{F}_W(x), y)$, using a mini-batch SGD strategy, to compute the gradients by backpropagation. Here $\mathcal{F}_W(x)$ denotes the forward pass function of the NN with learnable parameters W. In the training of QNNs, we employ the same procedure but the weights and biases are quantized during the forward pass of the training phase. In this way, the NN is trained with quantization and can achieve high accuracy despite the limited number of levels. This approach is often referred to as quantization-aware training in literature [13].

III. Our Proposed Method

Our layer-wise temperature-aware memory mapping consists of three main steps. In the first step, we determine the sensitivity of each NN layer to errors and use this information to identify the optimal mapping configuration of NN layer parameters to memory regions. Next, we improve the accuracy through error-aware training. Finally, we use heuristics to identify the power-efficient cooling configuration that meets the target accuracy during the cooling stage.

A. Profiling and Mapping Stage

The goal of this stage is to identify the best way to assign NN layers to different regions of on-chip memory, by taking into account their error sensitivity. We perform error sensitivity analysis of the layers through a per-layer approach. Specifically, we inject faults into each layer and measure the resulting accuracy. By repeating this process for all layers, we can identify the layer(s) that have the most impact on the overall output accuracy. This analysis can be carried out in $O(n)$ iterations of testing, where n is the number of NN layers.

After the per layer sensitivity analysis is done, we adopt a greedy approach to assign less error-tolerant layers to more reliable regions of on-chip memory. We assume that each memory region can accommodate at least one layer, and each layer can be mapped to only one region.

B. Error Training Stage

If the target accuracy cannot be achieved in the previous stage, we proceed to train the NN in the presence of temperature-induced errors. As discussed in Section II, the thermal gradient created by the MAC array on the on-chip memory results in different regions having different P_{error} values.

To adapt the NN to these errors and improve its accuracy, we first use the method described in the profiling stage to assign less error-tolerant layers to more reliable memory regions. Then, during the forward pass of the training procedure (similar to [9]), we inject the corresponding P_{error} values of the memory regions into the layers. This way, the NN can learn to better handle the errors and achieve higher accuracy.

C. Power-Efficient Region-Wise Active Cooling

On top of the two previous steps, we further use TEC devices as active on-chip cooling. We consider r regions (columns) in on-chip memory, each of which is covered with TECs, where we have control over each of the TEC regions, i.e. we can determine when and by how much power (cooling cost) should

TECs be turned on to dissipate the heat per region. We only consider active cooling that does not lead to Joule heating domination. This means that the cases where increasing the cooling cost leads an increase in temperature/probability of errors are disregarded. Therefore, minimizing the cooling cost under the different probability of errors is *monotonic* for each region.

Algorithm 1 shows our proposed method. We consider a $cooling\_cost_{r \times k}$ matrix, which includes k cooling power steps consumed for each of the r regions in increasing order. First, we use the most power-consuming active cooling configuration for all regions. The *cooling_config* array keeps track of the column index of the *cooling_cost* matrix for each region (Lines 2-5). Then, if the target accuracy fails to satisfy in this case, the algorithm ends with an error (Line 8). If not, the algorithm goes through a relaxation stage, where it tries to find a power-efficient cooling configuration by reducing the cooling power of each region to its previous power step.

In the relaxation stage (Lines 10-18), we start with a viable configuration, where every region is cooled down with the maximum cooling capacity and tagged as *eligible* for relaxation. Next, we choose a region for relaxation based on a heuristic (either LSF or LSEF which is discussed later) and decrease its cooling power by one step. If this new configuration can also meet the target accuracy, we continue finding more power-efficient configurations; otherwise, we backtrack to the previous configuration and the selected region is tagged as *not eligible* (Lines 15-16). Finally, when there is no eligible region for relaxation, the algorithm terminates and returns the cooling configuration, which shows which TECs should be switched on with what cooling power. The two heuristics that we consider in our work are as follows:

Largest Savings First (LSF): We select the region that provides the largest cost savings when its cooling power is reduced (or relaxed) by one step in the *cooling_cost* matrix. In other words, for region i ($1 \leq i \leq r$), if the current configuration is column j in the *cooling_cost* matrix, then the region with the largest ΔC is opted, where:

$$\Delta C = cooling\_cost[i,j] - cooling\_cost[i,j-1] \quad (2)$$

Largest Savings-Error ratio First (LSEF): We choose a region for relaxation that provides the largest cooling cost savings but with the least P_{error} coverage; in other words, the one that gives the largest cost-error ratio when relaxed to the previous step in the *cooling_cost* matrix is selected as:

$$\frac{\Delta C}{\Delta E} = \frac{cooling\_cost[i,j] - cooling\_cost[i,j-1]}{P_{error}[i,j-1] - P_{error}[i,j]} \quad (3)$$

Noteworthy, since we have at most rk iterations of relaxations, the complexity of the proposed Algorithm 1 is of $O(rk)$ iterations of testing.

D. Overview of our Proposed Algorithm

To conclude this section, we provide an overview of our proposed algorithm for temperature-aware memory mapping in Algorithm 2. The algorithm involves a profiling stage, followed by a mapping stage, and then potentially an error training stage, followed by an active cooling stage as a last resort. The profiling stage is performed using a per-layer sensitivity analysis, as described in Section III-A (Lines 1-5). Once the sensitivity analysis is complete, the layers are sorted in non-increasing order of sensitivity (Line 6), and then mapped to memory regions in a greedy manner, starting with the most reliable memory regions (Lines 7-12). If a layer can be accommodated in a memory region, it is assigned to that region; otherwise, the algorithm moves on to the next region (Lines 9-10). If the target accuracy is not met, the algorithm proceeds to the error training stage (Lines 13-15). In this stage, the model is trained with the probability of errors with the goal of improving accuracy. If further accuracy compensation is required, the algorithm proceeds to the active cooling stage (Lines 19-23), which is described in Algorithm 1. If active cooling is unsuccessful, the algorithm returns with no solution to achieving the target accuracy (Lines 24-26).

Algorithm 1 Power-Efficient Region-Wise Active Cooling

```
 1: procedure active_cooling (model)
 2: for i ← 1 to R do
 3:     cooling_config[i] ← k
 4:     eligible[i] ← 1
 5: end for
 6: accuracy ← test (model)
 7: if accuracy < target_accuracy then
 8:     return error
 9: end if
10: repeat
11:     Choose eligible region i for relaxation according to LSF or LSEF
12:     cooling_config[i] ← cooling_config[i] − 1
13:     accuracy ← test (model)
14:     if accuracy < target_accuracy then
15:         cooling_config[i] ← cooling_config[i] + 1
16:         eligible[i] ← 0
17:     end if
18: until ∀j, eligible[j] = 0
19: return cooling_config
20: end procedure
```

IV. Evaluation

A. Experiment Setup

In our experiments, we use four datasets, namely Fashion-MNIST, SVHN, CIFAR10, and Imagenette with the details given in Table I. We also employ VGG [14], MobileNetV2 [15], and ResNet [16] models adapted with their corresponding datasets. For Fashion-MNIST and SVHN, we run the Adam optimizer for 50 epochs, and for CIFAR10 and Imagenette we run it for 200 epochs with a batch size of 256 and an initial learning rate of 10^{-3} for all cases. To stabilize training, we decrease the learning rate by 50% after every 50 epochs for CIFAR10 and Imagenette, and after every second epoch for SVHN and Fashion-MNIST. To run our experiments, we use the PyTorch environment for NN training and evaluation. PyTorch allows Cpp-level access to the tensors using custom CUDA kernels, which we use for efficient quantization and fault injections. The bit-flip faults are injected into the weights and biases by creating random numbers for each bit to be flipped. If a random number is smaller than the P_{error}, the bit is flipped. In thermal multi-physics simulations, we use the commercial

Algorithm 2 Temperature-Aware Memory Mapping

```
for i ← 1 to n do
    fault_injection (layer[i], fixed_error_rate)
    accuracy ← test (model)
    sensitivity_list[i] ← 1 − accuracy
end for
Sort layers wrt sensitivity_list in non-increasing order
r ← 1
for i ← 1 to n do
    if map(region[r], layer[i]) = false then
        r ← r + 1
    end if
end for
accuracy ← test (model)
if accuracy < target_accuracy then
    train (model, P_error)
else
    return mapping
end if
accuracy ← test (model)
if accuracy ≥ target_accuracy then
    return mapping
else if active_cooling(model) ≠ error then
    return mapping with the cooling configuration
else
    print "no possible solution"
    return
end if
```

TABLE I: Datasets and models used for experiments.

Name (Model)	# Train	# Test	# Dim	# classes
Fashion-MNIST (VGG3)	60000	10000	(1,28,28)	10
SVHN (VGG7)	73257	26032	(3,32,32)	10
CIFAR10 (MobileNetV2)	50000	10000	(3,32,32)	10
Imagenette (ResNet18)	9470	3925	(3,64,64)	10

ANSYS tool (see Section II-A for modeling); 200W/cm^2 is considered for the power density of the MAC array, and 30W/cm^2 for on-chip memory and the auxiliary. Moreover, we assume a heat transfer coefficient of 100W/m^{2}K, which is the highest forced-convection of air under the maximum capability of fan-based cooling [6]. With this configuration, the total power of the TPU die reaches 72.5W in which MAC, SRAM, and auxiliary consume 50W, 7.5W, and 15W, respectively. In addition, we consider 3 equal regions (or columns) for the on-chip memory; from the farthest region to the MAC array (Region 1) to the nearest one (Region 3). Noteworthy, the number of regions can be determined based on the physical limitations of TECs and/or floorplan of NPU in addition to the thermal gradient of on-chip memory.

B. The Effect of Active Cooling on the Probability of Errors

In Fig. 2, the effect of using active on-chip TEC cooling on the P_{error} of the different memory regions is shown. The values for P_{error} in this figure, are the average of each region. Due to the spatial thermal gradient created by the MAC array through the on-chip memory, the temperature and the corresponding P_{error} for different memory regions have a considerable difference. Hence, the region in the immediate neighborhood of MAC (Region 3) is hotter, leading to a higher P_{error}. By paying more cooling cost, the P_{error} decreases for all the regions; however, it is noteworthy to note that beyond the values for the cooling cost in Fig. 2, since Joule heating

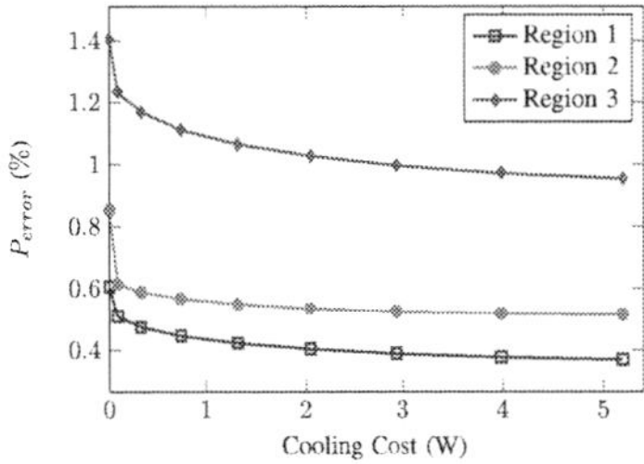

Fig. 2: The effect of active on-chip cooling on the probability of errors (P_{error}) for the 3 on-chip memory regions. Cooling cost is represented by the total power consumed by the TEC devices in Watt.

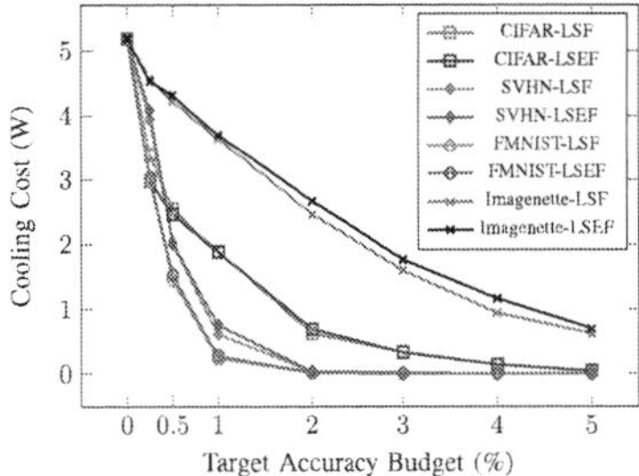

Fig. 3: The effect of different target accuracy budgets on the cooling cost using LSF and LSEF heuristics.

dominates and leads to higher temperatures, therefore it is not considered in the evaluations. This is also the reason why the trend in the figure starts with a steep decline and then the slope of the curves gradually plateaus.

C. The Effect of Target Accuracy on Cooling Cost

We consider the following set of target accuracy budgets: $\{0.25, 0.5, 1, 2, 3, 4, 5\}$ percentage below the maximum feasible accuracy achieved through maximum active cooling: 91.43% (SVHN), 90.43% (Fashion-MNIST), 80.46% (CIFAR10), and 56.36% (Imagenette). In Fig. 3, the two heuristics introduced in Section III-C for active cooling is compared for different target accuracies. Each data point in the figures corresponds to the average of 100 experiments. For Fashion-MNIST dataset, no active cooling is necessary for target accuracy budgets $\{3, 4, 5\}$%. Similarly, for SVHN, the previous stages of our proposed method can meet the accuracy demands for target accuracy budgets $\{4, 5\}$%. However, for CIFAR10 and Imagenette active cooling is essential for all target accuracies. On using the TEC devices efficiently, the greedy cost savings heuristic (LSF) performs slightly better than LSEF.

In Fig. 4, we present the accuracy results obtained with and without active cooling for different datasets, using the LSF heuristic. *Temperature-oblivious* in the figure represents the memory mapping that is done irrespective of the thermal profile of memory regions (similar to [10]). *Profiling* and *Error Training*, represent the two stages of our method.We consider two scenarios for active cooling: (1) with 1% target accuracy budget denoted as *Active Cooling*, and (2) with the maximum cooling cost (5.19W) denoted as *Max Cooling*. Finally, *No Error*, is given as a reference for the scenario where there are no errors in the memory.

For CIFAR10 dataset, our method can improve the accuracy from 21.6% in temperature-oblivious mapping to 79.8% in active cooling (1.8W), and above 80% through the maximum cooling. For Imagenette dataset, the accuracy cannot rise above 46% without using active cooling even under temperature-aware mapping and error training. For the Fashion-MNIST dataset, the values that are subject to errors are more error-tolerant compared to CIFAR10; even in temperature-oblivious mapping, the accuracy is as high as 74%. Through passing the first two stages of our method the accuracy reaches 88%. If the target accuracy is met until this point, there is no demand to incur active cooling; however, higher target accuracies are only viable via active cooling. Moreover, for SVHN dataset, while in the oblivious mapping the accuracy can be as low as 27%, only by consuming 0.62W, the accuracy can rise to over 90%.

On the Cost of Cooling Power Overhead: To estimate the obtained power saving via memory voltage scaling, we employ the state-of-the-art version of CACTI (i.e., FN-CACTI) [17]. We model in FN-CACTI, 24MB memory (mimicking the memory size in the TPUv1 chip) at two different voltages (0.7V and 0.3V). We then perform simulations for these two scenarios. An 83% reduction in on-chip memory power is observed, which is attributed to the exponential dependency of leakage power as well as the quadratic dependency of dynamic power on voltage. This, reduces the power consumed by on-chip SRAM by around 6.2W. Such obtained power saving compensates and cancels out the additional power cost needed for the TEC cooling even under the worst case in which 5.19W is requested.

On the Reliability Improvement: Note that operating SRAMs at lower voltage and temperature results in a significant improvement in reliability. To evaluate this improvement, we employ state-of-the-art physics-based aging models that are calibrated against industrial measurements for the FinFET technology [12] in order to accurately estimate the aging-induced ΔV_T. We consider two scenarios; (i) In the absence of our proposed cooling in which transistors operate at 0.7V, 97.5°C. (ii) In the presence of our proposed cooling in which transistors operate at 0.3V, 80.8°C. For a 10-year lifetime, the obtained results demonstrate that our cooling solution reduces the aging-induced ΔV_T from 34.32mV (obtained from scenario-i) down to merely 8.23mV (obtained from scenario-ii), which substantially improves (76%) the SRAM reliability for the entire lifetime.

V. Conclusion

Neural processing units are specifically designed to accelerate neural network applications. At the heart of these accelerators is a huge MAC array that speeds up the multiply-accumulate operations. In this work, we demonstrate that the power density of the MAC array has a significant impact on the nearby on-chip SRAM memory for Google's TPU. This effect creates a thermal gradient through the memory, making its regions sensitive to the different probability of errors. Superlattice thermoelectric cooling (TEC) as an active cooling can efficiently cool down the on-chip memory and compensate for the accuracy loss caused by the high power density of the MAC. The evaluations show that our temperature-aware memory mapping and active cooling method can be efficiently used to achieve higher target accuracies.

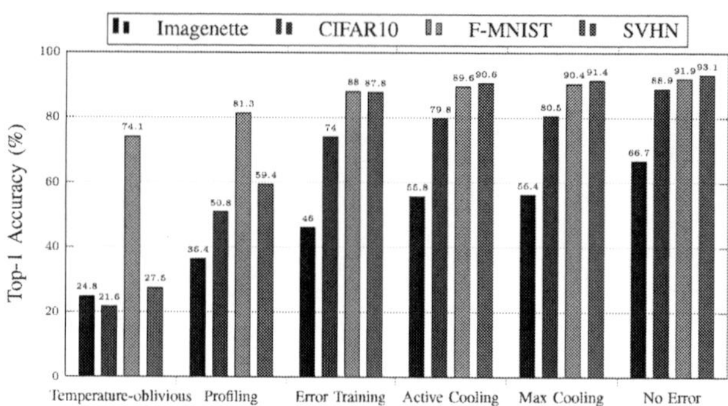

Fig. 4: Compensating accuracy loss at different stages of the proposed method for different datasets. The power consumption in active cooling for Imagenette, Cifar10, FMNIST, and SVHN are 3.7W, 1.8W, 0.24W, and 0.62W respectively.

Acknowledgment

This paper has been supported by Deutsche Forschungsgemeinshaft (DFG) within the project OneMemory (project number 405422836), the SFB 876 (project number 124020371), and project ACCROSS (428566201, AM 534/3-1).

References

[1] N. P. Jouppi *et al.*, "In-datacenter performance analysis of a tensor processing unit," in *Proc. of the 44th ISCA*, 2017, pp. 1–12.

[2] E. Talpes *et al.*, "Compute solution for tesla's full self-driving computer," *IEEE Micro*, vol. 40, no. 2, pp. 25–35, 2020.

[3] N. P. Jouppi *et al.*, "Ten lessons from three generations shaped google's TPUv4i: Industrial product," in *Proc. of the 48th ISCA*, 2021, pp. 1–14.

[4] H. Amrouch *et al.*, "Npu thermal management," *IEEE Trans. on CAD*, vol. 39, no. 11, pp. 3842–3855, 2020.

[5] S. H. Choday, M. S. Lundstrom, and K. Roy, "Prospects of thin-film thermoelectric devices for hot-spot cooling and on-chip energy harvesting," *IEEE Trans. on CPMT*, vol. 3, no. 12, pp. 2059–2067, 2013.

[6] I. Chowdhury *et al.*, "On-chip cooling by superlattice-based thin-film thermoelectrics," *Nature Nano.*, vol. 4, no. 4, pp. 235–238, 2009.

[7] S. Salamin *et al.*, "Impact of ncfet technology on eliminating the cooling cost and boosting the efficiency of google tpu," *IEEE Transactions on Computers*, 2021.

[8] B. Reagen *et al.*, "Ares: A framework for quantifying the resilience of deep neural networks," in *DATE*, 2018, pp. 1–6.

[9] S. Kim *et al.*, "Matic: Learning around errors for efficient low-voltage neural network accelerators," in *DATE*. IEEE, 2018, pp. 1–6.

[10] D. Stutz, N. Chandramoorthy, M. Hein, and B. Schiele, "Bit error robustness for energy-efficient dnn accelerators," *Proceedings of Machine Learning and Systems*, vol. 3, pp. 569–598, 2021.

[11] S. Koppula *et al.*, "Eden: Enabling energy-efficient, high-performance deep neural network inference using approximate dram," in *Proc. of the 52nd International Symposium on Microarchitecture*, 2019, pp. 166–181.

[12] S. Mahapatra and N. Parihar, "Modeling of nbti using bat framework: Dc-ac stress-recovery kinetics, material, and process dependence," *IEEE Trans. on Device and Materials Reliability*, vol. 20, no. 1, pp. 4–23, 2020.

[13] R. Krishnamoorthi, "Quantizing deep convolutional networks for efficient inference: A whitepaper," *arXiv preprint arXiv:1806.08342*, 2018.

[14] K. Simonyan and A. Zisserman, "Very deep convolutional networks for large-scale image recognition," *arXiv preprint arXiv:1409.1556*, 2014.

[15] M. Sandler, A. Howard, M. Zhu, A. Zhmoginov, and L.-C. Chen, "Mobilenetv2: Inverted residuals and linear bottlenecks," in *CVPR*, 2018.

[16] K. He, X. Zhang, S. Ren, and J. Sun, "Deep residual learning for image recognition," in *CVPR*, 2016.

[17] D. P. Ravipati *et al.*, "Fn-cacti: Advanced cacti for finfet and nc-finfet technologies," *IEEE Trans. on VLSI*, vol. 30, no. 3, pp. 339–352, 2022.

WeNet: Configurable Neural Network with Dynamic Weight-Enabling for Efficient Inference

Jingxiao Ma
School of Engineering
Brown University
Providence, Rhode Island, RI
jingxiao_ma@brown.edu

Sherief Reda
School of Engineering
Brown University
Providence, Rhode Island, RI
sherief_reda@brown.edu

***Abstract*—Deep Neural Networks (DNN) are widely deployed in resource-limited edge devices. Due to the limitation of computational resources, it is important to meet the timing and energy constraints while maintaining a high level of accuracy. To deploy the same DNN model on different edge devices, one challenge is to train a dynamic neural network with the flexibility of balancing the trade-off between accuracy and efficiency at runtime. In this paper, we present a novel methodology, dynamic Weight-enabling Network (WeNet), where the weights of neural network can be dynamically enabled or disabled to switch between different sub-networks, so that we are able to balance the trade-off between inference time, energy consumption and model accuracy. We extend the methodology to convolutional layers using group convolution and channel shuffling. We also propose a design space exploration approach to search for the optimal sub-network for different scenarios. We thoroughly evaluate our methodology using a number of DNN architectures on different hardware platforms, showing that WeNet provides a large number of energy-efficient operation modes, 73.2% of which provide better accuracy-efficiency trade-off compared to other methodologies.**

***Index Terms*—Deep learning, Energy-efficient application, Dynamic neural network.**

I. Introduction

Deep Neural Networks (DNN) are widely used in many applications, *e.g.* object detection, gesture recognition and augmented reality, which are often deployed on edge devices with limited computational resources. While it is possible to train DNNs on the cloud, inference on edge devices needs to meet timing and energy constraints. Different devices have different computing power, while other running tasks also limit the amount of available computational resources. To deploy the same DNN model in different scenarios, one solution is to train a *dynamic DNN* with *multiple configurations of operation modes*, where we are able to choose the most applicable configuration at runtime to meet the timing and energy constraints while optimizing accuracy.

The last few years have seen various methodologies for dynamic DNNs, which can be categorized into three classes: flexible width [1]–[3], flexible depth [4], [5] and flexible precision [6], [7]. In this paper, we propose a novel orientation, *flexible weight-enabling*, where connective weights between layers can be dynamically enabled or disabled. As Figure 1 illustrates, the entire model is first trained on server. For inference, the model can be dynamically configured to different sub-networks by enabling different set of weights to meet the timing and energy constraints of different devices. We name our methodology *dynamic Weight-enabling Network* (WeNet). The contributions of this paper are as follow.

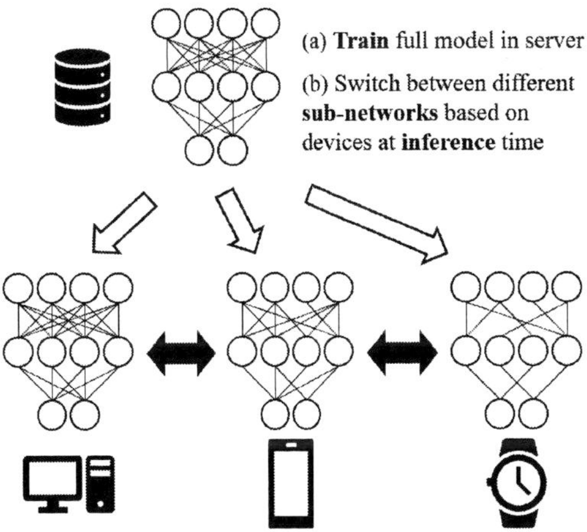

Fig. 1. Flexible weight-enabling methodology, where weights can be dynamically enabled to form different sub-networks for different hardware platforms.

- We introduce a novel dynamic DNN architecture, WeNet, that is able to dynamically enable different subsets of weights at runtime. We propose a weight-enabling pattern, that for each layer, the subset of weights forms a sub-network with several *independent groups*.
- We extend WeNet to convolutional layers by using *flexible group convolution* and *channel shuffling operation*.
- During training, random sub-networks are sampled at each iteration, where *switchable batch normalization* [2] is used. At inference time, we propose a *design space exploration* method to search optimal sub-networks and balance the trade-off between efficiency and accuracy.
- We evaluate WeNet using multiple DNN architectures, and measure inference time, energy consumption and accuracy with different configurations of sub-networks. By comparing against other dynamic DNNs, we demonstrate that WeNet provides better accuracy-efficiency trade-off.

The organization of this paper is as follow. In section II,

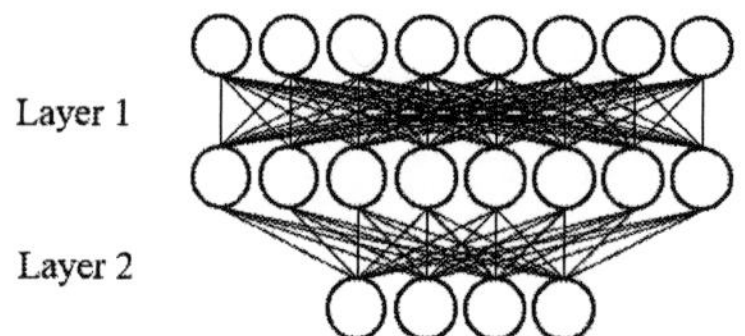

(a) Full network (1 channel)

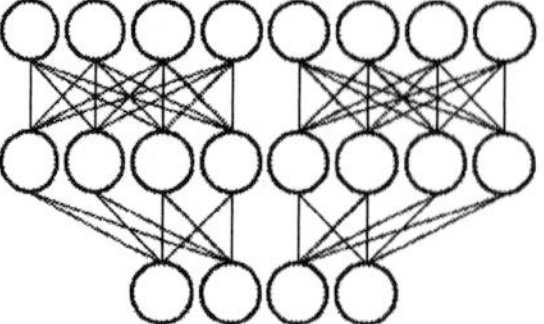
(b) 1/2-network (2 channels)

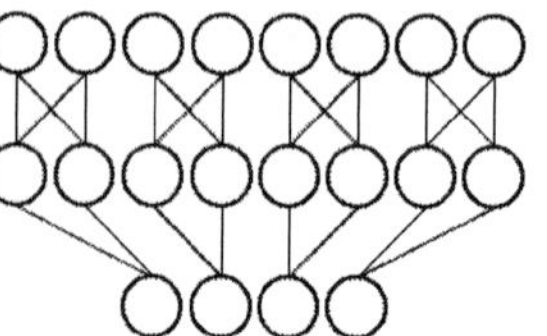
(c) 1/4-network (4 channels)

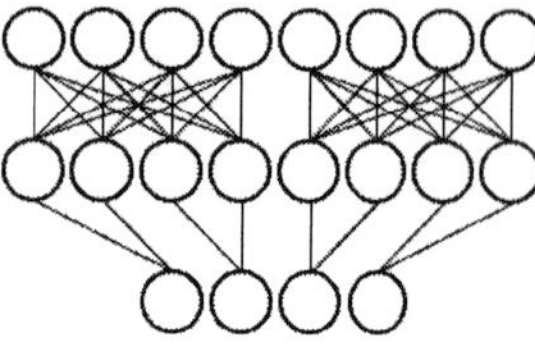
(d) Mixture of 1/2 and 1/4 networks

Fig. 2. Example of a WeNet on dense layers. (a) Full network: Enable all weights to restore original network with highest accuracy. (b) 1/2-network: Enable 1/2 weights and form 2 separated channels. (c) 1/4-network: Enable 1/4 weights and form 4 separated channels. (d) Combination of 1/2- and 1/4-network.

TABLE I
COMPARISON BETWEEN DIFFERENT DYNAMIC NETWORK METHODS

	Memory Footprint	Num. of Neurons	Special Device
Flex. Width	Reduced	Reduced	Not Needed
Flex. Depth	May Increased	Reduced	Not Needed
Flex. Precision	Reduced	Same	Needed
Flex. W.E.	Reduced	Same	Not Needed

we overview relevant previous works. In section III, we introduce WeNet and its training algorithms. We provide our experimental results in section IV. Finally, we summarize our conclusion and directions for future works in Section V.

II. PREVIOUS WORK

A number of methodologies for efficient DNNs have been proposed [8]–[10]. While these methodologies aim to shrink model size and reduce computational cost, none of them have the flexibility of adjusting model efficiency at runtime. The first design with such flexibility is "Big/Little" implementation [11], where two networks are trained and the "big" network is triggered only if the result from "small" network is not deemed confident enough. However, such methodology needs to store multiple models, and the latency for switching between different models is not negligible. Thus, a more efficient approach is to train *one single* network, with the flexibility to switch between different modes at runtime. Recent works on dynamic network can be categorized into three classes:

- **Flexible Width:** At runtime, the width of each layer can be shrunk to reduce computational cost, or expanded to increase accuracy. Tann *et al.* [1] propose one of the first runtime configurable DNN with flexible width, where DNN is trained incrementally at each width ratio. Slimmable neural network (SNN) [2] further improves the idea with switchable batch normalization. US-Nets [3] extends SNN to execute at arbitrary width ratio.
- **Flexible Depth:** Flexible depth means that the depth of DNN is adjustable at runtime. One common approach is to introduce early-exiting points, where the rest layers are ignored [5]. BranchyNet [4] proposes to add side branches to exiting points, where a forward pass can exit earlier from the main branch with higher confident inputs.
- **Flexible Precision:** Flexible precision means that the precision of operands, including both weights and inputs, can be adjusted dynamically at runtime. Pagliari *et al.* [6] find that many inputs do not need full precision to make accurate classification, and propose to reduce precision of operations when the confidence of input is high enough. SP-Nets [7] propose to train a multi-precision network using switchable batch normalization.

Table I compares three categories of dynamic networks, where all of them aim to reduce inference time and energy consumption. For *flexible width*, the number of neurons activated at each layer can be reduced at runtime, which leads to less memory footprint. For *flexible depth*, however, due to the additional side branches, memory footprint may even increase. Both *flexible width* and *flexible depth* reduce number of neurons, where crucial information may be lost and accuracy may drop significantly. For *flexible precision*, each weight uses less memory space by decreasing its floating-point precision, which reduces memory footprint. Since the number of neurons remains the same as original network, it is more likely to retain extracted features and remain high accuracy. But this class of dynamic networks cannot be used on any arbitrary devices, since the device has to support operations with multiple precision to exploit the efficiency of low-precision operands. Comparing to all three categories, we expect our methodology to reduce memory footprint for energy-efficient inference, keep the number of neurons unchanged for higher accuracy, while generalizing to all types of devices. Thus, as Table I shows, we propose *flexible weight-enabling* (Flex. W.E.), where different subsets of weights can be enabled dynamically at runtime.

III. PROPOSED METHODOLOGY

In this section, we propose the methodology of Weight-enabling Network (WeNet). As Figure 2 shows, WeNet enables different subsets of weights to dynamically switch between different sub-networks, where computational cost can be adjusted without changing the number of activated neurons. Such network architecture can be executed on any types of hardware platforms, though as discussed later, it will benefit even more with parallelism. After discussing WeNet and its extension to convolutional layers, we will introduce training algorithm and design space exploration method for optimizing the trade-off between efficiency and accuracy at inference time.

A. Dynamic Weight-enabling Network (WeNet)

WeNet dynamically balances between accuracy and efficiency by switching between different sub-networks, each of which enables a subset of weights, following a specific pattern as shown in Figure 2. Assume for a dense layer, input vector is $\mathbf{x}$ and output vector is $\mathbf{y}$. In a standard dense layer, every neuron in $\mathbf{y}$ is connected to every neuron in $\mathbf{x}$, which leads to $\ell(\mathbf{x}) \cdot \ell(\mathbf{y})$ weights, where $\ell(\mathbf{x})$ denotes the number of neurons in $\mathbf{x}$. In WeNet, we propose to divide both input $\mathbf{x}$ and output $\mathbf{y}$

into n groups, respectively, such that $\mathbf{x} = \mathbf{x}_1 \cup \mathbf{x}_2 \cup \ldots \cup \mathbf{x}_n$ and $\mathbf{y} = \mathbf{y}_1 \cup \mathbf{y}_2 \cup \ldots \cup \mathbf{y}_n$. To reduce the computational cost, instead of connecting all input and output neurons, only neurons in $\mathbf{y}_i$ are connected to neurons in $\mathbf{x}_i$, which forms a partially-connected sub-network. In other word, for each layer, $(\mathbf{x}_i, \mathbf{y}_i)$ pairs form *independent groups*. By adjusting the number of groups n at runtime, WeNet dynamically controls the number of enabled weights, and thus adjusts computational cost.

Notice that all *independent groups* can be executed in parallel. To optimize inference time and energy consumption, each layer is evenly divided, in other word, $\mathbf{x}_1, \mathbf{x}_2, \ldots, \mathbf{x}_n$ have same number of neurons. Evenly-divided layers have two benefits: (1) For layers with n groups, evenly division leads to the least number of weights as well as computational cost

$$O\left(\frac{\ell(\mathbf{x}) \cdot \ell(\mathbf{y})}{n}\right) \tag{1}$$

(2) Each group has the same amount of FLOPs, which further improves efficiency with parallel execution.

Figure 2(a) shows the full network. To improve efficiency, we may disable half of the weights and execute the $1/2$-network as Figure 2(b) shows. If faster inference time or lower energy consumption is expected, we may then switch to $1/4$-network shown in figure 2(c). As less weights are enabled during this process, WeNet loses more information, which leads to lower accuracy as trade-off. To further improve the flexibility, we may create more operating modes, where each layer may have individual weight-enabling ratio. Figure 2(d) shows a combination of $1/2$- and $1/4$-network. By enabling different subset of weights, WeNet dynamically switches between sub-networks to balance between efficiency and accuracy.

B. WeNet on Convolutional Layers

WeNet transforms standard dense layers into a set of switchable partially-connected sub-networks. Nowadays, the most widely used layers in DNNs are convolutional layers, which itself is a partially-connected layer. In this subsection, we extend the idea of WeNet to convolutional layers.

Similar to enabling subset of weights in dense layers as shown in Figure 2, in convolutional layer, WeNet dynamically enable *subset of kernels* between feature maps, which forms *independent groups of channels*, or in other word, using group convolution with the flexibility of switching number of groups. We consider a standard convolutional layer as the full network of convolutional WeNet. Assume that it takes an input tensor L_i of size $h_i \times w_i \times c_i$, and applies convolutional kernel $K \in \mathbb{R}^{k \times k \times c_i \times c_o}$ to produce an output tensor L_o of size $h_o \times w_o \times c_o$. The computational cost of standard convolution is

$$T_{full} = h_i \times w_i \times c_i \times c_o \times k \times k \tag{2}$$

Similar to WeNet on dense layer, input tensor L_i with c_i feature maps is divided *evenly* into n groups, each of which has c_i/n feature maps, such that $L_i = L_i^1 \cup L_i^2 \cup \ldots \cup L_i^n$. Output tensor L_o is also divided in same pattern $L_o = L_o^1 \cup L_o^2 \cup \ldots \cup L_o^n$. In standard convolution, there exist kernels that convolve any input feature map into any output feature map. In sub-network of WeNet, however, only kernels between

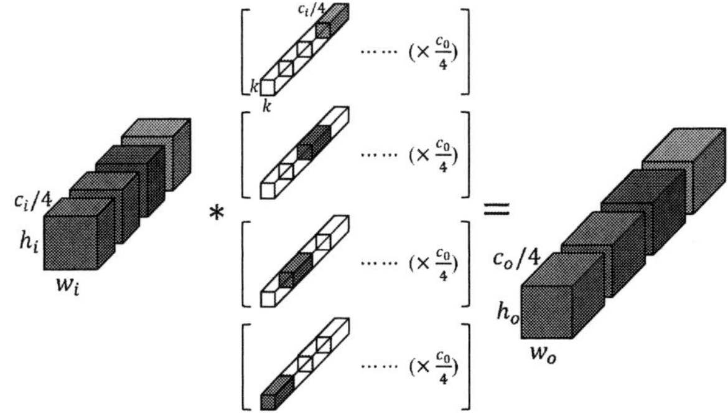

Fig. 3. $1/4$-network of convolutional layer

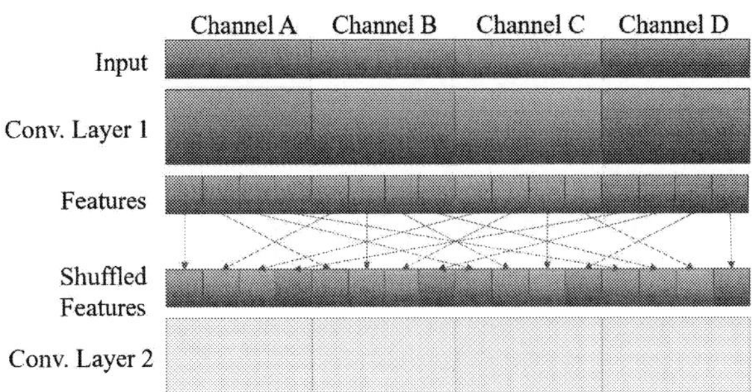

Fig. 4. Channel shuffling after $1/4$-network

feature maps of L_i^j and L_o^j is enabled. Figure 3 demonstrates the example of dividing input and output tensors into 4 groups, where input feature maps in L_i^j are only convolved to output feature maps in L_o^j, using only the kernels shown in same color. Since WeNet effectively forms 4 *separate groups of channels*, only $1/4$ of kernels in each filter are enabled. Thus, computation cost is reduced to $1/4$ of original cost T_{full}.

$$T_{1/4} = h_i \times w_i \times \frac{c_i \times c_o \times k \times k}{4} = \frac{T_{full}}{4} \tag{3}$$

Similar to dense layer, each group of channels is independent, which can be executed in parallel to further improve efficiency at inference time. However, on the other hand, each feature map can only receive inputs from its own group. Compared to standard convolution, feature maps in each group receive limited input information for features extractions, where accuracy may drop significantly. To mitigate the accuracy loss caused by *isolated* groups, we adopt a channel-shuffling operation proposed by ShuffleNet [9]. A channel-shuffling operation is performed between two convolutional layers. As Figure 4 shows, after the first convolutional layer, each group of features is first split, and then shuffled and merged to form new groups of features as input to the next convolutional layer. Although a standard convolution is replaced by several independent groups of channels, channel-shuffling allows each group to learn from the others, which significantly improves accuracy. Meanwhile, parallelism can still be used on each group to improve efficiency.

Implementation of channel-shuffling is straightforward. Suppose a convolutional layer of WeNet has *maximum* g independent groups (in experiment, we set $g = 16$), where each group has n feature maps. All $g \times n$ feature maps are

Algorithm 1: Train with random sampling and S-BN

```
Input  : Training set (X, Y) with features X and labels Y,
         Number of iterations n_iter, Number of sampled
         sub-network per iteration s, Loss function loss_fn,
         Optimizer opt
Output: WeNet M
1  Initialize WeNet M
2  Initialize S-BN layers for each weight-enabling pattern
3  for num_iter = 1, 2, ..., n_iter do
4      Get next batch of data and labels (x, y) from (X, Y)
5      Clear gradients opt.zero_grad()
6      Execute full network ŷ = M(x)
7      Compute loss loss = loss_fn(ŷ, y)
8      Accumulate gradient loss.backward()
9      Adjust S-BN to smallest sub-network
10     Execute smallest sub-network ŷ = M'(x)
11     Compute loss loss = loss_fn(ŷ, y)
12     Accumulate gradient loss.backward()
13     Randomly sample s − 2 sub-networks
14     for each sampled WeNET do
15         Adjust S-BN according to weight-enabling patterns
16         Forward training data ŷ = M'(x)
17         Compute loss loss = loss_fn(ŷ, y)
18         Accumulate gradient loss.backward()
19     end
20     Update gradients to weight opt.step()
21 end
22 return M
```

first reshaped into (g, n), which are then transposed to (n, g) and flattened as shuffled groups. Channel-shuffling operation only involves matrix transposing, where the computational cost is negligible compared to model inference.

C. Training WeNet with Switchable Batch Normalization

WeNet dynamically enables subset of weights or kernels to form sub-networks, where each layer may have different number of *independent groups*. Since it is impractical to enumerate and train all possible sub-networks, we propose to *randomly sample* sub-networks at each iteration.

We also need to reconsider the implementation of Batch Normalization (BN) layers, which normalize feature maps across each batch. Nowadays, BN is widely used to reduce internal covariate and stabilize training process. Using y and y' to denote inputs and outputs, computation of BN layers is defined as

$$y' = \gamma \cdot \frac{y - \mu}{\sqrt{\sigma^2 + \epsilon}} + \beta \tag{4}$$

where μ, σ^2 are means and variance of feature maps of current batch, γ, β are learnable variables. Since each layer has a number of possible weight-enabling pattern, where each neuron receives different inputs as Figure 2, μ and σ^2 are not consistent, which leads to inaccurate learning of γ and β. Yu *et al.* propose Switchable Batch Normalization (S-BN) in SNN [2], which privatizes μ, σ^2, γ, β settings for different pattern of the same layer. In other words, each weight-enabling pattern has its own customized BN layer, which is switchable according to the selected pattern. The number of learnable variables in BN layers is negligible compared to other weights and kernels. Notice that in WeNet, although

Algorithm 2: Design space exploration of WeNet

```
Input  : Full network of WeNet M, Accuracy threshold δ
Output: Optimal sub-network M'
1  M' = M // Keep track of most efficient sub-network so far
2  acc = Accuracy of M // Accuracy of current sub-network
3  while acc > δ do
4      for each layer i in M' that has < 16 groups do
5          Increase the number of groups in layer i as
             sub-network m_i
6          acc_i = Accuracy of m_i
7          t_i = Inference time of m_i
8          loss_i = t_i/acc_i
9      end
10     k = arg min_i loss_i
11     M' = m_k and acc = acc_k
12 end
13 return M'
```

there is a large number of possible sub-networks, for each layer the possible number of weight-enabling patterns is limited (in our experiment, maximum 16 independent groups). Thus, it is still efficient to use customized S-BN layers for all weight-enabling patterns in each layer. Algorithm 1 describes the process of training WeNet using S-BN. We initialize DNN with independent BN layers for each weight-enabling pattern (line 1-2). For each iteration, we accumulate gradients from full network (line 6-8), the smallest sub-network (line 9-12) and other $s - 2$ random sub-networks (line 14-19).

D. Design Space Exploration

After training with Algorithm 1, theoretically WeNet can be configured into operation modes corresponding to all possible sub-networks. But in practice, some sub-networks performs better in terms of efficiency-accuracy trade-off. Given an accuracy threshold, to determine the optimal operation mode on a specific hardware platform, we propose to explore the design space of possible sub-networks, using *design space exploration* (DSE) approach as Algorithm 2.

We start from the full network (line 1-2). In each iteration of DSE, we consider a number of candidate sub-networks by increasing the number of groups in one of the WeNet layers, respectively (line 5). We set an upper bound on number of groups (16 in experiment) to prevent accuracy dropping rapidly. As our objective is to minimizing computational cost (represented by inference time t) while keeping accuracy acc high, for each candidate sub-network i, we evaluate t_i and acc_i using targeted hardware device, and compute a loss using t_i/acc_i (line 8). By minimizing the loss in each iteration, we keep reducing computational cost while optimizing the trade-off against accuracy. When accuracy of current sub-network drops below threshold, DSE is finished.

Using DSE, we effectively explore a number of possible sub-networks. At each iteration, sub-network with smallest loss forms Pareto Frontier, which demonstrate the optimal trade-off between efficiency and accuracy.

IV. Experimental Results

In this section, we first describe our experiment setup. To highlight the benefits of channel-shuffling, we evaluate WeNet

TABLE II
COMPARISON OF TOP-1 ACCURACY AND INFERENCE TIME WITH AND WITHOUT CHANNEL-SHUFFLING OPERATIONS USING RESNET-50.

Sub-Network	Without Shuffling		With Shuffling	
	Acc.(%)	Time(ms)	Acc.(%)	Time(ms)
1/16	60.88	41.68	64.24 (+3.36)	43.41 (+1.73)
1/8	65.81	82.79	70.43 (+4.62)	85.62 (+2.38)
1/4	69.38	124.14	74.07 (+4.69)	127.35 (+3.21)
1/2	73.49	127.88	75.63 (+2.41)	132.17 (+4.29)
Full	75.99	198.62	76.07 (+0.08)	203.38 (+4.76)
Average	69.11	115.02	72.09 (+2.98)	118.38 (+3.36)

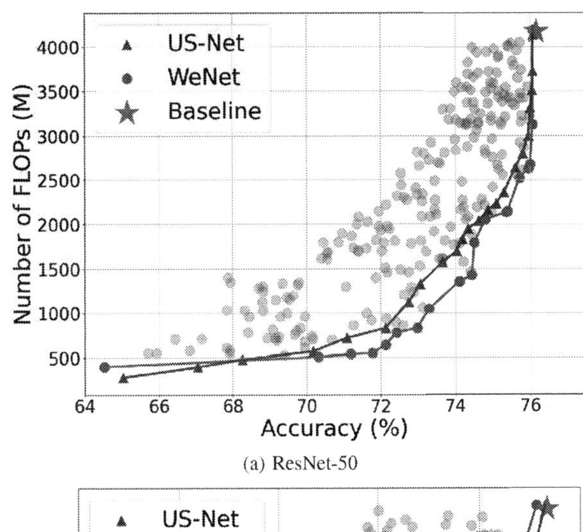

(a) ResNet-50

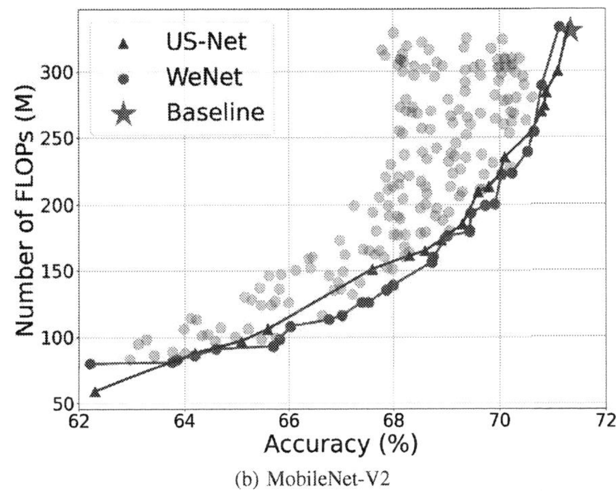

(b) MobileNet-V2

Fig. 5. Comparison between WeNet and US-Net [3]

with and without channel-shuffling operations. We also explore the design space of WeNet and compare optimal execution modes against other methodologies, showing that WeNet provides better trade-off. Finally, we measure inference time and energy consumption, showing the benefits on different types of real devices.

A. Experiment Setup

1) *Datasets and models:* We implement and evaluate WeNet using *ResNet-50* [12], *MobileNet-V2* [8] and *EfficientNet-B0* [10] on ImageNet classification problem.
2) *Software setup:* We implement weight-enabling operations with *group convolution* in *PyTorch*. We use default training settings for each benchmark, with one additional hyper-parameter $s = 20$, meaning that in each iteration we randomly sample 20 sub-networks.
3) *Hardware setup:* We measure inference time and energy consumption on the NVIDIA Jetson Nano board.

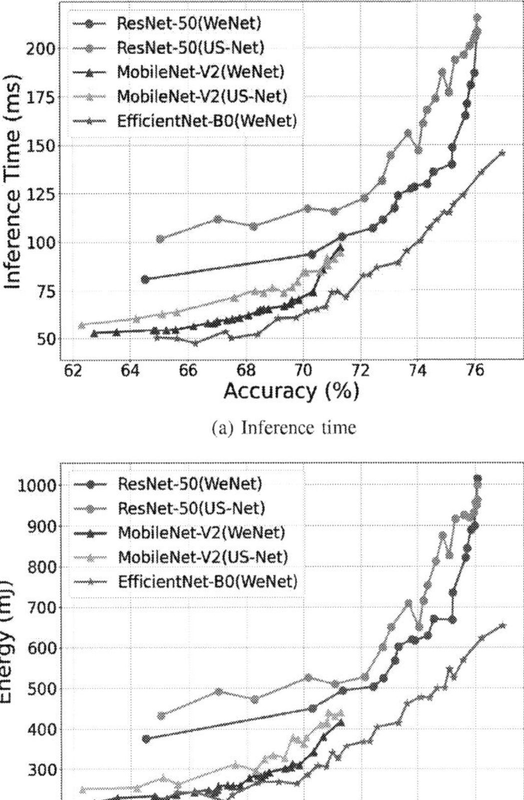

(a) Inference time

(b) Energy consumption

Fig. 6. Inference Time and Energy Consumption on Jetson Nano Board

B. Channel-Shuffling

In the first experiment, we demonstrate the benefit of channel-shuffling operation by evaluating five operating modes of WeNet with and without channel-shuffling. Table II shows the comparison result using ResNet-50, where each residual block has one channel-shuffling operation after the first 1×1 convolution layer. Channel-shuffling operation significantly improves accuracy for all five sub-networks. On average, channel-shuffling increases accuracy by 2.98%. Meanwhile, channel-shuffling introduces additional computational cost. On average, these operations use additional $3.36ms$ out of $118.38ms$ inference time. Compared to significant improvements on accuracy, increasing of inference time is negligible.

C. WeNet v.s. US-Net

In the second set of experiment, we compare WeNet against US-Net [3], which provides multiple operation modes by adjusting width ratio for each layer, using two benchmarks, i.e. *ResNet-50* [12] and *MobileNet-V2* [8]. After training WeNet

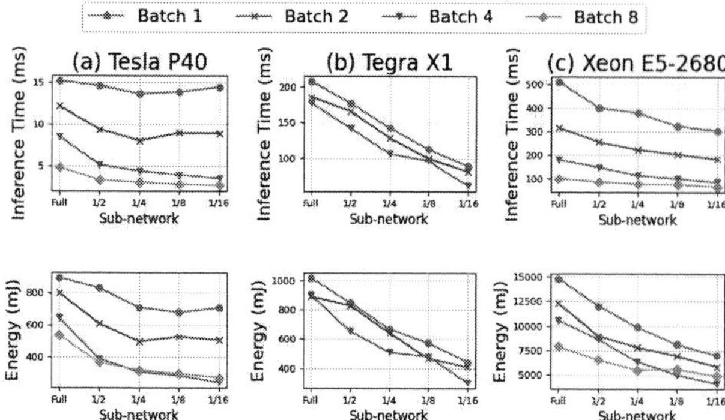

Fig. 7. Evaluation of *ResNet50* on three devices, with different batch size

using Algorithm 1, we explore the design space of WeNet operation modes using Algorithm 2, where computational cost is measured by number of Floating-Point Operations (FLOPs). Figure 5 shows the comparison results, where red star represents the original model. Compared to US-Nets, WeNet has less number of FLOPs in most accuracy range, which demonstrates the state-of-the-art performance. For ResNet-50, WeNet provides more efficient operating modes when accuracy is higher than 69%. For MobileNet-V2, WeNet performs better when accuracy is between 64% and 71%.

D. Inference Time and Energy Consumption

In the third set of experiment, we implement WeNet using all three benchmarks and plot operation modes on Pareto Frontier. Using Jetson Nano board, we evaluate inference time and energy consumption of these optimal points. Figure 6 shows the evaluation results, including US-Net in Section IV-C as comparison. As Figure 6a shows, for all three benchmarks, WeNet substantially saves inference time by trading-off small amount of accuracy. Compare to number of FLOPs in Figure 5, the improvement of inference time is more significant, which is always lower then US-Net. Since each layer of WeNet consists of *independent* groups, which can be executed in parallel, inference time can be further saved if there are idle threads (or warps). Thus, the benefit of inference time is more significant. As Figure 6b shows, WeNet also substantially improves energy consumption compared against US-Net.

E. Evaluation on Different Devices

Finally, in Figure 7, we measure inference time of models trained in Section IV-C using three devices, *i.e. Tesla P40* (GPU), *Tegra X1* (GPU) and *Xeon E5-2680* (CPU). High-performance GPUs (*Tesla P40*) are already well-optimized for tensor operations. Thus, on these devices, inference time of *full network* is already fast enough. Disabling part of weights does not accelerate inference time as expected, but does improve energy efficiency due to fewer FLOPs. If we increase batch size, inference time and energy consumption keep reducing, since there is still enough computational resources to do parallel computing for each batch. On the other hand, for low-performance devices (*Tegra X1*), all threads (or warps) are busy even if batch size is 1. In this case, increasing batch size does not make much difference. But inference time and energy consumption vary a lot according to sub-network, where disabling more weights improves efficiency a lot. Since CPU (*Xeon E5-2680*) is not designed specifically for tensor operations, the inference time and energy consumption is much higher compared to GPU. But disabling more weights still improves energy efficiency significantly.

V. Conclusion

In this paper, we proposed a novel dynamic network methodology, WeNet, which enables different subsets of weights on the fly to trade-off between accuracy and inference time. By enabling smaller subsets of weights, WeNet effectively forms a "sparser" sub-network with multiple separate groups of channels, where parallelism can be used to further improve efficiency. We also extended WeNet to convolutional layers using group convolution and channel shuffling, trained with switchable batch normalization and explored design space of possible sub-networks. By evaluating on different DNNs, we demonstrated that WeNet is able to optimize the trade-off between efficiency and accuracy.

References

[1] H. Tann, S. Hashemi, R. I. Bahar, and S. Reda, "Runtime configurable deep neural networks for energy-accuracy trade-off," in *2016 International Conference on Hardware/Software Codesign and System Synthesis (CODES+ ISSS)*. IEEE, 2016, pp. 1–10.

[2] J. Yu, L. Yang, N. Xu, J. Yang, and T. Huang, "Slimmable neural networks," *arXiv preprint arXiv:1812.08928*, 2018.

[3] J. Yu and T. S. Huang, "Universally slimmable networks and improved training techniques," in *Proceedings of the IEEE/CVF international conference on computer vision*, 2019, pp. 1803–1811.

[4] S. Teerapittayanon, B. McDanel, and H.-T. Kung, "Branchynet: Fast inference via early exiting from deep neural networks," in *2016 23rd International Conference on Pattern Recognition (ICPR)*. IEEE, 2016, pp. 2464–2469.

[5] S. Laskaridis, A. Kouris, and N. D. Lane, "Adaptive inference through early-exit networks: Design, challenges and directions," in *Proceedings of the 5th International Workshop on Embedded and Mobile Deep Learning*, 2021, pp. 1–6.

[6] D. J. Pagliari, E. Macii, and M. Poncino, "Dynamic bit-width reconfiguration for energy-efficient deep learning hardware," in *Proceedings of the International Symposium on Low Power Electronics and Design*, 2018, pp. 1–6.

[7] L. Guerra, B. Zhuang, I. Reid, and T. Drummond, "Switchable precision neural networks," *arXiv preprint arXiv:2002.02815*, 2020.

[8] M. Sandler, A. Howard, M. Zhu, A. Zhmoginov, and L.-C. Chen, "Mobilenetv2: Inverted residuals and linear bottlenecks," in *Proceedings of the IEEE conference on computer vision and pattern recognition*, 2018, pp. 4510–4520.

[9] X. Zhang, X. Zhou, M. Lin, and J. Sun, "Shufflenet: An extremely efficient convolutional neural network for mobile devices," in *Proceedings of the IEEE conference on computer vision and pattern recognition*, 2018, pp. 6848–6856.

[10] M. Tan and Q. Le, "Efficientnet: Rethinking model scaling for convolutional neural networks," in *International conference on machine learning*. PMLR, 2019, pp. 6105–6114.

[11] E. Park, D. Kim, S. Kim, Y.-D. Kim, G. Kim, S. Yoon, and S. Yoo, "Big/little deep neural network for ultra low power inference," in *2015 international conference on hardware/software codesign and system synthesis (codes+ isss)*. IEEE, 2015, pp. 124–132.

[12] K. He, X. Zhang, S. Ren, and J. Sun, "Deep residual learning for image recognition," in *Proceedings of the IEEE conference on computer vision and pattern recognition*, 2016, pp. 770–778.

Energy-Efficient Missing Data Recovery in Wearable Devices: A Novel Search-Based Approach

Dina Hussein*, Taha Belkhouja*, Ganapati Bhat, and Janardhan Rao Doppa
School of Electrical Engineering and Computer Science, Washington State University, Pullman, WA, 99164

***Abstract*—Wearable and internet of things (IoT) devices are transforming a number of high-impact applications. Machine learning (ML) algorithms on wearable devices assume that data from all sensors is available at runtime. However, one or more sensors may be unavailable at runtime due to malfunction, energy constraints or communication challenges. Loss of sensor data can potentially lead to severe degradation in application accuracy and quality of service. Commonly employed generative ML methods to recover missing data are not suitable for resource-constrained wearables because they incur significant memory, execution time, and energy overhead at runtime. In contrast to prior methods, this paper presents a novel search-based accuracy-preserving imputation (AIM) algorithm that obtains most likely imputation patterns of sensor data for each missing data scenario via *offline* analytics. Specifically, for each missing data condition, we store the most likely recovery patterns which preserve ML classifier-based application accuracy in a look up table and use it appropriately at runtime. The key insight behind AIM is that we do not need exact recovery of the missing data as long as the ML classifier-based application accuracy (e.g., health assessment) is preserved. To further improve the overall effectiveness of AIM, we train the ML classifiers to be robust to small errors in data recovery. Experiments on four diverse wearable sensor based time-series benchmarks demonstrate that AIM is able to maintain accuracy within 5% of the baseline with no missing data when one sensor is missing, and improves the overall accuracy by 15% compared to a state-of-the-art baseline. AIM achieves this improvement with negligible energy consumption overhead.**

I. Introduction

Wearable and internet of things (IoT) devices are enabling several challenging and high-impact applications such as health monitoring, rehabilitation, and fitness tracking [1–3]. They are also employed in mobile health applications including diagnosis of movement disorders such as essential tremor, and Parkinson's disease by monitoring the respective biomarkers [4, 5]. Wearable devices typically implement these applications by collecting data from various sensors mounted on the body and processing them using machine learning (ML) algorithms to make data-driven predictions and decisions.

One of the key assumptions made by wearable applications including the above-mentioned ones is that data from all sensors is available at runtime. However, data from one or more sensors might be missing during real-world, runtime usage. Sensor data may be missing due to energy limitations, user error, sensor malfunction, or data communication challenges [6, 7]. Missing data leads to significant degradation

*D. Hussein and T. Belkhouja contributed equally.

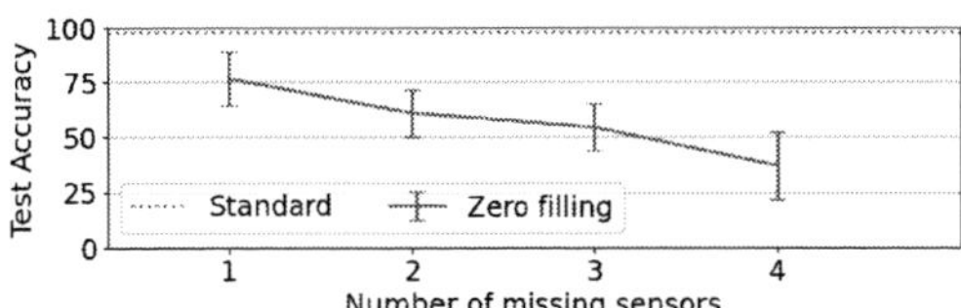

Fig. 1: Classification accuracy (mean and standard deviation) of a deployed ML classifier on Shoaib [8] dataset as a function of k missing sensors.

in the overall quality of applications since the underlying ML models are trained with the assumption that data from all sensors is available. Specifically, our experiments on real-world applications show that missing data from even a single sensor can degrade the accuracy of ML models by as much as 20%, as shown in Figure 1. Training multiple models for each combination of sensors is also not feasible due to increased memory overhead and context switching requirements at run-time. Therefore, there is a strong need to develop approaches that **1)** recover missing data at runtime to preserve the accuracy of a given ML-driven application; and **2)** have low-overhead in terms of execution time, power consumption, and memory footprint as wearable devices are resource-constrained.

Recent work has proposed imputation methods that aim to recover raw sensor data at runtine [9–12]. These approaches typically use statistical methods or deep generative networks. While imputation networks are able to recover accuracy loss, they suffer from high memory overhead since we have to store parameters for $2^n - 2$ possible missing data scenarios with n sensors on the wearable device. They also incur additional execution overhead to produce imputed data.

This paper presents a novel and energy-efficient search-based *Accuracy-Preserving Imputation (AIM)* approach to produce accuracy-preserving imputations for missing sensor data at runtime. The AIM method meets the desiderata of an effective solution for missing sensor data based on two *synergistic* principles. **First**, we do not need to recover the exact missing sensor data as long as the application accuracy is preserved. **Second**, we can train the ML model to be robust to small deviations from the exact sensor data, i.e., ability to make accurate predictions for sensor data with small perturbations. To instantiate the first principle, AIM starts with a set of possible scenarios for missing sensor data and formulates a search problem whose goal is to obtain the most likely imputation pattern for the missing sensors using the given training data with no missingness. When we run the search algorithm for each of the $2^n - 2$ missing data scenarios, we get a table of imputation patterns for each scenario. To

979-8-3503-1176-1/23 $31.00 © 2023 IEEE

instantiate the second principle, we train robust ML models using augmented data (i.e., perturbations of the clean and complete training data) to make accurate predictions when the imputed data deviates from the true data.

We validate the proposed AIM approach on four diverse wearable sensor-based time series benchmark datasets [8, 13–15]. For each of these applications, we enumerate all possible scenarios of missing sensor data and execute AIM for imputing the missing data. Our results demonstrate that for up to 2 missing sensors, AIM achieves accuracy within 5% of the upper-bound baseline with no missing data. This is a remarkable result because the likelihood of one or two sensors missing is higher than a larger number of sensors being unavailable in the real-world. The application accuracy with AIM drops with more than two missing sensors, however, it is still higher than the accuracy with no data recovery and generative imputation networks. We also compare AIM with a recent generative imputation network approach (GAIN) [12]. Compared to GAIN, our AIM approach achieves over 15% higher prediction accuracy on average with significantly lower energy and memory overhead. Specifically, our measurements on the Odroid-XU3 device [16] show that AIM enables 78–98% energy savings over GAIN, thus almost doubling the operating time of wearable devices.

Contributions: This paper makes the following contributions:

- Characterization of the accuracy loss when one or more sensors are unavailable in wearable applications.
- Novel and energy-efficient accuracy-preserving imputation approach called AIM to identify the most likely data patterns for imputation of missing sensor data.
- Experimental validation on four diverse wearable datasets to demonstrate that AIM enables reliable imputation of missing sensor data with minimal overhead.

II. Related Work

Wearable devices are being increasingly used in health applications [1–3, 17]. Integrating multiple sensors increases the likelihood of one or more sensors being unavailable due to energy constraints, user error, or communication challenges. Therefore, there is a strong need for energy-efficient and accuracy-preserving methods to recover missing data.

Recent work has proposed several methods to handle missing data in sensor based applications [9, 10, 12, 18, 19]. These approaches typically use statistical or deep generative methods to recover the missing data. Statistical methods, such as mean, median, or regression use available data around the missing instances to obtain an imputation [9, 10]. As such, they are suitable to handle isolated missing data instances where data around the missing samples is available. However, they are not suitable for long sequences of sensor unavailability with no reference data for statistical methods, which is the focus of this paper. Deep generative methods have been recently proposed to handle longer sequences of missing data [12, 20]. For instance, [12] employs a generative adversarial imputation network (GAIN) to impute the data. Specifically, the GAIN approach imputes the missing data conditioned upon observed

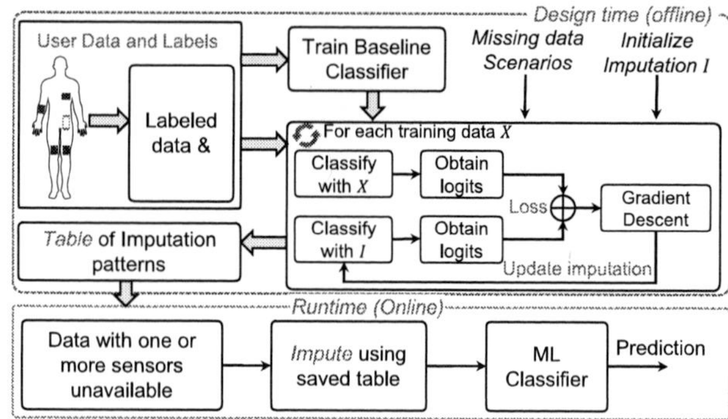

Fig. 2: Overview of the proposed accuracy-preserving imputation approach.

data. Similarly, the work in [20] employs deep auto-encoder models to impute EEG recordings from multiple patients and data collection days. However, the primary limitation of deep learning methods is the high memory overhead and the energy cost of performing imputation at runtime.

To precisely fill this gap in the current knowledge, this paper develops a novel and energy-efficient accuracy-preserving imputation method to handle missing sensor data at runtime.

III. Background and Problem Setup

This section first provides the background on wearable devices and introduces the missing sensor data problem.

A. *Wearable Devices Preliminaries*

We consider wearable systems with multiple sensors mounted on the body, as shown in Figure 2. The sensors are used to monitor physical and physiological parameters for applications including mobile health, which we use as the running example. Since the sensor data is continuous and streaming in nature, wearable devices perform the following steps to enable health assessment.

Data Segmentation: The streaming time series sensor data must be divided into equal sized windows for periodic health assessment and to provide fixed-sized inputs to the ML models. Given n sensors and T samples in each window, we denote the time-series sensor data with the variable $X \in \mathbb{R}^{n \times T}$.

Feature Generation and Classification: The time-series sensor data in each window $X \in \mathbb{R}^{n \times T}$ are fed into feature generation and classifier blocks to perform the health assessment. Labeled pairs of sensor data $X \in \mathbb{R}^{n \times T}$ and the class label y are used to train a classifier F_θ, where θ are the parameters for the classifier. At runtime, the sensor data and trained classifier are used to predict class labels of interest.

B. *Missing Sensor Data and Imputation Challenges*

Depending on the length of unavailability, we can classify the missing data patterns into two main categories as follows.

Random Missing Data: In this case, the sensor encounters isolated missing samples that are not clustered around any particular time instance. Prior work has proposed a number of approaches to handle random missing data [9, 10]. Random missing data is typically easier to handle since data around the missing instance is available for imputation. Therefore, we do not consider the random missing case and focus our attention on the more *challenging* block missing case described next.

Block Missing Data: Block missing data occurs when long sequences of sensor data are unavailable at runtime. The block missing data is challenging to recover since it does not contain any reference data for the missing sensors. Moreover, in a system with n sensors, we can have $2^n - 2$ possible combinations of block missing data.

C. Problem Setup

Let $X \in \mathbb{R}^{n \times T}$ be the time-series sensor data, where n is the total number of sensor channels and T is the number of samples in each input window. We denote every input $X \in \mathbb{R}^{n \times T}$ as $[X_1, \cdots, X_i, \cdots, X_n]$ where $X_i \in \mathbb{R}^T$ corresponds to a channel i of X. The class label for each window is denoted by y. We denote the set of training examples $\mathcal{D}$ as several input X and ground-truth output y pairs. Standard ML algorithms use the given data $\mathcal{D}$ to train a classifier F_θ which takes time-series X as input to predict the corresponding class label $\hat{y}$, where θ stands for parameters of the classifier.

During inference, the data from one or more sensors might be missing. Let $\{j\}_{0 \leq j \leq n}$ represent the subset of channels that are missing at runtime for a given input X. We denote the new input that has $\{j\}$ missing channels by $\tilde{X}_{\{j\}} \in \mathbb{R}^{n \times T}$:

$$\tilde{X}_{\{j\}} = \begin{cases} 0^T & \text{if } i \in \{j\} \\ X_i & \text{if } i \notin \{j\} \end{cases} \tag{1}$$

For example, for a human activity recognition (HAR) application, when one of the sensors goes missing, its three accelerometer channels data $\{j\} = \{1, 2, 3\}$ will have 0 value.

This missing data in the input results in misclassifications ($\hat{y} \neq y$) by the classifier F_θ as seen in Figure 1. Consequently, the overall performance of the classifier and quality of application service during deployment will degrade significantly.

IV. Search based Accuracy-Preserving Imputation

This section describes the accuracy-preserving imputation (AIM) algorithm to solve missing sensor data challenge for wearable applications. We first provide an overview of the AIM approach and list the two synergistic design principles behind AIM. Next, we describe the details of the algorithmic approaches which instantiate those two design principles.

Overview of AIM Approach: During the offline configuration of AIM, we perform the following two steps sequentially, as shown in Figure 2. First, we train a robust ML classifier F_θ that can make accurate predictions for small perturbations of time-series signals in the training data (i.e., no missingness). This step ensures accuracy even if there are small errors in the imputed values. Second, given the trained ML classifier F_θ, for each candidate missing sensors configuration, we execute a search algorithm to compute the most likely imputation pattern that preserves the accuracy of the classifier F_θ using the training data. The output of this step is a lookup table $\mathcal{I}$ that stores one imputation pattern for each missing configuration. At runtime, given an input X with some missing channels (one for each sensor), we impute the missing data using appropriate imputation pattern from the lookup table $\mathcal{I}$ (i.e., negligible overhead) and then use the classifier F_θ to make prediction.

Design Principles: AIM meets the desiderata of an effective solution for missing sensor data based on two synergistic principles. *1) Accuracy-preserving imputation:* there is no need to recover the exact missing sensor data as long as the accuracy of ML classifier is preserved. *2) Training robust classifiers:* training the ML classifier to make accurate predictions even with small deviations from the exact sensor data distribution will exhibit robustness to small errors in imputed data. Below we provide algorithms to instantiate these two principles.

A. Search Algorithm for Accuracy-Preserving Imputation

Intuition: During the offline training phase with no missing data, the ML classifier for wearable application achieves high accuracy on the given classification task. Since the inputs are multivariate time-series signals, the classifier relies on the information spread across n different channels to make accurate predictions. Prior work has shown that classifiers rely on a subset of critical channels to predict output labels [21]. This means that in case of missing channels, the input possesses enough information to allow the classifier to predict its true label. Therefore, we set our goal to find a recovery data pattern that lets the classifier predict the correct labels from the available channels. The recovery pattern pushes the classifier to predict the output label as if the data from all sensors is available. In summary, we can view our solution as the search for the most likely data pattern to impute the data for missing sensors for preserving the accuracy of the given ML classifier.

Algorithm: Formally, given a ML classifier F_θ, for any input $X = [X_1, \cdots, X_n]$ and a fixed set of missing channels $\{j\}$, we search for an imputation pattern $\mathcal{I}_{j \in \{j\}} \in \mathbb{R}^T$ s.t.:

$$\mathcal{I}_{\{j\}} = \begin{cases} \mathcal{I}_j & \text{if } i \in \{j\} \quad \text{and} \quad F_\theta(X) \approx F_\theta(\mathcal{I}_{\{j\}}) \\ X_i & \text{if } i \notin \{j\} \end{cases} \tag{2}$$

We note that $\mathcal{I}_{j \in \{j\}}$ *does not depend* on the available sensor data of the input X. Every imputation pattern is stored in a look up table indexed by the combination of the missing channels $\{j\}_{0 \leq j \leq n}$ where the samples are missing. We conduct this search for each missing sensor data configuration during design time (offline) where the goal is to find for every combination of $\{j\}$ missing channels, the corresponding imputation pattern $\mathcal{I}_{j \in \{j\}}$ that yields a prediction similar to the prediction on the original input without any missingness. Hence, at runtime, we use the stored lookup table to select the appropriate imputation pattern with negligible overhead.

We find imputation patterns based on a given set of missing channels $\{j\}_{0 \leq j \leq n}$ (missing configuration) that preserves the accuracy of classifier F_θ. We define our overall objective for the search of imputation pattern as shown below:

$$\text{Given } \{j\}: \text{ We find } \mathcal{I}_{\{j\}} \text{ s.t. } \forall X, F_\theta(X) \approx F_\theta(\mathcal{I}_{\{j\}}) \tag{3}$$

We compute the imputation pattern to fill values of the missing channels by solving the minimization problem below:

$$\min_{\mathcal{I}_{j \in \{j\}}} \mathcal{L}\Big(\text{Logits}(F_\theta(X)), \text{Logits}(F_\theta(\mathcal{I}_{\{j\}}))\Big) \tag{4}$$

Algorithm 1 AIM Search for accuracy-preserving imputation

Input: Training set $\mathcal{D} = \{X\}$; F_θ, pre-trained classifier on $\mathcal{D} = \{(X, y)\}$; $\{j\}$, missing sensors configuration; MAX_G, maximum iterations for gradient descent; MAX, maximum iterations over all inputs
Output: $\{\mathcal{I}_{j \in \{j\}}\}$, imputation pattern.

Random initialization of the set $\{\mathcal{I}_{j \in \{j\}}\}$
for i=1, $\cdots$, MAX **do**
  for each training example X **do**
    for i_G=1, $\cdots$, MAX_G **do**
      Compute the classifier's logits values: $\lg X$ = Logits($F_\theta(X)$)
      Compute the classifier's logits values: $\lg I$ = Logits($F_\theta(\mathcal{I}_{\{j\}})$)
      Estimate the loss $\mathcal{L}(\lg X, \lg I)$
      Estimate the gradient $\nabla_{\{\mathcal{I}_j\}}\mathcal{L}$ for $j \in \{j\}$
      Perform gradient descent and update $\{\mathcal{I}_j\}$ for $j \in \{j\}$
    end for
  end for
end for
return imputation pattern $\{\mathcal{I}_{j \in \{j\}}\}$ as per the requirements of Eq.2

The loss function $\mathcal{L}$ over the logits outcome of the classifier is used for accuracy-preserving imputation pattern search. The logits of a classifier are interpreted as the unnormalized predictions for each candidate class label and input time-series signal pair. The role of this loss function is to compute the similarity between the prediction outcomes of the classifier F_θ for the original input example X (no missingness) and the input with the imputed pattern $\mathcal{I}_{\{j\}}$. For example, $\mathcal{L}$ can be the Mean Squared Error between the logits values of both predictions $F_\theta(X)$ and $F_\theta(\mathcal{I}_{\{j\}})$. *When $\mathcal{L} \to 0$, accuracy of the classifier over imputed and original inputs will be similar.*

Algorithm 1 shows the pseudo-code of proposed search approach to compute accuracy-preserving imputation pattern $\mathcal{I}_{\{j\}}$. We set $MAX_G = 100$ to ensure that AIM can find the optimal imputation pattern per example X. Additionally, we set $MAX = 50$ to ensure that AIM optimizes the imputation pattern across the training data. The output of this algorithm is used to populate a look-up table $\mathcal{I}$ that maps the set of missing channels $\{j\}$ to the corresponding imputation pattern $\{\mathcal{I}_j\}$. Therefore, given a missing sensor data configuration $\{j\}$ and any input X at test time, we construct $\mathcal{I}_{\{j\}}$ as shown in Eq. 2 and employ the classifier F_θ to predict the class label after using $\mathcal{I}_{\{j\}}$ to impute the missing data in input X.

B. Training Robust Classifiers for Improved Effectiveness

Recall that we store *one* imputation pattern for each missingness configuration to preserve the accuracy and AIM's imputation strategy does not depend on the available sensor data of the given input X. As a result, AIM has negligible overhead, but ML classifier may not make correct predictions with generic imputation patterns for a small fraction of the input examples. Therefore, we propose to train ML classifiers to be robust to small errors in the imputed data.

The motivation to train robust ML classifiers is two-fold. First, the ML classifier is less sensitive to the natural noise in the training data. As a result, we will be able to find more robust imputation patterns using our search algorithm and robust ML classifier. Second, the ML classifier will be more robust to small errors in the imputed data at the runtime.

We train robust a ML classifier using data augmentation and propose to apply the recent framework of generating augmented data for time-series signals based on their statistical features [22, 23]. The key idea is to generate small perturbations over the original time-series signals in the training data which preserve the statistical features. To instantiate this framework for our specific use-case, we employ the following statistical features: mean absolute error, statistical average, and root mean square. Additionally, for HAR applications, we include the body acceleration feature due to its high relevance.

V. Experiments and Results

This section analyzes the performance of the proposed data recovery approach on four datasets along different dimensions.

A. Experimental Setup

1) Wearable Device Setup: We employ the Odroid-XU3 board [16] for sensor data processing, while noting that any low-power processor can be used. Odroid-XU3 contains four high-performance ARM Cortex-A15 and four low-power Cortex-A7 cores. We use the Odroid-XU3 to store the imputation table and measure the overhead on A7 cores.

2) Datasets: AIM is validated using four datasets described below. To validate the proposed missing data recovery approach, we vary the number of missing sensors for each dataset with n sensors from one to $n - 1$.

Shoaib et al. [8]: The Shoaib dataset includes three-axis accelerometer data for 10 users performing seven activities. The dataset has accelerometer sensors at five locations on the body: left pocket, right pocket, wrist, belt and upper arm.

PAMAP2 [13]: PAMAP2 is a HAR dataset that provides data from three accelerometers for five activities with nine users.

eRing [14]: eRing is a smart health dataset that uses a ring to capture data along four dimensions. The eRing dataset allows us to test the efficacy of AIM in gesture recognition settings.

SelfRegulationSCP1 (SR-SCP1) [15]: SR-SCP1 is a health monitoring dataset that includes EEG data from six channels. The data from EEG sensors is used to develop a control system to drive spelling devices for completely paralyzed patients.

3) Evaluation Metrics: We employ accuracy, memory, and energy consumption as evaluation metrics. Accuracy is used as a metric because accuracy is of utmost importance in health applications. Similarly, memory and energy are important for wearable devices due to resource constraints.

4) Classifier Representation: We use a 1-D convolutional neural network (CNN) as the classifier for all datasets. Specifically, we use a 1-D CNN with one conv. and max-pooling layers, and two fully connected layers with the ReLU activation and dropout value of 20%. We use the Adam optimizer [24] over 20 epochs for both standard and robust training.

B. Baseline Methods for Comparison

The proposed data recovery approach is compared against baseline approaches described below.

GAIN [12]: GAIN is a generative approach that recovers missing data as a function of the observed data. GAIN trains a deep neural network that takes the observed data and a mask specifying the missing time instances as input. The output of the generator is a data matrix that consists of imputed values.

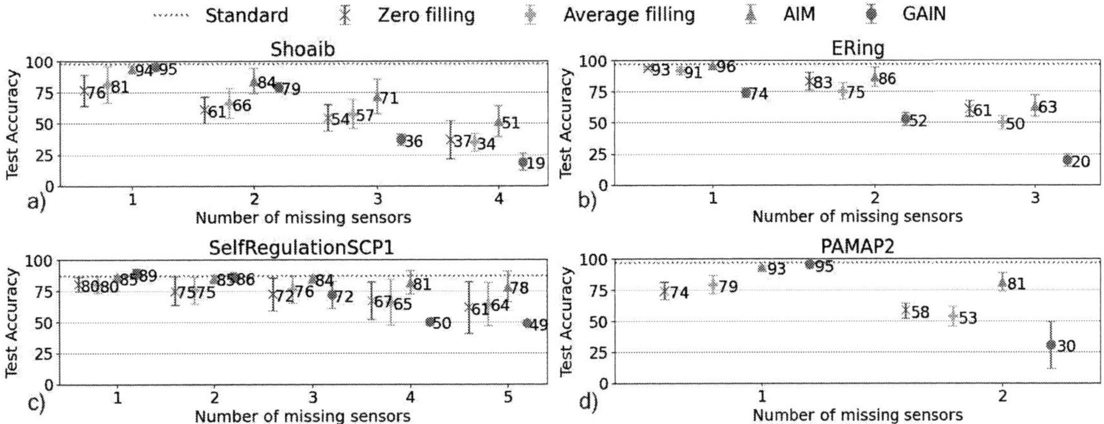

Fig. 3: Accuracy (Mean and standard deviation) of the robust-trained ML classifier via different imputation methods on all combinations of missing sensors.

One of the disadvantages of the GAIN approach is the high memory requirement for storage of the generator parameters and energy overhead for each imputation. Moreover, generative models for time-series data are challenging to train [25], which can affect their accuracy in complex tasks.

Zero Filling: In the absence of any data recovery algorithm, missing data will typically be filled with zeros. Therefore, we use it as one of the baselines for comparison. Filling missing values with previously observed data is not feasible since we assume the sensor is missing for the entire experiment.

Average Filling: Another realistic alternative for zero-filling is to fill the missing data with pre-determined values that represent the average case over the training data with no missingness. These values are determined by averaging sensor data across all training time-series signals.

C. Application Accuracy with Imputed Data

We start the experimental evaluation by analyzing the accuracy of the health applications under different missing data scenarios. For each dataset, we first train a classifier to perform the application tasks. Once the classifier is trained, we use it with the proposed search algorithm to find likely patterns of sensor data when one or more sensors are missing.

Figure 3 shows the comparison of accuracy for all four datasets. Each point on the figure shows the mean and standard deviation of the accuracy over all possible combinations of missing sensors. For example, in case of two missing sensors in the Shoaib dataset, we obtain the average and standard deviation over $\binom{5}{2}$ combinations of possible scenarios. We see that missing data with zero-filling or average-filling settings have a significant drop in accuracy. For example, for both HAR datasets, a single missing sensor results in more than 20% drop in average accuracy. In contrast, using the same classifier and missing data cases, AIM is able to efficiently recover the classification performance. Even when data from the entire window is missing, AIM is able to produce an average performance within 5% of the original accuracy for all datasets. AIM also succeeds in improving average performance of the classifiers on HAR datasets by 15% in the highly-unlikely case where almost all sensors are missing. Additionally, the confusion matrices in Figure 4 show that AIM

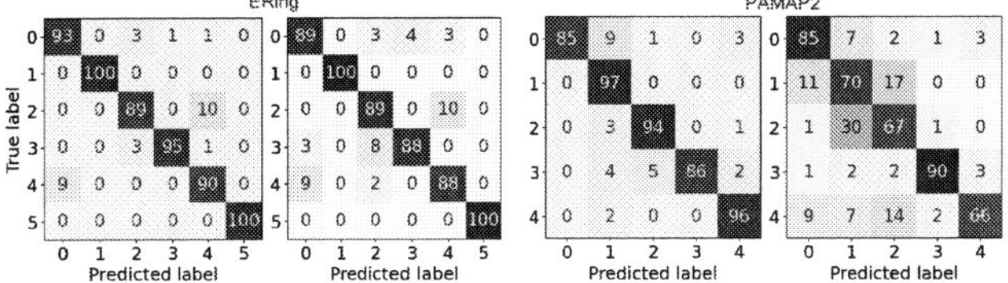

Fig. 4: Confusion matrix normalized over the true labels of the deployed classifier on ERing and PAMAP2 datasets using AIM (red) imputation methods in the event of a single missing sensor. The same performance using zero-filling (blue) is provided for reference.

improves the classification accuracy over different classes with equal importance in spite using a single imputation pattern.

Compared to the imputation provided by the baseline GAIN approach, AIM produces better results for most of the cases. Notably, GAIN fails to recover the original accuracy when more than one sensor is missing. GAIN has lower accuracy than zero-filling in some cases because GAIN is unable to follow real data accurately and incurs higher error. For PAMAP2, the average performance is reduced from 58% to 30% for the GAIN algorithm. In summary, the AIM approach is able to efficiently recover the data with low overhead, while the baseline GAIN approach is unable to recover the accuracy for more than one sensor missing and has a higher overhead.

D. Accuracy Improvement with Robust Classifiers

The AIM algorithm generates a pattern $\mathcal{I}_{\{j\}}$ to preserve the accuracy of the classifier in the case of $\{j\}$ missing input channels. Ideally, the generated pattern requires small adjustments to fit every input X to maintain $F_\theta(X) \approx F_\theta(\mathcal{I}_{\{j\}})$. To account for these adjustments, we use robust training for the ML classifier to overcome small errors in imputed data. Robust classifiers are also important because any health application must be able to handle small variations in data either due to natural disturbances or imputation. To this end, we compare the accuracy of the proposed robust classifiers with the standard classifier in Table I. Indeed, the table shows that robust training overcomes errors due to small imputation deviations by improving the average accuracy for majority of cases while reducing standard deviation. For example, SR-SCP1 has an increase of 6% in recovered accuracy with a

TABLE I: Classification accuracy of the imputed data of k missing sensors generated by AIM using different standard and robust training protocols. Table entries show mean with standard deviation in parantheses.

Dataset	k	Standard	Robust	Dataset	k	Standard	Robust
Shoaib	1	91 (2)	94 (1)	SR-SCP1	1	79 (8)	85 (1)
	2	80 (5)	84 (9)		2	79 (8)	85 (2)
	3	69 (13)	71 (13)		3	79 (8)	84 (2)
	4	50 (12)	51 (12)		4	81 (9)	81 (9)
ERing	1	95 (1)	96 (0)		5	73 (5)	78 (12)
	2	84 (8)	86 (7)	PAMAP2	1	92 (2)	93 (0)
	3	60 (2)	63 (8)		2	75 (3)	77 (11)

standard deviation of 1%. Overall, robust training is able to provide higher accuracy while reducing standard deviation.

E. Implementation Overhead

One of the primary advantages of AIM is low memory and energy overhead. Table II shows the memory overhead for AIM and GAIN, respectively. GAIN incurs high memory overhead to store the parameters of the generator network. In contrast, AIM has less than 1 MB memory overhead. AIM memory requirements are minimal even when the wearable device includes multiple health applications. The memory overhead for AIM can be further reduced by loading *only* the required imputation setting on detecting missing data.

Next, Figure 5(a) compares energy consumption of GAIN and AIM for all datasets. The energy consumption is obtained using power sensors on the Odroid-XU3 board. We see that AIM consumes less than 10 mJ per imputation while GAIN has significantly higher energy consumption. For instance, energy consumption for the SR-SCP1 dataset is close to 1 J for each imputation. Similarly, Figure 5(b) shows percentage energy savings achieved by AIM when compared to GAIN. The energy savings are close to 98% for all datasets except eRing. The eRing dataset has lower energy savings of about 74% since it has lower computation requirements for both GAIN and AIM, resulting in lower energy savings. In summary, the AIM approach provides superior performance over prior approaches while incurring significantly lower overhead.

TABLE II: Summary of memory overhead of AIM and GAIN approaches

Dataset	AIM Memory (MB)	GAIN Memory (MB)
Shoaib	0.180	25
PAMAP2	0.055	60
eRing	0.007	0.19
SR-SCP1	0.667	81

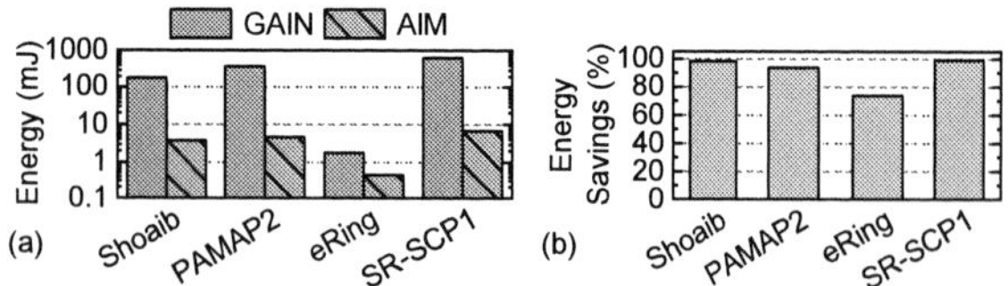

Fig. 5: a) Comparison of energy consumption for GAIN and AIM approach. The y-axis is shown in log scale to represent the large range of values. b) Energy savings achieved by AIM when compared to GAIN.

VI. Conclusion

Wearable devices are transforming a number of high-impact applications. However, they may suffer loss in the quality of service due to one or more sensors being unavailable at runtime. This paper presented a novel search-based algorithm that obtains most likely imputation patterns of sensor data for each missing data scenario via *offline* analytics. Experiments on four diverse wearable sensor based time-series benchmarks showed that the proposed approach is able to maintain accuracy within 5% of the ideal accuracy when the number of missing sensors is less than two, with negligible runtime overhead.

References

[1] A. Mosenia *et al.*, "Wearable Medical Sensor-Based System Design: A Survey," *IEEE TMSCS.*, vol. 3, no. 2, pp. 124–138, 2017.

[2] A. Limaye and T. Adegbija, "HERMIT: A Benchmark Suite for the Internet of Medical Things," *IEEE IoT J.*, vol. 5, no. 5, 2018.

[3] A. J. Espay *et al.*, "Technology in Parkinson's Disease: Challenges and Opportunities," *Movt. Disorders*, vol. 31, no. 9, pp. 1272–1282, 2016.

[4] P. Zappi *et al.*, "Activity Recognition from On-Body Sensors by Classifier Fusion: Sensor Scalability and Robustness," in *Proc. Int. Conf. on Intell. Sensors, Sensor Netw. and Info.*, 2007, pp. 281–286.

[5] H. Kim *et al.*, "Collaborative Classification for Daily Activity Recognition with a Smartwatch," in *Proc. SMC*, 2016, pp. 003 707–003 712.

[6] S. Liu *et al.*, "Handling Missing Sensors in Topology-Aware IoT Applications with Gated Graph Neural Network," *Proc. IMWUT*, vol. 4, no. 3, pp. 1–31, 2020.

[7] K. Kunze and P. Lukowicz, "Sensor Placement Variations in Wearable Activity Recognition," *IEEE Perv. Comput.*, vol. 13, no. 4, 2014.

[8] M. Shoaib *et al.*, "Fusion of Smartphone Motion Sensors for Physical Activity Recognition," *Sensors*, vol. 14, no. 6, pp. 10 146–10 176, 2014.

[9] I. M. Pires *et al.*, "Improving Human Activity Monitoring by Imputation of Missing Sensory Data: Experimental Study," *Future Internet*, vol. 12, no. 9, p. 155, 2020.

[10] T. De Waal, J. Pannekoek, and S. Scholtus, *Handbook of Statistical Data Editing and Imputation*. John Wiley & Sons, 2011, vol. 563.

[11] T. Hossain and S. Inoue, "A Comparative Study on Missing Data Handling Using Machine Learning for Human Activity Recognition," in *Proc. ICIEV and icIVPR*, 2019, pp. 124–129.

[12] J. Yoon, J. Jordon, and M. Schaar, "GAIN: Missing Data Imputation Using Generative Adversarial Nets," in *ICML*, 2018, pp. 5689–5698.

[13] A. Reiss and D. Stricker, "Introducing a New Benchmarked Dataset for Activity Monitoring," in *ISWC*, 2012, pp. 108–109.

[14] M. Wilhelm *et al.*, "eRing: Multiple Finger Gesture Recognition with One Ring Using an Electric Field," in *Proc. Int. Work. on Sensor-based Activity Recognition and Interaction*, 2015, pp. 1–6.

[15] N. Birbaumer *et al.*, "A Brain-Controlled Spelling Device for the Completely Paralyzed," *Nature*, pp. 297–298, 2001.

[16] Hardkernel. (2014) Odroid-xu3. https://www.hardkernel.com/shop/odroid-xu3/ Accessed 11/20/2020.

[17] G. Bhat, N. Tran, H. Shill, and U. Y. Ogras, "w-HAR: An Activity Recognition Dataset and Framework using Low-Power Wearable Devices," *Sensors*, vol. 20, no. 18, p. 5356, 2020.

[18] Z. Guo *et al.*, "A Data Imputation Method for Multivariate Time Series Based on Generative Adversarial Network," *Neurocomputing*, vol. 360, pp. 185–197, 2019.

[19] D. Hussein, A. Jain, and G. Bhat, "Robust Human Activity Recognition Using Generative Adversarial Imputation Networks," in *Proc. DATE*, 2022, pp. 84–87.

[20] S. Talukder *et al.*, "Deep Neural Imputation: A Framework for Recovering Incomplete Brain Recordings," *arXiv:2206.08094*, 2022.

[21] T. Belkhouja and J. R. Doppa, "Analyzing Deep Learning for Time-Series Data Through Adversarial Lens in Mobile and IoT Applications," *IEEE TCAD*, vol. 39, no. 11, pp. 3190–3201, 2020.

[22] ——, "Adversarial Framework with Certified Robustness for Time-Series Domain via Statistical Features," *JAIR*, 2022.

[23] D. Hussein *et al.*, "Reliable Machine Learning for Wearable Activity Monitoring: Novel Algorithms and Theoretical Guarantees," in *Proc. ICCAD*, 2022, pp. 1–9.

[24] D. P. Kingma and J. Ba, "Adam: A Method for Stochastic Optimization," in *The Int. Conf. on Learning Representations (Poster)*, 2015.

[25] E. Brophy, Z. Wang, Q. She, and T. Ward, "Generative Adversarial Networks in Time Series: A Survey and Taxonomy," *arXiv preprint arXiv:2107.11098*, 2021.

RecPIM: A PIM–Enabled DRAM–RRAM Hybrid Memory System For Recommendation Models

Heewoo Kim, Haojie Ye, Trevor Mudge, Ronald Dreslinski, and Nishil Talati
University of Michigan, Ann Arbor, MI, USA
{heewoo, yehaojie, tnm, rdreslin, talatin}@umich.edu

***Abstract*—The performance of modern recommendation models is limited because of the memory bandwidth-hungry embedding layer reductions. We propose RecPIM—a novel hybrid memory system with DRAM and RRAM with PIM capability. The performance of traditional RRAM PIM is limited by the latency of bit-serial computation. RecPIM presents a comprehensive optimization approach that includes access-pattern-aware mapping, compute complexity reduction, and selective PIM reduction to offset this computation latency. Our evaluation shows that RecPIM offers significant p erformance, e nergy, and EDP improvement of 2.6×, 1.7×, and 4.4×, on average, compared to a CPU baseline. We also co-design wear-leveling techniques and demonstrate a practical lifetime of more than 12 years.**

I. Introduction

The Deep Learning Recommendation Models (DLRM) are widely deployed in today's data centers to predict user preferences and deliver personalized advertisements [1], [2]. DLRM inference occupies more than 60% of AI inference cycles in commercial data centers [3]. The DLRM workload consists of a combination of *dense* Multi-Layer Perceptron (MLP) layers and *sparse* embedding layers. Out of these, the embedding layer operation takes a majority (around 80%) of DLRM execution time, and is bottlenecked by the **high memory bandwidth** requirement [1], [3]. Therefore, optimizing the embedding layer performance is essential to improving data center performance and energy consumption, leading to a significant reduction in computing's carbon footprint and Total Cost of Ownership (TCO).

Prior works employ DRAM–based Near Memory Processing (NMP) [1], [4] or frequent embedding partial sum caching [1], [5] to improve DLRM performance. Several emerging memory technologies (*e.g.,* RRAM and PCM) offer low cost–per–bit and improved technology scaling compared to DRAM. Given the ever-growing sizes of modern DLRM embedding tables, the performance of DLRM inference can be further improved using emerging memory technologies.

In this paper, we introduce RecPIM—a novel hybrid memory system design using DRAM and Resistive RAM (RRAM). RecPIM employs a massively parallel Processing-In-Memory (PIM) technique called Memristor Aided loGIC (MAGIC) for speeding up DLRM's sparse embedding layer computation in RRAM. DRAM, on the other hand, is used as the main memory. The core intuition behind this design choice is to exploit the high memory bandwidth available to RRAM PIM for reducing bandwidth–intensive embedding vectors. Despite high bandwidth availability, one of the **key design challenges** in achieving high performance is the bit–serial nature of MAGIC computation. The latency of PIM arithmetic operations is especially exacerbated in the case of floating point inputs [6], widespread in the DLRM workload. Therefore, a careful design of the DRAM–RRAM memory system is crucial to achieving high performance.

The **design goal** of RecPIM is to offset the cost of bit–serial PIM operations to achieve high performance. We also consider the performance-energy efficiency because even a tiny amount of energy efficiency improvement results in significant cost savings in data centers. To this end, RecPIM proposes three optimizations keeping the technological parameters of the proposed memory system in mind: 1) Access-Pattern-Aware Data Mapping, 2) Compute Complexity Reduction, and 3) Selective PIM Reduction. Access-Pattern-Aware Data Mapping presents a data mapping between DRAM and RRAM, and within RRAM arrays to optimize the overall throughput of embedding vector reductions. Compute Complexity Reduction improves the algorithmic efficiency to avoid redundant PIM computations. Selective PIM Reduction selectively offloads PIM computations whenever the benefits outweigh the bit–serial computation cost. This is the first work that optimizes MAGIC RRAM PIM for sparse and bandwidth-intensive embedding layer operation.

We evaluate the performance and energy consumption of RecPIM using nine real–world datasets from web service vendors. Our evaluation shows that RecPIM significantly improves the performance, energy consumption, and EDP (energy-delay product) of a commercial CPU baseline by 2.6×, 1.7×, and 4.4×, on average. A primary reason behind such significant improvements is a net 49% reduction in off-chip memory traffic due to PIM. We also compare the performance of RecPIM with a state-of-the-art NMP system called SPACE [1]. An iso-technology comparison shows that RecPIM outperforms SPACE by 1.6×, on average. To make the RecPIM technology appealing for commercial use, we demonstrate the lifetime of RecPIM to be more than 12 years. Below, we summarize our novel contributions.

- Identifying why naïve RRAM PIM cannot readily accelerate the embedding layer operation in DLRM inference.
- Designing optimizations including Access-Pattern-Aware Data Mapping, Compute Complexity Reduction, and Selective PIM Reduction to best speed up DLRM inference using massively parallel PIM operations in RRAM.

979-8-3503-1176-1/23 $31.00 © 2023 IEEE

- Proposing RecPIM—a hybrid DRAM–RRAM system design that significantly improves the performance of DLRM embedding layer reduction by 2.6×, on average.

II. Background and Motivation

A. Background: Deep Learning Recommendation Models

DLRM predicts and recommends items preferred by different users based on their attributes and previous user–item interactions [1], [2]. DLRM consists of MLP and embedding layers. MLP layers are used to process the dense and continuous features, such as user characteristics. The embedding layer is used to process sparse and categorical features, such as the users' previous selections.

The embedding layer typically employs several embedding tables for different item categories. The primary computation in this layer includes reading embedding vectors (consisting of a large vector of floating point numbers) from large embedding tables and performing element–wise floating–point additions (*i.e.,* reductions). Typical embedding tables have several million vectors, and each user reads 10s to 100s of these vectors at inference. The accessed indices are determined by each user's past interactions with items at runtime, which are not typically contiguous. As a result, several long embedding vectors are fetched that are far away from one another in the address space. Therefore, embedding vector reduction is a memory bandwidth intensive operation that takes up to 80% of DLRM inference time [3]. The **focus** of this paper is to optimize the memory bandwidth–hungry sparse embedding layer in DLRM.

B. Background: Processing-In-Memory (PIM) using RRAM

RRAM is an emerging non-volatile memory technology. RRAM stores the memory state using a resistance, as opposed to electrical charge in DRAM. Due to the resistive nature of RRAM cells, it is possible to compute logic operations within RRAM arrays. This paper uses RRAM to perform arithmetic operations using Memristor Aided loGIC (MAGIC) [7]. MAGIC enables the execution of bitwise logical NOR operations that can be conducted within the RRAM arrays. Because NOR is functionally complete, it is possible to execute any arithmetic/logical operation using a series of NOR gates.

There are two major advantages of PIM in RRAM using MAGIC NOR operations. First, MAGIC NOR can be performed within the RRAM memory arrays by applying voltages to bitlines/wordlines, without reading data outside [7]. This exposes massive data bandwidth for computation. Second, MAGIC executes NOR operations in parallel for all data present in bitlines/wordlines as well as different RRAM crossbars, enabling a massively parallel in-memory computation engine.

C. Motivation: Challenges of Adopting RRAM PIM for DLRM

Although the RRAM technology offers PIM capability with massive memory bandwidth available to computation and large opportunities of parallelism, simply employing RRAM PIM is not an attractive solution. The bit-serial nature of PIM results in long-latency arithmetic operations that require a large amount of parallelism to amortize the cost of bit–serial computation. For example, 1144 NOR cycles are required to add two 32–bit floating point numbers [6], [8]. This, in fact, results in a net performance slowdown compared to the CPU baseline (more results in Section V), obviating the performance benefits of PIM. Therefore, to meaningfully continue delivering high performance using PIM within RRAM, it is **critical** to rethink the RRAM system design. In what follows, we propose novel design optimizations to best employ a DRAM–RRAM hybrid memory system to significantly improve DLRM inference performance.

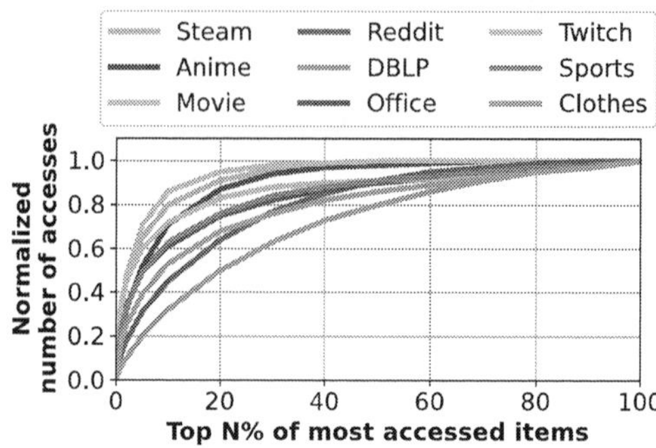

Fig. 1. Power-law distribution of DLRM datasets, where the top 30% most frequently accessed items take 84% of total item accesses, on average.

III. The Design of RecPIM

A. Access-Pattern-Aware Data Mapping

Motivation. While the RRAM–based MAGIC NOR operations can exploit abundant in-RRAM memory bandwidth by performing computation within the memory arrays, one of the key constraints of being able to exploit this bandwidth is *data co–location*. That is, the input operands to PIM computation must be present in the same RRAM memory arrays. If the inputs are present in different RRAM banks, it becomes necessary to move data within the memory, offseting the benefits expected from PIM. Furthermore, the sparse nature of embedding vector reductions exacerbates this problem as the input vectors can be physically distant in memory.

Opportunity. As a result of investigating the data access patterns in the DLRM workload, we find an opportunity to enable data co–location for PIM. Fig. 1 shows the power–law distribution of item access patterns by different users. The figure shows that *the top 30% most frequently accessed items are accessed 84% of the time*, on average.

Proposal. Motivated by the crucial necessity of data co–location in PIM and the skewed item access pattern in DLRM workloads, we propose a two–tier design optimization. First, we propose to sort the item IDs based on their frequencies. This is achieved by examining the past user–item interactions to find item access frequencies and re–labeling the item IDs in the decreasing order of their frequencies. Such design optimizations based on the runtime workload properties called *profile–guided optimizations* are widely deployed in today's data centers. Furthermore, as presented by a recent industrial work [9], re–labeling of items can be easily achieved using a remapping table, already used in DLRM workloads.

Second, we propose to map the top frequently accessed item

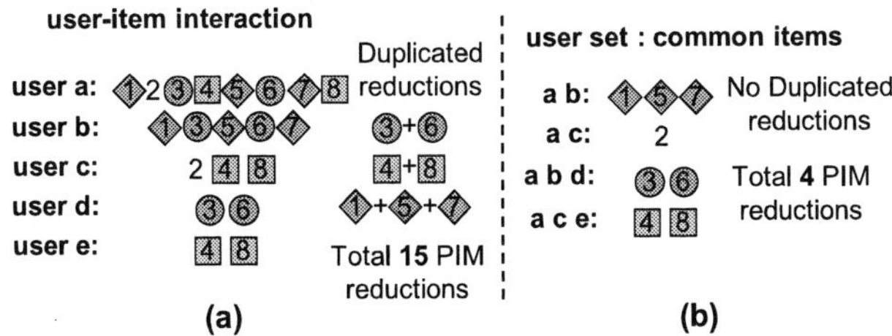

Fig. 2. (a) Before Compute Complexity Reduction and (b) After Compute Complexity Reduction. Duplicated reductions are removed, and the total number of PIM reduction decreases from 15 to 4.

embeddings to RRAM, and the rest to DRAM. The intuition behind this design decision is the reduction of the most frequently accessed embedding vectors using RRAM PIM, and the rest are reduced on the CPU. RecPIM maps each element of an embedding vector to one RRAM row. Different elements from the same vector are mapped to different RRAM arrays, allowing them to be used in parallel. Different memory rows host different embeddings sorted by their access frequencies. Mapping frequent embedding vectors to the same RRAM array increases the chance of data co-location for PIM.

Access-Pattern-Aware Data Mapping is processed offline, and the data mapped to physical memory arrays is used for inference during the runtime. This is possible because data center operators profile the user-item interactions offline (*e.g.*, a week) and use the updated system for inference (consecutive weeks).

B. Compute Complexity Reduction

Motivation. The sparse embedding layers in DLRM typically reduce a large number of embedding vectors for each user *individually*. Although the compute operation itself is simple (*i.e.*, floating point addition), element-wise reductions of hundreds of long vectors for each user result in large compute complexity.

Opportunity. By analyzing the item access patterns of different users in real-world DLRM models, we find that many different users reduce similar frequently accessed item embeddings. Because a small subset (30%) of items are accessed a majority (84%) of the time (see Fig. 1), it is intuitive that similar embeddings are reduced multiple times by different users. This workload behavior leads to redundant memory accesses and computation that can be avoided.

Proposal. To avoid redundant computations and improve the algorithmic complexity, we propose to reuse reduction results among multiple users. This is in stark contrast with how DLRMs currently operate, where the sparse features of different users are computed separately. Although RecPIM alters the operating principles, it does not impact the accuracy of DLRM.

The key idea behind this optimization is to identify the commonly reduced item embeddings across a set of users, and reuse their computation. The identification of common items is performed using an inverse-map of user-item interactions and applying a map-reduce process. Because RecPIM conducts this process at runtime it falls on the critical path of embedding reduction. Therefore, it is critical to minimize the latency of this logic. To this end, we find a *rich trade-off space* between the latency of common computation identification and the benefit of complexity reduction. While grouping a large number of users results in a greater complexity reduction, it incurs an increase in the latency of the identification process. Grouping a small number of users, on the other hand, reverses this trend. To effectively navigate this trade-off space, we empirically use a group size of 8 users to balance the cost and benefit of this method (not shown in detail due to space limitation). Furthermore, our experiments show that the latency of generating common item sets can be completely hidden by computing it during the PIM computation of the previous batch.

Fig. 2 shows a walk-through example of our proposal with five users a to e. A baseline design, where all users are reduced individually, requires 15 reduction operations, out of which, many are redundant as indicated in Fig. 2(a). Our proposal reduces this computation into 4 reductions by reusing common computation among multiple users (see Fig. 2(b)), resulting in a significant complexity reduction. Combining these partial reduction results are conducted in the CPU.

C. Selective PIM Reduction

Motivation. As discussed in Section II-C, one of the key design challenges is the slow, bit-serial nature of PIM computation. That is, each reduction operation takes hundreds of cycles to compute. Therefore, unless the cost of one PIM reduction is amortized by computing several reductions in parallel, RRAM PIM will result in inferior performance compared to the CPU. However, the number of reductions that can be performed in parallel depends on the workload.

Opportunity. It is possible to reduce a subset of embedding vectors stored in RRAM at the CPU as well, because the non-frequent item embeddings are read out of DRAM and reduced at the CPU (along with partial reductions from RRAM).

Proposal. To offset the performance slowdown because of PIM's bit-serial computation, we propose Selective PIM Reduction. The key idea of this optimization is to intelligently select between conducting reduction operations using PIM versus at the CPU. To this end, we build an analytical model to make this decision. The simplicity of this model is key to its practical utility in DLRM execution as this decision is taken at runtime. We define a threshold parameter N_{thres} as

$$N_{thres} = T_{PIM} \times BW_{mem}, \tag{1}$$

where T_{PIM} and BW_{mem} represent the latency of a PIM operation and the memory bandwidth, respectively. N_{thres} defines the minimum number of PIM reductions to compute in parallel to offset the cost of one bit-serial reduction. At runtime, the RRAM controller compares the number of parallel reductions to be performed within an RRAM array with N_{thres}. If the number of concurrent operations is greater than N_{thres}, then the RRAM controller offloads PIM commands. Otherwise, transferring data off-chip and reducing on the CPU is beneficial. Therefore, instead of the RRAM performing PIM

TABLE I
CHARACTERISTICS OF REAL-WORLD DATASETS USED FOR EVALUATION

Dataset (Idx+)	# Items	Avg. Pool. Factor	Freq. Item. (%)*
Steam (D1) [11]	10,978	67	10
Anime (D2) [12]	11,200	106	20
Movie (D3) [13]	26,744	143	20
Reddit (D4) [14]	232,965	492	40
DBLP (D5) [15]	540,459	62	30
Office (D6) [16]	598,943	43	10
Twitch (D7) [17]	739,991	31	5
Sports (D8) [16]	1,505,707	62	5
Clothes (D9) [16]	2,345,346	50	10
M1	twi+mov+ani+stm	342	10
M2	clo+off+dblp+ani	262	10
M3	spo+off+dblp+twi	187	10
M4	clo+spo+off+dblp	218	10

+ The index of each dataset for the results plotting.
*The proportion of top frequently accessed items that are stored in RRAM.

TABLE II
MODELED SYSTEM PARAMETERS

Emb. Table Characteristics	Precision	bfloat16 [10]
	Embedding Vector Dimension	64
	Batch / Sub-Batch Size	64 users / 8 users
Technology Specification	RRAM / DRAM capacity	4 GB [18] / 48GB
	RRAM Crossbar Size	1024×1024
	MAGIC NOR Latency	1.1 ns [6]
	bfloat16 PIM Reduction Latency	361 ns [6]
	N_{thres} (Section III-C)	58
	bfloat16 PIM Reduction Energy	86.9 pJ/reduction [6]
DDR4 interface	Mem Interface	DDR4-3200
	DDR4 Read Energy	40 pJ/bit

computation, the controller issues load requests to multiple embedding vectors to send the data to the CPU.

IV. METHODOLOGY

A. Datasets

We use various real-world datasets with different characteristics, as shown in Table I. The datasets are listed in the order of the number of items (# Items). The average pooling factor is the average item access per user, and the frequent item is the proportion of top frequently accessed items that are stored in RRAM. The proportion is determined heuristically for maximum performance. M1 to M4 are the combined real-world datasets to showcase the effectiveness of RecPIM for the real-world deployment. For each dataset, we randomly sample the data and assign 50% of the data to the training set and another 50% to the testing set. We get the frequently accessed item ranking from the training set and conduct the embedding table operation with the testing set.

When storing the testing set data in the memory, RRAM is used to store frequently used items, and DRAM to store non-frequently used items. In RRAM, one crossbar row stores one embedding vector element, and the rest of the columns of the same row are used for PIM intermediate results. RecPIM uses bfloat16 precision, a half-precision format widely used in DL applications and has no accuracy drop for DLRM [10].

B. RecPIM Configuration

RecPIM design configuration is divided into RRAM and DRAM. RRAM stores the top frequently accessed items, conducts parallel tree reductions, and transfers the partial reduction results to the CPU. These partial sums from RRAM are then reduced with the item embeddings read from DRAM at the CPU. DRAM stores the non-frequently used items and transfers them to the CPU in parallel with the PIM reduction. Table II presents the technology-specific parameters of both DRAM and RRAM based on prior works [6].

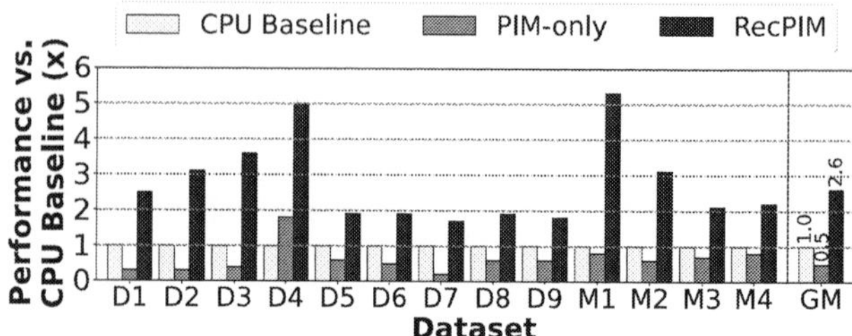

Fig. 3. Performance comparison of RecPIM with CPU and PIM–only.

C. Simulation Methodology

We build an in-house simulator to evaluate the performance, energy consumption, and lifetime of RecPIM. PIM reduction time and energy are modeled based on the Bitlet model [8] and FloatPIM model [6] (see Table II). Because the sparse embedding reduction workload in DLRM is heavily memory intensive [4], we ignore the host CPU and cache time/energy in our simulator. We have cross-validated our results with Intel i9-9900K CPU. To verify the functional correctness of our simulation infrastructure, we verify the embedding reduction results generated by our simulator and a traditional software implementation.

D. Baselines

CPU baseline. This baseline models a software implementation of DLRM used in today's data centers, and uses a DDR4 DRAM without PIM capability. All embedding vectors are read out of DRAM to the CPU for reduction.

PIM-only. This design stores all embedding vectors in the finite-size RRAM crossbars. This baseline fully reduces the embedding vectors using RRAM PIM without applying the proposed optimizations (Section III). This design exploits the row-level parallelism for tree-reduction and MAT-level parallelism for simultaneously reducing multiple vector elements. For large datasets, multiple RRAM chips are deployed. In contrast, RecPIM uses a single RRAM chip because RecPIM stores only the frequently used items in a RRAM chip.

SPACE [1]. This is a state-of-the-art domain-specific hybrid memory system design using DIMM and HBM-based memories. Similar to our RRAM design, both DIMM and HBM in SPACE are compute capable. SPACE further optimizes the architecture by caching the partial sums of a subset of highly frequent items. Because of technological idiosyncrasies, we compare RecPIM and SPACE using iso-technology baselines (*i.e.,* both RecPIM and SPACE with DIMM–HBM and DRAM–RRAM memory system designs).

V. EVALUATION RESULTS

A. Performance Analysis

RecPIM vs. CPU and PIM–only. Fig. 3 compares the performance of RecPIM with CPU and PIM–only baselines. The figure shows that the PIM–only baseline degrades the performance of a CPU baseline by 0.5×, on average. This

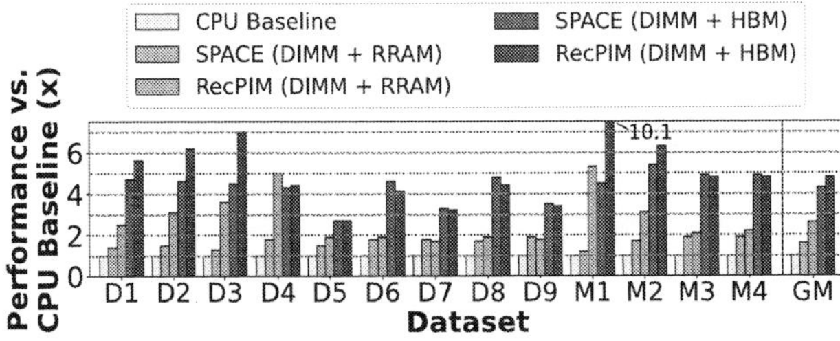

Fig. 4. Performance comparison of RecPIM with SPACE. The GM values are 1.0×, 1.6×, 2.6×, 4.3×, and 4.8×, respectively.

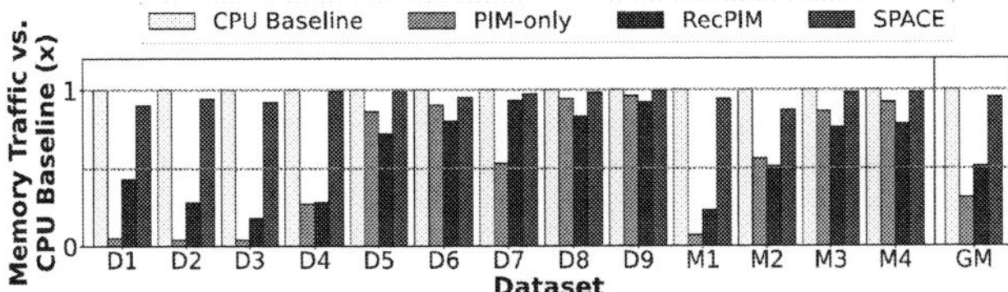

Fig. 5. Memory traffic comparison of RecPIM (GM: 0.51×) with CPU (GM: 1.0×), PIM–only (GM: 0.31×), and SPACE (GM: 0.95×).

is because the PIM–only design suffers from long latency of floating point arithmetic operations in reducing embedding vectors. This result validates our initial hypothesis that traditional RRAM PIM, by itself, is inadequate to improve DLRM performance.

Fig. 3 further shows that RecPIM significantly outperforms CPU and PIM–only baselines by 2.6× and 4.8×, on average. These high speedups are attributed to the proposed design optimizations that (a) co–locate PIM input data to reduce in-memory data transfers, (b) significantly reduce the computation complexity of the embedding layer, and (c) selectively execute PIM operations only when their benefits outweigh cost. *This result underscores the value of proposed RecPIM design optimizations that elevate the average performance of traditional RRAM PIM by 4.8×.*

RecPIM vs. SPACE [1]. Fig. 4 compares the performance of SPACE and RecPIM with a CPU baseline. For an iso-technology comparison, we model both SPACE and RecPIM using the same technology parameters (*i.e.*, both DIMM+HBM and DRAM+RRAM). The figure shows that RecPIM consistently outperforms a state-of-the-art hybrid memory system design SPACE by 1.6× (for DRAM+RRAM) and 1.1× (for DIMM+HBM), on average. *The superior performance of RecPIM compared to SPACE is due to exploiting massive PIM parallelism and improved memory traffic reduction (as discussed next).*

Memory traffic reduction. Fig. 5 compares the off–chip memory traffic reduction on PIM–only, RecPIM, and SPACE,

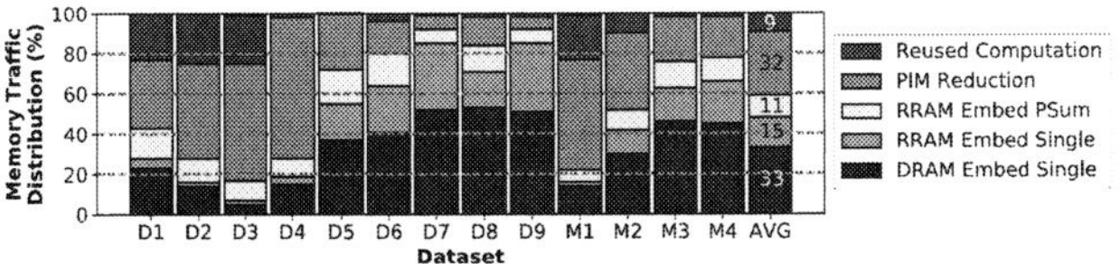

Fig. 6. Memory operation types in RecPIM. Reused Computation and PIM Reduction are the decreased memory traffic from computation reuse and PIM reduction, respectively. RRAM Embed PSum, RRAM Embed Single and DRAM Embed Single are memory traffic from partial sum, single embedding data from RRAM and DRAM, respectively.

normalized to a CPU baseline. By processing data within memory, RecPIM significantly reduces the memory traffic by 49%, on average. SPACE, on the other hand, only results in a 5% reduction in memory traffic that further corroborates our result in Fig. 4. Interestingly, the PIM–only baseline results in a greater memory traffic reduction, yet inferior performance, compared to RecPIM. Despite reducing more embedding vectors inside memory, the long latency of bit–serial PIM computation hurts the performance of PIM–only. Selective PIM Reduction in RecPIM carefully avoids PIM operations where parallelism cannot amortize the cost of bit–serial computation. *This interesting result shows that a more significant memory bandwidth reduction does not always result in optimal system performance, given long latency of PIM operations.*

Fig. 6 further shows the memory operation types in RecPIM normalized to off–chip data transfers in a CPU baseline. The figure shows that the computation reuse in RecPIM removes 9% of reductions, on average. Furthermore, RecPIM reduces 32% of data using PIM that saves off–chip data transfers. Examining individual workloads demonstrate a strong correlation between the fraction of reused computation plus PIM reductions with application speedup (see Fig. 3). The rest of the portions of the bars shows off–chip memory traffic in terms of 1) partial sum and 2) single embedding vector data transfer out of RRAM, and 3) single embedding vector data transfer out of DRAM. Off-chip memory traffic accounts for 59% of memory traffic (different than Fig. 5 as it computes geomean).

B. Energy Analysis

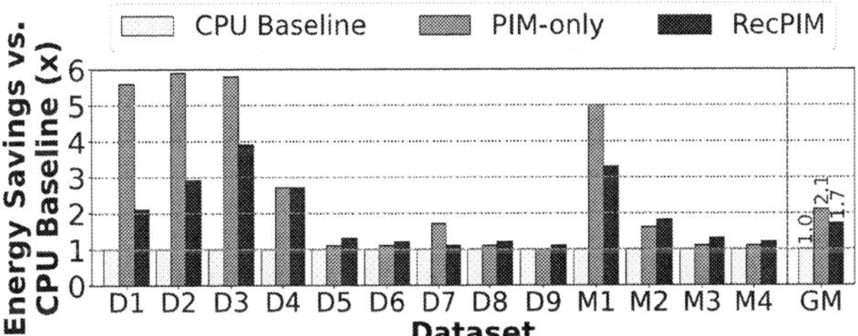

Fig. 7. Energy savings comparison of RecPIM with CPU and PIM-only. While RecPIM consumes more energy than the PIM-only design, RecPIM has 4.0× lower EDP than PIM-only because of superior performance.

As shown in Fig. 7, RecPIM achieves an average of 1.7× energy reduction compared to the CPU baseline. As CPU–memory data transfers consume a large amount of energy, RecPIM saves energy by processing data on the memory device. Furthermore, RecPIM also optimizes the energy savings by avoiding redundant PIM reductions (Section III-B). While the PIM–only baseline results in a greater energy reduction than RecPIM, this is mainly because of the greater memory traffic reduction (Fig. 5). RecPIM, on the other hand, significantly improves performance, and results in a net EDP reduction of 4.4× and 4.0× compared to the CPU and PIM–only baseline, respectively.

C. Sensitivity Study

N_{thres} in Selective PIM Reduction. We study how RecPIM performance changes for different values of N_{thres}.

First, we observe that $N_{thres} = 58$ is the optimal performance point. But even if we perturb N_{thres} by 50%, the performance degrades by less than 4%. Since the performance is not sensitive to the N_{thres}, it can be modified depending on the design goal. N_{thres} can be reduced to save memory traffic, or be increased to alleviate the wear-out of RRAM.

RRAM MAGIC NOR PIM time. We investigate how the RRAM technology parameter affects RecPIM performance. The chosen PIM latencies are $1.1ns$ [6], $3ns$, $5ns$, $7ns$, and $10ns$ [8]. The optimal value of N_{thres} increases from 58 to 525 and the average RecPIM performance changes from 2.6× to 1.7× as the PIM latency increases because additional parallelism is necessary to offset the longer PIM latency. The result shows that the Selective PIM Reduction optimization in RecPIM can adapt to tolerate longer PIM latencies to consistently deliver superior performance than a CPU baseline.

D. RecPIM Lifetime

We improve the lifetime of RecPIM and evaluate it for each dataset to show RecPIM's practicality. Without applying wear-leveling techniques, the baseline lifetime is 5-to-146 days when RRAM endurance is 10^{12} writes [19]. We adopt several wear-leveling techniques to spread out PIM computation across the memory fabric to improve the RRAM lifetime. First, we distribute the intermediate results of MAGIC NOR operations across many different bitlines in the same array. Second, we spread computations across the RRAM crossbar by modifying the data copying operation that aligns two embedding vectors in an RRAM wordline. Third, the RRAM controller employs a periodic data remapping from one crossbar array to another. These techniques improve the lifetime of RecPIM by 1633×, on average, to reach more than 12 years.

VI. Related Works

SPACE [1] uses HBM as a cache to store their frequently used items and the partial sum of frequently used items. The most N frequently used items are cached, and all pair-of-2 combinations of at most $sqrt(N)$ frequently used items are summed and cached. Using an iso–technology comparison, we show that RecPIM outperforms SPACE. MERCI [5] minimizes additional DRAM access by memoizing the frequently co-appearing items and their partial sums of arbitrary length. While MERCI employs an expensive pre–processing algorithm to find co–accessed items, RecPIM uses a lower complexity reduction algorithm that better scales to large embedding tables. TensorDIMM [2] is a custom DIMM module with near-memory processing cores. RecNMP [4] augments DIMMs with near–memory compute at the buffer chip. Unlike the above works, RecPIM improves the DLRM performance improvement with emerging memory technology.

VII. Conclusion

This paper introduced RecPIM, a hybrid DRAM–RRAM system with PIM–capable RRAM to accelerate the embedding layer performance in DLRM inference. To offset the long PIM computation latency of RRAM, RecPIM proposed novel optimizations that increased the spatial PIM operand locality, reduced the algorithmic complexity of embedding reductions, and selectively offloaded PIM computation. Our evaluation showed that RecPIM outperformed the CPU baseline by 2.6×, saved 1.7× energy, and improved EDP by 4.4×, on average. Moreover, RecPIM resulted in a net 49% reduction in off–chip memory traffic. We also showed that the lifetime of RecPIM is at least 12 years, demonstrating its practicality.

Acknowledgment

We thank the anonymous reviewers for their helpful feedback. The material is based on research sponsored by Air Force Research Laboratory (AFRL) and Defense Advanced Research Projects Agency (DARPA) under agreement number FA8650-18-2-7864. The U.S. Government is authorized to reproduce and distribute reprints for Governmental purposes notwithstanding any copyright notation thereon. The views and conclusions contained herein are those of the authors and should not be interpreted as necessarily representing the official policies or endorsements, either expressed or implied, of Air Force Research Laboratory (AFRL) and Defense Advanced Research Projects Agency (DARPA) or the U.S. Government.

References

[1] H. Kal *et al.*, "Space: locality-aware processing in heterogeneous memory for personalized recommendations," in *ISCA*, 2021, pp. 679–691.

[2] Y. Kwon *et al.*, "Tensordimm: A practical near-memory processing architecture for embeddings and tensor operations in deep learning," in *MICRO*, 2019, pp. 740–753.

[3] U. Gupta *et al.*, "The architectural implications of facebook's dnn-based personalized recommendation," in *HPCA*, 2020, pp. 488–501.

[4] L. Ke *et al.*, "Recnmp: Accelerating personalized recommendation with near-memory processing," in *ISCA*, 2020, pp. 790–803.

[5] Y. Lee *et al.*, "MERCI: efficient embedding reduction on commodity hardware via sub-query memoization," in *ASPLOS*, 2021, pp. 302–313.

[6] M. Imani *et al.*, "Floatpim: In-memory acceleration of deep neural network training with high precision," in *ISCA*, 2019, pp. 802–815.

[7] N. Talati *et al.*, "Logic design within memristive memories using memristor-aided loGIC (MAGIC)," *TNANO*, vol. 15, pp. 635–650, 2016.

[8] R. Ronen *et al.*, "The Bitlet model: A parameterized analytical model to compare PIM and CPU systems," *ACM JETC*, vol. 18, pp. 1–29, 2022.

[9] G. Sethi *et al.*, "RecShard: statistical feature-based memory optimization for industry-scale neural recommendation," in *ASPLOS*, 2022, pp. 344–358.

[10] D. Kalamkar *et al.*, "A study of BFLOAT16 for deep learning training," *arXiv preprint arXiv:1905.12322*, 2019.

[11] M. Wan and J. McAuley, "Item recommendation on monotonic behavior chains," in *RecSys*, 2018, pp. 86–94.

[12] "Anime Recommendations Database," https://www.kaggle.com/datasets/CooperUnion/anime-recommendations-database.

[13] F. M. Harper and J. A. Konstan, "The movielens datasets: History and context," *ACM TIIS*, vol. 5, pp. 1–19, 2015.

[14] W. Hamilton *et al.*, "Inductive representation learning on large graphs," *Advances in neural information processing systems*, vol. 30, 2017.

[15] R. Rossi and N. Ahmed, "The network data repository with interactive graph analytics and visualization," in *AAAI*, vol. 29, no. 1, 2015.

[16] J. Ni *et al.*, "Justifying recommendations using distantly-labeled reviews and fine-grained aspects," in *9th EMNLP-IJCNLP*, 2019, pp. 188–197.

[17] J. Rappaz *et al.*, "Recommendation on live-streaming platforms: Dynamic availability and repeat consumption," in *RecSys*, 2021, pp. 390–399.

[18] T.-Y. Liu *et al.*, "A 130.7 mm 2 2-layer 32Gb ReRAM memory device in 24nm technology," in *ISSCC*, 2013.

[19] M. Lanza *et al.*, "Recommended methods to study resistive switching devices," *Advanced Electronic Materials*, vol. 5, p. 1800143, 2019.

Weight-Aware Activation Mapping for Energy-Efficient Convolution on PIM Arrays

Kang Eun Jeon[1], Johnny Rhe[1], Hyeonsu Bang[2] and Jong Hwan Ko[1]
[1] Department of Electrical and Computer Engineering, Sungkyunkwan University, Suwon, Korea
[2] Department of Artificial Intelligence, Sungkyunkwan University, Suwon, Korea
{kejeon, djwhsdj, bhs1996, jhko}@skku.edu

***Abstract*—Convolutional weight mapping plays a stapling role in facilitating convolution operations on Processing-in-memory (PIM) architecture which is, at its essence, a matrix-vector multiplication (MVM) accelerator. Despite its importance, convolutional mapping methods are under-studied and existing mapping methods fail to exploit the sparse and redundant characteristics of heavily quantized convolutional weights, leading to low array utilization and ineffectual computations. To address these issues, this paper proposes a novel weight-aware activation mapping method where activations are mapped onto the memory cells instead of the weights. The proposed method significantly reduces the number of computing cycles by skipping zero-valued weights and merging those PIM array rows with the same weight values. Experimental results on ResNet-18 demonstrate that the proposed weight-aware activation mapping can achieve up to 90% energy saving and latency reduction compared to the conventional approaches.**

I. Introduction

The rapid development of high-performance and low-power electronics and computing technologies has been instrumental in the advancement and widespread adoption of deep learning techniques. Among these, Convolutional Neural Networks (CNN) have garnered significant attention owing to their outstanding performance in a wide range of computer vision applications. However, the inference of large CNN models on the traditional von Neumann architecture leads to significant energy consumption and latency due to the large amounts of data movement between the processor and memory. This phenomenon, commonly known as the *'memory wall'* problem, has become a major bottleneck in the performance and scalability of deep learning models/systems. To address this challenge, a new computing paradigm called *Processing-In-Memory* (PIM) architecture has emerged [1]. PIM architecture eliminates the need for unnecessary data movements between the processor and memory by performing matrix-vector multiplication (MVM) operations directly in the memory cells that contain the data of the deep learning model [2].

Processing a fully connected layer on a PIM array is a relatively straightforward task, as it primarily involves MVM operation. However, the same cannot be said for the convolutional layer/operation, which presents challenges due to its underpinning weight-sharing sliding kernel architecture and resulting translation-equivariant responses. The convolutional weight mapping method addresses these challenges by reshaping activation and weight tensors into a set of vectors and matrices, or *mappings*, such that convolution operation can be facilitated as an MVM operation. For instance, image to column (im2col) [3] performs matrix-vector multiplications (MVMs) by unrolling a kernel into a column of a PIM array and inputting an unrolled input feature map (IFM) corresponding to a single sliding window. However, the im2col mapping method often leads to low PIM array utilization, incurring additional computing cycles, increased energy consumption, and latency. To overcome its limitations, various high-utilization weight mapping methods such as shift and duplicate kernel (SDK) [4] mappings were proposed recently.

Although the previously proposed weight mapping methods promote higher utilization of the PIM array, they are indifferent to the inherently sparse and redundant nature of the convolutional weights. Indeed, up to 90% of multiplication operations will be with zero, hence ineffectual. To address such an issue, various PIM accelerators equipped with zero-skipping hardware features were proposed [5]. However, although the HW-based zero skipping can save energy by turning off unnecessary WLs, its latency may not benefit as much as the number of read cycles to generate the output feature map remains unchanged.

To address such an issue, this paper proposes a novel activation-aware mapping framework where activations are mapped onto the memory array instead of the weights. Doing so allows us to better exploit the beneficial qualities of the convolutional kernels, namely the sparsity, and redundancy. More specifically, zero-valued weights are skipped and repeated weights are merged to generate a smaller mapping and thereby reducing the number of computing cycles. Previously, there have been few attempts to map the activations on the PIM array [6]. However, previous approaches proposed an HW-based solution that may not be generic to a typical PIM architecture. To the best of our knowledge, this is the first attempt that proposes an activation mapping method that can be adopted to typical PIM hardware.

The key contributions of this work are summarized below.

- We propose a novel activation mapping framework that is adoptable by a typical PIM hardware architecture.
- We propose a novel row-merging technique that reduces the required number of rows in a mapping by merging rows with repeated values.
- We demonstrate the effectiveness of the proposed method through numerical experiments with ResNet-18, achieving up to 90% energy saving and latency reduction.

979-8-3503-1176-1/23 $31.00 © 2023 IEEE

II. Related Works and Motivations

A. Typical PIM Architecture

A PIM architecture comprises a memory array for computation and peripheral circuits for data conversion and storage, such as analog-to-digital converters (ADCs), wordline drivers, buffers, and control circuitry. An overview of a typical PIM architecture is shown in Fig. 1. Firstly, sliding windows of the input feature map (IFM) and the kernels are first transformed into a format compatible with MVM operations through a weight mapping method. The IFM windows are stored in the input buffer, whereas the kernels are loaded onto the PIM memory array. The memory array consists of memory cells connected to bitlines (BLs) and wordlines (WLs). When WLs are turned on simultaneously, the input vectors are fed into the memory array in a form of voltages. The current passing through each memory cell is determined by the accumulation of conductance of and voltage applied to each memory cell, thereby performing the MVM operation. Finally, the accumulated current is obtained by the ADC [7].

To ensure the model accuracy, multi-bit activations, and weights are used in the PIM array. There exist two major approaches to representing multi-bit inputs. The first approach is to use digital-to-analog converters (DAC) where the input is represented with multi-level voltages in an analog fashion. The second approach is to use bit-serial voltage input [8] where multi-bit inputs are sent sequentially to the WLs, bit-by-bit. In other words, to process input with b_i bit-precision, it would take b_i cycles, inducing $\times b_i$ more computing cycles compared to the DAC-based parallel input approach. Although the bit-serial method may take additional cycles for inference, it is often the preferred approach as DAC-based approaches are known to suffer from inaccuracies with the non-linear I-V relation of the RRAMs [7], [9]. In this paper, we evaluate our work assuming a bit-serial input structure as it allows us to compare various mapping methods with different input precision simply using its read and write cycles.

To represent multi-bit values in the memory array, many existing PIM implementations group multiple memory cells along the column to represent a single weight value [10], [11]. For example, to represent a weight with b_c bit-precision, b_c columns must be grouped together, assuming that a binary memory cell, such as SRAM, is being used. On the other hand, positive and negative values are represented through separated positive–negative weight placement [2], where positive and negative weights are placed in separate arrays and later subtracted. Therefore, $2N_cb_c$ PIM columns are required for a mapping with N_c columns and b_c bit precision.

B. Weight Mapping and Read/Write Cycles

A convolutional weight mapping method is a process of transforming the kernel and the corresponding IFM windows into a format compatible with the MVM operation. Image to column (im2col) [3] is the most well-known and widely adopted weight mapping method. As shown in Fig. 2 (a), im2col mapping first unrolls the IFM window into a vector.

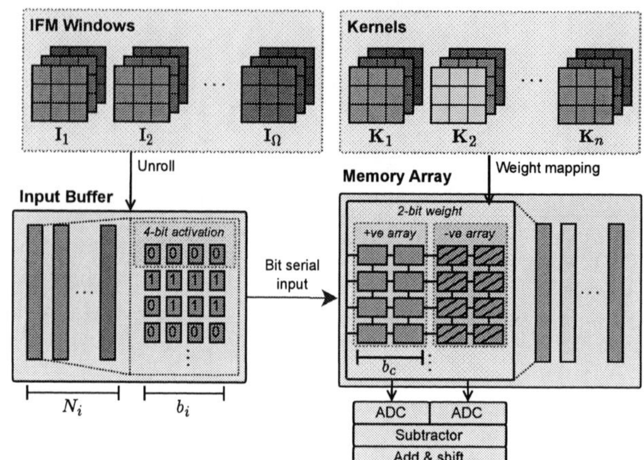

Fig. 1. An overview of a typical PIM architecture.

Similarly, each output channel of the kernel is unrolled and appended to form a mapping matrix. The numbers of rows and columns of the mapping are given by $N_r = mk^2$ and $N_c = n$ respectively. Here, k is the width/height of the kernel, m is the depth of the kernel, and n is the number of kernels.

Given N_r and N_c of the weight mappings, the numbers of read and write cycles, C and C_w respectively, can now be calculated. To this end, we first calculate array row (AR) and array column (AC) cycles, as proposed by Rhe et al. [12]. AR cycle, C_r, defines the number of arrays required to process the rows whereas AC cycle, C_c, the columns:

$$C_r = \left\lceil \frac{N_r}{h_s} \right\rceil, \quad C_c = \left\lceil \frac{2N_cb_c}{w_s} \right\rceil, \tag{1}$$

where b_c is the bit-precision of a single memory cell. Here, the numerator of C_c represents the number of columns required in a PIM array to host the mapping subject to the requirements for multi-bit and negative value representation.

Using the AR and AC cycles defined above, the read and write cycles, C and C_w respectively, are given by:

$$C = C_rC_cC_i, \quad C_w = N_mC_rC_c, \tag{2}$$

where $C_i = b_iN_i$ is the number of input cycles considering the bit-serial input structure of our assumed PIM architecture. N_m is the number of required/duplicated mappings per layer.

C. Zero-skipping Hardware Architecture

Besides optimizing the weight mapping method to generate smaller mapping, the zero-skipping technique takes advantage of the sparsity of the kernel to bypass ineffectual computations involving zeros. Yu et al. proposed an SRAM-based PIM architecture that implements HW-based zero-skipping of activations and weights with a dual 7T bitcell structure [5]. Kim et al. implemented a zero-skipping PIM hardware architecture, Z-PIM [6], where kernels are used as inputs over activations to better exploit the sparsity of the convolutional kernels. Z-PIM implements a unique circuitry to bypass computation when the inputted weight value is zero. Both studies were successful in achieving a high level of energy efficiency, owing to the sparsity of the convolutional kernels and IFMs.

However, there are several limitations associated with hardware-based zero-skipping methods. Fig. 2 (b) presents im2col activation mapping (AM) on a PIM array with HW-based zero-skipping capability. It can be seen that deactivating the WLs of zero-valued inputs does not necessarily result in the reduction of computing cycles. Although over half of the weights are 0, 4 computing cycles are still required, where only a small portion of the PIM array is actually being utilized. To address such issues, this paper proposes a novel weight-aware activation mapping framework that better leverages the zero-valued weights for accelerated computation, as shown in Fig. 2 (c). Here, we propose an activation mapping framework over weight mapping for two reasons. Firstly, since there are more zero-valued weights than zero-valued activations, using weights as input would allow more zero-skipping and therefore more performance gain. Secondly, if we were to use activations as input the weight mapping would have to be generated for every new input and therefore would be infeasible. On the other hand, since the weight values are already known, activation mapping reflecting the zero-valued weights can be generated offline. Although the proposed mapping framework requires more write cycles compared to the conventional approach, we discover that the read cycle reduction was large enough to nullify the impact of the additional write cycles. Besides, we also propose a novel row-merging technique that exploits redundant values in the kernel, as shown in Fig. 2 (d).

III. Weight-Aware Activation Mapping

To promote accelerated inference of the convolutional layers on a typical PIM architecture, here we propose weight-aware activation mapping framework that enables the exploitation of various beneficial characteristics of convolutional kernels, namely sparsity, and redundancy. To fully leverage the benefits of weight-aware activation mapping, two crucial conditions must be met. Firstly, the framework must be able to handle negative weight values as input to the PIM array. Secondly, it should maximize or at least retain the effectiveness of weight-aware techniques such as zero-skipping and row merging by preserving the original sparsity and maximizing the number of redundant values. By satisfying these conditions, the proposed framework can achieve optimal acceleration of convolutional layers on a PIM architecture.

However, facilitating activation mapping based on a typical PIM architecture faces several challenges. Unlike the traditional weight mapping methods where the positively valued activations are inputted into the PIM array, in an activation mapping, weights that contain both positive and negative values must be inputted into the network. The most obvious approach to handle negative input value is to simply quantize the weight, such that the minimum value is mapped to 0 and the maximum value is mapped to $2^{b_w}-1$, where b_w is the bit-precision of the weight. However, doing so would result in a loss of sparsity (i.e., zeros in the original kernels are not mapped to zero), greatly crippling the effectiveness of the zero-skipping technique.

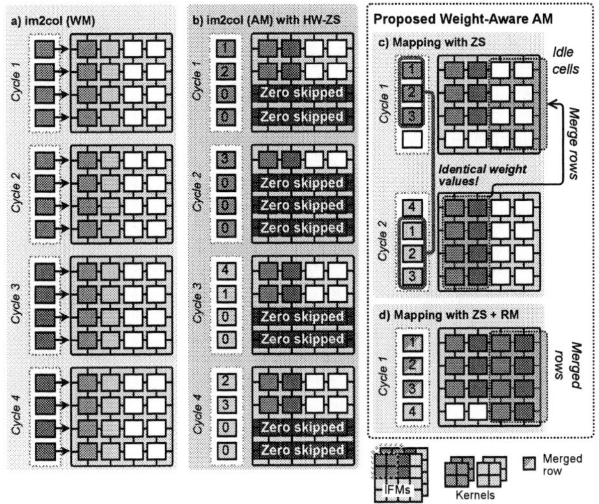

Fig. 2. Comparisons of the existing and proposed method: (a) im2col weight mapping (WM) method; (b) im2col activation mapping (AM) method with hardware-based zero-skipping (HW-ZS); (c) proposed AM with ZS; and (d) proposed AM with ZS and row-merging (RM).

Based on the above insights, this paper proposes an activation mapping framework where we exploit the positive and negative arrays, that already exist in most PIM architectures, to handle negative weight as inputs. More specifically, the activations are mapped to either the positive or the negative array depending on the sign of the weight inputs. On the contrary, the magnitude or absolute values of the weight are fed into the array. By doing so, we not only perfectly facilitate MAC operation between negative weight values and the activations but also promote weight-aware techniques. To facilitate this, we first propose sign-decomposed quantization method. Second, in order to amplify the ability to generate weight-aware mappings, we also proposed a novel weight-aware technique, coined row merging, to overcome the previously mentioned limitations of the zero-skipping technique.

A. Sign-decomposed Quantization

An overview of the process is shown in Fig. 3. Here, we first decompose an unrolled weight vector, $\mathbf{k}_j = [k_{j,1}, k_{j,2}, \cdots, k_{j,m}]$, into its magnitude and sign components:

$$\mathbf{k}_j = \text{sgn}(\mathbf{k}_j) \odot \text{abs}(\mathbf{k}_j), \tag{3}$$

where $\text{sgn}(\cdot)$ is a function returning a vector containing the signs of the input vector, $\text{abs}(\cdot)$ is an absolute value function, and $\odot$ is an element-wise multiplication operator. Then the magnitude vector is quantized in a uniform manner and its minimum value is forced to be 0, such that the sparsity of the original weights is preserved. Then the quantization function $Q(\cdot)$ and its scaling factor, S, is defined as

$$Q(\text{abs}(\mathbf{k}_j)) = \text{int}\left(\frac{\text{abs}(\mathbf{k}_j)}{S}\right), \quad S = \frac{\alpha}{2^{b_w}}, \tag{4}$$

where $\text{int}(\cdot)$ is a function that rounds the input to the nearest integer, α is the maximum value of the magnitude vector, $\text{abs}(\mathbf{k}_j)$, or the clipping range, and b_w is the bit precision

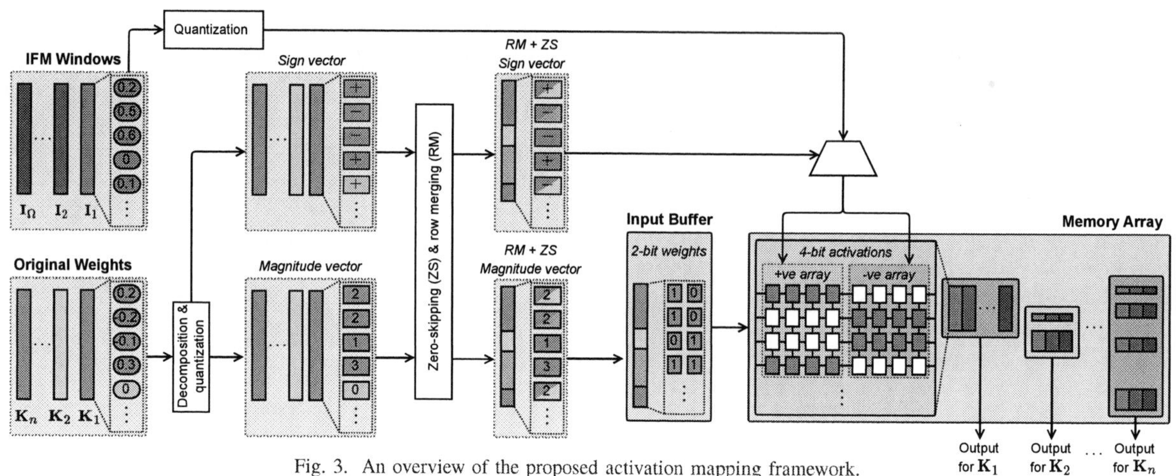

Fig. 3. An overview of the proposed activation mapping framework.

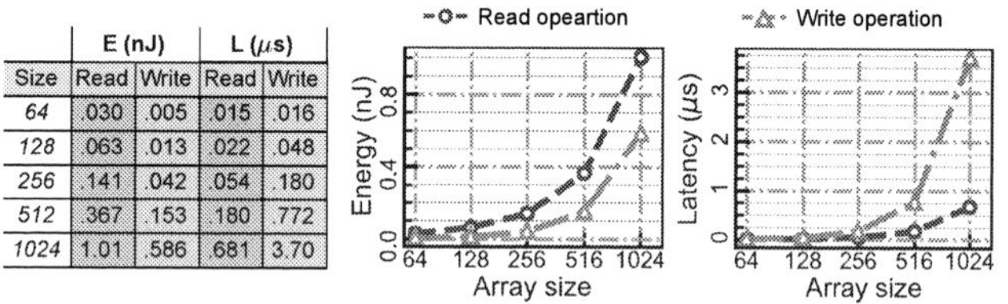

	E (nJ)		L (μs)	
Size	Read	Write	Read	Write
64	.030	.005	.015	.016
128	.063	.013	.022	.048
256	.141	.042	.054	.180
512	.367	.153	.180	.772
1024	1.01	.586	.681	3.70

Fig. 4. Energy consumption and latency of read/write operations simulated using NeuroSIM for varying array sizes.

of the input. Following the proposed quantization function, $2^{b_w+1} - 1$ uniformly quantized values can be represented.

With the proposed quantization method, we not only preserve the sparsity but also increase the number of redundant weight values; the decomposition process mapped all negative weight values to positive ones in the magnitude vector. Such qualities of the proposed quantization method ensure the effectiveness of the weight-aware techniques, promoting more zero-skipping and row-merging operations.

B. Row Merging Technique

Given the quantized magnitude vector from the previous section, weight-aware activation mapping can now be generated. Firstly, an im2col *activation* mapping is generated. Note that here we are dealing with activation mapping and not weight mapping. Unlike im2col weight mapping where kernels are unrolled into vectors to form the mapping, in the activation mapping counterpart, IFM windows are unrolled to form the mapping, and the weights are inputted.

Then the zero-skipping technique is first applied to the quantized magnitude vector of the kernel, $\text{abs}(\mathbf{k}_j)$, and its activation mapping that was generated in the previous step. Next, the row merging technique is applied to the resulting zero-skipped kernels and mappings. The mergeability of the two rows of the mappings is subject to two conditions. Firstly, the quantized magnitudes of the two weights corresponding to the two rows must be the same. Second, the specific cells that are needed for the merging should be idle as shown in Fig. 2. Note that the rows cannot be merged within a single mapping, because of the second condition. In other words, the weights from the same kernels cannot be merged.

IV. Experiments and Results

A. Experimental Setup

We evaluated the performance of the proposed weight-aware activation mapping framework on the ResNet-18 network. To generate the activation mappings, the kernels, and IFMs of the convolutional layers with kernel dimensions 3×3 were extracted from the network. Then the proposed quantization method was applied post-training. To evaluate the energy consumption and latency of the proposed mapping method, we calculated the related parameters for read and write operations from NeuroSIM [2], one of the most widely adopted simulators for PIM accelerators. The energy consumption and latency for a single cycle and a single sub-array were calculated for varying dimensions, ranging from 64×64 to 1024×1024. The details of the calculated values are presented in Fig. 4. Here, we have modified the simulation parameters to SRAM devices and to simulate inference only. In this paper, we only experiment with SRAM devices as NeuroSIM does not support write energy simulation for RRAM devices. Moreover, since the proposed mapping method involves a large number of write operations, the mapping method is better suited for SRAM devices that can facilitate the write operations with less energy cost. Also, RRAMs are known to suffer from endurance issues after repeated write operations. Besides the above changes, all other parameters were set to default values.

B. The Number of Read and Write Cycles

Here, we first analyze the performance of three different mapping methods, namely the im2col weight mapping (WM) method denoted by 'im2col (WM)', the im2col activation mapping (AM) with zero-skipping (ZS) denoted by 'ZS (AM)', and lastly, the im2col AM with ZS and row merging denoted by 'RM + ZS (AM)'. Fig 5 presents the number of read and write cycle, C and C_w respectively, calculated for ResNet-18 inference using array size of 256×256 and weight bit precision of 1 bit. In terms of read cycle, it can be seen that the weight-aware activation mappings greatly outperform the classical im2col weight mapping method. The outcome was foreseeable given the highly sparse nature of the convolutional kernels.

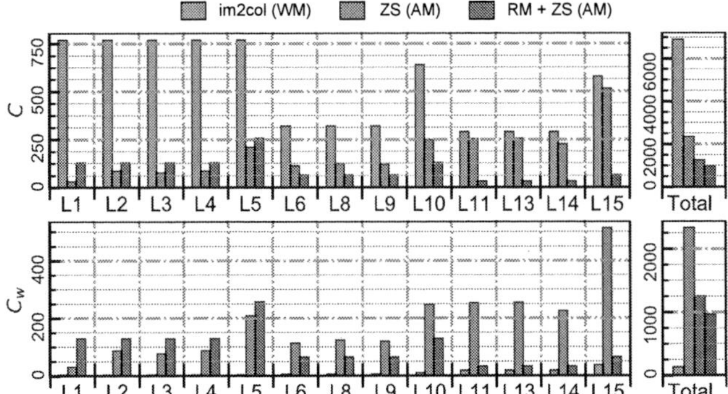

Fig. 5. Read and write cycles for 256×256 array and 1-bit weight precision.

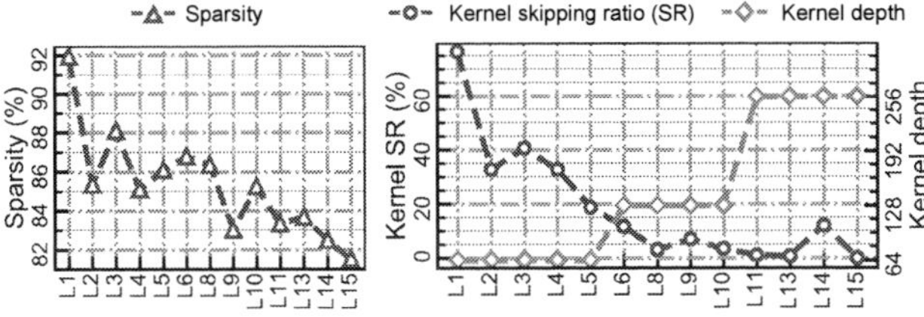

Fig. 6. Sparsity of ResNet-18 layers and corresponding kernel skipping ratio.

Comparing the ZS and RM + ZS methods, we can observe that the ZS method performs much better in the first 5 layers of the network. Its superior performance in the earlier layers is largely attributed to the fact that the earlier layers tend to be more sparse than the later layers. To support this claim, the sparsity of the kernels from each layer is shown in Fig. 6. We can see that the sparsity of the kernel reaches over 90% in the first layer, L1, and averages to around 88% in the first five layers, L1 to L5.

On the other hand, kernel skipping ratio (SR) is another factor that makes zero-skipping extremely effective in the first few layers. When the sparsity of the kernel is sufficiently high, there are cases where a kernel is populated with just 0 valued weights. Hence, the entire kernel can be skipped greatly reducing the number of read cycles. The kernel SR refers to the percentage of kernels that are skipped due to the just previously mentioned phenomenon. Fig. 6 plots kernel SR along with kernel depth. We can see that the kernel SR reaches up to 70% in the first layer and averages at around 40% in the first 4 layers. Here, the high kernel SR is attributed to not only high sparsity but also shallow kernel depth which makes it more probable for the kernel skipping to occur.

However, as we traverse into deeper layers, both sparsity and kernel SR, two crucial factors for the zero-skipping technique's effectiveness, quickly diminish. Starting from layer L6, we can see that activation mapping adopting the proposed row-merging technique starts to shine. Compared to the ZS (AM), RM + ZS (AM) reduces the number of cycles by around 50% from L6 to L10, and by a whopping 80% from L11 to L15. The experimental results demonstrate how the row-merging technique could mitigate the under-performance of the zero-skipping technique in the deeper layers, where sparsity and kernel depth conditions are not favorable.

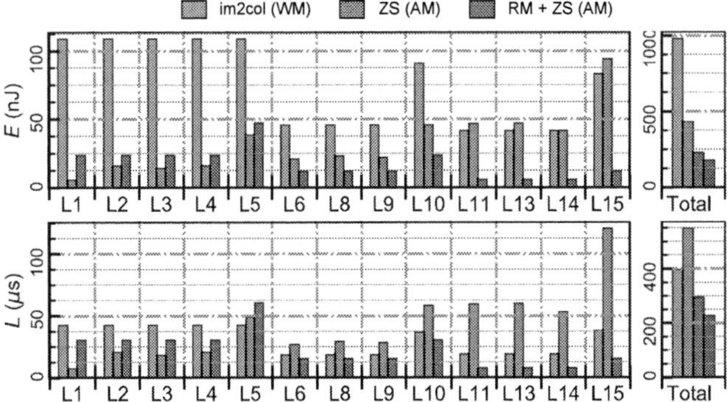

Fig. 7. Energy and latency for 256×256 array and 1-bit weight precision.

A similar trend can be observed with the number of write cycles, C_w, where ZS (AM) outperforms the RM + ZS (AM) in the earlier layers, but not in the later layers. This is because there are more kernels in the deeper layer, and, for ZS (AM), an activation mapping must be generated for every kernel. For example, in layer L1, since there are only 64 kernels, among which 80% are skipped, around 12 mappings are generated and written onto the PIM array. Whereas in layer L5, there are 256 kernels, among which less than 1% are skipped. Hence, the number of write operations required in the deeper layers is significantly larger than those in the first few layers. On the other hand, the proposed row-merging technique works around the issue as it only has a single merged kernel and the corresponding activation mapping. By doing so, the row-merging technique generates an activation mapping that has a much higher array utilization than that of the ZS (AM), consequently greatly reducing the number of write cycles.

To summarize, Fig. 5 also shows the total number of read and write cycles for the inference of the entire network on the far right. It can be seen that the proposed row-merging technique achieves less number of read cycles than the other two mapping methods. Here, a red bar is added which represents a hybrid mapping method that can choose between the three mappings to minimize its read cycle. The hybrid method utilized ZS (AM) mapping in the first 5 layers and RM + ZS (AM) in the remaining layers to achieve the least number of read cycles. Whereas for the number of write cycles, the RM + ZS (AM) requires fewer write operations than ZS (AM), but still requires significantly more operations than the im2col (WM) method. However, since the energy consumption of a single read operation is larger than that of the write operation, as shown in Fig. 4, we are willing to invest some more write cycle for a severalfold return on the read cycle reduction.

C. Energy Consumption and Latency

To better illustrate the benefits of the proposed method, here we translate our previous experimental results of read and write cycles to energy consumption and latency, as shown in Fig. 7. It can be observed that the RM +ZS (AM) method can achieve lower energy consumption and latency against the two other mapping methods, whereas the hybrid method achieves the least in both criteria. It is noteworthy that ZS (AM)

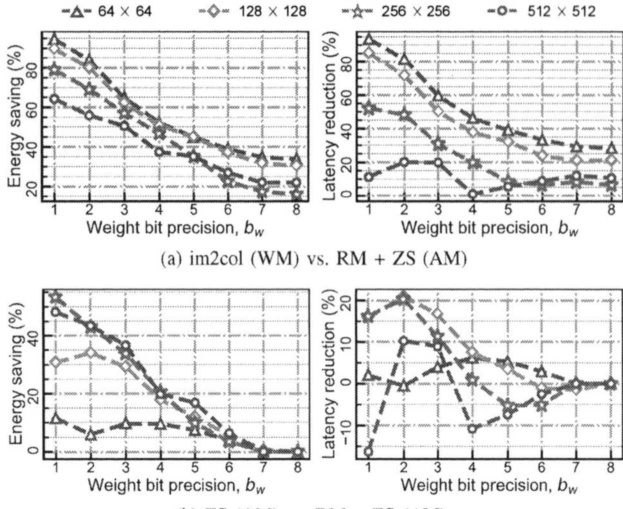

(a) im2col (WM) vs. RM + ZS (AM)

(b) ZS (AM) vs. RM + ZS (AM)

Fig. 8. Energy saving and latency reduction of the proposed framework plotted against varying array size and weight bit precision.

method achieves better energy consumption compared to the im2col (WM) method, but performs worse in terms of latency. This is because the ZS (AM) method involves more write operations, which is cheaper energy-wise but more expensive in latency. With the proposed row-merging technique used together with the zero-skipping technique, as shown in the previous subsection, we can achieve a much lower write cycle than ZS (AM). Hence, we can achieve better performance in terms of both energy and latency. Once again, we demonstrate how the proposed row-merging technique can complement the shortcomings of the zero-skipping technique.

Now we investigated the impact of varying weight bit precision and array size on the energy saving and latency reduction performance. Fig. 8 (a) presents the performance gain when ZS + RM (AM) method is compared with the im2col (AM) method, whereas (b) when ZS + RM (AM) is compared with ZS (AM). In Fig. 8 (a), we can see that the ZS + RM achieves energy saving and latency reduction in all scenarios. Among them, the method achieves the most energy saving and latency reduction when using a small array size of 64×64 and low bit precision of 1-bit. This is because the proposed method will generate mappings consisting of few rows, often less than a hundred rows. Therefore, its utilization is maximized with smaller arrays. Whereas lower bit precision promoted more zero-skipping and row-merging operations, achieving higher energy saving and latency reduction.

On the other hand, when compared with ZS (AM) method, as shown in Fig. 8 (b), the proposed method achieves the best performance with the array size 256×256. This is because the ZS (AM) generates mappings that use even fewer rows than that of the ZS + RM (AM). Hence, the performance of the ZS (AM) method also increased with a smaller array size. Whereas the performance of RM + ZS (AM) is maximized with smaller bit precision because the number of mergeable rows decreases with higher bit precision. Based on this result, we find that it is most optimal to use the proposed method with the array size of either 256×256 or 128×128 and bit precision less than 2. Based on the recommended configurations, the RM + ZS (AM) achieves up to 80% energy saving and latency reduction against the traditional weight mapping methods. Whereas against ZS (AM) method, 60% energy saving and 20% latency reduction can be achieved.

V. Conclusion

Processing-In-Memory (PIM) architectures have emerged as a promising solution to drive various power-hungry deep learning systems. However, processing a convolutional layer on a PIM array remained a challenge. This paper proposes a novel weight-aware activation mapping framework that enables the exploitation of the sparsity and redundancy of the convolutional kernels. Moreover, we proposed a novel row-merging technique that can complement various shortcomings of the zero-skipping technique. We demonstrate the effectiveness of the proposed method through experiments with the ResNet-18 network architecture, achieving up to 90% energy savings and latency reduction compared to the conventional im2col weight mapping method.

Acknowledgement

This work was partly supported by the National Research Foundation (NRF) grants (RS-2023-00251438, 2022R1F1A1074142, 2022R1A4A3032913) and Institute of Information and Communication Technology Planning & Evaluation (IITP) grants (IITP-2019-0-00421, IITP-2020-0-00821, IITP-2021-0-02052, IITP-2021-0-02068), and Samsung Electronics Co., Ltd (IO230404-05747-01).

References

[1] K. Roy *et al.*, "Towards spike-based machine intelligence with neuromorphic computing," *Nature*, 2019.

[2] P.-Y. Chen *et al.*, "NeuroSim: A circuit-level macro model for benchmarking neuro-inspired architectures in online learning," *IEEE Trans. Comput.-Aided Design Integr. Circuits Syst.*, 2018.

[3] K. Yanai *et al.*, "Efficient mobile implementation of a CNN-based object recognition system," in *Proc. 24th ACM Int. Conf. on Multimedia*, 2016.

[4] Y. Zhang *et al.*, "Efficient and robust RRAM-based convolutional weight mapping with shifted and duplicated kernel," *IEEE Trans. Comput.-Aided Design Integr. Circuits Syst.*, vol. 40, no. 2, 2020.

[5] C. Yu *et al.*, "A zero-skipping reconfigurable SRAM in-memory computing macro with binary-searching ADC," in *IEEE 51st European Solid-State Device Research Conf. (ESSDERC)*, 2021.

[6] J.-H. Kim *et al.*, "Z-PIM: A sparsity-aware processing-in-memory architecture with fully variable weight bit-precision for energy-efficient deep neural networks," *IEEE J. Solid-State Circuits*, vol. 56, no. 4, 2021.

[7] X. Peng *et al.*, "Optimizing weight mapping and data flow for convolutional neural networks on processing-in-memory architectures," *IEEE Trans. Circuits Syst. I*, 2020.

[8] S. Okumura *et al.*, "A Ternary Based Bit Scalable, 8.80 TOPS/W CNN accelerator with Many-core Processing-in-memory Architecture with 896K synapses/mm2." *2019 Symp. on VLSI Circuits*, 2019.

[9] P.-Y. Chen *et al.*, "Technology-design co-optimization of resistive crosspoint array for accelerating learning algorithms on chip," in *2015 Design, Automation & Test in Europe Conference & Exhibition (DATE)*, 2015.

[10] E. Lee *et al.*, "A charge-domain scalable-weight in-memory computing macro with dual-SRAM architecture for precision-scalable DNN accelerators," *IEEE Trans. Circuits Syst. I*, vol. 68, no. 8, 2021.

[11] A. Jaiswal *et al.*, "8T SRAM cell as a multibit dot-product engine for beyond Von Neumann computing," *IEEE Trans. VLSI Syst.*, 2019.

[12] J. Rhe *et al.*, "VWC-SDK: Convolutional weight mapping using shifted and duplicated kernel with variable windows and channels," *IEEE Trans. Emerg. Sel. Topics Circuits Syst.*, vol. 12, no. 2, 2022.

Teleport: A High-Performance ShiftNet Hardware Accelerator with Fused Layer Computation

Hyunmin Kim
Department of Electronic Engineering
Sogang University
Seoul, Republic of Korea
hyunminkim@sogang.ac.kr

Sungju Ryu*
Department of System Semiconductor Engineering
Sogang University
Seoul, Republic of Korea
sungju@sogang.ac.kr

***Abstract*—In this paper, we introduce a high-performance ShiftNet-optimized hardware accelerator called Teleport. ShiftNet replaces the standard convolutional layers with zero-flop-based shift convolution and pointwise convolution to reduce the number of computations. However, previous hardware acceleration approaches do not support the shift convolution, and hence they mapped the shift operation to the 3×3 convolution, and thereby the shift layer still shows the same number of computations as the conventional convolutional layers. To mitigate such a limitation, we first fuse the shift and convolutional layers without modifying the original configuration of the ShiftNets, and the fused computations are accelerated using a custom address translator, a systolic loader, and a systolic array. Our work improved the performance by 6.1-103× over the previous hardware acceleration approach on the ShiftNet benchmark.**

***Index Terms*—Neural network, hardware accelerator, shift convolution, systolic array, artificial intelligence.**

I. Introduction

As the neural networks are rapidly evolving, many customers from all round the world have adopted the neural networks in various applications. Among the various neural networks, convolutional neural network (CNN) has been used a lot in our daily lives. For example, object detection and image filtering are very popular tasks on personal mobile devices, and they are also often utilized in recognition tasks on autonomous driving and driving assistance control tasks. Such a CNN inference requires a fast response time because it is related to both customer satisfaction and safety of the drivers, so edge computing of the CNN is preferred over datacenter inference methods.

Recent CNN models consist of a larger number of layers to obtain high inference accuracy than the previous models, which leads to the increased number of computational complexity. However, the computation of more accurate but complex models on edge devices is difficult to achieve real-time processing, and hence we need to compress the network models while maintaining the number of layer/weight parameters. One of the approaches is to use shift convolution where standard convolution is partly replaced with the shift operations, thereby reducing the model complexity [1]. In the shift convolution, spatial domain convolution becomes shift operation which accounts for zero-flops. The shift convolution mitigates the number of multiply-accumulate (MAC) operations, but conventional hardware does not support the zero-flop shift computation. As a result, the shift convolution is first converted to tensor multiplication, and it is finally computed in the same manner as the previous convolution. Thus, shift convolution on the conventional hardware does not mitigate the large number of MAC computations. GPU-based shift convolution acceleration was introduced in a previous work [2], but it computes the shift operations based on the software-based approach.

In this paper, we present a design method for the shift convolution-tailored hardware accelerator (Teleport) with a slight logic overhead. By doing so, we exploit the advantage of the zero-flop-based shift convolution. We first fuse the shift and pointwise convolutional layers without any changes of conventional networks and map the fused layer to the computing units. Using custom address translator logic, we can achieve high throughput of ShiftNets.

The rest of this paper is organized as follows. Section II briefly explains the shift convolution and the systolic array. Our Teleport architecture including custom address translator logic is introduced in Section III. The evaluation of the Teleport architecture is performed in Section IV, and we discuss the additional points in Section V. We finally conclude this paper in Section VI.

II. Preliminaries

A. Shift Convolution

The main concept of the shift convolution is that the channel-wise input feature map is shifted in a single direction (Fig. 1a-b). In other words, input features in each channel are independently shifted depending on the shift direction dedicated to each channel. Such a channel-wise shift operation is actually similar to the depthwise convolution. As the MobileNets decompose the conventional standard convolution into the depthwise convolution and the pointwise convolution [3], the ShiftNets [1] decompose the standard convolution into the shift convolution and the pointwise convolution. The shift convolution is a weight-free and zero-flop-based convolution, because it does not convolve the input feature map by weights

*Corresponding Author

979-8-3503-1176-1/23 $31.00 © 2023 IEEE

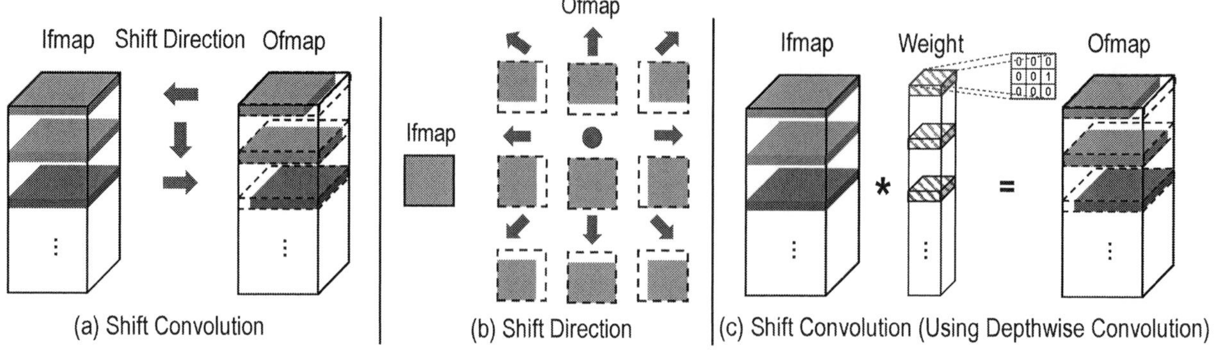

Fig. 1. (a) A simple illustration on Shift Convolution [1] and (b) generated output features depending on the shift direction. (c) Shift convolution on the conventional hardware.

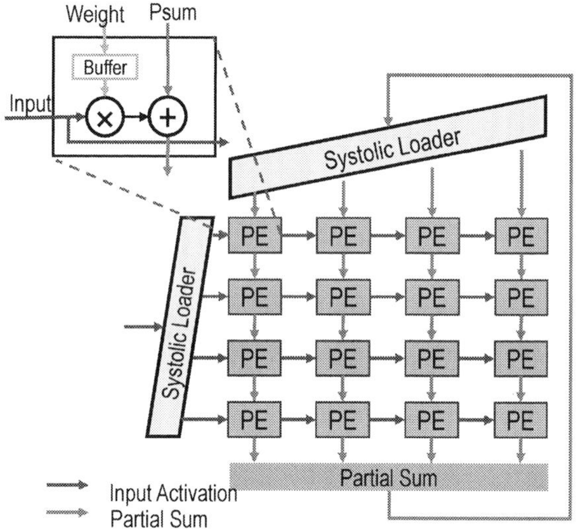

Fig. 2. Baseline systolic array architecture [4], [5].

and it replaces the depthwise convolution with simple shift operations.

B. Systolic Array

Systolic array is one of the well known approaches to perform matrix multiplication. Google TPU also adopts the systolic array to accelerate both datacenter workloads [4] and edge computation [5] with a weight stationary dataflow. The weight stationary-based systolic array consists of PEs for the MAC computation and input/psum loaders for the pipelined systolic array operation. Fig. 2 shows the operation of the systolic array. First, inputs and weights are fetched from the SRAM buffer, and they are fed to the systolic array. Weights are pre-stored in the weight buffer. Second, inputs are loaded to the systolic array through the loader. Then, inputs from the loader are sequentially sent to the PEs in a pipelined manner. The psum from the previous PE stage is added by the multiplication result in the corresponding PE stage, and the MAC result is sent to the following PE stage.

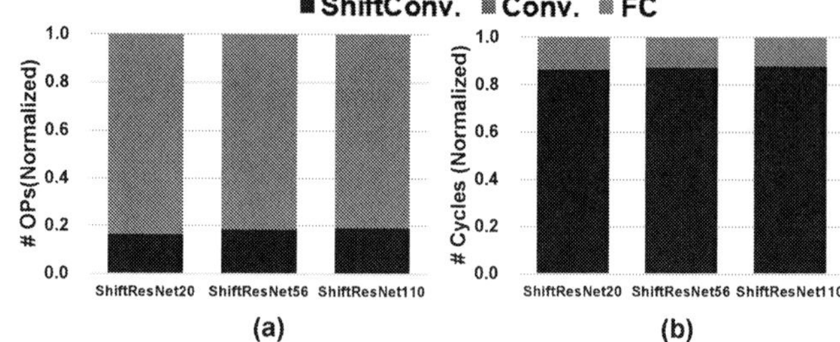

Fig. 3. (a) The number of operations on ShiftResNets. (b) The breakdown of clock cycles when computing the ShiftResNets on the conventional systolic array-based hardware [4].

C. Previous Shift Convolution Methods

For the fast inference task of the ShiftNet, previous works have tried to use a commercial GPU [1], [2], [6], [7] and a custom hardware accelerator [8].

The authors of the original ShiftNet paper [1] developed the PyTorch implementation [6] using their CUDA kernel for the shift operation [7]. However, commercial GPU microarchitecture does not support the zero-flop-based shift operation. Therefore, the CUDA kernel converts the shift operation into the depthwise convolution with the 3×3 filter.

Furthermore, previous neural network hardware accelerators [4], [5] also do not support the shift convolution. When computing the ShiftNets on the conventional accelerator, the shift operation must be converted to the 3×3 depthwise convolution (Fig. 1c). Fig. 3 shows the number of MAC operations for the ShiftNets and the breakdown of the computing clock cycles. As shown in the Fig. 3a, the shift convolutional layer only accounts for 16-19%. However, the shift convolutional layer (86-88%) on the systolic array dominates the other layers (Fig. 3b), because the depthwise convolution only uses a single weight filter (limited input reuse pattern) and mapping the depthwise convolution on the matrix multiplication leads to extremely low MAC utilization [9], [10]. By doing so, such schemes show the same throughput as the depthwise convolution-based model, and it does not have any improvements in the throughput even when using the shift convolution.

An FPGA implementation of the shift convolution was

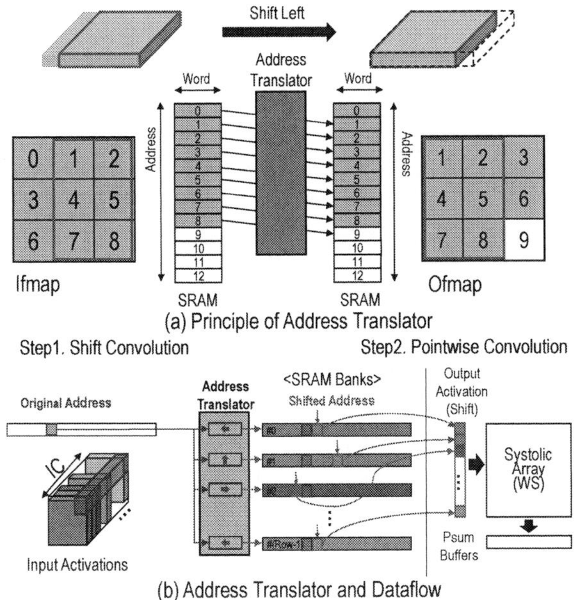

Fig. 4. (a) The operating principle of the address translator. (b) Proposed address translator and dataflow.

introduced [8], and it used a dedicated computational unit for the shift operation. In the shift unit, they designed the line buffer to relax the latency of loading inputs for the shift operation and tried to accelerate the ShiftNet using the extra computing block. However, such an approach suffers serious throughput degradation. First, the maximum throughput of the shift unit is 1 output per a clock cycle as written in the original paper due to their custom line buffering scheme. Second, their 1×1 conv unit array replaces previous 32×32 weights with new 32×32 weights in every computation as written in the paper, which significantly degrades the multiplier array utilization.

III. Proposed Architecture: Teleport

Our main contribution is to implement the ShiftNet-optimized hardware architecture. In this Section, we propose the detailed design method of the address translator, its peripheral logic including input/psum loaders, and network mapping scheme.

A. Hardware Design of Address Translator

Our approach is to compute the zero-flop-based shift convolution on the proposed custom digital logic. The definition of the shift operation is to move the features stored in the activation memory to another memory address. However, such an approach requires relocating the features by reading it from memory and writing it to a different address. Instead, we aim to mitigate the burden of repetitive memory read and write operations by designing a custom digital logic called an address translator. By doing so, we do not need to replace the shift convolution with depthwise convolution [6], [7] and software-level implementation [2], and our method does not suffer the performance degradation [8], as explained in the previous Section II-C. In this Subsection, we propose a detailed implementation method of the custom address translator circuit.

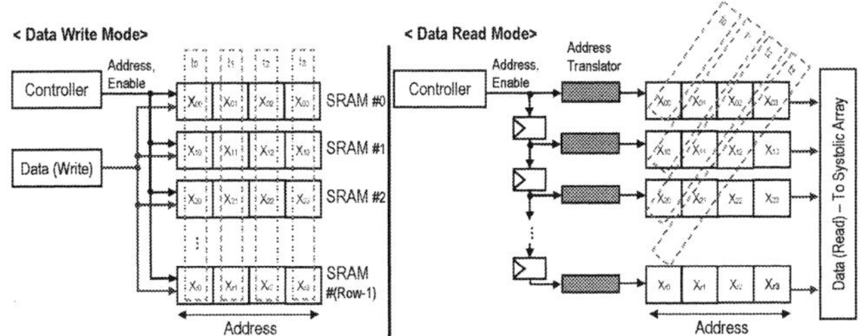

Fig. 5. Read and write operations of memory banks using the low-cost systolic loader.

Fig. 4a illustrates the principle of the address translator for the shift convolution. This example shows the shift left operation, and other directional shift operations can be easily realized in the same manner. Instead of relocating the feature address, our address translator modifies the original input feature address to the shifted output feature address. As a result, extra instructions for the address relocation are not necessary, and we can access the effective relocated address by simply translating the input feature address.

By definition of shift convolution, each channel of the input feature map is shifted independently in different shift directions. Hence, an address translation logic is dedicated to each channel of the input feature map. Fig. 4b explains the implementation of our custom address translator. First, each input SRAM bank is dedicated to an input channel. In other words, the consecutive input activations in the spatial domain are packed in an SRAM bank, and input activations in the different input channels are stored in the different SRAM banks. Second, an original address of the input activation is changed into the shifted address in the address translation logic. In this stage, the shift operation is replaced by the simple address translation. The address translation logic dedicated to each SRAM bank determines the destination shifted address depending on pre-defined shift parameters. Finally, the input SRAM outputs the activation located at the shift address of each SRAM bank.

B. Low-Cost Systolic Loader

Considering that the PEs in the systolic array compute the MAC operations in the pipelined manner as explained in Section II-B, the systolic array usually requires a systolic data loader. One of the simple approaches is to use the flip-flop array or multiple first-in-first-out (FIFO) registers to adjust the pipeline latency for the input/psum data fed to the systolic array. There are a lot of methods to implement the loader, but we introduce a low-cost systolic loader (Fig. 5). The key idea behind the low-cost systolic loader is to inject the delay in the address and enable signals using flip-flops. Such a design still uses the flip-flops, but it significantly reduces the number of flip-flops compared to the vanilla systolic loader design method and multiple-FIFO-based design approach. The

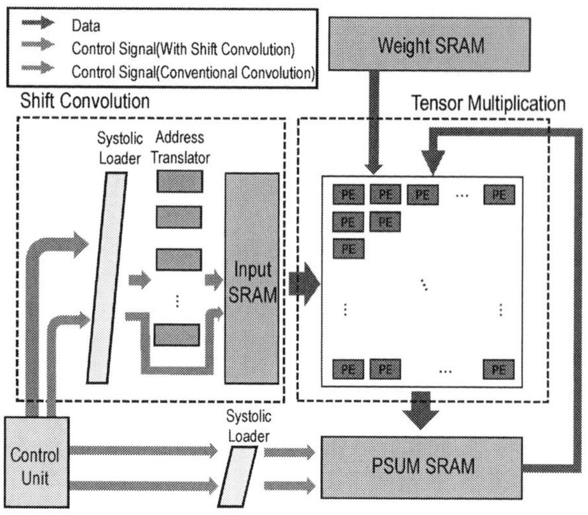

Fig. 6. Top-level illustration of proposed Teleport architecture.

delayed address and enable signals are then sent to the address translator. The translated address for each input channel is sent to each SRAM bank for the memory read operation. In the case of memory write operation, the write data does not care about the systolic loader because the memory write is not related to the backend systolic array operation. As a result, the write operation is performed in the same manner as the original SRAM write operation.

C. Top-Level Architecture

Fig 6 shows the top-level illustration of our Teleport accelerator. Our architecture includes the systolic array with 16×16 PEs, 32KB of input activation SRAM, 32KB of weight SRAM, 25KB of psum SRAM, an address translator for the shift convolution, and the low-cost systolic loader. Our architecture is based on Google's weight stationary-based systolic array which is widely used in the edge devices [5] and the datacenter hardware [4]. Only additional component is our address translator logic. The baseline systolic array also use the data loader for the systolic array, but the previous work does not explain the detailed information for the loader, and hence we designed a custom low-cost systolic loader as explained in the previous Section III-B. The shift parameter for each input channel is generated from the control unit, and it is sent to each address translation logic dedicated to each channel before the computation. Considering that most of the components are not modified compared to the baseline, it is also possible to run the conventional CNN models by simply bypassing the address translator.

D. Network Mapping and Dataflow

In this Subsection, we explain the network mapping method and dataflow for the ShiftNets on our Teleport hardware. The neural network is first trained, and model parameters are extracted using neural network frameworks (Fig. 7). Input activations and weights are fetched from the DRAM, and they are stored in on-chip SRAM. For the computation, the weights are loaded from the weight SRAM and pre-stored in the flip-flop buffers in the PE array, which is the same as the conventional weight stationary-based systolic array. The controller generates the SRAM address and the enable signal for the input feature. Considering that the shift convolutional layer always be followed by the standard (or mostly pointwise) convolutional layer, the consequent two layers are merged and they are simultaneously handled on our hardware (Fig. 7). The shift convolution is performed at the address translator, and the 1×1 convolutional layer is computed at the systolic array (Fig. 4b). Other layers outside the convolution-shift-convolution (CSC) module are consumed by the systolic array, which is the same as the conventional systolic array operation. The ReLU function is simply achieved by a very simple logic to check if the psum is a positive number. There are many ways to compute batch normalization layers and other activation functions using a custom digital computing block or a host processor, which is orthogonal to our main contribution. Therefore, we skip the detailed explanation of the orthogonal extra components in this work.

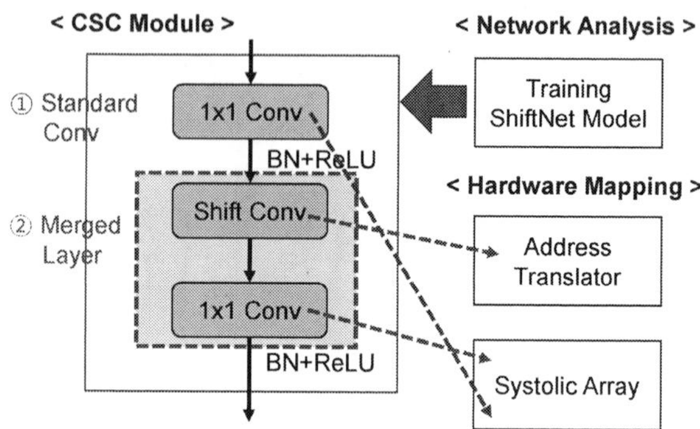

Fig. 7. Mapping the ShiftNet on the Teleport hardware. The ShiftNet is implemented based on the convolution-shift-convolution (CSC) module [1].

IV. Results

A. Experimental Setup

We evaluate the proposed Teleport hardware architecture in this Section. We designed the behavioral model of our hardware in the Verilog HDL, and the hardware was synthesized in a gate-level design using a 28nm CMOS standard cell library, targeting 500MHz. Synopsys Design Compiler and PrimePower were used for the synthesis and area/power analysis. Timing closure was also checked using Synopsys PrimeTime. The analysis on the throughput and the energy consumption was performed on our custom architecture simulator. In the evaluation, we adopted the LPDDR3 model parameters [11]. We compared our Teleport architecture with two custom hardware baselines: 1) Weight stationary-based systolic array [4], [5] where shift operation is not supported and depthwise convolution replaces the shift convolution (Base_WS in the Figures) and 2) Previous FPGA implementation with dedicated shift unit [8] (Base_FPGA in the Figures). We used a 16×16 systolic array, and design parameters including SRAM capacity are set to the values as explained in Section III-C. We used ShiftResNet benchmarks [6] which were provided by the authors of the ShiftNet original paper [1].

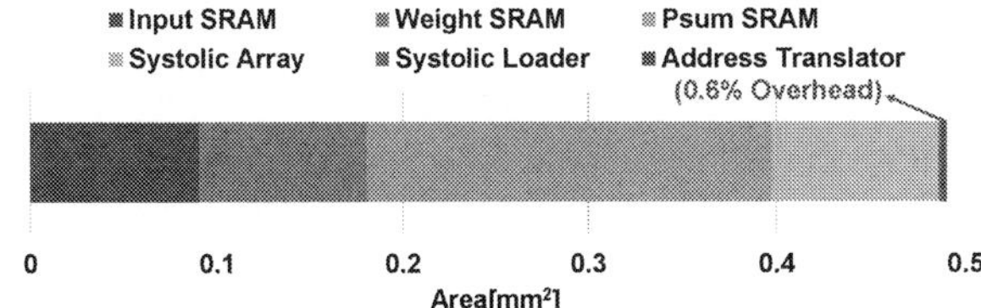

Fig. 8. The area breakdown of the proposed Teleport architecture. The additional overhead over the conventional systolic array baseline [4], [5] is only 0.6%.

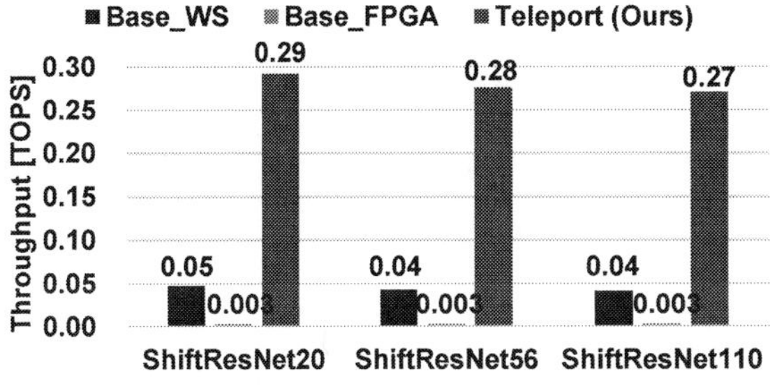

Fig. 9. The comparison of the throughput on the baseline and proposed hardware accelerators when computing ShiftResNets.

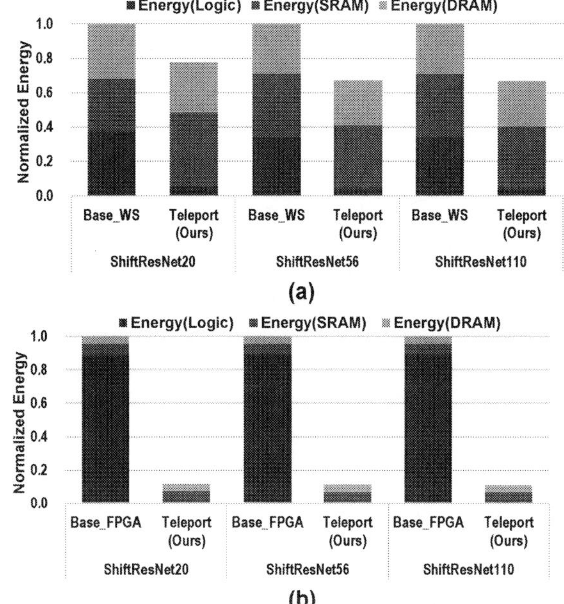

Fig. 10. The comparison of the energy consumption on the baseline and proposed hardware accelerators when computing ShiftResNets. (a) The comparison with Base_WS. (b) The comparison with Base_FPGA.

B. Results

1) Area: Fig. 8 shows the detailed area information of our Teleport hardware accelerator. As described in the previous Section III-C, the main components of our top-level architecture consist of the memory arrays (input SRAM, weight SRAM, psum SRAM), the systolic array for the tensor multiplication, the systolic loader to support the pipelined design of the systolic array, and the additional address translator for the fast computation of the zero-flop-based shift convolution. The memory arrays are 81% of the design area, which is the largest part of the total area. The systolic array for the computation is 18%, and the systolic loader is 0.3%. The address translator, the area overhead over the vanilla systolic array baseline (Base_WS) [4], [5], is only 0.6% which is negligible.

2) Throughput: Fig. 9 shows the throughput of the baseline and our Teleport architectures. As explained in the Section II-C, the operations of the shift convolution is only a small part of the entire task. However, conventional architectures do not support the shift convolution, and hence the shift convolution must be converted to the 3×3 depthwise convolution for the inference task on the conventional hardware. Meanwhile, the depthwise convolution on the conventional matrix multiplication units shows very low MAC array utilization, which was also described in the Section II-C. The throughput of the Base_WS shows 0.04-0.05 TOPS. Teleport architecture shows much larger throughputs (0.27-0.29 TOPS) than the Base_WS. Meanwhile, as explained in the Section II-C, the Base_FPGA suffers serious performance degradation on the 1×1 conv unit due to the weight reloading for the every multiplication stages and due to the low throughput of the shift unit as directly explained in the original paper [8]. Therefore, it only shows 0.003 TOPS for the ShiftResNets. As a result, our Teleport architecture improved the throughput by 6.1-6.5$\times$ and 96-103$\times$ compared to Base_WS and Base_FPGA baselines.

3) Energy Consumption: Fig. 10 shows the energy consumption of the baseline and our Teleport architecture on the ShiftResNet benchmarks. Base_WS consumes a large number of clock cycles due to the low utilization on the shift convolutional layer, and hence the energy consumption of the logic is much larger than our Teleport (Fig. 10a). Meanwhile, the energy consumption of the SRAM is the dominant part among the on-chip components in the Teleport architecture. The definition of the weight stationary-based systolic array used by baseline and our designs does not exploit the PE level convolutional reuse. The input features are unrolled through the `im2col` computation, and the tensor multiplication is only performed in the computing array. In this case, the same inputs are repeatedly fetched from the SRAM to realize the `im2col` computation, which leads to an increase in the energy consumption in the SRAM. In this evaluation stage, our Teleport reduced the energy consumption by 22-33% compared to the Base_WS design. However, if the PE-level convolutional reuse is further optimized which is beyond our main contribution, the energy consumption on the SRAM can be decreased in both baseline and our designs. In the SRAM-optimized case, the improvement of the energy efficiency on the Teleport over the Base_WS can be much higher than the current experimental condition.

On the other hand, the Base_FPGA experiences a significant degradation of PE utilization due to the dataflow of the 1×1 conv unit as described above. Hence, a large amount of logic energy becomes wasted due to the low throughput on the computational unit. As a result, our Teleport shows 88-89% reduction in the overall energy consumption compared to the Base_FPGA design.

TABLE I
THE RESOURCE UTILIZATION ON ZCU104 FPGA EVALUATION BOARD.

Design ZU7EV (ZCU104)	CLB LUTs (230000)	CLB REGs (460800)	BRAMs (312)
Teleport	22537	10445	24

4) FPGA Implementation: We furthermore verified our Teleport architecture on the FPGA chip. For the implementation, we used a Xilinx ZCU104 evaluation board with the ZU7EV FPGA chip. The target clock frequency of the FPGA chip was set to 100MHz, and the built-in BRAM arrays were used for the input/weight/psum SRAM buffers. Table I shows the resource utilization on the FPGA chip. Our Teleport architecture consumed 9.8% of the configurable logic block (CLB) LUTs to construct the combinational logic, 2.3% of the CLB registers (REGs) for the sequential elements such as the flip-flop, and 7.7% of the BRAMs for on-chip memory arrays. Total on-chip power of the FPGA chip was 1.651W with the power report in the Xilinx Vivado tool.

V. DISCUSSION

A. Standard CNNs on Teleport

Except for the additional component for the shift convolution, most of our design is based on the conventional systolic array. Hence, conventional standard CNN models are also seamlessly handled on our Teleport accelerator by skipping and bypassing the address translator (Fig. 6).

B. ShiftResNets on MobileNet Acclerator

Previous hardware architectures do not support the computation of the ShiftNets, and hence the shift convolution must be mapped into the depthwise convolution (Section II-C). In the case, one of the possible ways to compute the shift convolution is to accelerate the ShiftResNets on the custom MobileNet Accelerator [9]. However, such an approach still cannot maintain the zero-flop-based characteristics of the shift convolution thereby leading to the extra 3×3 convolutional layers. Furthermore, the channel stationary dataflow for the MobileNet Accelerator increases chip area due to the distributed placement of the memory arrays as described in the paper, thereby degrading the throughput/area. As a result, comparison to the MobileNet-optimized hardware is not required and time consuming.

C. Accuracy of the fused layer computation

Our fused layer computation does not modify the original layer configuration of the ShiftNets, and our approach is only to place the fused shift/pointwise convolutions on the same temporal stage of the top-level architecture using separated blocks including the address translator for the shift convolution and the systolic array for the pointwise convolution. As a result, this method does not affect the accuracy of the original implementation of the ResNet benchmark.

VI. CONCLUSION

We introduced a high-performance ShiftNet-optimized hardware accelerator called Teleport. Compared to the previous works where zero-flop-based shift convolution is replaced by depthwise convolution or software-level low-throughput shift operations are implemented, we designed a custom address translator and a low-cost systolic loader which enable the fast computation of the ShiftNet workloads. By fusing the shift/pointwise convolutional layers and making a single fused layer without changing the original layer information of the ShiftNets, our Teleport seamlessly achieved improved performance by 6.1-103× over the previous hardware acceleration approach on the ShiftNet benchmark.

ACKNOWLEDGMENT

This work was partly supported by Institute of Information & communications Technology Planning Evaluation (IITP) grant funded by the Korea government(MSIT) (No.2022-0-00266, Development of Ultra-Low Power Low-Bit Precision Mixed-mode SRAM PIM, 50%) and the National Research Foundation of Korea(NRF) grant funded by the Korea government. (MSIT) (NRF-2022R1F1A1070414, 50%). The EDA tool was supported by the IC Design Education Center(IDEC), Korea.

REFERENCES

[1] B. Wu, A. Wan, X. Yue, P. Jin, S. Zhao, N. Golmant, A. Gholaminejad, J. Gonzalez, and K. Keutzer, "Shift: A zero flop, zero parameter alternative to spatial convolutions," in *Proceedings of the IEEE conference on computer vision and pattern recognition*, 2018, pp. 9127–9135.

[2] H. Zhong, X. Liu, Y. He, and Y. Ma, "Shift-based primitives for efficient convolutional neural networks," *arXiv preprint arXiv:1809.08458*, 2018.

[3] A. G. Howard, M. Zhu, B. Chen, D. Kalenichenko, W. Wang, T. Weyand, M. Andreetto, and H. Adam, "Mobilenets: Efficient convolutional neural networks for mobile vision applications," *arXiv preprint arXiv:1704.04861*, 2017.

[4] N. P. Jouppi, C. Young, N. Patil, D. Patterson, G. Agrawal, R. Bajwa, S. Bates, S. Bhatia, N. Boden, A. Borchers *et al.*, "In-datacenter performance analysis of a tensor processing unit," in *Proceedings of the 44th annual international symposium on computer architecture*, 2017, pp. 1–12.

[5] GoogleLLC. Edge tpu compiler: Parameter data caching. [Online]. Available: https://coral.ai/docs/edgetpu/compiler/#parameter-data-caching

[6] A. Wan. Shiftresnet. [Online]. Available: https://github.com/alvinwan/shiftresnet-cifar

[7] P. Jin. Shift operation cuda implementation. [Online]. Available: https://github.com/peterhj/shiftnet_cuda_v2

[8] Y. Yang, Q. Huang, B. Wu, T. Zhang, L. Ma, G. Gambardella, M. Blott, L. Lavagno, K. Vissers, J. Wawrzynek *et al.*, "Synetgy: Algorithm-hardware co-design for convnet accelerators on embedded fpgas," in *Proceedings of the 2019 ACM/SIGDA international symposium on field-programmable gate arrays*, 2019, pp. 23–32.

[9] S. Ryu, Y. Oh, and J.-J. Kim, "Mobileware: A high-performance mobilenet accelerator with channel stationary dataflow," in *2021 IEEE/ACM International Conference On Computer Aided Design (ICCAD)*. IEEE, 2021, pp. 1–9.

[10] J.-S. Park, C. Park, S. Kwon, H.-S. Kim, T. Jeon, Y. Kang, H. Lee, D. Lee, J. Kim, Y. Lee *et al.*, "A multi-mode 8k-mac hw-utilization-aware neural processing unit with a unified multi-precision datapath in 4nm flagship mobile soc," in *2022 IEEE International Solid-State Circuits Conference (ISSCC)*, vol. 65. IEEE, 2022, pp. 246–248.

[11] M. Gao, J. Pu, X. Yang, M. Horowitz, and C. Kozyrakis, "Tetris: Scalable and efficient neural network acceleration with 3d memory," in *Proceedings of the Twenty-Second International Conference on Architectural Support for Programming Languages and Operating Systems*, 2017, pp. 751–764.

Energy-Efficient ReRAM-based ML Training via Mixed Pruning and Reconfigurable ADC

Chukwufumnanya Ogbogu[1], Mohapatra Soumen[1], Biresh Kumar Joardar[2], Janardhan Rao Doppa[1], Deuk Heo[1], Krishnendu Chakrabarty[3], Partha Pratim Pande[1]. [1]School of EECS Washington State University, Pullman WA, USA. [2]University of Houston, Houston TX, USA, [3]Arizona State University, Tempe AZ, USA.

Abstract— **Machine learning (ML) models have gained prominence in solving real-world tasks. However, implementing ML models is both compute- and memory-intensive. Domain-specific architectures such as Resistive Random Access Memory (ReRAM)-based Processing-in-Memory (PIM) platforms have been proposed to efficiently accelerate ML training and inference. However, existing ML workloads require a high amount of area and power for training. A major contributor to the area and power overheads is the Analog-to-Digital Converter (ADC). In this work, we propose a mixed pruning technique along with a novel reconfigurable ADC design to improve the power consumption profile. Overall, the pruned model with the reconfigurable ADC achieves ~50% reduction in power for training compared to existing state-of-the-art ReRAM-based architectures.**

Keywords—ReRAM, CNN training, Pruning, ADC

I. Introduction

Training machine learning (ML) models at the edge (on-chip training or on end-user devices) can address many pressing challenges associated with data privacy/security, increase the accessibility of ML applications to different parts of the world by reducing the dependence on communication fabric and the cloud infrastructure, and meet the real-time requirements of AR/VR applications. However, existing edge platforms do not have sufficient computing capabilities to support complex ML tasks such as training deep Convolutional Neural Networks (CNNs). ReRAM-based processing-in-memory (PIM) offers high-performance yet energy-efficient computing platforms for on-chip CNN training. This is due to their inherent capability to perform energy-efficient and high-throughput matrix-vector multiplication (MVM). Despite these advantages, deep CNNs with multiple layers require high power for training.

Recent work has proposed quantization and pruning to reduce the storage and power overheads of implementing deep CNNs on ReRAM-based PIM platforms [1] [2]. However, they are mostly targeted toward CNN inferencing and are not suited for training at the edge. Unlike inferencing, we also need to store the intermediate activations (in addition to the weights) during training. Simply pruning/quantizing weights does not help to reduce the storage and power overhead for activations. Moreover, existing techniques do not adequately reduce the number of ReRAM peripherals (e.g., ADCs). The ADCs in particular contribute between 50-70% of the power in a ReRAM-based PIM accelerator [3] [4]. Since, the power cost of ADCs scales exponentially with their precision, reducing the precision of ADCs without compromising the accuracy is key to saving power.

In this work, we achieve this by proposing a mixed-weight pruning technique that maximally prunes weights. In addition to the mixed pruning technique, we leverage dropout to further reduce the number of activations that must be stored in memory. The mixed pruning method involves a coarse-grained structured weight pruning where the granularity varies from filters to channels to index of the CNN weights. This is followed by fine-grained unstructured pruning. Hence, we refer to this mixed pruning method as coarse-to-fine-grained pruning (**C2F**). As we show later, C2F also prunes some activations. Dropout is also used as a mechanism to sparsify activations further. The C2F pruning along with Dropout enables the reduction of both the number and precision of required ADCs.

As we show later, different CNN layers achieve different levels of sparsity using C2F pruning and Dropout; this necessitates variable ADC precisions, which is not possible with conventional ADC designs. Hence, to support the mixed weight pruning (structured + unstructured) and Dropout, we propose to use a reconfigurable ADC. In this work, we propose a reconfigurable time-based ADC design [5]. The reconfigurability of the ADC stems from its ability to reconfigure the number of bits (precision) by disabling the lower LSB bits of the ADC. Hence, this approach enables a flexible and power-efficient design for ReRAM-based architectures with varying ADC requirements. The main contributions of this paper are:

- We propose a weight pruning technique called C2F, which along with dropout, sparsifies both weights and activations thereby reducing the power required for CNN training on ReRAM-based PIM architectures.
- We propose a reconfigurable ADC to efficiently support the C2F pruned CNN model during training to achieve greater power savings
- Experimental results indicate that the proposed solution achieves significant power reduction compared to existing architectures for CNN training.

II. Related Prior Work

ReRAM crossbars are well-suited for performing efficient MVM operations, which are predominant in CNN workloads [6]. Recent work has proposed ReRAM-based architectures for CNN inference. However, these architectures cannot efficiently support CNN training especially as the CNNs grow larger. Some ReRAM-based PIM architectures for CNN training have been proposed [7] [8]. However, they require a high bandwidth memory hierarchy for off-chip activation storage, which results in high latency and power costs [3].

Pruning is a popular technique for removing redundant weights and effectively reducing the computation and memory costs of CNN training and inferencing. Unstructured pruning does not result in a commensurate area or power savings compared to the amount of sparsity. Recent work has proposed

This work was supported, in part by the US National Science Foundation (NSF) under grants CNS-1955353, and CNS-1955196.

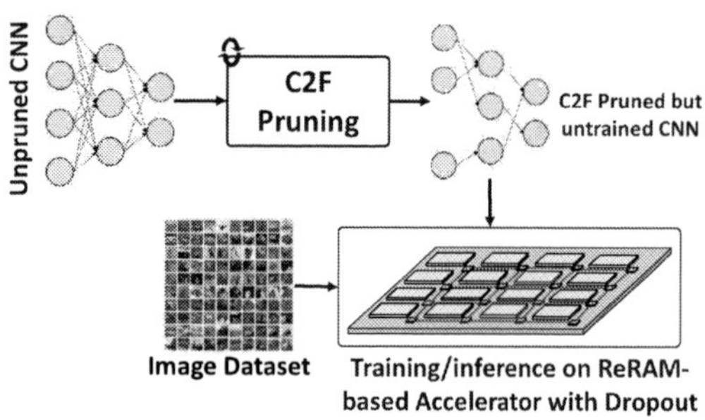

Fig. 1: Overview of the C2F pruning enabled energy-efficient deep CNN training/inference on ReRAM-based PIM platforms.

crossbar-aware pruning methodologies that remove weights in regular (or structured) shapes to enable more hardware savings [4] [9] [10]. These existing pruning techniques when applied naively to ReRAM-based architectures, prune out only the CNN model weights and do not sparsify activations during training. Existing pruning strategies also do not reduce the hardware overhead due to peripheral circuit components such as ADCs [4] [1]. As a result, ADCs with high power and area overheads are still required for ReRAM crossbars even after pruning.

Recent work has proposed ADC-aware optimization techniques for ReRAM crossbars [4] [11] [2]. However, they assume lower but uniform ADC precision requirements across CNN layers. This can lead to accuracy loss when low-bit ADCs are used for precision-critical CNN layers as we show later. Recently, a load balancing optimization technique for CNN inferencing was proposed, which leverages a reconfigurable ADC design for ReRAM-based accelerators [12]. However, the load balancing-based method is also focused on inferencing and is not suitable for the input-dependent nature of activations during CNN training. Therefore, it is important to explore peripheral-circuit-aware techniques that efficiently accelerate CNN training on ReRAM-based architectures without any accuracy loss. In this work, we address these limitations by proposing the C2F pruning technique with Dropout, which along with the reconfigurable ADCs reduces power.

III. Joint Weights and Activations Pruning

In this section, we discuss the important features of the coarse- to fine-grained pruning technique (C2F). Also, we complement C2F with dropout specifically for activation pruning. Overall, we aim to enable training CNN models at the edge. CNN training requires storing both weights and activations. The C2F pruning requires several training iterations to generate the pruned model. Hence, we implement the C2F pruning offline (i.e., on GPUs) to first obtain the pruned but untrained CNN model as shown in Fig. 1. However, activations are input-dependent and cannot be implemented offline. Hence, we implement the dropout-enabled activation pruning online during the training with the pruned CNN model on the ReRAM-based architecture. We implement Dropout using Linear Feedback Shift Registers (LFSRs), which require little area and power. Next, we present more details about sparsifying both weights and activations.

A. Coarse- to Fine-grained (C2F) Pruning Technique

As mentioned earlier, existing pruning methods can be categorized as either structured or unstructured. Unstructured pruning can remove more weight than the structured counterpart, but it is oblivious to the underlying crossbar structure. Hence, it often does not translate to a significant area or power savings [4]. On the other hand, structured pruning does not achieve the same level of sparsity as the unstructured counterpart, but it saves more area and power than unstructured pruning. C2F synergistically combines both structured and unstructured pruning and hence is more effective.

The C2F technique achieves this by adopting an iterative magnitude pruning approach to find a highly sparse and trainable model following the Lottery Ticket Hypothesis (LTP) [13]. Figs. 2 (a)-(d) show the overall C2F pruning process. C2F prunes (in order) (a) filter-wise, to reduce the number of columns across multiple ReRAM crossbars required for storing weights, (b) channel-wise to ensure that columns within a ReRAM crossbar are reduced, (c) index-wise, which prunes entries along rows of a crossbar, and (d) element-wise (finest pruning) which prunes individual weights in an unstructured fashion to maximize sparsity. The filter-, channel-, and index-wise pruning constitute the structured part of the C2F pruning. C2F prioritizes filter-wise pruning as it reduces both the weights and activations. As shown in Fig 2(b), pruning out an entire filter from a CNN layer ensures that an entire column is pruned in all four crossbars. This produces sparse output (activations) from the crossbars. However, the relatively coarse granularity of pruning does not lead to significant sparsity without accuracy loss. Once C2F fails to prune filter-wise without significant accuracy drop, it switches to channel-wise, followed by index-wise, and then finally unstructured pruning (Fig. 2(c)-(d)) to maximize sparsity with less than 1% accuracy drop. As mentioned earlier, the iterative C2F pruning process is implemented on conventional computing platforms such as GPU/CPU offline. This pruning phase is a one-time process.

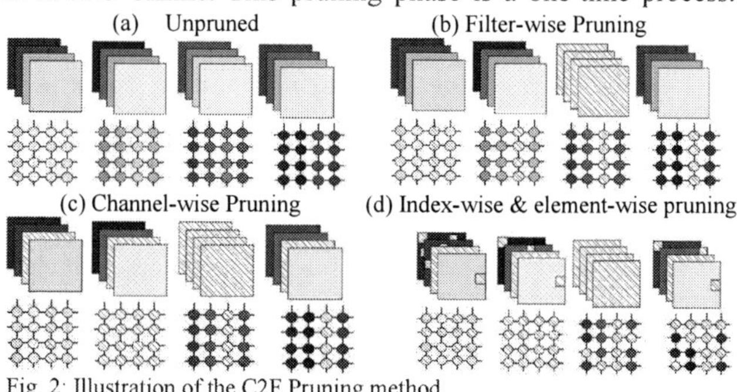

Fig. 2: Illustration of the C2F Pruning method.

Algorithm 1. C2F Pruning

Input: Unpruned CNN model, pruning percentage p
Output: C2F Pruned CNN model
Algorithm:

Initialize: $\theta^l \leftarrow \theta_{initial}$;
While *itr* < *N* and no *accuracy_drop* **do**
  Train for *E* epochs
  Prune $p\%$ of filters based on magnitude ($|\theta^l|$)
  If *new_accuracy* < *Baseline_accuracy* **do**
    Undo last pruning step
    Switch to finer pruning strategy (Channel-wise -> Index-wise -> element-wise pruning)
  Reinitialize remaining **weights** with $\theta_{initial}$
Return *Hardware-friendly C2F Pruned Model*

Once the pruned model is obtained, it can be reused since LTP-enabled pruned models are largely dataset-agnostic [14]. Here, we deploy the previously pruned model to the ReRAM accelerator as shown in Fig. 1 for future training tasks with any dataset [14], thereby amortizing the cost (time/energy) for the offline C2F pruning itself.

Algorithm 1 presents the high-level details of the C2F pruning strategy. We start by initializing the CNN weights (θ) using Xavier or Kaiming initialization (line 1) [7], then we train the CNN for $\boldsymbol{E}$ epochs. Next, we first prune the p% of the low magnitude CNN weights using coarse filter-wise pruning, and progressively switch to finer punning strategies (channel-, index-, and element-wise pruning), with <1% accuracy loss (lines 4-7). Finally, the C2F pruned (but untrained) model can then be used to train on resource-constrained edge devices with little area and power.

B. Activation Pruning using Dropout

Weight pruning does not automatically result in activation sparsity. Only the filter-wise pruning in C2F introduces some sparsity in the activations. To further sparsify the activations, we incorporate the dropout method on the activations. Dropout introduces unstructured sparsity without loss of accuracy. Dropout is a well-known regularization technique for CNN training which solves the overfitting problem and improves the model's generalizability to unseen data [15]. Dropout randomly prunes (temporarily) output activations of layers, which can be leveraged to save power as well. Traditionally, dropout is applied to the activations of the last few fully connected layers of CNNs only [15]. However, activation sizes typically decrease as the input passes through the various CNN layers. The last few CNN layers have a few activations. Hence, dropout only on these layers does not result in any hardware savings.

In reality, the initial layers of the CNN produce most of the activations stored on-chip during the pipelined training of CNNs on the ReRAM architectures [7]. For example, in VGG11 the output activations of the first two layers make up more than 70% of the total activations required for training as we show later. This happens because the initial layers process larger-sized activations. In addition, due to the pipelined execution of CNN training on ReRAM-based platforms, the memory requirement is especially high for the first few layers [6] [7]. The memory required for activation storage of a given CNN layer at depth (l) for pipelined training is multiplied by a factor of $(2(L-l)+1)$ (where L = number of CNN layers) [7]. As is clear from the equation, the initial layers need much more storage for activations compared to the later layers. Hence, in this work, we apply dropout on these first few layers to significantly sparsify activations. As typically done in practice, we also incorporate dropout on the last fully connected layers to prevent overfitting and improve prediction accuracy. However, adding dropout to the last few layers does not help with reducing activation storage significantly.

Other activation pruning methods such as magnitude-based activation pruning can also be used too. However, we show later that magnitude activation pruning leads to a significant accuracy drop during training. Similarly, applying dropout to every CNN layer also leads to significant accuracy degradation. Therefore, to effectively reduce the number of activations stored on the ReRAM crossbars during CNN training without accuracy loss, dropout should be applied to a few initial layers only. This reduces the ADC precision requirements for ReRAM crossbars that store the activations of the first few layers.

C. ADC Design for C2F Pruned CNNs on ReRAM Platforms

An ADC is necessary to interface between the ReRAM crossbar and the digital peripherals. The minimum ADC precision ($ADC_{precision}$) required for a ReRAM crossbar is determined by the number of rows activated (r), the number of input bits processed per cycle (v), and the number of weight or activation bits (w) stored per ReRAM cell. The required ADC precision (following [6]) is computed as follows [6]:

$$ADC_{precision} = v + w + \log_2(r) - 2 \tag{1}$$

Following equation (1), a ReRAM crossbar of size 128×128 (r = 128), with 1-bit input per cycle (v = 1), and 2-bits stored per cell (w = 2), would require an 8-bit ADC in an unpruned scenario. It is well known that large ADCs (e.g., 8-bit) are power-hungry and this problem must be addressed. Besides the precision, having too many ADCs incurs significant power overhead. A single ADC is often shared by multiple ReRAM columns (one ADC is shared by 128 columns as in [6]). Therefore, it follows that the number of ADCs required for c ReRAM columns is given by: $\lceil 128/c \rceil$.

The structured pruning part of C2F removes (prunes) all the weights mapped to selected columns while leaving all the weights in other columns intact, i.e., structured pruning reduces the value of c only. This enables us to use fewer ADCs. However, it does not affect the precision of ADCs. Hence, we will have a few, but high precision ADCs with structured pruning. On the other hand, unstructured pruning removes weights from every column but may not prune all the weights in a column i.e., unstructured pruning reduces the effective value of r. This allows us to use ADCs with lower precision. However, it does not reduce the number of ADCs required. Hence, unstructured pruning results in many but lower precision ADCs. C2F combines the benefits of both these pruning techniques and maximizes power savings by reducing both the number and size of ADCs.

In addition, dropout can be seen as a form of unstructured pruning. Hence, we can further reduce the precision of ADCs for activations following equation (1). For example, if only 8 out of 128 weights or activations are left on a column after C2F pruning and dropout, this effectively activates only 8 rows (r = 8). Thus, a 4-bit ADC can be used instead of an 8-bit ADC (as in the unpruned model) for the computation. This is a non-trivial reduction in peripheral power as the ADC power increases rapidly with the bit precision [3]. The power consumption increases by approximately 1.5-2× for every additional bit of resolution. Overall, C2F pruning combined with dropout can reduce both the precision and the number of ADCs required.

However, the amount of pruning varies with each CNN layer. The first few layers in the CNN have few weights that cannot be easily pruned, while the later convolution layers can be pruned to a greater extent [13]. As a result, C2F (and other pruning techniques) achieve different levels of sparsity for different CNN layers. This necessitates ADCs with different precisions for each layer. Hence, we propose to incorporate a

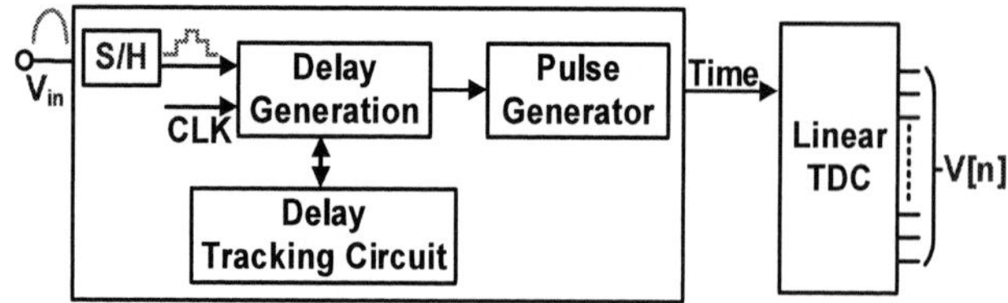

reconfigurable ADC design (discussed next), that can enable different precisions as needed.

D. Time-Based ADC Design for ReRAM Crossbars.

Existing Successive Approximation Register (SAR) and Flash ADC architectures have been adopted for the ReRAM-based accelerators [3]. SAR ADCs are commonly used for ReRAM crossbars due to their relatively simple structure and low area/power overhead compared to Flash ADCs [3]. However, SAR ADCs do not scale with technology, due to their analog sub-block circuits. On the other hand, Flash ADCs provide higher conversion speed, but they suffer from limited precision, which is not suitable for CNN training.

Time-based ADCs offer superior time resolution and scalability, making them attractive for designing complex PIM systems [16] [17]. The time-based ADC design is composed of two primary blocks, namely the voltage-to-time converter (VTC) and the time-to-digital converter (TDC). The analog input signal is sampled and processed by the VTC to produce an output pulse width that corresponds to the input voltage. This timed signal is subsequently converted to the digital domain by the TDC, as illustrated in Fig. 3. In this context, a timed signal implies a time-delayed signal where differently delayed clock signals are generated based on the input voltage.

The proposed ADC features a new VTC with a low-power consumption, process, and temperature-insensitive design, high linearity, and dynamic range [16]. The VTC consists of three main components: a sampler, a delay generation circuit with self-tracking, and a pulse generator. The sampler generates a discrete-time voltage signal from the input voltage, which is then used by the delay generation circuit to produce a delayed pulse width modulation (PWM) signal. The proposed VTC design, with its delay generation circuit featuring self-tracking, achieves superior linearity. This makes it highly effective, regardless of temperature or other environmental factors. We reduce this ADC architecture's dynamic power consumption by limiting the ramp signal, making it a power-efficient solution. Furthermore, the delay increases linearly with the sampled input voltages, ensuring accuracy and stability in detecting input signals. Incorporating a self-tracking circuit and limiting the ramp makes the proposed ADC architecture a highly effective solution for ReRAM-based architectures.

IV. Experimental Evaluation

A. Experimental Setup

We evaluate the C2F pruning method on the CIFAR-10 dataset using diverse CNN models: VGG-11 (V11), VGG-19 (V19), ResNet-18 (R18), ResNet-34 (R34), and GoogleNet (GN) [13]. The C2F pruning is implemented on an NVIDIA Titan GPU with 24GB of memory. The C2F pruning is offline and it results in a pruned but untrained CNN, which is then mapped to a ReRAM-based PIM architecture and trained with dropout for 50 epochs and a learning rate of 0.01 as an example. In this work, we simulate the on-chip training process on the ReRAM-based architecture using NeuroSim [18]. We follow the hierarchical ReRAM tile configuration presented in [6]. Each ReRAM tile consists of multiple 128×128 crossbar arrays that can be used for storage and computation with both weights and activations [6]. The weights and activations are stored using 16-bit fixed-point precision. Each tile contains a 16-bit reconfigurable LFSR operating at 1GHz for implementing dropout, which consumes less than 1% of the ReRAM tile area and power [19]. We summarize the hardware specifications of the ReRAM-based PIM architecture in Table I.

TABLE I. ReRAM-based PIM Architecture Specifications

16 PEs, 4 tiles per PE	
ReRAM Tile	96 Time-based ADCs, 128×96 DACs (1-bit), 96 crossbars, 128×128 crossbar array, 2-bit/cell resolution, 0.34W, 0.38 mm^2 [19] 1 Programable LFSR (16-bit)

TABLE II. Time-Based ADC Power Requirements

$ADC_{precision}$	3-bits	4-bits	5-bits	6-bits	7-bits	8-bits
Power (mW)	0.18	0.20	0.24	0.31	0.46	0.77

ADC Power and Area: The power consumption of the proposed time-based reconfigurable ADC is shown in Table II for various bit precisions obtained through Cadence simulations. The proposed Time-based reconfigurable ADC operates at a 1.2GHz sampling frequency and occupies an area of around 0.0013mm^2 in the TSMC-28nm technology node. The VTC's size and power consumption are independent of the number of bits, as only the number of TDC increases based on the number of bits and its area. Therefore, the power behavior is not linear with respect to the number of bits. The proposed VTC architecture, along with the constant field scaling and time interleaving, offers a promising solution for reducing the power consumption of ReRAM-based architectures while maintaining high accuracy and performance.

Baseline Pruning: To ensure a fair and thorough evaluation, we consider the unpruned CNN model as a baseline (BL). We benchmark the performance of the C2F pruned models trained with dropout, with two existing techniques. We consider the existing standard LTP method as the representative unstructured pruning method [13], and a recently proposed crossbar-aware structured pruning (CSP) technique [10]. The CSP method utilizes a multi-group LASSO algorithm to prune groups of weights that would otherwise be mapped along a column in a ReRAM crossbar. We implement iterative pruning (as shown in Algorithm 1) in all the methods to ensure that maximum sparsity can be achieved without incurring significant accuracy loss.

B. Accuracy of C2F Pruned Models with Dropout

As mentioned earlier, we must prune weights and activations as they both must be stored on-chip for training. C2F inherently prunes weights and some activations. We further increase the sparsity on the activations using Dropout to reduce ADC precision and number; this step leads to high power savings. Dropout is applied to the fully connected layers to improve inferencing accuracy. However, the last fully connected layers generate fewer activations than the initial convolution layers. Hence, having dropout only on the last few layers is not very useful for reducing storage requirements.

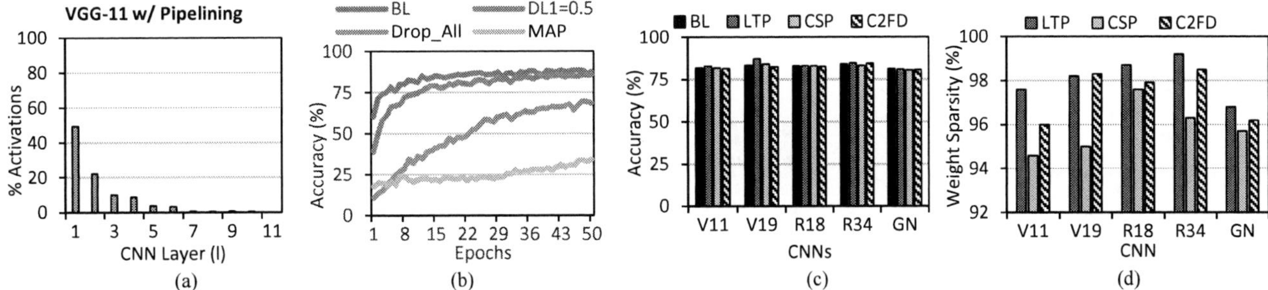

Fig. 4: (a) Distribution of activations across multiple VGG11 layers during pipelined training, (b) prediction accuracy of VGG-11 C2F pruned model with dropout on the first layer (D_{L1}), all layers (*Drop_All*), magnitude pruning (MAP) compared to the baseline (BL), (c) Overall accuracy of pruned CNNs using LTP, CSP, & C2FD compared to the unpruned model BL, (d) average sparsity of pruned CNNs using LTP, CSP, & C2FD.

Fig 4(a) shows the distribution of activations across layers during the pipelined training of VGG11 on the ReRAM-based architecture. As mentioned earlier in Section IIIB, the first layer of VGG-11 generates ~50% of the activations during pipelined training on the ReRAM-based architecture, while the last few layers generate less than 2%, as shown in Fig. 4(a). Hence, simply applying dropout to just the last few fully connect layers only is insufficient. Applying dropout to all layers leads to diminishing returns as the majority of the activation storage is used to store the data from the first CNN layer. Hence, we apply a dropout ratio of 0.5 on the activations of the first layer (D_{L1}=0.5) while training the C2F pruned model on the ReRAM-based architecture. We set the dropout ratio in the first layer to be 0.5 (D_{L1}=0.5), as this is sufficient to ensure a 1-bit reduction in the ADC precision requirement.

We compare the accuracy of this selective Dropout with two other activation pruning methods (a) pruning activations using simple magnitude-based pruning (MAP) and (b) a Dropout of 50% is incorporated in all CNN layers (*Drop_All*). Both MAP and *Drop_All* include C2F for weight pruning. Fig. 4(b) shows the accuracy of the C2F pruned model trained on the ReRAM-based architecture with different activation pruning methods incorporated. As shown in Fig.4(b), pruning activations using MAP causes the CNN model to train poorly (<30% accuracy). Moreover, implementing MAP on-chip requires sorting of the activations based on their magnitude, which requires additional hardware overhead. It is also evident that *Drop_All* leads to poorer training performance compared to the baseline unpruned model (BL). Hence, in this work, we use dropout in the first layer (D_{L1}=0.5), which is a simple and effective method to prune activations without drastic accuracy degradation. Following existing work, we also apply dropout on the last fully connected layers to prevent overfitting and ensure no accuracy loss [15]. C2F reduces both the number of weights and activations and Dropout reduces the number of activations further. Overall, this leads to significantly lower energy requirements. Henceforth, we refer to the C2F pruned model with dropout as **C2FD**. In all the subsequent analyses, we employ C2FD as the overall pruning method.

Fig. 4(c) and Fig. 4(d), compare the effectiveness of the C2FD in terms of the prediction accuracy and the achievable sparsity respectively with other pruning methods. The CNN models are pruned using each pruning technique (LTP, CSP, & C2F), and then we map the pruned CNNs to the ReRAM-based platform for the training evaluation. The goal of the iterative pruning approach (Algorithm 1) is to find the sparsest model that can be trained from scratch with minimal accuracy loss (<1%). As shown in Fig. 4(c), the C2FD pruned model achieves comparable accuracy with other pruning methods for all CNN models. In Fig 4(d), we show the percentage of weights pruned (sparsity) using each method (UP, CSP, & C2F). LTP achieves the highest sparsity (99.20%) due to its unstructured nature, while CSP achieves the least sparsity because of its structured pruning (94.60%). C2FD combines both unstructured and structured pruning to achieve high sparsity (98.40%) as shown in Fig 4(d). Moreover, as discussed earlier, in addition to pruning the weights, C2FD also prunes activations due to its filter-wise pruning and dropout. For example, in V19, C2FD achieves an average activation sparsity of 53.27%, while LTP and CSP achieve a lower activation sparsity of 30.25% and 33.41% respectively. Next, we present the power savings when training the C2FD-enabled CNN on the ReRAM-based PIM.

C. Overall Power Analysis

It is well known that the high precision ADC peripheral circuits in ReRAM-based PIM architectures contribute significantly to the overall chip power consumption [3] [6]. Hence, the reconfigurable ADC design proposed in this work enables optimizations that focus on reducing the bit precision of ADCs, thereby reducing the overall power consumption. As discussed earlier, the C2F pruning approach leads to varying sparsity levels across different CNN layers as shown in Fig. 5(a). Highly sparse CNN layers tend to have fewer weights per ReRAM crossbar column left on average after pruning (smaller r in eqn. (1)). Hence, the ADCs required for such layers can be reconfigured to use a lower bit precision to save power. In Fig. 5(a), we show the minimum ADC requirements and per-layer sparsity of the weights in the C2FD pruned VGG-19 CNN as an example. Here, we observe that the initial and final layers have less sparsity (i.e., more weights remaining), and hence require high-precision ADCs (8 bits). Meanwhile, the intermediate layers are significantly pruned and thus require low-precision ADCs (4-6 bits). This necessitates a reconfigurable ADC design for the ReRAM-based architecture as proposed in this work.

Fig. 5(b) compares the power consumption of two types of ReRAM-based PIM architectures: with uniform 8-bit ADC design, and the proposed architecture with reconfigurable ADC design. Here, we map the VGG19 pruned model obtained using the offline pruning techniques (LTP, CSP, & C2FD) to both architectures. We observe that for each pruned model, the reconfigurable ADC design inherently consumes less power compared to the all 8-bit ADC counterpart. Moreover, the C2FD

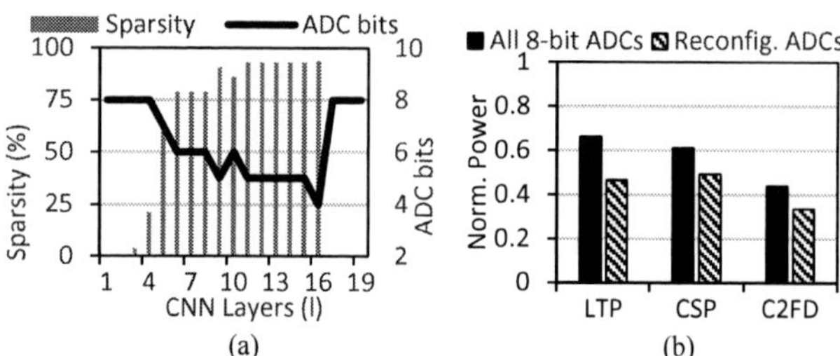

Fig. 5: (a) ADC requirements and sparsity distribution for C2F pruned VGG19, (b) Power consumption for LTP, CSP and C2FD pruned VGG-19 mapped to an all 8-bit ADC design vs the proposed Reconfigurable ADC design, all normalized w.r.t. the unpruned VGG-19 CNN model.

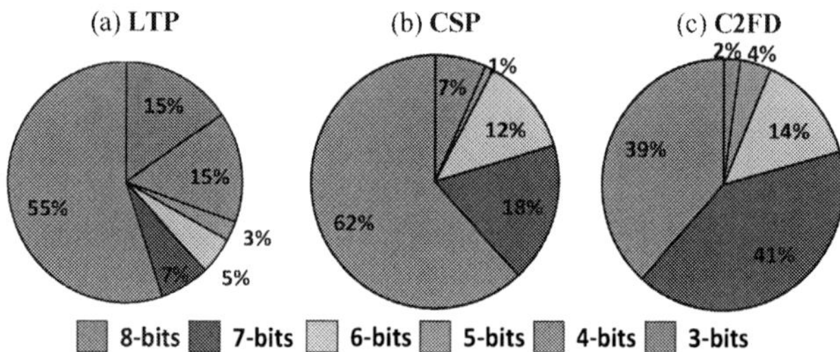

Fig. 6: Breakdown of ADC requirements for ResNet18 after pruning using; (a) unstructured pruning (LTP), (b) Crossbar-aware structured pruning (CSP) (c) the proposed coarse- to -fine- grained pruning with dropout (C2FD).

pruned model consumes the least power (<40%) when it is mapped to the reconfigurable ADC design.

Fig. 6 shows the breakdown of the ADC precision requirements when the ResNet-18 CNN is pruned using all three methods (LTP, CSP, & C2FD) as an example. Each pruned version of ResNet-18 is mapped onto the ReRAM architecture with reconfigurable ADCs. As seen in Fig. 6, the LTP and CSP pruned models require more high-precision ADCs compared to C2FD. For example, 8-bit ADCs constitute 55% and 62% of the total number of ADCs for LTP and CSP respectively, while for C2FD, this number is only 39%. This is because the finer-grained pruning in C2FD reduces the number of elements left in each ReRAM crossbar column, thereby reducing r (as in eqn. (1)) and the ADC precision. Moreover, the dropout incorporated in C2FD further reduces the number of 8-bit ADCs from 48% in C2F only to 39% as we show in Fig. 6(c). Hence, the C2FD-enabled pruned model requires fewer 8-bit ADCs and more lower precision ADCs (7bits, 6bits, etc.). This precision reduction leads to overall power savings.

We evaluate the overall power consumption of the ReRAM-based architecture with reconfigurable ADC design. Fig. 7 presents the overall power consumption of the C2FD pruned model compared to other pruning methods for the different models considered in this work. The C2FD pruned model achieves ~50% reduction in power consumption compared to the unpruned model (BL). Moreover, the C2FD pruned model consumes ~10% and ~18% less power compared to the CSP and LTP methods respectively. Overall, the C2FD pruning enables significant power savings on the ReRAM-based architecture with the proposed reconfigurable ADC design.

V. Conclusion

In this work, we demonstrate that coarse-to-fine weight pruning and dropout-enabled activation pruning enable us to reduce both the number and precision of power-hungry ADC circuits in ReRAM-based architectures for training CNNs. By incorporating a reconfigurable ADC architecture, we can reduce the ADC requirements (both number and precision) significantly without any noticeable loss in model accuracy while training the CNN models. This approach effectively reduces the overall on-chip power consumption by ~50% compared to the unpruned counterpart.

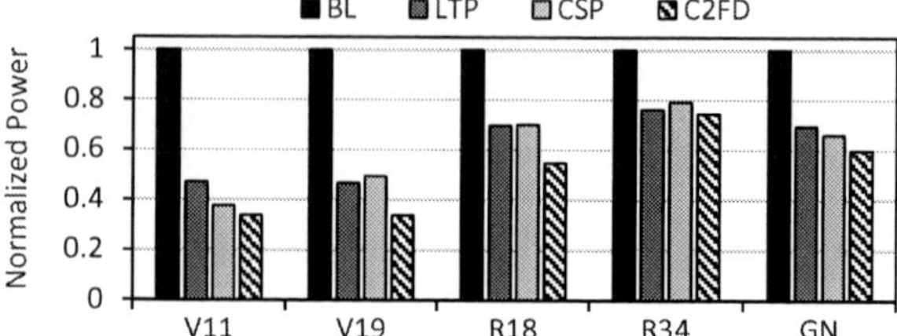

Fig. 7: Overall power consumption for each CNN normalized with respect to their unpruned versions (BL).

Refrences

[1] G. Yuan et al., "FORMS: Fine-grained Polarized ReRAM-based In-situ Computation for Mixed-signal DNN Accelerator," in *ISCA*, 2021.

[2] S. Huang et.al, "Mixed Precision Quantization for ReRAM-based DNN Inference Accelerators," in *ASP-DAC*, 2021.

[3] K. Roy et al., "In-Memory Computing in Emerging Memory Technologies for Machine Learning: An Overview," in *(DAC)*, 2020.

[4] G. Yuan et. al, "TinyADC: Peripheral Circuit-aware Weight Pruning Framework for Mixed-signal DNN Accelerators," in *DATE*, 2021.

[5] S. Mohapatra et al., "Low-Power Process and Temperature-Invariant Constant Slope-and-Swing Ramp-Based Phase Interpolator," *IEEE JSSC*, 2023.

[6] A. Shafiee et al., "ISAAC: A Convolutional Neural Network Accelerator with In-Situ Analog Arithmetic in Crossbars Ali," in *ISCA*, 2016.

[7] L. Song, X. Qian, L. Hai and Y. Chen, "PipeLayer: A Pipelined ReRAM-Based Accelerator for Deep Learning," in *IEEE HPCA*, 2017.

[8] A. Ankit et. al, "PANTHER: A Programmable Architecture for Neural Network Training Harnessing Energy-Efficient ReRAM," *IEEE Transactions on Computers*, 2020.

[9] C. Ogbogu et. al, "Accelerating Large-Scale Graph Neural Network Training on Crossbar Diet," *IEEE TCAD*, 2022.

[10] J. Meng et al., "Structured Pruning of RRAM Crossbars for Efficient In-Memory Computing Acceleration of Deep Neural Networks," *IEEE Transactions on Circuits and Systems II: Express Briefs*, vol. 68, no. 5, pp. 1576-1580, 2021.

[11] Y. He et. al, "InfoX: An Energy-Efficient ReRAM Accelerator Design with Information-Lossless Low-Bit ADCs," in *IEEE DAC*, 2022.

[12] D. Kim et. al, "SAMBA : Sparsity Aware In-Memory Computing Based Machine Learning Accelerator," *IEEE Transactions on Computers* , 2023.

[13] J. Frankle and M. Carbin, "The lottery ticket hypothesis: Finding sparse, trainable neural networks," in *ICLR*, 2019.

[14] A. Morcos, Y. Haonan, M. Paganini and Y. Tian, "One ticket to win them all: Generalizing lottery ticket initializations across datasets and optimizers," NeurIPS, 2019.

[15] N. Srivastava et. al, "Dropout: A simple way to prevent neural networks from overfitting," *JMLR*, 2014.

[16] K. Ohhata, "A 2.3-mW, 1-GHz, 8-Bit Fully Time-Based Two-Step ADC Using a High-Linearity Dynamic VTC," *IEEE JSSC*, 2019.

[17] M. Zhang et.al, "A 20GS/s 8b Time-Interleaved Time-Domain ADC with Input-Independent Background Timing Skew Calibration," in *Symposium on VLSI Circuits*, 2021.

[18] X. Peng et al., "DNN+NeuroSim V2.0: An end-to-end benchmarking framework for compute-in-memory accelerators for on-chip training," *arXiv:2003.06471*, 2020.

[19] A. Arka et al., "DARe: DropLayer-Aware Manycore ReRAM architecture for Training Graph Neural Networks," in *ICCAD*, 2021.

Digital Implementation of On-Chip Hebbian Learning for Oscillatory Neural Network

Edgar Luhulima[1], Madeleine Abernot[2], Federico Corradi[1], and Aida Todri-Sanial[1,2] (a.todri.sanial@tue.nl)
[1]*Eindhoven Univ. of Technology*, Eindhoven, Netherlands, [2]*LIRMM, Univ. of Montpellier, CNRS*, Montpellier, France

***Abstract*—This work proposes a digital implementation of an Oscillatory Neural Network (ONN) in a Field-Programmable Gate Array (FPGA), demonstrating excellent associative memory capabilities. This work goes beyond previous implementations by enabling on-chip learning directly in the FPGA. More specifically, we implement on-chip Hebbian learning, and we compare three different design strategies. The first s trategy t akes a dvantage of a System-on-Chip (SoC) composed of a Processing System (PS) and Programmable Logic resources (PL) to integrate Hebbian learning in PS. The two other strategies implement the Hebbian learning directly in PL. We compare the three different design strategies on a digit recognition task in terms of accuracy, utilization, execution time, and maximum frequency. We show that implementing Hebbian learning in PL gives more advantages in terms of resource utilization and latency than implementing Hebbian in PS with several orders of magnitude because the weight matrix computation is performed in hardware. Moreover, we develop an application interface to demonstrate the pattern learning and recognition capabilities of our digital ONN implementation.**

Keywords: *Artificial intelligence, auto-associative memory, pattern recognition, oscillatory neural network, FPGA implementation, Hebbian learning*

I. Introduction

In the past few years, there has been a rising trend of employing machine learning instead of traditional computing to perform tasks that previously seemed impossible to be performed by conventional machines. For example, machine learning technique as Artificial Neural Networks (ANN) has been used to perform object recognition [1], character recognition [2], and even speech recognition [3]. Different techniques have been explored to keep up with the growing constraints, especially with embedded applications. One of these constraints is power. One approach to employ a neural network with low power that suits embedded devices is Oscillatory Neural Network (ONN) [4]. ONN is a network of coupled oscillators with unique phase and frequency dynamics that can be used to perform low-power parallel computation [5, 6]. ONN has good associative memory capability [7] such that it can memorize patterns and retrieve them from corrupted input information. Also, different solutions has been explored to implement ONN in hardware, in analog [8], mixed analog and digital [9], or even fully-digital [10]. This paper focuses only on the digital implementation of the ONN from [10]. The current digital implementation is implemented in an FPGA. This digital implementation showcases a good associative memory capability of the ONN. However, the learning part of this ONN is still performed off-chip, meaning the weights are hard-coded every time the ONN is trained.

This work was supported by the European Union's Horizon 2020 research and innovation program, EU H2020 NEURONN (www.neuronn.eu) project under Grant 871501 and Horizon EU research and innovation program, Horizon EU PHASTRAC (https://phastrac.eu) project under Grant no. 101092096.

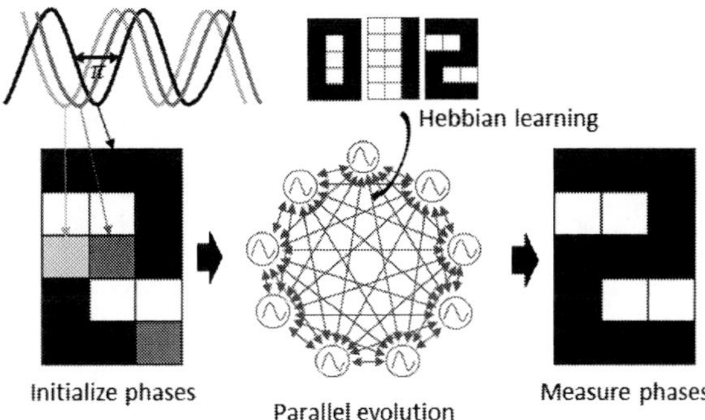

Fig. 1: Oscillatory Neural Network computing paradigm.

Recently, authors in [11], proposed a solution to enable on-chip learning, with the digital ONN taking advantage of the processing system (PS) of a Zynq processor. They performed on-chip learning with Hebbian and Storkey learning rules on a 15-neuron ONN configured for digit recognition. In this paper, we propose two other solutions to implement on-chip Hebbian learning [12] using the Programmable Logic (PL) resources of the Pynq FPGA board, and we compare the three solutions in terms of resource utilization. Finally, we also propose an application interface to demonstrate the pattern learning and recognition performances of our digital ONN solutions on a digit recognition application.

This paper is organized as follows. Section II describes the ONN computing paradigm, its auto-associative memory properties, and the digital implementation on FPGA. Then, Section III explains the different design approaches to implement on-chip Hebbian learning digitally. After, Section IV highlights the measurement results which validate each design, and compares the three design solutions. Section IV also presents the application demonstration on digit recognition. Finally, Section V discusses the advantages and limitations of the different solutions.

II. Oscillatory Neural Networks

A. Computing paradigm

Oscillatory Neural Networks (ONNs) [13, 14, 15] are brain-inspired computing models emulating neural oscillations from the brain. In ONNs, each neuron is an oscillator coupled with synaptic elements [16]. In this work, we consider phase-based ONNs which encode information in the phase relationship between oscillators. For example, if we consider bipolar information $\{-1, 1\}$, a $\{-1\}$ represents an oscillator with 0° phase, while a $\{+1\}$ represents an oscillator with 180° phase. Phase-computing ONNs use the natural phase synchronization behavior of coupled oscillators to compute in parallel. Thus, using phase computing allows for fast and parallel computation while possibly reducing the voltage amplitude and so limiting the power consumption [5], making ONN attractive for edge AI. During training, couplings between oscillators are configured depending on the task to solve. Then, inference starts with the initialization of each oscillator with the input phase information. Then, thanks to coupling, oscillators interact among them and phases evolve in time until stabilization. Reading the stable oscillator's phases gives the ONN output solution, see Figure 1.

B. Auto-associative memory

ONN configured with fully-connected architecture using unsupervised learning rules is known to perform auto-associative memory or pattern recognition tasks [7], like in Hopfield Neural Networks (HNNs) [17]. In this case, the network memorizes patterns in its coupling using some unsupervised learning rules such that when the network is initialized with a corrupted input pattern, it will evolve and stabilize to one of the memorized patterns. The main learning algorithm used to configure ONN or HNN for pattern recognition is the unsupervised Hebbian learning rule [12]. Hebbian configures the synaptic weights W_{ij} between neuron i and neuron j following:

$$W_{ij}^{p} = \sum_{p} \sigma_i^p \sigma_j^p \tag{1}$$

with $W_{ii} = 0$ and p the number of memorized or training patterns σ. There are other unsupervised learning rules which can be used to train ONN for pattern recognition and give better capacity, meaning being able to learn and retrieve more patterns. However, Hebbian is the simplest algorithm, and so the easiest to implement. Thus, in this paper, we only consider the Hebbian learning rule.

C. Digital ONN implementation

In this work, we focus on enabling on-chip learning for an ONN implemented on FPGA. A first fully-digital ONN design was introduced in [10] without the on-chip learning capability. In [10] each neuron is a phase-controlled digital oscillator allowing 16 phase stages, and each synapse is a 5-bits signed register. Then, a first solution to perform on-chip learning with the previous digital ONN design was introduced in [11] proposing to take advantage of the PS of a Zynq processor to implement Hebbian and Storkey learning rules in a 15-neuron ONN configured for digits recognition. In this work, we study two other solutions to enable on-chip Hebbian learning with the digital ONN on FPGA and compare the scalability of each solution.

III. Designs and Methods

A. Design exploration

The weights for the ONN determine the coupling between two distinct neurons. In the current implementation, these weights are computed off-chip using the Hebbian learning rule, see Figure 2. This learning rule requires arranging the learning patterns into a column vector. Each element of the vector is bipolar (-1/1.) Multiple learning patterns translate to multiple column vectors, which are then appended to form a matrix. The matrix rows are equal to the number of neurons in the ONN. The matrix is multiplied with its transpose, and the diagonal elements of the product are set to 0, see Equation (1). The resulting square matrix defines the weights for the ONN of which the size is determined by the number of neurons. Figure 2 and Equation (1) illustrate the computation of the weights. The term Hebbian learning and weights computation are used interchangeably in this paper.

The weights are embedded into the hardware description code of the ONN which is then programmed into the FPGA. This limits the flexibility of modifying the learning patterns, therefore the weights, of the ONN after the FPGA is programmed. It prevents the possibility of on-chip learning. This chapter describes the different design variations that were explored in order to realize the implementation

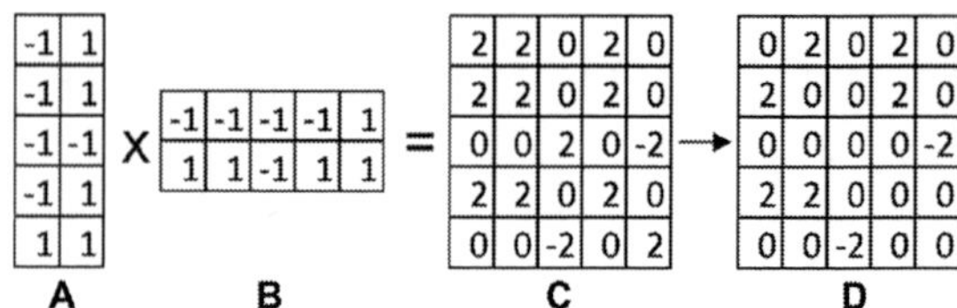

Fig. 2: Hebbian learning rule in the form of matrix multiplication. **(A)** Matrix of 2 learning patterns for 5 neurons. **(B)** Transposed matrix. **(C)** Square matrix product of matrix multiplication. **(D)** Weights with diagonal elements equal to 0.

of on-chip Hebbian learning. This allows for the computation of the weights for the ONN to be executed on chip. In this paper, we discuss three designs. These designs make use of the two units in modern FPGAs, the PS and the PL. Figure 3 illustrates the block diagram of each design variation. The first design, introduced in [11], incorporates Hebbian learning, indicated as the Hebb block in the diagram, as a part of the application software running on the PS. The second and third design integrate Hebbian learning as a hardware block in the PL.

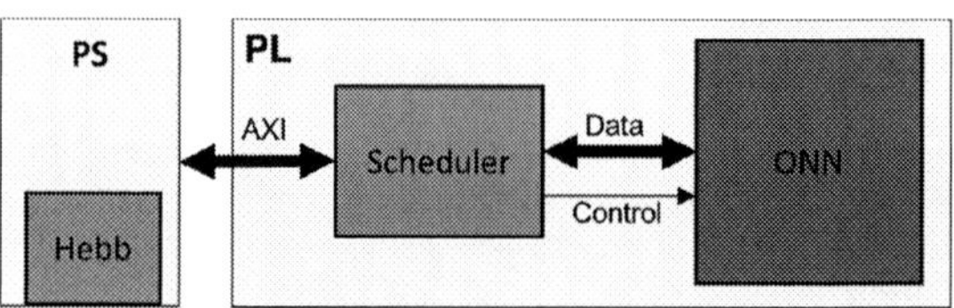

(a) Hebbian learning in PS

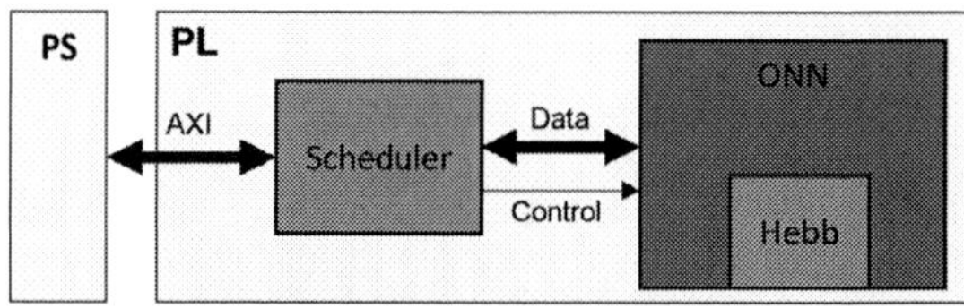

(b) Hebbian learning in PL

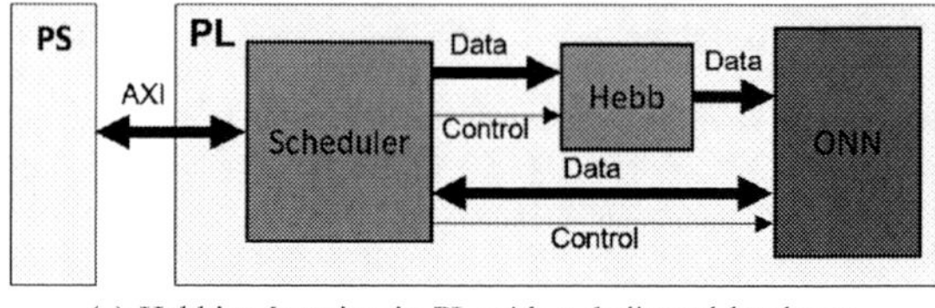

(c) Hebbian learning in PL with a dedicated hardware

Fig. 3: Design variations of the on-chip Hebbian learning

1) ONN with Hebbian learning in PS: The Hebbian learning in PS method, from [11], is designed as a function that computes the weights for the ONN in the application software. It is depicted in Figure 3a. The application software applies the Hebbian learning rule to each learning vector sequentially and accumulates the result in a weight matrix. This is a different technique than the one used in off-chip Hebbian learning. Instead of forming a matrix from different learning vectors, learning vectors are processed sequentially to produce a weight matrix. This process requires a multiplication between a vector with its transpose and an accumulation with previous results. After all learning vectors are processed, the final weight matrix is transmitted from PS to the Scheduler block in PL through the AXI interface. The Scheduler is a control unit for the ONN block. It bridges the communication between the application software and the ONN. It writes the weights to the ONN during training. During inference, it writes the test pattern to the ONN and reads back the inference result, which is then transmitted to the application software. While this design is a solution for on-chip learning, the weights computation is still executed in software which is relatively slow compared to hardware. To maximize the benefit of implementation in an FPGA, we explored other designs that directly exploit PL hardware during the learning process.

2) ONN with Hebbian learning in PL: The Hebbian learning in PL integrates the learning or weights computation as hardware to the ONN block as shown in Figure 3b. Since the learning is executed in hardware, the performance in terms of execution time should improve. The weights computation is added to the hardware description language of the ONN. In contrast to the Hebbian learning in PS, the application software sequentially transmits the learning vectors instead of the weight matrix to the Scheduler. The Scheduler, which controls the ONN, writes the vectors to the ONN during training. The processing of the vectors is consequently executed sequentially in the ONN block. It multiplies a vector with its transpose, then the result is accumulated. During inference, it writes the test pattern to the ONN block and reads back the inference result. The drawback of this design is that it lacks parallelism since the multiplication and accumulation for each element of the vectors are performed serially. It also utilizes more resources in the FPGA in comparison to the previous design because of the addition of the learning block as hardware.

3) ONN with Hebbian learning in PL with a dedicated hardware: The last design uses a separate dedicated hardware for Hebbian learning, as shown in Figure 3c, whose purpose is to add parallelism to the weight matrix computation. This dedicated hardware, represented by the Hebb block in the diagram, is a multiply and accumulate (MAC) unit. During training, the Scheduler receives the learning vectors from the application software and writes them to the MAC unit. The MAC unit has two data inputs, A and B. It is illustrated in Figure 4 for a learning vector of size 5. Elements of the learning vector are pushed serially into the multiplier through input A. The elements of the transposed vector are pushed in parallel to input B. As an improvement to the previous design, the multiplication result for a single column of the weight matrix is computed in a single clock cycle. This result will be accumulated with the previous results in the accumulator. The accumulator also acts as a weights buffer for the ONN. The Scheduler, similar to previous designs, writes the test pattern to the ONN, reads back the inference result and transmits it to the application software during inference. Because some computations are parallelized, this design should yield a better execution time for learning than the previous design.

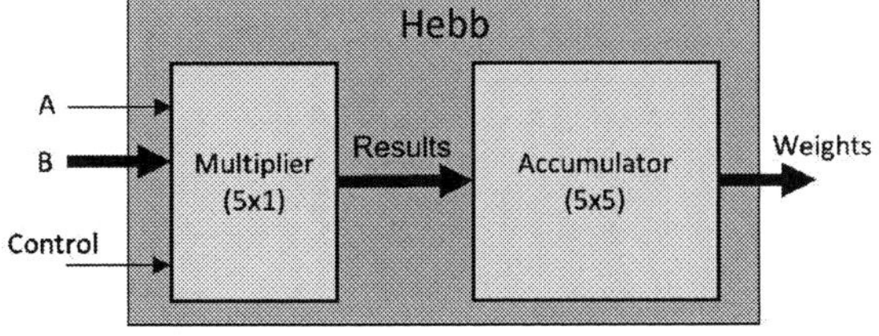

Fig. 4: MAC unit for a learning vector of size 5

B. FPGA Implementation

As a proof of concept, all three designs are implemented and tested on a PYNQ-Z2 board, which integrates a Dual ARM Cortex-A9 processor with 85K of programmable logic cells and 630 KB of block RAM. The board is supported by a Jupyter-based framework and Python APIs. These APIs provide access to the low-level control of the hardware on PYNQ. They also allow the overlay, which configures the architecture of the FPGA, to be loaded through a Python interface. Xilinx's Vivado is used in the development of the digital design and the generation of bitstream. A pattern, in the form of numerical digits, is chosen to showcase the associative memory capability of the ONN. A 5x3 ONN is set up on the FPGA and trained to learn digit 0, 1, and 2. Then a fuzzy digit is presented to the ONN during inference. A set of fuzzy digits is divided into 3 groups. Each group corresponds to each learning digit. In this setup, a learning pattern or digit consists of 15 black-and-white pixels represented in bipolar values. On the other hand, a fuzzy digit can contain a fractional value between -1 and 1 to indicate a grayscale pixel. This grayscale pixel represents a corrupted pixel in this application. Hamming distance (HD) is used as the metric to determine the fuzziness of a digit. The HD value determines how many pixels deviate from its corresponding learning digit. Figure 5 illustrates the fuzzy digits with different HDs.

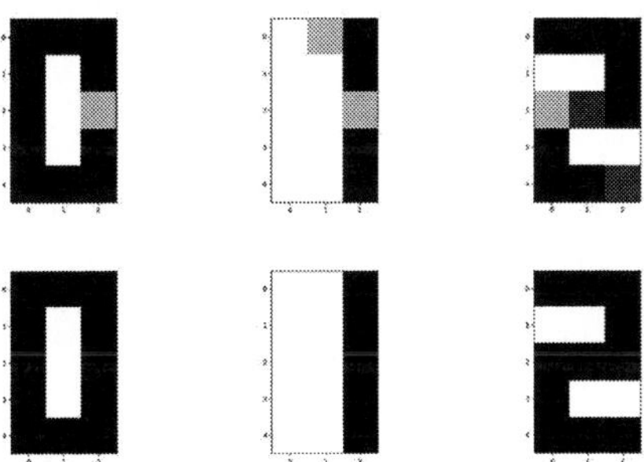

Fig. 5: Fuzzy digits 0,1,2 with HD = 1,2,3 respectively, and their corresponding learning digit.

The application software loads the overlay and controls the communication between PS and PL. The communication between PS and PL depends on the design. For Hebbian learning in PS, the software sends the computed weight matrix and the fuzzy digits to PL. For the other designs, the software sends the learning and fuzzy digits to PL. In all three designs, the software carries the retrieval of the inference result from PL. The user-defined learning and fuzzy digits are written in the software code. This software is mainly used for testing and for comparing the designs. In the demo, which is later described in section IV, a user interface feature is added to the software, allowing users to define the learning and fuzzy digits in real time. The Scheduler handles every data transfer to and from PL. State machines in the Scheduler handles the data transfer with the AXI, Hebb, and ONN blocks. The clock frequency for these blocks can be configured by modifying the clock divider in the Scheduler. To compare the learning and inference execution time between designs, a clock cycle counter is included in the Scheduler. This counter counts the number of clock cycles taken for performing the learning and inference.

IV. RESULTS

A. Design comparison and test

Several design metrics are used to compare these designs. They consist of error rate, utilization, execution time and maximum clock frequency. The three designs are implemented on the PYNQ board for tests and measurements. How the design metrics are measured and the results are discussed in this chapter.

1) Error rate: The accuracy of all three implementations can be determined by their error rates. The error rate can help identify the correlation between the number of learning digits and the ability of the ONN to recognize fuzzy digits. The error rate is calculated as the number of errors divided by the number of fuzzy digits. An

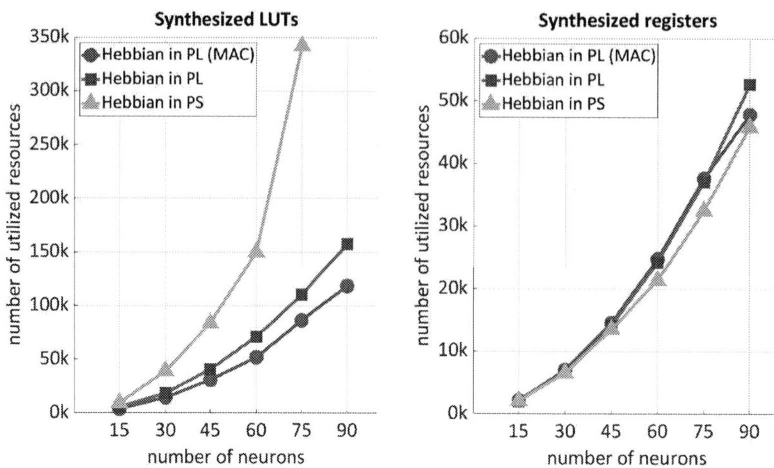

Fig. 6: Resource utilization of the 3 implementations

error is defined as when a fuzzy digit is incorrectly recognized by the ONN. The test results of the three implementations show that they produce the same error rate across different combinations of learning digits and HDs. The test was performed on a 5x3 (15 neurons) ONN. Table I shows the error rate of all three on-chip Hebbian learning implementations. The different combinations of 2 and 3 learning digits of digit 0,1,2 were tested. Each digit has 15 randomly generated and distinct fuzzy numbers for each HD. Thus, 45 fuzzy digits are used in total. HD = 1,2,3 are used in this test.

TABLE I: Error rate of all three on-chip Hebbian learning implementations for 5x3 ONN

Learning digits	HD	#Fuzzy digits	Errors	Error rate (%)
0,1	1	30	0	0
0,1	2	30	0	0
0,1	3	30	0	0
0,2	1	30	0	0
0,2	2	30	1	3.34
0,2	3	30	4	13.34
1,2	1	30	0	0
1,2	2	30	0	0
1,2	3	30	0	0
0,1,2	1	45	3	6.67
0,1,2	2	45	4	8.89
0,1,2	3	45	11	24.44

The result shows that the error rates of 2 learning digits, with respect to the HD, are lower than 3 learning digits. Thus, the error rate increases with the number of learning digits. It also increases with the HD.

2) Utilization: The size of the FPGA implementation is determined by the number of utilized resources. The resource utilization of the three implementations are compared by observing their number of synthesized LUTs and registers in Vivado. Figure 6 shows the number of utilized resources for LUTs and registers vs the number of neurons in the ONN. For Hebbian learning in PS, only the size of the ONN block is observed. For Hebbian learning in PL and PL (MAC), the total size of the ONN and Hebb block is observed. The Scheduler block is not included in this observation because the size is insignificant compared to other blocks and it grows linearly with the number of neurons. The number of neurons is increased to see how the number of utilized resources grows. The result shows that the number of synthesized LUTs for Hebbian learning in PL and PL (MAC) has a polynomial growth. This is because as the number of neurons increases, the weight matrix increases by a factor of n^2. The discrepancy in the number of synthesized LUTs between Hebbian in PL and Hebbian in PL (MAC) is because the weight matrix computation in Hebbian in PL (MAC) is pipelined. It removes the need to finish the multiplication before performing the accumulation. Thus, reducing the number of LUTs. The number of synthesized LUTs for Hebbian in PS grows almost exponentially. This is because the processing of the weight matrix increases by a factor of n^2 in the ONN block. The processing requires a large portion of the LUTs because it needs to sequentially parse the incoming weight matrix data. The numbers of synthesized registers for all 3 implementations show relatively the same growth and no significant discrepancies. The maximum number of neurons that can be implemented on the PYNQ board for Hebbian in PL (MAC) is 60. For Hebbian in PS and PL is 30. This is limited by the number of available LUTs on PYNQ. It can be concluded that Hebbian in PL (MAC) is a better choice among the other two for an application that requires a large number of neurons.

3) Execution time and maximum clock frequency: Execution time and maximum clock frequency are important metrics, especially for an application requiring a real-time constraint. Therefore, we must determine the execution time for the learning and inference and the maximum implementation frequency. This test was performed on a 5x3 ONN. The average learning period for Hebbian in PS is measured by measuring the average time span (μs) to learn a digit. In Hebbian in PL and PL (MAC), this is performed by using a counter in the Scheduler. This counter counts the number of clock cycles between the moment it is triggered and stopped. The Scheduler triggers the counter when the learning starts and stops it when the learning is done. The average inference period for all three implementations was also measured using a counter. The maximum clock frequency is determined by evaluating the timing analysis in Vivado. Table II shows the execution time and maximum clock frequency of the three designs. Hebbian in PL and PL (MAC) spend almost the same clock cycles for inference. Hebbian in PS is a little bit faster because the ONN block is more optimized. The average learning speed for Hebbian in PL (MAC), independent of their maximum clock frequencies, is 8x faster than Hebbian in PL. This is due to the parallelism in the weight matrix computation. At their maximum frequencies of 65 and 41 MHz, Hebbian in PL and PL (MAC) can perform learning 280x and 1500x faster than Hebbian in PS. In conclusion, Hebbian in PL and PL (MAC) are faster than Hebbian in

TABLE II: Execution time and maximum clock frequency for 5x3 ONN

Design	Average learning period (clock cycles)	Average inference period (clock cycles)	Maximum clock frequency (MHz)
Hebbian learning in PS	700 (*)	157	32.5
Hebbian learning in PL	162	218	65
Hebbian learning in PL (MAC)	19	217	41

(*) in μs instead of clock cycles because the learning period is measured in the application software.

PS because the weight matrix computation is performed in hardware and they have higher maximum clock frequencies. Thus, they are a better choice for an application that requires low latency.

B. Pattern recognition demo

The pattern recognition demo aims to showcase the associative memory capability with a user-defined pattern in real time. This demo enables users to create their own pattern and verify if the ONN can recognize it from a fuzzy representation. The demo is performed using the Jupyter platform, which runs a demo application from the ARM core of the PYNQ board. A 5x3 ONN is used for the demo with the Hebbian in PL (MAC) design. Figure 7 illustrates the complete demo setup with a PYNQ-Z2 board and the user interface. The user interface consists of 3 grid boxes namely the Learning, Input, and Output pattern grid boxes. In the Learning pattern grid box, a user can create the desired pattern. This pattern is comprised of black and white pixels. The Learn button sends the pattern to the ONN for learning. The Reset ONN button resets the weights in the ONN. The Input pattern grid box is used to generate the fuzzy pattern. The fuzzy pattern is comprised of black, white and grayscale pixels. The Send button sends the fuzzy pattern for inference. Finally, the inference result will be displayed in the Output pattern grid box.

V. Discussion

This chapter discusses the advantages and limitations of the different learning method designs of digital ONN on FPGA. All three designs show the same performance in terms of accuracy for 5x3 ONN. The test shows that the accuracy can be improved by limiting the number of learning patterns. Another way to improve the accuracy is to resort to a more advanced learning rule. The results show that Hebbian in PL and PL (MAC) yield a better resource utilization trend of synthesized LUTs than Hebbian in PS. This is due to the handling of weight matrix data in the ONN block of Hebbian in PS which consumes a lot of resources. The number of utilized resources can be minimized by optimizing the data parsing in the ONN block. Hebbian in PL (MAC) is slightly better than Hebbian in PL as it can offer more neurons i.e. larger ONN size. For the current implementation in PYNQ board, the maximum number of neurons Hebbian in PL (MAC) is 60 and Hebbian in PL is 30. For both designs, optimization on the Hebb and ONN block can help improve the number of resource utilization for example by adding more pipelines. Hebbian in PL and PL (MAC) are again better than Hebbian in PS in terms of overall execution time and maximum clock frequency. The bottleneck of the overall execution time of Hebbian PS lies on the Hebb block which is implemented in software. It can be concluded that the MAC unit significantly improves the average learning period in hardware. The average inference period, for Hebbian in PL and PL (MAC), can be improved by optimizing the ONN block to have more parallelism on the processing of the inference data. This test shows that implementing Hebbian learning in hardware gives more advantage in terms of execution time or speed. One of the limitations of the Hebb learning designs is the memory size which determines the maximum number of stored patterns. The memory size is determined by the number of encoding bits of the weight. Each weight is encoded in a 5-bit signed integer for Hebbian in PS and PL, and a 6-bit signed integer for Hebbian in PL (MAC). Since the weight is computed off the bipolar learning vector elements by the Hebbian learning rule, the memory size can be formulated as memory size = $\frac{2^n}{2}$ to calculate the memory size. Variable n represents the number of encoding bits. The term in the denominator is a result of using signed integer. The number of learning patterns is used as the unit for the memory size.

In this case, the Hebbian in PS and PL can only store a maximum of 16 patterns. While the Hebbian in PL (MAC) can store up to 32 patterns. The memory size grows exponentially with the number of encoding bits. Increasing the number of encoding bits will have an impact on the number of utilized resources. Increasing the number of stored patterns above the limit will result to a loss of information due to overflow on the signed integer.

VI. Conclusion

In summary, this paper provides three unique designs for on-chip Hebbian learning in a digital implementation on FPGA. They serve a purpose which is to make an on-chip learning possible. These designs make use of the PS and PL units in an FPGA. The first design incorporates the Hebbian learning to the PS. The second design integrates the Hebbian learning into a hardware block in PL. The last design also integrates the Hebbian learning into a hardware block but it provides more parallelism, as it uses a dedicated MAC unit for the weight matrix computation. Several design metrics are used to compare these designs. In terms of error rate, they show an identical performance. The error rate increases with the number of learning digits and Hamming distance. Hebbian in PL (MAC) utilizes the least LUTs amongst the three. The utilization of Hebbian in PL and PL (MAC) has a polynomial growth as the number of neurons increases. While the utilization of Hebbian in PL grows almost exponentially. It is due to the processing of the incoming weight matrix data in the ONN block. The maximum number of neurons that can be implemented on PYNQ-Z2 board for Hebbian in PL (MAC) is 60, while for Hebbian in PL and PS is 30. Thus, Hebbian in PL (MAC) is a better choice for an application that requires a large number of neurons. Hebbian in PL and PL (MAC) performs learning and inference faster than Hebbian in PS, resulting in a better choice for an application that requires low latency.

References

[1] J. Redmon et al. "You Only Look Once: Unified, Real-Time Object Detection". In: *2016 IEEE Conference on Computer Vision and Pattern Recognition (CVPR)*. 2016, pp. 779–788. DOI: 10.1109/CVPR.2016.91.

[2] L.D. Jackel et al. "A neural network approach to hand-print character recognition". In: *COMPCON Spring '91 Digest of Papers*. 1991, pp. 472–475. DOI: 10.1109/CMPCON.1991.128851.

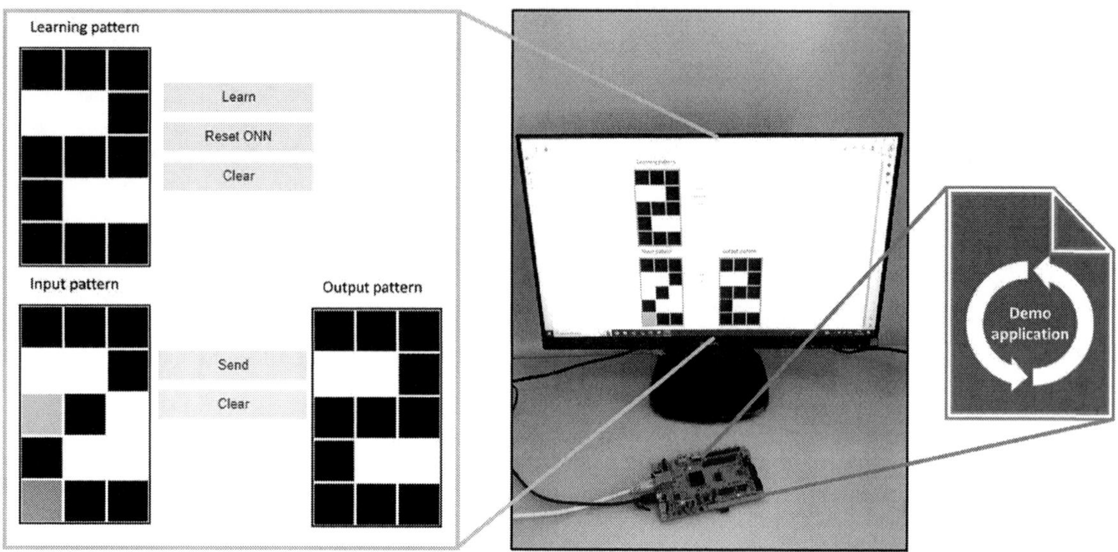

Fig. 7: Pattern recognition demo running on a PYNQ-Z2 board with a user interface.

[3] W. Xiong et al. "The microsoft 2016 conversational speech recognition system". In: *2017 IEEE ICASSP*. 2017, pp. 5255–5259. DOI: 10.1109/ICASSP.2017.7953159.

[4] A. Raychowdhury et al. "Computing with networks of oscillatory dynamical systems". In: *Proceedings of the IEEE* 107.1 (2019), pp. 73–89. DOI: 10.1109/jproc.2018.2878854.

[5] C. Delacour et al. "Energy-Performance Assessment of Oscillatory Neural Networks Based on VO $_2$ Devices for Future Edge AI Computing". In: *IEEE Transactions on Neural Networks and Learning Systems* (2023), pp. 1–14. DOI: 10.1109/TNNLS.2023.3238473.

[6] A. Todri-Sanial et al. "How Frequency Injection Locking Can Train Oscillatory Neural Networks to Compute in Phase". In: *IEEE Transactions on Neural Networks and Learning Systems* 33.5 (2022), pp. 1996–2009. DOI: 10.1109/TNNLS.2021.3107771.

[7] F.C. Hoppensteadt and E.M. Izhikevich. "Associative memory of weakly connected oscillators". In: *Proceedings of ICNN'97*. Vol. 2. Houston, TX, USA: IEEE, 1997, pp. 1135–1138. ISBN: 978-0-7803-4122-7. DOI: 10.1109/ICNN.1997.616190.

[8] R. Shi et al. "On the design of phase locked loop oscillatory neural networks: Mitigation of transmission delay effects". In: *2016 IJCNN*. 2016, pp. 2039–2046. DOI: 10.1109/IJCNN.2016.7727450.

[9] Th. Jackson, S. Pagliarini, and L. Pileggi. "An Oscillatory Neural Network with Programmable Resistive Synapses in 28 Nm CMOS". In: *2018 IEEE ICRC*. McLean, VA, USA, Nov. 2018, pp. 1–7. DOI: 10.1109/ICRC.2018.8638600.

[10] M. Abernot et al. "Digital implementation of oscillatory neural network for image recognition applications". In: *Frontiers in Neuroscience* 15 (2021). DOI: 10.3389/fnins.2021.713054.

[11] M. Abernot, Th. Gil, and A. Todri-Sanial. "On-Chip Learning with a 15-neuron Digital Oscillatory Neural Network Implemented on ZYNQ Processor". In: *Proceedings of the ICONS 2022*. ICONS '22. New York, NY, USA: Association for Computing Machinery, Sept. 2022, pp. 1–4. DOI: 10.1145/3546790.3546822. URL: https://doi.org/10.1145/3546790.3546822.

[12] R.G.M Morris. "D.O. Hebb: The Organization of Behavior, Wiley: New York; 1949". In: *Brain Research Bulletin* 50.5-6 (1999), p. 437. DOI: 10.1016/s0361-9230(99)00182-3.

[13] N. Shukla et al. "Ultra low power coupled oscillator arrays for computer vision applications". In: *2016 IEEE Symposium on VLSI Technology*. June 2016, pp. 1–2. DOI: 10.1109/VLSIT.2016.7573439.

[14] G. Csaba and W. Porod. "Coupled oscillators for computing: A review and perspective". en. In: *Applied Physics Reviews* 7.1 (Mar. 2020), p. 011302. DOI: 10.1063/1.5120412.

[15] C. Delacour et al. "Oscillatory Neural Networks for Edge AI Computing". In: July 2021, pp. 326–331. DOI: 10.1109/ISVLSI51109.2021.00066.

[16] C. Delacour and A. Todri-Sanial. "Mapping Hebbian Learning Rules to Coupling Resistances for Oscillatory Neural Networks". In: *Frontiers in Neuroscience* 15 (2021). ISSN: 1662-453X.

[17] J. J. Hopfield. "Neurons with graded response have collective computational properties like those of two-state neurons." en. In: *Proceedings of the National Academy of Sciences* 81.10 (May 1984), pp. 3088–3092. DOI: 10.1073/pnas.81.10.3088.

PAIRS: Pruning-AIded Row-Skipping for SDK-Based Convolutional Weight Mapping in Processing-In-Memory Architectures

Johnny Rhe*, Kang Eun Jeon*, and Jong Hwan Ko†
*Department of Electrical and Computer Engineering, Sungkyunkwan University, Suwon, South Korea
†College of Information and Communication Engineering, Sungkyunkwan University, Suwon, South Korea
{djwhsdj, kejeon, jhko}@skku.edu

***Abstract*—Processing-in-memory (PIM) architecture is becoming a promising candidate for convolutional neural network (CNN) inference. A recent weight mapping method called shift and duplicate kernel (SDK) improves the utilization by the deployment of shifting the same kernels into idle columns. However, this method inevitably generates idle cells with an irregular distribution, which limits reducing the size of the weight matrix. To effectively compress the weight matrix in the PIM array, prior works have introduced a row-wise pruning scheme, one of the structured weight pruning schemes, that aims to skip the operation on a row by zeroing out all weight in the specific row (we call it row-skipping). However, due to the deployment of shifting kernels, SDK mapping complicates zeroing out all the weight in the same row. To address this issue, we propose pruning-aided row-skipping (PAIRS) that effectively reduces the number of rows of convolutional weights that are mapped with SDK mapping. By pairing the SDK mapping-aware pruning pattern design and row-wise pruning, PAIRS achieves a higher row-skipping ratio. In comparison to pruning methods, PAIRS achieves up to 1.95× rows skipped and 4× higher compression rate with similar or even better inference accuracy.**

***Index Terms*—convolutional neural network, processing in memory, weight mapping and pattern-based pruning.**

I. Introduction

Convolutional neural networks (CNNs) are being widely applied in various mobile computer vision applications owing to the advancement of digital accelerators. However, CNN accelerators based on the conventional Von Neumann architecture require massive memory accesses to fetch or export data, which leads to huge latency and energy overheads [1]. Accordingly, the processing-in-memory (PIM) architecture that performs CNN computations on memory arrays has emerged as an efficient alternative to digital accelerators. As most of the PIM-based computing energy is consumed by the digital-to-analog (DA) and analog-to-digital (AD) conversions at every computing cycle [2], reducing the number of computing cycles is critical for energy-efficient CNN inference. As the number of computing cycles required can vary significantly depending on how the weights are mapped onto the PIM array, optimizing a convolutional weight mapping method is critical for efficient in-memory CNN inference. One of the basic mapping techniques is image to column (im2col) [3], which unrolls the 3D-shaped kernel into the column of the PIM array. It then performs matrix-vector multiplications (MVMs)

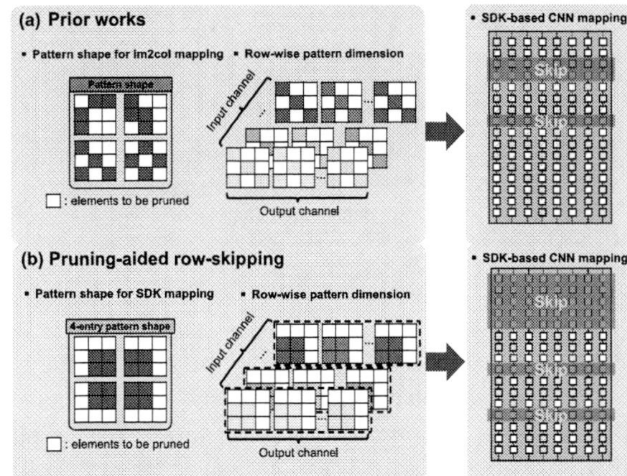

Fig. 1. Illustration shows skipping the rows using pattern-based pruning with (a) prior works; (b) proposed one in the PIM array.

with the unrolled input data, and the accumulated output can be obtained through each column. However, it utilizes only a small portion of the PIM array if the unrolled convolutinoal weight matrix is smaller than the given array size. To resolve this issue, Zhang et al. [4] proposed a novel weight mapping technique called shifted and duplicated kernel (SDK) mapping. By shifting the same kernels and deploying them into idle columns, it provides higher array utilization than im2col. However, kernel shifting inevitably generates idle cells with an irregular distribution, which limits reducing the size of the weight matrix mapped into the PIM array.

One of the powerful techniques for compressing the weight matrix is weight pruning [5], [6], which forces certain weight elements to be zeros. To effectively compress the weight matrix in the PIM array, many studies have proposed row-wise weight pruning methods [7]–[9]. By zeroing out all the weights in a specific row, row-wise pruning aims to skip operations on that row (we call it row-skipping). Please note that all of such methods are targeted for the im2col mapping method, where row-skipping with row-wise pruning is trivial as it maps the same index of weight element on the same row. However, SDK mapping deploys the same kernel on multiple columns by shifting them, zeroing out all the weights in the same row can be complicated. Therefore, directly applying prior im2col-based pruning methods on SDK mapping yields limited row-

979-8-3503-1176-1/23 $31.00 © 2023 IEEE

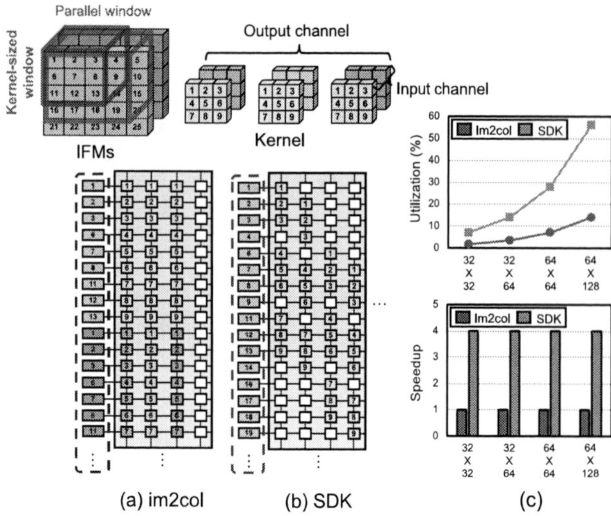

Fig. 2. Mapping methods for computing the convolutional weights in the PIM array: (a) im2col; (b) SDK. (c) shows the speedup and utilization of im2col and SDK mapping methods. The speedup is normalized to the number of computing cycles of im2col. The X-axis means input channel × output channel, where the kernel size is 3×3, and we assumed that there is a single PIM array which size is 512×512.

skipping performance (Fig. 1 (a)). Above all, to date, no work has proposed row-wise weight pruning for skipping the rows of the SDK-mapped convolutional weights.

In this paper, we propose pruning-aided row-skipping (PAIRS) that effectively reduces the number of rows of convolutional weights that are mapped with SDK mapping. To the best of our knowledge, this is the first work that pairs row-wise pruning and SDK mapping-aware pruning pattern design for a higher row-skipping ratio (Fig. 1 (b)). For a given CNN and a PIM array, PAIRS determines the optimal pruning pattern that maximizes the row-skipping ratio, depending on the mapping structure and the number of elements to be pruned. The simulation with various CNNs and sub-arrays shows that PAIRS achieves a significant reduction in the used rows as well as the compression rate. In comparison to prior pruning methods, PAIRS achieves up to 1.95× rows skipped and 4× higher compression rate with similar or even better inference accuracy.

II. Background and Motivation

A. Convolutional Weight Mapping on PIM arrays

In PIM-based CNN inference, sliding windows of the input feature maps (IFMs) are supplied to the PIM array rows, which are then multiplied with the kernel weights mapped to the array cells to generate a part of output feature maps (OFMs) summed at each column. To perform inference of large IFMs, the PIM array computes multiple cycles by loading different parts of windows in the IFMs. Every computing cycle requires digital-to-analog (DA) and analog-to-digital (AD) conversions, which are estimated to be 98% of total energy [2]. Thus, it is required to reduce the number of data conversions between DA/AD converters for minimizing inference energy and improving the latency. Consequently, several studies [3], [4] proposed PIM-based convolutional weight mapping methods to reduce the computing cycles. Based on [10], the computing cycle with a single PIM array can be expressed as follows:

$$C_t = C_r C_w, \quad C_w = \left\lceil \frac{W}{S} \right\rceil, \tag{1}$$

where S is the size of the sub-array, C_r is the read cycle, C_w is the write cycle when using a single array, W is the weight matrix, and $\lceil \cdot \rceil$ is the ceil function. The read cycle is a cycle to load input data through the PIM array rows. The write cycle is a cycle to map the weights of the convolutional layer into the single PIM array, and it is affected by the size of both the weight matrix and sub-array.

Im2col. Image to column (im2col) [3] is a conventional weight mapping scheme that maps each kernel set with multiple input channels (ICs) (which size is $K \times K \times IC$, where K is the kernel size) into a column of the PIM array and output channels (OCs) into other columns, as illustrated in Fig. 2(a). In a row, the identical location of weight elements is mapped into the columns. A kernel-sized window is then loaded into the PIM array rows to convolve with the weights, obtaining an output through PIM array columns at every computing cycle. However, this mapping generates idle memory cells depending on the size of the kernel mapped into the PIM array, which causes lower utilization and more computing cycles.

SDK. For higher array utilization, shift and duplicate kernel (SDK) [4] reuses both the input data and weights by shifting duplicated kernels into multiple columns, as Fig. 2(b) demonstrates. Unlike the im2col mapping method, other weight elements can be mapped in a row. A set of kernel-sized windows called a parallel window (PW) (which size is $PW \times PW \times IC$, where PW is the PW size) is supplied to the rows to convolve with duplicated kernels, generating multiple outputs at each computation cycle. As a result, this mapping method requires fewer computing cycles than the im2col one. Based on higher utilization, the computing speed of the SDK mapping method can be improved, as Fig. 2(c) shows. Therefore, the SDK mapping method is an advanced im2col mapping method not only to solve the challenge of low utilization but also to improve the inference speed.

B. Weight Pruning Methods

Weight pruning [5], [6] is a common strategy for reducing the computation memory requirements of deep neural networks by removing unnecessary connections. Most of the conventional pruning approaches are non-structured pruning, which irregularly removes unimportant weight elements [5]. Structured pruning [6] prunes weight matrices with regularity such as filters and channels, generally with the purpose of preserving hardware-friendly structural regularity. One of the prior studies [11] introduced pattern-based pruning that uses pre-defined kernel-wise patterns ($K \times K \times 1 \times 1$), and it focused on the remaining four un-pruned weight elements since a 4-entry kernel is extremely favorable to SIMD architectures.

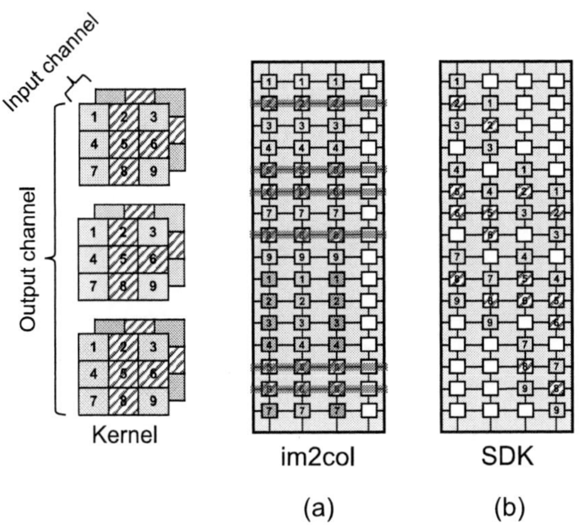

Fig. 3. Comparison of row-skipping with a row-wise pruning with the (a) im2col; (b) SDK mapping method in the 5-entry pattern, where the PW is 4×4, and the red line is a skipped row by the pattern.

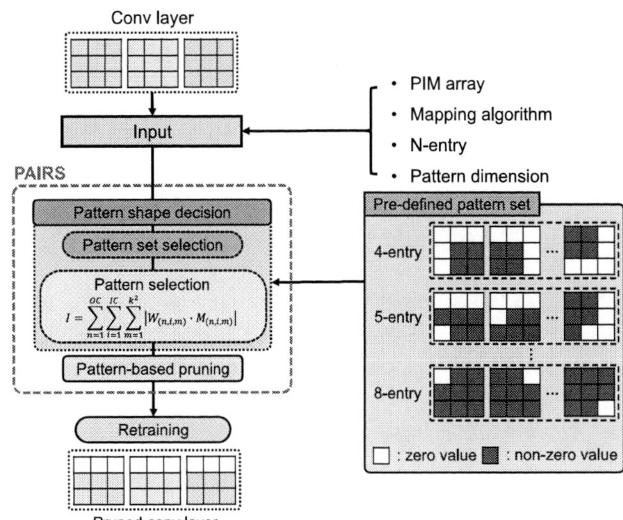

Fig. 4. Process of pruning-aided row-skipping.

Recently, several studies [7], [8], [12], [13] have proposed a structured weight pruning scheme for the PIM architecture. For compressing the weight matrix to map into the PIM array, the column-wise pruning strategy [8], [13] prunes OC to exploit structural column sparsity of the weight matrix. To skip the row, the row-wise pruning strategy [7]–[9] prunes the same index of weight element across total filters ($1 \times 1 \times 1 \times OC$). Based on the pattern-based pruning, Yu et al. [14] obtained the pattern from the pre-trained model calculating the importance with the largest-magnitude weights. Wang et al. [15] used the probability density function to select the candidate pattern with the largest probability from the irregularly pruned model. However, prior works proposed their pruning methods considering only the im2col mapping method. To extend weight pruning towards the SDK mapping, Rhe et al. [12] used column-wise weight pruning by pruning the residual channels that lead to low utilization. To date, however, no work has proposed the row-wise weight pruning scheme for the SDK mapping method.

C. Requirements of Row-Wise Pruning on SDK Mapping

A common property of CNNs is an abundance of zero-valued weights [16]. By exploiting the weight sparsity, we can skip the mapping of the row when all the cells in a specific row are assigned zero-valued weights. Since the im2col mapping method maps the same index of weight element into a row, row-skipping can be easily performed through row-wise weight pruning. As Fig. 3(a) shows, for row-wise pruning on im2col mapping, the number of skipped rows is determined by the number of pruned elements in the kernel. However, when using the SDK mapping method, the deployment of the shifted and duplicated kernel generates an irregular distribution of the weights, which complicates zeroing out weight elements in the same row (Fig. 2(b)). For instance, all the rows have at least one non-zero weight element in spite of pruning the same kernel elements as in im2col, as shown in Fig 3(b). Thus, in this case, none of the rows can be skipped when the SDK mapping is used, although four weight elements in the 3×3 kernel are pruned.

Based on this observation, we can infer that row skipping on SDK mapping requires different elements on each kernel to be pruned. Then, more important question here is which elements in different kernels should be pruned to maximize the row-skipping performance and minimize the accuracy loss. When a certain row is fully occupied, four elements among nine kernel weights have to be pruned to skip the row, resulting in the loss of accuracy. On the other hand, pruning one element can skip a row when the row is mapped with only one element (as in the top and bottom rows in Fig. 3(b)). In addition, as SDK mapping shifts the same kernel elements on different columns and rows, pruning a specific kernel element can affect the weight occupation (and row-skipping performance) on multiple rows (e.g., pruning the element '1' affects the weight occupation of both first and second rows). Therefore, designing the optimal pattern shape for the SDK mapping method is critical for achieving the best performance in terms of the row-efficient weight compression and the inference accuracy.

III. Pruning-Aided Row-Skipping

A. Basic Concept of Proposed Method

To reduce the number of used rows and computing cycles of SDK-based CNN inference on PIM arrays, we propose pruning-aided row-skipping (PAIRS) that pairs SDK mapping-aware pruning pattern design and row-wise pruning to facilitate row-skipping. As Fig. 4 shows, PAIRS prunes the weights of a given CNN using the proposed pruning pattern that maximizes the row-skipping and minimizes the accuracy loss under various conditions such as the PIM array size, the PW size of SDK mapping, the number of entries to be pruned, and row-wise pruning dimension. Then, among the pre-defined pattern set that maximizes the row-skipping for the N-entry, PAIRS selects the appropriate pattern shape by considering the

TABLE I
CHARACTERISTICS OF SELECTED PATTERNS FOR PAIRS

N-entry	#Patterns		#Rows skipped
	Total	Selected	
8	${}_9C_1 = 9$	4 (44.5%)	1
7	${}_9C_2 = 36$	14 (38.9%)	2
6	${}_9C_3 = 84$	4 (4.8%)	PW_h
5	${}_9C_4 = 126$	16 (12.7%)	$PW_h + 1$
4	${}_9C_5 = 126$	4 (3.2%)	$2PW_h - 1$

importance of each weight element. After pruning the CNN weights according to the determined pattern dimension and shape, PAIRS retrains the network to recover the accuracy.

B. SDK Mapping-Aware Pruning Pattern Design

The SDK mapping method overlaps the input data and shifts the duplicated kernels, as shown in Fig. 2(b) and Fig. 5(a). As the kernels are shifted, different weight elements can be mapped to different rows. As a result, pruning a certain weight element may not lead to row-skipping. Thus, when the SDK mapping method is used, the number of rows skipped is dependent on the pattern shape as well as the number of pruned entries, as illustrated in Fig. 6. The figure also shows that the SDK mapping method skips fewer rows than the im2col mapping method for certain patterns, and even fails to perform the row-skipping. This is because a row can be skipped only when all the different weight elements mapped to the same row are zeros. Also note that, for each entry, there are patterns (marked by a red star) that can skip most of the rows. Therefore, to maximize the row-skipping for the SDK mapping method, we need to use these patterns when pruning the kernel weights.

Fig. 5(b) shows the set of pattern shapes for each number of entries that maximize row-skipping assuming the 4×4 PW as in Fig. 5(a). It should be noted that we explain our proposed SDK mapping-ware pattern design in the 3×3 kernel, but the main process of finding a pattern shape in the SDK mapping method is equally valid regardless of the kernel and PW size. As the weight elements at the edge of the kernel ([a, c, g, i]) are used only once during the kernel sliding over the PW, one row can be skipped by making one of the elements zero. Therefore, one of the kernel elements [a, c, g, i] should be selected when making 8-entry patterns. In the 7-entry pattern, two rows can be skipped by pruning two of the kernel edge elements [a, c, g, i]. Furthermore, pruning one of the elements [a, c, g, i] and a neighboring element also leads to 2-row skipping since they can simultaneously remove the overlapped weight elements generated by the kernel sliding. In the 6-entry pattern, four rows that are the width or height of the PW can be skipped by pruning the kernel elements at each side of the kernel. In order to further prune the weight, the edge or adjacent element toward the height or width direction must be selected except for the center element ([e]).

Table I shows the ratio of the patterns suggested by PAIRS among all possible pattern combinations, and the corresponding number of rows skipped for each of the N-entry patterns.

TABLE II
INFORMATION ON NETWORKS AND PWS

PIM array	Layer	Image ($I \times I$)	Kernel ($K^2 \times IC \times OC$)	PW ($PW^2 \times IC \times OC$)
ResNet-20				
512×512	1	32×32	3×3×16×16	5×5×16×16
	2	16×16	3×3×32×32	4×4×32×32
	3	8×8	3×3×64×64	4×4×64×64
WRN16-4				
2048×2048	1	32×32	3×3×16×16	5×5×16×16
	2	16×16	3×3×32×32	4×4×32×32
	3	8×8	3×3×64×64	4×4×64×64

TABLE III
A SUMMARY OF THE TRAINING HYPER-PARAMETERS.

Hyper-parameter	ResNet-20	WRN16-4
Dataset	CIFAR-10 [17]	CIFAR-100 [17]
# of Training/re-training epochs	40/100	
Learning rate	0.01 $\sim$ 0.0001	
Scheduler cycle	{25, 50, 75}	
Batch size	512	
Weight decay	0.0005	
Optimizer	Adam	
Weight bit-precision.	2	

The pattern of PAIRS maximizes the performance of row-skipping when using the SDK mapping method. Note that we can skip more rows as we prune more elements (as N decrease), where the number of rows skipped is determined by the PW size when N is lower than 7. In 6- and 4-entry patterns, only under 5% patterns in total combinations can skip the maximum rows.

IV. EXPERIMENTAL RESULT

A. Experimental Settings

Softeware. To evaluate the performance of PAIRS, we used ResNet-20 [18] and WRN16-4 [19], and the PW sizes were generated by SDK algorithm [4] as in Table. II. The networks were trained by ADMM (Alternating Direction Methods of Multipliers) [20] to select a proper pattern, and the topology of the networks and hyperparameters are shown in Table. III. We adopted the quantization function used in [21] to train the weights using Pytorch, and we did not quantize the activation. During retraining, both networks were trained with a cosine annealing strategy for optimal accuracy. We conducted three trials of experiments with three random seeds to obtain the averaged results. Codes are available at the following address (https://github.com/djwhsdj/PAIRS).

Hardware. We assumed that the PIM array can represent the multi-bit weight precision using the multi-column structure [22]. It should be noted that the purpose of using the PIM array sizes in Table. II is to generate the PW according to the convolutional layer based on the SDK mapping algorithm [4], and we assumed various sizes of sub-array to compare the compression rate.

Pattern and its dimension. To compare the performance, we used the pattern of PatDNN [11] to compare the performance of row-skipping. Also, we implemented another counterpart called *Random*, which generates the pattern randomly. The purpose of adopting the pattern of PatDNN and Random is

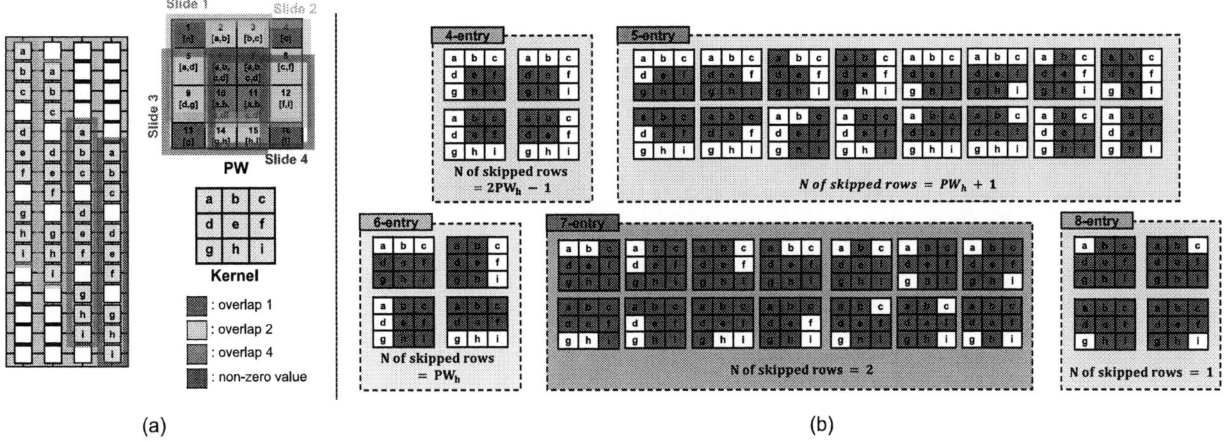

Fig. 5. The proposed patterns with various entries for the SDK mapping in the 3×3 kernel.

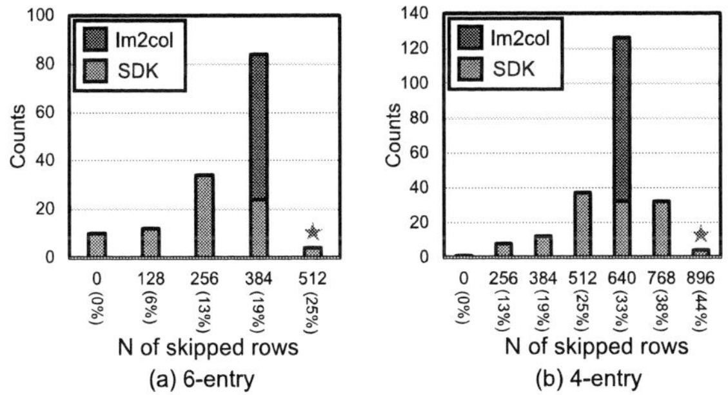

Fig. 6. The histogram of rows skipped with the row-wise dimension in total pattern combinations, where the PW size is 4×4, and the filter size is 3×3×128×128. The X-axis is the number of rows skipped and their percentage. The red star represents the patterns that can skip the maximum number of rows.

to generate the N-entry pattern shape. For a fair comparison in the PIM array, we applied the row-wise pruning [7]–[9] to the N-entry pattern shape. As a result, by pairing the pattern shape and row-wise pruning, we can compare the row-skipping in the PIM array between the prior and our SDK mapping-aware pruning pattern designs.

Pattern selection For selecting the proper row-wise pruning pattern to the kernel, we use the importance of the weights, which is calculated by using the magnitude [11] of the dot-product between the pattern and the weights as follows:

$$I = \sum_{n=1}^{OC} \sum_{i=1}^{IC} \sum_{m=1}^{k^2} |W_{(n,i,m)} \cdot M_{(n,i,m)}|, \tag{2}$$

where I is an importance; M is the mask that has a zero value.

B. Row-Skipping Performance and Accuracy Analysis

To evaluate the performance of our proposed pattern design, we compared the number of rows skipped, its percentage compared to No pruning, and model accuracy according to the N-entry, and the results are shown in Table IV. The proposed pattern design achieves the highest performance of row-skipping regardless of the networks. For example, in Layer 1 of WRN16-4 and the 4-entry, PAIRS can further skip the rows by 1.93× and 1.29× compared to Random and PatDNN, respectively. It can be observed that generally, PAIRS achieves the highest performance of row-skipping with similar accuracy to other patterns. Therefore, it can be concluded that PAIRS can compress the row-efficient weight matrix of the SDK mapping with a reasonable accuracy drop.

TABLE IV
A PERFORMANCE OF ROW-SKIPPING AND ACCURACY

N-entry	Pattern +Row-wise [7]	♯Skipped rows Layer1	Layer2	Layer3	Accuracy [%]
ResNet-20 on Cifar-10					
No pruning: 88.78					
8	Random	5 (1.3 %)	6 (1.2 %)	25 (2.4 %)	88.13 (-0.65)
	PatDNN [11]	6 (1.5 %)	10 (1.0 %)	13 (1.3 %)	88.17 (-0.61)
	PAIRS	**16 (4.0%)**	**32 (6.3 %)**	**64 (6.3 %)**	**88.33 (-0.45)**
7	Random	13 (3.3 %)	30 (5.9 %)	72 (7.0 %)	87.68 (-1.10)
	PatDNN [11]	17 (4.3 %)	51 (10.0 %)	77 (7.5 %)	87.54 (-1.24)
	PAIRS	**32 (8.0 %)**	**64 (12.5 %)**	**128 (12.5 %)**	**87.75 (-1.03)**
6	Random	32 (8.0 %)	74 (14.5 %)	120 (11.7 %)	**86.91 (-1.50)**
	PatDNN [11]	54 (13.5 %)	81 (15.8 %)	166 (16.2 %)	87.28 (-1.87)
	PAIRS	**80 (20.0 %)**	**128 (25.0 %)**	**256 (25.0 %)**	87.01 (-1.77)
5	Random	46 (11.5 %)	104 (20.3 %)	179 (17.5 %)	**87.12 (-1.68)**
	PatDNN [11]	83 (20.1 %)	143 (27.9 %)	259 (25.3 %)	86.66 (-2.12)
	PAIRS	**96 (24.0 %)**	**160 (31.3 %)**	**320 (31.3 %)**	86.72 (-2.06)
4	Random	83 (20.1 %)	166 (32.4 %)	328 (32.0 %)	**86.79 (-1.96)**
	PatDNN [11]	112 (28.0 %)	192 (37.5 %)	384 (37.5 %)	86.82 (-1.99)
	PAIRS	**144 (36.0 %)**	**224 (43.8 %)**	**448 (43.8 %)**	86.17 (-2.61)
WRN16-4 on Cifar-100					
No pruning: 59.51					
8	Random	14 (0.8 %)	43 (2.1 %)	114 (2.8 %)	**59.93 (+0.42)**
	PatDNN [11]	28 (1.8 %)	0 (0.0 %)	85 (2.1 %)	58.88 (-0.63)
	PAIRS	**64 (4.0 %)**	**128 (6.3 %)**	**256 (6.3 %)**	59.05 (-0.42)
7	Random	71 (4.4 %)	128 (6.3 %)	313 (7.6 %)	58.22 (-1.29)
	PatDNN [11]	114 (7.1 %)	142 (6.9 %)	341 (8.3 %)	58.40 (-1.11)
	PAIRS	**128 (8.0 %)**	**256 (12.5 %)**	**512 (12.5 %)**	**58.65 (-0.84)**
6	Random	114 (7.1 %)	356 (17.4 %)	514 (12.5 %)	57.49 (-2.02)
	PatDNN [11]	114 (7.1 %)	284 (13.9 %)	599 (14.6 %)	**58.94 (-0.57)**
	PAIRS	**320 (20.0 %)**	**512 (25.0 %)**	**1024 (25.0 %)**	58.67 (-0.84)
5	Random	220 (18.7 %)	313 (15.3 %)	768 (18.8 %)	57.25 (-2.26)
	PatDNN [11]	192 (12.0 %)	483 (23.6 %)	882 (21.5 %)	58.07 (-1.44)
	PAIRS	**384 (24.0 %)**	**640 (31.3 %)**	**1280 (31.3 %)**	**58.66 (-0.85)**
4	Random	299 (18.7 %)	583 (28.5 %)	1109 (27.1 %)	55.35 (-4.16)
	PatDNN [11]	448 (28.0 %)	768 (37.5 %)	1536 (37.5 %)	**57.15 (-2.36)**
	PAIRS	**576 (36.0 %)**	**896 (43.8 %)**	**1792 (43.8 %)**	56.88 (-2.63)

Based on Table IV, we compared the number of total rows skipped when the 1-entry weight element is pruned, as shown in Fig. 7. With the 6-entry pattern, PAIRS further skips the rows 1.95× and 1.64× compared to Random and PatDNN in ResNet-20 and 1.89× and 1.86× in WRN16-4, respectively. As N-entry decreases in the given kernel size, the pattern shape is similar to our proposed pattern design, and thus, the performance of others is close to that of PAIRS.

To evaluate the row-efficient weight compression, we com-

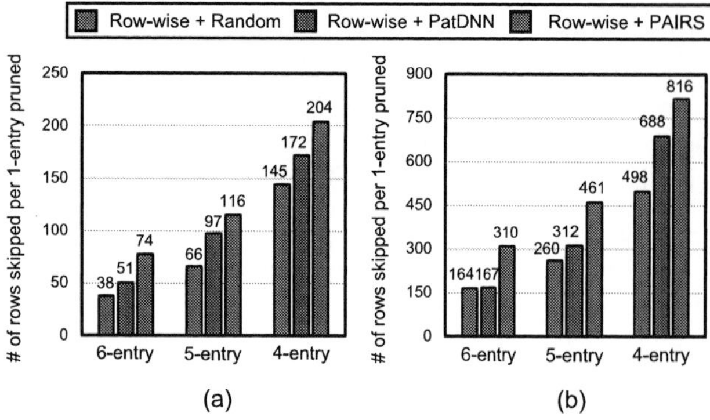

Fig. 7. Comparison of the number of total rows skipped across the convolutional layers per 1-entry pruned in (a) ResNet-20; (b) WRN16-4.

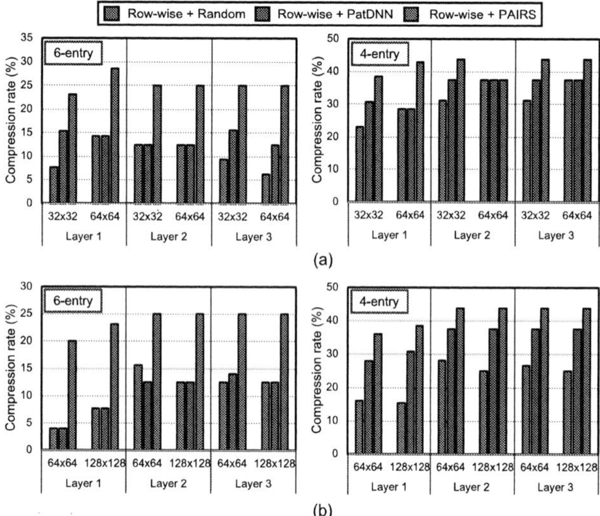

Fig. 8. Comparison of the compression rate with different sub-array sizes in (a) ResNet-20; (b) WRN16-4.

pared a metric called a compression rate that is used in the prior work [13], and the results are shown in Fig. 8. The purpose of this metric is to compare the number of required sub-arrays for mapping the shrunken weight matrix into the PIM array. Generally, based on the maximum row-skipping, PAIRS achieves a higher compression rate than others regardless of the networks. However, with other layers and sub-array sizes, PAIRS shows better performance than others. Furthermore, in Layer 1 of WRN16-4 and the 6-entry, the compression rate gap between PAIRS and others is about $4\times$ and $3\times$ when the 64×64 and 128×128 sub-arrays are used, respectively.

V. Conclusion

This paper specifies the problem of row-wise pattern pruning when the SDK-based mapping method is used, which blocks row-skipping to compress the weight matrix due to the presence of non-zero values in a row. To address this limitation, we proposed PAIRS that performs row-skipping by combining the SDK mapping-aware pruning pattern design and row-wise weight pruning. Based on PAIRS, we can compress the SDK-based weight matrix for an efficient PIM-based CNN inference. In comparison to other patterns, PAIRS achieves up to $1.95\times$ row-skipping ratio and $4\times$ higher compression rate with similar or even better inference accuracy.

Acknowledgment

This work was partly supported by the National Research Foundation (NRF) grants (RS-2023-00251438, 2022R1F1A1074142, 2022R1A4A3032913) and Institute of Information and Communication Technology Planning & Evaluation (IITP) grants (IITP-2019-0-00421, IITP-2020-0-00821, IITP-2021-0-02052, IITP-2021-0-02068), and Samsung Electronics Co., Ltd (IO230404-05747-01).

References

[1] Z. Guo *et al.*, "Spintronics for energy-efficient computing: An overview and outlook," *Proceedings of the IEEE*, 2021.
[2] L. Xia *et al.*, "Switched by input: Power efficient structure for RRAM-based convolutional neural network," in *DAC*, 2016.
[3] K. Yanai *et al.*, "Efficient mobile implementation of a CNN-based object recognition system," in *ACM MM*, 2016.
[4] Y. Zhang *et al.*, "Efficient and robust RRAM-based convolutional weight mapping with shifted and duplicated kernel," *TCAD*, 2020.
[5] S. Han *et al.*, "Learning both weights and connections for efficient neural network," *NIPS*, 2015.
[6] Y. He *et al.*, "Channel pruning for accelerating very deep neural networks," in *ICCV*, 2017.
[7] B. K. Joardar *et al.*, "ReaLPrune: ReRAM crossbar-aware lottery ticket pruned cnns," *arXiv preprint arXiv:2111.09272*, 2021.
[8] C. Chu *et al.*, "PIM-Prune: Fine-grain DCNN pruning for crossbar-based process-in-memory architecture," in *DAC*, 2020.
[9] G. Yuan *et al.*, "An ultra-efficient memristor-based DNN framework with structured weight pruning and quantization using admm," in *ISLPED*, 2019.
[10] L. Song *et al.*, "Pipelayer: A pipelined ReRAM-based accelerator for deep learning," in *HPCA*, 2017.
[11] W. Niu *et al.*, "PatDNN: Achieving real-time DNN execution on mobile devices with pattern-based weight pruning," in *Proceedings of the Twenty-Fifth International Conference on Architectural Support for Programming Languages and Operating Systems*, 2020.
[12] J. Rhe *et al.*, "VWC-SDK: Convolutional weight mapping using shifted and duplicated kernel with variable windows and channels," *JETCAS*, 2022.
[13] S. Yang *et al.*, "AUTO-PRUNE: automated DNN pruning and mapping for ReRAM-based accelerator," in *Proceedings of the ACM International Conference on Supercomputing*, 2021.
[14] S. Yu, L. Zhang, J. Wang, J. Yue, Z. Yuan, X. Li, H. Yang, and Y. Liu, "High area/energy efficiency RRAM CNN accelerator with pattern-pruning-based weight mapping scheme," in *NVMSA*, 2021.
[15] J. Wang, S. Yu, J. Yue, Z. Yuan, Z. Yuan, H. Yang, X. Li, and Y. Liu, "High pe utilization CNN accelerator with channel fusion supporting pattern-compressed sparse neural networks," in *DAC*, 2020.
[16] S. Han *et al.*, "Deep compression: Compressing deep neural networks with pruning, trained quantization and huffman coding," *arXiv*, 2015.
[17] A. Krizhevsky *et al.*, "Learning multiple layers of features from tiny images," 2009.
[18] K. He *et al.*, "Deep residual learning for image recognition," in *CVPR*, 2016.
[19] S. Zagoruyko *et al.*, "Wide residual networks," *arXiv preprint arXiv:1605.07146*, 2016.
[20] S. Boyd *et al.*, "Distributed optimization and statistical learning via the alternating direction method of multipliers," *Foundations and Trends® in Machine learning*, 2011.
[21] H. Yu *et al.*, "Any-precision deep neural networks," in *AAAI*, 2021.
[22] E. Lee *et al.*, "A charge-domain scalable-weight in-memory computing macro with dual-SRAM architecture for precision-scalable DNN accelerators," *TCAS-I*, 2021.

A Fully-Integrated Energy-Scalable Transformer Accelerator Supporting Adaptive Model Configuration and Word Elimination for Language Understanding on Edge Devices

Zexi Ji*
MIT
Cambridge, MA, USA

Hanrui Wang*
MIT
Cambridge, MA, USA

Miaorong Wang
MIT
Cambridge, MA, USA

Win-San Khwa
TSMC Corporate Research
Hsinchu, Taiwan

Meng-Fan Chang
TSMC Corporate Research
Hsinchu, Taiwan

Song Han
MIT
Cambridge, MA, USA

Anantha P. Chandrakasan
MIT
Cambridge, MA, USA

**Equally contributing authors*

***Abstract*—Efficient natural language processing on the edge is needed to interpret voice commands, which have become a standard way to interact with devices around us. Due to the tight power and compute constraints of edge devices, it is important to adapt the computation to the hardware conditions. We present a Transformer accelerator with a variable-depth adder tree to support different model dimensions, a SuperTransformer model from which SubTransformers of various sizes can be sampled enabling adaptive model configuration, and a dedicated word elimination unit to prune redundant tokens. We achieve up to 6.9× scalability in network latency and energy between the largest and smallest SubTransformers, under the same operating conditions. Word elimination can reduce network energy by 16%, with a 14.5% drop in F1 score. At 0.68V and 80MHz, processing a 32-length input with our custom 2-layer Transformer model for intent detection and slot filling takes 0.61ms and 1.6μJ.**

***Index Terms*—hardware accelerators, machine learning, natural language processing, transformers**

I. Introduction

There are close to 15 billion connected devices in the internet of things (IoT) today and many of them require efficient natural language processing (NLP) algorithms to interpret voice commands. Attention-based Transformer models have replaced recurrent neural networks (RNNs) as the predominant model for NLP applications due to parallel input processing and the attention mechanism being able to capture both short and long-range relations, resulting in faster training and increased accuracy on a variety of NLP tasks. However, existing mainstream models (e.g. BERT, GPT) are way too large for edge devices, with the largest model (Switch Transformer) containing over a trillion parameters (Fig. 1). For simple NLP tasks on the edge (e.g. smart watches, home assistants), tiny custom Transformer models can achieve good accuracy, while being much more suitable for constrained hardware [1].

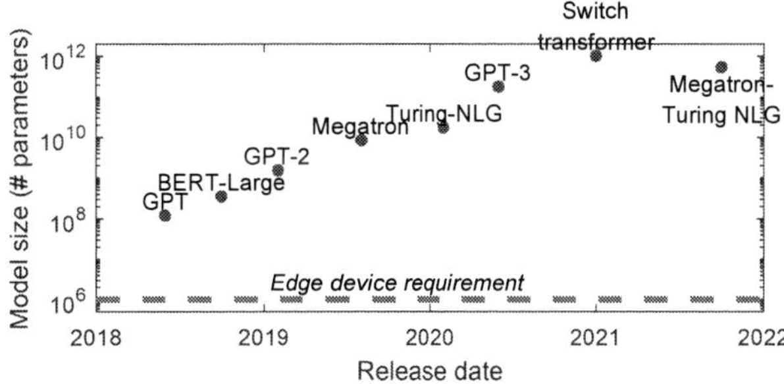

Fig. 1. Growth in model size of recent popular language models.

There are two main challenges when deploying lightweight NLP models on edge devices (Fig. 2). Firstly, hardware constraints can fluctuate based on battery level, latency requirements, availability of compute resources, and accuracy tolerance. Effectively adapting to these varying conditions typically requires multiple models of different sizes. For instance, when the device is less constrained (i.e. high battery level, relaxed latency demands etc.), we may use a large model. On the other hand, under more constrained conditions, we may opt for a small model to reduce latency and energy consumption. But storing multiple models incurs a significant memory overhead, potentially exceeding the on-chip memory capacity and requiring us to load different models from external memory. This additional data transfer would diminish or even negate the energy and latency benefits from using multiple models. Secondly, sentences usually contain redundant words that contribute little to the overall understanding and may potentially be skipped during the majority of the processing. Conventional models spend an equal amount of time processing each word, leading to unnecessary computation [2].

Our paper addresses these challenges with an energy-

979-8-3503-1176-1/23 $31.00 © 2023 IEEE

scalable Transformer accelerator targeting small IoT devices with three key features: 1) a variable-depth MAC adder tree to support different model dimensions; 2) adaptive model configuration using a custom SuperTransformer model to generate models of various sizes, while only taking up the memory footprint of a single full model; and 3) a comparator-based word elimination unit to progressively remove unimportant words from the sentence, reducing computation.

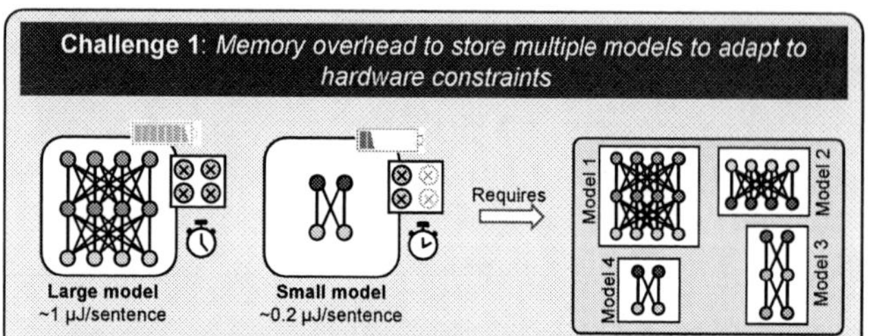

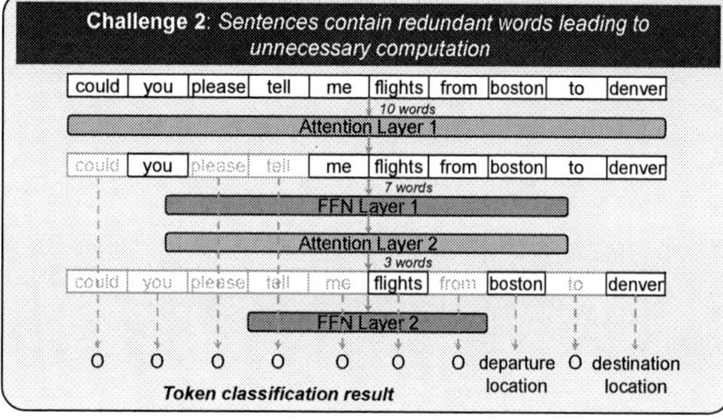

Fig. 2. Challenges of deploying lightweight models on edge devices.

II. Background

A. Transformer models

Transformer models are used to process sequence data (e.g. sentences) with a basic structure shown in Fig. 3. Inputs are first passed through an embedding layer, where each input is mapped to an embedding vector of length d_{emb}. A positional encoding vector is then added to each embedding vector to represent the location of each input. This is distinct from RNNs where the positional information is inherent in the sequential processing, whereas Transformers process all inputs in parallel.

The position-encoded inputs are passed through a series of encoder layers. Each layer consists of a multi-head attention mechanism followed by a feed-forward network (FFN, i.e. fully-connected layer). In the attention mechanism, each input is split along the embedding dimension into several heads and projected to a query, key, and value vector of dimension d_{QKV}/N_{head}. A similarity score (typically dot product) is computed between every pair of query and key vectors and normalized by softmax to generate the attention probability matrix, which gives a measure of the relevance of each pair of inputs. This matrix is multiplied by the value matrix to obtain a new representation of the inputs. Results from the different heads are combined through a fully-connected layer. This is followed by two FFNs, first mapping to dimension d_{FFN}, then back to d_{emb} to act as input to the next layer. The final encoder layer can be connected to either a classification head or a language model head, depending on the nature of the task.

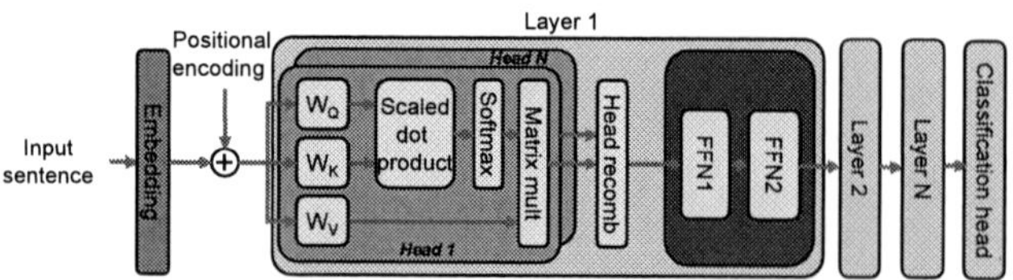

Fig. 3. Basic structure of transformer model.

B. Intent detection and slot filling task

We focus on the intent detection and slot filling task, which is representative of the complexity required for processing voice commands. This is a classification task in which the overall intent of a sentence is determined (i.e. the command) and relevant fields for the intent are identified by labeling each word with an appropriate slot. Fig. 4 shows an example with the desired intent being requesting flight information. The relevant words are "Boston" as the departure location and "Orlando" as the destination location. The remaining words are labeled as O for "outside." The intent is obtained by appending a <*CLS*> token (for "classification") to the beginning of the sentence, which is later passed through an intent head after the final encoder layer (while all the other words get processed by a slot head).

The ATIS (Airline Travel Information Systems) and SNIPS datasets are common benchmarks [3] [4]. ATIS consists of air travel related sentences with 26 different intent classes and 129 slots. SNIPS features more everyday tasks such as asking for the weather and playing music. It contains 10 different intents and 39 slots.

For evaluation on this task, we look at the intent accuracy and slot F1 score. The F1 score is the harmonic mean of the precision and recall, with the positive class being associated with non-O labels. This is preferred over token-wise slot accuracy due to the high proportion of O labels. Simply

	Intent	*Slots*									
Input sentence	<CLS>	i	need	a	flight	that	goes	from	boston	to	orlando
Ground Truth	flight info	O	O	O	O	O	O	O	departure location	O	destination location
Prediction	flight info	O	O	O	departure location	O	O	O	departure time	O	destination location
Predicted Non-O	-	✗	✗	✗	✓	✗	✗	✗	✓	✗	✓
Correctly predicted Non-O	-	✗	✗	✗	✗	✗	✗	✗	✗	✗	✓

$$\text{Precision} = \frac{\#(\text{correctly predicted non-O tokens})}{\#(\text{predicted non-O tokens})} = \frac{1}{3}$$

$$\text{Recall} = \frac{\#(\text{correctly predicted non-O tokens})}{\#(\text{ground truth non-O tokens})} = \frac{1}{2}$$

$$\text{F1} = \frac{2(\text{Precision} \times \text{Recall})}{\text{Precision} + \text{Recall}} = \frac{2}{5}$$

$$\text{Accuracy} = \frac{\#(\text{correctly predicted tokens})}{\#(\text{all tokens})} = \frac{8}{10}$$

Fig. 4. Example of intent detection and slot filling task with calculation of evaluation metrics.

labeling all tokens as O would result in a high accuracy, making it a misleading measure of model performance.

C. Existing work

Early works targeting Transformers were mainly architectures focused on accelerating the attention mechanism [5]. More recently, several accelerators for Transformers have emerged, including both digital and compute-in-memory implementations. They take advantage of techniques such as approximate computing [6], sparsity [7] [8], and varying the amount of computation based on input complexity [9]. All these works focus on larger networks (e.g. BERT) and achieving high throughput, but may not be optimized for running smaller networks that are more appropriate for edge applications.

III. Design Features

Our design combines algorithmic and circuit innovations to create a low-power scalable solution for performing NLP inference on the edge. It can efficiently support different model configurations with a variable-depth adder tree and uses a SuperTransformer model from which submodels of varying sizes can be generated through weight sampling. This enables adapting the model based on changing hardware conditions without the overhead of storing multiple distinct models. A comparator-based word elimination unit with programmable threshold provides another way to trade off energy and model performance.

Fig. 5 shows the system architecture of the proposed design. A 16-bit external memory interface facilitates transfer of model weights, loading the input data, and outputting the final results. The entire model and inputs (up to 32 words) are stored on-chip (in the W/B/I-Mem) so external data accesses are not required during operation. The remaining memory is divided into groups dedicated to storing certain intermediate values. Each memory bank is connected to its own power switch, allowing it to be dynamically connected to a lower retention voltage when inactive to minimize leakage. Data is transferred between the memory and processing core through three read channels and one write channel. Each read channel has access to a unique set of memory groups to reduce the amount of multiplexing required. The processing core consists of a reconfigurable 64-way MAC for matrix multiplication, 16-way L1-norm for computing attention score, word elimination unit, softmax for generating attention probabilities, and layer normalization unit. The model configuration is supplied through a serial interface.

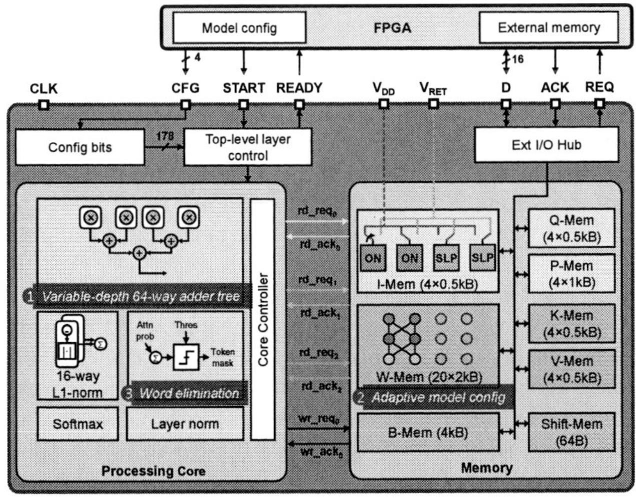

Fig. 5. System architecture of our proposed transformer accelerator.

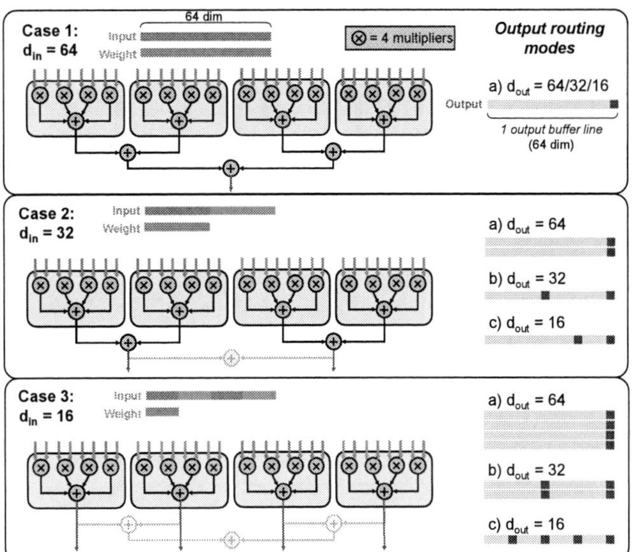

Fig. 6. Variable-depth adder tree and output routing modes to support different model dimensions.

A. Variable Depth Adder Tree

We design a variable-depth 64-way adder tree with seven output routing modes to easily support different model dimensions in the hardware (Fig. 6). The input dimension (d_{in}) determines the configuration of the adder tree and weight read pattern, while the output dimension (d_{out}) determines the spacing between consecutive words during writeback. Each dimension can be one of {16, 32, 64}. We use an output stationary dataflow, computing each output in a single cycle. The outputs of the 64 multipliers are combined through a 3-stage adder tree: the first stage adds 16 products together, while the remaining stages add two values at a time. Outputs can be taken at any point in the adder tree to obtain the desired sum, which are then routed to the output buffer. In most cases, only one output buffer line needs to be activated, but up to four lines may be required depending on the configuration. Fig. 7 shows the latency breakdown for three models where all dimensions are equal. The flexible compute unit maintains the throughput across different configurations, as shown by the latency scaling with $d_{in} \times d_{out}$ for most layers. The compute units can also be configured as 64 independent MAC units for softmax, layer normalization, and transposed matrix multiplication.

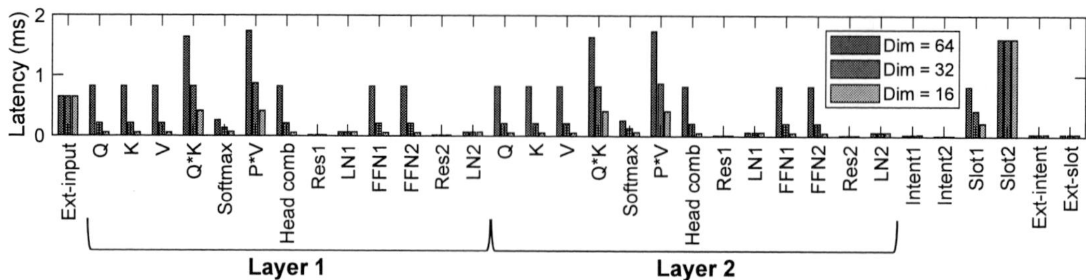

Fig. 7. Latency breakdown with $d_{emb} = d_{QKV} = d_{FFN}$.

B. SuperTransformer Model

The reconfigurable MAC unit is used in conjunction with a Super-Sub-Transformer framework, in which models of different sizes share parameters [10]. This enables energy scalability by varying the model size, but with the storage requirement of only a single full-sized model (Fig. 8). The SuperTransformer is the largest model in the design space and a SubTransformer is a submodel that shares all its parameters with the SuperTransformer. The SuperTransformer has an elastic layer number (n_L), embedding dimension (d_{emb}), QKV dimension (d_{QKV}), and feed-forward network hidden dimension (d_{FFN}).

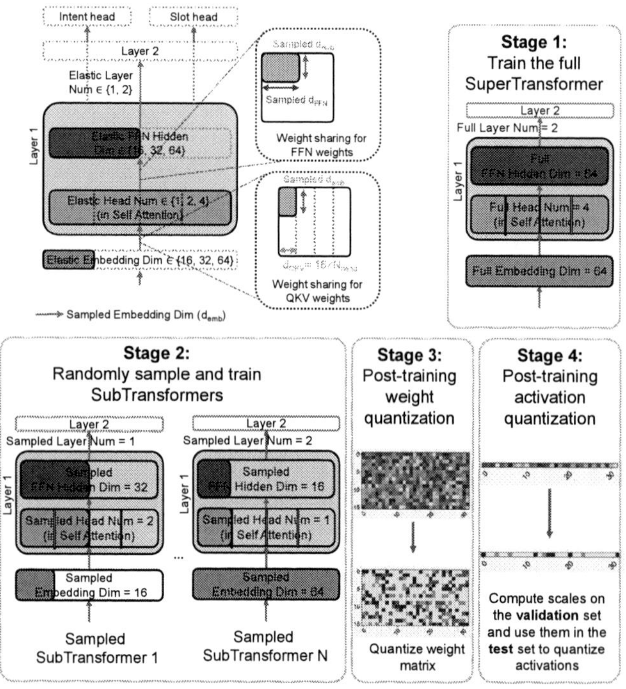

Fig. 8. SuperTransformer architecture and training pipeline.

To ensure each SubTransformer can perform the task properly, we uniformly sample a SubTransformer during each training step and update its parameters. After enough steps, all SubTransformers can be individually used. Our custom SuperTransformer model has 2 layers with d_{emb}, d_{QKV}, and d_{FFN} all equal to 64. For each SubTransformer, n_L can be 1 or 2, and each dimension can be chosen from 16, 32, 64. This gives us 54 (2×3^3) SubTransformers with the footprint of one full model. Practically, not all submodels may be used as several may be of similar size, with varying performance. Software evaluation can determine which submodel has the best accuracy for a given model size. Though we train a small model for our design, this framework can be applied to larger models as well, so long as the number of training cycles is large enough such that each submodel is sampled a sufficient number of times during training. We quantize weights to 4b and activations to 8b. L1-norm is used instead of dot-product attention to reduce multiplications, without accuracy loss.

The SuperTransformer model is stored entirely on-chip in the weight memory, which comprises the majority of the on-chip SRAM. Due to the output stationary dataflow, new weights are read every cycle. Thus, the weight read access energy is a significant fraction of the total energy consumed by the memory. In order to minimize this energy, the weight memory is divided into an array of 5×4 memory banks (5 rows, 4 columns). Only one row of memory banks is active at a time, and the number of columns that need to be active is determined by the sublayer d_{in}.

C. Word Elimination Unit

Besides adjusting model size, we can adjust how many words to process in each sentence. Motivated by the high redundancy in human language, we include word elimination to remove unimportant words. The elimination threshold is tunable, offering another way to trade off accuracy and energy. It is performed after computing attention probabilities in the softmax layer (Fig. 9). Each row of the attention matrix provides a normalized measure of the impact of a particular word to every other word in the sentence. The column-wise sum provides a proxy for the overall importance of a word. The larger the sum, the larger the overall impact. A threshold can be applied on the sum to generate a word mask indicating which words will be eliminated. Compared to other implementations which employ a top-k engine [2] [8], thresholding is more hardware-friendly, but provides less control over the elimination ratio. The word elimination unit consists of 32 parallel accumulators and comparators. During the computation of the attention output ($P*V$), multiplications

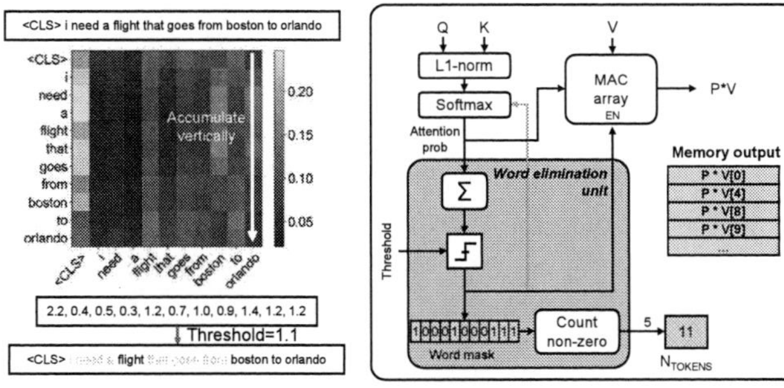

Fig. 9. Detailed operation of word elimination unit.

and writeback are skipped for words for which the word mask is 0. The resulting output memory has no gaps associated with removed words, allowing subsequent computations to be performed without regard to the word mask. Only the register storing the sentence length needs to be updated.

IV. Measurement Results

The design was taped out in TSMC 28nm HPC+ technology and occupies 0.9mm×0.9mm. The die micrograph and test setup are shown in Fig. 10. An FPGA was used to interface between the chip and the PC and the supply voltage and current were measured by Keithley sourcemeters, from which the power consumption was computed. The chip can operate up to 80MHz at 0.68V, consuming 2.62mW. Under these conditions, processing a 32-length input on the full model takes 0.61ms and 1.6μJ (or 20.5 inferences/s/MHz). For our target application, it was not necessary to have a latency much less than 1 ms. Because of this, we set a moderate target maximum frequency during the design phase to ease the effort during synthesis and place-and-route, rather than pushing to the maximum achievable frequency of the technology node. The peak efficiency point occurs at 0.56V and 20MHz, where the energy efficiency reaches 3.8 TOPS/W. A combination of

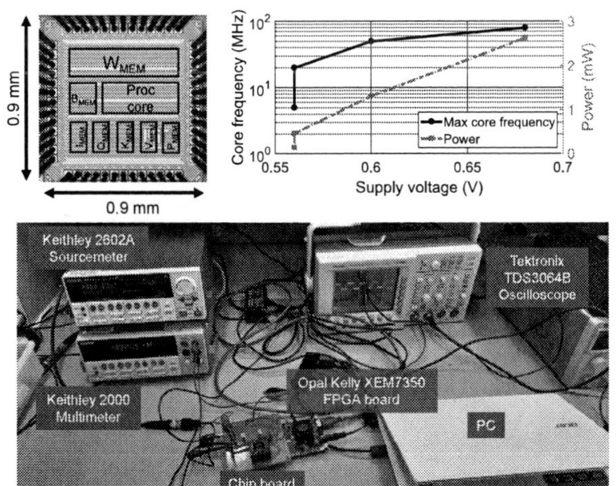

Fig. 10. Die micrograph, voltage-frequency scaling characteristics, and test setup.

SRAM sleep mode and power switches to connect inactive memory banks to a lower retention voltage reduces SRAM leakage by 11×, translating to a 56% reduction in the overall leakage.

Fig. 11 shows the tradeoff between model accuracy and network energy for the ATIS and SNIPS datasets. The accuracy is most sensitive to changes in n_L and d_{emb}, rather than d_{QKV} and d_{FFN}. In general, a two layer SubTransformer performs better than a single layer SubTransformer of similar size. The F1 score remains fairly high, even for submodels less than 1/4 the size of the full model (down to about $(n_L, d_{emb}, d_{QKV}, d_{FFN}) = (2, 32, 16, 16)$). This may be a result of some transfer learning occurring during the training process between the SuperTransformer and SubTransformers from the model sampling. Intent accuracy remains high across configurations. The difference in network energy between the largest and smallest SubTransformers is 5.8× and 6.9× for ATIS and SNIPS, respectively, demonstrating a wide range for scalability.

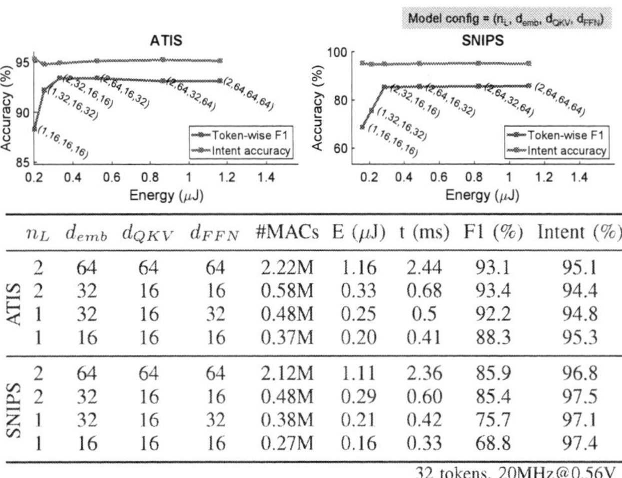

	n_L	d_{emb}	d_{QKV}	d_{FFN}	#MACs	E (μJ)	t (ms)	F1 (%)	Intent (%)
ATIS	2	64	64	64	2.22M	1.16	2.44	93.1	95.1
	2	32	16	16	0.58M	0.33	0.68	93.4	94.4
	1	32	16	32	0.48M	0.25	0.5	92.2	94.8
	1	16	16	16	0.37M	0.20	0.41	88.3	95.3
SNIPS	2	64	64	64	2.12M	1.11	2.36	85.9	96.8
	2	32	16	16	0.48M	0.29	0.60	85.4	97.5
	1	32	16	32	0.38M	0.21	0.42	75.7	97.1
	1	16	16	16	0.27M	0.16	0.33	68.8	97.4

32 tokens, 20MHz@0.56V

Fig. 11. Model accuracy for various configurations on the SNIPS and ATIS datasets.

Fig. 12 shows the effect of the overall elimination ratio on the accuracy for the full model on the ATIS dataset. When the overall elimination ratio is increased to 0.2, the F1 score drops by 14.5%, while the energy and latency are reduced by 16%. The accuracy degradation is mainly due to the relatively short length of test inputs (average 16 words), providing fewer opportunities for word elimination. For tasks with longer inputs, less accuracy loss would be expected at higher elimination ratios, due to increased redundancy.

Compared to other transformer accelerators in Table I, ours is the only one that targets lightweight networks that can fit entirely on-chip. This allows it to achieve much lower power and smaller area, making it more suitable for edge devices with limited compute and energy resources. Our peak energy efficiency is generally lower because we do not take as much advantage of sparsity. The performance of BERT-base on a scaled up (12×) version of our design is estimated based on

TABLE I
COMPARISON WITH OTHER WORKS.

	This work	**[6]**	**[7]**	**[8]**	**[9]**
Target application	Transformer	Transformer	Transformer	Multimodal Transformer	Transformer
Implementation	Digital	Digital	Digital CIM	Digital CIM	Digital
Technology (nm)	28	28	28	28	12
Max frequency (MHz)	80	510	240	275	717
SRAM (kB)	56	336	192	128 (CIM)/192 (buffer)	647
Area (mm^2)	0.81	6.82	6.83	6.83	4.6
Power (mW)	0.14–2.6	12–272	27–118	30–153	9–122
Energy efficiency (TOPS/W)	2.6–3.8	1.91–27.56	3.1–20.5	12.1–101.1	3.0–18.1
Word elimination support	Yes	No	No	Yes	No
Model	Custom (ATIS)	GPT-2	BERT (base)	ViLBERT	BERT (base)
Model on-chip	Yes	No	No	No	No
Network latency (ms)	0.61–9.76 (247[1])	-	1684	268–925	682
Network energy (μJ/token)	0.036–0.05 (43.8[1])	-	15.6	2.24–7.72	4000

[1]Estimated performance of BERT-base on a scaled up (12×) version of our design operating at 0.56V and 20 MHz

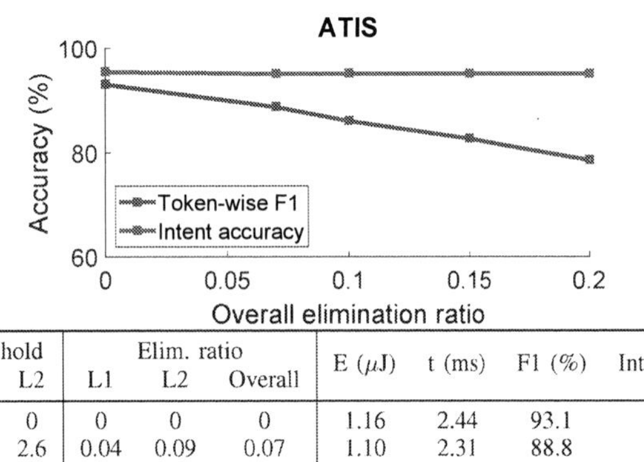

Threshold L1	Threshold L2	Elim. ratio L1	Elim. ratio L2	Elim. ratio Overall	E (μJ)	t (ms)	F1 (%)	Intent (%)
0	0	0	0	0	1.16	2.44	93.1	95.1
0.6	2.6	0.04	0.09	0.07	1.10	2.31	88.8	95.1
0.65	2.8	0.06	0.14	0.1	1.04	2.17	86.1	95.2
0.7	3.0	0.09	0.20	0.15	1.01	2.10	82.7	95.1
0.75	3.1	0.14	0.26	0.2	0.98	2.05	78.6	95.5

ATIS dataset, 32 tokens, 20MHz@0.56V

Fig. 12. Effect of word elimination on the full model accuracy with the ATIS dataset.

the number of operations. Latency and energy are close to other designs. However, our scaled design would have more on-chip memory than most of the other works, reducing the amount of required external memory accesses, which is not captured in the reported numbers.

V. Conclusion

We demonstrate an energy-scalable Transformer accelerator for edge devices. It supports 5.8× (for ATIS) to 6.9× (for SNIPS) scalability in the network latency and energy with a small memory footprint by adopting a custom-trained SuperTransformer model and having reconfigurable hardware to support various model dimensions. The word elimination unit can reduce energy by up to 16%, with a 14.5% drop in F1 score.

Acknowledgment

The authors would like to thank TSMC for funding and the TSMC University Shuttle Program for tapeout support.

References

[1] D. Wu, L. Ding, F. Lu, and J. Xie, "SlotRefine: A Fast Non-Autoregressive Model for Joint Intent Detection and Slot Filling," in *Proceedings of the 2020 Conference on Empirical Methods in Natural Language Processing (EMNLP)*, (Online), pp. 1932–1937, Association for Computational Linguistics, Nov. 2020.

[2] H. Wang, Z. Zhang, and S. Han, "SpAtten: Efficient Sparse Attention Architecture with Cascade Token and Head Pruning," in *2021 IEEE International Symposium on High-Performance Computer Architecture (HPCA)*, (Los Alamitos, CA, USA), pp. 97–110, IEEE Computer Society, mar 2021.

[3] C. T. Hemphill, J. J. Godfrey, and G. R. Doddington, "The ATIS Spoken Language Systems Pilot Corpus," in *Speech and Natural Language: Proceedings of a Workshop Held at Hidden Valley, Pennsylvania*, 1990.

[4] A. Coucke, A. Saade, A. Ball, T. Bluche, A. Caulier, D. Leroy, C. Doumouro, T. Gisselbrecht, F. Caltagirone, T. Lavril, M. Primet, and J. Dureau, "Snips Voice Platform: an embedded Spoken Language Understanding system for private-by-design voice interfaces," *CoRR*, vol. abs/1805.10190, 2018.

[5] T. J. Ham, S. J. Jung, S. Kim, Y. H. Oh, Y. Park, Y. Song, J.-H. Park, S. Lee, K. Park, J. W. Lee, and D.-K. Jeong, "A^3: Accelerating Attention Mechanisms in Neural Networks with Approximation," in *2020 IEEE International Symposium on High Performance Computer Architecture (HPCA)*, pp. 328–341, 2020.

[6] Y. Wang, Y. Qin, D. Deng, J. Wei, Y. Zhou, Y. Fan, T. Chen, H. Sun, L. Liu, S. Wei, and S. Yin, "A 28nm 27.5TOPS/W Approximate-Computing-Based Transformer Processor with Asymptotic Sparsity Speculating and Out-of-Order Computing," in *International Solid- State Circuits Conference (ISSCC)*, vol. 65, pp. 1–3, 2022.

[7] F. Tu, Z. Wu, Y. Wang, L. Liang, L. Liu, Y. Ding, L. Liu, S. Wei, Y. Xie, and S. Yin, "A 28nm 15.59μJ/Token Full-Digital Bitline-Transpose CIM-Based Sparse Transformer Accelerator with Pipeline/Parallel Reconfigurable Modes," *International Solid- State Circuits Conference (ISSCC)*, vol. 65, pp. 466–468, 2022.

[8] F. Tu, Z. Wu, Y. Wang, W. Wu, L. Liu, Y. Hu, S. Wei, and S. Yin, "MulTCIM: A 28nm 2.24μJ/Token Attention-Token-Bit Hybrid Sparse Digital CIM-Based Accelerator for Multimodal Transformers," pp. 248–250, 2023.

[9] T. Tambe, J. Zhang, C. Hooper, T. Jia, P. N. Whatmough, J. Zuckerman, M. C. D. Santos, E. J. Loscalzo, D. Giri, K. Shepard, L. Carloni, A. Rush, D. Brooks, and G.-Y. Wei, "A 12nm 18.1TFLOPs/W Sparse Transformer Processor with Entropy-Based Early Exit, Mixed-Precision Predication and Fine-Grained Power Management," *International Solid-State Circuits Conference (ISSCC)*, pp. 342–344, 2023.

[10] H. Wang, Z. Wu, Z. Liu, H. Cai, L. Zhu, C. Gan, and S. Han, "HAT: Hardware-Aware Transformers for Efficient Natural Language Processing," in *Annual Conference of the Association for Computational Linguistics*, 2020.

Learning from Output Transitions: A Chosen Challenge Strategy for ML Attacks on PUFs

Chia-Chih Lin
Graduate Institute of Electrical Engineering
National Taiwan University
Taipei, Taiwan.
cclin@arbor.ee.ntu.edu.tw

Ming-Syan Chen
Graduate Institute of Electrical Engineering
National Taiwan University
Taipei, Taiwan.
mschen@ntu.edu.tw

***Abstract*—The susceptibility of many strong Physically Unclonable Functions (PUFs) to machine learning (ML)-based modeling attacks is a significant challenge in hardware security. To evaluate the predictability of a PUF, the Hamming Distance Test (HDT) was introduced to measure the probability of output transitions. Poorer HDT values indicate that the attacker can predict some responses better than random guessing. However, existing work has not fully explored the integration of the HDT property for enhancing ML attacks. This paper proposes a chosen challenge strategy combining the HDT property with ML algorithms. The proposed strategy, the Differential Chosen Challenge Attack (DCCA), efficiently models a PUF by forcing the ML algorithm to learn from output transitions. Experimental results show at most 50% of Challenge-Response Paris (CRPs) are reduced compared with conventional ML attacks when attacking XOR Arbiter PUFs (XOR APUF) and Interpose PUFs (IPUFs). Furthermore, the proposed method efficiently l earns t he X OR trigger-based Adversarial APUFs (AAPUF) that conventional ML attacks are hard to model.**

***Index Terms*—Physically unclonable function, machine learning attacks, Hamming Distance Test, chosen challenge attack**

I. Introduction

A Physically Unclonable Function (PUF) is a physical system that uses its physical properties to map an external stimulus, called a challenge, to its output, called a response. PUFs have drawn much interest recently due to their resource-efficient properties for resource-limited devices. Depending on the scale of challenge-response pairs (CRPs), PUFs can be divided into strong and weak PUFs. Strong PUFs provide exponentially large CRPs, which are suitable for authentication. However, previous studies have shown that most strong PUFs are vulnerable to machine learning (ML)-based modeling attacks [1]. Although numerous ML-resistant PUFs have been proposed, the competition between attacks and defenses is still ongoing.

To evaluate whether a newly developed PUF is unpredictable, the Hamming Distance Test (HDT) [2] is a commonly used scheme. The HDT measures the average output transition probability between two challenges with a specified Hamming distance. A poor HDT value implies that an adversary can predict the response of an unseen challenge better than random guessing, given the received CRP and the mismatch pattern vector. Although statistical attacks based on HDT are considered inefficient compared with conventional ML attacks [2], it is essential to note that ML algorithms can exploit the information leaked by the PUF's output transition probabilities and may be underestimated. Thus, how to utilize the output transition information to model a target PUF efficiently is worth exploring.

Existing ML attacks against PUF can be roughly divided into two categories: conventional ML attacks and hyperparameter search [1], [3], [4], and integration of prior knowledge into the algorithm [5]–[8]. The former focused on exploring a function set to fit the data better, whereas the latter combined properties analyzed from a PUF or a particular group of PUFs with the ML algorithm. However, most ML attacks only consider training models using random CRPs, which may rapidly increase the required number of CRPs as the complexity of PUF increases. Some studies analyzed how to select informative training CRPs for ML models to reduce the number of required CRPs [9]–[11]. Nevertheless, the potential of using the output transition of PUF to select CRPs has not been thoroughly investigated.

This paper proposes a novel chosen challenge strategy, Differential Chosen Challenge Attack (DCCA), to improve the efficiency of ML attacks. The proposed DCCA chooses CRPs with differential patterns, CRP_{diff}, and combines them with random CRPs CRP_r as training data. The CRP_{diff} reveals the output transition of PUFs, whereas the CRP_r mitigates data mismatch during training. We evaluated the DCCA with deep learning-based ML algorithms on XOR Arbiter PUF (XOR PUF) [12], Interpose PUF (IPUF) [13], and XOR-triggered Adversarial APUF (AAPUF) [14]. Experimental results show that the proposed DCCA reduces the number of CRPs by 25%-50% when attacking XOR PUFs and IPUFs compared with traditional ML attacks. Besides, the DCCA efficiently attacks XOR-triggered AAPUFs, which are difficult for traditional ML attacks. Moreover, we show that directly using CRPs generated from HDT or manually selecting mismatch patterns based on poor HDT values will decrease the learning efficiency of ML algorithms. We list our contributions below:

This work was supported in part by the Ministry of Science and Technology, Taiwan, under Grant MOST 111-2221-E-002-135-MY3.

- To our best knowledge, this is the first work that systematically explores the output transitions of PUF to improve ML attacks in terms of data utility.
- We show that HDT-based chosen challenge strategies would mislead the ML algorithms, resulting in more training CRPs than conventional ML attacks.
- We propose a novel challenge strategy, Differential Chosen Challenge Attacks (DCCA), which focuses on learning the PUF's output transition by adding Hamming distance constraints to the training CRP set.
- Compared with baselines, our proposed method reduces the number of CRPs by 25%-50% when learning XOR PUFs [12] and IPUFs [13]. Moreover, the proposed DCCA can efficiently learn XOR-triggered AAPUFs [14] using approximately 10^5 CRPs, which is hard to learn by random CRPs.

The rest of the article is organized as follows: Section II describes the threat model, APUF and its variants, and related works. The proposed DCCA method is illustrated in Section III. Section IV evaluates the performances of XOR APUFs, IPUFs, and AAPUFs under the proposed attacks. Finally, Section V concludes the findings of this article.

II. Background

A. Threat Model

We adopted standard Strong PUF attack scenarios [1], [7]. Adversaries are restricted to non-invasive attacks. The PUF is generally assumed to be public and unprotected. The communication channel is insecure and noiseless. Adversaries can collect a subset of CRPs by eavesdropping on the channel or sending arbitrary challenges. The structures of the PUFs are known to adversaries, but the internal parameters are unknown.

B. Arbiter PUF and its variants

Arbiter PUF consists of cascaded MUX pairs and a flip-flop as an arbiter at the end of the MUX pairs. Due to each mux pair's intrinsic path delay variations, APUF can provide an exponentially large challenge-response space. The response of APUF can be expressed as a linear additive model [15]. Hence, APUF is well known to be vulnerable to machine learning algorithms due to its simplicity [1]. Consequently, many APUF variants have been proposed to resist machine learning attacks.

One approach to improving ML resistance is XORing the responses of multiple APUFs, as suggested in [12], also known as XOR Arbiter PUF (XOR PUF). XOR PUF is believed to be secure when the number of APUFs is sufficiently large [1]. However, increasing the number of APUFs cannot effectively counter reliability-based ML attacks [5]. Therefore, Nguyen et al. proposed the Interpose PUF (IPUF), consisting of two XOR PUFs, against classic and reliability-based ML attacks [13]. On the other hand, the recently proposed adversarial APUF (AAPUF) [14] exploits the concept of adversarial attacks against ML algorithms. The author proposed three types of triggers using only a few logic gates to flip a subset of APUF's responses. Interestingly, even though the trigger functions are simple, the performance of AAPUFs is promising, especially for XOR triggers. We illustrate the APUF, XOR PUF, IPUF, and AAPUF structures in Fig.1.

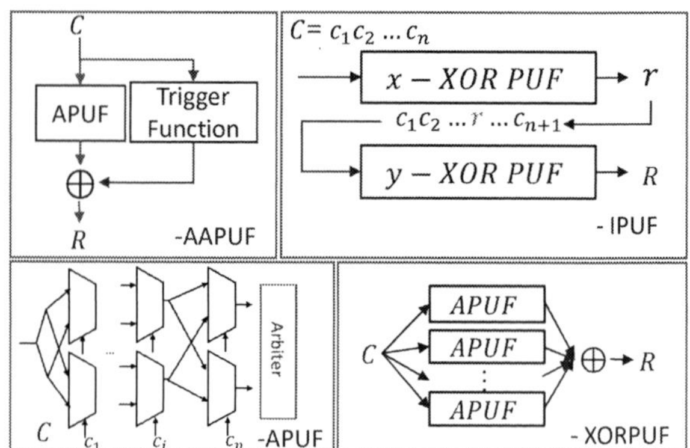

Fig. 1: Structures of the APUF and its variants.

C. Related Work

1) Hamming Distance Test: The Hamming Distance Test (HDT) was proposed in [16] to measure the output transition probability given challenge pairs with a specified Hamming distance. Specifically, a PUF is said to satisfy the $HDT(t)$ when

$$HDT(t) = Pr[(f(\mathbf{c}) \oplus f(\mathbf{c} \oplus \mathbf{p})) = 1] = 1/2, \qquad (1)$$

where $\mathbf{c}$ is a challenge vector, $\mathbf{p}$ is a mismatch pattern vector with a Hamming weight equal to t, i.e., $HW(\mathbf{p}) = t$, $\mathbf{c}, \mathbf{p} \in \{0,1\}^n$, and $f : \{0,1\}^n \rightarrow \{0,1\}$ is the PUF instance. The $HDT(1)$ is also known as the Strict Avalanche Criterion (SAC). A later extended version, $HDT(p,t)$, where the mismatch pattern p is specified, was proposed in [2]. To measure the HDT, a set of challenges $\mathbf{c}$ is randomly generated, and then XORs with each mismatch pattern vector $\mathbf{p}$ in the pattern vector set. Next, the output transition probability is calculated by averaging $f(\mathbf{c}) \oplus f(\mathbf{c} \oplus \mathbf{p})$. An adversary can choose several mismatch pattern vectors with poor HDT values to attack the PUF, as suggested in [2]. However, we will discuss in Section III-A that either directly using the CRPs generated by the HDT test or using CRPs with poor HDT values as the training data are inefficient for ML algorithms.

2) ML attacks incorporating prior knowledge: Several studies have combined additional information with ML algorithms. For instance, reliability-based ML attacks exploited the observation that the response of an APUF becomes unreliable when the delay difference is close to zero [5], [6]. Besides, some studies used divide-and-conquer approaches to learn the sub-components of complex PUFs separately [7], [8].

While most studies assumed challenges are generated randomly, Several works pointed out that not all challenges contribute equally during model training [9]–[11]. Wen et al. applied active learning to decrease CRPs for learning APUFs [9], but the suitability of active learning for complex PUFs remains uncertain [17]. Zhu et al. proposed sample essentiality based on sample-hyperplane relationship [11], which is limited to PUFs with known formulas. Lin et al. addressed the issue similarly in [18] by heuristically creating small challenge perturbations, but frequent CRP removal incurs significant computational costs as the required quantity increases. Ganji et al. [10] increased the efficiency of ML algorithms by using combination rules of influential challenge bits, but finding these rules requires additional statistical analysis. Unlike previous works, our proposed method selects CRPs based on the output transition of PUF as training data and does not require prior statistical analysis.

III. Methodology

A. The Proposed Differential Chosen-Challenge Attacks

Algorithm 1: Generating a differential challenge set

DCSG ($c_{init.}, N$);
Input : An initial n-bit randomly sampled challenge $c_{init.} \in \{0,1\}^n$; Number of required CRPs N
Output: Differential challenge set C_{diff}, where $C_{diff} \in \{0,1\}^{N \times n}$
$C_{diff} \leftarrow \{c_{init.}\}$
Pattern set $P \leftarrow \emptyset$
$N_{rest} \leftarrow N$
$i \leftarrow 1$
while $N_{rest} > 0$ **do**
  Create a set P^i consisting of all possible combinations of pattern vectors. $P^i = \{p^i_j \mid HW(p^i_j) = i, 1 \leq i \leq \binom{n}{i}\}$
  if $|P^i| \leq N_{rest}$ **then**
    $P \leftarrow P \cup P^i$
    $N_{rest} \leftarrow N_{rest} - |P^i|$
    $i \leftarrow i + 1$
  else
    Randomly choose N_{rest} pattern vectors, P^i_{select}, from P^i
    $P \leftarrow P \cup P^i_{select}$
    $N_{rest} \leftarrow 0$
foreach $p_i \in P$ **do**
  $c' \leftarrow c_{init} \oplus p_i$
  add c' to C_{diff}
shuffle(C_{diff})
return C_{diff};

The Hamming Distance Test first generates a set of random challenges, denoted as C_r, then applies each mismatch pattern p in set P to each challenge in C_r to measure the average output transition. That is, $C_r^{'} = \{c_i \oplus p_j \mid \forall c_i \in C_r, \forall p_i \in P\}$. Let $C_{hdt} = \{C_r, C_r^{'}\}$ be the challenges used in the Hamming Distance Test, and CRP_{hdt} be the CRPs obtained from a PUF instance. Intuitively, the adversary can analyze the HDT property of the targeted PUF and select some CRPs with poor HDT values in CRP_{hdt} as training data, as suggested in [2]. However, in addition to the computation cost brought by the

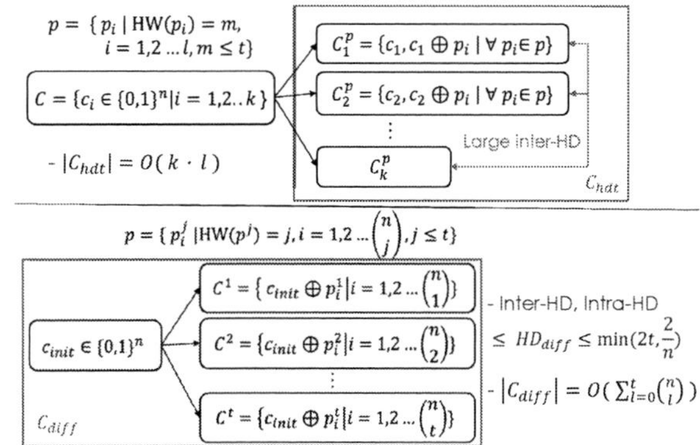

Fig. 2: An illustration of the proposed strategy compared with CRPs used in the Hamming Distance Test.

HDT process, there are two main reasons why this method is inefficient for ML algorithms. (1) Since randomly generating challenges is usually the premise of PUF design, most PUFs use the randomness of the challenge itself as one of the sources of entropy. The randomness introduced by C_r may increase the learning complexity of the ML algorithms. (2) Manually selected patterns may introduce unnecessary bias that misleads the ML algorithm. Therefore, we consider another approach that forces ML algorithms to focus on the output transition.

We generated a differential challenge set C_{diff} from only one randomly sampled challenge c_{init} to address the first problem. This implies that many pattern vectors are needed for the required training data. Next, the second problem is how to select pattern vectors. Instead of manually selecting, we exhaustively enumerated all combinations of pattern vectors whose Hamming weight is less than t. Specifically, the pattern set P_{diff} is defined as $P^i_{diff} \in P_{diff}$, $P^i_{diff} = \{p^i_j \mid HW(p^i_j) = i, 1 \leq j \leq \binom{n}{i}\}$, and $i \leq t$, where $HW(\cdot)$ is the Hamming weight and n is the length of a challenge. Thus, the order of the $|P_{diff}|$ is $\mathcal{O}(\sum_{i=1}^{t} \binom{n}{i})$. Since $\binom{n}{t}$ may provide more pattern vectors than needed, the remaining ones are randomly sampled from $\binom{n}{t}$. Consequently, a differential CRP set CRP_{diff} is obtained by $CRP_{diff} = PUF(C_{diff})$ and $C_{diff} = \{c_i \mid c_{init} \oplus p_i, \forall p_i \in P_{diff}\}$. Algorithm 1 depicts the detailed process of generating the differential challenge set C_{diff}.

Assume $\max(HW(P_{diff})) = t$ be the maximum Hamming weight of P_{diff}, the intra-Hamming distance in C_{diff} is restricted to $HD(c_i, c_j) \leq min(2t, 2/n)$, $c_i, c_j \in C_{diff}$. On the other hand, if the C_{hdt} is grouped by the challenges in the C_r, the inter-Hamming distance between groups is not restricted. As a result, ML algorithms may struggle to concentrate on patterns while they are being trained. We illustrate the concept in Fig.2.

B. Mitigate the Data Mismatch

The CRP set CRP_{diff} is obtained by applying the C_{diff} to a targeted APUF. Although CRP_{diff} forces the ML algorithms to focus on output transition, data bias is also introduced to ML algorithms. When the distribution of training data

TABLE I: Parameters for DNN Models in the experiments

DNN params.	XORPUF		IPUF		XOR-based AAPUF	
	4-X	5-X	(3,3)	(4,4)	4-bit	8-bit
No. of hidden layers	3	3	3	3	2	2
Avg. nodes / layer	50	200	50	100	50	50
(Hidden)Activation	tanh	tanh	relu	relu	relu	relu

deviates from that of testing data, the data mismatch problem is inevitable. To address the issue, a random challenge set is generated, and the corresponding CRP set CRP_r is obtained from the targeted PUF. Afterward, the CRP_r is combined with the biased CRP set C_{diff} to reduce the distribution gap between the training and testing sets. Depending on the learning complexity, the data's scale, and the model's size, the ratio r of biased to unbiased CRP sets could be different. Therefore, r is a tunable hyperparameter for the training set. Intuitively, the more unbiased CRPs in a training set, the less likely it is to obtain a biased model. However, adding more unbiased CRPs also increases the learning complexity. The adversary could adjust r to balance the tradeoff based on the performance of a trained model. We argue that adding $r\%$ random CRPs still forces the ML algorithm to be aware of the $1 - r\%$ CRP_{diff}, which still differs from the CRP_{hdt}.

On the other hand, an additional random CRP set CRP_t is collected as a hold-out test set to ensure the trained model is unbiased. CRP_t is only used to evaluate the ML model and fine-tune training hyperparameters. Thus, it is unseen during the training phase of the model. The experiment results in this paper are reported based on the hold-out test CRP set CRP_t.

C. Deep Learning Attacks

This study considers deep neural networks (DNN) as the machine learning algorithm for adversaries. The structure of the DNN model generally follows previous work [3]. The input of the DNN model is the parity vector obtained from the challenge. Several fully connected layers are used as hidden layers. A one-bit output layer with a sigmoid activation function predicts the response. The optimizer is the ADAM optimizer, and the loss function is a binary cross-entropy loss. The train-test split ratio is set to 9:1. We list the model parameters in Table I for each targeted PUF.

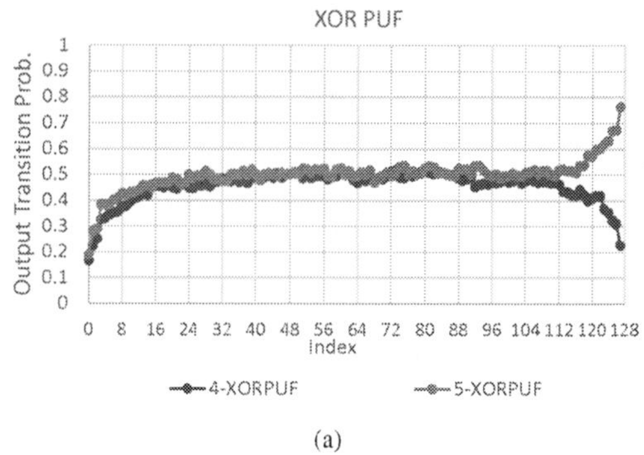

(a)

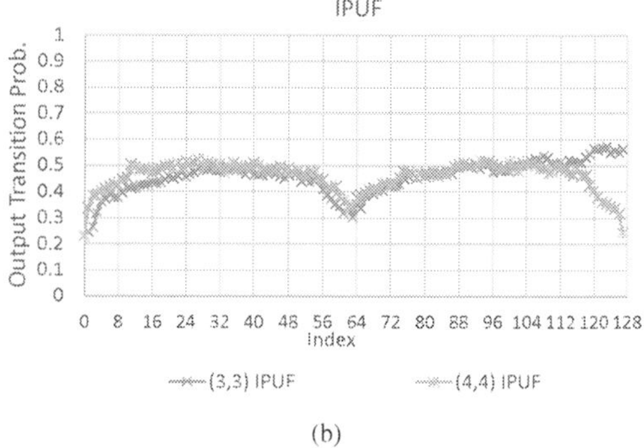

(b)

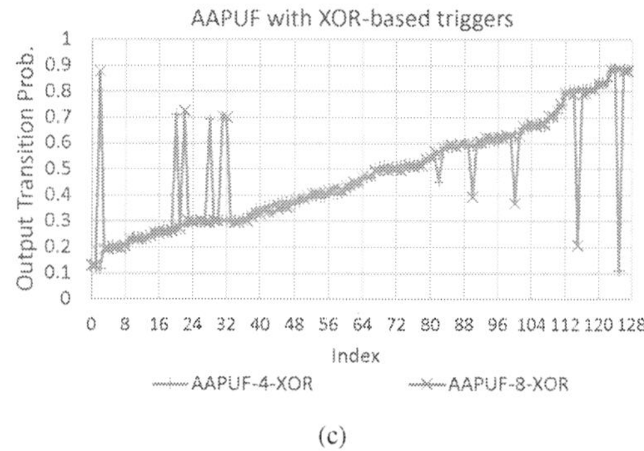

(c)

Fig. 3: The SAC properties ($HDT(1)$) of three APUF-based PUFs: (a) XOR PUFs, (b) Interpose PUFs, and (c) XOR-triggered AAPUFs

IV. Experimental Results

Previous research has shown that CRPs collected from simulated APUFs perform very similarly to those from silicon APUFs [19]. Because XOR PUF, IPUF, and AAPUF consist of APUFs and digital functions, we simulated 128-bit PUFs and conducted experiments in Python [20]. In the experiment, the hyperparameters of the DNN model are the same for each targeted PUF, as listed in Table I. The experimental results are reported using the hold-out test set.

A. Output Transition Probability

We followed the algorithm in [2] and analyzed $HDT(1)$ to investigate the predictability of 128-bit XOR PUFs, IPUFs, and XOR-triggered AAPUFs. Ten thousand randomly generated CRPs were used to obtain the output transition probability whenever a single challenge bit flips, i.e., the Strict Avalanche Criterion (SAC) property. The position of the interpose bit of IPUFs is $n/2 = 64$, as suggested in their work [13]. The challenge indices of the trigger functions were uniformly selected for AAPUFs. The trigger signals are $T_4 = c_{20} \oplus c_{28} \oplus c_{82} \oplus c_{125}$ and $T_8 = c_2 \oplus c_{22} \oplus c_{31} \oplus c_{32} \oplus c_{72} \oplus c_{90} \oplus c_{100} \oplus c_{115}$ for two AAPUFs with 4-bit and 8-bit XOR trigger functions, respectively, where c_i denotes the i-th challenge bit in a given challenge. The output transition probabilities are shown in Fig.3.

An HDT value is considered poor if it is close to 0 or 1.

As observed in Fig. 3, all Arbiter-based PUFs have poor HDT values at challenge indices 0, 1, 127, and 128. Furthermore, IPUFs exhibit a deviation from 0.5 at the interpose bit position. XOR-triggered AAPUFs present significant fluctuations at the trigger bit positions. These results are consistent with findings from previous studies [2], [13], [21]. Intuitively, an adversary can exploit these poor mismatch patterns to model a PUF, as a similar idea is presented in [2]. To evaluate the performance of DCCA, we use mismatch pattern vectors with poor HDT values as one of the baselines. Specifically, let $p_i \in P$ denote a mismatch pattern vector where $HW(p_i) = 1$ and the "1" only appears at index i. We use $p_0, p_1, p_{126}, p_{127}$ for all PUFs, and an extra vector p_{64} for IPUFs. For AAPUFs, vectors at all trigger indices, $p_{20}, p_{28}, p_{82}, p_{125}$, and $p_2, p_{22}...p_{115}$, are used additionally. We denote the challenge set consisting of randomly generated challenges C_r and patterned challenges $C_r \oplus P$ as C_{hdt}, and the corresponding CRP set is CRP_{hdt}.

B. Performance of the Proposed Chosen-Challenge Strategy

To better understand the performance of the proposed DCCA, three chosen challenge strategies were evaluated as baselines. The first one is a CRP set CRP_{rnd} where challenges are randomly generated. Model training with CRP_{rnd} is known as conventional ML attacks. The second baseline is CRP_{hdt}, which is defined in the previous subsection. The last CRP set is CRP_{hdt-s} where the challenges are created similarly to CRP_{diff} in Algorithm 1 except for using 1000 randomly sampled challenges instead of only one initial challenge. Compared with CRP_{hdt} that manually selects pattern vectors, the CRP_{hdt-s} uniformly samples pattern vectors with a constraint $HW(p) \leq t$, where t is set to $t = 4$. The CRP set generated from the proposed method is denoted as $CRP_{dcca} = \{CRP_{diff}, CRP_r\}$, where CRP_{diff} is generated by Algorithm 1 and CRP_r is randomly sampled to mitigate the data mismatch. The proportions of CRP_r in CRP_{dcca} for learning XOR PUFs, IPUFs, and AAPUFs are 50%, 30%, and 30%, respectively.

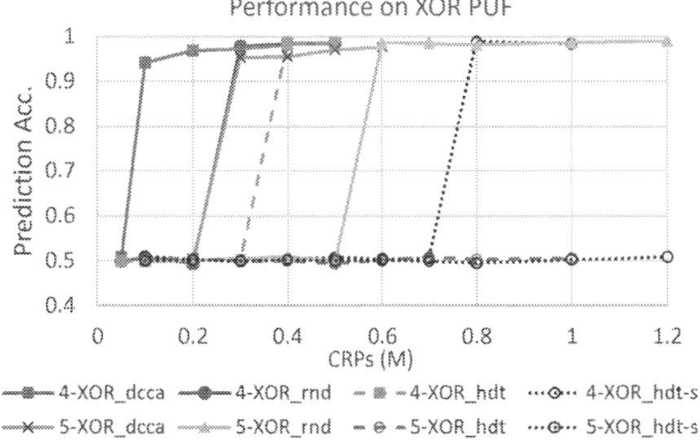

Fig. 4: Performance of the DCCA on XOR PUFs.

Fig.4 shows the result of the ML attacks on XOR PUFs with different chosen challenge strategies. Conventional ML attacks can model the 4-XOR PUFs with 97.84% accuracy using 300k training CRPs, while the DCCA can achieve the same accuracy with 33.3% fewer training CRPs. A 50% reduction in the required training CRPs is observed for the 5-XOR PUF, indicating that the ML model benefits more with DCCA-based CRPs when the learning complexity is higher. By contrast, CRP_{hdt} and CRP_{hdt-s} mislead the ML algorithm and perform worse than conventional ML attacks. Moreover, CRP_{hdt-s} fails to model the 5-XOR PUF, which means that, compared with CRP_{hdt}, increasing the number of pattern vectors does not enhance the learning efficiency of the ML algorithm.

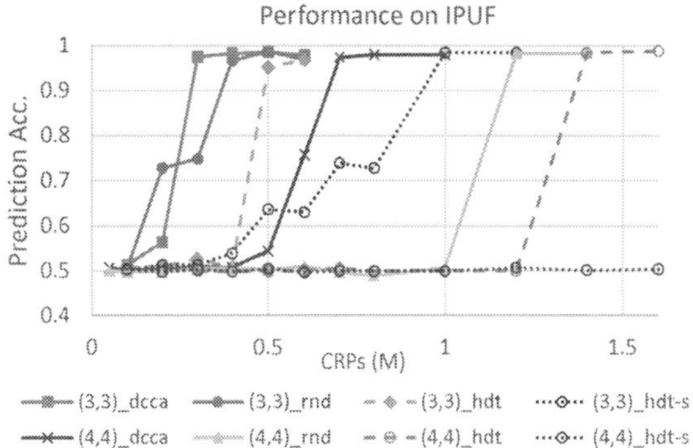

Fig. 5: Performance of the DCCA on IPUFs.

To investigate the performance of the proposed DCCA on IPUFs, (3,3)-IPUF and (4,4)-IPUF were evaluated, and the results are shown in Fig. 5. Since (3,3)-IPUF is relatively easy to learn, the reduction of required training CRPs is only 25% compared with the conventional ML attack, i.e., the CRP_{rnd}. Besides, the performance of the CRP_{dcca} is worse than the CRP_{rnd} when using 200k training CRPs, showing that the CRP_{diff} might mislead ML algorithms when the targeted PUF is easy to learn with the CRP_{rnd}. By contrast, 41.67% of the required training CRPs are reduced when attacking the (4,4)-IPUF. The results reveal that ML algorithms can benefit from additional information in CRP_{diff} when dealing with higher learning complexity. On the other hand, both CRP_{hdt} and CRP_{hdt-s} perform worse than the conventional ML attacks, indicating using the CRPs created from HDT as training data is inefficient.

Finally, we evaluated the performance of the DCCA on XOR-triggered AAPUFs, as shown in Fig. 6. Interestingly, the XOR trigger functions become easy to learn by using the proposed DCCA, while all other chosen challenge strategies fail to model the AAPUFs. Besides, even if the number of triggered bits for the AAPUF is doubled, the number of required training CRPs for the DCCA increases linearly. The result shows that the CRP_{diff} successfully forces the ML algorithm to focus on the mismatch patterns.

C. The Ratio of Biased and Unbiased Data in Training Sets

While the proposed chosen-challenge strategy reduces learning complexity and helps machine learning algorithms converge efficiently, data bias is also introduced to the ML

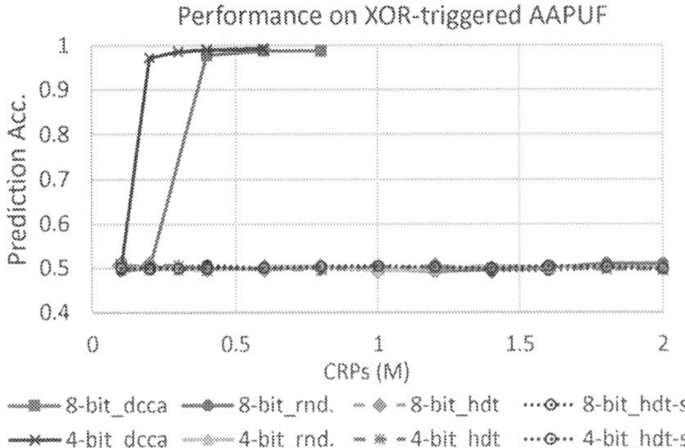

Fig. 6: Performance of the DCCA on AAPUFs.

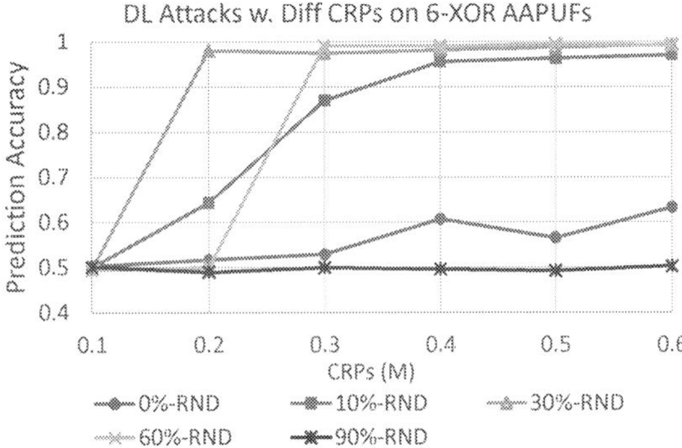

Fig. 7: Impact of different ratios of random CRPs in DCCA. $x\%$-RND denotes that the training CRP set contains $x\%$ randomly generated CRPs.

model. To investigate the tradeoff between data mismatch and learning complexity, we varied the ratio of unbiased CRPs, i.e., randomly generated CRPs CRP_r, in CRP_{dcca}. The result is shown in Fig. 7 where CRP_{dcca} is evaluated on a 6-bit XOR-based AAPUF. When there are no unbiased CRPs in the training sets, the DNN model performs perfectly on the training set but is close to random guesses in the testing set. As unbiased CRPs increase, the DNN model begins to perceive and learn from unbiased data. The results illustrate that when the data volume is limited, varying the ratio between biased and unbiased data might improve the model's performance.

V. Conclusion

This paper proposes a chosen challenge strategy-based ML attack, DCCA, to explore the information leakage from output transitions. By learning output transition patterns and data distribution from the selected training CRPs, the proposed DCCA improves the data utility of the ML algorithm. To evaluate the performance of the DCCA, DNN-based ML attacks on XOR PUFs, IPUFs, and AAPUFs were conducted. The experimental results show that compared with conventional ML attacks, the DCCA reduces the required training CRPs by 25%-50% when attacking XOR PUFs and IPUFs. Besides, the DCCA significantly decreases the learning complexity of AAPUFs with XOR-based triggers, which is hard to learn for conventional ML attacks. The experimental result also demonstrated that using CRPs created from the HDT test as training data is inefficient. Manually selecting CRPs with poor HDT values might also mislead the ML algorithm. Finally, we analyzed the impact of different ratios of randomly generated CRPs in the proposed DCCA. The results show that adding a small proportion of unbiased CRPs to training sets alleviates the data mismatch problem.

References

[1] U. Rührmair, F. Sehnke, J. Sölter, G. Dror, S. Devadas, and J. Schmidhuber, "Modeling attacks on physical unclonable functions," in *CCS*, 2010, p. 237–249.

[2] P. H. Nguyen, D. P. Sahoo, R. S. Chakraborty, and D. Mukhopadhyay, "Security analysis of arbiter puf and its lightweight compositions under predictability test," *ACM TODAES*, vol. 22, no. 2, Dec. 2016.

[3] P. Santikellur, A. Bhattacharyay, and R. S. Chakraborty, "Deep learning based model building attacks on arbiter puf compositions," *IACR Cryptol. ePrint Arch*, 2019.

[4] N. Wisiol, B. Thapaliya, K. T. Mursi, J.-P. Seifert, and Y. Zhuang, "Neural network modeling attacks on arbiter-puf-based designs," *IEEE TIFS*, vol. 17, pp. 2719–2731, 2022.

[5] G. T. Becker, "The gap between promise and reality: On the insecurity of xor arbiter pufs," in *CHES*, 2015, pp. 535–555.

[6] J. Tobisch, A. Aghaie, and G. T. Becker, "Combining optimization objectives: New machine-learning attacks on strong pufs," Cryptology ePrint Archive, Paper 2020/957, 2020.

[7] N. Wisiol, C. Mühl, N. Pirnay, P. H. Nguyen, M. Margraf, J.-P. Seifert, M. van Dijk, and U. Rührmair, "Splitting the interpose puf: A novel modeling attack strategy," *TCHES*, vol. 2020, no. 3, p. 97–120, 2020.

[8] D. P. Sahoo, P. H. Nguyen, D. Mukhopadhyay, and R. S. Chakraborty, "A case of lightweight puf constructions: Cryptanalysis and machine learning attacks," *IEEE TCAD*, vol. 34, no. 8, pp. 1334–1343, 2015.

[9] Y. Wen and Y. Lao, "Puf modeling attack using active learning," in *Proc. of IEEE ISCAS*, 2018, pp. 1–5.

[10] F. Ganji, S. Tajik, F. Fäßler, and J.-P. Seifert, "Strong machine learning attack against pufs with no mathematical model," in *CHES*. Springer-Verlag, 2016, p. 391–411.

[11] S. Zhu, Y. Tang, J. Zheng, Y. Cao, H. Wang, Y. Huang, and M. Margraf, "Sample essentiality and its application to modeling attacks on arbiter pufs," *ACM TECS*, vol. 18, no. 5, pp. 42:1–42:25, 2019.

[12] G. E. Suh and S. Devadas, "Physical unclonable functions for device authentication and secret key generation," in *Proc. of ACM/IEEE DAC*, 2007, pp. 9–14.

[13] P. H. Nguyen, D. P. Sahoo, C. Jin, K. Mahmood, U. Rührmair, and M. van Dijk, "The interpose puf: Secure puf design against state-of-the-art machine learning attacks," *CHES*, no. 4, p. 243–290, 2019.

[14] S. Wang, Y. Chen, and K. S. Li, "Adversarial attack against modeling attack on pufs," in *Proc. of ACM/IEEE DAC*, 2019, pp. 1–6.

[15] Daihyun Lim, J. W. Lee, B. Gassend, G. E. Suh, M. van Dijk, and S. Devadas, "Extracting secret keys from integrated circuits," *IEEE TVLSI SYST*, vol. 13, no. 10, pp. 1200–1205, 2005.

[16] M. Majzoobi, F. Koushanfar, and M. Potkonjak, "Testing techniques for hardware security," in *2008 IEEE International Test Conference*, 2008, pp. 1–10.

[17] B. Settles, "From theories to queries: Active learning in practice," in *Active Learning and Experimental Design workshop In conjunction with AISTATS 2010*, ser. PMLR, vol. 16, 2011, pp. 1–18.

[18] C.-C. Lin and M.-S. Chen, "Enhancing reliability and security: A configurable poisoning puf against modeling attacks," *IEEE TCAD*, vol. 41, no. 11, pp. 4301–4312, 2022.

[19] U. Rührmair, J. Sölter, F. Sehnke, X. Xu, A. Mahmoud, V. Stoyanova, G. Dror, J. Schmidhuber, W. Burleson, and S. Devadas, "Puf modeling attacks on simulated and silicon data," *IEEE TIFS*, vol. 8, no. 11, pp. 1876–1891, 2013.

[20] "https://github.com/d06921014/islped2023/."

[21] C.-C. Lin and M.-S. Chen, "Attack is the best defense: A multi-mode poisoning puf against machine learning attacks," in *PAKDD*. Springer, 2021, pp. 176–187.

Efficient Machine Learning on Encrypted Data using Hyperdimensional Computing

Yujin Nam*, Minxuan Zhou*, Saransh Gupta†, Gabrielle De Micheli*
Rosario Cammarota‡, Chris Wilkerson‡, Daniele Micciancio*, Tajana Rosing*
*Department of Computer Science and Engineering, UC San Diego, La Jolla, USA
† IBM Research, Santa Clara, USA, ‡ Intel Labs, Santa Clara, USA
{yujinnam, miz087, gdemicheli, dmicciancio, tajana}@ucsd.edu
saransh@ibm.com, {rosario.cammarota, chris.wilkerson}@intel.com

Abstract—**Fully Homomorphic Encryption (FHE) enables arbitrary computations on encrypted data without decryption, thus protecting data in cloud computing scenarios. However, FHE adoption has been slow due to the significant c omputation and memory overhead it introduces. This becomes particularly challenging for end-to-end processes, including training and inference, for conventional neural networks on FHE-encrypted data. Additionally, machine learning tasks require a high throughput system due to data-level parallelism. However, existing FHE accelerators only utilize a single SoC, disregarding the importance of scalability. In this work, we address these challenges through two key innovations. First, at an algorithmic level, we combine hyperdimensional Computing (HDC) with FHE. The machine learning formulation based on HDC, a brain-inspired model, provides lightweight operations that are inherently well-suited for FHE computation. Consequently, FHE-HD has significantly lower complexity while maintaining comparable accuracy to the state-of-the-art. Second, we propose an efficient a nd scalable FHE system for FHE-based machine learning. The proposed system adopts a novel interconnect network between multiple FHE accelerators, along with an automated scheduling and data allocation framework to optimize throughput and hardware utilization. We evaluate the value of the proposed FHE-HD system on the MNIST dataset and demonstrate that the expected training time is 4.7 times faster compared to state-of-the-art MLP training. Furthermore, our system framework exhibits up to 38.2 times speedup and 13.8 times energy efficiency improvement over the baseline scalable FHE systems that use the conventional data-parallel processing flow.**

I. Introduction

Fully Homomorphic Encryption (FHE) is a family of encryption methods that enables computations on encrypted data. It allows data owners to securely outsource their models to servers without compromising data privacy. However, FHE faces challenges when it comes to easy deployment in production due to the significant p erformance degradation compared to computations on plaintext data [1]. This limitation restricts the usability of FHE in various machine learning scenarios, particularly during the training procedure. Previous research [2] has demonstrated that training a simple neural network on the MNIST dataset [3] using FHE requires over two months. Since many machine learning applications necessitate larger networks and datasets to achieve high accuracy, training with FHE becomes impractical. In addition to algorithm-level optimization, hardware acceleration plays a critical role in FHE programs because conventional systems fail to provide sufficient compute throughput and memory bandwidth for operations involving large polynomials (e.g., polynomials of degree greater than 32K). Although existing ASIC accelerators already promise to offer at least four orders of magnitude speedup compared to conventional systems [1], accelerating machine learning tasks on large datasets that require high throughput to exploit data-level parallelism remains challenging. Unfortunately, there is currently no prior work focusing on a high-throughput and scalable FHE acceleration system.

In this work, we explore the combination of FHE with hyperdimensional Computing (HDC), a lightweight machine learning approach that has demonstrated promising performance and accuracy across a wide range of applications [4]. HDC operates by mapping data into a high-dimensional representation, leveraging this representation to enable learning capabilities. Compared to conventional training methods used in NN-based models (such as back-propagation), HDC training is considerably simpler and composed of friendly operations for FHE. Additionally, HDC models exhibit noise tolerance, which is crucial considering that FHE training introduces errors to the model. As a result, HDC offers advantages over NN-based models for FHE training. We propose both algorithmic and hardware optimizations for HDC based on FHE. We propose the first implementation of FHE for end-to-end HDC using the widely used CKKS FHE scheme [5], [6]. The CKKS scheme is commonly employed for learning tasks due to its support for real (and complex) number arithmetic and SIMD-style operations through vector encryption. Utilizing CKKS for HDC poses several challenges. Firstly, HDC requires several non-linear operations that are not directly supported by the limited operations of CKKS. Secondly, CKKS introduces approximation errors that can impact the training accuracy of HDC. To address these challenges, we introduce an efficient and noise-tolerant FHE-friendly HDC algorithm based on the CKKS scheme. In addition, we propose a novel system architecture consisting of multiple FHE ASIC chips and inter-ASIC connections. To optimize system throughput and memory utilization, we present an automated scheduling and data allocation framework. The contributions of this work are as follows:

- To the best of our knowledge, we provide the first FHE-based end-to-end HDC classification, covering all steps of HDC - encoding, training, inference, and retraining.

979-8-3503-1176-1/23 $31.00 © 2023 IEEE

- We introduce an efficient FHE-friendly HDC model with good accuracy. The proposed model exhibits 4.7 times faster training compared to state-of-the-art FHE-based MLP models, while achieving over 1000 times faster inference than FHE-based RNN.
- We propose a novel FHE system with an automated scheduling and data allocation for high-throughput FHE learning. We demonstrate the efficiency of the proposed method on a scalable FHE system with state-of-the-art FHE accelerators, showing that the proposed system achieves up to 38.2 times speedup and 13.8 times energy efficiency improvement over baseline systems with conventional data-parallel processing flow.

II. Background and Motivation

A. Fully Homomorphic Encryption

In this work, we focus on the CKKS scheme [5]–[7] which supports processing in real numbers and SIMD packing.

1) Basics of CKKS

A ciphertext in the CKKS scheme is given as a tuple $c = (b, a) \in R_Q^2$, where the polynomial rings considered are $R = \mathbb{Z}[X]/(X^N + 1)$ and $R_Q = R/QR$. For a given security parameter λ, CKKS sets the ring size N and a ciphertext modulus Q. A ciphertext c encrypts a vector $\mathbf{m}$ in $\mathbb{C}^{N/2}$ of up to $N/2$ real (or complex) elements. Applying residue number system (RNS), a large integer Q_L can be decomposed into smaller primes q_i where $i \in [0, L]$ and $q_0 q_1 ... q_L = Q_L$. RNS-CKKS is then a leveled homomorphic encryption method where each (ciphertext) multiplication consumes one level (or depth), thus reducing the ciphertext modulus from Q_l to Q_{l-1}. When level 0 is reached, no further multiplications are possible. Bootstrapping [7] then allows to raise the modulus to further continue with operations.

Each homomorphic operation works in SIMD-style over the vector elements. The homomorphic operations supported by the CKKS scheme can be explained as follows, where $\circ$ denotes element-wise multiplication and pk is a public key:

- HomAdd (c_0, c_1) = $\texttt{Enc}_{pk}(\mathbf{m}_0 + \mathbf{m}_1)$
- HomMult (c_0, c_1) = $\texttt{Enc}_{pk}(\mathbf{m}_0 \circ \mathbf{m}_1)$
- HomRot $(c_0, k) = \texttt{Enc}_{pk}((m_k, m_{k+1}, ..., m_{k-1}))$ circularly rotates the message by k slots.

2) On the security of CKKS

As this paper focuses mostly on efficiency and performance, we refer the readers to [8], [9] for discussions on the security of the CKKS scheme.

B. Hyperdimensional Computing (HDC)

HDC based classification first maps the input data into hypervectors (high dimensional vectors). The rest of the steps in HDC - training, inference and retraining - use hypervectors.

1) Encoding

The encoding procedure converts an input vector F with f features into a hypervector HV of dimension D. We used a commonly used encoding, random projection with nonlinear activation [10]. A hypervector HV can be evaluated as $HV = \sigma(A \cdot F)$, where A is a randomly chosen matrix in $\{-1, 1\}^{D \times f}$ and $\sigma()$ applies a chosen nonlinear function to each element of the input vector.

2) Training

After all the training data are mapped to high-dimensional space, training can be done by adding up the hypervectors that belong to the same class. We call the generated hypervectors as class hypervectors $classHV_i$.

3) Inference

To infer a query, the input is first mapped to a query hypervector using the same encoding method. The result label is then decided as $\texttt{argmax}(\delta(queryHV, classHV_i))$, where $\delta(\cdot,\cdot)$ is a similarity metric between two hypervectors. We used cosine similarity as it is commonly used.

4) Retraining

Retraining improves the accuracy of a model. When a train hypervector $trainHV$ with the correct label c infers to a wrong label w, class hypervectors are updated as follows with learning rate r:

$$classHV_c \leftarrow classHV_c + r \cdot trainHV$$
$$classHV_w \leftarrow classHV_w - r \cdot trainHV$$

C. FHE Accelerator

There have been various accelerator proposals for the CKKS FHE scheme [1], [11], [12]. These accelerators achieve significant performance improvement over conventional architectures by exploiting high-throughput function units, large on-chip storage, high-bandwidth memory, and algorithmic optimization. However, existing work focuses on single SoC, ignoring the scalability of FHE acceleration for applications with high data-level parallelism, like machine learning. There exist several challenges in designing a scalable FHE system, including 1) interconnect network to keep a high bandwidth when increasing FHE accelerators, 2) operation scheduling on different accelerators, and 3) data allocation on the distributed memory. Section IV introduces the details of our scalable FHE system.

III. FHE-HD Algorithm

We propose FHE-friendly algorithm for HDC classification, FHE-HD. The overall algorithm is described in Figure 1.

A. Data Encoding

The data is first encoded into high dimensional space. We used random projection followed by an activation function.

1) Random Projection

Random projection multiplies a random projection matrix $A = \{-1, 1\}^{D \times f}$ with a feature vector F of size f. The random projection matrix does not need to be encrypted as it is generated randomly and contains no sensitive information. A naive method would be making D different plaintext vectors and evaluating dot projects between an encrypted vector F and plaintexts. However, this is inefficient as a dot product generates only one element of the result and it wastes the majority of available slots. Also because D is large, many plaintext vectors and dot product operations are required.

We propose an efficient method for random projection that takes advantage of every slots in a ciphertext while requiring

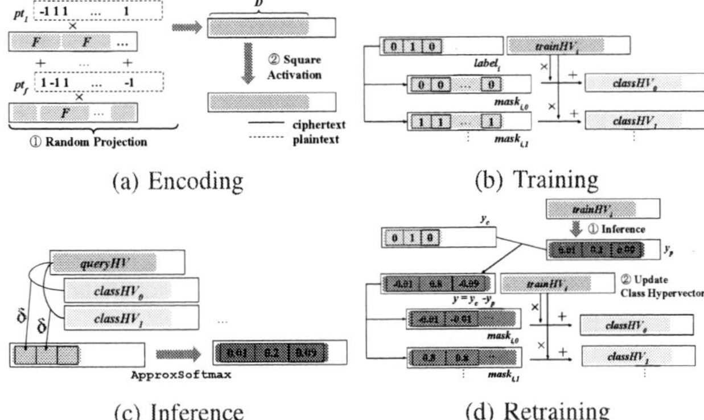

(a) Encoding (b) Training (c) Inference (d) Retraining

Fig. 1: **FHE-HD algorithm.**

minimal number of operations. The random projection process can be carried out by encrypting the repetitions of vector F using every available slots in the ciphertext c. We evaluate rotations of the ciphertext, $\texttt{HomRot}(c, i)$, $i = 0, \dots f - 1$. Then we multiply each rotated ciphertext with a random plaintext vector and add up all the results. This process is described in Figure 1a. We further reduced the number of rotations using the baby-step-giant-step method [13].

2) Activation Function

The CKKS scheme can only evaluate linear functions because it supports limited arithmetic operations (e.g., addition and multiplication). Therefore, to evaluate a nonlinear function, a polynomial approximation of those functions is needed. Previous works approximated activation functions with low-degree polynomials, like sigmoid [14] or ReLu [15]. Other works [16] replaced activation functions with square activation, i.e., $f(x) = x^2$. Previous work [10] used cos and $sign$ for activation in HDC. To minimize the depth consumption of the activation, we used square activation. We empirically found that square activation has similar accuracy to other nonlinear activations.

B. Label Encoding

For data privacy, the labels should be encrypted. Labels for classification are typically integers from 0 to $l - 1$, where l is the total number of labels. However, directly encrypting these integer values into ciphertexts incurs comparisons between integer labels with high computational costs. We utilized one-hot encoding for the labels, where a label i is represented as a vector of length l whose the i-th element is 1 and others are 0. Then, using the SIMD style encryption of CKKS, we encrypted this one-hot encoded vector as one ciphertext. Later in the training process, we use the encrypted labels to create mask ciphertexts, which eliminates the need for comparison.

C. Training

Once training data are mapped to hypervectors, we need to add up the train hypervectors for each class. We create mask ciphertexts by extending each slot of a one-hot encoded label ciphertext. This refers to copying a value from one slot to a encrypted hypervector's D slots, using plaintext multiplication, rotations and additions. Our model is then trained by creating a j-th class hypervector performing $\sum_i mask_{i,j} \times trainHV_i$, where $mask_{i,j}$ is a mask generated by copying the j-th slot of the i-th data label. Due to the inability to execute different branches based on encrypted values, certain redundant operations with masks are introduced. Despite this, the proposed method avoids costly comparison operations that are not inherently supported by the CKKS scheme.

D. Inference

We replaced `argmax` with `ApproxSoftmax`, which enabled us to compare values while avoiding expensive comparison operations. For the similarity metric, we used cosine similarity: $\delta(queryHV, classHV_i) = \langle queryHV, classHV_i \rangle / \| classHV_i \|$. Inferring a query includes computing a dot product, an inverse of square root, and `ApproxSoftmax`. Dot product can be computed easily using homomorphic multiplication, rotations and additions.

1) Approximate Inverse and Inverse of Square Root

The inverse function $f(x) = \frac{1}{x}$ and the inverse of square root function $f(x) = \frac{1}{\sqrt{x}}$ can be used for division and division by l_2-norm of class hypervectors. Because CKKS does not support nonlinear operations, we need to use an approximation approach. We used Goldschmidt's iterative method [17] for both functions.

2) Approximate Softmax

We proposed a hybrid approximation of `argmax` and `softmax`, in place of the `argmax` function of the initial HDC algorithm. For CNN models, Lee et al. [15] approximated `softmax` function. In the region of [-1,1], they used a degree-13 polynomial approximation of the exponent function. Cheon et al. [18] proposed `argmax` (or `MaxIdx`) procedure, which iteratively approximates $\frac{a_j^k}{a_1^k + \cdots + a_n^k}$ for large k. Both works used Goldschmidt's method for division operation.

We empirically verified that the `softmax` approximation from Lee et al. alone is insufficient to provide precision for the retraining procedure. On the other hand, Cheon et al. iteratively perform inverse operations, which uses up multiplication depths. As a result, we suggest a hybrid approach, `ApproxSoftmax`, that combines two earlier research. First, we evaluate the approximation of $e^{x/\lambda}$, using λ to match the approximation region. Repeating square operations for t times, we can evaluate $(e^{x/\lambda})^{2^t}$. We applied division $(a_j / (n \cdot \sum_{i=1}^{n} a_i))$ from Cheon et al. once in between the squaring to prevent the overflow. The result is obtained by dividing each value by the sum of the values.

3) Packed Inference

We made full use of SIMD operation by packing multiple hypervectors into one ciphertext and evaluating the dot product for cosine similarity. We also inferred multiple queries simultaneously by placing their cosine similarities in one ciphertext.

E. Retraining

As stated in II-B4, the retraining is carried out by adjusting the class hypervectors with inference results. We can retrain the model using the predicted label and the ground truth label, which is an one-hot encoded label. To elaborate, let us assume that y_c is the ground truth label, y_p is the predicted label, and $y = y_c - y_p$. By performing

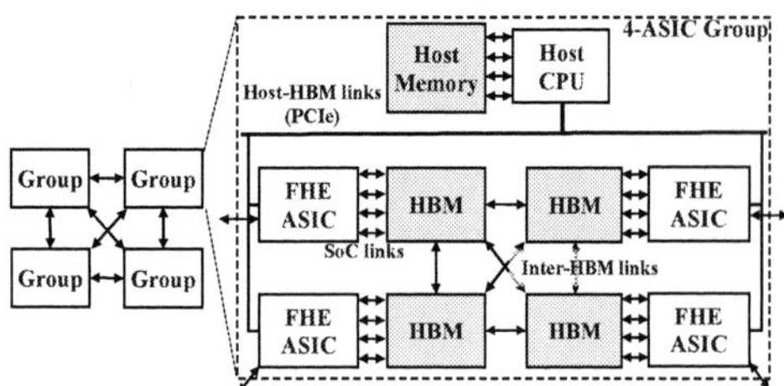

Fig. 2: **The scalable FHE system.**

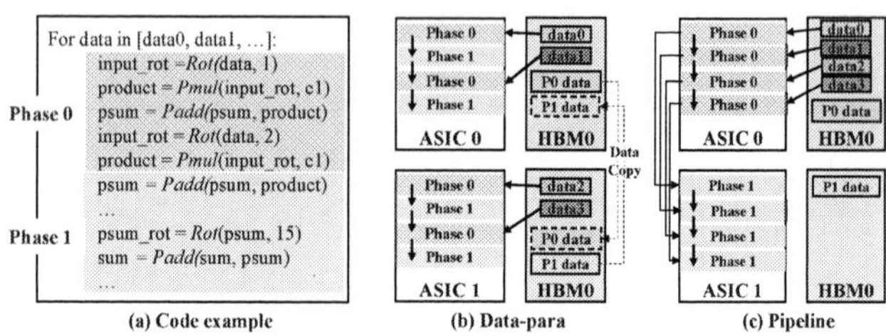

Fig. 3: **Scheduling and data allocation schemes.**

$classHV_j \leftarrow classHV_j + r \cdot y[j] \cdot trainHV_i$ for every possible j, we can retrain the class hypervectors. Note that multiplying by a single slot $y[j]$ can be achieved through the same mask generation process of the training step. Because we used `ApproxSoftmax` for inference, the prediction result y_p is a vector of values in $[0, 1]$ rather than 0 or 1. This approach updates each class hypervector while taking into account their similarity with the query, as opposed to only updating two class hypervectors. Comparing this method to the original HDC algorithm, we confirmed that it still provides acceptable accuracy.

F. Computaion-Comunication Trade-Off

We can consider different scenarios to reduce the server side computation. For example, encoding can be done in the client side in plaintext and the server can receive encrypted hypervectors. Also every training masks can be created on the client and sent to the server. This would raise the communication cost as more ciphertexts need to be sent while improving the computation efficiency on the server side by skipping the encoding or mask generation process.

IV. Scalable FHE-HD System

Machine learning tasks, including inference and training, feature high data-level parallelism. Thus, the scalability of the underlying system is critical to the end-to-end performance of machine learning tasks. In this section, we first introduce a scalable FHE system with FHE accelerators. Then, we propose several operation scheduling and data allocation schemes that can be exploited by the runtime to achieve the best performance for different FHE-HD phases.

A. Scalable FHE System

Figure 2 shows the scalable FHE systems, consisting of multiple FHE ASIC chips. Each FHE ASIC is connected to a High-Bandwidth Memory (HBM). For the scalable system, we connect all FHE ASICs in a dragonfly interconnect, similar to previous near-memory acceleration [19]. Therefore, all FHE ASICs are organized into multiple 4-ASIC groups, where four ASICs in each group are fully connected through the inter-memory links. Even though each ASIC can assess its local HBM through fast links, the ASIC HBM cannot hold the extremely large FHE-HD data, including the encrypted dataset and FHE keys. In this case, each group contains a host CPU and a host memory. The host stores the whole FHE-HD data and transfers the required data to ASIC HBM. The scalable system supports various types of FHE ASIC, which are treated as black boxes. Our contribution lies in the overall system design with efficient operation scheduling and data allocation schemes. In the following subsections, we introduce different methods of operation scheduling and data allocation that can fully utilize the scalable FHE system. Furthermore, we propose an automated framework to optimize the scheduling and data allocation in large-scale FHE systems for HDC.

B. Operation Scheduling

The operation scheduling scheme indicates the way of distributing FHE-HD applications across all FHE ASICs in the system, critical to the system's efficiency. A straightforward operation scheduling is data-parallel, as shown in Figure 3(b) for an example code (Figure 3(a)). In this example, the algorithm processes different data points with identical operations, which is the common pattern in HD inference, training, and encoding phases. The data-parallel scheduling distributes the computation of different data points to different ASICs. In this case, each ASIC processes the end-to-end process for a subset of encrypted data points as well as FHE. The advantage of data-parallel scheduling is we can balance the workloads on different FHE ASICs. However, data-parallel scheduling requires each ASIC to load all FHE materials throughout the end-to-end process, significantly increasing the memory footprint, and hence increasing the host-ASIC data loading. The host-ASIC is slow due to the limited bandwidth of the host-ASIC link that is based on PCIE.

An alternative scheduling scheme is pipelining which allocates operations of phases to different FHE ASICs, shown in Figure 3(c). In the pipeline scheduling, each ASIC avoids loading all FHE materials to its local HBM. However, the pipeline scheduling requires data transfers between different FHE ASIC HBM. Furthermore, pipeline scheduling needs to carefully balance the memory footprint as well as operation latency among different ASICs. In general, the pipeline scheduling may suffer from significant inefficiency due to the overwhelming inter-ASIC transfers and unbalanced pipeline.

C. Data Allocation

Another problem of running FHE-HD applications on scalable FHE systems is data allocation where we need to determine how to allocate data in different ASIC HBMs. The naive way is to treat all ASIC HBMs as a unified memory pool, where only one copy for each data can be loaded. For example, in Figure 3(b), P0 data is only stored in HBM0 so ASIC1 needs to load P0 data from ASIC0 to its on-chip memory (i.e., scratchpad) if needed. Such a no-copy method can minimize the memory footprint. However, it introduces a lot of inter-ASIC data transfers that can significantly slow down the system. Another data allocation method is to maintain multiple copies of data in the AISCs that need the data. In this case, each ASIC can access its local HBM for all needed data. However, the data copy introduces coherence issues as well as increases the memory footprint, increasing the inter-ASIC and host-ASIC transfers. We propose a hybrid data allocation scheme that only allows the data copy for read-only data with a high reuse rate. For example, all the key-switching keys, including homomorphic rotation and multiplication, are read-only and highly reusable among different ASICs. With the hybrid data allocation, we can avoid a large number of inter-ASIC and host-ASIC data transfers while maintaining a relatively low memory footprint.

D. System Support

We implement an offline exploration framework to determine the best combination of scheduling and data allocation for each FHE-HD step. In addition to the high-level exploration, the framework optimizes the detailed scheme for pipeline and hybrid data allocation. For the pipeline scheduling, we need to optimize the pipeline phase segmentation to balance the overall latency. For a given program on one data point, consisting of a sequence of FHE instructions, we use a random search with trimming to approximately optimize the pipeline. For each searched pipeline, we run a simulation to determine its goodness. For hybrid data allocation, our framework checks all data in the input program and determines which data can be duplicated. Our framework runs in the offline stage and will not affect the production runtime for FHE-HD applications because the optimized processing can be used for any new data points without re-optimization.

V. Results

A. Experimental Setup

We implemented our framework using OpenFHE [20] library. Our implementation was tested with Intel Core i7-8700K processor and 64GB memory. We evaluated our algorithm and hardware design on MNIST [3] and ISOLET [21] datasets with hypervectors of size 8192. We chose a minimum power of 2 hypervector dimensionality that gives an acceptable accuracy to maximize the utilization of packing method. For CKKS parameters, we used uniform ternary secret distribution following the Standard [22], ring dimension $N = 2^{16}$, ciphertext prime $logQ = 1770$ and 53-bit scaling factor, which satisfies $\lambda = 128$-bit security [23]. We used the iterative bootstrapping of the library for extra precision of our workload. We used ARK [1], which is the state-of-the-art FHE accelerator, as the FHE ASIC in our scalable system. Specifically, each ARK ASIC has a 2048-lane processing pipeline for FHE operations and 512MB on-chip scratchpads to buffer the data. Each ARK ASIC is connected to two HBM2 stacks which have a total of 16GB capacity and 512GB/s bandwidth. The host-ASIC and inter-ASIC connection follows the dragonfly interconnect network, similar to previous near-data processing acceleration [19]. We assume a total 32GB/s PCIe bandwidth (shared by all hosts) and 40GB/s inter-ASIC bandwidth for each link. For the latency, energy, and power evaluation, we use the reported values from previous work [1], [19] for different

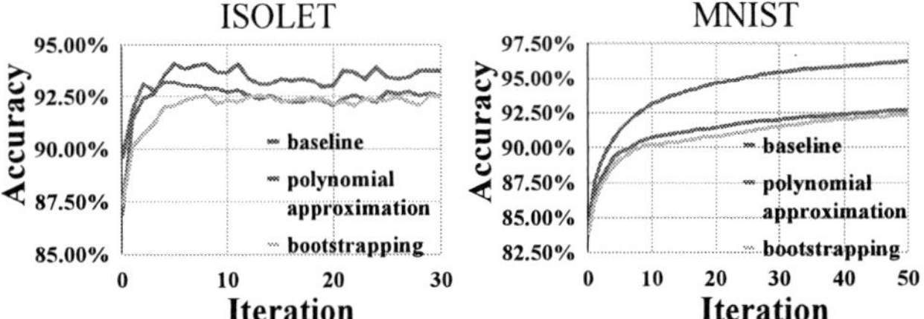

Fig. 4: **Model accuracy by retraining epochs. The baseline HDC is the initial algorithm described in Section II-B with square activation. "polynomial approximation" uses proposed approximations, and "bootstrapping" adds bootstrapping error to the "polynomial approximation".**

TABLE I: **Amortized latency of each step per data point.**

Latency (s) on 48-thread CPU			
Encoding	Training	Retraining	Inference
ISOLET / MNIST 4.67 / 5.76	6.28	16.59	10.31

operations. For evaluation, we implement an in-house simulator that can simulate instruction traces of FHE operations. We implement all hardware components and corresponding mechanisms to accurately calculate the number of operations, including multiplications, NTTs, scratchpad load/store, HBM load/store, and inter-ASIC communication.

B. Accuracy

The accuracy degradation of the model was evaluated in plaintext. We empirically measured the magnitude of bootstrapping error and simulated it. Figure 4 shows the model's accuracy by retraining epochs. For the ISOLET dataset, the model only lost around 1% accuracy compared to the baseline HDC algorithm with both polynomial approximation and bootstrapping errors. Because of the noise robustness of the HDC algorithm, we were able to achieve minimal accuracy degradation with FHE-HD algorithm. For the MNIST dataset, the accuracy loss was about 3% with polynomial approximation and 3.8% with bootstrapping error. Most of the accuracy loss was due to polynomial approximation rather than the bootstrapping error. This can be improved by a better approximation of `argmax`. As we are focusing on the efficiency of the task in this paper, we leave it as further work.

C. Latency

Table I reports the latency of each step in our HDC model. Note that the encoding latency differs as it depends on $\sqrt{f}$ in random projection. Other operations are the same because they operate in the same high dimensional space. Table II compares our training latency with previous FHE-based training models, which train the entire model without other techniques like transfer learning. Compared to the previous MLP works, our HDC trains 4.7× and 5.8× faster. We assumed that the data encoding and mask generation for training was done by the client. But even when assuming they are done by the server, it only increases the total training time by 1.5 days. Also compared to the FHE-based RNN model which takes 49 minutes to inference an image, our work takes 2.58 seconds.

TABLE II: **Latency comparison with previous works using MNIST dataset on CPU (48 threads). Inference latency is an amortized result.**

Work	#epoch	Latency		
		1-epoch	Total training	Inference
MLP 1 [2]	5	14 days	69.9 days	-
MLP 2 [24]	5	17.4 days	86.8 days	-
FHE-HD	5	2.9 days	15 days	2.58 s
RNN [25]	-	-	-	49 min

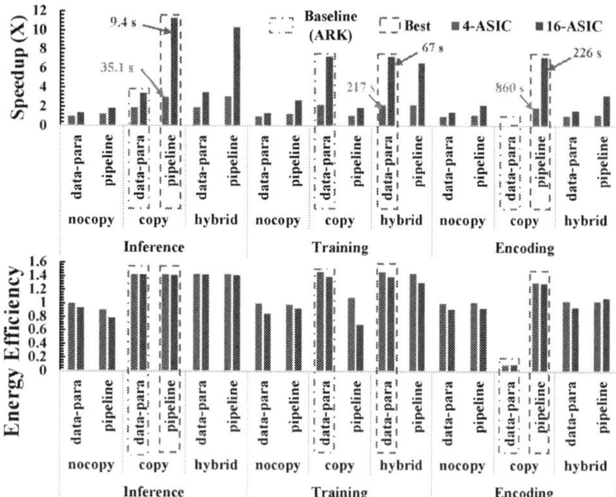

Fig. 5: **The performance and energy efficiency of various hardware configurations. The data-para with a "copy" data allocation is the conventional way of utilizing multiple accelerators for data-parallel execution, which is similar to the straightforward usage of multiple ARK chips [1].**

D. Hardware Acceleration

Figure 5 shows the performance and energy efficiency results of FHE-HD hardware acceleration using various scheduling and data allocation schemes. In the 4-ASIC system, the best scheme, explored by our framework, is 3.1×, 2.2×, and 1.9× faster while consuming 1.4×, 1.5×, and 1.3× less energy than the naive data-parallel processing with no-copy data allocation on HDC inference, training, and encoding respectively. We also observe good scalability when we increase the number of ASICs from 4 to 16. Specifically, the best scheme on the 16-ASIC system is 3.7×, 3.3×, and 3.8× faster than the 4-ASIC system, achieving almost linear speedup with the number of ASICs. As compared to the baseline (data-para copy (ARK) [1]), the best solution achieves 1.6×, 1.1×, and 38.2× speedup. The reason for the extremely slow encoding of the baseline is encoding requires a large number of rotation keys, where the copy-based data allocation significantly increases the memory footprint, thus introducing large host-ASIC communication overhead. Overall, the proposed system can finish FHE-HD inference, training, and encoding on 60,000 MNIST data points in 35.1 (9.4) s, 217 (67) s, and 860 (226) s on the 4-ASIC (16-ASIC). As for energy, the proposed system consumes up to 13.8× less energy than the baseline system.

VI. CONCLUSION

We proposed an efficient algorithm-hardware optimization, FHE-HD, for FHE-based HDC classification. First, we suggested the first FHE-based HDC model using polynomial approximation and packing method. Our model shows 4.7× faster training and over 1000× faster inference compared to the state-of-the-art works. Furthermore, we suggested a scalable FHE accelerator system considering the scheduling and data allocation of HD workload, which achieved up to 38.2× speedup compared to a state-of-the-art.

ACKNOWLEDGMENT

This work was supported in part by PRISM and CoCoSys, centers in JUMP 2.0, an SRC program sponsored by DARPA, SRC Global Research Collaboration (GRC) grants, and NSF grants #1826967, #1911095, #2003279, #2052809, #2112665, #2112167, and #2100237.

REFERENCES

[1] J. Kim *et al.*, "Ark: Fully homomorphic encryption accelerator with runtime data generation and inter-operation key reuse," in *MICRO '22*.

[2] Q. Lou *et al.*, "Glyph: Fast and accurately training deep neural networks on encrypted data," *NeurIPS '20*, vol. 33, pp. 9193–9202, 2020.

[3] Y. LeCun, "The mnist database of handwritten digits," *http://yann. lecun. com/exdb/mnist/*, 1998.

[4] M. Imani *et al.*, "Revisiting hyperdimensional learning for fpga and low-power architectures," in *HPCA '21*.

[5] J. H. Cheon *et al.*, "Homomorphic encryption for arithmetic of approximate numbers," in *Asiacrypt '17*. Springer, 2017, pp. 409–437.

[6] J. H. Cheon *et al.*, "Bootstrapping for approximate homomorphic encryption," in *Eurocrypt '18*. Springer, 2018, pp. 360–384.

[7] K. Han and D. Ki, "Better bootstrapping for approximate homomorphic encryption," in *CT-RSA'20*.

[8] B. Li and D. Micciancio, "On the security of homomorphic encryption on approximate numbers," in *EUROCRYPT '21*. Springer.

[9] B. Li *et al.*, "Securing approximate homomorphic encryption using differential privacy," in *CRYPTO '22*. Springer, 2022, pp. 560–589.

[10] Z. Zou *et al.*, "Manihd: Efficient hyper-dimensional learning using manifold trainable encoder," in *DATE '21*. IEEE, 2021, pp. 850–855.

[11] N. Samardzic *et al.*, "Craterlake: a hardware accelerator for efficient unbounded computation on encrypted data," in *ISCA '22*, 2022.

[12] R. Cammarota, "Intel HERACLES: homomorphic encryption revolutionary accelerator with correctness for learning-oriented end-to-end solutions," in *CCSW*, 2022.

[13] S. Halevi and V. Shoup, "Faster homomorphic linear transformations in helib," in *CRYPTO '18*. Springer, 2018, pp. 93–120.

[14] K. Han *et al.*, "Logistic regression on homomorphic encrypted data at scale," in *AAAI'19*.

[15] J.-W. Lee *et al.*, "Privacy-preserving machine learning with fully homomorphic encryption for deep neural network," *IEEE Access*, 2022.

[16] A. Brutzkus *et al.*, "Low latency privacy preserving inference," in *ICML '19*.

[17] R. E. Goldschmidt, "Applications of division by convergence," Ph.D. dissertation, Massachusetts Institute of Technology, 1964.

[18] J. H. Cheon *et al.*, "Numerical method for comparison on homomorphically encrypted numbers," in *Asiacrypt '19*.

[19] J. Ahn *et al.*, "A scalable processing-in-memory accelerator for parallel graph processing," in *ISCA '15*, 2015, pp. 105–117.

[20] A. A. Badawi *et al.*, "Openfhe: Open-source fully homomorphic encryption library," Cryptology ePrint Archive, Paper 2022/915, 2022.

[21] D. Dua and C. Graff, "Uci machine learning repository," 2017.

[22] M. Albrecht *et al.*, "Homomorphic encryption security standard," HomomorphicEncryption.org, Toronto, Canada, Tech. Rep., November 2018.

[23] B. R. Curtis and R. Player, "On the feasibility and impact of standardising sparse-secret lwe parameter sets for homomorphic encryption," Cryptology ePrint Archive, Paper 2019/1148.

[24] K. Nandakumar *et al.*, "Towards deep neural network training on encrypted data," in *CVPR Workshops*, June 2019.

[25] J. Jang *et al.*, "Privacy-preserving deep sequential model with matrix homomorphic encryption," in *ASIA CCS '22*, 2022, pp. 377–391.

AUTHOR INDEX